Photoelectron spectroscopy and the future of surface analysis

Burlington House, London, UK and online

20–22 April 2022

FARADAY DISCUSSIONS

Volume 236, 2022

ROYAL SOCIETY OF CHEMISTRY

The **Faraday Division** of the Royal Society of Chemistry, previously the Faraday Society, was founded in 1903 to promote the study of sciences lying between chemistry, physics and biology.

Faraday Discussions (Print ISSN 1359-6640, Electronic ISSN 1364-5498) is published 8 times a year by the Royal Society of Chemistry, Thomas Graham House, Science Park, Milton Road, Cambridge, UK CB4 0WF.

Volume 236 ISBN 978-1-83916-693-8

2022 annual subscription price: print+electronic £1133 US $1994; electronic only £1079, US $1899.
Customers in Canada will be subject to a surcharge to cover GST. Customers in the EU subscribing to the electronic version only will be charged VAT.

All orders, with cheques made payable to the Royal Society of Chemistry, should be sent to the Royal Society of Chemistry Order Department, Royal Society of Chemistry, Thomas Graham House, Science Park, Milton Road, Cambridge, CB4 0WF, UK
Tel +44 (0)1223 432398; E-mail **orders@rsc.org**

If you take an institutional subscription to any Royal Society of Chemistry journal you are entitled to free, site-wide web access to that journal. You can arrange access via Internet Protocol (IP) address at **www.rsc.org/ip**

Customers should make payments by cheque in sterling payable on a UK clearing bank or in US dollars payable on a US clearing bank.

Printed in the UK

MIX
Paper from responsible sources
FSC® C013604

ROYAL SOCIETY OF CHEMISTRY

Faraday Discussions

Faraday Discussions are unique international discussion meetings that focus on rapidly developing areas of chemistry and its interfaces with other scientific disciplines.

Information for Authors

∞ The paper used in this publication meets the requirements of ANSI/NISO Z39.48–1992 (Permanence of Paper).

Registered charity number: 207890

Photoelectron spectroscopy and the future of surface analysis

Faraday Discussions

www.rsc.org/faraday_d

A General Discussion on Photoelectron spectroscopy and the future of surface analysis was held in London, UK on the 20th, 21st and 22nd of April 2022.

The Royal Society of Chemistry is the world's leading chemistry community. Through our high impact journals and publications we connect the world with the chemical sciences and invest the profits back into the chemistry community.

CONTENTS

ISSN 1359-6640; ISBN 978-1-83916-693-8

Cover
See G. Galli *et al.*, *Faraday Discuss.*, 2022, **236**, 352–363.

The electronic properties of salt water under pressure are probed by computing photoelectron spectra.

Image reproduced by permission of Giulia Galli from *Faraday Discuss.*, 2022, **236**, 352–363.

INTRODUCTORY LECTURE

PAPERS AND DISCUSSIONS

CONCLUDING REMARKS

ADDITIONAL INFORMATION

Faraday Discussions

ROYAL SOCIETY OF CHEMISTRY

PAPER

Spiers Memorial Lecture: prospects for photoelectron spectroscopy

Wendy R. Flavell *

Received 3rd April 2022, Accepted 18th May 2022
DOI: 10.1039/d2fd00071g

An overview is presented of recent advances in photoelectron spectroscopy, focussing on advances in *in situ* and time-resolved measurements, and in extending the sampling depth of the technique. The future prospects for each are considered. Progress in the last few years has been remarkable, aided by substantial advances in photon sources, environmental cells and in analyser and detection technology. It is now possible to use the technique at ambient pressures and above, and to routinely probe interfaces buried many tens of nanometres below the surface in a non-destructive way. Our capabilities in time-resolved photoemission of the solid state have extended so far as to enable exploration of the fundamental timescale of the photoemission process itself.

1. Introduction and some basic principles

Einstein famously proposed a direct link between the kinetic energy (E_{KE}) of a photoelectron emitted from a solid and the energy of the photon absorbed ($h\nu$) in 1905,[1] an achievement for which he was awarded the Nobel Prize in 1921:

$$E_{\mathrm{KE}} = h\nu - E_{\mathrm{BE}} - \Phi, \tag{1}$$

where E_{BE} is the binding energy relative to the Fermi energy and Φ is the work-function (in fact, that of the spectrometer, as the sample is in direct electrical contact with the solid sample). Historically, photoemission in solids has been treated by the 'three-step model', involving initial optical excitation, followed by transport of the photoelectron to the solid, then transmission through the surface with escape into the vacuum.[2] The optical excitation is regarded as taking place instantaneously, with no change of momentum; we will return to this point in Section 4.5, as our experimental capabilities have advanced so far in the last few years that we can start to examine this assumption.

The simple relationship between the measured kinetic energy and the binding energy in the solid (eqn (1)) immediately defines photoelectron spectroscopy as a technique sensitive to chemical state. The demonstration that the binding

Department of Physics and Astronomy, Photon Science Institute and Henry Royce Institute, School of Natural Sciences, The University of Manchester, Manchester, M13 9PL, UK. E-mail: wendy.flavell@manchester.ac.uk

energy of an electron in a solid depends on its chemical state was made by Siegbahn (who received the Nobel Prize in 1981);[3] his work led to the acronym 'ESCA' (electron spectroscopy for chemical analysis), although a chemical shift (in X-ray absorption) was first measured by Robinson in 1930.[4] In nearly a century since, X-ray photoelectron spectroscopy (XPS) has become ubiquitously used as a 'go-to' technique for chemical analysis and is implemented in many thousands of labs around the world. The development of photoemission excited by UV-sources (ultraviolet photoelectron spectroscopy or UPS), primarily as a technique for measuring the energy- and momentum-dependent filled bandstructure of solids, using angle-resolved photoelectron spectroscopy (ARPES, Fig. 1),[5] proceeded in parallel by Spicer and others.[6] More recently, the gap between the two photon energy regimes has been plugged by synchrotron sources, and photoelectron spectroscopy, implemented in multiple ways, is now one of the most important and direct techniques for measuring the chemical and electronic structure of solids.

A large number of excellent reviews has been written[5,7] so here we focus on recent developments in the technique that promise to be of key importance to the future. A characteristic of the technique is that the exiting photoelectrons are of relatively low energy and are charged. This has historically led to interlinked difficulties (described below) that have imposed some limits on the applications of photoemission. A hallmark of the last decade is that significant progress has been made in tackling these issues.

Firstly, when a large group of photoelectrons leaves a solid at the same time (for example if excited with a short and intense pulse from a free-electron laser (FEL)), they interact with one another, causing space charge effects in the vacuum above the solid. This, until the last few years, slowed the progress in developing ultrafast time-resolved photoemission from solids. This roadblock has been largely removed very recently with the advent of new time-of-flight detectors and higher repetition-rate sources. The rapid recent progress in this field is reviewed in Section 4.

A related issue is that the low energy and charge of the photoelectrons means that they interact strongly with any medium they pass through. Photoelectron

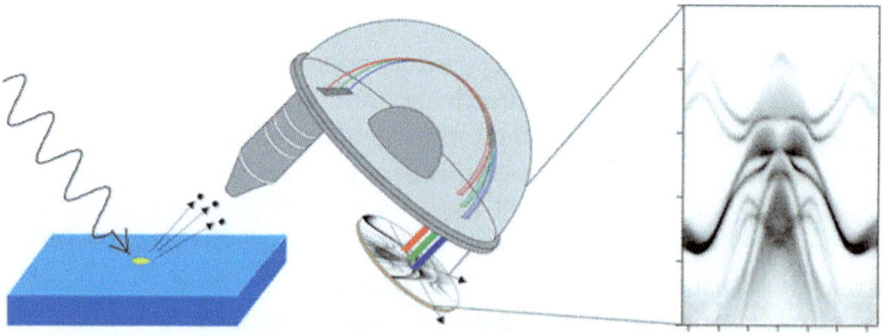

Fig. 1 Schematic diagram of angle-resolved photoelectron spectroscopy (ARPES) using a conventional hemispherical electron energy analyser. By measuring the intensity of the photoelectrons as a function of their kinetic energy and emission angle, ARPES gives access to the electron-removal spectral function in the crystal momentum space, allowing band dispersions to be obtained (right). Reprinted with permission from ref. 5. Copyright (2021) American Chemical Society.

spectroscopy measurements have historically been carried out in ultrahigh vacuum (UHV), partly to allow the photoemitted electrons to reach the electron energy analyser without attenuation by an intervening gas. This leaves the technique open to the criticism that results cannot be obtained under realistic pressures (for example in studies of catalysts or thin film growth) or in a liquid (for example in studies of electrocatalysts). Here again, astonishing progress has been made in stretching the capabilities of near-ambient pressure XPS (NAP-XPS) so that in the last few years, several measurements have been reported at ambient pressure, and substantial progress has been made in our capability to carry out real-time measurements of cells and devices under working conditions. This progress in NAP-XPS is summarised in Section 2.

A further consequence of the strong interaction of charged low-energy photoelectrons with matter is that they are very strongly attenuated by inelastic scattering as they leave the solid. This leads to the hallmark surface sensitivity of photoelectron spectroscopy, which in many ways is a boon, but can also be a limitation, as results from it are often taken to apply to the bulk of the material. The initial peak intensity, I_0, of a photoelectron signal is attenuated with depth, d, as

$$I = I_0 \exp\left(-\frac{d}{L}\right), \tag{2}$$

where L is the electron effective attenuation length. This is related to the inelastic mean free path (IMFP, λ) of photoelectrons, and for materials where elastic scattering has a negligible influence, $L = \lambda$. The sampling depth for photoelectron spectroscopy may be estimated as

$$d_S = 3\lambda \cos \theta, \tag{3}$$

where θ is the emission angle of photoelectrons leaving the sample with respect to the surface normal. The photoemitted intensity is diminished to 5% for electrons originating from this depth according to eqn (3) with $\theta = 0$ (normal emission). For traditional XPS (X-rays with energies 200–1500 eV), d_S is in the range of 1–10 nm.

The sampling depth as defined in eqn (3) applies to the use of primary photoelectron peaks, *i.e.*, those electrons which have escaped the surface to detection without inelastic scattering. These 'elastic' peaks are used for chemical state determination (as chemical state changes lead to chemical shifts in core level electron binding energies), and for the measurement of the distribution of elements near the surface. Probing beyond the elastic limit can be achieved by modelling the inelastic (secondary electron) background associated with each core level excitation at kinetic energies (KEs) lower than the primary peak.[8–12] The shape of this background is dependent on the depth distribution below the surface,[13] and this background signal remains significant even after the signals from the core level peaks have been attenuated by overlayers. This may be used to yield information from depths much larger than 3λ.[8] The mean energy loss is approximately linear with the distance the electron has travelled, meaning that extracting a depth distribution requires measurement of the inelastic background over several hundred eV.[13] This technique can be applied to surface-sensitive XPS with photon energies of around 1.5 keV. However, in conventional XPS, the inelastic background from a single core level is often obscured by other core level peaks and Auger lines present in the spectrum, meaning that the energy range over which the inelastic background is measured may be limited. The information depth has been found to be $\sim 8\lambda$ (perhaps some tens of nm) when the background is measured over a range of ~100 eV to the low kinetic energy of the primary peak, as might be typical in XPS.[14] There are a great many applications (for example to device heterostructures) where this is simply too short to be useful, and there is therefore a strong incentive to extend the sampling depth of XPS. Recent progress in extending the sampling depth *via* hard X-ray photoelectron spectroscopy (HAXPES) and X-ray standing wave techniques has been remarkable, and has also aided the realisation of ambient-pressure XPS. This progress is reviewed in Section 3.

This paper serves as an introduction to the 2022 *Faraday Discussion* meeting 'Photoelectron spectroscopy and the future of surface analysis'. I have framed it around three of the key themes of the meeting, developments in *in situ* techniques (based on NAP-XPS), in measurements of buried interfaces by non-destructive techniques, and in time-resolved photoemission. I present a personal view, so the paper is not exhaustive and cannot fully reflect the diversity and wealth of discussions that took place during the meeting. Far from being impacted by the pandemic, there has been quite astonishing progress in all three areas in the last few years alone, which is reflected in the papers presented at this meeting.

2. *In situ* methods: developments in near-ambient pressure XPS

2.1. Background

Without doubt, one of the most important developments in photoemission in the last decade has been the significant increase in the availability of near-ambient pressure XPS instruments, and their widespread use. The important applications are well-represented by the submissions to this meeting, and include heterogeneous catalysis (https://doi.org/10.1039/D1FD00120E), including water splitting[15] electrochemistry (https://doi.org/10.1039/D1FD00114K),[16] thin film growth, for example by atomic-layer deposition,[17] gas sensing[18] and corrosion and degradation.[19] A large number of excellent reviews has been published.[20–23]

The low IMFP of low-energy photoelectrons, which gives XPS its surface sensitivity, also of course limits the IMFP of electrons in gases (in the case of 100 eV electrons travelling through 1 mbar of water vapour, to about 1 mm).[20] This is short compared to the typical distance between the sample and the electron energy analyser in XPS, so in order to carry out XPS under non-vacuum conditions, typically differential pumping stages are used that reduce the pathlength of the electrons through the gas phase, together with scattering of the electrons and signal attenuation. Near-ambient pressure- (NAP-)XPS (also called ambient-pressure- (AP-)XPS) was developed initially by the Siegbahn group in Uppsala as long ago as the 1970's,[24,25] but suffered from low count rates. The technique did not start to become deployed in a widespread way until the early 2000's when NAP-XPS instruments were developed at high flux synchrotron sources and were combined with a new generation of high transmission analysers (the first of these at the Lawrence Berkeley National Laboratory (LBNL) in the late 1990's).[26] Further technical developments, in particular by SPECS Surface Nano Analysis GmbH (with LBNL and the Fritz Haber Institute in Berlin), based on the SPECS Phoibos 150 NAP analyser, have seen an explosion of activity in the last decade, and NAP-XPS is now widely available at both synchrotron sources and in commercial lab-based instruments.[20,23]

Most NAP-XPS instruments utilise an electron energy analyser with multiple (typically four) pumping stages that reduce the pressure from (typically) a few mbar in the sample environment to $<10^{-7}$ mbar in the hemispherical analyser, with the aperture nozzle of the first pumping stage held typically around 1 mm from the sample. Different NAP-XPS systems realise the sample environment in different ways (Fig. 2),[20] and in the last few years, this has led to specialised commercial instruments with rather different applications. In the simplest experiments, the whole of the vacuum chamber that isolates the sample from the laboratory environment is exposed to the gas atmosphere during data taking (Fig. 2(a)). This method is in use, for example at the TEMPO beamline at SOLEIL, Paris and at beamlines 9.3.2 and 11.0.2 at the Advanced Light Source (ALS), Berkeley. The main disadvantage of this approach is the time taken to regain ultra-high vacuum (UHV) if UHV XPS measurements are required (which

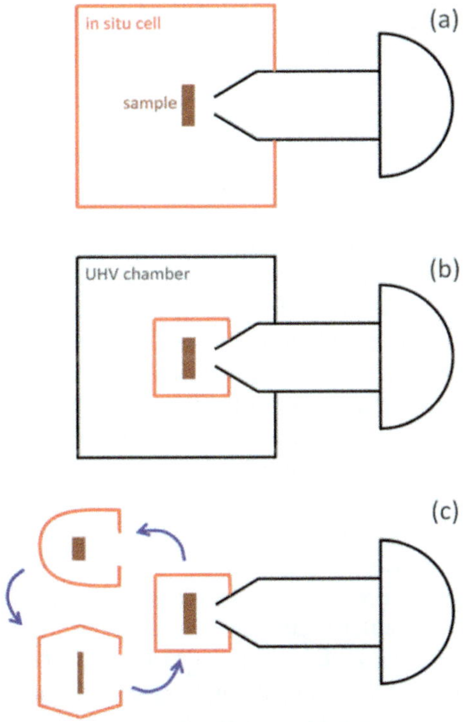

Fig. 2 Schematic diagram of the main arrangements of *in situ* measurement cells used in NAP-XPS systems. (a) The analysis chamber/*in situ* cell is part of a vacuum system (often with load lock and preparation chamber). The whole chamber is exposed to gases during APXPS experiments. (b) The *in situ* cell is located inside a larger vacuum chamber. Only the *in situ* cell is exposed to gases during APXPS measurements. (c) Exchangeable *in situ* cells, tailored to a particular experiment, are attached to the analyser. Reproduced from ref. 20. Copyright © 2013 Author(s).

normally involves bake-out), and also the possibility of cross-contamination between experiments. However, these disadvantages may be turned to spectacular advantage if the instrument can be dedicated to one purpose (*e.g.* wet environments), and if UHV experiments are not required, as discussed further in Section 2.6.

In instruments where other sample analyses (including UHV XPS) and complex sample preparation are needed, it is more typical to use an environmental or *in situ* cell, which is placed inside the analyser chamber and sealed from it during the exposure of the sample to gases (Fig. 2(b)). This allows UHV to be regained more rapidly between experiments (a particular consideration for synchrotron experiments), but adds to the complexities of sample manipulation, as the sample must be manoeuvred in and out of the cell. In a refinement of this approach, called the 'Lund' or cell-in-cell concept (developed by Lund University and SPECS Surface Nano Analysis GmbH), the nozzle that is normally attached to the analyser aperture is incorporated into the cell, so that the analyser collection angle is not constrained by it during UHV measurements.[27] The 'cell-in-cell' concept has been implemented at the SPECIES and HIPPIE beamlines at MAX IV.[28,29] It is also possible to make use of a set of interchangeable sample cells, each of which can be devoted to one purpose (*e.g.* corrosive environments, liquid samples, electrochemical cells *etc.*, Fig. 2(c)). This again adds to experimental complexity, but improves flexibility. This type of arrangement is available, for example, at the VERSOX beamline at Diamond Light Source and at HIPPIE and SPECIES (MAX IV).

2.2. Applications in heterogeneous catalysis

A number of excellent reviews of NAP-XPS have noted the preponderance of applications to heterogeneous gas-phase catalysis,[20,23] and these applications are well-represented at this meeting. Even though new applications of NAP-XPS are emerging rapidly, in 2020, surface science and catalysis still accounted for around 50% of all NAP-XPS publications. In heterogeneous catalysis, studies of model reactions, such as the CO oxidation reaction, accounted for 45% of NAP-XPS research at this point.[23] In these studies, it has become clear that under NAP-XPS conditions, the reaction is typically mass-transfer limited (for example in CO oxidation, all CO that reaches the surface is instantaneously oxidised to CO_2, and there is a CO depletion layer at the surface that extends for some mm above the surface).[23] Under these conditions, the total conversion rate is determined by the rate of gas diffusion through the depletion layer.[23] Planar laser-induced fluorescence has been used to explore the significant differences between the CO_2 concentration measured above the sample and that measured by mass spectrometry at the gas outlet.[30] Although there is now an increasing number of studies of oxide catalysts (https://doi.org/10.1039/D1FD00120E),[31] there is a preponderance of studies of metal single crystal surfaces (including curved crystals). These have been well-reviewed,[20,23] and have been used to elucidate the differing catalytic activity of different metals (*e.g.* Pt(111) *vs.* Ir(111) in CO oxidation,[32] and in the preferential and partial oxidation of CO/H_2 mixtures[23]) and of different metal surfaces, highlighting the catalytic activity of stepped and faceted surfaces (*e.g.* in studies of a Pt curved crystal in the same reaction[33]).

The study of the reactions of larger molecules, particularly carbon-containing ones, is of huge importance in the production of fine chemicals and pharmaceuticals. However, the study of these reactions by NAP-XPS remains challenging because of issues of rapid surface contamination, complexity of the mechanism, and the low vapour pressure of the reactants. In addition, the binding energies of the gas and surface-phase components often overlap, necessitating parallel UHV studies. As an example, a recent study of the Sonogashira cross-coupling reaction (a C–C bond-formation reaction between iodo- or chlorobenzene and phenylacetylene to form diphenylacetylene) confirmed earlier UHV results that Au(111) catalyses the reaction at moderate temperature and pressure.[34] However, rapid inactivation due to the formation of a carbonaceous species was observed at higher temperature and pressure,[34] illustrating the importance of NAP-XPS in providing more industrially relevant conditions. Although, as we discuss in Section 2.7, it has recently become possible to carry out NAP-XPS at pressures of 1 bar and above of a few mbar, and there remain significant challenges to solve in studying real catalytic reactions of large molecules.

2.3. Electrochemistry and electrocatalysis

NAP-XPS now contributes powerfully to the study of electrocatalysis and electrochemical devices, such as batteries, fuel cells and supercapacitors. Although such studies only began around 2008,[36] the capacity of NAP-XPS for *in situ* and *operando* measurements has led to a rapid expansion in these studies, which by 2020 accounted for about 6% of NAP-XPS publications.[23] The first devices studied in this way were solid oxide electrochemical cells (SOCs), which typically operate in an atmosphere of gaseous fuel at temperatures > 700 °C, precluding the application of most surface analytical tools.[36–39] The first experiments were carried out at the ALS, and used a specially designed SOC in which a Au–ceria working electrode and a Pt counter electrode were deposited on a single crystal yttria-stabilised zirconia (YSZ) electrolyte disc.[36,37] This was designed with all the components located on the same side of the electrolyte disc so that all the cell components were exposed to the gas atmosphere, and were accessible to NAP-XPS. As well as providing chemically specific information about each element of the cell under *in situ* conditions (at *ca.* 750 °C, in a pressure of 1.3 mbar (1 : 1H_2/H_2O)) while the cell was working, this study showed that local electrical potential changes at the surface can be determined from the changes in the KE of the core level photoelectron peaks.[36,37] The authors were also able to show that the active electrochemical region on ceria extends up to 150 mm away from the current collector, with significant changes in the equilibrium Ce^{3+}/Ce^{4+} ratio observed throughout this region.[37] The first measurement of individual overpotentials across an interface was made using a Ni/YSZ/Pt SOC;[38] in the water splitting reaction, and it was used to show that H_2O splitting is faster than H_2 oxidation on Ni, while the converse is true on Pt.[38]

Clearly a particular target for NAP-XPS is to understand the oxidation and reduction half-reactions in lithium-ion and lithium–air batteries. These pose particular challenges, not least the chemical complexity, leading to parasitic side reactions, but also the use of a liquid electrolyte in many conventional cells. Currently most NAP-XPS uses soft X-rays (with energies <2 keV), which limits the probing depth to less than about 5 nm below the sample surface. This in general allows access to gas/solid or liquid/solid interfaces but buried interfaces (liquid/

solid or solid/solid) are more problematic. It means that the liquid electrolyte surrounding the immersed electrode must be stable (to evaporation) and less than a few nm thick.[40] As a result, some of the first NAP-XPS studies were carried out instead on solid state Li-ion/O_2 cells.[41] These allowed the Li–O_2 reactions to be probed in the absence of binder and liquid electrolyte, and for the first time revealed reversible Li^+ intercalation into a $Li_xV_2O_5$ electrode *in situ* under applied potential in *ca.* 0.5 mbar O_2.[41] This general approach has been extended to Na–O_2 model batteries, where it showed the formation of Na_2CO_3 as the main discharge product of the cell.[42]

In the years following the initial measurements, there has been a strong drive to towards developing methods to access buried interfaces (particularly the liquid/solid interface) with NAP-XPS, so that more realistic devices with liquid electrolytes may be studied under real working (*operando*) conditions with a temporal resolution high enough to detect metastable states.[43] Two main threads have been pursued: the development of liquid cells for XPS,[16,44–52] and the use of harder X-rays with energies of a few keV (often called 'tender' X-rays) producing photoelectrons with a higher IMPF, and thus a larger sampling depth (Section 3).[53,54] The latter also facilitates the use of standing wave techniques (Section 3.4) combined with NAP-XPS in a technique known as standing-wave ambient-pressure photoemission spectroscopy (SWAPPS).[53] The use of tender X-rays has also made possible the study of electrode/electrolyte interfaces created *via* the meniscus or 'dip and pull' approach.[54] This has been recently implemented, for example, at the HIPPIE beamline at MAX IV (Sweden)[28] and at the SpAnTeX beamline at BESSY II.[58]

In the last decade, there has been significant development of cells that use ultrathin membranes of graphene, graphene oxide or silicon,[16,44–51,59,60] which separate the sample volume from the vacuum (Fig. 3). If these are sufficiently thin, the emerging photoelectrons can be emitted through the membrane into the electron energy analyser. The approach has been implemented in flow cell geometries with both liquid and vapour flows, and used, for example, to monitor the electrodeposition of an oxidised cobalt layer from a $CoSO_4$ solution onto the graphene membrane, anchored by $Co(CO)_x$ species.[16] *In situ* electrochemical cells using proton exchange membranes (PEMs) have been implemented in both 2-electrode and 3-electrode (flow) manifestations to probe the oxygen evolution reaction (OER) during water electrolysis over Pt- and Ir-based electrodes, revealing the working oxidation states of both anions and cations.[48,52,56]

Current research on electrochemical systems is clearly benefitting from a number of simultaneous technical developments such as those described above; particularly important future directions are the development of liquid-jet sources[22] and the rapid development of capability to perform XPS at pressures of 1 bar and above.[35,62–64] These developments are discussed further in Section 2.7.

2.4. Thin film growth

In contrast to many processes in gas-phase heterogeneous catalysis, the surface chemistry and mechanisms of thin film growth remain largely unknown.[65] In the last decade, the potential of NAP-XPS to contribute to the understanding of commercially important thin film growth processes has been realised, and although there remain rather few published studies, investments are being made

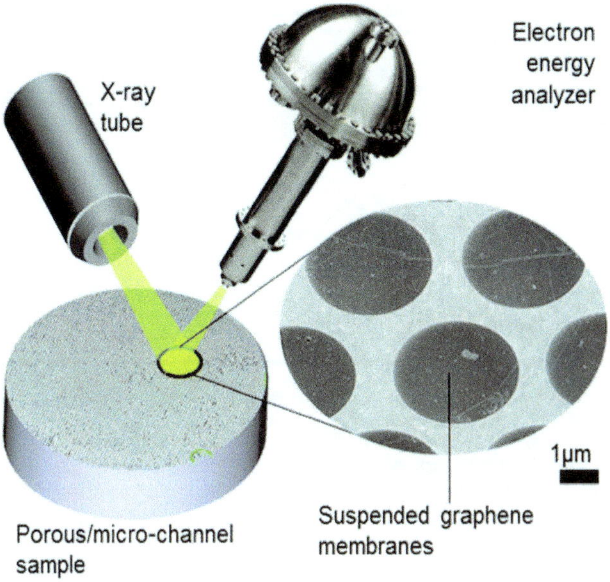

Fig. 3 Schematic illustration of the collection of XPS through graphene; in this example, a microporous substrate that contains a high density of individual micro-volumes is separated from the vacuum by electron-transparent graphene membranes. The sample is impregnated with the analyte. The concept is implemented at the ESCA Microscopy beamline at Elettra (Trieste).[61] Reproduced from ref. 51 with permission from the Royal Society of Chemistry.

to support these challenging experiments at a number of synchrotron sources, including at the SPECIES and HIPPIE beamlines at MAX-IV.[66,67] The pressure used in atomic layer deposition (ALD) (and the useful range in some chemical vapour deposition (CVD) processes) is typically in the 10^{-3} to 20 mbar range, which is well-matched to NAP-XPS.[23] However, there are also formidable challenges. ALD makes use of dedicated reactor geometries for self-limiting film growth, using short vapour pulses of each precursor in alternation, with purging of the sample volume with an inert gas between every precursor pulse – so NAP-XPS instruments of the types shown in Fig. 2(b) and (c) are needed, with a small cell volume. Because the precursors are designed to stick well to surfaces and decompose easily, there are serious issues with the contamination of succeeding experiments, and X-rays can induce surface reactions that dominate the experiment.[23] It is nevertheless possible to obtain completely new insights. An example is the use of NAP-XPS to probe the growth of TiO_2 on $RuO_2(110)$ by ALD using tetrakis(dimethylamido) titanium (TDMAT) and water in alternate half-cycles of the reaction.[17] NAP-XPS was recorded in real time during repeated dosing of the surface with first TDMAT and then water at 60 °C and pressures up to 0.1 mbar (Fig. 4).[23]

Prior to these experiments, the proposed reaction mechanism was a simple Ti ligand displacement by surface hydroxyl to bind the TDMAT to the surface, followed by hydrolysis of the remaining ligands by water, involving no change the oxidation state of Ti(IV).[68] However, a much more complex picture was revealed by NAP-XPS. In the first half cycle, evidence of Ti(II), (III) and (IV) is observed, together

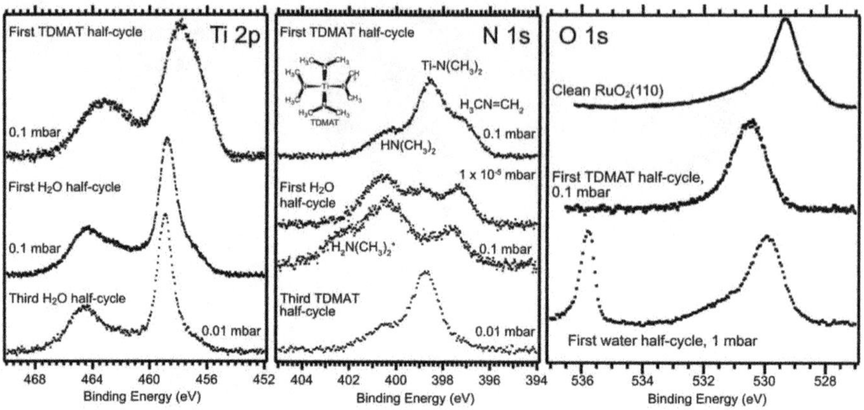

Fig. 4 NAP-XPS study of TiO_2 ALD from TDMAT and H_2O on $RuO_2(110)$: Ti 2p, N 1s and O 1s spectra acquired during the initial cycles of ALD. The sample temperature was 110 °C, and a photon energy of 580 eV was used in the recording of all spectra. Reprinted with permission from ref. 17 and ref. 23. Copyright (2016) American Chemical Society.

with a much more complex N 1s chemistry than expected. In addition to the expected surface $-N(CH_3)_2$ ligands from the precursor, dimethylamine ($HN(CH_3)_2$) and methylene imine ($CH_3N{=}CH_2$) are observed (Fig. 4).[17,23] These species persist during the water half cycle, while the $-N(CH_3)_2$ ligands react with water, and dimethylammonium ions are formed. It has been proposed that the reaction mechanism in the initial TDMAT half-cycle is dominated by a β-hydride elimination reaction involving the reduction of Ti(IV).[17,23] The expected ligand exchange mechanism appears to take over in subsequent cycles; the imine peak is removed and a hydroxyl peak appears in the O 1s spectrum during the first water half-cycle (Fig. 4).[17,23]

Although in the example discussed above, spectra were recorded in real time during ALD, they were not recorded during typical reactor operating conditions. As the typical pulse durations in ALD are on the ms to s timescale,[69] it is clear that this will usually not allow enough time to accumulate NAP-XPS with sufficient statistics. It is necessary then to develop a 'snapshot' mode capability, where data from many repeated cycles may be summed at the same temporal delay between pulse and spectrum (akin to those developed for some forms of time-resolved pump–probe photoelectron spectroscopy, discussed in Section 4.4).[66] This can then be used to study 'steady-state' ALD, when the chemistry is identical for a large number of successive pulses. This capability was realised for a test reaction (CO oxidation on Pd(100)) in 2021.[67] Because the mass flow controllers that switch the gas pulses in the cell are generally distant from the surface, gas switching may not be a suitable signal for triggering the averaging. In this demonstration, the work-function-induced binding energy shift of the gas-phase O 1s peak as the gas environment around the sample changes from CO_2 (at the start of a CO pulse) to O_2 (at the end of the pulse) was used, and recognised using image recognition software.[67] It was possible to record data at frame rates of 6–17 Hz, giving a time resolution of the order of 150–50 ms, limited by the software and frame rate of the detector camera.[67] Clearly, with a delay-line or similar detector, the time resolution can be increased substantially if there is a sufficiently large trigger signal. Indeed, a demonstration of sub-ms time resolution in gas-pulsing experiments was published in 2021.[70] This utilised gated detection based on a fast camera, combined with short gas pulses generated using a fast valve that generates pulses with a rising edge of a few 100 μs, and internal pressures in the mbar range.[70] The fast camera shutter was triggered by the valve operation. The development of time-resolved NAP-XPS measurements is an exciting area of current development that is discussed further in Section 2.7.

2.5. Material degradation in humid environments

There still remains a 'pressure-gap' in NAP-XPS between the conditions of most experiments (a few mbar) and atmospheric pressure and the pressures used in real heterogeneous catalysis. This is rapidly being closed, as we describe in Section 2.7. Nevertheless, there is a number of fields where the 'few mbar' pressure range that is already readily covered in NAP-XPS is ideal – an example is the thin film growth studies described above. A further important application is to problems of material degradation in wet environments,[15] and the surface passivation necessary to avoid it. A pressure of 9 mbar of water vapour (easily achieved in NAP-XPS) is equivalent to an atmospheric relative humidity (RH) of *ca.* 30% at a standard temperature of 25 °C, a fairly typical ambient atmosphere in temperate climates. This means that realistic water pressures can be achieved within the controlled environment of the spectrometer.

Organic–inorganic hybrid perovskites, exemplified by the 'MAPI' (methylammonium lead iodide, $CH_3NH_3PbI_3$) perovskite, have shown immense promise as cheap light harvesters for a new generation of solar cells, with device efficiencies rising in the last 9 years to more than 25% – rivalling that of the conventional material, silicon.[71,72] However, the exploitation of these materials is currently prevented by their instability when exposed to water and air, which leads to rapid degradation. Understanding the mechanism of this reaction may allow researchers to identify passivation routes. The prototypical perovskite MAPI is particularly unstable.[73,74] The key difficulties in understanding the degradation mechanism of MAPI are then the difficulty in determining the surface chemistry of the first few atomic layers of the pristine material without any exposure to ambient atmospheric conditions, and, conversely, the difficulty in measuring this same surface chemistry in real time under exposure to realistic environments. Both problems were addressed simultaneously by utilising NAP-XPS to investigate the real-time degradation of an *in situ*-deposited MAPI film, exposed to 9 mbar water vapour, which provides a realistic model of the partial pressure of water vapour in the ambient environment (Fig. 5).[19] A thin film of MAPI was grown onto a $SrTiO_3(100)$ substrate *in situ* in the spectrometer using vapour deposition.[75] The use of a film thickness of only a few nm meant that peak intensities could be reliably quantified by referencing to the attenuation of signals from the underlying substrate. When exposed to water vapour (Fig. 5), complete decomposition of the methylammonium cation of the perovskite was observed, evidenced by the complete loss of the N 1s signal, and the evolution of C : N stoichiometry (Fig. 5).[19] This occurred concomitantly with the collapse of the lattice of PbI_6 octahedra of the perovskite structure to form PbI_2, evidenced by changes in the Pb 4f and I 3d binding energies and a change in the Pb : I stoichiometry (Fig. 5).[19] The data

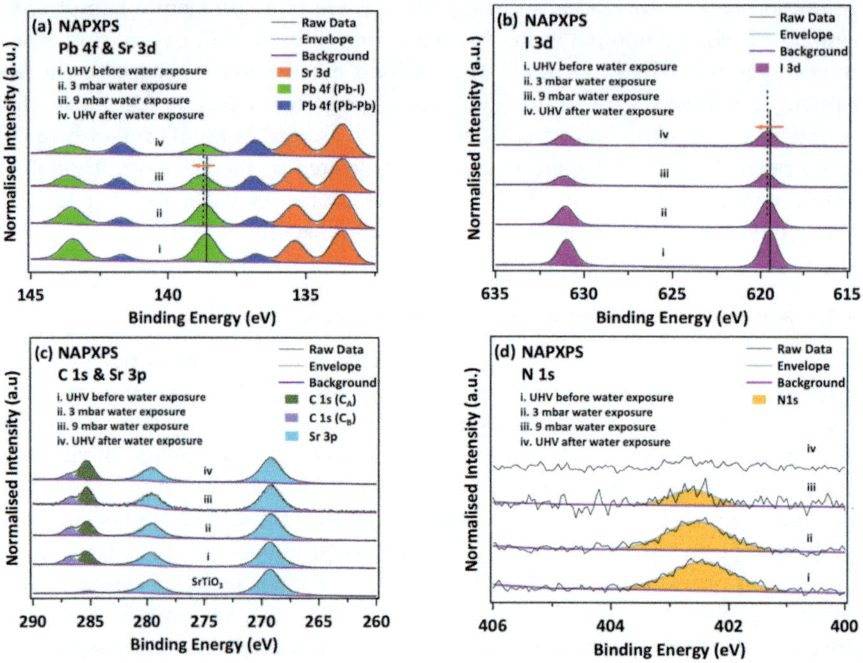

Fig. 5 Degradation of *in situ*-grown MAPI monitored by NAP-XPS: (a) Pb 4f & Sr 3d, (b) I 3d, (c) C 1s & Sr 3p, and (d) N 1s core levels before/during/after water exposure. C 1s (C_A) and C 1s (C_B) refer to adventitious carbon and to carbon associated with the perovskite methylammonium ion/C=O/C–OH, respectively. All spectra are normalised to the corresponding integrated area of the Sr $3d_{5/2}$ peak. Reproduced from ref. 20. Copyright © 2017 Author(s).

indicate a degradation pathway by reaction with water, which involves the complete loss of nitrogen from the film without the initial formation of methylammonium iodide, consistent with a degradation reaction of the type:

$$CH_3NH_3PbI_3(s) \xrightarrow{H_2O} (\text{–}CH_2\text{–})(s) + NH_3(g) + HI(g) + PbI_2(s). \tag{4}$$

2.6. Natural and biological samples

A simple NAP-XPS spectrometer of the type shown in Fig. 2(a) has some serendipitous advantages in the study of materials with little vacuum compatibility. Typically, these instruments operate at a significantly higher working pressure than conventional XPS, which can lead to much reduced sample pump-down times, and can remove the need for a load lock as the sample may be directly introduced into the analytical chamber of the instrument. This reduces the time needed for sample loading, and increases the capability for high throughput.[76] A perhaps more significant advantage is the in-built charge compensation that NAP-XPS provides, because the incident X-rays ionize the gas molecules around the sample, creating cations (together with photoelectrons, secondary and Auger electrons). These efficiently compensate the surface charging of insulating molecules.[77–79] The amount of charge compensation can be tuned by changing the

X-ray flux or the pressure and type of gas in the analysis chamber.[78] In the last 5 years, these advantages have been turned to good use in a new generation of high-throughput, simple NAP-XPS instruments exemplified by the SPECS EnviroESCA.[78] These instruments are suitable for the routine and fast measurement of a wide range of materials that are insulating and/or outgas in UHV, such as zeolites and polymers, biological and other liquid and wet samples (including nanoparticle suspensions) and consumer goods, such as foods and cosmetics.[76,80] This approach has made XPS easy and accessible to new communities. It has been possible, for example, to use NAP-XPS to study *Escherichia coli* in the hydrated state,[81] and to show that freeze drying (which has been commonly used to study biological systems by XPS[82]) alters the composition of carbon species at the surface.[81] In later studies of the bacterium *Pseudomonas fluorescens*, it has been shown that this difference can be minimised by using fast freezing in cryo-XPS.[83] Recently, the first NAP-XPS studies of radiation damage in fully hydrated double-strand DNA have been published.[84] DNA damage by ionising radiation is of fundamental importance to cancer therapy, but is difficult to study by XPS in realistic environments. XPS in vacuum, nitrogen and water was used to deconvolute direct DNA damage, caused by low energy electrons, from the indirect effects caused by the ionisation of water molecules to form reactive oxygen species, such as $OH^{\bullet}$ and H_2O^+. The latter initiate indirect damage processes which lead to deoxyribose and nucleobase modification and base release. In the presence of water, the authors show that these processes predominate over the strand breakage, which is the dominant type of damage in dry DNA, although damage of all types increases when water is introduced.[84] Thus, it is clear that NAP-XPS is beginning to make powerful contributions to resolving long-standing controversies in biological systems.

2.7. Future directions for NAP-XPS

As noted above, there is currently a strong drive to close the pressure gap completely in NAP-XPS, and to make possible studies at atmospheric pressure and above[35,62–64] (finally allowing the 'N' in the acronym to be dropped without any shade of exaggeration!) This is coupled with a strong drive to use harder X-rays, and to implement the technique in the regime of hard X-ray photoelectron spectroscopy (HAXPES). HAXPES is discussed in more detail in Section 3, but here we note that these two drivers are coupled; the use of higher kinetic energies reduces the scattering of photoelectrons in the gas phase, allowing higher pressures to be used. The use of X-rays in the few keV energy range, coupled with a very small analyser aperture placed very close to the sample, can then allow AP-XPS at atmospheric pressure, for example as implemented on BL36XU at SPring-8, using a 30 μm aperture at a working distance of 60 μm from the sample.[85] Of course, the use of higher kinetic energies also increases the inelastic mean free path in the solid. Surface sensitivity can then be enhanced by using grazing angles of incidence of *ca.* 1–5° relative to the surface (close to the condition for total external reflection).[58,62,85] In this geometry, the footprint of the X-ray beam on the surface becomes significantly elongated in the plane of incidence. In the POLARIS high-pressure HAXPES instrument recently developed for beamline P22 at PETRA III, this has led to the development of an analyser aperture made up of a linear array of evenly spaced μm-sized holes, matching the beam footprint on the sample.[63,64]

In this system, the sample-to-aperture distance is a few tens of μm, and gas dosing directed at the sample is incorporated into the aperture front cone. This leads to a very localised high-pressure pocket at the sample, allowing for pressures exceeding 1 bar to be achieved locally at the sample, while maintaining the main chamber of the spectrometer at a pressure of a few mbar.[62] This has been used to measure CO oxidation on Pd(100) in the 'light-off' regime (when CO_2 begins to be produced) in pressures of up to 1 bar for the first time.[35] The measurements indicate that under the conditions of the experiment, the system is not in the mass-transfer-limited regime immediately after light-off, allowing the measurement to be performed in the kinetic regime. The variation with temperature of the intensity of the C 1s and O 1s signals from the working catalyst allowed the extraction of an apparent activation energy.[35] In this arrangement, there is no separation between the sample and the main chamber, but effectively a 'virtual' cell is achieved by the localised high-pressure pocket. An alternative approach to closing the pressure gap is to further develop the *in situ* cells discussed in Section 2.3. In a development of the graphene membrane technology for electrochemical cells discussed earlier, AP-XPS at 1 atm has been achieved by separating the vacuum and high pressure environments by a silicon nitride grid with an array of 1 μm-diameter holes coated by bilayer graphene.[64]

A further strong area of development in the last decade has been in studies of the liquid/vapour interface – partly driven by the work in electrocatalysis described above (Section 2.3), but also by studies of water and heterogeneous reactions at liquid–vapour surfaces, such as those of atmospheric aerosols.[86] One driver here is climate change, as about one third of anthropogenically generated CO_2 is sequestered at the air/water interface.[87] With the exception of ionic liquids (which have sufficiently low vapour pressures at room temperature that they can be studied in UHV), these studies pose significant challenges, which have been recently reviewed,[22,88] but include issues of charge referencing, vapour pressure, contamination and beam damage. The development of *in situ* cells with thin membranes has allowed some significant advances, for example the study of dispersed nanoparticles in solution.[89] However, in the main the approaches to preparing a liquid surface fall into two categories: fast flowing jet or droplet sources suitable for short (ms) reaction times, and the use of a bulk reservoir of liquid inside the vacuum chamber generated, for example, *via* a wet wire, Langmuir trough,[22] deliquesced salt,[90,91] supported droplet (or offset droplet for studies of the liquid/solid interface[92]) or dip-and-pull approach.[22,54] The latter requires the experiment to be done in the presence of an equilibrium vapour pressure (to avoid evaporation), while fast moving jets can be used in the vacuum environment as the jet or droplet train moves through the chamber at a speed of many tens of m s^{-1}, so the interaction time is brief.[22] Jet and droplet sources can also be used in fast mixing modes, for example where two identical laminar jets collide just inside the measurement chamber, forming a stable liquid sheet known as a 'flatjet', and so far implemented primarily in X-ray absorption spectroscopy.[93,94] The solutions can also be fast-mixed just prior to entering the analysis chamber, an approach used in studies of the aqueous carbonate system to mix HCl and $NaHCO_3$, generating carbonic acid (H_2CO_3) *in situ*.[86] Depth-profiling the mixed jets by varying the photon energy was used to reveal surface partitioning of CO_3^{2-} over HCO_3^{2-}.[86] The effects of organic surfactants may also be probed, as in recent studies of octanoic acid at the water–gas interface.[95] This experiment exploited the geometry of liquid-jet experiments to extract photoelectron angular distributions as a means of characterising the orientation and relative position of the surfactant molecules at the interface.[95] These experiments are not without their difficulties and a number of studies have outlined the care that must be taken in interpreting these measurements;[95,96] an example is the distortion of the spectra by the quasi-elastic scattering of low kinetic energy electrons recently highlighted in measurements from liquid water made using the liquid jet sources at the PLEIADES and DESIRS beamlines at SOLEIL.[96] However, as liquid jet sources have now been implemented in a significant number of labs, this area is burgeoning. Although many studies have been made of water, the approach has been extended to other liquids. A particularly elegant example from BESSY II is studies of solvated electrons in liquid ammonia microjets.[97] Here, increasing amounts of Li, Na or K were introduced into the jets, and the transition from a blue, dilute electrolyte solution of paired electrons to a delocalised bronze metallic solution with a strong Fermi edge was tracked.[97]

As noted in Section 2.4, there is currently significant work devoted to improving the time-resolution of NAP-XPS. A number of spectrometers incorporating delay-line detectors have been developed in the last decade, implemented for example at beamline 11.0.2 at the ALS[98] and at the SpAnTeX beamline at BESSY II.[58] These detectors in principle have a precision of *ca.* 150 ps, as the detector records the arrival time and position of every electron at the exit plane of the analyser. However, the ultimate time resolution in such systems is limited by the spread in the times of flight of the photoelectrons in the analyser to values on the order of 1 ns.[98] The spectra are typically collected by repeated scans of a narrow binding energy window, so in reality the time resolution is limited by the signal level and the duty cycle of the repeated process. However, 30 ns snapshot capability with viable signal has been demonstrated.[58] Methods for accessing faster timescales, including using free-electron laser (FEL) and high-harmonic generation (HHG) light sources, have been recently reviewed.[99] Time-resolved photoemission is discussed further in Section 4.

In the next few years, we are likely to see the very rapid development of multi-modal experiments incorporating NAP-XPS.[100] The grazing incidence geometry used with tender X-rays allows for the incorporation of a number of additional techniques that use a reflection geometry. As an example, a schematic layout of an instrument under construction at the Fritz Haber Institute in Berlin is shown in Fig. 6.[22] This is designed to probe surfactant chemistry at the aqueous solution vapour interface. The samples will be prepared in a Langmuir trough, located inside the spectrometer. A micro-focussed X-ray source at grazing incidence will provide 3 different photon energies, allowing depth-profiling in NAP-XPS, and is combined with 2 other techniques in reflection geometry – reflection absorption infrared spectroscopy, RAIRS (providing optical characterisation) and grazing incidence small angle X-ray scattering, GISAXS (providing information on the morphology and ordering of the surfactant layer, supplemented by Brewster angle microscopy). X-ray fluorescence, measured with a fluorescence yield (FY) detector, will be used to measure the bulk composition, and an in-vacuum force sensor will be used to measure the surface tension as a function of the compression of the surfactant layer.[22]

In general, there is a strong push to combine NAP-XPS with structural techniques and/or lateral resolution. In addition to grazing incidence X-ray

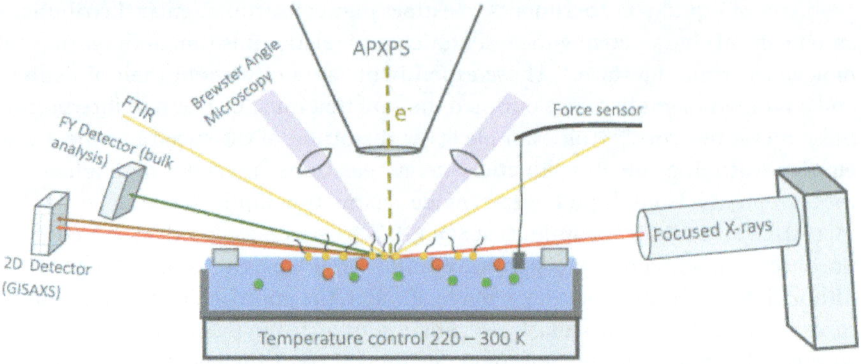

Fig. 6 Schematic layout of a combined AP-XPS–Langmuir trough instrument under construction at the Fritz Haber Institute in Berlin. Published in ref. 22. Copyright © 2021 Author(s).

scattering,[101] examples include its combination with SWAPPS[53,55–57] (Section 2.7), where there are now prospects to extend the technique to studies at elevated temperature,[57] and near-edge X-ray absorption fine structure (NEXAFS).[102] Combination with scanning photoelectron microscopy (SPEM) adds lateral imaging capability to the picture, and is implemented with NAP-XPS, for example, at the ESCA Microscopy beamline at Elettra (Trieste), offering sub-μm resolution.[61,103,104] The use of a 3D delay-line detector with a lateral resolution lens at the SpAnTeX beamline at BESSY II[58] has opened up the possibility of combined time- and lateral resolution in NAP-XPS with tender X-rays.

3. Buried interfaces

3.1. Developments in HAXPES

Over the last decades, XPS has become embedded and ubiquitously used as a core technique for materials characterisation. However, as the emerging photoelectrons have a photoelectron inelastic mean free pathlength, λ, of only a few nm, there are some dangers in probing only the surface. Surface contamination, degradation, and segregation effects are very significant over these probing depths. Indeed, as a consequence of the absence of the crystal potential at the surface, the equilibrium electronic structure and composition of the surface is expected to differ from that of the bulk. Thus, although conventional XPS is very commonly applied to materials, the information obtained about the bulk is at best an inference and at worst meaningless. There is now a large number of advanced functional materials (such as the MAPI perovskite photovoltaic) where the surface is sufficiently unstable as to render conventional XPS useless, and an increasing number of 2D layered materials (such as van der Waals heterostructures) where it is necessary to determine the layer-by-layer composition in order to tune the electronic properties. There are numerous technological applications (for example to device heterostructures and passivating coatings) where there is a need to study interfaces buried some tens of nm below the surface. For some decades, there has thus been a recognition of the desirability of extending the probing depth of XPS.[105]

At electron kinetic energies larger than around 100 eV, the electron mean free pathlength, λ, scales roughly with $E^{1/2}$ (where E is the photoelectron kinetic energy),[106] so one obvious way to achieve this is to use a higher X-ray energy source to increase the energy of the photoelectrons leaving the sample. Hard X-ray photoelectron spectroscopy (HAXPES), which typically uses X-rays with energies *ca.* 3–10 keV, has developed from the first experiments in 2001–2002 to be an important area of research, as illustrated by the rapid increase in HAXPES publications and citations to them.[105] One significant drawback that has slowed development is the rapid drop in photoelectron cross sections as the photon energy is increased (Fig. 7).[107,108] Until recently, any form of lab-based HAXPES has utilised conventional low-intensity, unfocussed X-ray sources (typically of the twin-anode type using 5.41 keV Cr Kα or 2.9 keV Ag Lα sources), making many HAXPES experiments inviable in reality as unrealistic counting times are needed. The need for a brilliant photon source, however, led to rapid development of the technique at synchrotron facilities in the last 20 years (such as the GALAXIES beamline at SOLEIL, I09 at Diamond, 9.3.1 at ALS and no fewer than 12 beamlines at SPring-8), which in general became rapidly oversubscribed. In the last few years, there has been a significant effort by manufacturers to alleviate this situation by providing higher flux HAXPES sources with high transmission analysers in order to make high-throughput lab-based HAXPES feasible. The capability to perform such spectroscopy in the laboratory environment with sufficient flux and sensitivity to be able to make rapid measurements is a step change in photoemission spectroscopy. The widespread deployment of these instruments could see HAXPES become as ubiquitously used in the next 20 years as XPS is now.

The new generation of high flux lab-based HAXPES instruments is primarily based on Cr Kα and Ga Kα sources. A high flux, monochromated and focussed Cr

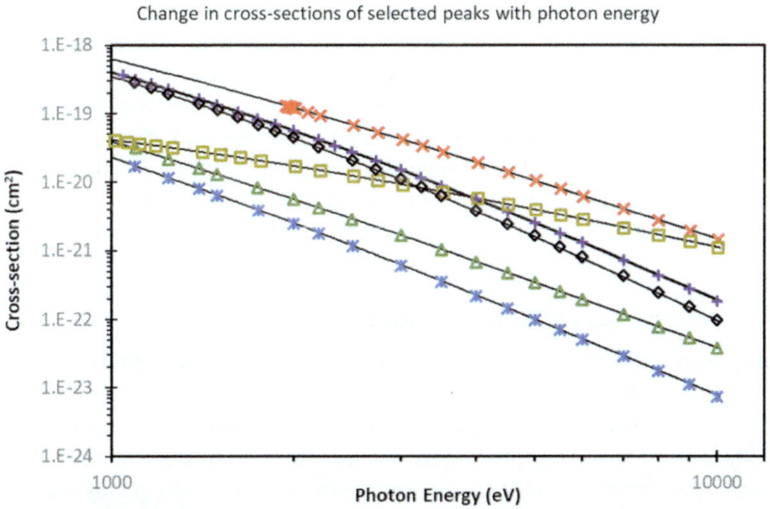

Fig. 7 Cross section values from Sabbatucci and Salvat (ref. 107) for some selected subshells, along with the associated model (lines), depicted on a log–log plot. Values from the database are shown as points, with descriptive curves from ref. 108 shown as lines. Markers correspond to subshells as follows: (△ – C 1s); (× – Si 1s); (∗ – Si $2p_{3/2}$); (+ – Ag $3d_{5/2}$); (□ – Au 4s); (◇ – Au $4f_{7/2}$). Reproduced from ref. 108.

Kα source was first combined with a high transmission electron energy analyser around 2010.[109,110] This has now been implemented in a number of commercially available sources, including the Physical Electronics Quantes (ULVAC-PHI) system.[111] More recently, Scienta Omicron GmbH launched the HAXPES-lab instrument.[112] This system uses a 9.25 keV metal jet X-ray source (Excillum)[113,114] coupled with a bespoke monochromator and a high-transmission EW-4000 electron energy analyser. The focussed X-ray source (with 50 μm spot size) has an intensity three orders of magnitude greater than that of traditional twin-anode X-ray sources, sufficient to obviate the core-level sensitivity issue. The result is that HAXPES measurements can be made on similar timescales to conventional Al Kα XPS (https://doi.org/10.1039/D2FD00021K). Lab-based HAXPES measurements of light elements such as lithium, boron, carbon and nitrogen are made more feasible (https://doi.org/10.1039/D2FD00021K).

Applications of HAXPES are diverse, ranging from the identification of buried oxygen in Cu catalysts for CO_2 reduction[115] to characterising the likely outcomes of nuclear accidents.[116] In particular, HAXPES has proved strikingly powerful in locating and characterising layers and buried interfaces,[117] for example in devices (such as prototype metal-oxide-semiconductor field effect transistor devices, https://doi.org/10.1039/D1FD00110H[118] and spintronic materials[119]) and interphases, for example the solid electrolyte interphase in battery materials.[120–122] In both cases, there are significant advantages in making measurements in a more realistic environment in a working device or cell, moving towards *in operando* conditions. Although much of this work has been carried out with synchrotron sources, the new generation of lab-based sources make this work viable in the laboratory. As an example, in this meeting, lab-based HAXPES of a prototypical oxide memory device taken during resistive switching is presented. The chemical modifications under bias and the potential distribution at the interface between a Ti electrode and a film of $La_{0.8}Sr_{0.2}MnO_3$ is investigated (https://doi.org/10.1039/D1FD00110H). In the case of battery materials, the difficulties in approaching *operando* conditions are perhaps more significant, and much of the initial work was carried out with *ex situ* electrochemical cycling, with the material subsequently recovered from the cells for examination.[120,121] More recently, however, it has become possible to obtain HAXPES from prototype cells during charging and discharging *in situ* inside the UHV chamber. An example, from BL28XU at SPring-8, is the observation of reversible changes in the Co 2p and O 1s spectra during *in situ* cycling of a $LiCoO_2$ thin film cell.[122]

3.2. Depth-profiling using the inelastic background

Probing beyond the elastic limit by modelling the inelastic background associated with each core level excitation is proving very productive in HAXPES experiments.[8–13,123–125] This is because, compared with XPS, rather few Auger lines are typically present in the regions of interest, so often the inelastic background can be measured over a very wide kinetic energy range. This is especially true for the deep core levels of heavier atoms, such as in the example shown in Fig. 8,[126] where the Ir 3d core level at *ca.* 2000 eV binding energy is used. This shows 9.25 keV HAXPES from a thin (18 nm-thick) layer of metal–organic complex ($Ir(ppy)_2(acac)$, $C_{27}H_{23}IrN_2O_2$) buried at various depths of up to 200 nm in organic material (Irganox1010 ($C_{73}H_{108}O_{12}$)).[126] The core level peaks from Ir are strongly attenuated

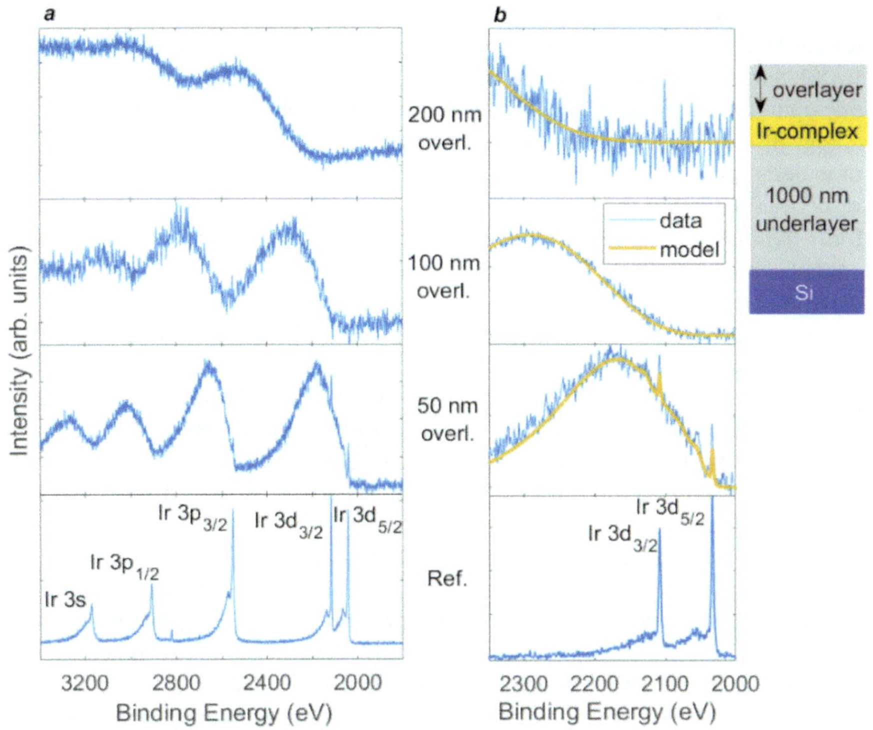

Fig. 8 9.25 keV HAXPES for (a) Ir 3s/3p/3d region, and (b) Ir 3d with inelastic background modelling for (from bottom upwards, as labelled) an $Ir(ppy)_2(acac)$ layer without an overlayer, and with overlayers of 50, 100, and 200 nm Irganox1010. The layers were deposited on a 1000 nm Irganox1010 underlayer on top of a silicon substrate. Yellow lines show inelastic background modelling using the QUASES-Generate software package (ref. 127) used to obtain the overlayer thickness. The ordering of the layers is illustrated in the inset diagram (not to scale). Reproduced from ref. 126. Copyright © 2021 Author(s).

by the 50 nm overlayer, and are completely lost for the 100 and 200 nm layers. However, the inelastic background shows a striking depth-dependent oscillation to lower kinetic energy (higher binding energy) from each core level peak. These features shift as the overlayer thickness is increased, and it can be seen that the onset of the inelastic background from each Ir core level shifts to lower kinetic energy as the Ir photoelectrons travel through more overlayer material, and therefore lose more energy to inelastic scattering, before escaping the surface to detection. Modelling of this background, done here using the QUASES-Generate software package,[127] generates overlayer thicknesses within an error of ±5% for the 50 and 100 nm overlayers, and ±10% for the thicker 200 nm overlayer.[126] Use of the inelastic background has been demonstrated to enable information to be extracted from an *inelastic* sampling depth of up to 20λ (potentially corresponding to more than hundreds of nm).[126]

3.3. Future directions for HAXPES

3.3.1. HAXPES metrology. Compared with XPS, HAXPES is still in its infancy. If lab-based systems are to become as ubiquitously used as XPS, then there is still a significant amount of metrological work to do before the technique can be used for routine quantification with confidence. Even the energy scale calibration of HAXPES instruments still requires attention as, to date, energy scale calibrations rely upon lower energy XPS reference points which require validation for higher energy instruments. In order to allow quantitative measurements of the amount of material present, two key parameters must be quantified: the intensity/energy response function of the spectrometer, and the relative electron intensity emitted from different materials under a given instrumental setup (geometry, photon energy, *etc.*). In general, there is work to do on both aspects. The relative sensitivity factors (RSFs) for the core levels of each element are not necessarily well-established for new lab-based HAXPES sources, although a number of sets of calculated photoionisation cross sections are available.[107,128–130] Initial work to produce RSFs has been carried out for both Ag Lα[131] and Ga Kα,[126] and a method of calculating reasonably accurate theoretical RSFs applicable to all instrument geometries and photon energies in the range 1.5–10 keV has been proposed.[108] The geometry chosen for HAXPES experiments also needs special consideration, as it is necessary to consider both the effect of X-ray polarisation, and the applicability of the dipole approximation within HAXPES. The divergence from a dipole emission distribution becomes more significant as the X-ray energy increases,[129] and is significant in the tender X-ray range. Thus, for unpolarised light, it is not possible to straightforwardly choose a simple 'magic angle' geometry as in XPS where photoelectrons can be collected with no angular distribution effects.[132] However, it is possible to eliminate both dipole and non-dipole terms using a part-polarised beam and an angle of 90° between the vectors connecting the X-ray source and the sample, and the sample and the analyser[132] (fortuitously close to the geometry of some commercial instruments).

A further metrological aspect currently receiving attention is the extent to which species identification *via* the Auger parameter can be extended into the HAXPES regime. As was first recognised for XPS by Wagner, the sum of the binding energies of a strong core level photoelectron line and the kinetic energy of a sharp core–core Auger transition from the same element is a characteristic quantity which can be used to diagnose the local chemical state of the atom.[133] The advent of commercial HAXPES instruments which combine traditional low energy XPS sources (such as Al Kα) with a higher energy source (such as Cr Kα or Ga Kα) means that it is now possible to determine the Auger parameter for some very common elements that have high energy KLL transitions (such as Si, Al and Ti) without the use of a synchrotron.[134] If the photoelectron and Auger lines are chosen carefully, it is possible to choose pairs with similar probing depth, adding to the accuracy of the analysis.[134]

3.3.2. HAXPES instrumentation. As discussed in Section 2.7, an important area of current focus is the development of HAXPES combined with NAP-XPS or AP-XPS. We saw that the use of higher kinetic energies reduces the scattering of photoelectrons in the gas phase allowing higher pressures to be used, so helps in the drive to close the 'pressure gap'. We can also expect to see much further development of *operando* studies, for example the *in situ* cycling of battery materials under realistic working conditions and timeframes.

In general, the other rapidly developing areas in HAXPES are strikingly similar to those commented on above for NAP-XPS, and have been recently reviewed.[105] There is a strong push to combine HAXPES with structural techniques and/or lateral and time resolution. Time-resolved experiments are discussed in detail in Section 4, but it is worth noting here that sub-100 ps HAXPES has been achieved at both synchrotron[135] and free-electron laser[136] sources. As we shall see below, developments in time-resolved spectroscopy have occurred simultaneously with advances in full-field imaging momentum microscopes operating using time-of-flight.[137–139] Although developed primarily for its valence band mapping capability, the momentum microscope can be used directly for its diffraction capability in the hard X-ray regime,[140] leading to the rapid development of hard X-ray photoelectron diffraction combined with HAXPES.[105,141] The use of harder X-rays in HAXPES also allows for the possibility of extracting structural information using the X-ray standing wave that is set up above and below the surface as the incoming and outgoing beam interference, discussed below.

3.4. Probing buried interfaces using X-ray standing waves (XSW)

The X-ray standing wave (XSW) or SW-XPS (standing wave X-ray photoelectron spectroscopy) technique uses the interference between the incident and a reflected X-ray beam to add depth resolution to XPS, allowing buried interfaces (for example in superlattices) to be probed non-destructively.[143] An XSW is formed in the region of spatial overlap between the waves, which results in a periodic modulation of the X-ray intensity perpendicular to the reflecting planes (normally vertically through the sample). Inelastic scattering from the atoms within the XSW acts as the probe, and is measured by X-ray fluorescence (XRF) or photoelectron yield (typically for light atoms). Varying the period of the XSW then allows successive atoms to be brought into regions of high and low XSW electric field amplitude, causing periodic modulation in the XRF or photoelectron yield. The experiment has been implemented in two main modes; in the standard rocking-curve method, the X-rays are incident at the Bragg angle of the surface plane, giving a strong reflected wave at the Bragg condition for the superlattice:

$$\lambda_{\text{X-ray}} = 2d_{\text{SL}} \sin \theta_{\text{B}} \tag{5}$$

where $\lambda_{\text{X-ray}}$ is the X-ray wavelength, θ_{B} is the Bragg angle of the reflecting plane and d_{SL} is the superlattice period. At this condition, the period of the XSW matches d_{SL} of the sample, and the phase of the XSW may then be tuned by rotating the incidence angle around θ_{B} (moving through the rocking curve, Fig. 9(a)) or by tuning the beam energy.[144] By moving the standing wave in space in this way, information is obtained about the location of the emitters along the direction perpendicular to the Bragg planes. The structural accuracy of the technique is striking – estimated to be around 1–10% of d_{SL},[144,145] allowing deeply buried interfaces to be probed with few-Ångstrom precision.[144] In an alternative implementation, the sample is held at glancing incidence angles approaching zero (where close to total external reflection of the X-rays may be achieved, sometimes called near-total reflection).[146]

XSW measurements have been typically implemented at synchrotron sources, where the energy tunability can be used to match the XSW to the superlattice/multilayer period. The rocking curve method allows the signal from the buried interface to dominate over the surface signal when the surface is in a region of low electric field amplitude of the XSW. Nevertheless, the photoelectron IMFP still poses a limitation which may be alleviated to some extent by implementing the

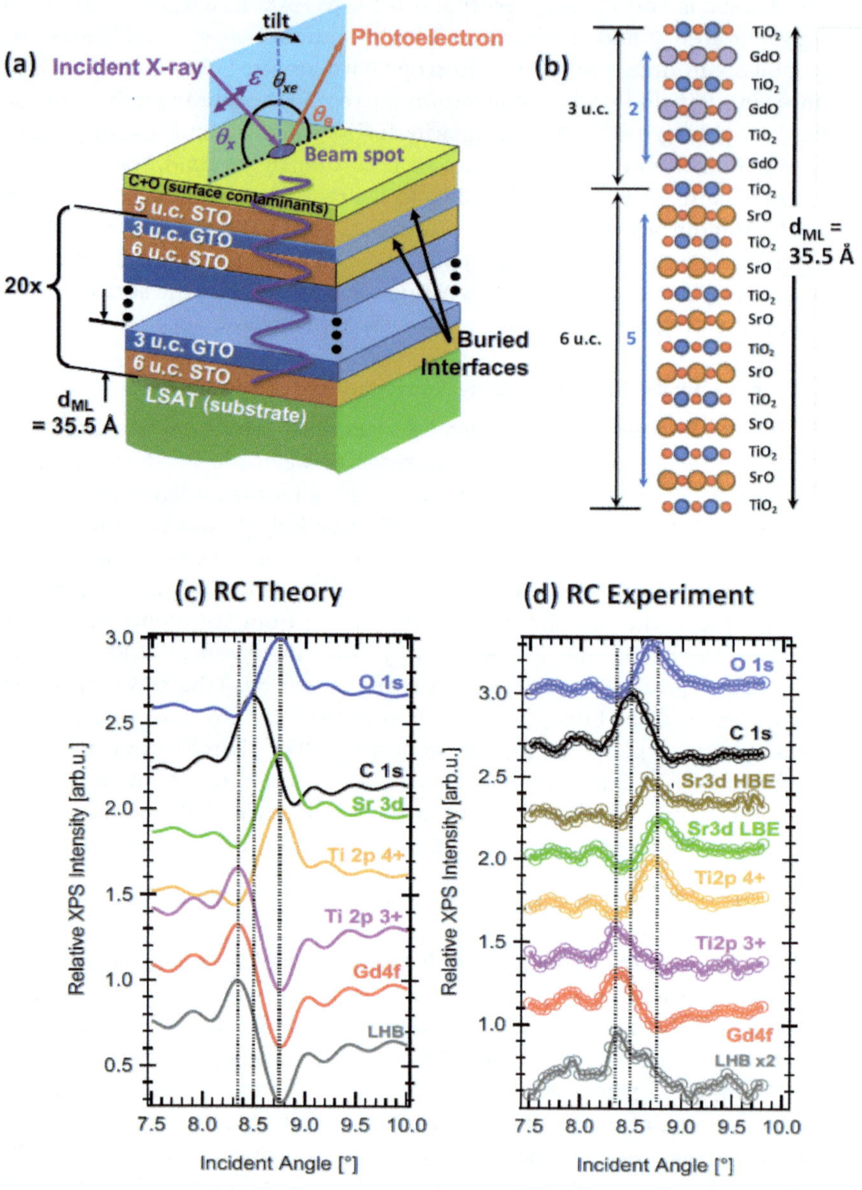

Fig. 9 (a) The experimental geometry of the XSW rocking curve method applied to a heterostructure of $SrTiO_3$ and $GdTiO_3$ layers, with various key parameters defined. The angle of incidence, θ_x is tuned in the plane of reflection through the Bragg angle, changing the period of the XSW shown in purple. θ_{xe} is fixed. Tilt in the perpendicular plane can be used to obtain angle-resolved photoemission data. (b) A layer-by-layer drawing of a single $GdTiO_3/SrTiO_3$ bilayer in the sample, with 3 unit cells (u.c.) of $GdTiO_3$ (GTO) and 6 u.c. of $SrTiO_3$ (STO). (c) Calculated RCs for various core levels and the lower Hubbard band. (d) Experimental RCs, taken with excitation at 1181 eV, chosen just below the Gd M_5 resonance, for the same core levels and the lower Hubbard band. The low binding energy and high binding energy Sr 3d signals are shown separately. Experimental data are from the ALS. Adapted with permission from ref. 142. Copyright 2016 the American Physical Society.

approach combined with HAXPES at soft-X-ray beamlines, allowing deeply buried layers to be probed. In the last ten years, this approach has been very productively implemented at a number of synchrotron beamlines, including the Galaxies beamline at SOLEIL (and at lower energies, Cassiopee),[147] beamline BL15XU at SPring-8,[142,148] BLP09 at PETRA III[142,148] and beamline 9.3.1 at the ALS.[142,146,148] In parallel with these developments, the theoretical background that allows the rocking curves to be simulated in order to fit experiment has been developed and made available,[143,149] and measurements have often been combined with resonant photoemission or complemented by real-space imaging, such as by high angular dark-field scanning tunnelling electron microscopy.[144] The result is that the chemical structure, oxidation state, interdiffusion or roughness at interfaces,[144] and even the built-in potential,[147] can now be measured with few-ångstrom resolution. This has been particularly powerful in studies of various types of oxide superlattices, for example in characterising the polarisation-induced charge density accumulation at the interface between a ferroelectric and a Mott insulator,[146] in characterising multilayer magnetic tunnel junctions,[144] and (by isolating the binding energy shifts due to individual multilayers) in determining the form of the built-in potential in a $LaCrO_3/SrTiO_3$ superlattice.[147] Fig. 9 shows an example where standing wave techniques in the hard and soft and X-ray regime are used with resonant photoemission and electronic structure calculation to characterise the two-dimensional electron gas at the interface between two oxides, $SrTiO_3$ and $GdTiO_3$.[142]

The structural sensitivity and accuracy of the XSW technique has also been demonstrated using a fs pulsed source of photons from a FEL, FLASH at DESY in Hamburg.[145] Photoelectron yield was measured from Si/Mo multilayers terminated with surface SiO_2 layers of various thicknesses, allowing determination of the oxide layer thickness to an accuracy of a few percent.[145] This opens up the prospect of combining the fs temporal resolution of a FEL with a high level of structural accuracy, in order to better understand the structural dynamics of surfaces. Because XSW does not require lateral long-range order (unlike, for example, low energy electron diffraction), the technique could provide a route to measuring the structural evolution occurring during chemical reactions at surfaces, significantly enhancing our understanding of heterogeneous catalysis. Recent developments in time-resolved photoemission at FEL and other light sources are reviewed in the next section.

4. Time-resolved surface analysis

4.1. Introduction

The study of the 'real-time' dynamics of charge carriers at material interfaces is becoming ever more important as next generation technologies increasingly access nanoscale dimensions. An example is solar harvesting technologies, where a clear understanding of the dynamics of photoexcited carriers is of key fundamental importance. Similarly, in catalysis, there is a need to underpin our understanding of reaction pathways by acquiring data on the timescales of bond making and bond breaking, so that we may improve our understanding of the transition state and the way it is modified by the catalyst. The last 20 years in particular has been characterised by a drive to complement our often detailed understanding of static structure and chemical composition by an improved understanding of dynamics of various types. Surface scientists have been in the vanguard of this campaign, driven by the fundamental importance of the surface to problems in passivation, catalysis and device design.

In fact, the primary processes of interest to physical scientists occur over a wide range of timescales (Table 1), and so require a range of different experimental arrangements for their measurement. One common feature of many such experiments is that they operate on the 'pump–probe' principle, where a pulse of (typically) electromagnetic radiation induces the process of interest, and a second pulse, with a known but variable time delay from the first, probes the dynamics of the system. Clearly, the shortest dynamics that can be measured are then limited by the pulse length of the pump pulse, so typical synchrotron pulses (of some tens of ps or longer) may only be used to probe the longer timescales listed in Table 1, such as exciton lifetimes. Studies of molecular vibrations and bond breaking (in the 10–100 fs range) have been addressed primarily using high power pulsed table-top lasers or FELs, which can attain pulse lengths of a few fs routinely. Time-resolved photoemission has emerged as an important technique in all these ranges, using lab-based high-harmonic generation (HHG) sources,[137] at FELs such as FLASH,[138] the linac coherent light source (LCLS) at Stanford[150] SACLA at SPring-8 [151]and FERMI at Elettra,[152] and at synchrotrons including SOLEIL (https://doi.org/10.1039/D1FD00107H), ALS[153] and SPring-8.[135]

4.2. Attosecond and few-femtosecond dynamics probed using the 'core-hole clock' method

At the shortest ultrafast times, attosecond (as) pulse lengths can now be achieved using lab-based tabletop lasers, but this is still far from routine, and studies have so far focussed on gas-phase studies where Auger decay and valence electron motion have been observed in real time (for reviews, see *e.g.* ref. 154–157) and

Table 1 Approximate timescales for some processes of interest to physical scientists

Timescale	Process	Providing access to
as	Fundamental electron motion	Wave packet dynamics
10s as	Photoemission from valence and low-lying core levels	Correlation dynamics and relative photoemission delays
100s as to few fs	Core hole lifetime	Core-hole clock spectroscopy
Few fs	Nuclear motion	Nuclear dynamics
<10 fs	Some charge transfers	Charge transfer into some semiconductors
10–100s fs	Vibrations, bond-breaking	Reaction dynamics
100s fs	Photoexcited quasiparticle relaxation	Electron dynamics
100s fs	Spin and lattice changes	Spin dynamics
1 ps	Electron-phonon thermalisation	Charge-density-wave melting
Few ps	Cooper-pair formation in superconductors	Superconducting dynamics
>50 ps	Spin-lattice equilibration	Magnetisation dynamics
10s ps to ms	Exciton lifetimes in semiconductors	Charge dynamics
10s ns	Domain wall motion	Magnetisation switching
100s ns to s	Formation of reaction products	Reaction kinetics

wavepacket dynamics are probed on timescales where the nuclei are not effectively frozen, as assumed by the Franck–Condon principle. In the solid state, the space charge effect in ultrafast photoelectron generation has proved a huge challenge, and, as we shall see in Sections 4.3 and 4.5, significant progress enabling solid-state attosecond (or even fs) photoemission has only been made very recently. By contrast, research using all-optical 'photon-in, photon-out' techniques (such as fs and attosecond XAS) has progressed more rapidly.[155,156] To date, an alternative photoemission-based approach, the 'core-hole clock' method, has been used more widely to probe ultrafast dynamics in solid state systems, particularly for molecules adsorbed at surfaces.

The core-hole clock method uses the core hole created by photoelectron emission as an internal clock to detect changes in electronic configuration that occur before it decays. This is typically implemented at soft X-ray (often called 'tender X-ray') synchrotron beamlines, using photon energies in the range of 2–10 keV. The core holes created by this radiation typically have a lifetime on the order of a few fs, and sometimes even hundreds of attoseconds, allowing (in particular) the detection of some ultrafast charge transfer processes that occur faster than this (or on similar timescales).[158,159] The technique, as applied to surface adsorbates, is typically implemented by a careful comparison of X-ray absorption (XAS) and resonant photoemission (or resonant photoelectron spectroscopy, RPES) spectra, aided by the high resolution now available at soft X-ray synchrotron beamlines. RPES monitors decay events of the core excited state that involve the photoexcited electron (Fig. 10). If a transfer of the photoexcited electron (say out of a surface-adsorbed molecule) takes place on a timescale that competes with the normal decay of the core hole, this modifies the RPES intensities (Fig. 10). The RPES intensities can then be compared with XAS intensities to estimate the electron transfer time.[158]

A seminal early demonstration of this technique was its application to dye-sensitised solar cells, where measurements at the nitrogen K-edge were used to show that the transfer of a photoexcited electron from a surface-adsorbed bi-isonicotinic acid molecule (a model for the ligands of the dye[160]) into a TiO_2 substrate occurs on an estimated timescale of <3 fs.[161] At the time of this experiment, the available time-resolved laser techniques had shown only that the upper time limit was of the order of 20–100 fs.[162] This was the first study to demonstrate that the electron transfer from an aromatic adsorbate to a semiconductor can occur in only a few fs, and so competes effectively with loss processes, such as the intramolecular thermalisation of excited states.[161] In 2008, this approach was extended to the full so-called 'N3' dye molecule (*cis*-bis(isothiocyanato)bis(2,2′-bipyridyl-4,4′-dicarboxylato)-ruthenium(II)), deposited intact onto TiO_2 by electrospray deposition.[163] This showed that electron injection from the N3 dye into the substrate takes place in <16 fs.[163] By focussing on the creation of short-lived holes, where the initial and final states are in the same electronic shell, the resolution has been improved into the attosecond regime. Measurements on the $c(4 \times 2)$–S/Ru(0001) surface (using the S LLM decay) showed that electron transfer from an adsorbed S atom to the Ru surface takes place in around 320 as.[164] The approach has now been implemented at tender X-ray HAXPES/XAS beamlines (for example at the GALAXIES beamline at SOLEIL), allowing the comparison between resonant Auger emission and resonant inelastic X-ray scattering (RIXS).[159] Tender X-ray beamlines provide access to a wider range of K-edges of important elements,

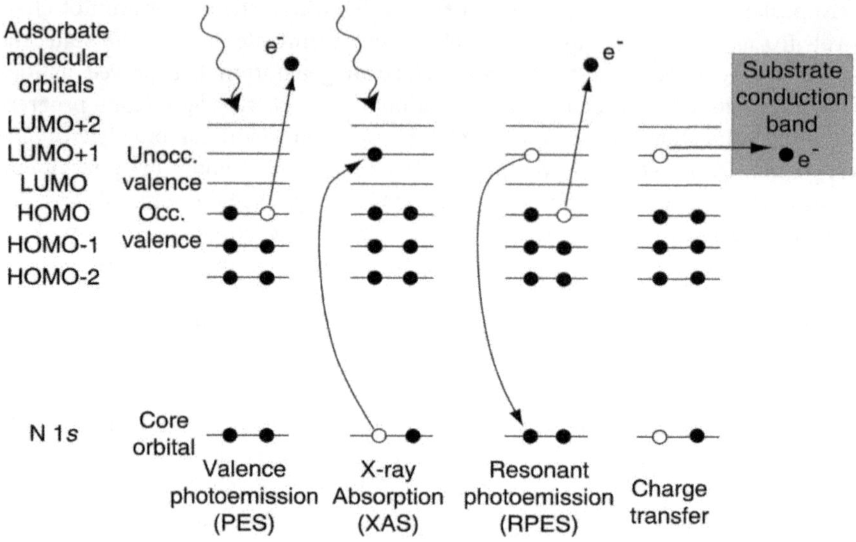

Fig. 10 Left to right: photoelectron spectroscopy (PES), X-ray absorption spectroscopy (XAS) and resonant photoelectron spectroscopy (RPES) applied to a nitrogen-containing molecule. The electronic states are indicated by lines. PES measures the occupied and XAS the local unoccupied density-of-states. RPES monitors the decay of the excited state. Far right: If the molecular orbital couples to the substrate density-of-states, an electron transfer may be possible that competes with RPES. HOMO, highest occupied molecular orbital; LUMO, lowest unoccupied molecular orbital. Here, an electron transfer is energetically possible from LUMO+1 and LUMO+2, but not from the LUMO. If the depicted electron charge transfer occurs, the RPES is reduced in spectral intensity. Copyright Joachim Schnadt, 2022, private communication, reproduced with permission.

including those of S and P. Recently studies at the S K-edge have been used to probe charge transfer in organic heterojunctions,[165] in MoS_2)[166] and SnS_2, using polarised light to probe the effects of in-plane and out-of-plane excitation.[167] Charge transfer times as short as tens of attoseconds have been determined.[166]

4.3. Towards reaction dynamics and molecular movies – femtosecond pump–probe measurements

On slightly slower, fs timescales, the concept of making a 'molecular movie' showing atoms and molecules moving and reacting in real time has been a long-held goal, as this would revolutionise our understanding of reaction dynamics. The ability to probe species on surfaces at these timescales took a major leap forward with the development in the last 15 years of FELs that can deliver ultra-short (few fs) soft X-ray pulses, combined with atom-specific spectroscopic tools that can monitor the evolution of the electronic structure.[139,151,168,169] The implementation of these measurements in solid state photoemission has not been without its difficulties (associated with space charge issues, discussed further below), and in general, until very recently most progress had been made with 'photon-in-photon-out' approaches that do not change the electron count of the sample.

In fact, the first direct time-resolved measurement of the transition state (TS) region during a surface reaction made using an X-ray FEL (XFEL) employed X-ray emission spectroscopy (XES) and XAS applied to the oxidation of CO on a Ru(0001) substrate.[170] Here, an optical laser pump pulse was used to excite the substrate electrons, with energy transfer leading to a rapid increase in adsorbate-substrate vibrational excitation (in several hundred fs), sufficient to drive the CO oxidation. The LCLS was used to provide synchronised X-ray pulses probing the O K edge. Within about 800 fs, new electronic states were detected due to the TS region en route to the formation of CO_2. Remarkably, about 10% of the CO was observed to populate the TS region within around 1 ps, the majority of which dissociated back to O and CO.[170] This ability to probe transient species close to the TS gives completely new insights into the electronic states of reacting molecules at surfaces and provides essential input to theory. Further experiments in the last few years have allowed the timescales of electron excitation, phonon thermalization, and the excitation of various adsorbate modes to be distinguished.[171,172] An example is recent studies using the seeded FEL facility at FERMI at Elcttra, using variably polarised soft X-ray pulses to excite the C K-edge to probe CO dynamics at the Ru surface (Fig. 11).[173] As the C K-edge is subject to less vibrational broadening than the O K-edge, this allows the evolution of the spectral features to be distinguished more clearly. This has revealed the surface diffusion of excited CO molecules to highly coordinated sites in the first ps, before a fraction enter the precursor state. In this state, the molecules are free to rotate isotropically for several ps before being desorbed or readsorbed, illustrated in Fig. 11.[173]

Although these experiments are undoubtedly challenging, the future looks exciting for studies of surfaces and solid-state systems *via* time-resolved XAS. While FELs typically deliver few-fs pulses, the delivery of attosecond pulses with good coherence and sufficiently high repetition rate has been demonstrated.[174] Meanwhile, the development of lab-based HHG sources has continued apace, and isolated soft X-ray attosecond pulses with a broad bandwidth suitable for XAS measurements have become available. This has allowed the sub-cycle optoelectronic response in a range of quantum solids to be probed.[155] A recent example is studies of the transition-metal dichalcogenide TiS_2, where attosecond Ti $L_{2,3}$

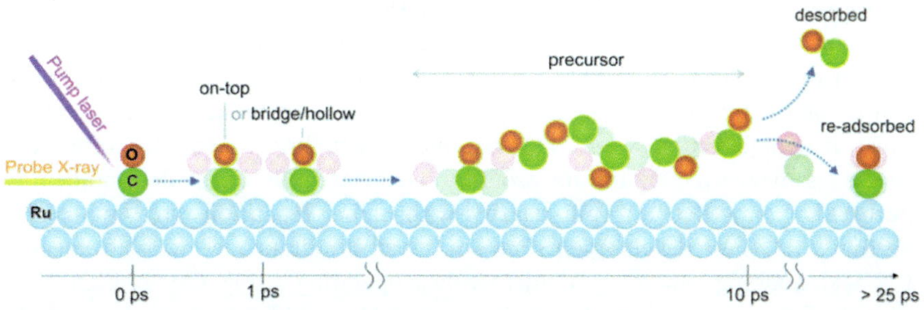

Fig. 11 Dynamic evolution as function of pump–probe delay for CO chemisorbed on Ru(0001) following optical laser excitation. After laser irradiation of Ru(0001), CO molecules are excited and diffuse from the on-top sites towards more highly coordinated sites, followed by isotropically rotating in a precursor state. A fraction of CO in the precursor state desorbs while the majority re-adsorb back onto the surface. Reproduced from ref. 173 with permission from the PCCP Owner Societies.

XANES is used to probe the 3d-state dynamics, in the presence of a near-IR optical control field (bridging the bandgap). This shows that localization of electron density around the Ti atoms occurs within a fraction of the cycle of the optical control field.[155]

Turning to photoemission, in the last few decades, fs time-resolved photoemission excited with optical lasers has become well-established (typically using as a pump an infrared (1.5 eV) output from a Ti:sapphire laser with a pulse length of a few tens of fs). For one-photon photoemission, the probe pulse energy must at least exceed the sample work function, requiring a probe pulse energy of typically ≥6 eV. This can be readily provided by cascaded frequency upconversion in non-linear crystals (as in *e.g.* ref. 175–177), although it does mean that only low binding energy states may be probed, and the low tunability of the source can mean that only states around the Brillouin zone centre can be probed in most materials. This approach has been used to probe a variety of phenomena in condensed matter physics and chemistry, such as charge-density wave melting,[176] the momentum dependence of Cooper pair dynamics in high temperature superconductors,[175,178] insulator–metal transitions in Mott insulators,[179] the non-thermal carrier dynamics in graphite,[180] and the ultrafast electron dynamics at the surface of the topological insulator Bi_2Se_3.[181]

These laser-excited photoemission studies are limited by the lack of chemical specificity that arises as the core levels cannot be accessed. There has been a strong drive in recent years to provide extreme ultraviolet (XUV) or soft X-ray pulses suitable for photoemission measurements from solids. Rapid developments have occurred along two main fronts, the development of HHG sources that provide sufficient flux in the few tens of eV energy range,[182] and FEL sources with sufficiently reproducible pulse trains.[139] However, until very recently, the great potential for time-resolved photoemission from solids using these sources has been hampered by vacuum space charge effects, which arise when too many photoelectrons are emitted from the same ultrashort high intensity photon pulse. This causes broadening and shifting of the spectral features. The effect was noted, in the case of FEL radiation, in studies of W(110) using FLASH (Hamburg) as early as 2008,[169] and, by controlling the fluence, some measurements of CDW melting in 1T-TaS_2 were possible.[168,183] However, the developmental work needed to find ways of bringing vacuum space charge effects under control[139,168,184–188] has delayed the widespread deployment of both FEL and HHG sources in time-resolved photoemission from the solid state. In order to control these effects, it is necessary to control the number of emitted photoelectrons per pulse,[139] which is obviously counterproductive to the spectral signal-to-noise. It is thus necessary to use the most efficient detection schemes in combination with high repetition-rate sources. In the last few years, significant progress has been made in the development of both FEL[139] and HHG[189,190] pulse trains with higher repetition rate. As noted in Section 3.3.2, these developments have occurred simultaneously with advances in full-field imaging momentum microscopes operating using time-of-flight.[137–139] This has occurred alongside an improved theoretical understanding of the problem,[151,187] and experimental developments such as a retarding the electrostatic front lens, which suppresses the pump-induced slow photoelectrons.[186] These developments have transformed the prospects for solid state photoemission using FEL and HHG sources, and made possible solid-state time- and angle-resolved photoemission with combined time-resolved momentum microscopy (tr-ARPES/

tr-MM, Fig. 12).[139] As full-field photoelectron diffraction is now demonstrated in static experiments, there is a realistic prospect that tr-ARPES/tr-MM can be used with fs core-level excitation to simultaneously probe how the both the electronic states and the coherent lattice motion evolve after a pump pulse, by exploiting photoelectron diffraction and X-ray standing waves.[138,145,191]

These recent developments are enabling a new generation of time-resolved photoemission experiments, giving unprecedented insight into ultrafast dynamics in the excited state, ranging from the nonequilibrium band structure of WSe_2 (ref. 139) to momentum-space imaging of transiently excited unoccupied molecular orbitals in surface-adsorbed organic molecules.[137] An example is the use of an IR pump pulse and soft X-ray probe pulse at FLASH to measure the ultrafast linewidth broadening of the C 1s core level of graphene, illustrated in Fig. 12.[139] This is used to show that the broadening is caused by an exchange of energy and momentum between the outgoing photoelectron and the hot electron gas, rather than by vibrational excitation. The momentum change of the outgoing photoelectrons leads to a small change in the time-resolved photoelectron diffraction pattern measured using a momentum microscope.

4.4. Slower dynamics; probing carrier recombination dynamics and reaction kinetics

As illustrated in Table 1, a great many dynamic processes, particularly those of interest to chemists, occur on timescales of the order of some tens of ps or longer,

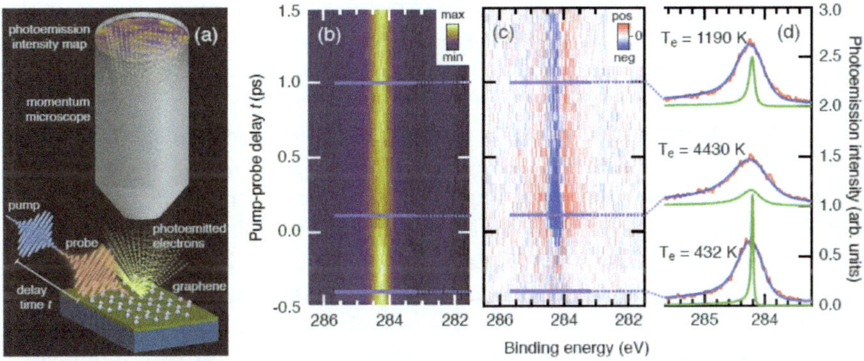

Fig. 12 Time-resolved momentum microscopy of graphene. (a) Schematic diagram of the experiment. The time-resolved photoemission intensity is measured from the graphene C 1s core level and valence band by combining an infrared pump pulse with a soft X-ray probe pulse from FLASH. The electrons are detected over a wide angular range using a momentum microscope. (b) C 1s photoemission intensity as a function of the pump–probe delay and binding energy. (c) Same as in (b) but with the average spectrum before excitation subtracted. Red and blue indicate an increase and decrease in electron counts, respectively. (d) Selected spectra (cuts at specific time delays) fitted with a model described in ref. 138 (blue line). The temperature-dependent asymmetry kernel for each fit is shown in green, showing the way in which the low temperature asymmetry due to electron–hole pair excitations around the Fermi energy is reduced as energy gains become possible at high electronic temperature. The electronic temperatures resulting from the fit are noted close to the spectra. Adapted with permission from ref. 138. Copyright 2021 by the American Physical Society.

and do not require the use of an ultrashort pulsed source for their investigation. Here, the highly reproducible, lower intensity and high repetition rate source provided by a synchrotron is highly suitable, as it largely avoids the space charge issues encountered in photoemission with ultrashort pulses from FEL and HHG sources. In the last decade or so, various types of 'hybrid' or 'low α-' mode operation (where both continuous and few-ps pulsed electron bunch trains are available in the same machine fill) have been developed at several synchrotron sources (including SOLEIL, SPring-8 and BESSY II) with the aim of providing access to dynamic measurements.[192–194] Photoemission measurements using these sources have focussed on probing carrier recombination dynamics at surfaces.

Carrier dynamics at the surfaces of semiconductors may be probed by pump–probe spectroscopy using time-resolved surface photovoltage (SPV) spectroscopy.[195–197] The space charge layer at the surface of semiconductors where the Fermi level is pinned by surface states within the band gap leads to band bending in this region. For an n-type semiconductor with a depletion layer at the surface, the bands bend upwards, and downwards in a p-type semiconductor. A laser is used to promote carriers across the band gap, and, in the depletion region of an n-type semiconductor, electrons and holes migrate away from and towards the surface, respectively (and *vice versa* in a p-type semiconductor, Fig. 13).[198] This additional electric field reduces the band bending, creating a shift in the surface photovoltage which shifts the energy of the valence band maximum in the space charge region (in n-type semiconductors to a higher BE and to a lower BE in p-type).[197] The SPV shift relaxes once all carriers have recombined and the surface potential is returned to its original value, on timescales that vary from typically a few μs to seconds or longer in the case of persistent photoconductivity.[195] In the period before this occurs, the SPV shift can be measured with photoemission as a function of time, either as a change in the position of the valence band maximum (VBM), or a change in the binding energy position of a core level peak.[197,199,200] In the last 30 years, the technique has been implemented in various ways, including with both continuous (multibunch) and pulsed (single bunch or hybrid mode) synchrotron X-ray sources as the probe beam.[193,196,197,201,202] In the latter case, the X-ray probe pulses are synchronised to the initial photoexcitation, and used to excite core or valence level photoelectrons (using XPS). By varying the time delay between the laser pump and XPS probe, the recombination dynamics are monitored over time as the carriers recombine and the band bending and binding energies return to equilibrium.[196,197,201,202] This allows the recombination dynamics at the surface to be inferred (https://doi.org/10.1039/D1FD00107H).[196,197,201,202] Such measurements have been implemented at a number of world synchrotron sources, one of the earliest being the TEMPO beamline at SOLEIL.[192,194,203] As well as measurements of recombination at semiconductor surfaces (primarily Si) (https://doi.org/10.1039/D1FD00107H),[196,197,201,202] charge injection across interfaces where there is a surface depletion layer has also been monitored, for example from gold nanoparticles into TiO_2 [153] or from light-absorbing quantum dots into ZnO.[204] In combination with time-resolved magneto-optical Kerr effect measurements, pump–probe photoemission of this type has also been used to study magnetisation dynamics, such as the antiferromagnetic–ferromagnetic

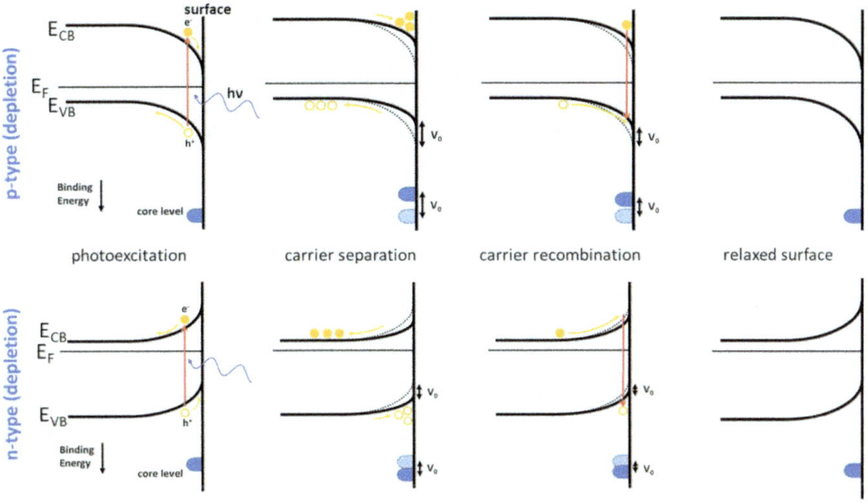

Fig. 13 Schematic band diagrams of the surface photovoltage effect in (top) a p-type semiconductor material with a hole depletion layer at the surface and (bottom) an n-type semiconductor material with an electron depletion layer at the surface. The change in the band bending is indicated by the SPV shift (V_0), and the positions of the core levels relative to their equilibrium positions are indicated. Reproduced from ref. 198. Copyright © 2021 Author(s).

transition in FeRh films, which occurs within 100 ps [205] and electron relaxation dynamics over several hundreds of ps in the half metallic $La_{1-x}Sr_xMnO_3$.[135]

Laser-pump X-ray-probe SPV experiments using pulsed lasers, where the pump laser is synchronized to individual X-ray pulses from a synchrotron, are limited by the time window available, determined by the synchrotron pulse length and repetition rate. For synchrotron excitation, this limits the time resolution to (typically) some tens of ps and the maximum time window to some hundreds of ns, often less. This means that for processes occurring on slower timescales, the observed SPV transient is influenced by the residual SPV induced by preceding laser pump pulses.[197] If separate experiments can be carried out using as a probe both the normal 'multibunch' train from the synchrotron and the isolated pulse of hybrid mode, then it becomes possible to start to deconvolute processes occurring on ps timescales from longer-lived effects. An example is work carried out at the ALOISA beamline at Elettra, where a lattice distortion occurring in 2H-$MoTe_2$ on sub-ns timescales is distinguished from a SPV shift persisting for μs (https://doi.org/10.1039/D1FD00105A). The use of angle-resolved time-of-flight (ARToF) analysers with two-dimensional delay-line detectors (such as at BESSY II[206] and at the BL07LSU beamline at SPring-8 [193]) has allowed synchronisation of the pump laser pulse with hybrid mode pulses in multiple synchrotron periods, enabling SPV measurements extending to μs timescales, for example in studies of Si(111),[207] ZnO(0001),[208] $TiO_2(110)$[209] and phthalocyanine/fullerene layers adsorbed on TiO_2.[210]

A different approach is necessary to enable SPV measurements over much longer timescales (up to ms or longer), which may be required where the electron dynamics are intrinsically very slow, for example due to persistent photoconductivity caused by carrier trapping.[197,204] This has been enabled at the TEMPO beamline at SOLEIL, where pump–probe measurements are now possible over timescales ranging from 50 ps to ms or longer. Fast (ps to ns) measurements use a pulsed fs laser in combination with hybrid mode synchrotron radiation.[192] Dynamics with slower characteristic times are conveniently monitored using a CW, rather than a pulsed laser. This is modulated using the output of a signal generator, where the period and duty cycle are easily controlled over wide ranges, and a fast XPS detection system (with a two-dimensional delay-line detector) is used to monitor XPS spectra at small time intervals over this modulation period.[197,211,212] The access that this provides to a wide range of timescales is very convenient in systems where small changes in the chemistry may result in a change of orders of magnitude in the carrier decay rates.

One such example is in colloidal quantum dot (CQD) solids, made up of an ordered arrays of quantum dots passivated with ligands. These solids take on bulk material properties, as the wave functions of adjacent CQDs overlap,[213] coupling them together.[214] This makes them fascinating subjects for the study of the transition from localised to itinerant properties. It also provides potential for controlling bulk properties, including the semiconductor doping level, carrier concentration, mobility, and lifetimes, through control of the CQD material, size, shape and surface chemistry.[214] They have a range of optoelectronic devices ranging from solar cells[215,216] to photodetectors.[217–219] Tailoring the band alignment and band bending at the junctions between CQD solids and oxides, metal cathodes or other CQD solids has become important for device design.[220,221] For example, in CQD photovoltaics, this has led to more efficient carrier separation and extraction.[222] It is thus important to develop a clear understanding of how band bending occurs in these solids. Time-resolved SPV spectroscopy has been used to show that band bending is intrinsic to CQD solids and occurs at the CQD solid–vacuum interface (Fig. 14).[198,223] It is observed in the presence and in the absence of detectable surface oxidation, and for a number of ligand types and synthesis routes. The timescale of the photoexcited SPV dynamics is found to depend dramatically on the surface chemistry of the film. If large amounts of oxygen-containing contaminants are present in the film (as in the example in Fig. 14(b)), the rates of carrier recombination decrease by 6 orders of magnitude compared to those of a film with no observable contamination (Fig. 14(c)).[198] This is consistent with the hypothesis that dynamic deep traps are formed when photoexcited carriers react with contaminants, decreasing the mobility of the photoexcited carriers.[224,225]

Finally in this section, we note that the advent of high transmission electron energy analysers (often with two- or three-dimensional delay-line detectors) means they may conveniently be used in 'snapshot' mode to record in real time the evolution of core level peaks over the comparatively long timescales of 10^3 s or longer that may be needed for kinetic analysis. Some examples are measurements of the adsorption kinetics of water on the Si(001)-2 $\times$ 1 surface,[226] of ammonia on the hydroxylated Si(001)-2 $\times$ 1 surface[227] and the role of adsorbed atomic oxygen in the insertion of Pd ions into the 2H-TPP porphyrin.[228]

4.5. Future directions in time-resolved experiments

As we have noted in previous sections, there is in general a very strong push towards multi-modal analysis, and to improve the time resolution of both NAP-

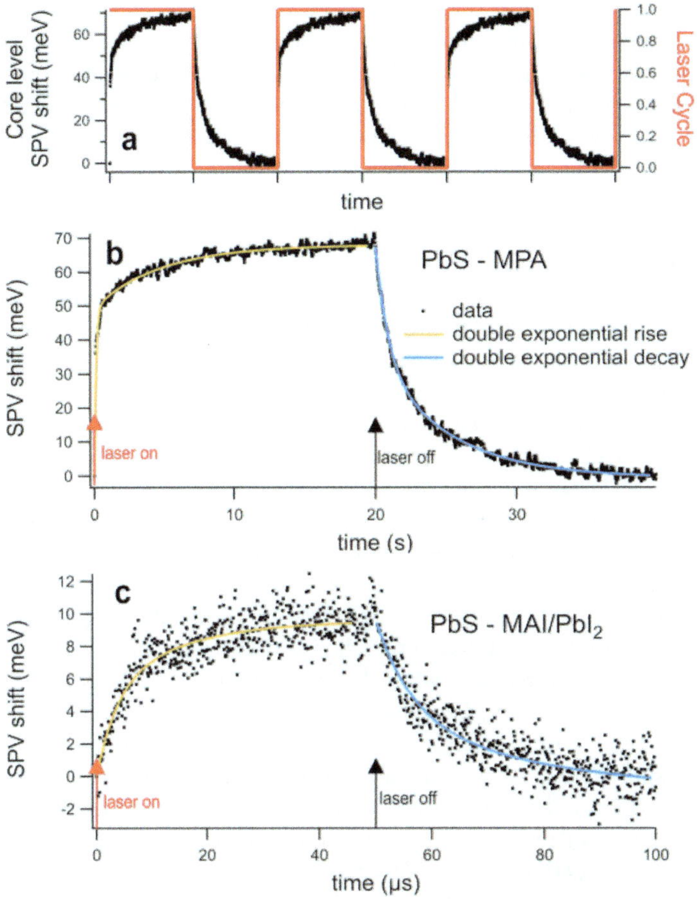

Fig. 14 (a) Time-resolved SPV shift of the Pb 4f core level in a PbS CQD film with the CW laser 'on–off' cycle superimposed; (b) a film treated with 3-MPA ligands under illumination by a $\lambda = 375$ nm CW laser modulated on a 50% duty cycle with a period of 40 s; (c) a film with a quasi-epitaxial MAI/PbI_2 ligand shell, with the CW laser modulated on a 50% duty cycle with a period of 100 μs. The X-ray photon energy was 475 eV, corresponding to a sampling depth of 1.4 nm.[223] The rise and decay of the pump-induced SPV shift are fitted using double exponential functions (yellow and blue lines, respectively). Reproduced from ref. 198. Copyright © 2021 Author(s).

XPS (discussed in Section 2.7) and HAXPES (discussed in Section 3.3.2). It is also clear that the advent of time-resolved momentum microscopy, together with higher repetition rate pulse trains, is transforming the prospects for fs time-resolved solid-state photoemission in greatly alleviating the problems associated with sample charging in the solid state. The drive for shorter and shorter pulses, allowing the fastest dynamics to be probed, will continue. However, measurements in the attosecond regime remain challenging. One fundamental issue affecting attosecond photoemission is that it has proved very hard to attain the few meV resolution that is routinely attainable in static photoemission, as the use of an attosecond pulse intrinsically implies a broad bandwidth of up to a few tens of eV.[229] Since 2000, two major experimental advances in attosecond metrology have made such photoemission possible; attosecond streaking, using isolated attosecond pulses and RABBIT (reconstruction of attosecond harmonic beating by interference of two-photon transitions), using attosecond pulse trains. A detailed description can be found in a number of reviews,[154,156,157] but both exploit the sub-cycle temporal structure of HHG radiation, and its synchronisation with the driving field provided by the intense visible/near IR laser pulses used to create it, to analyse the evolution of the photoelectron spectra with the varying time delay between the two optical fields. Both techniques have been used extensively to measure attosecond photoemission delays from gases, providing insight into electron dynamics within the atom.[154,156,157,230] By contrast, solid-state attosecond photoemission has proved much more challenging, largely because the space charge effects we discussed in the context of fs spectroscopy are significantly more problematic in the case of shorter attosecond pulses.[231] As in fs photoemission, this has accelerated the drive to produce pulse trains with high repetition rates which allow for a high average photoelectron flux while releasing only a relatively low number of photoelectrons per pulse. Recently, high flux MHz HHG sources have become available,[156,232] enabling measurements from solids. In the photoemission process, the outgoing photoelectron wavepacket forms in a few tens of attoseconds, so these studies tell us a great deal about the photoemission process itself,[233] requiring us to revise our understanding of the process. For example, studies of WSe_2 have shown that electrons with low angular momentum, l, are emitted first, while those with high l are delayed.[234] This effect, due to electron correlation, is well-established in studies of atoms,[230,235] but is neglected in models of solid-state photoemission, which historically have treated photoemission as an instantaneous single-particle phenomenon.[230,234] More recently, a high flux 18.4 MHz HHG source has been used to enable a study of attosecond intra-valence band dynamics in W(110), showing that sp-band photoemission is delayed by almost 40 as compared with emission from the d-band.[231] It seems clear that in the immediate future we are set to learn more than we ever thought possible about the photoemission process itself.

5. Outlook and conclusions

With the obvious pun fully intended, the future is bright for surface analysis by photoelectron spectroscopy. In the last decade, we have seen the development of a number of truly disruptive technologies. These have included the development of lab-based, high intensity tender X-ray sources, and high repetition rate HHG sources. The deployment of three-dimensional delay-line detectors in full-field imaging momentum microscopes has transformed the prospects for ultrafast time- and angle-resolved photoemission from solids. In some cases, an advance in one area has aided another – for example, the rapid uptake of HAXPES has made it easier to achieve ambient pressure XPS measurements, and the geometries of these experiments lend themselves to multimodal approaches with the incorporation of other spectroscopies. The clever design of sample environments has meant that we are just beginning to be able to make genuinely *operando* measurements in real time on working systems. There is a strong drive to harness these capabilities together and to provide all-encompassing measurements with lateral resolution and time resolution (and also spin resolution, a topic which I have not discussed, but where there is also huge progress[5,236]). Advances in our theoretical understanding have supported these developments – for example in developing our understanding of the vacuum space charge effect in time-resolved photoemission and of the inelastic background in HAXPES. In turn, experimental advances, for example the observation of photoemission delay in the solid state, are a salutary demonstration to us that our understanding of photoelectron emission from surfaces is not complete. It is not an over-statement to say that we will learn more in the next decade than we currently know about how real systems behave in real time. Certainly, it seems clear that today's postgraduate students inherit an exciting and burgeoning field (which will be worth the investment needed to learn an exponentially rising collection of acronyms!).

Data availability

This review brought together existing research data obtained upon request and subject to licence restrictions from a number of different sources. No data were created or analysed in this review.

Conflicts of interest

There are no conflicts to declare.

Acknowledgements

I am grateful to all my co-workers for the studies reported in this paper that come from our own work. I also thank a number of colleagues, including Joachim Schnadt, Slavomir Nemšák, Davide Curcio, Ben Spencer, Pip Clark, Andrew Thomas and Chun-Ren Ke, for providing or editing original figures. Aspects of the research described in this paper have been supported by the EPSRC, including the Henry Royce Institute for Advanced Materials (funded through the EPSRC grants EP/R00661X/1, EP/P025021/1 and EP/P025498/1) and EP/K009710. I acknowledge SOLEIL for providing access to synchrotron radiation facilities. For the purpose of open access, the author has applied a Creative Commons Attribution (CC BY) licence (where permitted by UKRI, 'Open Government Licence' or 'Creative Commons Attribution No-derivatives (CC BY-ND) licence may be stated instead) to any Author Accepted Manuscript version arising.

References

1 A. Einstein, *Ann. Phys.*, 1905, **322**, 132–148.
2 C. N. Berglund and W. E. Spicer, *Phys. Rev.*, 1964, **136**, A1030–A1044.
3 S. Hagström, C. Nordling and K. Siegbahn, *Z. Phys.*, 1964, **178**, 439–444.
4 H. R. Robinson and C. L. Young, *London, Edinburgh Dublin Philos. Mag. J. Sci.*, 1930, **10**, 71–75.
5 P. D. C. King, S. Picozzi, R. G. Egdell and G. Panaccione, *Chem. Rev.*, 2021, **121**, 2816–2856.
6 W. E. Spicer, *CRC Crit. Rev. Solid State Sci.*, 1976, **6**, 317–336.
7 S. Hüfner, *Photoelectron Spectroscopy: Principles and Applications, Springer Series in Solid State Sciences*, Springer, Berlin, 3rd edn, 2003, vol. 82.

8 Y.-T. Cui, S. Tougaard, H. Oji, J.-Y. Son, Y. Sakamoto, T. Matsumoto, A. Yang, O. Sakata, H. Song and I. Hirosawa, *J. Appl. Phys.*, 2017, **121**, 225307.

9 C. Zborowski, O. Renault, A. Torres, Y. Yamashita, G. Grenet and S. Tougaard, *Appl. Surf. Sci.*, 2018, **432**, 60–70.

10 P. Risterucci, O. Renault, C. Zborowski, D. Bertrand, A. Torres, J.-P. Rueff, D. Ceolin, G. Grenet and S. Tougaard, *Appl. Surf. Sci.*, 2017, **402**, 78–85.

11 C. Zborowski, O. Renault, A. Torres, C. Guedj, Y. Yamashita, S. Ueda, G. Grenet and S. Tougaard, *J. Appl. Phys.*, 2018, **124**, 085115.

12 C. Zborowski and S. Tougaard, *Surf. Interface Anal.*, 2019, **51**, 857–873.

13 S. Tougaard, *J. Surf. Anal.*, 2017, **24**, 107–114.

14 S. Tougaard, *Surf. Interface Anal.*, 1998, **26**, 249–269.

15 P.-M. Deleuze, H. Magnan, A. Barbier, M. Silly, B. Domenichini and C. Dupont, *J. Phys. Chem. Lett.*, 2021, **12**, 11520–11527.

16 J. J. Velasco-Velez, V. Pfeifer, M. Hävecker, R. S. Weatherup, R. Arrigo, C.-H. Chuang, E. Stotz, G. Weinberg, M. Salmeron, R. Schlögl and A. Knop-Gericke, *Angew. Chem., Int. Ed.*, 2015, **54**, 14554–14558.

17 A. R. Head, S. Chaudhary, G. Olivieri, F. Bournel, J. N. Andersen, F. Rochet, J.-J. Gallet and J. Schnadt, *J. Phys. Chem. C*, 2016, **120**, 243–251.

18 M. Vorokhta, I. Khalakhan, M. Vondráček, D. Tomeček, M. Vorokhta, E. Marešová, J. Nováková, J. Vlček, P. Fitl, M. Novotný, P. Hozák, J. Lančok, M. Vrňata, I. Matolínová and V. Matolín, *Surf. Sci.*, 2018, **677**, 284–290.

19 J. C.-R. Ke, A. S. Walton, D. J. Lewis, A. Tedstone, P. O'Brien, A. G. Thomas and W. R. Flavell, *Chem. Commun.*, 2017, **53**, 5231–5234.

20 D. E. Starr, Z. Liu, M. Hävecker, A. Knop-Gericke and H. Bluhm, *Chem. Soc. Rev.*, 2013, **42**, 5833–5857.

21 *Ambient Pressure Spectroscopy in Complex Chemical Environments*, ed. A. R. Head, S. Nemšák and B. Eren, American Chemical Society, 2021, vol. 1396, ACS Symposium Series.

22 R. Dupuy, C. Richter, B. Winter, G. Meijer, R. Schlögl and H. Bluhm, *J. Chem. Phys.*, 2021, **154**, 060901.

23 J. Schnadt, J. Knudsen and N. Johansson, *J. Phys.: Condens. Matter*, 2020, **32**, 413003.

24 H. Siegbahn and K. Siegbahn, *J. Electron Spectrosc. Relat. Phenom.*, 1973, **2**, 319–325.

25 H. Siegbahn, *J. Phys. Chem.*, 1985, **89**, 897–909.

26 D. F. Ogletree, H. Bluhm, G. Lebedev, C. S. Fadley, Z. Hussain and M. Salmeron, *Rev. Sci. Instrum.*, 2002, **73**, 3872–3877.

27 J. Knudsen, J. N. Andersen and J. Schnadt, *Surf. Sci.*, 2016, **646**, 160–169.

28 S. Zhu, M. Scardamaglia, J. Kundsen, R. Sankari, H. Tarawneh, R. Temperton, L. Pickworth, F. Cavalca, C. Wang, H. Tissot, J. Weissenrieder, B. Hagman, J. Gustafson, S. Kaya, F. Lindgren, I. Källquist, J. Maibach, M. Hahlin, V. Boix, T. Gallo, F. Rehman, G. D'Acunto, J. Schnadt and A. Shavorskiy, *J. Synchrotron Radiat.*, 2021, **28**, 624–636.

29 E. Kokkonen, F. Lopes da Silva, M.-H. Mikkelã, N. Johansson, S.-W. Huang, J.-M. Lee, M. Andersson, A. Bartalesi, B. N. Reinecke, K. Handrup, H. Tarawneh, R. Sankari, J. Knudsen, J. Schnadt, C. Såthe and S. Urpelainen, *J. Synchrotron Radiat.*, 2021, **28**, 588–601.

30 J. Zhou, S. Blomberg, J. Gustafson, E. Lundgren and J. Zetterberg, *Catalysts*, 2017, **7**, 29.

31 X. Garcia, L. Soler, N. J. Divins, X. Vendrell, I. Serrano, I. Lucentini, J. Prat, E. Solano, M. Tallarida, C. Escudero and J. Llorca, *Catalysts*, 2020, **10**, 286.

32 N. Johansson, M. Andersen, Y. Monya, J. N. Andersen, H. Kondoh, J. Schnadt and J. Knudsen, *J. Phys.: Condens. Matter*, 2017, **29**, 444002.

33 F. Schiller, M. Ilyn, V. Pérez-Dieste, C. Escudero, C. Huck-Iriart, N. Ruiz del Arbol, B. Hagman, L. R. Merte, F. Bertram, M. Shipilin, S. Blomberg, J. Gustafson, E. Lundgren and J. E. Ortega, *J. Am. Chem. Soc.*, 2018, **140**, 16245–16252.

34 N. Johansson, S. Sisodiya, P. Shayesteh, S. Chaudhary, J. N. Andersen, J. Knudsen, O. F. Wendt and J. Schnadt, *J. Phys.: Condens. Matter*, 2017, **29**, 444005.

35 S. Blomberg, U. Hejral, M. Shipilin, S. Albertin, H. Karlsson, C. Hulteberg, P. Lömker, C. Goodwin, D. Degerman, J. Gustafson, C. Schlueter, A. Nilsson, E. Lundgren and P. Amann, *ACS Catal.*, 2021, **11**, 9128–9135.

36 S. C. DeCaluwe, G. S. Jackson, R. Farrow, A. McDaniel, F. E. Gabaly, K. McCarty, S. Nie, M. Linne, H. Bluhm, J. Newberg, Z. Liu and Z. Hussain, *ECS Trans.*, 2009, **16**, 253.

37 C. Zhang, M. E. Grass, A. H. McDaniel, S. C. DeCaluwe, F. E. Gabaly, Z. Liu, K. F. McCarty, R. L. Farrow, M. A. Linne, Z. Hussain, G. S. Jackson, H. Bluhm and B. W. Eichhorn, *Nat. Mater.*, 2010, **9**, 944–949.

38 F. E. Gabaly, M. Grass, A. H. McDaniel, R. L. Farrow, M. A. Linne, Z. Hussain, H. Bluhm, Z. Liu and K. F. McCarty, *Phys. Chem. Chem. Phys.*, 2010, **12**, 12138–12145.

39 C. Zhang, M. E. Grass, Y. Yu, K. J. Gaskell, S. C. DeCaluwe, R. Chang, G. S. Jackson, Z. Hussain, H. Bluhm, B. W. Eichhorn and Z. Liu, *ACS Catal.*, 2012, **2**, 2297–2304.

40 E. J. Crumlin, H. Bluhm and Z. Liu, *J. Electron Spectrosc. Relat. Phenom.*, 2013, **190**, 84–92.

41 Y.-C. Lu, E. J. Crumlin, G. M. Veith, J. R. Harding, E. Mutoro, L. Baggetto, N. J. Dudney, Z. Liu and Y. Shao-Horn, *Sci. Rep.*, 2012, **2**, 715.

42 B. Mao, Y. Dai, J. Cai, Q. Li, C. Jiang, Y. Li, J. Xie and Z. Liu, *Top. Catal.*, 2018, **61**, 2123–2128.

43 S. Narayanan, J. S. Gibson, J. Aspinall, R. S. Weatherup and M. Pasta, *Curr. Opin. Solid State Mater. Sci.*, 2022, **26**, 100978.

44 C. H. Wu, R. S. Weatherup and M. B. Salmeron, *Phys. Chem. Chem. Phys.*, 2015, **17**, 30229–30239.

45 A. Kolmakov, L. Gregoratti, M. Kiskinova and S. Günther, *Top. Catal.*, 2016, **59**, 448–468.

46 R. S. Weatherup, *Top. Catal.*, 2018, **61**, 2085–2102.

47 R. S. Weatherup, B. Eren, Y. Hao, H. Bluhm and M. B. Salmeron, *J. Phys. Chem. Lett.*, 2016, **7**, 1622–1627.

48 V. Streibel, M. Hävecker, Y. Yi, J. J. Velasco Vélez, K. Skorupska, E. Stotz, A. Knop-Gericke, R. Schlögl and R. Arrigo, *Top. Catal.*, 2018, **61**, 2064–2084.

49 A. Kolmakov, D. A. Dikin, L. J. Cote, J. Huang, M. K. Abyaneh, M. Amati, L. Gregoratti, S. Günther and M. Kiskinova, *Nat. Nanotechnol.*, 2011, **6**, 651–657.

50 T. Masuda, H. Yoshikawa, H. Noguchi, T. Kawasaki, M. Kobata, K. Kobayashi and K. Uosaki, *Appl. Phys. Lett.*, 2013, **103**, 111605.

51 J. Kraus, R. Reichelt, S. Günther, L. Gregoratti, M. Amati, M. Kiskinova, A. Yulaev, I. Vlassiouk and A. Kolmakov, *Nanoscale*, 2014, **6**, 14394–14403.
52 R. Arrigo, M. Hävecker, M. E. Schuster, C. Ranjan, E. Stotz, A. Knop-Gericke and R. Schlögl, *Angew. Chem., Int. Ed.*, 2013, **52**, 11660–11664.
53 O. Karslıoğlu, S. Nemšák, I. Zegkinoglou, A. Shavorskiy, M. Hartl, F. Salmassi, E. M. Gullikson, M. L. Ng, C. Rameshan, B. Rude, D. Bianculli, A. A. Cordones, S. Axnanda, E. J. Crumlin, P. N. Ross, C. M. Schneider, Z. Hussain, Z. Liu, C. S. Fadley and H. Bluhm, *Faraday Discuss.*, 2015, **180**, 35–53.
54 S. Axnanda, E. J. Crumlin, B. Mao, S. Rani, R. Chang, P. G. Karlsson, M. O. M. Edwards, M. Lundqvist, R. Moberg, P. Ross, Z. Hussain and Z. Liu, *Sci. Rep.*, 2015, **5**, 9788.
55 S. Nemšák, A. Shavorskiy, O. Karslioglu, I. Zegkinoglou, A. Rattanachata, C. S. Conlon, A. Keqi, P. K. Greene, E. C. Burks, F. Salmassi, E. M. Gullikson, S.-H. Yang, K. Liu, H. Bluhm and C. S. Fadley, *Nat. Commun.*, 2014, **5**, 5441.
56 Y. Takagi, H. Wang, Y. Uemura, T. Nakamura, L. Yu, O. Sekizawa, T. Uruga, M. Tada, G. Samjeské, Y. Iwasawa and T. Yokoyama, *Phys. Chem. Chem. Phys.*, 2017, **19**, 6013–6021.
57 O. Karslıoğlu, L. Trotochaud, F. Salmassi, E. M. Gullikson, A. Shavorskiy, S. Nemšák and H. Bluhm, *J. Vac. Sci. Technol., A*, 2022, **40**, 013207.
58 M. Favaro, P. C. J. Clark, M. J. Sear, M. Johansson, S. Maehl, R. van de Krol and D. E. Starr, *Surf. Sci.*, 2021, **713**, 121903.
59 D. M. Itkis, J. J. Velasco-Velez, A. Knop-Gericke, A. Vyalikh, M. V. Avdeev and L. V. Yashina, *ChemElectroChem*, 2015, **2**, 1427–1445.
60 E. S. Jones, J. E. N. Swallow and R. S. Weatherup, in *Ambient Pressure Spectroscopy in Complex Chemical Environments*, American Chemical Society, 2021, vol. 1396, pp. 175–218.
61 M. Amati, V. Bonanni, L. Braglia, F. Genuzio, L. Gregoratti, M. Kiskinova, A. Kolmakov, A. Locatelli, E. Magnano, A. A. Matruglio, T. O. Menteş, S. Nappini, P. Torelli and P. Zeller, *J. Electron Spectrosc. Relat. Phenom.*, 2019, 146902.
62 P. Amann, D. Degerman, M.-T. Lee, J. D. Alexander, M. Shipilin, H.-Y. Wang, F. Cavalca, M. Weston, J. Gladh, M. Blom, M. Björkhage, P. Löfgren, C. Schlueter, P. Loemker, K. Ederer, W. Drube, H. Noei, J. Zehetner, H. Wentzel, J. Åhlund and A. Nilsson, *Rev. Sci. Instrum.*, 2019, **90**, 103102.
63 C. Schlueter, A. Gloskovskii, K. Ederer, I. Schostak, S. Piec, I. Sarkar, Yu. Matveyev, P. Lömker, M. Sing, R. Claessen, C. Wiemann, C. M. Schneider, K. Medjanik, G. Schönhense, P. Amann, A. Nilsson and W. Drube, *AIP Conf. Proc.*, 2019, **2054**, 040010.
64 J. J. Velasco-Vélez, V. Pfeifer, M. Hävecker, R. Wang, A. Centeno, A. Zurutuza, G. Algara-Siller, E. Stotz, K. Skorupska, D. Teschner, P. Kube, P. Braeuninger-Weimer, S. Hofmann, R. Schlögl and A. Knop-Gericke, *Rev. Sci. Instrum.*, 2016, **87**, 053121.
65 F. Zaera, *Coord. Chem. Rev.*, 2013, **257**, 3177–3191.
66 E. A. Redekop, N. Johansson, E. Kokkonen, S. Urpelainen, F. Lopes da Silva, M. Kaipio, H.-E. Nieminen, F. Rehman, V. Miikkulainen, M. Ritala and U. Olsbye, *Rev. Sci. Instrum.*, 2021, **92**, 044101.
67 J. Knudsen, T. Gallo, V. Boix, M. D. Strømsheim, G. D'Acunto, C. Goodwin, H. Wallander, S. Zhu, M. Soldemo, P. Lömker, F. Cavalca, M. Scardamaglia,

D. Degerman, A. Nilsson, P. Amann, A. Shavorskiy and J. Schnadt, *Nat. Commun.*, 2021, **12**, 6117.
68 Q. Xie, Y.-L. Jiang, C. Detavernier, D. Deduytsche, R. L. Van Meirhaeghe, G.-P. Ru, B.-Z. Li and X.-P. Qu, *J. Appl. Phys.*, 2007, **102**, 083521.
69 V. Miikkulainen, M. Leskelä, M. Ritala and R. L. Puurunen, *J. Appl. Phys.*, 2013, **113**, 021301.
70 A. Shavorskiy, G. D'Acunto, V. Boix de la Cruz, M. Scardamaglia, S. Zhu, R. H. Temperton, J. Schnadt and J. Knudsen, *ACS Appl. Mater. Interfaces*, 2021, **13**, 47629–47641.
71 J. Jeong, M. Kim, J. Seo, H. Lu, P. Ahlawat, A. Mishra, Y. Yang, M. A. Hope, F. T. Eickemeyer, M. Kim, Y. J. Yoon, I. W. Choi, B. P. Darwich, S. J. Choi, Y. Jo, J. H. Lee, B. Walker, S. M. Zakeeruddin, L. Emsley, U. Rothlisberger, A. Hagfeldt, D. S. Kim, M. Grätzel and J. Y. Kim, *Nature*, 2021, **592**, 381–385.
72 H. Min, D. Y. Lee, J. Kim, G. Kim, K. S. Lee, J. Kim, M. J. Paik, Y. K. Kim, K. S. Kim, M. G. Kim, T. J. Shin and S. Il Seok, *Nature*, 2021, **598**, 444–450.
73 B. Conings, J. Drijkoningen, N. Gauquelin, A. Babayigit, J. D'Haen, L. D'Olieslaeger, A. Ethirajan, J. Verbeeck, J. Manca, E. Mosconi, F. D. Angelis and H.-G. Boyen, *Adv. Energy Mater.*, 2015, **5**, 1500477.
74 R. K. Misra, S. Aharon, B. Li, D. Mogilyansky, I. Visoly-Fisher, L. Etgar and E. A. Katz, *J. Phys. Chem. Lett.*, 2015, **6**, 326–330.
75 M. Liu, M. B. Johnston and H. J. Snaith, *Nature*, 2013, **501**, 395–398.
76 D. I. Patel, T. Roychowdhury, V. Jain, D. Shah, T. G. Avval, S. Chatterjee, S. Bahr, P. Dietrich, M. Meyer, A. Thißen and M. R. Linford, *Surf. Sci. Spectra*, 2019, **26**, 016801.
77 M. Salmeron and R. Schlögl, *Surf. Sci. Rep.*, 2008, **63**, 169–199.
78 P. M. Dietrich, S. Bahr, T. Yamamoto, M. Meyer and A. Thissen, *J. Electron Spectrosc. Relat. Phenom.*, 2019, **231**, 118–126.
79 H. Bluhm, *J. Electron Spectrosc. Relat. Phenom.*, 2010, **177**, 71–84.
80 M. Kjærvik, A. Hermanns, P. Dietrich, A. Thissen, S. Bahr, B. Ritter, E. Kemnitz and W. E. S. Unger, *J. Phys.: Condens. Matter*, 2017, **29**, 474002.
81 M. Kjærvik, K. Schwibbert, P. Dietrich, A. Thissen and W. E. S. Unger, *Surf. Interface Anal.*, 2018, **50**, 996–1000.
82 P. G. Rouxhet and M. J. Genet, *Surf. Interface Anal.*, 2011, **43**, 1453–1470.
83 M. Kjærvik, M. Ramstedt, K. Schwibbert, P. M. Dietrich and W. E. S. Unger, *Front. Chem.*, 2021, **9**, 666161.
84 M. B. Hahn, P. M. Dietrich and J. Radnik, *Commun. Chem.*, 2021, **4**, 1–8.
85 Y. Takagi, T. Nakamura, L. Yu, S. Chaveanghong, O. Sekizawa, T. Sakata, T. Uruga, M. Tada, Y. Iwasawa and T. Yokoyama, *Appl. Phys. Express*, 2017, **10**, 076603.
86 R. K. Lam, J. W. Smith, A. M. Rizzuto, O. Karslıoğlu, H. Bluhm and R. J. Saykally, *J. Chem. Phys.*, 2017, **146**, 094703.
87 C. L. Sabine, R. A. Feely, N. Gruber, R. M. Key, K. Lee, J. L. Bullister, R. Wanninkhof, C. S. Wong, D. W. R. Wallace, B. Tilbrook, F. J. Millero, T.-H. Peng, A. Kozyr, T. Ono and A. F. Rios, *Science*, 2004, **305**, 367–371.
88 C. Richter, R. Dupuy and H. Bluhm, in *Ambient Pressure Spectroscopy in Complex Chemical Environments*, American Chemical Society, 2021, vol. 1396, pp. 39–66.
89 L. Nguyen, P. P. Tao, H. Liu, M. Al-Hada, M. Amati, H. Sezen, L. Gregoratti, Y. Tang, S. D. House and F. F. Tao, *Langmuir*, 2018, **34**, 9606–9616.

90 H. Tissot, G. Olivieri, J.-J. Gallet, F. Bournel, M. G. Silly, F. Sirotti and F. Rochet, *J. Phys. Chem. C*, 2015, **119**, 9253–9259.
91 H. Tissot, J.-J. Gallet, F. Bournel, G. Olivieri, M. G. Silly, F. Sirotti, A. Boucly and F. Rochet, *Top. Catal.*, 2016, **59**, 605–620.
92 S. G. Booth, A. M. Tripathi, I. Strashnov, R. A. W. Dryfe and A. S. Walton, *J. Phys.: Condens. Matter*, 2017, **29**, 454001.
93 M. Ekimova, W. Quevedo, M. Faubel, P. Wernet and E. T. J. Nibbering, *Struct. Dyn.*, 2015, **2**, 054301.
94 M. Kubin, M. Guo, M. Ekimova, M. L. Baker, T. Kroll, E. Källman, J. Kern, V. K. Yachandra, J. Yano, E. T. J. Nibbering, M. Lundberg and P. Wernet, *Inorg. Chem.*, 2018, **57**, 5449–5462.
95 R. Dupuy, J. Filser, C. Richter, R. Seidel, F. Trinter, T. Buttersack, C. Nicolas, J. Bozek, U. Hergenhahn, H. Oberhofer, B. Winter, K. Reuter and H. Bluhm, *Phys. Chem. Chem. Phys.*, 2022, **24**, 4796–4808.
96 S. Malerz, F. Trinter, U. Hergenhahn, A. Ghrist, H. Ali, C. Nicolas, C.-M. Saak, C. Richter, S. Hartweg, L. Nahon, C. Lee, C. Goy, D. M. Neumark, G. Meijer, I. Wilkinson, B. Winter and S. Thürmer, *Phys. Chem. Chem. Phys.*, 2021, **23**, 8246–8260.
97 T. Buttersack, P. E. Mason, R. S. McMullen, H. C. Schewe, T. Martinek, K. Brezina, M. Crhan, A. Gomez, D. Hein, G. Wartner, R. Seidel, H. Ali, S. Thürmer, O. Marsalek, B. Winter, S. E. Bradforth and P. Jungwirth, *Science*, 2020, **368**, 1086–1091.
98 A. Shavorskiy, S. Neppl, D. S. Slaughter, J. P. Cryan, K. R. Siefermann, F. Weise, M.-F. Lin, C. Bacellar, M. P. Ziemkiewicz, I. Zegkinoglou, M. W. Fraund, C. Khurmi, M. P. Hertlein, T. W. Wright, N. Huse, R. W. Schoenlein, T. Tyliszczak, G. Coslovich, J. Robinson, R. A. Kaindl, B. S. Rude, A. Ölsner, S. Mähl, H. Bluhm and O. Gessner, *Rev. Sci. Instrum.*, 2014, **85**, 093102.
99 A. Shavorskiy, J. Schnadt and J. Knudsen, in *Ambient Pressure Spectroscopy in Complex Chemical Environments*, American Chemical Society, 2021, vol. 1396, pp. 219–248.
100 H. Kersell, L. Falling, A. Shavorskiy and S. Nemšák, in *Ambient Pressure Spectroscopy in Complex Chemical Environments*, American Chemical Society, 2021, vol. 1396, pp. 333–358.
101 H. Kersell, P. Chen, H. Martins, Q. Lu, F. Brausse, B.-H. Liu, M. Blum, S. Roy, B. Rude, A. Kilcoyne, H. Bluhm and S. Nemšák, *Rev. Sci. Instrum.*, 2021, **92**, 044102.
102 J. Cole, Z. Henderson, A. G. Thomas, C. L. Compeán-González, A. J. Greer, C. Hardacre, F. Venturini, W. Q. Garzon, P. Ferrer, D. C. Grinter, G. Held and K. L. Syres, *J. Phys. Chem. C*, 2021, **125**, 22778–22785.
103 M. Scardamaglia, C. Struzzi, A. Zakharov, N. Reckinger, P. Zeller, M. Amati and L. Gregoratti, *ACS Appl. Mater. Interfaces*, 2019, **11**, 29448–29457.
104 M. Amati, L. Gregoratti, P. Zeller, M. Greiner, M. Scardamaglia, B. Junker, T. Ruß, U. Weimar, N. Barsan, M. Favaro, A. Alharbi, I. J. T. Jensen, A. Ali and B. D. Belle, *J. Phys. D: Appl. Phys.*, 2021, **54**, 204004.
105 C. Kalha, N. K. Fernando, P. Bhatt, F. O. L. Johansson, A. Lindblad, H. Rensmo, L. Z. Medina, R. Lindblad, S. Siol, L. P. H. Jeurgens, C. Cancellieri, K. Rossnagel, K. Medjanik, G. Schönhense, M. Simon,

A. X. Gray, S. Nemšák, P. Lömker, C. Schlueter and A. Regoutz, *J. Phys.: Condens. Matter*, 2021, **33**, 233001.
106 M. P. Seah and W. A. Dench, *Surf. Interface Anal.*, 1979, **1**, 2–11.
107 L. Sabbatucci and F. Salvat, *Radiat. Phys. Chem.*, 2016, **121**, 122–140.
108 D. J. H. Cant, B. F. Spencer, W. R. Flavell and A. G. Shard, *Surf. Interface Anal.*, 2022, **54**(4), 442–454.
109 K. Kobayashi, M. Kobata and H. Iwai, *J. Electron Spectrosc. Relat. Phenom.*, 2013, **190**, 210–221.
110 M. Kobata, I. Píš, H. Iwai, H. Yamazui, H. Takahashi, M. Suzuki, H. Matsuda, H. Daimon and K. Kobayashi, *Anal. Sci.*, 2010, **26**, 227–232.
111 O. Renault, E. Martinez, C. Zborowski, J. Mann, R. Inoue, J. Newman and K. Watanabe, *Surf. Interface Anal.*, 2018, **50**, 1158–1162.
112 A. Regoutz, M. Mascheck, T. Wiell, S. K. Eriksson, C. Liljenberg, K. Tetzner, B. A. D. Williamson, D. O. Scanlon and P. Palmgren, *Rev. Sci. Instrum.*, 2018, **89**, 73105.
113 M. Otendal, T. Tuohimaa, U. Vogt and H. M. Hertz, *Rev. Sci. Instrum.*, 2008, **79**, 016102.
114 M. Wansleben, C. Zech, C. Streeck, J. Weser, C. Genzel, B. Beckhoff and R. Mainz, *J. Anal. At. Spectrom.*, 2019, **34**, 1497–1502.
115 H.-Y. Wang, M. Soldemo, D. Degerman, P. Lömker, C. Schlueter, A. Nilsson and P. Amann, *Angew. Chem., Int. Ed.*, 2022, **61**, e202111021.
116 T. Okane, M. Kobata, I. Sato, K. Kobayashi, M. Osaka and H. Yamagami, *Nucl. Eng. Des.*, 2016, **297**, 251–256.
117 H.-Q. Wang, J. Xu, X. Lin, Y. Li, J. Kang and J.-C. Zheng, *Light: Sci. Appl.*, 2021, **10**, 153.
118 G. Panaccione and K. Kobayashi, *Surf. Sci.*, 2012, **606**, 125–129.
119 T. Pincelli, V. Lollobrigida, F. Borgatti, A. Regoutz, B. Gobaut, C. Schlueter, T.-L. Lee, D. J. Payne, M. Oura, K. Tamasaku, A. Y. Petrov, P. Graziosi, F. M. Granozio, M. Cavallini, G. Vinai, R. Ciprian, C. H. Back, G. Rossi, M. Taguchi, H. Daimon, G. van der Laan and G. Panaccione, *Nat. Commun.*, 2017, **8**, 16051.
120 G. Assat, D. Foix, C. Delacourt, A. Iadecola, R. Dedryvère and J.-M. Tarascon, *Nat. Commun.*, 2017, **8**, 2219.
121 R. Dubey, J. Sastre, C. Cancellieri, F. Okur, A. Forster, L. Pompizii, A. Priebe, Y. E. Romanyuk, L. P. H. Jeurgens, M. V. Kovalenko and K. V. Kravchyk, *Adv. Energy Mater.*, 2021, **11**, 2102086.
122 H. Kiuchi, K. Hikima, K. Shimizu, R. Kanno, F. Toshiharu and E. Matsubara, *Electrochem. Commun.*, 2020, **118**, 106790.
123 P. Risterucci, O. Renault, E. Martinez, B. Detlefs, V. Delaye, J. Zegenhagen, C. Gaumer, G. Grenet and S. Tougaard, *Appl. Phys. Lett.*, 2014, **104**, 051608.
124 C. Zborowski and S. Tougaard, *Surf. Interface Anal.*, 2022, **54**, 433–441.
125 S. Tougaard, *J. Vac. Sci. Technol., A*, 2021, **39**, 011201.
126 B. F. Spencer, S. Maniyarasu, B. P. Reed, D. J. H. Cant, R. Ahumada-Lazo, A. G. Thomas, C. A. Muryn, M. Maschek, S. K. Eriksson, T. Wiell, T.-L. Lee, S. Tougaard, A. G. Shard and W. R. Flavell, *Appl. Surf. Sci.*, 2021, **541**, 148635.
127 Quases-Tougaard - QUASES, https://www.quases.com/products/quases-tougaard/, accessed March 11, 2022.
128 J. H. Scofield, *Theoretical Photoionization Cross Sections from 1 to 1500 keV*, U.S. Atomic Energy Commission, 1973.

129 M. B. Trzhaskovskaya and V. G. Yarzhemsky, *At. Data Nucl. Data Tables*, 2018, **119**, 99–174.
130 M. B. Trzhaskovskaya and V. G. Yarzhemsky, *At. Data Nucl. Data Tables*, 2019, **129–130**, 101280.
131 A. G. Shard, J. D. P. Counsell, D. J. H. Cant, E. F. Smith, P. Navabpour, X. Zhang and C. J. Blomfield, *Surf. Interface Anal.*, 2019, **51**, 763–773.
132 D. J. H. Cant, B. P. Reed, B. F. Spencer, W. R. Flavell and A. G. Shard, unpublished work.
133 C. D. Wagner, *J. Electron Spectrosc. Relat. Phenom.*, 1977, **10**, 305–315.
134 S. Siol, J. Mann, J. Newman, T. Miyayama, K. Watanabe, P. Schmutz, C. Cancellieri and L. P. H. Jeurgens, *Surf. Interface Anal.*, 2020, **52**, 802–810.
135 T. Pincelli, R. Cucini, A. Verna, F. Borgatti, M. Oura, K. Tamasaku, H. Osawa, T.-L. Lee, C. Schlueter, S. Günther, C. H. Back, M. Dell'Angela, R. Ciprian, P. Orgiani, A. Petrov, F. Sirotti, V. A. Dediu, I. Bergenti, P. Graziosi, F. Miletto Granozio, Y. Tanaka, M. Taguchi, H. Daimon, J. Fujii, G. Rossi and G. Panaccione, *Phys. Rev. B*, 2019, **100**, 045118.
136 L.-P. Oloff, M. Oura, K. Rossnagel, A. Chainani, M. Matsunami, R. Eguchi, T. Kiss, Y. Nakatani, T. Yamaguchi, J. Miyawaki, M. Taguchi, K. Yamagami, T. Togashi, T. Katayama, K. Ogawa, M. Yabashi and T. Ishikawa, *New J. Phys.*, 2014, **16**, 123045.
137 R. Wallauer, M. Raths, K. Stallberg, L. Münster, D. Brandstetter, X. Yang, J. Güdde, P. Puschnig, S. Soubatch, C. Kumpf, F. C. Bocquet, F. S. Tautz and U. Höfer, *Science*, 2021, **371**, 1056–1059.
138 D. Curcio, S. Pakdel, K. Volckaert, J. A. Miwa, S. Ulstrup, N. Lanatà, M. Bianchi, D. Kutnyakhov, F. Pressacco, G. Brenner, S. Dziarzhytski, H. Redlin, S. Y. Agustsson, K. Medjanik, D. Vasilyev, H.-J. Elmers, G. Schönhense, C. Tusche, Y.-J. Chen, F. Speck, T. Seyller, K. Bühlmann, R. Gort, F. Diekmann, K. Rossnagel, Y. Acremann, J. Dęmsar, W. Wurth, D. Lizzit, L. Bignardi, P. Lacovig, S. Lizzit, C. E. Sanders and P. Hofmann, *Phys. Rev. B*, 2021, **104**, L161104.
139 D. Kutnyakhov, R. P. Xian, M. Dendzik, M. Heber, F. Pressacco, S. Y. Agustsson, L. Wenthaus, H. Meyer, S. Gieschen, G. Mercurio, A. Benz, K. Bühlman, S. Däster, R. Gort, D. Curcio, K. Volckaert, M. Bianchi, Ch. Sanders, J. A. Miwa, S. Ulstrup, A. Oelsner, C. Tusche, Y.-J. Chen, D. Vasilyev, K. Medjanik, G. Brenner, S. Dziarzhytski, H. Redlin, B. Manschwetus, S. Dong, J. Hauer, L. Rettig, F. Diekmann, K. Rossnagel, J. Demsar, H.-J. Elmers, Ph. Hofmann, R. Ernstorfer, G. Schönhense, Y. Acremann and W. Wurth, *Rev. Sci. Instrum.*, 2020, **91**, 013109.
140 O. Fedchenko, A. Winkelmann, K. Medjanik, S. Babenkov, D. Vasilyev, S. Chernov, C. Schlueter, A. Gloskovskii, Y. Matveyev, W. Drube, B. Schönhense, H. J. Elmers and G. Schönhense, *New J. Phys.*, 2019, **21**, 113031.
141 K. Medjanik, S. V. Babenkov, S. Chernov, D. Vasilyev, B. Schönhense, C. Schlueter, A. Gloskovskii, Y. Matveyev, W. Drube, H. J. Elmers and G. Schoenhense, *J. Synchrotron Radiat.*, 2019, **26**, 1996–2012.
142 S. Nemšák, G. Conti, A. X. Gray, G. K. Palsson, C. Conlon, D. Eiteneer, A. Keqi, A. Rattanachata, A. Y. Saw, A. Bostwick, L. Moreschini, E. Rotenberg, V. N. Strocov, M. Kobayashi, T. Schmitt, W. Stolte, S. Ueda, K. Kobayashi, A. Gloskovskii, W. Drube, C. A. Jackson, P. Moetakef, A. Janotti, L. Bjaalie,

B. Himmetoglu, C. G. Van de Walle, S. Borek, J. Minar, J. Braun, H. Ebert, L. Plucinski, J. B. Kortright, C. M. Schneider, L. Balents, F. M. F. de Groot, S. Stemmer and C. S. Fadley, *Phys. Rev. B*, 2016, **93**, 245103.

143 S.-H. Yang, A. X. Gray, A. M. Kaiser, B. S. Mun, B. C. Sell, J. B. Kortright and C. S. Fadley, *J. Appl. Phys.*, 2013, **113**, 073513.

144 C. S. Conlon, G. Conti, S. Nemšák, G. Palsson, R. Moubah, C.-T. Kuo, M. Gehlmann, J. Ciston, J. Rault, J.-P. Rueff, F. Salmassi, W. Stolte, A. Rattanachata, S.-C. Lin, A. Keqi, A. Saw, B. Hjörvarsson and C. S. Fadley, *J. Appl. Phys.*, 2019, **126**, 075305.

145 G. Mercurio, I. A. Makhotkin, I. Milov, Y. Y. Kim, I. A. Zaluzhnyy, S. Dziarzhytski, L. Wenthaus, I. A. Vartanyants and W. Wurth, *New J. Phys.*, 2019, **21**, 033031.

146 M. Marinova, J. E. Rault, A. Gloter, S. Nemsak, G. K. Palsson, J.-P. Rueff, C. S. Fadley, C. Carrétéro, H. Yamada, K. March, V. Garcia, S. Fusil, A. Barthélémy, O. Stéphan, C. Colliex and M. Bibes, *Nano Lett.*, 2015, **15**, 2533–2541.

147 S.-C. Lin, C.-T. Kuo, R. B. Comes, J. E. Rault, J.-P. Rueff, S. Nemšák, A. Taleb, J. B. Kortright, J. Meyer-Ilse, E. Gullikson, P. V. Sushko, S. R. Spurgeon, M. Gehlmann, M. E. Bowden, L. Plucinski, S. A. Chambers and C. S. Fadley, *Phys. Rev. B*, 2018, **98**, 165124.

148 G. Conti, A. M. Kaiser, A. X. Gray, S. Nemšák, G. K. Pálsson, J. Son, P. Moetakef, A. Janotti, L. Bjaalie, C. S. Conlon, D. Eiteneer, A. A. Greer, A. Keqi, A. Rattanachata, A. Y. Saw, A. Bostwick, W. C. Stolte, A. Gloskovskii, W. Drube, S. Ueda, M. Kobata, K. Kobayashi, C. G. Van de Walle, S. Stemmer, C. M. Schneider and C. S. Fadley, *J. Appl. Phys.*, 2013, **113**, 143704.

149 O. Karslıoğlu, M. Gehlmann, J. Müller, S. Nemšák, J. A. Sethian, A. Kaduwela, H. Bluhm and C. Fadley, *J. Electron Spectrosc. Relat. Phenom.*, 2019, **230**, 10–20.

150 K. R. Siefermann, C. D. Pemmaraju, S. Neppl, A. Shavorskiy, A. A. Cordones, J. Vura-Weis, D. S. Slaughter, F. P. Sturm, F. Weise, H. Bluhm, M. L. Strader, H. Cho, M.-F. Lin, C. Bacellar, C. Khurmi, J. Guo, G. Coslovich, J. S. Robinson, R. A. Kaindl, R. W. Schoenlein, A. Belkacem, D. M. Neumark, S. R. Leone, D. Nordlund, H. Ogasawara, O. Krupin, J. J. Turner, W. F. Schlotter, M. R. Holmes, M. Messerschmidt, M. P. Minitti, S. Gul, J. Z. Zhang, N. Huse, D. Prendergast and O. Gessner, *J. Phys. Chem. Lett.*, 2014, **5**, 2753–2759.

151 L.-P. Oloff, A. Chainani, M. Matsunami, K. Takahashi, T. Togashi, H. Osawa, K. Hanff, A. Quer, R. Matsushita, R. Shiraishi, M. Nagashima, A. Kimura, K. Matsuishi, M. Yabashi, Y. Tanaka, G. Rossi, T. Ishikawa, K. Rossnagel and M. Oura, *Sci. Rep.*, 2016, **6**, 35087.

152 M. Fushitani, S. T. Pratt, D. You, S. Saito, Y. Luo, K. Ueda, H. Fujise, A. Hishikawa, H. Ibrahim, F. Légaré, P. Johnsson, J. Peschel, E. R. Simpson, A. Olofsson, J. Mauritsson, P. A. Carpeggiani, P. K. Maroju, M. Moioli, D. Ertel, R. Shah, G. Sansone, T. Csizmadia, M. Dumergue, N. G. Harshitha, S. Kühn, C. Callegari, O. Plekan, M. Di Fraia, M. B. Danailov, A. Demidovich, L. Giannessi, L. Raimondi, M. Zangrando, G. De Ninno, P. R. Ribič and K. C. Prince, *J. Chem. Phys.*, 2021, **154**, 144305.

153 M. Borgwardt, J. Mahl, F. Roth, L. Wenthaus, F. Brauße, M. Blum, K. Schwarzburg, G. Liu, F. M. Toma and O. Gessner, *J. Phys. Chem. Lett.*, 2020, **11**, 5476–5481.
154 M. Nisoli, P. Decleva, F. Calegari, A. Palacios and F. Martín, *Chem. Rev.*, 2017, **117**, 10760–10825.
155 B. Buades, A. Picón, E. Berger, I. León, N. Di Palo, S. L. Cousin, C. Cocchi, E. Pellegrin, J. H. Martin, S. Mañas-Valero, E. Coronado, T. Danz, C. Draxl, M. Uemoto, K. Yabana, M. Schultze, S. Wall, M. Zürch and J. Biegert, *Appl. Phys. Rev.*, 2021, **8**, 011408.
156 J. Li, J. Lu, A. Chew, S. Han, J. Li, Y. Wu, H. Wang, S. Ghimire and Z. Chang, *Nat. Commun.*, 2020, **11**, 2748.
157 F. Krausz and M. I. Stockman, *Nat. Photonics*, 2014, **8**, 205–213.
158 P. A. Brühwiler, O. Karis and N. Mårtensson, *Rev. Mod. Phys.*, 2002, **74**, 703–740.
159 M. N. Piancastelli, G. Goldsztejn, T. Marchenko, R. Guillemin, R. K. Kushawaha, L. Journel, S. Carniato, J.-P. Rueff, D. Céolin and M. Simon, *J. Phys. B: At., Mol. Opt. Phys.*, 2014, **47**, 124031.
160 A. G. Thomas, W. R. Flavell, C. Chatwin, S. Rayner, D. Tsoutsou, A. R. Kumarasinghe, D. Brete, T. K. Johal, S. Patel and J. Purton, *Surf. Sci.*, 2005, **592**, 159–168.
161 J. Schnadt, P. A. Brühwiler, L. Patthey, J. N. O'Shea, S. Södergren, M. Odelius, R. Ahuja, O. Karis, M. Bässler, P. Persson, H. Siegbahn, S. Lunell and N. Mårtensson, *Nature*, 2002, **418**, 620–623.
162 T. A. Heimer, E. J. Heilweil, C. A. Bignozzi and G. J. Meyer, *J. Phys. Chem. A*, 2000, **104**, 4256–4262.
163 L. C. Mayor, J. Ben Taylor, G. Magnano, A. Rienzo, C. J. Satterley, J. N. O'Shea and J. Schnadt, *J. Chem. Phys.*, 2008, **129**, 114701.
164 A. Föhlisch, P. Feulner, F. Hennies, A. Fink, D. Menzel, D. Sanchez-Portal, P. M. Echenique and W. Wurth, *Nature*, 2005, **436**, 373–376.
165 F. O. L. Johansson, M. Ivanović, S. Svanström, U. B. Cappel, H. Peisert, T. Chassé and A. Lindblad, *J. Phys. Chem. C*, 2018, **122**, 12605–12614.
166 F. O. L. Johansson, U. B. Cappel, M. Fondell, Y. Han, M. Gorgoi, K. Leifer and A. Lindblad, *Phys. Chem. Chem. Phys.*, 2020, **22**, 10335–10342.
167 F. O. L. Johansson, X. Chen, O. Eriksson, B. Sanyal and A. Lindblad, *Phys. Rev. B*, 2020, **102**, 035165.
168 S. Hellmann, C. Sohrt, M. Beye, T. Rohwer, F. Sorgenfrei, M. Marczynski-Bühlow, M. Kalläne, H. Redlin, F. Hennies, M. Bauer, A. Föhlisch, L. Kipp, W. Wurth and K. Rossnagel, *New J. Phys.*, 2012, **14**, 013062.
169 A. Pietzsch, A. Föhlisch, M. Beye, M. Deppe, F. Hennies, M. Nagasono, E. Suljoti, W. Wurth, C. Gahl, K. Döbrich and A. Melnikov, *New J. Phys.*, 2008, **10**, 033004.
170 H. Öström, H. Öberg, H. Xin, J. LaRue, M. Beye, M. Dell'Angela, J. Gladh, M. L. Ng, J. A. Sellberg, S. Kaya, G. Mercurio, D. Nordlund, M. Hantschmann, F. Hieke, D. Kühn, W. F. Schlotter, G. L. Dakovski, J. J. Turner, M. P. Minitti, A. Mitra, S. P. Moeller, A. Föhlisch, M. Wolf, W. Wurth, M. Persson, J. K. Nørskov, F. Abild-Pedersen, H. Ogasawara, L. G. M. Pettersson and A. Nilsson, *Science*, 2015, **347**, 978–982.
171 M. Beye, H. Öberg, H. Xin, G. L. Dakovski, M. Dell'Angela, A. Föhlisch, J. Gladh, M. Hantschmann, F. Hieke, S. Kaya, D. Kühn, J. LaRue,

G. Mercurio, M. P. Minitti, A. Mitra, S. P. Moeller, M. L. Ng, A. Nilsson, D. Nordlund, J. Nørskov, H. Öström, H. Ogasawara, M. Persson, W. F. Schlotter, J. A. Sellberg, M. Wolf, F. Abild-Pedersen, L. G. M. Pettersson and W. Wurth, *J. Phys. Chem. Lett.*, 2016, **7**, 3647–3651.

172 E. Diesen, H.-Y. Wang, S. Schreck, M. Weston, H. Ogasawara, J. LaRue, F. Perakis, M. Dell'Angela, F. Capotondi, L. Giannessi, E. Pedersoli, D. Naumenko, I. Nikolov, L. Raimondi, C. Spezzani, M. Beye, F. Cavalca, B. Liu, J. Gladh, S. Koroidov, P. S. Miedema, R. Costantini, T. F. Heinz, F. Abild-Pedersen, J. Voss, A. C. Luntz and A. Nilsson, *Phys. Rev. Lett.*, 2021, **127**, 016802.

173 H.-Y. Wang, S. Schreck, M. Weston, C. Liu, H. Ogasawara, J. LaRue, F. Perakis, M. Dell'Angela, F. Capotondi, L. Giannessi, E. Pedersoli, D. Naumenko, I. Nikolov, L. Raimondi, C. Spezzani, M. Beye, F. Cavalca, B. Liu, J. Gladh, S. Koroidov, P. S. Miedema, R. Costantini, L. G. M. Pettersson and A. Nilsson, *Phys. Chem. Chem. Phys.*, 2020, **22**, 2677–2684.

174 J. Duris, S. Li, T. Driver, E. G. Champenois, J. P. MacArthur, A. A. Lutman, Z. Zhang, P. Rosenberger, J. W. Aldrich, R. Coffee, G. Coslovich, F.-J. Decker, J. M. Glownia, G. Hartmann, W. Helml, A. Kamalov, J. Knurr, J. Krzywinski, M.-F. Lin, J. P. Marangos, M. Nantel, A. Natan, J. T. O'Neal, N. Shivaram, P. Walter, A. L. Wang, J. J. Welch, T. J. A. Wolf, J. Z. Xu, M. F. Kling, P. H. Bucksbaum, A. Zholents, Z. Huang, J. P. Cryan and A. Marinelli, *Nat. Photonics*, 2020, **14**, 30–36.

175 C. L. Smallwood, J. P. Hinton, C. Jozwiak, W. Zhang, J. D. Koralek, H. Eisaki, D.-H. Lee, J. Orenstein and A. Lanzara, *Science*, 2012, **336**, 1137–1139.

176 F. Schmitt, P. S. Kirchmann, U. Bovensiepen, R. G. Moore, L. Rettig, M. Krenz, J.-H. Chu, N. Ru, L. Perfetti, D. H. Lu, M. Wolf, I. R. Fisher and Z.-X. Shen, *Science*, 2008, **321**, 1649–1652.

177 A. Gauthier, J. A. Sobota, N. Gauthier, K.-J. Xu, H. Pfau, C. R. Rotundu, Z.-X. Shen and P. S. Kirchmann, *J. Appl. Phys.*, 2020, **128**, 093101.

178 J. Graf, C. Jozwiak, C. L. Smallwood, H. Eisaki, R. A. Kaindl, D.-H. Lee and A. Lanzara, *Nat. Phys.*, 2011, **7**, 805–809.

179 L. Perfetti, P. A. Loukakos, M. Lisowski, U. Bovensiepen, H. Berger, S. Biermann, P. S. Cornaglia, A. Georges and M. Wolf, *Phys. Rev. Lett.*, 2006, **97**, 067402.

180 G. Rohde, A. Stange, A. Müller, M. Behrendt, L.-P. Oloff, K. Hanff, T. J. Albert, P. Hein, K. Rossnagel and M. Bauer, *Phys. Rev. Lett.*, 2018, **121**, 256401.

181 Y. H. Wang, D. Hsieh, D. Pilon, L. Fu, D. R. Gardner, Y. S. Lee and N. Gedik, *Phys. Rev. Lett.*, 2011, **107**, 207602.

182 E. J. Sie, T. Rohwer, C. Lee and N. Gedik, *Nat. Commun.*, 2019, **10**, 3535.

183 S. Hellmann, M. Beye, C. Sohrt, T. Rohwer, F. Sorgenfrei, H. Redlin, M. Kalläne, M. Marczynski-Bühlow, F. Hennies, M. Bauer, A. Föhlisch, L. Kipp, W. Wurth and K. Rossnagel, *Phys. Rev. Lett.*, 2010, **105**, 187401.

184 S. Hellmann, K. Rossnagel, M. Marczynski-Bühlow and L. Kipp, *Phys. Rev. B: Condens. Matter Mater. Phys.*, 2009, **79**, 035402.

185 J. Graf, S. Hellmann, C. Jozwiak, C. L. Smallwood, Z. Hussain, R. A. Kaindl, L. Kipp, K. Rossnagel and A. Lanzara, *J. Appl. Phys.*, 2010, **107**, 014912.

186 G. Schönhense, D. Kutnyakhov, F. Pressacco, M. Heber, N. Wind, S. Y. Agustsson, S. Babenkov, D. Vasilyev, O. Fedchenko, S. Chernov, L. Rettig, B. Schönhense, L. Wenthaus, G. Brenner, S. Dziarzhytski,

S. Palutke, S. K. Mahatha, N. Schirmel, H. Redlin, B. Manschwetus, I. Hartl, Yu. Matveyev, A. Gloskovskii, C. Schlueter, V. Shokeen, H. Duerr, T. K. Allison, M. Beye, K. Rossnagel, H. J. Elmers and K. Medjanik, *Rev. Sci. Instrum.*, 2021, **92**, 053703.

187 L.-P. Oloff, K. Hanff, A. Stange, G. Rohde, F. Diekmann, M. Bauer and K. Rossnagel, *J. Appl. Phys.*, 2016, **119**, 225106.

188 M. Dell'Angela, T. Anniyev, M. Beye, R. Coffee, A. Föhlisch, J. Gladh, S. Kaya, T. Katayama, O. Krupin, A. Nilsson, D. Nordlund, W. F. Schlotter, J. A. Sellberg, F. Sorgenfrei, J. J. Turner, H. Öström, H. Ogasawara, M. Wolf and W. Wurth, *Struct. Dyn.*, 2015, **2**, 025101.

189 C. W. Nicholson, A. Lücke, W. G. Schmidt, M. Puppin, L. Rettig, R. Ernstorfer and M. Wolf, *Science*, 2018, **362**, 821–825.

190 A. K. Mills, S. Zhdanovich, M. X. Na, F. Boschini, E. Razzoli, M. Michiardi, A. Sheyerman, M. Schneider, T. J. Hammond, V. Süss, C. Felser, A. Damascelli and D. J. Jones, *Rev. Sci. Instrum.*, 2019, **90**, 083001.

191 S. Gerber, S.-L. Yang, D. Zhu, H. Soifer, J. A. Sobota, S. Rebec, J. J. Lee, T. Jia, B. Moritz, C. Jia, A. Gauthier, Y. Li, D. Leuenberger, Y. Zhang, L. Chaix, W. Li, H. Jang, J.-S. Lee, M. Yi, G. L. Dakovski, S. Song, J. M. Glownia, S. Nelson, K. W. Kim, Y.-D. Chuang, Z. Hussain, R. G. Moore, T. P. Devereaux, W.-S. Lee, P. S. Kirchmann and Z.-X. Shen, *Science*, 2017, **357**, 71–75.

192 N. Bergeard, M. G. Silly, D. Krizmancic, C. Chauvet, M. Guzzo, J. P. Ricaud, M. Izquierdo, L. Stebel, P. Pittana, R. Sergo, G. Cautero, G. Dufour, F. Rochet and F. Sirotti, *J. Synchrotron Radiat.*, 2011, **18**, 245–250.

193 M. Ogawa, S. Yamamoto, Y. Kousa, F. Nakamura, R. Yukawa, A. Fukushima, A. Harasawa, H. Kondoh, Y. Tanaka, A. Kakizaki and I. Matsuda, *Rev. Sci. Instrum.*, 2012, **83**, 023109.

194 M. G. Silly, T. Ferté, M. A. Tordeux, D. Pierucci, N. Beaulieu, C. Chauvet, F. Pressacco, F. Sirotti, H. Popescu, V. Lopez-Flores, M. Tortarolo, M. Sacchi, N. Jaouen, P. Hollander, J. P. Ricaud, N. Bergeard, C. Boeglin, B. Tudu, R. Delaunay, J. Luning, G. Malinowski, M. Hehn, C. Baumier, F. Fortuna, D. Krizmancic, L. Stebel, R. Sergo and G. Cautero, *J. Synchrotron Radiat.*, 2017, **24**, 886–897.

195 L. Kronik and Y. Shapira, *Surf. Sci. Rep.*, 1999, **37**, 1–206.

196 J. P. Long, H. R. Sadeghi, J. C. Rife and M. N. Kabler, *Phys. Rev. Lett.*, 1990, **64**, 1158–1161.

197 B. F. Spencer, D. M. Graham, S. J. O. Hardman, E. A. Seddon, M. J. Cliffe, K. L. Syres, A. G. Thomas, S. K. Stubbs, F. Sirotti, M. G. Silly, P. F. Kirkham, A. R. Kumarasinghe, G. J. Hirst, A. J. Moss, S. F. Hill, D. A. Shaw, S. Chattopadhyay and W. R. Flavell, *Phys. Rev. B: Condens. Matter Mater. Phys.*, 2013, **88**, 195301.

198 P. C. J. Clark, N. K. Lewis, J. C.-R. Ke, R. Ahumada-Lazo, Q. Chen, D. C. J. Neo, E. A. Gaulding, G. F. Pach, I. Pis, M. G. Silly and W. R. Flavell, *Nanoscale*, 2021, **13**, 17793–17806.

199 B. F. Spencer, M. J. Cliffe, D. M. Graham, S. J. O. Hardman, E. a Seddon, K. L. Syres, A. G. Thomas, F. Sirotti, M. G. Silly, J. Akhtar, P. O'Brien, S. M. Fairclough, J. M. Smith, S. Chattopadhyay and W. R. Flavell, *Faraday Discuss.*, 2014, **171**, 275–298.

200 S. Tanaka, S. D. Moré, K. Takahashi and M. Kamada, *J. Phys. Soc. Jpn.*, 2003, **72**, 659–663.

201 M. Marsi, R. Belkhou, C. Grupp, G. Panaccione, A. Taleb-Ibrahimi, L. Nahon, D. Garzella, D. Nutarelli, E. Renault, R. Roux, M. E. Couprie and M. Billardon, *Phys. Rev. B: Condens. Matter*, 2000, **61**, R5070–R5073.
202 W. Widdra, D. Bröcker, T. Gießel, I. V. Hertel, W. Krüger, A. Liero, F. Noack, V. Petrov, D. Pop, P. M. Schmidt, R. Weber, I. Will and B. Winter, *Surf. Sci.*, 2003, **543**, 87–94.
203 F. Polack, M. Silly, C. Chauvet, B. Lagarde, N. Bergeard, M. Izquierdo, O. Chubar, D. Krizmancic, M. Ribbens, J. -P. Duval, C. Basset, S. Kubsky and F. Sirotti, *AIP Conf. Proc.*, 2010, **1234**, 185–188.
204 B. F. Spencer, M. A. Leontiadou, P. C. J. Clark, A. I. Williamson, M. G. Silly, F. Sirotti, S. M. Fairclough, S. C. E. Tsang, D. C. J. Neo, H. E. Assender, A. A. R. Watt and W. R. Flavell, *Appl. Phys. Lett.*, 2016, **108**, 091603.
205 F. Pressacco, V. Uhlíř, M. Gatti, A. Nicolaou, A. Bendounan, J. A. Arregi, S. K. K. Patel, E. E. Fullerton, D. Krizmancic and F. Sirotti, *Struct. Dyn.*, 2018, **5**, 034501.
206 A. Vollmer, R. Ovsyannikov, M. Gorgoi, S. Krause, M. Oehzelt, A. Lindblad, N. Mårtensson, S. Svensson, P. Karlsson, M. Lundvuist, T. Schmeiler, J. Pflaum and N. Koch, *J. Electron Spectrosc. Relat. Phenom.*, 2012, **185**, 55–60.
207 M. Ogawa, S. Yamamoto, R. Yukawa, R. Hobara, C.-H. Lin, R.-Y. Liu, S.-J. Tang and I. Matsuda, *Phys. Rev. B: Condens. Matter Mater. Phys.*, 2013, **87**, 235308.
208 R. Yukawa, S. Yamamoto, K. Ozawa, M. Emori, M. Ogawa, Sh. Yamamoto, K. Fujikawa, R. Hobara, S. Kitagawa, H. Daimon, H. Sakama and I. Matsuda, *Appl. Phys. Lett.*, 2014, **105**, 151602.
209 K. Ozawa, M. Emori, S. Yamamoto, R. Yukawa, S. Yamamoto, R. Hobara, K. Fujikawa, H. Sakama and I. Matsuda, *J. Phys. Chem. Lett.*, 2014, **5**, 1953–1957.
210 K. Ozawa, S. Yamamoto, M. D'angelo, Y. Natsui, N. Terashima, K. Mase and I. Matsuda, *J. Phys. Chem.C*, 2019, **123**, 4388–4395.
211 B. F. Spencer, M. J. Cliffe, D. M. Graham, S. J. O. Hardman, E. A. Seddon, K. L. Syres, A. G. Thomas, F. Sirotti, M. G. Silly, J. Akhtar, P. O'Brien, S. M. Fairclough, J. M. Smith, S. Chattopadhyay and W. R. Flavell, *Surf. Sci.*, 2015, **641**, 320–325.
212 H. Magnan, P. M. Deleuze, J. Brehin, T. Plays, D. Stanescu, W. R. Flavell, M. G. Silly, B. Domenichini and A. Barbier, *J. Phys. Chem. C*, 2020, **124**, 10315–10323.
213 S. Koley, J. Cui, Y. E. Panfil and U. Banin, *Acc. Chem. Res.*, 2021, **54**, 1178–1188.
214 C. R. Kagan and C. B. Murray, *Nat. Nanotechnol.*, 2015, **10**, 1013–1026.
215 E. M. Sanehira, A. R. Marshall, J. A. Christians, S. P. Harvey, P. N. Ciesielski, L. M. Wheeler, P. Schulz, L. Y. Lin, M. C. Beard and J. M. Luther, *Sci. Adv.*, 2017, **3**, eaao4204.
216 M. Hao, Y. Bai, S. Zeiske, L. Ren, J. Liu, Y. Yuan, N. Zarrabi, N. Cheng, M. Ghasemi, P. Chen, M. Lyu, D. He, J. H. Yun, Y. Du, Y. Wang, S. Ding, A. Armin, P. Meredith, G. Liu, H. M. Cheng and L. Wang, *Nat. Energy*, 2020, **5**, 79–88.
217 A. Maulu, P. J. Rodríguez Cantó, J. Navarro Arenas, R. Abargues, J. F. Sanchez-Royo, R. García Calzada and J. P. Martínez-Pastor, *RSC Adv.*, 2016, **6**, 80201–80212.

218 A. Yousefiamin, N. A. Killilea, M. Sytnyk, P. Maisch, K. C. Tam, H. J. Egelhaaf, S. Langner, T. Stubhan, C. J. Brabec, T. Rejek, M. Halik, K. Poulsen, J. Niehaus, A. Köck and W. Heiss, *ACS Nano*, 2019, **13**, 2389–2397.
219 I. Ramiro, O. Özdemir, S. Christodoulou, S. Gupta, M. Dalmases, I. Torre and G. Konstantatos, *Nano Lett.*, 2020, **20**, 1003–1008.
220 A. G. Pattantyus-Abraham, I. J. Kramer, A. R. Barkhouse, X. Wang, G. Konstantatos, R. Debnath, L. Levina, I. Raabe, M. K. Nazeeruddin, M. Grätzel and E. H. Sargent, *ACS Nano*, 2010, **4**, 3374–3380.
221 N. Oh, B. H. Kim, S.-Y. Cho, S. Nam, S. P. Rogers, Y. Jiang, J. C. Flanagan, Y. Zhai, J.-H. Kim, J. Lee, Y. Yu, Y. K. Cho, G. Hur, J. Zhang, P. Trefonas, J. A. Rogers and M. Shim, *Science*, 2017, **355**, 616–619.
222 C.-H. M. Chuang, P. R. Brown, V. Bulović and M. G. Bawendi, *Nat. Mater.*, 2014, **13**, 796–801.
223 P. C. J. Clark, H. Radtke, A. Pengpad, A. I. Williamson, B. F. Spencer, J. O. Hardman, M. A. Leontiadou, D. C. J. Neo, S. M. Fairclough, A. A. R. Watt, I. Pis, S. Nappini, F. Bondino, E. Magnano, K. Schulte, M. Silly, F. Sirotti and W. R. Flavell, *Nanoscale*, 2017, **9**, 6056–6067.
224 Y. Zhang, Q. Chen, A. P. Alivisatos and M. Salmeron, *Nano Lett.*, 2015, **15**, 4647–4663.
225 L. Hu, S. Huang, R. Patterson and J. E. Halpert, *J. Mater. Chem. C*, 2019, **7**, 4497–4502.
226 D. Pierucci, J.-J. Gallet, F. Bournel, F. Sirotti, M. G. Silly, H. Tissot, A. Naitabdi and F. Rochet, *J. Phys. Chem. C*, 2016, **120**, 21631–21641.
227 L. Pérez Ramírez, J.-J. Gallet, F. Bournel, F. Lim, S. Carniato, F. Rochet, O. V. Yazyev, A. Pasquarello, E. Magnano and F. Bondino, *J. Phys. Chem. A*, 2020, **124**, 5378–5388.
228 F. Armillotta, E. D'Incecco, M. Corva, M. Stredansky, J.-J. Gallet, F. Bournel, A. Goldoni, A. Morgante, E. Vesselli and A. Verdini, *Angew. Chem., Int. Ed.*, 2021, **60**, 25988–25993.
229 E. Goulielmakis, M. Schultze, M. Hofstetter, V. S. Yakovlev, J. Gagnon, M. Uiberacker, A. L. Aquila, E. M. Gullikson, D. T. Attwood, R. Kienberger, F. Krausz and U. Kleineberg, *Science*, 2008, **320**, 1614–1617.
230 M. Ossiander, F. Siegrist, V. Shirvanyan, R. Pazourek, A. Sommer, T. Latka, A. Guggenmos, S. Nagele, J. Feist, J. Burgdörfer, R. Kienberger and M. Schultze, *Nat. Phys.*, 2017, **13**, 280–285.
231 S. Heinrich, T. Saule, M. Högner, Y. Cui, V. S. Yakovlev, I. Pupeza and U. Kleineberg, *Nat. Commun.*, 2021, **12**, 3404.
232 T. Saule, S. Heinrich, J. Schötz, N. Lilienfein, M. Högner, O. deVries, M. Plötner, J. Weitenberg, D. Esser, J. Schulte, P. Russbueldt, J. Limpert, M. F. Kling, U. Kleineberg and I. Pupeza, *Nat. Commun.*, 2019, **10**, 458.
233 V. S. Yakovlev and N. Karpowicz, *Science*, 2017, **357**, 1239–1240.
234 F. Siek, S. Neb, P. Bartz, M. Hensen, C. Strüber, S. Fiechter, M. Torrent-Sucarrat, V. M. Silkin, E. E. Krasovskii, N. M. Kabachnik, S. Fritzsche, R. D. Muiño, P. M. Echenique, A. K. Kazansky, N. Müller, W. Pfeiffer and U. Heinzmann, *Science*, 2017, **357**, 1274–1277.
235 M. Schultze, M. Fieß, N. Karpowicz, J. Gagnon, M. Korbman, M. Hofstetter, S. Neppl, A. L. Cavalieri, Y. Komninos, Th. Mercouris, C. A. Nicolaides, R. Pazourek, S. Nagele, J. Feist, J. Burgdörfer, A. M. Azzeer, R. Ernstorfer,

R. Kienberger, U. Kleineberg, E. Goulielmakis, F. Krausz and V. S. Yakovlev, *Science*, 2010, **328**, 1658–1662.
236 Y.-J. Chen, M. Hoffmann, B. Zimmermann, G. Bihlmayer, S. Blügel, C. M. Schneider and C. Tusche, *Commun. Phys.*, 2021, **4**, 1–6.

Faraday Discussions

PAPER

Understanding methanol dissociative adsorption and oxidation on amorphous oxide films†

Sri Krishna Murthy Padavala, [a] Kateryna Artyushkova,[b] Shannon W. Boettcher, [c] Slavomír Nemšák[d] and Kelsey A. Stoerzinger [*ae]

Received 29th November 2021, Accepted 20th January 2022
DOI: 10.1039/d1fd00109d

Interactions between a transition metal (oxide) catalyst and a support can tailor the number and nature of active sites, for instance in the methanol oxidation reaction. We here use ambient pressure X-ray photoelectron spectroscopy (AP-XPS) to identify and compare the surface adsorbates that form on amorphous metal oxide films that maximize such interactions. Considering $Al_{(1-x)}M_xO_y$ (M = Fe or Mn) films at a range of methanol : oxygen gas ratios and temperatures, we find that the redox-active transition metal site (characterized by methoxy formation) dominates dissociative methanol adsorption, while basic oxygen sites (characterized by carbonate formation) play a lesser role. Product detection, however, indicates complete oxidation to carbon dioxide and water with partial oxidation products (dimethyl ether) comprising a minor species. Comparing the intensity of methoxy and hydroxyl features at a fixed XPS chemical shift suggests methanol deprotonation during adsorption in oxygen rich conditions for high transition metal content. However, increasing methanol partial pressure and lower metal site density may promote oxygen vacancy formation and the dehydroxylation pathway, supported by a nominal reduction in the oxidation state of iron sites. These findings illustrate that AP-XPS and mass spectrometry together are powerful tools in understanding metal–support interactions, quantifying and probing the nature of catalytic active sites, and considering the link between electronic structure of materials and their catalytic activity.

[a]*School of Chemical, Biological and Environmental Engineering, Oregon State University, Corvallis, Oregon 97331, USA. E-mail: kelsey.stoerzinger@oregonstate.edu*

[b]*Physical Electronics Inc., Chanhassen, MN, 55317, USA*

[c]*Department of Chemistry & Biochemistry and the Materials Science Institute, University of Oregon, Eugene, Oregon 97403, USA*

[d]*Advanced Light Source, Lawrence Berkeley National Laboratory, 1 Cyclotron Road, Berkeley, CA 94720, USA*

[e]*Physical and Computational Sciences Directorate, Pacific Northwest National Laboratory, Richland, Washington 99354, USA*

† Electronic supplementary information (ESI) available. See DOI: 10.1039/d1fd00109d

1. Introduction

Methanol (CH_3OH) is a versatile and flexible chemical that can be used both as a fuel source and chemical feedstock.[1,2] CH_3OH oxidation involves a series of oxidation and dehydration reactions[3] with numerous potential products depending on the reaction pathway. For example, partial oxidation of CH_3OH can yield formaldehyde (CH_2O), dehydration and C–O coupling produces dimethyl ether $(CH_3)_2O$, and complete oxidation yields water (H_2O) and carbon dioxide (CO_2). The methanol oxidation reaction (MOR) can be used as a probe to quantify the density of active sites on oxide surfaces,[4,5] and due to its diverse potential products and pathways, help ascertain the nature of these active sites.[4] Considering a broad range of metal oxides, CH_2O is formed on redox-active sites, $(CH_3)_2O$ is formed on Brønsted acid sites, and CO_2 is formed on basic sites.[4]

The MOR has been studied on transition metals (*e.g.* V, Cu, Mn, Fe, Mo, Pd, Au) and their oxides with and without supports.[6–13] Supports not only disperse the catalyst as small particles with high surface area, but also can modify the activity of the catalyst and influence its selectivity towards the different products of MOR. Considering the common support material of γ-alumina, previous reports have shown that this material alone will oxidize CH_3OH to $(CH_3)_2O$ in oxygen rich environments at temperatures <275 °C, above which carbon monoxide (CO) begins to form. By incorporating manganese and iron oxide catalysts onto γ-alumina supports, $(CH_3)_2O$ remained the primary oxidation product but formed at lower temperatures, and above 200 °C complete oxidation to CO_2 dominated.[14] As oxide supports can carry their own activity towards MOR, the activity of and product distribution from supported catalysts thus depends on the catalyst loading.[9]

In order to maximize effects localized to the catalyst–support interface, the interface density can be increased by dispersion of small catalyst nanoparticles. The extension of this is the incorporation of transition metals into a support host directly, thus maximizing electronic interactions and potential dual-active site effects. Herein, we prepare amorphous metal oxides (AMOs) by embedding the transition metal ion into the support, resulting in uniform interaction between "support" and "catalyst" metal centers over the surface of AMOs. Such AMOs find use in many fields ranging from electronics to heterogeneous catalysis.[15–17] The electronic structure of these materials can be tailored for a given application, influencing for example their corrosion-resistance properties[18] and electrochemical activity.[19]

Alumina's acid/base characteristics, which maximize dispersion of transition metal (oxides) on supports, facilitate the incorporation of a wide range of transition metal cations at high concentrations in AMOs.[17] The addition of iron (Fe) and manganese (Mn), here model catalyst centers for the MOR, leads to new filled electronic states in alumina-based AMOs at the top of the valence band.[20] In our previous work, we investigated the influence of these new electronic states on the surface acidity, basicity, and the hydroxylation behavior in a humid environment.[21] In the current work, we study the influence of these new electronic states on the surface reactivity towards CH_3OH.

We have used ambient pressure X-ray photoelectron spectroscopy (AP-XPS) to investigate CH_3OH oxidation on $Al_{(1-x)}M_xO_y$ (M = Fe, Mn) AMO films across temperatures and methanol : oxygen gas ratios. For the compositions considered here, Fe and Mn are fully dispersed without evidence of phase segregation and the formation of transition metal oxide clusters.[17] The nominal oxidation state of Fe and Mn in the films is +3, and the basicity of the films increases with Fe content. Both methoxy (here referred to as OCH_3) and carbonate (here termed CO_3) species are observed on the surface, where the increase in OCH_3 coverage with increase in transition metal content suggests that transition metal sites are active for CH_3OH oxidation. The $Al_{(1-x)}Fe_xO_y$ and $Al_{(1-x)}Mn_xO_y$ oxide films studied here had fewer basic sites (characterized by CO_3 formation) compared to acidic and redox-active sites (characterized by OCH_3 formation). Consideration of OCH_3 and hydroxyl (here termed OH) features in the O 1s and C 1s core levels suggests that the chemical shift for and/or intensity of OH varies with the methanol : oxygen gas ratio and AMO composition.

2. Experimental

2.1 Thin-film preparation

Precursor solutions were prepared by mixing different ratios of 1 M aqueous solutions of transition metal nitrate salts, $Mn(NO_3)_2$ and $Fe(NO_3)_3$ and 1 M aqueous solution of $Al(NO_3)_x(OH)_{3-x}$ clusters, prepared by bulk electrolysis using a procedure from the literature and described previously.[17] The solution was spin coated onto a polished Si(100) substrate previously coated with Ti and Ir by electron-beam evaporation to mitigate charging, and annealed at 450 °C for 30 min.[20] The thin films obtained were atomically smooth with a root-mean-square roughness (R_{rms}) < 1 nm as determined by atomic force microscopy. The films are referred to by their metal ratios, *i.e.* a film with 80% Al and 20% Fe is called as Al8Fe2.

2.2 AP-XPS

AP-XPS was performed at the beamline 9.3.2 at the Advanced Light Source.[22] Films were placed atop a BN heating element on a sample holder containing no Pt, and grounded through a thermocouple pressed onto the film surface and electrically isolated from the holder with a ceramic. Films were cleaned by UV ozone prior to loading into the sample chamber, and cleaned *in situ* by heating to 300 °C in an atmosphere of 100–150 mTorr of O_2. Following cleaning, films were then cooled to room temperature under 100 mTorr O_2. Films were characterized at room temperature and during a temperature ramp for 10 : 100, 50 : 100, and 100 : 50 mTorr methanol : oxygen gas. Fig. S1† shows example survey spectra collected at 690 eV for 50 mTorr CH_3OH and 100 mTorr oxygen at 25 °C for the Al8Fe2 and Al4Fe6 films depicting the expected features. The incident photon energy was calibrated at room temperature by cross-referencing O 1s (690 eV, 750 eV) and Al 2p (490 eV, 350 eV) core levels in duplicate films previously calibrated to the Au 4f (84 eV) of an auxiliary reference,[21] where the resultant O 1s and Al 2p binding energy (BE) is dependent on composition as discussed in Section 3.2. Subsequent measurements at elevated temperatures were calibrated to the Al 2p core level as an internal standard. The spectra were fit with CasaXPS using a Shirley-type background subtraction. The identified features were fit using a line shape with 70% Gaussian and 30% Lorentzian character with average fitted values shown in

Tables S1 and S2.† The OH + OCH_3 feature in the O 1s core level was constrained at 1.65 eV above the bulk O 1s BE based on previous measurements made for duplicate films in water,[21] unless stated otherwise.

3. Results and discussion

3.1 Feature identification

We first describe the features observed in each core level, considering the example case of 50 mTorr CH_3OH and 100 mTorr oxygen at room temperature.

3.1.1 O 1s core level. For the oxygen core level, four distinct features, shown in Fig. 1a, were observed across all samples and conditions. Depth profiling established that the feature at ~531 eV is due to the bulk of the film, while the higher BE shoulder is due to adsorbates, with integrated intensity ratio of shoulder/bulk as 0.19 and 0.18 in Fig. 1 probed at 690 eV (shallow inelastic mean free path (IMFP) ~ 5.8 Å (ref. 23)) and 750 eV (deep IMFP ~ 6.6 Å (ref. 23)) incident photon energy (IPE). This shoulder at 1.65 eV above the bulk is attributed in part to OH species based on previous measurements of duplicate samples in water.[21] Any oxidized carbon species on the surface will also have similar O 1s BEs, and are primarily attributed to OCH_3 species based on the C 1s spectra (Fig. 1b). The relative sensitivity factor (RSF) for O 1s (690 eV) to the C 1s (490 eV) was determined to be 1.08 by measuring gaseous CO_2. The features at ~535 eV (ref. 6) and ~538 eV (ref. 6) are attributed to gas phase CH_3OH and oxygen, respectively, with their relative intensities consistent with gas-phase partial pressures and experimentally measured O : C RSF. Table S1† shows the average BEs and the full width half maximum (FWHM) of the features observed in the oxygen core level for the films across different methanol : oxygen gas ratios at different temperatures. Example room temperature oxygen core level spectra in Fig. S2† depict the changes in feature BE and intensity with film composition and gas ratio.

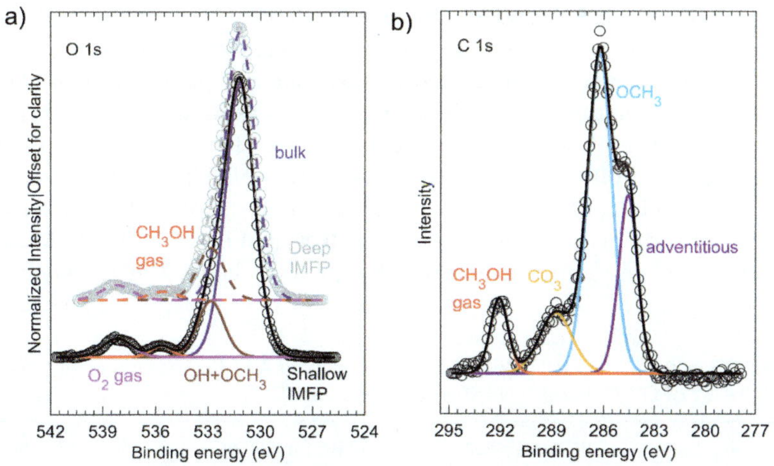

Fig. 1 (a) O 1s and (b) C 1s for Al8Fe2 in 50 mTorr of CH_3OH and 100 mTorr of oxygen at 25 °C. (a) O 1s probed at an incident photon energy (IPE) of 690 eV (shallow inelastic mean free path (IMFP) ~ 5.8 Å,[23] solid line) and 750 eV (deep IMFP ~ 6.6 Å,[23] dashed line). Features at the surface (OH + OCH_3) are less intense when probed with a greater IPE. (b) C 1s core level (490 eV IPE) depicting the different features observed.

3.1.2 C 1s core level. For the carbon core level, four distinct features at ~284.8 eV, ~286.1 eV, ~288.6 eV and ~292 eV were observed as shown in Fig. 1b. The feature at ~284.8 eV is adventitious carbon (CH_x).[24] The feature at 286.1 eV is at BE consistent with C–O or C≡O bonds, potentially from a OCH_3 species or CO.[25] Although both can be intermediates in MOR, we assign this feature to OCH_3 as it is the predominant intermediate for the MOR on transition metals.[4,7,26] The feature at 288.6 eV arises from CO_3 (ref. 27) and the feature at 292 eV from gas phase CH_3OH. A small amount of K was observed on the Al4Fe6 sample (Fig. S3†); based on tabulated cross sections, the abundance of K is ~10× lower than its 2p intensity ratio with C 1s at 490 eV. Table S2† shows the average BE and FWHM of the features observed in the carbon core level for the films across different methanol : oxygen gas ratios at different temperatures. Example room temperature carbon core level spectra in Fig. S2† depict the changes in feature BE and intensity with film composition and gas ratio.

3.1.3 Metal core levels. For Al8Fe2, four distinct features at ~74 eV, ~62 eV, ~66 eV and ~56 eV were observed as shown in Fig. S4.† The feature at ~74 eV is assigned to the Al 2p core level and that at ~56 eV to the Fe 3p core level, whose BE is referred to here as that from a single Gaussian–Lorentzian feature. The features at ~62 eV and ~66 eV are assigned to the Ir 4f doublet from the conductive substrate coating. Comparing across IPEs with varied mean free paths, the Ir layer remains buried under the $Al_{(1-x)}M_xO_y$ film (Fig. S5†). A similar spectrum was obtained for the Al4Fe6 film as well. For the sample with Mn content, a feature at ~47 eV was observed (Fig. S4†) instead of ~56 eV, which is assigned to the Mn 3p.

3.2 Electronic structure of the amorphous oxide films

We first consider the electronic structure of the Al8Fe2, Al4Fe6, and Al6Mn4 films by comparing their metal and oxygen core levels, as well as the valence band (VB). In all films, fitting of the Fe 3p and Mn 3p (Fig. S6 and S7†) core levels gives a nominal oxidation state of +3.[20] The oxidation state of the Fe 3p can be assessed using its FWHM (Fig. S7†) and BE (discussed in detail in Section 3.3.4). While the oxidation state of Fe 3p is predominantly +3 for the different gas compositions and temperatures, small changes in oxidation state with gas composition and in some cases temperature (consistent with VB spectra as shown in Fig. S8†) for Fe-containing films are discussed in more detail in Section 3.3.4. The VB edge, fit with a step down function and noted at half height, is highest in BE—furthest from the Fermi level—for Al8Fe2 (Table 1 and Fig. S9†), but shifts approximately ~1 eV lower with increasing Fe content in Al4Fe6, and is similar for Al6Mn4. For other methanol : oxygen gas ratios, the VB edge is similar for Al8Fe2 and Al6Mn4,

Table 1 BE of the bulk O 1s, Al 2p, M 3p and VB edge for Al8Fe2, Al4Fe6 and Al6Mn4 films for 50 mTorr of CH_3OH and 100 mTorr of oxygen at 25 °C

Core level	BE for Al8Fe2 (eV)	BE for Al4Fe6 (eV)	BE for Al6Mn4 (eV)
O 1s (bulk)	531.4	530.9	531.2
Al 2p	74.2	74	74.3
M 3p	56.1	55.7	48.8
VB edge	3.7	2.7	2.5

but decreases with increase in Fe content in Al4Fe6 (Fig. S9†). The use of a conductive Ir underlayer beneath the 5 nm AMO film, grounding of the AMO film surface through the thermocouple, and presence of gas phase photoelectrons mitigate potential charging effects. For example, the as measured BE of the bulk O 1s (750 and 690 eV IPE) and Al 2p (690, 490, and 350 eV IPE) of Al8Fe2 in 10 : 100 mTorr methanol : oxygen decreased only ~0.2 eV upon heating from room temperature to 275 °C, suggesting minimal potential charging effects.

Similarly, both the bulk O 1s feature and the Fe 3p shift to ~0.5 eV and ~0.4 eV lower BE with increasing Fe content (Fig. S9†). This could include effects from a shift in the rigid electronic structure (lesser charging) with Fe substitution, given the shift of VB edge towards the Fermi level, however we note that at the same IPE, the VB edge shifts by 1 eV while the Al 2p shifts by only 0.2 eV. Thus, a lower bulk O 1s BE may also indicate a greater basicity (electron density) on O, though surprising given the more covalent nature of Fe–O *versus* Al–O bonds based on electronegativity differences. In addition, the bulk O 1s feature shifts to lower BE for more reducing methanol : oxygen gas ratios with increasing temperature (Fig. S10†).

3.3 Adsorbate formation and mechanistic implications

We next consider the surface speciation and its relation to AMO composition and electronic structure. On oxides, CH_3OH dissociatively adsorbs to form CH_3O groups either by deprotonation at a metal site or dehydroxylation at an oxygen site. The metal site (M) is considered redox-active ($O + M + CH_3OH \rightarrow CH_3O\text{–}M + H\text{–}O$), whereas a surface O–H group can give up a proton (acidic character) to dehydrate CH_3OH in this process ($H\text{–}O + CH_3OH \rightarrow CH_3\text{–}O + H_2O$).[4] OCH_3 groups can subsequently deprotonate to form CH_2O, or C–O coupling can occur between a CH_3 group and CH_3O–M to form $(CH_3)_2O$. Adsorbed CO_3 is expected to form at basic sites as a precursor to CO_2 desorption ($O_2C\text{–}O \rightarrow O + CO_2$). We consider the relative abundance of OCH_3 and CO_3 surface species as obtained by the fitting of the C 1s spectra for different methanol : oxygen gas ratios (noted in mTorr) and temperature to understand the nature of AMO active sites for MOR.

3.3.1 Formation of adsorbates across compositions and conditions. Changes in film composition can affect the formation of OCH_3 groups. The intensity of the OCH_3 (C 1s) normalized to that of the bulk oxide (O 1s) is compared across Fe-containing samples (Fig. 2). If the intensity of bulk O is not attenuated appreciably by other species, such as adventitious carbon, this ratio is proportional to elemental coverage.[28] As expected with an increase in thermal energy, the OCH_3/bulk ratio decreases with increase in temperature. We note, however that the steeper decrease for Al4Fe6 for temperatures > 150 °C is commensurate with observed conversion to CO_2 and H_2O (Fig. S11†). The OCH_3 coverage is greater for Al4Fe6 compared to Al8Fe2 for all methanol : oxygen gas ratios, indicating that Fe is active for OCH_3 formation. While the OCH_3/bulk ratio at room temperature increases with the partial pressure of CH_3OH in the chamber, indicating sub-monolayer concentration on the surface, this increase is smallest on Al8Fe2 (and negligible comparing 50 : 100 and 100 : 50 methanol : oxygen gas ratios), further confirming the role of Fe in OCH_3 formation. The drop in OCH_3/bulk ratio with increasing temperature is larger for Al4Fe6 (77% for 50 : 100 methanol : oxygen gas ratio) than Al8Fe2 (46%) which may indicate that while

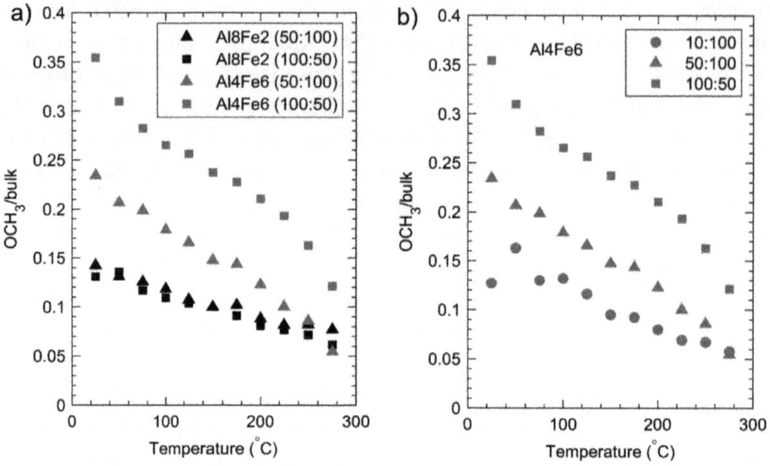

Fig. 2 Ratio of OCH_3 (C 1s) to the bulk (O 1s) with temperature for (a) Al8Fe2 and Al4Fe6 for 50 : 100 and 100 : 50 methanol : oxygen gas ratios and (b) Al4Fe6 for different gas ratios, where the OCH_3 coverage increases with CH_3OH partial pressure in the reactant gas. In the above figure, 10 : 100, 50 : 100 and 100 : 50 denote the pressures of methanol : oxygen in mTorr.

substitution of more acidic Fe for Al increases the number of sites for OCH_3 formation, it also reduces the strength of their binding or facilitates their oxidative removal.

To assess the coverage of CO_3 (expected at basic sites as a precursor to CO_2 desorption), its intensity is calculated relative to the bulk. The CO_3/bulk ratio is comparable on both Fe-containing films (Fig. 3 and S12†). The shift in bulk O 1s to lower BE from Al8Fe2 to Al4Fe6 is consistent with an increase in basicity of the O^{2-} sites upon incorporation of Fe^{3+} into the Al^{3+}-oxide host (Table 1 and Fig. S9†). The comparable CO_3/bulk ratio for these two films may thus suggest basic sites are not the primary contributor to CH_3OH oxidation under these conditions, assessed more in the following section. The CO_3/bulk ratio decreases with temperature for all film compositions (Fig. S12†). Generally, the amount of

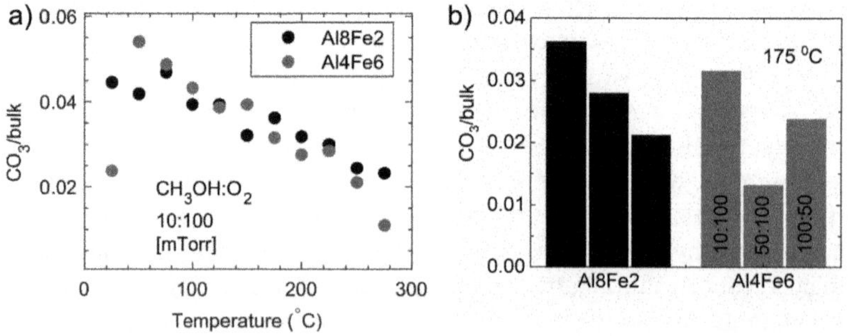

Fig. 3 Ratio of the CO_3 (C 1s) to bulk (O 1s) for (a) Al8Fe2 and Al4Fe6 for 10 mTorr of CH_3OH and 100 mTorr of oxygen with temperature and (b) Al8Fe2 and Al4Fe6 at 175 °C for noted methanol : oxygen gas pressures [mTorr].

CO_3 on the surface also decreases with decreasing partial pressure of oxygen as shown in Fig. 3, as would be expected given its production from complete CH_3OH oxidation at a stoichiometric 3 : 2 methanol : oxygen ratio. A notable exception is Al4Fe6 in 100 : 50 mTorr methanol : oxygen, which has greater CO_3 coverage compared to 50 : 100 mTorr methanol : oxygen.

3.3.2 Nature of active site. To better understand the nature of the active site and pathway for CH_3OH oxidation, we consider the relative coverage of CO_3 and OCH_3 species (Fig. 4). The CO_3/OCH_3 ratio is similar (0.3–0.4) across films in 10 : 100 mTorr of methanol : oxygen.

The CO_3/OCH_3 ratio tells the ratio of the number of basic (CO_3 forming) to acidic and redox-active (OCH_3 forming) sites. For all films considered here, the number of basic sites is more than two times lower than the total number of acidic and redox-active sites. The CO_3/OCH_3 ratio decreases with increase in CH_3OH partial pressure indicating that the formation of CO_3 (completely oxidized product) decreases as oxygen partial pressure is decreased, also evident from considering the CO_3/bulk oxide ratio (Fig. 3 and S12†). For amorphous Al4Fe6, we observe complete oxidation products of CO_2 and H_2O with a mass spectrometer for both 10 : 100 and 50 : 100 methanol : oxygen ratios (Fig. S10†) starting at 150 °C, but not $(CH_3)_2O$ at any temperatures. For amorphous Al6Mn4 in 50 : 100 methanol : oxygen, CO_2 and H_2O are produced starting at 200 °C, with small amounts of $(CH_3)_2O$ at lower temperatures (Fig. S11†), although we note that the quantification of evolved gases in the spectrometer is not calibrated. More notable complete oxidation products on Al4Fe6, compared to Al6Mn4, is consistent with both the lower CO_3/OCH_3 ratio (Fig. 4) and the concurrent greater reduction in adsorbed OCH_3 as temperature is increased (Fig. 2).

3.3.3 Mode of methanol dissociation on the films. We next examine the intensity ratio of the OCH_3 and OH features to ascertain the mode of CH_3OH activation on the surface. In the O 1s, these features cannot be distinguished and are fit as a single peak. In the C 1s, a distinct OCH_3 feature is fit. As the experimentally measured RSF of O : C is close to 1, taking the ratio of the OH + OCH_3

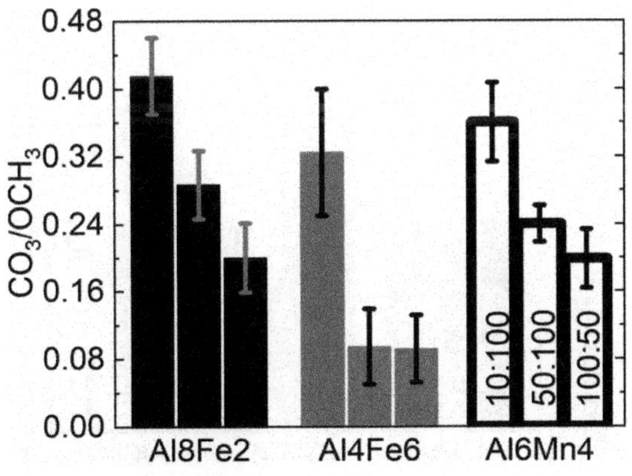

Fig. 4 Average ratio of CO_3 (C 1s) to OCH_3 (C 1s) species taken across temperatures for the Al8Fe2, Al4Fe6 and Al6Mn4 films. The relative abundance of CO_3 decreases with increasing the CH_3OH partial pressure in the reactant gas. In the above figure, 10 : 100, 50 : 100 and 100 : 50 denote the pressures of methanol : oxygen in mTorr. Error bars are from the standard deviation obtained across temperatures.

feature in the O 1s (fit at a chemical shift of 1.65 eV from the bulk) to the OCH_3 feature in the C 1s (approximately 1.6 eV relative to adventitious carbon), here referred to as the O : C ratio (Fig. 5), readily approximates the abundance ratio of these adsorbates, provided they are observable at the fit chemical shift. We note that the lower abundance of CO_3 on the surface, which is similar across samples and conditions, is not fit in the O 1s and would result in a roughly consistent offset to this ratio. CH_3OH has been previously proposed[5] to adsorb dissociatively by forming surface OCH_3 groups with the oxygen atom bonding to the metal site (CH_3O–M) and the proton from the alcoholic group bonding to a nearby lattice oxygen site forming O–H on the surface. An O : C ratio greater than 1 is consistent with both OCH_3 and OH groups giving rise to distinct XPS chemical shifts in the O 1s. The reduction in O : C ratio with increasing CH_3OH partial pressure for Al4Fe6 suggests some O species may not give rise to distinct chemical shifts (or a smaller shift poorly fit in the fixed binding energy offset employed here), or the removal of OH groups (probably as H_2O) from the surface, with lattice oxygen replenished by O_2 gas. The value of the fit O : C ratio is greater when the BE of the O 1s OH + OCH_3 feature floats freely, however the ratio for Al4Fe6 remains higher than that for Al8Fe2 at low CH_3OH partial pressure and comparable under reducing conditions (Fig. S13†). Al8Fe2 maintains comparable O : C ratio with CH_3OH partial pressure, however the Al6Mn4 O : C ratio decreases with increase in methanol partial pressure, just as for Al4Fe6, with relaxed fit.

The higher O : C ratio for Al4Fe6 at low CH_3OH pressures may indicate that the increased basicity of O sites upon Fe incorporation results in robust O 1s XPS chemical shifts at low OCH_3 coverage but less so as Fe sites become saturated. This is consistent with Al8Fe2, with constant OCH_3 coverage indicating site saturation, having a low O : C ratio regardless of CH_3OH partial pressure.

3.3.4 Changes in Fe electronic structure. To further assess the role of transition metal in activating CH_3OH, we consider potential changes in Fe valence. The BE of Fe 3p remains similar with temperature but decreases with increase in

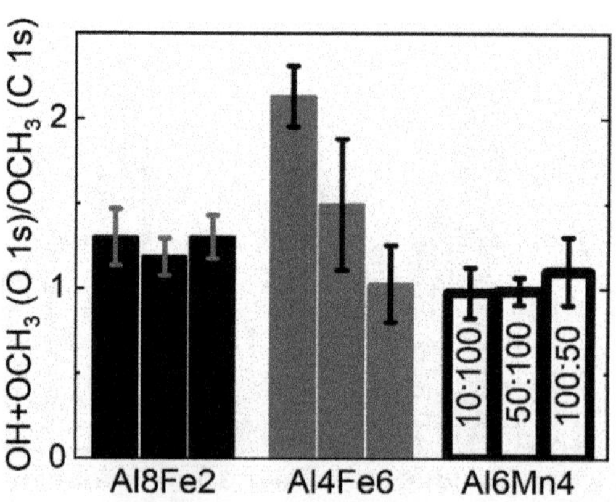

Fig. 5 Average across temperatures for the ratio of OH + OCH_3 feature (O 1s) to OCH_3 feature (C 1s) for different methanol : oxygen gas pressures for the Al8Fe2, Al4Fe6 and Al6Mn4 films. In the above figure, 10 : 100, 50 : 100 and 100 : 50 denote the pressures of methanol : oxygen in mTorr. Error bars are from the standard deviation obtained across temperatures.

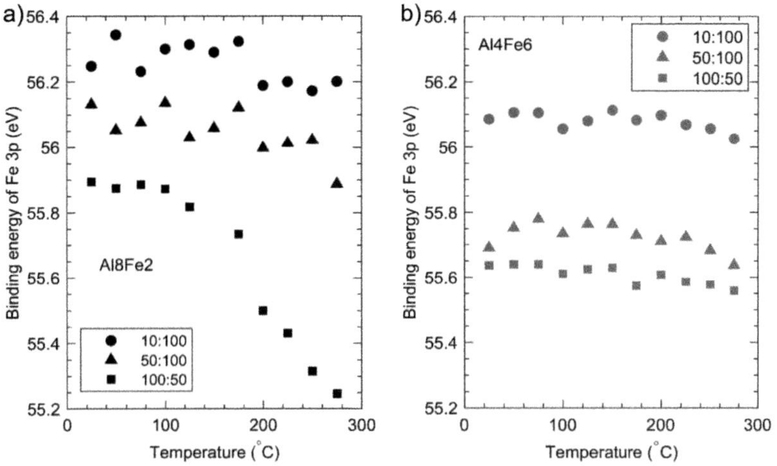

Fig. 6 BE of Fe 3p with temperature for (a) Al8Fe2 and (b) Al4Fe6 for different methanol : oxygen gas ratios. In the above figure, 10 : 100, 50 : 100 and 100 : 50 denote the pressure of methanol : oxygen in mTorr.

CH_3OH partial pressure in the reactant gas (Fig. 6 and S9†). Considering the negligible change in OCH_3 coverage with increase in partial pressure of CH_3OH for Al8Fe2, suggesting saturation of active sites, this decrease may result from the formation of oxygen vacancies under more reducing conditions. Oxygen vacancies may similarly contribute to the observed Fe 3p shift to lower BE in Al4Fe6, however the increase in OCH_3 coverage on Al4Fe6 with CH_3OH pressure might also result in Fe reduction. In contrast to other conditions, the Fe 3p BE of Al8Fe2 in a 100 : 50 methanol : oxygen gas ratio decreases with increasing temperature, commensurate with a new feature below the VB edge consistent with the presence of Fe^{2+} (Fig. S8†). Observable reduction of Fe in Al8Fe2, but not Al4Fe6, might arise from a larger proportion of the Fe sites (presumed saturated with OCH_3 species on Al8Fe2) driving CH_3OH (partial) oxidation with increasing temperature.

4. Discussion

In this study, we have incorporated transition metals (Fe, Mn) into an amorphous Al oxide host as a model of highly dispersed catalyst/support for the MOR. Cations with d^5 (Fe^{3+}) and d^4 (Mn^{3+}) electronic configuration add new electronic states at top of the VB edge and within the bandgap of alumina.[17,20] Further, the average coordination number of the Fe^{3+} and Mn^{3+} in the films considered is similar to that of alumina, ~4.5, indicating dispersed substitution for Al within the amorphous host without phase segregation. In these films, the interaction between "catalyst" (Fe, Mn ions) and "support" (Al oxide) is extended uniformly across the surface, as evidenced by the shift in VB maximum and bulk oxygen BE towards the Fermi level with Fe content (Fig. S7†). However, this increase in oxygen basicity did not translate to increased coverage of CO_3 on the surface under MOR conditions, suggesting that oxygen sites do not dominate CH_3OH activation. With low metal content, the coverage of OCH_3 was independent of CH_3OH partial pressure, suggesting saturation of active sites. In contrast, OCH_3 coverage increased with CH_3OH partial pressure at higher metal content, consistent with CH_3OH activation primarily occurring at these redox-active sites whose oxidation state lowered with OCH_3 adsorption.

Previous reports have incorporated manganese and iron oxide catalysts onto γ-alumina supports, which lowered the temperature of MOR compared to the support alone in an approximately 10 : 100 methanol : oxygen ratio.[14] At low temperatures, $(CH_3)_2O$ was the primary oxidation product, while complete oxidation to CO_2 dominated above 200 °C. For AMO films, despite the greater coverage of OCH_3 compared to CO_3 species, little to no $(CH_3)_2O$ is observed on Al4Fe6, with small amounts forming at temperatures as low as ~50 °C on Al6Mn4. Formation of $(CH_3)_2O$ requires C–O coupling between two adjacent $(O)CH_3$ groups, which may be limited by the dispersion of active transition metal sites in the solid-substitution amorphous films considered here. In contrast, complete oxidation to CO_2 and H_2O is observed at 150 °C—lower in AMOs than discrete supported catalysts by 50 °C—indicating that dispersing metal sites in amorphous oxide hosts is a promising route to maximize support effects in MOR.

5. Conclusions

We have employed AP-XPS to study the role of transition metal and its concentration on the dissociative adsorption and oxidation of methanol on amorphous $Al_xM_{(1-x)}O_y$ films, where M = Fe or Mn. Alumina's acid/base characteristics, which maximize dispersion of transition metal (oxides) on supports, facilitate their incorporation at high concentrations in AMOs. We here find that this dispersion in an amorphous framework reduces the temperature of complete methanol oxidation compared to supported MO_x/Al_2O_3 systems in literature and promotes complete oxidation over C–O coupling. The transition metal site (Fe, Mn) was redox-active, leading to the formation of OCH_3 groups whose coverage increases with transition metal content, indicating that the transition metal sites are active for the MOR. By comparison, the basic O^{2-} sites formed adsorbed CO_3 species with roughly comparable coverages across samples and conditions, present at roughly half the concentration of OCH_3 groups. Oxygen vacancy formation is supported by slight reductions in Fe valence (though remaining primarily +3) with increase in methanol partial pressure for the AMO films. These findings illustrate the nature of active sites for the MOR on AMO films and suggest dispersion of transition metals within an amorphous matrix offers new handles to tailor activity and selectivity in oxidation reactions.

Conflicts of interest

There are no conflicts to declare.

Acknowledgements

Acknowledgment is made to the donors of The American Chemical Society Petroleum Research Fund for partial support of this work. K. A. S. acknowledges support from Oregon State University as a Callahan Faculty Scholar. Film deposition was supported by the NSF CCI, The Center for Sustainable Materials

Chemistry (CHE-1606982). This research used resources of the Advanced Light Source, which is a DOE Office of Science User Facility under contract no. DE-AC02-05CH11231.

References

1 G. Iaquaniello, G. Centi, A. Salladini and E. Palo, in *Methanol*, ed. A. Basile and F. Dalena, Elsevier, 2018, pp. 595–612, DOI: 10.1016/B978-0-444-63903-5.00022-4.

2 W.-C. Liu, J. Baek and G. A. Somorjai, *Top. Catal.*, 2018, **61**, 530–541.

3 J. M. Tatibouët, *Appl. Catal., A*, 1997, **148**, 213–252.

4 M. Badlani and I. E. Wachs, *Catal. Lett.*, 2001, **75**, 137–149.

5 L. J. Burcham, L. E. Briand and I. E. Wachs, *Langmuir*, 2001, **17**, 6164–6174.

6 H. Bluhm, M. Hävecker, A. Knop-Gericke, E. Kleimenov, R. Schlögl, D. Teschner, V. I. Bukhtiyarov, D. F. Ogletree and M. Salmeron, *J. Phys. Chem. B*, 2004, **108**, 14340–14347.

7 M. Bowker, E. K. Gibson, I. P. Silverwood and C. Brookes, *Faraday Discuss.*, 2016, **188**, 387–398.

8 J. Döbler, M. Pritzsche and J. Sauer, *J. Am. Chem. Soc.*, 2005, **127**, 10861–10868.

9 H. Hu and I. E. Wachs, *J. Phys. Chem.*, 1995, **99**, 10911–10922.

10 K. Kähler, M. C. Holz, M. Rohe, A. C. van Veen and M. Muhler, *J. Catal.*, 2013, **299**, 162–170.

11 T. Kropp, J. Paier and J. Sauer, *J. Am. Chem. Soc.*, 2014, **136**, 14616–14625.

12 S. Oh, Y. K. Kim, C. H. Jung, W. H. Doh and J. Y. Park, *Chem. Commun.*, 2018, **54**, 8174–8177.

13 x. Zhang, X. Chen, Y. Liu and M. Guo, *Water, Air, Soil Pollut.*, 2020, **231**, 277.

14 U. S. Ozkan, R. F. Kueller and E. Moctezuma, *Ind. Eng. Chem. Res.*, 1990, **29**, 1136–1142.

15 H.-J. Freund and G. Pacchioni, *Chem. Soc. Rev.*, 2008, **37**, 2224–2242.

16 B. R. Goldsmith, B. Peters, J. K. Johnson, B. C. Gates and S. L. Scott, *ACS Catal.*, 2017, 7, 7543–7557.

17 M. G. Kast, E. A. Cochran, L. J. Enman, G. Mitchson, J. Ditto, C. Siefe, P. N. Plassmeyer, A. L. Greenaway, D. C. Johnson, C. J. Page and S. W. Boettcher, *J. Am. Chem. Soc.*, 2016, **138**, 16800–16808.

18 M. Aykol and K. A. Persson, *ACS Appl. Mater. Interfaces*, 2018, **10**, 3039–3045.

19 C. Zhang, R. D. Fagan, R. D. L. Smith, S. A. Moore, C. P. Berlinguette and S. Trudel, *J. Mater. Chem. A*, 2015, **3**, 756–761.

20 L. J. Enman, M. G. Kast, E. A. Cochran, E. Pledger, M. B. Stevens and S. W. Boettcher, *J. Phys. Chem. C*, 2018, **122**, 13691–13704.

21 K. A. Stoerzinger, L. J. Enman, E. A. Cochran, J. T. Diulus, R. T. Frederick, K. Artyushkova, E. J. Crumlin, G. S. Herman and S. W. Boettcher, *J. Phys. Chem. C*, 2019, **123**, 27048–27054.

22 M. E. Grass, P. G. Karlsson, F. Aksoy, M. Lundqvist, B. Wannberg, B. S. Mun, Z. Hussain and Z. Liu, *Rev. Sci. Instrum.*, 2010, **81**, 053106.

23 A. Akkerman, T. Boutboul, A. Breskin, R. Chechik, A. Gibrekhterman and Y. Lifshitz, *Phys. Status Solidi B*, 1996, **198**, 769–784.

24 E. R. Mysak, J. D. Smith, P. D. Ashby, J. T. Newberg, K. R. Wilson and H. Bluhm, *Phys. Chem. Chem. Phys.*, 2011, **13**, 7554–7564.

25 A. S. Duke, R. P. Galhenage, S. A. Tenney, S. C. Ammal, A. Heyden, P. Sutter and D. A. Chen, *J. Phys. Chem. C*, 2015, **119**, 23082–23093.
26 C. Cao and K. L. Hohn, *Appl. Catal., A*, 2009, **354**, 26–32.
27 D. J. Miller, M. C. Biesinger and N. S. McIntyre, *Surf. Interface Anal.*, 2002, **33**, 299–305.
28 E. McCafferty and J. P. Wightman, *Surf. Interface Anal.*, 1998, **26**, 549–564.

Faraday Discussions

PAPER

Oxygen relocation during HfO_2 ALD on InAs

Giulio D'Acunto,[ab] Esko Kokkonen,[c] Payam Shayesteh,[ab] Virginia Boix,[ab] Foqia Rehman,[ab] Zohreh Mosahebfard,[abe] Erik Lind,[bd] Joachim Schnadt[abc] and Rainer Timm[*ab]

Received 6th December 2021, Accepted 19th January 2022
DOI: 10.1039/d1fd00116g

Atomic layer deposition (ALD) is one of the backbones for today's electronic device fabrication. A critical property of ALD is the layer-by-layer growth, which gives rise to the atomic-scale accuracy. However, the growth rate – or growth per cycle – can differ significantly depending on the type of system, molecules used, and several other experimental parameters. Typically, ALD growth rates are constant in subsequent ALD cycles, making ALD an outstanding deposition technique. However, contrary to this steady-state – when the ALD process can be entirely decoupled from the substrate on which the material is grown – the deposition's early stage does not appear to follow the same kinetics, chemistry, and growth rate. Instead, it is to a large extent determined by the surface composition of the substrate. Here, we present evidence of oxygen relocation from the substrate-based oxide, either the thermal or native oxide of InAs, to the overlayer of HfO_2 in the initial ALD phase. This phenomenon enables control of the thickness of the initial ALD layer by controlling the surface conditions of the substrate prior to ALD. On the other hand, we observe a complete removal of the native oxide from InAs already during the first ALD half-cycle, even if the thickness of the oxide layer exceeds one monolayer, together with a self-limiting thickness of the ALD layer of a maximum of one monolayer of HfO_2. These observations not only highlight several limitations of the widely used ligand exchange model, but they also give promise for a better control of the industrially important self-cleaning effect of III–V semiconductors, which is crucial for future generation high-speed MOS.

[a]Division of Synchrotron Radiation Research, Department of Physics, Lund University, 22100 Lund, Sweden. E-mail: Rainer.Timm@sljus.lu.se
[b]NanoLund Center for Nanoscience, Lund University, 22100 Lund, Sweden
[c]MAX IV Laboratory, Lund University, 22100 Lund, Sweden
[d]Department of Electrical and Information Technology, Lund University, 22100 Lund, Sweden
[e]Department of Physics, University of Siegen, 57072 Siegen, Germany

Introduction

The future development of semiconductor technology is intimately connected to the development of faster and smaller devices, with a reduced power consumption. The research efforts towards this goal are intensive, not least in relationship to the metal oxide semiconductor field-effect transistor (MOSFET). One of the most significant limitations of present MOSFET technology is the use of silicon, which, although cheap and easy to produce, can hardly reach the required speed of device operation needed to achieve the development goals defined by the International Technology Roadmap for Semiconductors.[1] A first step towards the replacement of Si materials has been made by replacing the native silicon oxide with a high-κ oxide such as HfO_2, allowing drastically smaller gate sizes.[2–4] Substituting the silicon channel material with a III–V semiconductor has often been suggested as another promising step towards faster MOSFET devices.[2,5] Among the III–V semiconductors, InAs plays a particularly important role, since it has a narrow band gap compared to that of silicon and an electron mobility that is around 20 times higher than that of silicon;[6] these properties make InAs particularly suitable for use in new-generation high-speed metal oxide semiconductor transistors, especially in radio frequency applications.[2,3,7]

One of the challenges that must be overcome before InAs/HfO_2 MOSFET devices can enter a broad market is to obtain a clean interface between the two materials. The quality of this interface plays an essential role, since defects related to the oxide of the substrate InAs hinder the overall device performance. Atomic layer deposition (ALD) is a deposition technique that potentially can provide a defect-free and conformal layer of HfO_2 over the entire surface of InAs; especially as it comes along with the so-called "self-cleaning" effect, which comprises the removal of the unwanted native oxide from InAs together with the deposition of the desired HfO_2. This "self-cleaning" or "clean-up" effect has been reported for native oxides of typical III–V surfaces including InAs, GaAs, GaSb, and mixed alloys of these materials[8–13] upon the ALD of Al_2O_3 and HfO_2, performed with trimethylaluminum (TMA) and tetrakis(DMA^-)hafnium (TDMAHf, where DMA^- stands for the dimethylamido ligands), respectively. However, many questions about the efficiency of the self-cleaning and the surface chemical reactions during the ALD process behind this are still unsolved, leaving substantial room for improvement of the resulting III–V/high-κ interface.

ALD, with enormous potential in a large variety of material-dependent technologies,[14–17] is based on the self-limiting reaction of gaseous precursors with the surface of a substrate. The ALD of transition metal oxides, such as HfO_2, uses two precursors, a metal and an oxygen one. After one cycle of ALD, defined by self-limiting reactions of the two precursors with the surface in two separate half-cycles, ideally one monolayer of the chosen material is obtained. The self-limiting deposition step, which is the basis of the layer-by-layer growth in ALD and which ensures control and uniformity of the ALD layers, is often described by an ideal reaction mechanism. In reality, typically more or less severe deviations from the ideal mechanism occur.[18,19] For example, the standard model of the ALD of high-κ oxide is based on a ligand exchange reaction mechanism, which does not consider the surface as an active element. This assumption is in contrast to the experimentally observed and strongly appreciated self-cleaning effect.

One should expect the influence of the substrate to decrease with an increasing number of ALD cycles. This is also reflected by the thickness of the ALD layer which, according to the ideal ALD mechanism, should grow linearly with increasing numbers of ALD cycles. Numerous studies already show that this linearity only exists from a certain number of cycles onward, while the initial ALD cycles behave differently and can be characterized by either higher or lower growth rates. For more details, we suggest the review by Sønsteby *et al.*[20] The number of cycles after which the overlayer growth is completely decoupled from the surface can vary strongly, depending on experimental parameters and especially the precursor pulse length. Importantly, the surface chemistry that defines the final interface between the substrate and the ALD-grown overlayer typically occurs during the initial half-cycles of ALD,[8,12,13] *i.e.* before the decoupling and, thus, before the ideal ALD models gain validity. Hence, it is important to obtain better insight into the details of the initial ALD process, where the influence of the surface plays a significant role.

Here, we investigate eight InAs samples prepared with varying oxide thicknesses, ranging from a few Å to a few nanometres, by ambient pressure X-ray photoelectron spectroscopy (AP-XPS), a method which allows the real-time monitoring of the deposition process.[8,21,22] We evaluate the influence of the surface on the composition and thickness of the overlayer of HfO_2 already formed during the first half cycle of ALD performed with TDMAHf. We find evidence that the interaction with the TDMAHf molecules guarantees perfect self-cleaning, *i.e.* the removal of almost all InAs oxide irrespective of the initial oxide thickness. The interface is sharp and defect-free and characterized by the presence of a thin In–O–Hf layer. The oxygen contained in the original InAs oxide is used to form HfO_2 already in the first ALD half-cycle. We find that the thickness of the thus formed HfO_2 layer appears to be self-limited to one monolayer. Interestingly, we also find that TDMAHf does not interact with the clean InAs surface, but only with its oxide.

The results show that the ligand exchange model becomes valid only after the initial ALD half-cycle, and they provide answers to previously open questions about the role of oxygen atoms during the self-cleaning effect. Furthermore, the observation of a sharp, defect-free InAs/high-κ interface is of high relevance for future MOSFET applications. Our findings suggest that the quality of the interface will depend strongly on the ALD parameters, especially a sufficiently long first half cycle, rather than the exact thickness and chemical composition of the initial InAs oxide.

Experimental

The experiments were performed at the AP-XPS end stations of the SPECIES beamline[23] of the MAX IV Laboratory in Lund, Sweden, and of the I511/SPECIES beamlines at the previous MAX-lab[24–26] synchrotron radiation facility. Eight samples of InAs were treated equally in the cleaning process as follows (*cf.* Table 1): the samples were etched in 37% HCl : H_2O 1 : 1 for 30 s to remove the original native oxide and other contamination and then rinsed in ultrapure water for 60 s. The result of this first treatment step is a clean InAs (100) surface, which after the water rinse and subsequent contact to air is covered with a thin native oxide again. After that, the samples were mounted onto a stainless-steel sample holder and transferred to the ultrahigh vacuum (UHV) preparation chamber of the

Table 1 Cleaning treatments performed on each sample. A, B, C and D are referred to as "thermal oxide InAs" in this paper. E, F, G and H are referred to as "native oxide InAs"

A	B	C	D	E, F, G and H
Hydrogen cleaned	Hydrogen cleaned + 0.05 L O_2	Hydrogen cleaned + 6.00 L O_2	Hydrogen cleaned + 10.00 L O_2	Water bath

beamline. Samples E, F, G and H did not receive any other treatment, while samples A, B, C and D were heated up to around 360 °C and exposed to atomic hydrogen for 20 minutes, obtained using a thermal cracker (from MBE Komponenten) operated at about 1700 °C and a H_2 pressure of 2×10^{-6} mbar. This process produced four clean InAs surfaces without the presence of oxygen or carbon (as inspected by XPS). A more detailed study of the atomic hydrogen cleaning procedure can be found in ref. 27. No other treatment was done for one of the samples (here called sample A). For the other three samples, exposure to oxygen was performed in the preparation chamber at a sample temperature of 320 °C: 0.05 L (sample B), 6 L (sample C) and 10 L (sample D). The cleaning treatments performed on each sample are summarised in Table 1.

The first HfO_2 ALD half-cycle on all InAs(100) samples was performed with TDMAHf in the reaction cell of the AP-XPS chamber with the sample heated to 200–220 °C. The TDMAHf container (acquired from Strem Chemicals, Inc.) was attached to the gas system of the setup and heated to about 80 °C in order to increase the vapour pressure. The reaction cell was operated in constant flow mode, so that the TDMAHf was delivered at a constant pressure of about 10^{-3} mbar, measured by a gauge installed on the outlet of the gas line system. No carrier gases were used.

The InAs surfaces were first characterized in UHV with XPS. Then, time-resolved AP-XPS experiments were performed during the TDMAHf deposition; finally, another characterization data set was measured after the precursor gases were evacuated at a pressure of around 10^{-6} mbar. The spectra acquired before and after each half cycle were collected at 675 eV and 350 eV photon energy with a total instrumental broadening of around 0.35 eV for samples A, B, C and D. For the other samples (E, F, G and H), photon energies of 300 eV, 330 eV and 900 eV were used for the In 4d/Hf 4f and As 3d core levels and overview spectra, respectively, with a total instrumental broadening of around 0.50 eV. The sequences of In 4d, Hf 4f, and As 3d core-level spectra recorded during TDMAHf exposure at $\sim 10^{-3}$ mbar were measured with a photon energy of 675 eV. The time resolved spectra were measured in the fixed mode of the electron energy analyser in a sequence that took around six seconds to complete. The X-ray photoelectron (XP) spectra were curve-fitted using the IGOR Pro software by Wavemetrics, Inc. The binding energies were calibrated to the As $3d_{5/2}$ bulk component set to 41.0 eV.[28–31] Shirley backgrounds were removed before the fit. The intensities were normalized to that of the As 3d bulk component.

The thickness evaluation of the thermal oxide was carried out using an overlayer model explained elsewhere.[8] The cross sections, inelastic mean free paths, and atom densities were estimated from literature values.[8]

Results

In this section, we will first present the results on InAs samples covered by a thin thermal oxide layer and then on those with a slightly thicker native oxide layer. Finally, we will discuss the role of the oxide layers during the first half-cycle of ALD. The thermal and native oxides have a similar chemical composition and differ mainly in their thickness. The growth rate and stoichiometry of the thermal oxides can be controlled rather well, which is not the case for the native oxides. Moreover, as previously shown by our group,[21] the interface between the InAs surface and the high-κ oxide deposited by ALD is improved in terms of the interfacial defect densities, Fermi level unpinning and lower capacitance frequency dispersion when the ALD is carried out on an InAs surface covered by a thermal oxide layer in comparison to a native oxide-covered surface. In the present experiment, we have investigated both types of oxide for a practical reason: growing a thick thermal oxide layer is difficult, as is achieving thin native oxide layers. Since the goal of the current experiment is to compare thin and thick oxide layers, we adopt thermal oxidation of InAs to achieve submonolayer oxides and native oxidation to achieve thick oxide layers.

Thermal and native oxides of InAs, before TDMAHf deposition

We start by describing the state of the InAs surface and the effect of the treatments on the composition and thickness of the grown oxide layer. The surfaces do not exhibit any considerable C or N contamination, as can be seen from the survey spectra of samples A, C and G, shown in Fig. 1. High-resolution spectra of the As 3d and In 4d core levels before TDMAHf deposition are reported in Fig. 2. The As 3d XP spectrum of sample A, acquired at a photon energy of 350 eV, is shown in Fig. 2a. The As 3d spectrum is characterized by a doublet centred at 41.0 eV, and no component due to As oxide is visible. The corresponding In 4d core level (*cf.*

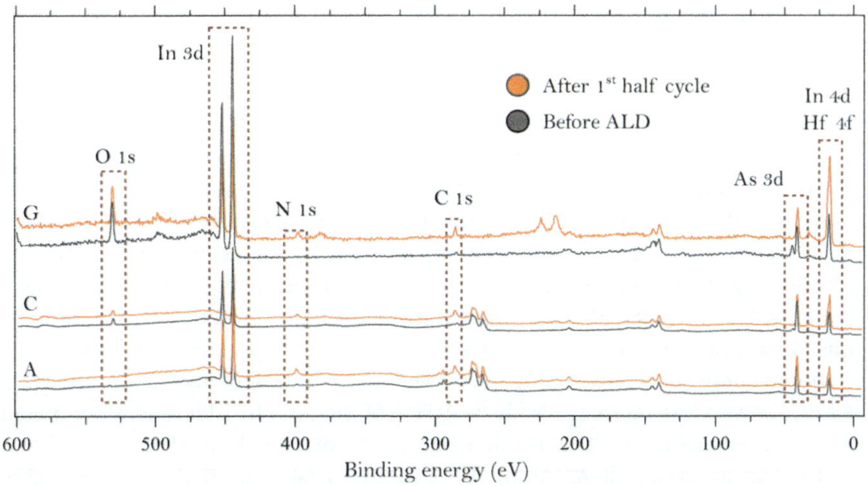

Fig. 1 Survey spectra collected on three of the InAs samples using photon energies of 675 eV (samples A and C) and 900 eV (sample G).

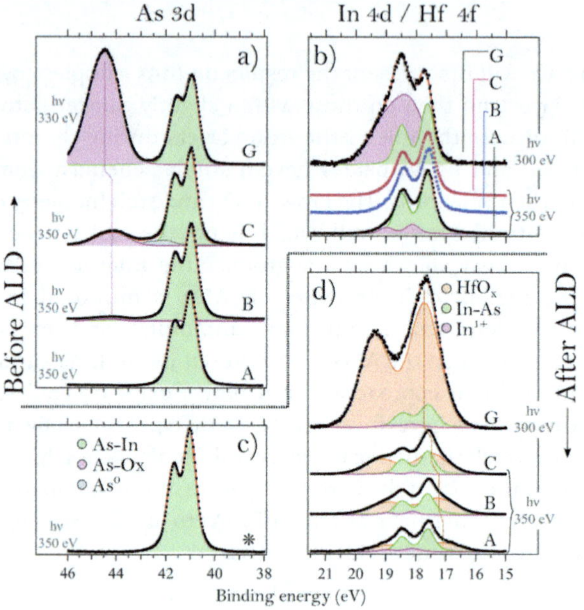

Fig. 2 Core-level spectra before and after TDMAHf deposition. (a) As 3d spectra of samples A, B, C and G acquired before TDMAHf deposition. The green doublet is associated with As bonded to In, while the broad purple doublet is associated with As bonded to O. (b) In 4d spectra of samples A, B, C and G before TDMAHf deposition. The green doublet is associated with In bonded to As, while the broad purple doublet is associated with In bonded to O. For samples B and C, the fit is not shown since the line shape is identical to that of sample A. (c) As 3d spectrum of sample C after the first ALD half-cycle. Since the spectra of all other samples look the same as that of sample C, only this particular spectrum is shown. (d) In 4d and Hf 4f spectra acquired after TDMAHf deposition. Green fits indicate the In 4d bulk component of InAs (In bonded to As), purple (small) its oxide, and orange (large) the HfO_x component. The vertical dotted line highlights the binding energy shift of the Hf 4f core level. (a–d) The photon energy used for measuring the spectra on samples A, B and C was 350 eV, while 330 and 300 eV were used for sample G. The spectra of sample G-H were collected at MAX-lab with a lower energy resolution (0.50 eV) in comparison to the spectra of samples A, B, C and D acquired at the MAX IV Laboratory (0.35 eV).

Fig. 2b), measured at a photon energy of 350 eV, is characterized by a dominant doublet at 17.6 eV, corresponding to In bonded to As, and a small shoulder towards high binding energy associated with In oxide (found at 18.1 eV; +0.5 eV from the bulk component); these values are in line with values found elsewhere in the literature.[8,12,32] The hydrogen cleaning process should result in an InAs surface without any oxide;[21] however, clearly this is not the case, but a thin In oxide layer remains. This unexpected behaviour confirms the high stability of the In oxide, as was already observed in our previous studies.[8,21] The In 4d core level acquired at a photon energy of 675 eV, *i.e.* in a somewhat more bulk-sensitive mode, shown in Fig. 1, shows a less intensive oxide peak (8% of the total In 4d intensity compared to 13% at $h\nu = 350$ eV). This implies that the In oxide is located at the surface.

In contrast to sample A, sample B – an InAs surface that had been exposed to 0.05 L of oxygen gas – shows a small As oxide component in the As 3d spectrum (*cf.* Fig. 2a). The component is located at 44.2 eV, *i.e.* at +3.2 eV higher binding energy than the InAs bulk component, in agreement with literature values.[8,10,29,33–34] The lineshape of the In 4d line is identical to that of sample A (see Fig. 2b).

Sample C – an InAs surface exposed to 6 L O_2 – differs from samples A and B, not only in terms of the expected more intense As oxide component due to the higher exposure to oxygen gas, but also by a small component at 42.0 eV, +1.0 eV above the bulk component (Fig. 2a). We assign this component to arsenic in the zero oxidation state, *i.e.* to As antisite defects or elemental As. This component is unexpected, since all samples were cut from the same InAs wafer, and sample C is the only one with an As^0 component. Its contribution to the overall As intensity is less than 5%, and its origin is likely associated with a non-perfect homogeneity of the wafer. The line shape of the In 4d spectrum of sample C is identical to the one of sample A (*cf.* Fig. 2b).

Lastly, the As 3d and In 4d core levels of the sample exposed to 10 L of oxygen (sample D) exhibit the main As 3d component centred at 41.0 eV and its oxide at 44.0 eV, +3.0 eV above the bulk component. As for the other samples, the In 4d bulk component is found at 17.6 eV and its oxide at 18.1 eV, +0.5 eV from the bulk component; however, sample D is the only InAs sample with a thin layer of thermal oxide that shows slightly higher intensity (with respect to the bulk) of the In oxide.

From a comparison of all the examined thermal oxide samples, it is clear that the exposure time to O_2 changes only the relative intensities of the As oxide and bulk components without changing the oxidation states of the surface. Indeed, we found that all As 3d oxide components are found at (+3.1 ± 0.1) eV from the InAs bulk component and the In 4d oxide ones at (+0.5 ± 0.1) eV above the bulk component. We also note that the relative binding energies of the oxide peak with respect to the bulk suggest that the As and In oxidation states are As^{3+} and In^+, as in As_2O_3 and In_2O. Interestingly, it appears as if exposure to O_2 doses below 10 L mainly leads to oxidation of the As atoms, while the surface In atoms are left unchanged, including a bulk InAs component and a small amount of In oxide. Finally, the thickness of the thermal oxide layers can be evaluated from the intensity ratio between the oxide and bulk components.[8] This is reported in Fig. 5 (first four points of the graph).

Turning to the samples with a thick, uniform native oxide layer, we find a common oxide stoichiometry for all of them; they only differ in the thickness of the oxide layer (see the Discussion section). The binding energies of the bulk and oxide components are found at values in excellent agreement with those found for samples A–D, except for the As 3d oxide. As is seen in Fig. 2a, the As oxide component of sample G is slightly shifted towards a higher binding energy with respect to the As oxide component of sample C. InAs oxides can contain As atoms in two different oxidation states, As^{3+} and As^{5+}. The As 3d component of As^{5+} is found at a 1–2 eV higher binding energy[12,33] than that of As^{3+}. The shift of the As oxide peak of sample G towards a higher binding energy amounts to around +0.3 eV, and we attribute it to the presence of a small amount of As^{5+} suboxides, not observed for the other samples. For simplicity, we fitted the data with a single doublet only that includes the contribution of both the As^{3+} and As^{5+} oxidation states. The presence of an As^{5+} suboxide in the native oxide of sample G is not completely unexpected; we have observed it previously,[21] and it is due to the lower homogeneity of the native oxide as compared to that of the thermal oxide layer.

To conclude this section, we find that the main difference between the analysed samples is the relative intensity of the oxide component with respect to the bulk peak, which has to be attributed to a different thickness of the oxide layer. Moreover, only the amount of As oxide differs for the samples with a thermal oxide.

Thermal and native oxides of InAs, during and after TDMAHf deposition

The core-level spectra collected after exposure of sample C to TDMAHf are shown in Fig. 2c and d. Additionally, in Fig. 4 we show the As 3d, In 4d and Hf 4f core level spectra acquired during the exposure, at a pressure of around 10^{-3} mbar. From Fig. 2C, it is seen that the As oxide of sample C gets fully removed by the interaction with TDMAHf. The observation is the same for all eight samples (not shown for the other seven samples), independent of the original quantity of the As oxide and its nature. The In 4d region (Fig. 2d) is more difficult to interpret. Here, a broad peak – the Hf 4f component – appears and overlaps with the In 4d peak. From a curve fit of this component we find the following Hf 4f binding energies: 17.1, 17.2, 17.4 and 17.5 eV for samples A, B, C and D, respectively, while it is 17.6 eV for all the other samples (E, F, G and H). Moreover, a small component of In oxide is visible in the In 4d/Hf 4f region for all examined samples. This is associated with an interface oxide, with the In oxide acting as As–In–O–Hf bridging bonds between bulk InAs and the HfO_x[8]. It is important to highlight that all remaining In oxides get fully removed during the first TDMAHf deposition, independent of the original amount, similar to the behaviour of the As oxide. Thus, we can confirm here a perfect self-cleaning of all investigated samples already during the first ALD half-cycle, since the only remaining O atoms found afterwards are those needed to form the InAs–HfO_2 interface.

The time evolution of the exposure can be evaluated from time-resolved As 3d, In 4d, and Hf 4f ambient pressure XP spectra taken during TDMAHf exposure. They are shown for sample C in Fig. 4. Here, the origin of the timescale (0 min) is defined as the beginning of the data acquisition; the green dotted line, around 3 minutes, defines the moment when the TDMAHf exposure starts. The higher photon energy of 675 eV used in the acquisition of the time-resolved spectra, as compared to the As 3d and In 4d/Hf 4f spectra shown in Fig. 2, makes the measurements somewhat more bulk-sensitive and therefore less sensitive to the As oxide signal. The time-resolved data show clearly that the removal of the As oxide is rapid and that it coincides with the growth of HfO_x in the Hf 4f signal. Hence, the spectra serve as a confirmation for the incorporation of the oxygen from the oxide of the InAs surface in the HfO_2 layer.

Turning back to Fig. 2, we note that the H4 4f binding energies vary between the samples. The variation can be understood from the dependency of the binding energy on the chemical environment of the Hf ions. Recently, we found that the first reaction step during the early stage of HfO_2 ALD by TDMAHf is a physisorption event, after which the Hf ion is still bound to all four ligands,[8,35] indicated by a lower binding energy. The interaction with the surface is mediated by a lone pair of the nitrogen atoms. Later in the deposition, the Hf ion is partially bound to oxygen atoms (from the surface) and to some of its DMA^- ligands, coming along with a shift to a higher binding energy. However, the binding energy of the Hf 4f line in stoichiometric HfO_2 is found to be even higher.[36–38] Therefore, we suppose that the Hf 4f binding energy observed here is a function of the average number of retained ligands: the larger the number of ligands, the lower the Hf 4f binding energy. This implies that with the increasing thickness of the original thermal oxide (samples A to D), a decreasing average number of ligands of the Hf precursor molecule is still retained after the first deposition.

It is also of interest to consider the molecular adsorbate species that appear on the surface upon interaction with TDMAHf. These species are easily identified

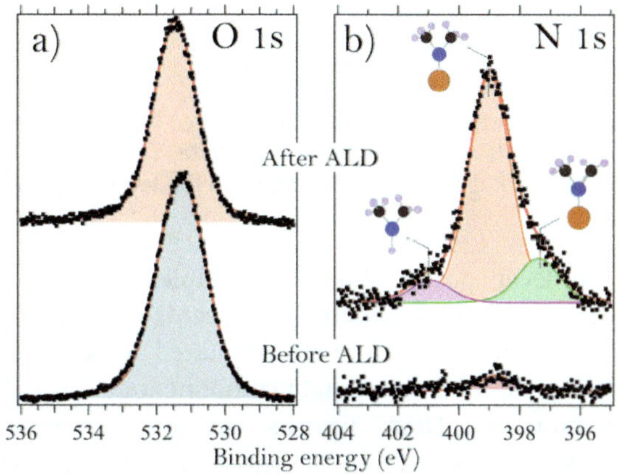

Fig. 3 (a) O 1s core level spectra of sample G acquired at 660 eV photon energy before (bottom) and after (top) the first half ALD half cycle. (b) N 1s spectrum acquired before (bottom) and after (top) the first ALD half-cycle on sample B. The photon energy was 675 eV. The three fitted components are assigned to different ligands, as indicated by their sketched structures, where blue (black, grey, orange) circles correspond to N (C, H, Hf) atoms, respectively.

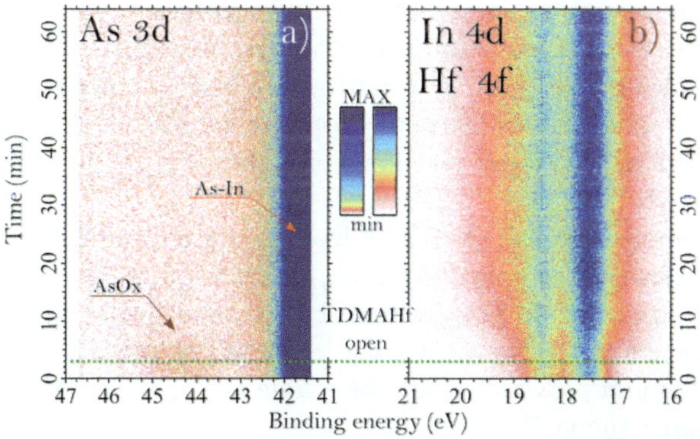

Fig. 4 Series of (a) As 3d and (b) In 4d/Hf 4f ambient pressure XP spectra measured during the first half-cycle of ALD using a photon energy of 675 eV. The *x* axes represent the binding energy scale and the *y* axes show the acquisition time. The green dotted line indicates when the valve to the TDMAHf precursor was opened. The pressure after exposure start was approximately 10^{-3} mbar. The intensity is represented by a colour gradient from white (minimum) to blue (maximum). The colour scale is logarithmic in (a) so that the less intense As 3d oxide peak at around 44 eV binding energy is enhanced. It is linear in (b).

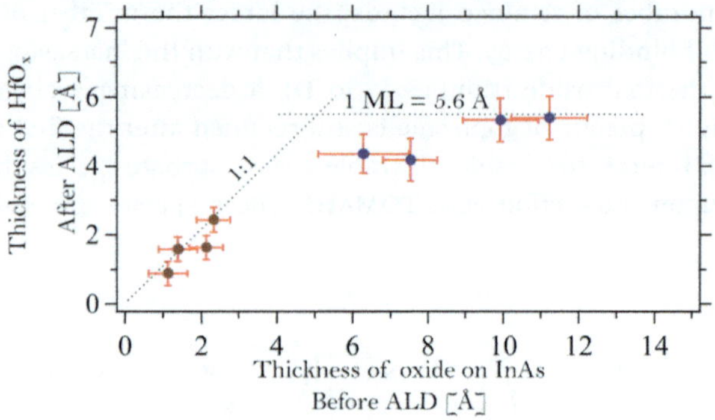

Fig. 5 Correlation graph between the oxide thickness of InAs prior to the deposition and the thickness of the HfO_x layer after the first half cycle. The samples are named alphabetically following their oxide thickness starting from the left (sample A).

from the N 1s core level spectra, shown for sample B in Fig. 3b. The spectrum shows a broad peak with two shoulders, one towards high and the other towards low binding energies. It is therefore deconvoluted with three identical Voigt profiles. For all samples, we find the main component at (399.0 $\pm$ 0.1) eV. This is due to photoemission from DMA^- ligands. The shoulders at 400.8 eV (+1.8 eV from the main peak) and 397.3 eV (−1.7 eV) are assigned to dimethylamine (DMA) and methyl methylene imine (MMI).[35,39,40] The relative intensities of the DMA^-, DMA and MMI components vary somewhat between the samples, but in all cases the DMA^- component is dominant with around 70% of the overall intensity, while the MMI and DMA peaks account for approximately 25% and 5% of the total N 1s intensity, respectively.

From a consideration of the In and As core levels, we find that the surface-adsorbed DMA^- is most likely bound to Hf: no shift or extra components are visible in these core levels. MMI, together with an equal amount of DMA, is formed in elimination reactions of DMA^- β-hydrides.[39,41] However, the areas subtended by the peaks of MMI and DMA are dissimilar, and this suggests a partial desorption of DMA, justified by the high temperature used during deposition. Alternatively, DMA could, in principle, also be formed in a ligand exchange reaction. This latter reaction would, however, require the presence of surface hydroxyls.[35,39]

A careful analysis of the O 1s core level before and after the deposition of TDMAHf, as shown for sample G in Fig. 3a, results in one symmetric peak. Oxygen bonded to As, In, or Hf (in HfO_2) cannot be distinguished since it results in the same binding energy. However, we can exclude the presence of any relevant amount of –OH groups, which would be visible as a separate O 1s component at a higher binding energy.[10]

Discussion

The functioning of ALD is based on the self-limiting interaction of gas-phase precursors with the surface onto which a thin layer of material should be deposited. Typically, the ALD of transition metal oxides is explained by invoking a ligand exchange reaction between the surface and the precursor molecules. The surface provides the active sites for this reaction in the form of surface hydroxyl groups, while the InAs substrate itself does not get involved. However, none of the examined InAs samples accommodate any hydroxyls. This is exemplified in Fig. 2a, which shows the O 1s core level spectrum of sample G. Clearly, the symmetric peak of O 1s does not exhibit any presence of –OH groups, which would lead to the appearance of a high-binding energy feature. Hence, we find that the standard ligand exchange reaction mechanism is not applicable to the initial ALD on the oxides of InAs.

Instead, the role of the surface oxide needs to be taken into account. It is seen in Fig. 1 and 2c, d that the surface oxide has almost completely vanished after the deposition of TDMAHf; actually, from the time-resolved data in Fig. 4 it is apparent that the removal took place during the first few moments of TDMAHf exposure. At the same time, the O 1s peak does not change drastically in intensity: for example, the O 1s line of sample G, displayed in Fig. 3a, loses only around (18 $\pm$ 5)% of its intensity. The trend is the same for all investigated samples with a native oxide layer. This evidence suggests strongly that the oxygen atoms are transferred from the InAs oxide to the chemisorption layer: HfO_2 is formed already in the first half-cycle of ALD.

As shown in Fig. 4, it is clearly possible to see how the component associated with arsenic oxide decreases after about 10 minutes from the exposure of the surface to the molecules and, with a slightly lower rate, the component associated with Hf 4f increases. The same behaviour is also observed for the time evolution APXPS measurement on sample D[35] (not shown here), moreover, for that sample we observe a rate of reduction for the In 4d oxide peak similar to the one of the Hf 4f.

The phenomenon of oxygen transport and oxide formation without an external source is not included in any of the standard ALD models; instead, these models assume a static surface with O (or rather hydroxyl) sites to which the TDMAHf precursor binds after splitting off a ligand. Only the subsequent water deposition half-cycle provides additional O sites in the shape of hydroxyl groups, according to the models. Since the oxide thicknesses in our study vary strongly between the thermal and native oxide samples, we now consider the details of the oxygen transfer separately for the two types of oxides.

InAs with a thermal oxide: samples A, B, C and D

A model structure containing two different overlayers was used to estimate the thickness of the interface and HfO_x layers on the InAs samples after TDMAHf deposition: the bulk InAs sample was assumed to be covered by an In_2O interface layer, which in turn is capped by the HfO_x layer. This model neglects the presence of molecular nitrogen and carbon materials on top of the surface. However, we consider that the DMA, DMA^-, and MMI surface species containing N and C are always located at the outermost surface. Their influence on the attenuation of the Hf 4f, As 3d, and In 4d signals from different layers underneath is then identical, and can, therefore, be ignored in the layer thickness calculation. In the same way, the thickness of the thermal oxide layer on the InAs substrates prior to TDMAHf deposition was calculated. The calculations deliver an average thickness of 1.1, 1.4, 2.1 and 2.3 Å for samples A, B, C and D, respectively. Hence, the thermal oxide does not continuously cover the entire InAs surface, but rather consists of islands. Assuming that a monolayer of the oxide has a thickness of around 5 Å, estimated by an average of In_2O and As_3O_5 oxides,[42,43] this translates into surface coverages of 22%, 28%, 42% and 46%, respectively. Since the XP spectra were collected in several spots of the sample and no differences between the different spots were detected, we conclude that the oxide islands are significantly smaller than the size of the beam spot on the sample ($100 \times 100\ \mu m^2$) and that they are distributed homogeneously on the sample surface.

As is shown in Fig. 5 (dark red points at low oxide thickness) we find an excellent one-to-one correlation between the thickness of the final HfO_x layer and that of the initial thermal oxide layer. A linear fit of the form $T_{HfO_x} = \alpha T_{InAsO_x}$ was performed and a coefficient $\alpha = 0.95 \pm 0.10$ is found. In the formula, T is the thickness of the HfO_x and the one of the oxide layer on InAs. Clearly, the amount of TDMAHf precursor material that interacts with the surface depends strictly on the quantity of thermal oxide prior to the deposition. The figure also shows that TDMAHf will interact only with the thermal oxide, but not with the clean InAs – otherwise, no correlation should be visible between the thickness of the initial InAs oxide layer and that of the final HfO_2 layer, and we would expect to see thicker HfO_2 layers also for samples with very thin thermal oxide layers.

The indium oxide component and the In 4d line shape, as shown in Fig. 2b, turn out to be the same prior to ALD for samples A, B and C. The main difference between them is the As oxide content. Interestingly, the amount of InO_x is reduced upon TDMAHf deposition by only 35% (from 12% to 8% of the overall In 4d signal), which shows that almost the entire indium oxide present before deposition was needed to form the interface between the InAs and the HfO_x. The As oxide, on the other hand, gets removed by 100% upon the initial ALD half-cycle.

For sample D, we also performed another half-cycle, executed with water vapour at 10^{-1} mbar. Interestingly, the InAs bulk component does not get re-oxidized during water deposition.[35] This suggests that an initial Hf oxide coverage of around 50% is enough to produce a clean and sharp interface between HfO_2 and InAs, which is uniform all along the surface and protected from re-oxidation during the water half-cycle.

Furthermore, after water deposition, a distinct hydroxyl peak was found in the O 1s spectra of sample D.[35] This is another strong indication for the validity of the ligand exchange model after the first ALD cycle, even though we demonstrate here the limitation of the model during the initial ALD cycle.

InAs with a native oxide: samples E, F, G and H

The thickness of the initial native oxide layer on samples E, F, G, and H is estimated to be 6.3, 7.5, 9.9, and 11.2 Å, respectively. The thickness of the HfO_2 layer after interaction with TDMAHf is also shown in Fig. 5 (blue data points). For these samples, the initial oxide thickness amounts to more than a monolayer. In contrast, the final HfO_2 thickness appears to be limited to one monolayer. This behaviour contrasts with that of samples A to D. The fact that no As oxide is left after TDMAHf deposition requires that part of the oxygen atoms leave the surface in a parallel reaction. However, the exact mechanism involved is not clear, since our time-resolved *in situ* studies show that the rate of formation of HfO_x and those of removal of the arsenic and indium oxide species are the same,[8] independently on the thickness of the oxide layer on InAs. Further studies on the nature of the oxide layer and the formed interface, as well as the exact mechanism of the oxygen relocation, *via* structural characterization based *e.g.* on surface diffraction and theoretical models, are needed here. Nevertheless, the experimental observation that the thickness of the HfO_2 layer after TDMAHf deposition is limited to one monolayer is important, as it confirms the self-limitation of ALD reactions, one of the basic ALD principles, even for the initial half-cycle.

Conclusions

In conclusion, we report an oxygen transition from the oxide of InAs to the HfO_x deposited by ALD using TDMAHf. Eight InAs samples, prepared with different oxide thicknesses, are analysed with *in situ* XPS. We find a linear correlation between the thickness of InAs oxide before deposition and the thickness of HfO_x at the end of the first half cycle of ALD, self-limited though to a maximum of one monolayer of HfO_x. Moreover, we observe that a 50% surface coverage of HfO_x after the first half-cycle is enough to form a uniform interface which will, after further ALD cycles, result in a sharp and native oxide-free interface between the InAs substrate and the HfO_2 layer.

In contrast to the widely accepted ligand exchange model, we reveal that no surface hydroxyl groups are needed to initiate the ALD reaction of HfO_2 formation on InAs.[39] Instead, InAs oxide reacts readily with TDMAHf in the first half-cycle, irrespective of whether the oxide is native or thermal in nature, while pure InAs is inert with respect to an interaction with the TDMAHf ALD precursor.

Even though the surface structure of the InAs substrate prior to ALD and possible pre-treatments does influence the thickness of the HfO_x layer after the first ALD half-cycle, which is not included in the ideal ALD models, we want to highlight that we observe an apparently perfect self-cleaning upon the first half-cycle over a broad range of initial InAs substrate configurations, if the initial deposition process is given sufficient time. This result is highly promising for the further development of high-speed, InAs-based MOSFET devices with improved III–V/high-κ interface quality.

Author contributions

G. D'A.: conceptualization, methodology, software, validation, formal analysis, investigation, writing – original draft, writing – review & editing, visualization. P. S., E. K., V. B., F. R. and N. M.: investigation, writing – review & editing. E. L., J. S., and R. T.: conceptualization, methodology, validation, formal analysis, investigation, writing – original draft, writing – review & editing, visualization.

Conflicts of interest

There are no conflicts to declare.

Acknowledgements

We acknowledge the MAX IV Laboratory for time on beamline SPECIES under Proposal 20200396. Research conducted at the MAX IV Laboratory, a Swedish national user facility, is supported by Vetenskapsrådet (Swedish Research

Council) under contract 2018-07152, Vinnova (Swedish Governmental Agency for Innovation Systems) under contract 2018-04969 and Formas under contract 2019-02496. The staff at MAX IV are gratefully acknowledged for assistance in the beamtimes. We acknowledge project funding from Vetenskapsrådet, grant no. 2016-6186, 2017-03871 and 2017-4108. This work was further supported by the NanoLund Center for Nanoscience at Lund University.

References

1 M. Graef, *International Roadmap for Devices and Systems - More than Moore - IRDS*, https://irds.ieee.org/editions/2020.

2 P. D. Ye, G. D. Wilk, J. Kwo, B. Yang, H. J. L. Gossmann, M. Frei, S. N. G. Chu, J. P. Mannaerts, M. Sergent, M. Hong, K. K. Ng and J. Bude, *IEEE Electron Device Lett.*, 2003, **24**, 209–211.

3 J. A. Del Alamo, *Nature*, 2011, **479**, 317–323.

4 H. Ko, K. Takei, R. Kapadia, S. Chuang, H. Fang, P. W. Leu, K. Ganapathi, E. Plis, H. S. Kim, S. Y. Chen, M. Madsen, A. C. Ford, Y. L. Chueh, S. Krishna, S. Salahuddin and A. Javey, *Nature*, 2010, **468**, 286–289.

5 J. Wu, A. S. Babadi, D. Jacobsson, J. Colvin, S. Yngman, R. Timm, E. Lind and L. E. Wernersson, *Nano Lett.*, 2016, **16**, 2418–2425.

6 S. M. Sze and K. K. Ng, in *Physics of Semiconductor Devices*, John Wiley & Sons, Inc., 2006, pp. 789–789.

7 K. Zhang, G. R. Zhu, Y. C. Kong, and T. S. Chen, *2018 Int. Conf. Microw. Millim. Wave Technol. ICMMT 2018 - Proc.*, 2018, vol. 39, pp. 179–180.

8 G. D'Acunto, A. Troian, E. Kokkonen, F. Rehman, Y. P. Liu, S. Yngman, Z. Yong, S. R. McKibbin, T. Gallo, E. Lind, J. Schnadt and R. Timm, *ACS Appl. Electron. Mater.*, 2020, **2**, 3915–3922.

9 R. Timm, A. Fian, M. Hjort, C. Thelander, E. Lind, J. N. Andersen, L. E. Wernersson and A. Mikkelsen, *Appl. Phys. Lett.*, 2010, **97**, 132904.

10 R. Timm, A. R. Head, S. Yngman, J. V. Knutsson, M. Hjort, S. R. McKibbin, A. Troian, O. Persson, S. Urpelainen, J. Knudsen, J. Schnadt and A. Mikkelsen, *Nat. Commun.*, 2018, **9**, 1412.

11 L. Ye and T. Gougousi, *Appl. Phys. Lett.*, 2014, **105**, 121604.

12 A. P. Kirk, M. Milojevic, J. Kim and R. M. Wallace, *Appl. Phys. Lett.*, 2010, **96**, 202905.

13 C. L. Hinkle, A. M. Sonnet, E. M. Vogel, S. McDonnell, G. J. Hughes, M. Milojevic, B. Lee, F. S. Aguirre-Tostado, K. J. Choi, H. C. Kim, J. Kim and R. M. Wallace, *Appl. Phys. Lett.*, 2008, **92**, 071901.

14 S. M. George, *Chem. Rev.*, 2010, **110**, 111–131.

15 J. A. Raiford, S. T. Oyakhire and S. F. Bent, *Energy Environ. Sci.*, 2020, **13**, 1997–2023.

16 A. S. Asundi, J. A. Raiford and S. F. Bent, *ACS Energy Lett.*, 2019, **4**, 908–925.

17 J. Plutnar and M. Pumera, *Small*, 2021, **17**, 2102088.

18 F. Zaera, *Coord. Chem. Rev.*, 2013, **257**, 3177–3191.

19 N. E. Richey, C. De Paula and S. F. Bent, *J. Chem. Phys.*, 2020, **152**, 040902.

20 H. H. Sønsteby, A. Yanguas-Gil and J. W. Elam, *J. Vac. Sci. Technol., A*, 2020, **38**, 020804.

21 A. Troian, J. V. Knutsson, S. R. McKibbin, S. Yngman, A. S. Babadi, L. E. Wernersson, A. Mikkelsen and R. Timm, *AIP Adv.*, 2018, **8**, 125227.

22 J. Schnadt, J. Knudsen and N. Johansson, *J. Phys.: Condens. Matter*, 2020, **32**, 413003.

23 E. Kokkonen, F. L. Da Silva, M. H. Mikkela, N. Johansson, S. W. Huang, J. M. Lee, M. Andersson, A. Bartalesi, B. N. Reinecke, K. Handrup, H. Tarawneh, R. Sankari, J. Knudsen, J. Schnadt, C. Sathe and S. Urpelainen, *J. Synchrotron Radiat.*, 2021, **28**, 588–601.

24 S. Urpelainen, C. Säthe, W. Grizolli, M. Agåker, A. R. Head, M. Andersson, S. W. Huang, B. N. Jensen, E. Wallén, H. Tarawneh, R. Sankari, R. Nyholm, M. Lindberg, P. Sjöblom, N. Johansson, B. N. Reinecke, M. A. Arman, L. R. Merte, J. Knudsen, J. Schnadt, J. N. Andersen and F. Hennies, *J. Synchrotron Radiat.*, 2017, **24**, 344–353.

25 J. Schnadt, J. Knudsen, J. N. Andersen, H. Siegbahn, A. Pietzsch, F. Hennies, N. Johansson, N. Mårtensson, G. Öhrwall, S. Bahr, S. Mähl and O. Schaff, *J. Synchrotron Radiat.*, 2012, **19**, 701–704.

26 R. Denecke, P. Väterlein, M. Bässler, N. Wassdahl, S. Butorin, A. Nilsson, J. E. Rubensson, J. Nordgren, N. Mårtensson and R. Nyholm, *J. Electron Spectrosc. Relat. Phenom.*, 1999, **101–103**, 971–977.

27 M. Hjort, J. V. Knutsson, B. Mandl, K. Deppert, E. Lundgren, R. Timm and A. Mikkelsen, *Nanoscale*, 2015, **7**, 9998–10004.

28 G. Hollinger, R. Skheyta-Kabbani and M. Gendry, *Phys. Rev. B*, 1994, **49**, 11159–11167.

29 D. M. Poirier and J. H. Weaver, *Surf. Sci. Spectra*, 1993, **2**, 224–231.

30 D. Y. Petrovykh, M. J. Yang and L. J. Whitman, *Surf. Sci.*, 2003, **523**, 231–240.

31 L. Olsson, L. Ilver, J. Kanski, P. Nilsson, C. Andersson and U. Karlsson, *Phys. Rev. B: Condens. Matter Mater. Phys.*, 1996, **53**, 4734–4740.

32 B. Brennan and G. Hughes, *J. Appl. Phys.*, 2010, **108**, 053516.

33 R. Timm, M. Hjort, A. Fian, C. Thelander, E. Lind, J. N. Andersen, L. E. Wernersson and A. Mikkelsen, *Microelectron. Eng.*, 2011, **88**, 1091–1094.

34 L. Ley, R. A. Pollak, F. R. McFeely, S. P. Kowalczyk and D. A. Shirley, *Phys. Rev. B*, 1974, **9**, 600–621.

35 G. D'Acunto, P. Shayesteh, E. Kokkonen, V. Boix, F. Rehman, N. Mosahebfard, E. Lind, J. Schnadt and R. Timm, in *Manuscr*, 2022.

36 O. Renault, D. Samour, J. F. Damlencourt, D. Blin, F. Martin, S. Marthon, N. T. Barrett and P. Besson, *Appl. Phys. Lett.*, 2002, **81**, 3627–3629.

37 J. C. Lee, S. J. Oh, M. Cho, C. S. Hwang and R. Jung, *Appl. Phys. Lett.*, 2004, **84**, 1305–1307.

38 G. Aygun and I. Yildiz, *J. Appl. Phys.*, 2009, **106**, 014312.

39 G. D'Acunto, R. Tsyshevsky, P. Shayesteh, J.-J. Gallet, F. Bournel, F. Rochet, I. Pinsard, R. Timm, A. R. Head, M. Kukla and J. Schnadt, In *Manuscr*, 2022.

40 A. R. Head, S. Chaudhary, G. Olivieri, F. Bournel, J. N. Andersen, F. Rochet, J. J. Gallet and J. Schnadt, *J. Phys. Chem. C*, 2016, **120**, 243–251.

41 A. R. Head, S. Chaudhary, G. Olivieri, F. Bournel, J. N. Andersen, F. Rochet, J. J. Gallet and J. Schnadt, *J. Phys. Chem. C*, 2016, **120**, 243–251.

42 P. Villars, *Inorganic Solid Phases, SpringerMaterials (online database)*, https://materials.springer.com/isp/crystallographic/docs/sd_1714120.

43 L. Liu, C. C. He, J. Zeng, Y. H. Peng, W. Y. Chen, Y. J. Zhao and X. B. Yang, *J. Phys. Chem. C*, 2021, **125**, 7077–7085.

Faraday Discussions

PAPER

Localized X-ray photoelectron impedance spectroscopy (LoXPIS) for capturing charge dynamics of an ionic liquid electrolyte within an energy storage device

Mustafa Başaran,[a] Erdinc Oz, [b] Said Ergoktas,[c] Coskun Kocabas,[c] Burak Ulgut, [b] Askin Kocabas*[a] and Sefik Suzer *[b]

Received 29th November 2021, Accepted 12th January 2022
DOI: 10.1039/d1fd00102g

Many electrochemical devices are based on the fundamental process of ion migration and accumulation on surfaces. Complex interplay of molecular properties of ions and device dimensions control the entire process and define the overall dynamics of the system. Particularly, for ionic liquid-based electrolytes it is often not clear which property, and to what extent, contributes to the overall performance of the device. Herein we use X-ray photoelectron spectroscopy (XPS) while the device is under electrical bias. Such a procedure reveals localized electrical potential developments, through binding energy shifts of the atomic core levels, in a chemically specific fashion. Combining it with square-wave AC modulation, the information can also be extended to time domain, and we investigate devices configured as a coplanar capacitor, with an ionic liquid as the electrolyte, in macro-dimensions. Our analysis reveals that a nonlinear voltage profile across the device emerges from spatially non-uniform electrical double layer formation on electrode surfaces. Interestingly the coplanar capacitor has an extremely slow time response which is particularly controlled by IL film thickness. XPS measurements can capture the ion dynamics in the tens of seconds to microseconds range, and reveal that ionic motion is all over the device, including on metallic electrode regions. This behavior can only be attributed to motion in more than one dimension. The ion dynamics can also be faithfully simulated by using a modified PNP equation, taking into account steric effects, and device dimensions. XPS measurements on two devices with different dimensions corroborated and validated the simulation results. The present results propose a new experimental approach and provide new insights into the dynamics of ions across electrochemical devices.

[a]Department of Physics, Koc University, Istanbul, 34450, Turkey. E-mail: akocabas@ku.edu.tr
[b]Department of Chemistry, Bilkent University, Ankara, 06800, Turkey. E-mail: suzer@fen.bilkent.edu.tr
[c]Department of Materials, University of Manchester, Manchester M13 9PL, UK

Introduction

Motion of ions within electrified electrodes that have liquid electrolytes in between is the most important process controlling the performance of many electrochemical systems, spanning a wide range of micro- and macro-devices, used for investigation of bio-systems,[1,2] batteries,[3] fuel-cells,[4] solar-cells[5] and super-capacitors,[6] *etc.* Ionic liquids have recently been added to the conventional list of electrolytes, and have quickly assumed a promising role, due to their convenient physical and chemical properties, such as low-volatility, high conductivity, larger operating potential window, and low toxicity.[7–10] The most important phenomenon affecting the ionic motion or transport in all of these systems has long been advocated to be the formation of the Electrical Double-Layer (EDL) for both faradaic and non-faradaic processes, since ions at the interface of the polarized/charged surfaces can and do rearrange to screen this charge.[11] The resulting concentration and, therefore, the electrical potential profile at the electrode/electrolyte interface further triggers a multitude of processes controlling the kinetics and thermodynamics of the electrochemical systems.[12] Therefore, investigation of static and dynamic properties of the EDL has been a focus both experimentally and theoretically dating back to Helmholtz approaching 17 decades ago.[13]

Recent progress has shed further light on the time dependency, particularly the nonlinear response of the thin double layer in a parallel plate configuration, where Bazant *et al.* identified different time scales controlled by geometric and ionic diffusion properties of the chemical make-up within such devices.[14–16] Most of these approaches have been developed and tested for small electrical potentials and relatively dilute aqueous solutions, while extension to higher voltages and highly concentrated electrolytes, as in the case of ionic liquids, still remains a challenging task due to the high interaction energy and long-range coulombic forces. Accordingly, the same group had modified their theoretical approaches and developed a comprehensive model (Bazant–Storey–Kornyshev, aka BSK theory) including steric interactions, overcrowding and under-screening dynamics of RTILs.[17–19] Alternatively, the pattern formation frameworks particularly mesoscale phase separation principles have been recently implemented to reveal the clustering behavior of ILs.[20–22]

All of these new concepts have also been tested using electro-chemical,[23–27] microscopic,[28–30] terahertz imaging,[31] and X-ray scattering techniques,[32,33] strongly guided and/or corroborated by modeling/simulation studies.[34,35] Use of electrochemical force microscopy has given intriguing new input into the dynamics of charge screening at the solid–aqueous interfaces by identifying multiple time constants and relatively larger length scales of up to 10 microns.[36,37] But again, such distances are still too small to represent real-life electrochemical devices, moreover, information not only at the interfaces but in regions covering the entire device are needed.

Recent findings clearly point out that dynamics of the charging processes in devices with ionic liquids as electrolytes differ widely from conventional ones exhibiting multiple time and length scales, spanning from very fast, Debye-layer charging (10^{-9} s) to ultra-slow, bifurcation-like (10^{4} s) processes under static or time-varying electrical stresses.[38–45] Although these methods illuminate the electrostatic phenomena at the interface, the link to the underlying physics/chemistry is missing. Deeper understanding of the detailed charge diffusion dynamics and evolution of double layer formation across millimeter-large device structures is still elusive. In order to give a different perspective, we combine X-ray photoelectron spectroscopy (XPS) with AC modulation and computational models, covering steric effects and large-scale device configurations.

XPS is a favorable chemical analysis tool and gives quantitative information with high surface sensitivity, but only with a limited spatial resolution, on orders of 50–100 microns for the lab-based instruments. On the other hand, an underutilized advantage of XPS is the capability to faithfully reflect the local electrical potential profiles, vital for further advancement of various applications of electrochemical systems. Under an electrical bias, core level photoelectrons measured by XPS undergo binding energy shifts which reflect the local potential (V) on investigated surfaces in a chemically specific fashion. Here, the limit of the energy sensitivity of the common lab-based instruments is about 100 meV, and the technique is also relatively non-invasive. This is related to the expected IR drop due to the low currents related to the photoelectron emission process on orders of ~1 nA, which translates to a negligibly small energy shift of ~1 meV for an ionic liquid film having ~1 MΩ resistance, and not measurable by XPS. In one of the first studies of ionic liquids, a drop of ionic liquid was placed on an inclined electrode.[46] Their experimental set-up enabled formation of a liquid film thin enough to probe the IL/electrode interface such that the authors were able to follow the EDL processes by tracing the shifts in the core-level binding energies under external potential bias. Whereas, application of DC potentials during XPS data acquisition provides steady state information, our group has previously shown that AC excitation is indispensable for investigation of the dynamics of electrical potential developments.[47–52]

In this study, we report on using XPS to directly probe certain features of the diffuse charge dynamics of a sample in the form of a coplanar-plate capacitor geometry, having an IL as the electrolyte, while imposing a square-wave (AC) potential excitation with frequencies ranging from mHz to kHz at room temperature. Note that for the commonly used parallel plate capacitor, the probing XPS beam would not have direct access to the electrical double layer. Using a coplanar capacitor configuration instead, we have the opportunity to spatially extend the beam and probe the effects of screening the applied potential through ion movement across a large electrolyte medium, since the ionic motion materializes in all regions of the device. Square-wave, rather than conventional electrochemical sine-wave modulation is preferred, since it causes almost no distortion of the XPS peaks.

Supported with numerical simulations, particularly by taking into account the 2-dimensional nature of ionic motion, our analyses consist of XPS data gathering in the forms of areal and line scans at different but constant frequencies for giving spatial information throughout the extended length scales.[47,48,53] Our approach provides new insights into potential developments, particularly on spatial charge dynamics on devices, at the positions they occur. Mapping the 2-D spatiotemporal dynamics of IL is an important and timely approach to tap into the complex non-equilibrium and pattern formation phenomena which have attracted significant scientific interest recently.[21,22,54–56]

Experimental

N,*N*-Diethyl-*N*-methyl-*N*-(2-methoxyethyl)ammonium bis(trifluoromethanesulfonyl)imide (DEME-TFSI) was purchased from Sigma-Aldrich and used without any purification. DEME-TFSI is chosen as the ionic liquid medium, since it has a relatively large electro-chemical window, and has a strong F 1s signal (6 F atoms per anion) for better data processing. The coplanar capacitor device configuration used in this study consists of two Au electrodes deposited onto a porous polyethylene membrane (PEM, from Gelon LIB), which are used as the source and the drain electrodes, and a 5 μL ionic liquid (DEME-TFSI) is soaked into the uncovered membrane in between, acting as a continuous liquid film as the electrolyte medium. One of the electrodes is used as a working electrode to apply the external bias while the other one, the counter electrode is always kept at ground potential.

A Thermo Fisher K-alpha X-ray photoelectron spectrometer with monochromatized X-rays of 1486.6 eV is used for all XPS measurements. Core level spectra were collected using 50 eV analyzer pass energy, but for time-resolved XPS measurements, the fast data gathering, the snapshot mode with 150 eV pass

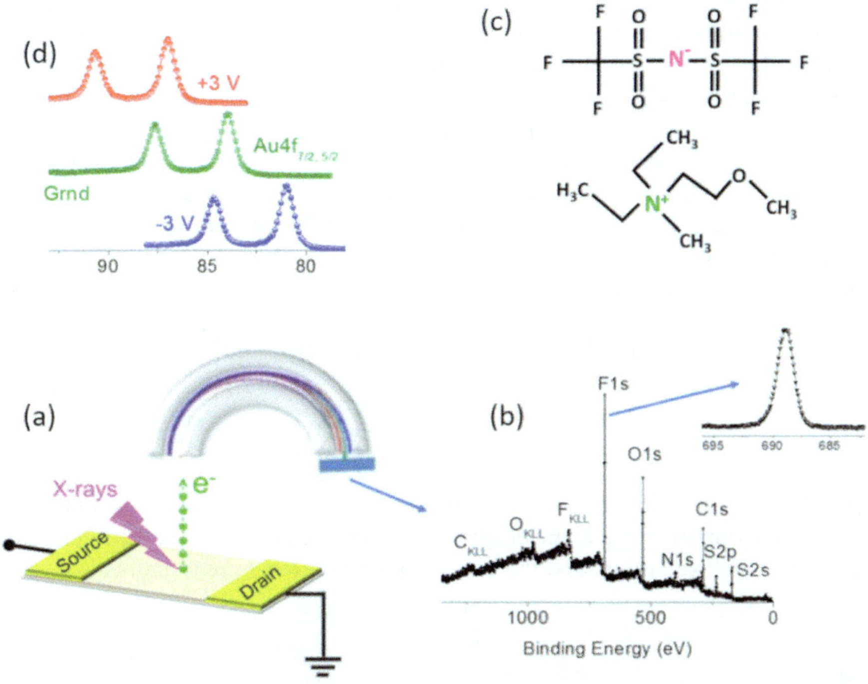

Fig. 1 (a) Schematics of the co-planar capacitor for probing the dynamics of the charging of the ionic liquid medium by the XPS probe beam, under *operando* conditions, where both current and peak positions are recorded simultaneously. The function generator is connected to the source electrode to impose a DC, or an AC (SQW), excitation with a fixed or variable frequency. (b) Survey XP spectrum of the IL and F 1s region representative of the anion. (c) Chemical formula of the IL. (d) Au 4f XP spectra, recorded at the source electrode and under −3, 0, and +3 V DC bias.

energy is used instead of the scanning mode. X-Ray spot size is set to 50 μm for line scan and 100 μm for single point measurements. Pressure in the analyzing chamber is always kept under 10^{-8} mbar. Fitting procedure is performed using the Avantage software equipped with the instrument. All the fits reported in this work have been carried out using a symmetrical Gaussian/Lorentzian product function with ratios of 30/70 after Shirley background subtraction. A schematic representation of the device with the ionic liquid and the gold electrodes is shown in Fig. 1(a). An XP survey spectrum of the IL, together with a scanned spectrum of the F 1s region, is depicted in Fig. 1(b). The chemical formula of the IL is displayed in Fig. 1(c). Fig. 1(d) shows XP spectra of the Au 4f spin–orbit doublet, recorded at the source electrode and under −3, 0, and +3 V DC bias.

Simulations

To gain a more quantitative representation of charge dynamics, we performed Finite Element Method (FEM) simulations based on Poisson–Nernst–Planck (PNP) equations.[21,22] Conventional PNP equations can only capture the drift-diffusion dynamics of dilute electrolytes. However, in IL due to finite size effects of ions, these equations require precise modifications to take into account the essential steric effects. The other important feature we must also consider is the diffusion process of ionic liquids. An IL is a solvent-free, molten salt and it maintains uniform total ion density ($p + n = 2c_0$). Unlike dilute ions in a solvent, due to this constraint, diffusion dynamics of an IL are driven by the inter-diffusion process which can be implemented by following the Onsager framework ($\mathbf{J} \sim pn\nabla(\mu)$), where μ is the chemical potential.[21,22]

Several models have recently been developed for RTILs. Specifically, we restrict ourselves to a Chan–Hillard–Poisson type framework to include all these steric effects, coulombic interactions and the Onsager diffusion process.[20–22] We used COMSOL Multiphysics for modeling. Following Gavish and Yochelis,[20] the free energy can be written as

$$\mathcal{F} = \int k_\mathrm{B}T[p \ln p + n \ln n] + \left[q(p-n)\phi - \frac{\varepsilon}{2}|\nabla\phi|^2\right]$$

$$+\frac{1}{c_0}\left[\frac{\beta np}{4} + \frac{\varepsilon_0\kappa^2}{4}\left(|\nabla p|^2 + |\nabla n|^2\right)\right]$$

p and n are cation and anion molar concentrations, $c_0 = 3.3$ M is the initial bulk ion concentration and ϕ is the electric potential. Further, the modified Poisson–Nernst–Planck equations and the corresponding currents are given below;

$$\frac{\partial p}{\partial t} = -\nabla J$$

$$\mathbf{J} = \frac{Dpn}{k_\mathrm{B}Tc_0}\nabla\left(k_\mathrm{B}T \ln\frac{p}{n} + 2ze\phi - \frac{\beta(c_\mathrm{ref} - p)}{2c_0} - \varepsilon_0\kappa^2\nabla^2 p\right)$$

$$\varepsilon\nabla^2\phi = -q(p-n)$$

where, $D_i \cong 1 \times 10^{-11}$ m^2 s^{-1} is the isotropic diffusion coefficient,[56] q is the elementary charge and $\varepsilon \sim 20$ is the dielectric constant of the IL.[57] We assume an equal ion diameter $a \cong 0.4$–0.6 nm which sets the maximum ion concentration to $2c_0 \cong 6.6$ M through the ionic steric interactions.

Results and discussion

(I) XPS measurements

(I-a) Measuring the localized equilibrium voltage distributions. The Au $4f_{7/2}$ peak has a binding energy of 84.0 eV when the system is grounded but shifts to 87.0 and 81.0 eV when biased with +3 and −3 V, respectively, and no corresponding shifts are observable for the same spectrum recorded on the grounded drain electrode. When both of the electrodes are grounded, the F 1s peak appears at the tabulated binding energy position of 688.5 eV, and is expected to shift to a higher (lower) position under positive (negative) bias respectively. However, contrary to the trivial 3.0 eV shifts of the electrode Au 4f peaks [see Fig. 1(d), 2(a) and (c)], the F 1s peak undergoes smaller shifts due to screening of the electrode potential, hence gives pertinent information both under DC and AC actuations, related to static and dynamic screening, respectively, operative within the entire probed system. Such a situation is shown in Fig. 2, for our coplanar capacitor,

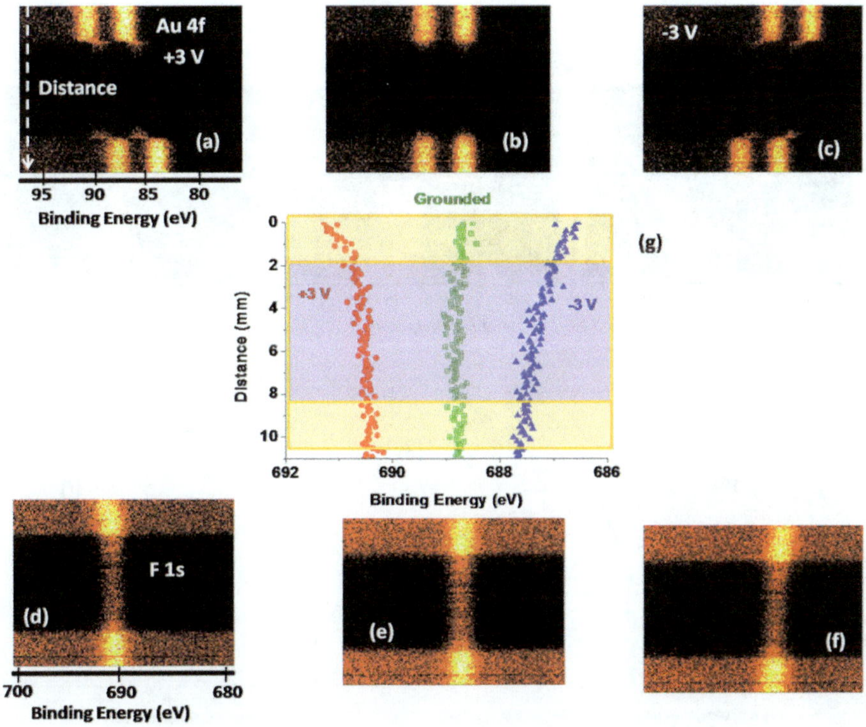

Fig. 2 Recorded line-scan XP spectra under +3, 0 and −3 V DC bias, respectively. (a)–(c) Au 4f region representative of the metal electrodes. (d)–(f) F 1s region representative of the IL. (g) Extracted F 1s binding energy variations.

where line scan spectra of the Au 4f and also F 1s peaks are displayed along the central line of the capacitor.

The line starts on the source electrode, which is ~2 mm wide and continues along the entire ~6 mm electrolyte film and ends after spanning the grounded electrode. Note that the electrodes are also partially covered with a thin layer of the liquid electrolyte, since the F 1s signal is also measurable in that region. Screening effects at the quasi-steady-state (>500 s) are readily seen, as exemplified by the measured F 1s peak as a function of the position of the probe beam on and across the electrodes, in Fig. 2(d) and (f). The F 1s peak undergoes shifts less than the applied voltage, due to the effective screening through the double layer formation on the electrodes, hence can be used as the reflection of the local electrical potential within the device.

(I-b) Measuring the temporal and localized voltage distributions. To capture the temporal variations across the device, we recorded the F 1s region in the snap-shot mode with 1 second time-intervals at a low frequency of 10 mHz (100 spectra in one cycle), as depicted in Fig. 3, at the three positions indicated.

It is seen from Fig. 3(a) and (e), when the positive cycle begins, the F 1s shifts instantaneously by almost +3 eV (688.5 + 3 = 691.5 eV) and gets shifted down as the screening takes place, with an approximate time constant of 4 seconds. During the negative cycle [Fig. 3(a) and (f)], the corresponding value is 685.5 eV and the time constant is more than twice that of the positive one. Moreover, response of the F 1s on the other electrode is quite different. In addition, there are

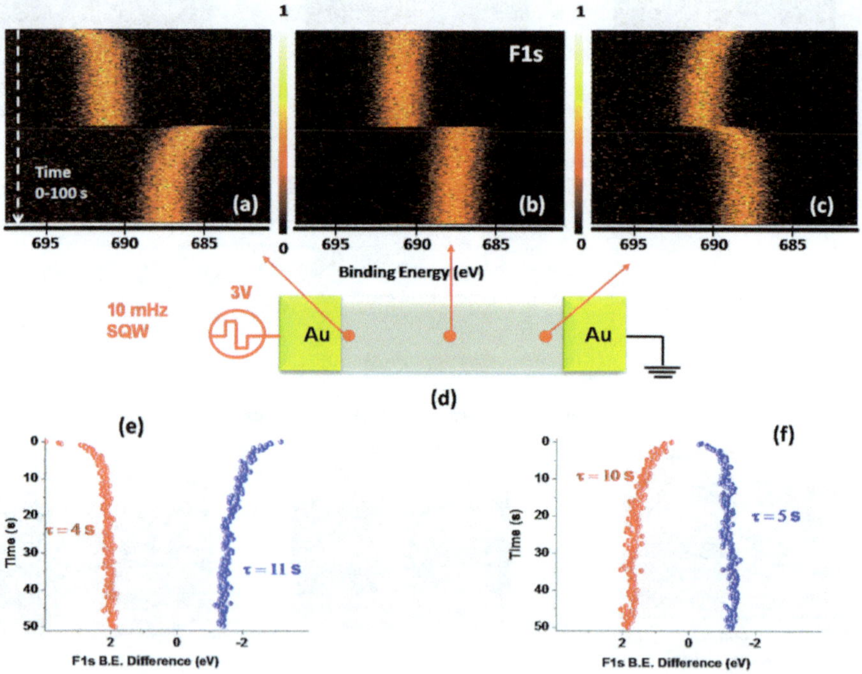

Fig. 3 (a)–(c) F 1s XP spectra of the IL recorded in the snap-shot mode at three different positions of the capacitor. (d) Schematics of the device. (e) and (f) Variation of the differences in the F 1s B.E. as a function of time, extracted from (a) and (c), respectively.

no measurable temporal shifts in the F 1s peak position in the middle of the device, in perfect harmony with the near-electrode measurements.

The most logical hypothesis explaining this observation is that the varying and transient voltage levels across the device are gradually approaching to the equilibrium distribution (~1.5 V). These results indicate that the response time of the device is slow enough to observe the detailed voltage transitions across the capacitor using XPS measurements. Note in passing that ionic motion is evident all over the device, and is controlled presumably by a combination of different electrochemical processes, and can easily be captured by XPS measurements.

These results bring out several interesting features which differ from those of a simple parallel plate electrolytic capacitor with blocking metal electrodes, such as: (i) very slow electrical response time, (ii) presence of asymmetric equilibrium voltage distribution, (iii) non-uniform and nonconforming (due to effective screening) electrical potential distribution on the surface of IL moieties and across the metal electrodes. Naively, we can claim that these observations are originating from the interplay of steric effects, geometrical constraints and the IL's droplet coverage of the large gold electrodes. Using the computational model we will further focus on these distinct features to reveal the detailed factors controlling these responses.

(I-c) Measuring the transient and localized voltage distributions. As a next step we focused on frequency response of our coplanar capacitor. We have been using a procedure to extract time-dependent variations of localized electrical potential developments, by recording XP spectra under square-wave AC modulation, which introduces a controllable time-window while causing minimum distortion to the

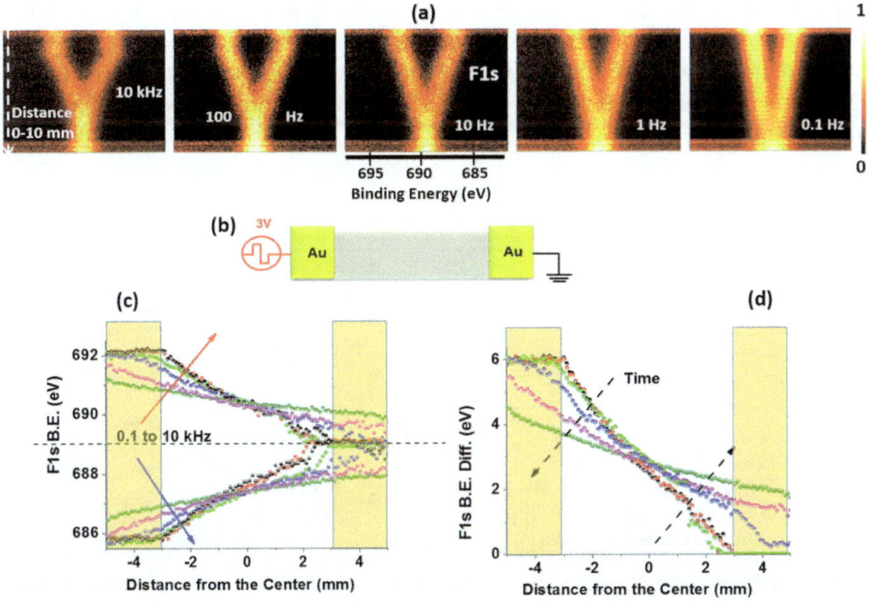

Fig. 4 (a) XP spectra of the F 1s region of the IL across the coplanar capacitor, recorded under various 3 V SQW excitations. (b) Schematics of the device. (c) Extracted BEs, and (d) their differences as a function of the distance across the coplanar-capacitor.

line-shape as depicted in Fig. 4(a). The format of the spectra is in the forms of line-scans along the central line of the capacitor, and is similar to those given in Fig. 2. However, now both the positive and the negative cycles are depicted simultaneously, spanning a wide frequency range of 5 decades (0.1 Hz to 10 kHz), reflecting the time-dependent local potential developments along the entire capacitor.

(I-d) Transient and localized voltage distributions in two different devices. As was mentioned in the Introduction section, for a parallel plate capacitor, Bazant, through simulations predicted three time scales, which are: $\tau_D = \lambda_d^2/D$, $\tau_c = \lambda_d L/D$ and $\tau_L = L^2/D$, where λ_d is the Debye length, $2L$ is the distance between electrodes and D is the diffusion constant. These are the time constants associated with local charge redistribution (EDL), diffusion across the cell, and the diffuse-charge dynamics, respectively. To check the validity of this prediction, we have also carried out similar measurements on another co-planar capacitor device which has only half of the IL channel width of $2L = 3$ mm. The results are depicted in Fig. 5, together with the first device ($2L = 6$ mm) reported in previous sections, yielding more or less the same local potential screening, except for a faster response, probably attributable to the higher electrical field imposed. The largest potential screenings are measured right near the two interfaces of the metal electrodes with the IL film, but screening throughout the metal electrodes is also significant. We can obtain information regarding: (i) the quantity of the screened charges, and also (ii) their dynamics, using the extracted screened potentials (−2.4 and −1.9 V for these two devices, both in 10 seconds).

The screening potential developed between the center of the device ($L = 0$) and the interface ($L = 3$ mm) can in turn be related to the ionic charges through the following equation:[11]

$$q = 4\pi\varepsilon\varepsilon_0 L \times \Phi = 4\pi \times 20 \times 8.85 \times 10^{-12}\frac{F}{m} \times 0.003 \text{ m} \times (2.4 \text{ V}) = 6 \times 10^{-13} \text{ C}$$

where a dielectric constant of 20 is used for the IL.[57] This formula tells us that an ion concentration difference of $\sim 4 \times 10^6$ develops within 10 seconds (between 0.1 Hz and 10 kHz).

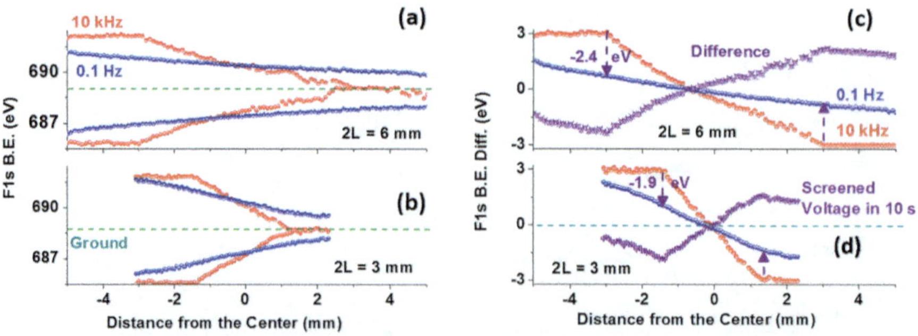

Fig. 5 Measured binding energy of the F 1s peak recorded with 10 kHz and 0.1 Hz 3 V SQW modulation across the two coplanar capacitors: (a) $2L = 6$ mm and (b) $2L = 3$ mm, respectively. The corresponding and extracted B.E. differences between the two frequencies together with effective screened local electrical potentials for the two devices are given in (c) and (d).

If we use the diffusion constant of $5 \times 10^{-11}\ m^2\ s^{-1}$, we would get a diffusion length of only [$(5 \times 10^{-11}\ m^2\ s^{-1} \times 10\ s)^{0.5}$ =] 0.02 mm for the time interval of 10 s. Alternatively, we can use the viscosity coefficient of the IL ($\eta = \sim 50$ cP),[58] under the applied potential of 3 V to estimate the drift velocity ($\bar{u}$) using the following formula:[11]

$$\bar{u} = E \times e/6\pi\eta r = (2.4\ \mathrm{V}/0.003\ \mathrm{m}) \times (1.6 \times 10^{-19}\ \mathrm{C})/(6\pi \times 50\ \mathrm{cP} \times 0.2\ \mathrm{nm}) = 7 \times 10^{-6}\ \mathrm{m\ s^{-1}}$$

where, r is the ion radius (~0.2 nm). This formula yields ~0.07 mm for the time interval of 10 s. Both of these two naive approaches yield more than 2 orders of magnitude smaller ion movements when compared with our XPS measurements. Hence, other experimental and/or geometrical parameters must be in effect to explain the results.

(II) Simulations

(II-a) Numerical simulations of charge dynamics of a parallel plate capacitor. To validate the numerical simulations, we first focused on the 1D parallel plate capacitor configuration. As an initial step we assessed the time response of the capacitor and contribution of steric effects and clustering dynamics to this charging process. We transiently applied a voltage step (1 V) to one electrode and grounded the other one. Fig. 4 summarizes the time responses of the IL based capacitor. Although the initial voltage distribution across the capacitor is linear between 0 to 1 V, at the steady state, through the two equivalent EDL formation it drops the voltage to the equilibrium level of 0.5 V. As ions accumulate on the electrodes, this initially linear voltage distribution converges to a flat equilibrium profile. Depending on the analyzed points on the device the local voltage shows up and down shifts towards this equilibrium level. Note that the center of the device shows a flat response, this is because this point is already at the equilibrium condition. As a second step we probed the time response of the system. The results of the simulations are depicted in Fig. 6.

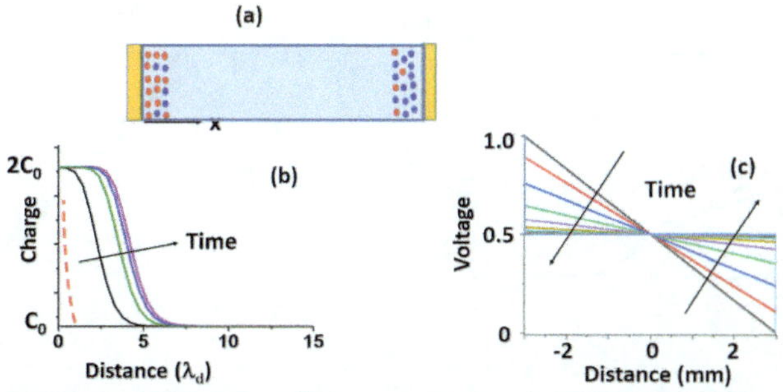

Fig. 6 Schematic representation of: (a) the ion distribution near both electrodes. (b) The ion distribution near the electrodes for the parallel plate capacitor and changes in the ion distribution near the electrified electrode as a function of time, after imposing a steric limit. Red dotted curve shows the voltage response in the absence of any steric limit. (c) Corresponding local voltage changes as a function of time along the perpendicular line to the plates.

As expected, the steric effects dramatically decrease the relaxation time of the capacitor from 200 to 2 ms, since the steric interactions significantly extend the width of the electric double layer and decrease the capacitance of the electrode surface. In the absence of a steric limit, the conventional Debye screening length is about $\lambda = 0.1$ nm, but length scale and accumulated charge densities around the electrodes are not realistic. In our simulation, the finite size of the ion leads to the formation of a much larger condensed layer of $\lambda > 1$ nm. This extension and the decrease in the capacitance eventually decrease the response of the system. The intuition behind this process can be simply grasped by analyzing a typical response time of a parallel plate capacitor which is defined by the *RC* time constant. Capacitance *C* is controlled by the electric double layer formation on the surfaces and resistance *R* is set by ion conduction across the device.

When we compare the experimental and simulation results of time responses, we find that the experimental results are much slower than can be predicted by the numerical calculations, $\tau_{\text{exp}} = 4\ \text{s} \gg \tau_{\text{sim}} = 2$ ms. Moreover, as we noted earlier, the other surprising feature we observed in XPS measurements is the extension of the voltage modulations deep into the electrode surfaces. These observations support the idea that 2-dimensional effects play a significant role in defining overall response of the system. Motivated by these observations and in order to account for these effects, we focused on simulating a more realistic configuration of the coplanar capacitor.

(II-b) Simulations of charge dynamics of a co-planar capacitor. The main difference between parallel and coplanar configurations is occurrence of time-dependent and non-uniform electric fields on the electrodes. To observe the

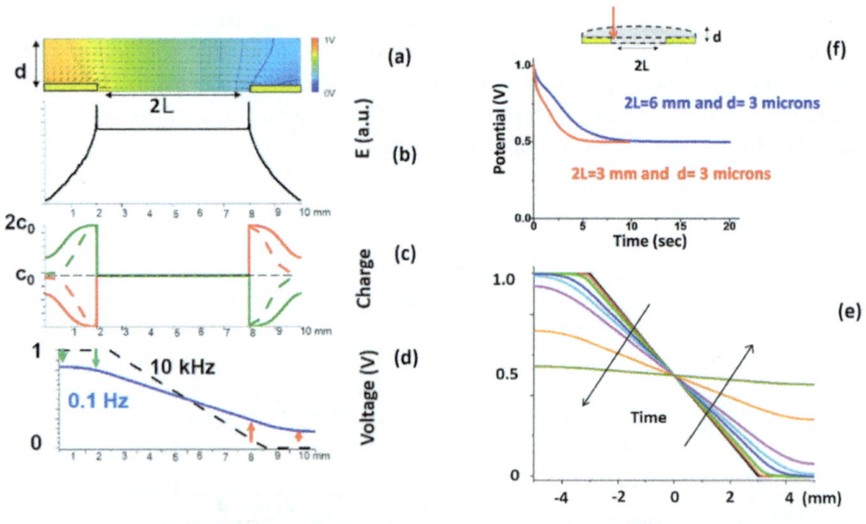

Fig. 7 (a) A schematic representation of 2D coplanar capacitor configuration, together with the color coded voltage profiles. (b) Magnitude of electric field on the surface of the electrodes under a suddenly applied voltage within 10 kHz and 0.1 Hz time-windows (*i.e.* 1/10 charging time). (c) Charge densities over the electrodes. (d) Voltage profiles at the designated two frequencies, and (e) at more frequencies. (f) Temporal variations of the extracted potential changes for the two devices near the electrified electrode as shown by the red arrow.

influence of this electric field profile on charge dynamics we performed 2-dimensional FEM simulations. Fig. 7 shows typical results of 2-dimensional cross-sectional view of electric field vectors, charge profiles and voltage distributions across the device. The magnitude of the electric field decays across the electrode surface. The main impact of this electric field profile is the emergence of a transient non-uniform electric double layer. Our simulations show that the drift of the charges and screening of the voltage on electrode surfaces significantly depend on the magnitude of the electric field. The non-uniform electric field eventually gives rise to non-uniform EDL layer formation across the surface of the electrodes. Due to steric effects the width of this condensed layer gradually grows.

In turn, the non-uniform and extended electrical double layer abruptly drops the electrode voltage and results also in a non-uniform voltage distribution. Although the double layer is buried under the IL film, the voltage profile on the electrodes can be measured on the surface of the IL droplet. This is simply observed by following the equipotential lines across the liquid film. Fig. 7(d) shows the simulated voltage profile on the droplet surface. This profile is dictated by the transient non-uniform EDL on electrode surfaces and can also be captured by XPS measurements. Moreover, the dynamics of this EDL can be monitored indirectly by measuring surface voltage. At a later stage, a non-uniform EDL gradually converges to a uniform distribution, and eventually reaches the equilibrium voltage across the device.

(II-c) Effects of the device dimensions. Through our simulation efforts, we have also noticed that the time response of the coplanar capacitor is particularly controlled by the thickness of the IL layer. This thickness together with the device length simply tunes the ion conduction of the system and sets the *RC* time

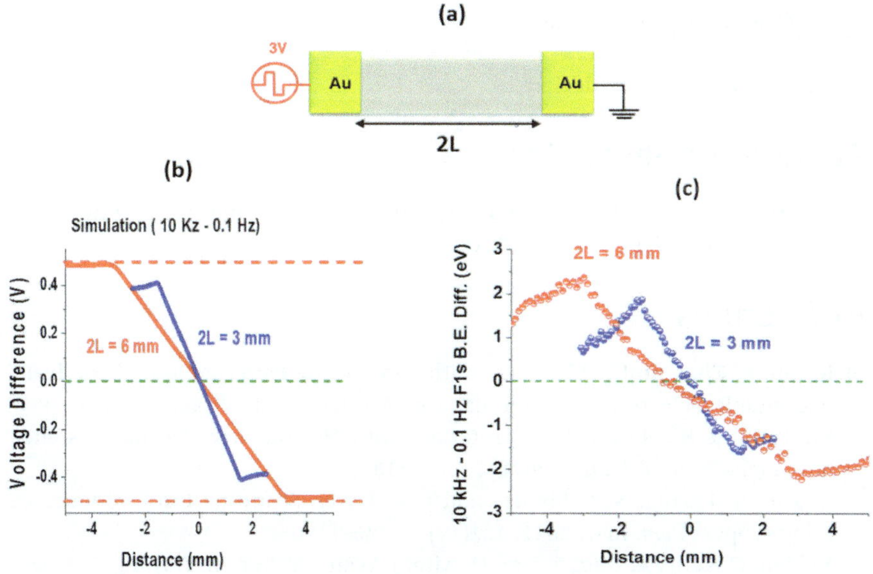

Fig. 8 (a) Schematics of the co-planar capacitor. Localized potential screening across the two co-planar capacitors obtained *via*: (b) simulations and (c) XPS measurements, for the two different devices, respectively.

constant. Our simulation results showed that a $d = 3$ μm thick and $2L = 6$ mm long IL layer provides $\tau_{sim} = 5$ s response time which is close to our experimental measurements. Moreover, for a similar but shorter device ($2L = 3$ mm) the same time constant is reduced to ~2.5 s, again not too far off from the experimental data which are presented in Fig. 7(f) and 8.

The overall time constant of the device ($\tau = RC$) can now be formulated to scale linearly with the device length and inversely with the ionic liquid film thickness ($\tau \sim L/d$). This is in harmony with Bazant's predictions formulated for a parallel-plate capacitor, except for the very important finding that IL film thickness needs to be taken into account for 2-D ionic motion, as is the case for our co-planar capacitor.

Conclusions

XPS under square-wave modulation is utilized to tap into the dynamics of localized charge accumulation and dissipation of a co-planar capacitor device, with an ionic liquid film as the electrolyte in between two gold electrodes in macro-dimensions. The results revealed two important findings: (i) dynamics of the charging/discharging processes are very slow, and (ii) temporal voltage variations are also measurable on electrode surfaces. In order to have deeper understanding, the COMSOL modelling software was used considering both 1- and 2-dimensional ionic motion. Simulation results were very successful in reproducing both temporal and spatial voltage variations across entire devices, measured by XPS. The simulation results also brought up a new understanding that the device time constant ($\tau = RC$) is found to scale linearly with the device length and inversely with the ionic liquid film thickness ($\tau \sim L/d$). This is in harmony with theoretical predictions, except for the new finding that IL film thickness has to be included.

Conflicts of interest

There are no conflicts to declare.

Acknowledgements

This work is supported by Grant No.: 118Z902 of the Scientific and Technological Research Council of Turkey (TUBITAK).

References

1 R. Liu, S. Zhao and J. Liu, From Lithographically Patternable to Genetically Patternable Electronic Materials for Miniaturized, Scalable, and Soft Implantable Bioelectronics to Interface with Nervous and Cardiac Systems, *ACS Appl. Electron. Mater.*, 2021, **3**, 101–118.

2 C. Zhu, K. Huang, N. P. Siepser and L. A. Baker, Scanning Ion Conductance Microscopy, *Chem. Rev.*, 2021, **121**(19), 11726–11768.

3 Y. Tian, G. Zeng, A. Rutt, T. Shi, H. Kim, J. Wang, Y. Sun, B. Ouyang, T. Chen, Z. Lun, Z. Rong, K. Persson and G. Ceder, Promises and Challenges of Next-Generation "Beyond Li-ion" Batteries for Electric Vehicles and Grid Decarbonization, *Chem. Rev.*, 2021, **121**(3), 1623–1669.

4 K. Chen, D.-Y. Yang, G. Huang and X.-B. Zhang, Lithium–Air Batteries: Air-Electrochemistry and Anode Stabilization, *Acc. Chem. Res.*, 2021, **54**(3), 632–641.

5 B. Selvaraj, G. Shanmugam, S. Kamaraj, A. Gunasekeran and A. Sambandam, Effect of 1-Substituted 2-(Pyridin-2-yl)-1*H*-Benzo[*d*]imidazole Ligand-Coordinated Copper and Cobalt Complex Redox Electrolytes on Performance of Ru(II) Dye-Based Dye-Sensitized Solar Cells, *Inorg. Chem.*, 2021, **60**(3), 1937–1947.

6 A. Eftekhari, Supercapacitors utilising ionic liquids, *Energy Storage Mater.*, 2017, **9**, 47–69.

7 B. B. Hansen, *et al.*, Deep Eutectic Solvents: A Review of Fundamentals and Applications, *Chem. Rev.*, 2021, **121**(3), 1232–1285.

8 K. S. Egorova, E. G. Gordeev and V. P. Ananikov, Biological Activity of Ionic Liquids and Their Application in Pharmaceutics and Medicine, *Chem. Rev.*, 2017, **117**(10), 7132–7189.

9 M. Watanabe, M. L. Thomas, S. Zhang, K. Ueno, T. Yasuda and K. Dokko, Application of Ionic Liquids to Energy Storage and Conversion Materials and Devices, *Chem. Rev.*, 2017, **117**(10), 7190–7239.

10 T. Welton, Ionic liquids: a brief history, *Biophys. Rev.*, 2018, **10**(3), 691–706.

11 A. J. Bard and L. R. Faulkner, *Electrochemical Methods: Fundamentals and Applications*, Wiley, 2000.

12 E. J. F. Dickinson and R. G. Compton, Influence of the diffuse double layer on steady-state voltammetry, *J. Electroanal. Chem.*, 2011, **661**(1), 198–212.

13 https://en.wikipedia.org/wiki/Double_layer_(surface_science).

14 M. Z. Bazant, K. Thornton and A. Ajdari, Diffuse-charge dynamics in electrochemical systems, *Phys. Rev. E: Stat., Nonlinear, Soft Matter Phys.*, 2004, **70**(2), 021506.

15 M. S. Kilic, M. Z. Bazant and A. Ajdari, Steric effects in the dynamics of electrolytes at large applied voltages. I. Double-layer charging, *Phys. Rev. E: Stat., Nonlinear, Soft Matter Phys.*, 2007, **75**(2), 021502.

16 M. S. Kilic, M. Z. Bazant and A. Ajdari, Steric effects in the dynamics of electrolytes at large applied voltages. II. Modified Poisson-Nernst-Planck equations, *Phys. Rev. E: Stat., Nonlinear, Soft Matter Phys.*, 2007, **75**(2), 021503.

17 M. Z. Bazant, B. D. Storey and A. A. Kornyshev, Double Layer in Ionic Liquids: Overscreening versus Crowding, *Phys. Rev. Lett.*, 2011, **106**(4), 046102.

18 B. D. Storey and M. Z. Bazant, Effects of electrostatic correlations on electrokinetic phenomena, *Phys. Rev. E: Stat., Nonlinear, Soft Matter Phys.*, 2012, **86**(5), 056303.

19 M. V. Fedorov and A. A. Kornyshev, Ionic Liquids at Electrified Interfaces, *Chem. Rev.*, 2014, **114**(5), 2978–3036.

20 N. Gavish and A. Yochelis, Theory of Phase Separation and Polarization for Pure Ionic Liquids, *J. Phys. Chem. Lett.*, 2016, **7**, 1121–1126.

21 S. Bier, N. Gavish, H. Uecker and A. Yochelis, From bulk self assembly to electrical diffuse layer in a continuum approach for ionic liquids: the impact of anion and cation size asymmetry, *Phys. Rev. E*, 2017, **95**, 060201.

22 N. Gavish, D. Elad and A. Yochelis, From Solvent-Free to Dilute Electrolytes: Essential Components for a Continuum Theory, *J. Phys. Chem. Lett.*, 2018, **9**, 36–42.

23 V. Lockett, R. Sedev, J. Ralston, M. Horne and T. Rodopoulos, Differential Capacitance of the Electrical Double Layer in Imidazolium-Based Ionic Liquids: Influence of Potential, Cation Size, and Temperature, *J. Phys. Chem. C*, 2008, **112**(19), 7486–7495.

24 V. Lockett, M. Horne, R. Sedev, T. Rodopoulos and J. Ralston, Differential capacitance of the double layer at the electrode/ionic liquids interface, *Phys. Chem. Chem. Phys.*, 2010, **12**(39), 12499–12512.

25 M. Druschler, B. Huber and B. Roling, On Capacitive Processes at the Interface between 1-Ethy-3-methylimidazolium tris(pentafluoroethyl) trifluorophosphate and Au(111), *J. Phys. Chem. C*, 2011, **115**, 6802–6808.

26 J. Zheng, S. S. Moganty, P. C. Goonetilleke, R. E. Baltus and D. Roy, A Comparative Study of the Electrochemical Characteristics of [$Emim^+$][BF_4^-] and [$Bmim^+$][BF_4^-] Ionic Liquids at the Surfaces of Carbon Nanotube and Glassy Carbon Electrodes, *J. Phys. Chem. C*, 2011, **115**(15), 7527–7537.

27 M. Jitvisate and J. R. T. Seddon, Direct Measurements of the Differential Capacitance of Solvent-Free and Dilute Ionic Liquids, *J. Phys. Chem. Lett.*, 2018, **9**, 126–131.

28 R. Hayes, N. Borisenko, M. K. Tam, P. C. Howlett, F. Endres and R. Atkin, Double Layer Structure of Ionic Liquids at the Au(111) Electrode Interface: An Atomic Force Microscopy Investigation, *J. Phys. Chem. C*, 2011, **115**(14), 6855–6863.

29 R. Atkin, N. Borisenko, M. Drüschler, F. Endres, R. Hayes, B. Huber and B. Roling, Structure and dynamics of the interfacial layer between ionic liquids and electrode materials, *J. Mol. Liq.*, 2014, **192**, 44–54.

30 R. Hayes, G. G. Warr and R. Atkin, Structure and Nanostructure in Ionic Liquids, *Chem. Rev.*, 2015, **115**(13), 6357–6426.

31 I. Baylam, O. Balci, N. Kakenov, C. Kocabas and A. Sennaroglu, Femtosecond pulse generation with voltage-controlled graphene saturable absorber, *Opt. Lett.*, 2014, **39**, 5180.

32 A. Uysal, H. Zhou, G. Feng, S. S. Lee, S. Li, P. T. Cummings, P. F. Fulvio, S. Dai, J. K. McDonough, Y. Gogotsi and P. Fenter, Interfacial ionic 'liquids': connecting static and dynamic structures, *J. Phys.: Condens. Matter*, 2015, **27**(3), 032101.

33 M. Chu, M. Miller, T. Douglas and P. Dutta, Ultraslow Dynamics at a Charged Silicon–Ionic Liquid Interface Revealed by X-ray Reflectivity, *J. Phys. Chem. C*, 2017, **121**(7), 3841–3845.

34 D. Bedrov, J.-P. Piquemel, A. D. MacKerell, B. Roux and C. Schroder, Molecular Dynamics Simulations of Ionic Liquids and Electrolytes Using Polarizable Force Fields, *Chem. Rev.*, 2019, **119**(13), 7940–7995.

35 Y.-L. Wang, B. Li, S. Sarman, F. Mocci, Z.-Y. Lu, J. Yuan, A. Laaksonen and M. D. Fayer, Microstructural and Dynamical Heterogeneities in Ionic Liquids, *Chem. Rev.*, 2020, **120**(13), 5798–5877.

36 L. Collins, S. Jesse, J. I. Kilpatrick, A. Tselev, O. Varenyk, M. B. Okatan, S. A. L. Weber, A. Kumar, N. Balke, S. V. Kalinin and B. J. Rodriguez, Probing charge screening dynamics and electrochemical processes at the solid–liquid interface with electrochemical force microscopy, *Nat. Commun.*, 2014, **5**(1), 3871.

37 L. Collins, J. I. Kilpatrick, S. V. Kalinin and B. J. Rodriguez, Towards nanoscale electrical measurements in liquid by advanced KPFM techniques: a review, *Rep. Prog. Phys.*, 2018, **81**(8), 086101.
38 M. A. Gebbie, M. Valtiner, X. Banquy, E. T. Fox, W. A. Henderson and J. N. Israelachvili, Ionic liquids behave as dilute electrolyte solutions, *Proc. Natl. Acad. Sci. U. S. A.*, 2013, **110**(24), 9674–9679.
39 M. A. Gebbie, A. M. Smith, H. A. Dobbs, A. A. Lee, G. G. Warr, X. Banquy, M. Valtiner, M. W. Rutland, J. N. Israelachvili, S. Perkin and R. Atkin, Long range electrostatic forces in ionic liquids, *Chem. Commun.*, 2017, **53**(7), 1214–1224.
40 A. A. Lee, C. S. Perez-Martinez, A. M. Smith and S. Perkin, Underscreening in concentrated electrolytes, *Faraday Discuss.*, 2017, **199**, 239–259.
41 L. Fillaud, T. Petenzi, J. Pallu, B. Piro, G. Mattana and V. Noel, Switchable Hydrogel-gated Organic Field-Effect Transistors, *Langmuir*, 2018, **34**, 3686–3693.
42 C. S. Perez-Martinez and S. Perkin, Surface forces generated by the action of electric fields across liquid films, *Soft Matter*, 2019, **15**, 4255–4265.
43 Y. Wang, R. Jarosova, G. M. Swain and G. J. Blanchard, Characterizing the Magnitude and Structure-Dependence of Free Charge Density Gradients in Room-Temperature Ionic Liquids, *Langmuir*, 2020, **36**, 3038–3045.
44 L. Richter, P. J. Zuk, P. Szymczak, J. Paczesny, K. M. Bak, T. Szymborski, P. Garstecki, H. A. Stone, R. Holyst and C. Drummond, Ions in an AC Electric Field: Strong Long-Range Repulsion between Oppositely Charged Surfaces, *Phys. Rev. Lett.*, 2020, **125**, 056001.
45 M. Han, H. Kim, C. Leal, M. Negrito, J. D. Batteas and R. M. Espinosa-Marzal, Insight into the Electrical Double Layer of Ionic Liquids Revealed through Its Temporal Evolution, *Adv. Mater. Interfaces*, 2020, **7**, 2001313.
46 D. Weingarth, A. Foelske-Schmitz, A. Wokaun and R. Kötz, In situ electrochemical XPS study of the Pt/[EMIM][BF_4] system, *Electrochem. Commun.*, 2011, **13**(6), 619–622.
47 M. T. Camci, P. Aydogan, B. Ulgut, C. Kocabas and S. Suzer, XPS enables visualization of electrode potential screening in an ionic liquid medium with temporal- and lateral-resolution, *Phys. Chem. Chem. Phys.*, 2016, **18**(41), 28434–28440.
48 P. Aydogan Gokturk, B. Ulgut and S. Suzer, AC Electrowetting Modulation of Low-Volatile Liquids Probed by XPS: Dipolar vs. Ionic Screening, *Langmuir*, 2019, **35**(9), 3319–3326.
49 C. B. Uzundal, O. Sahin, P. Aydogan Gokturk, H. Wu, F. Mugele, B. Ulgut and S. Suzer, X-ray Photoelectron Spectroscopy with Electrical Modulation Can Be Used to Probe Electrical Properties of Liquids and Their Interfaces at Different Stages, *Langmuir*, 2019, **35**(52), 16989–16999.
50 P. Aydogan Gokturk, B. Ulgut and S. Suzer, DC Electrowetting of Nonaqueous Liquid Revisited by XPS, *Langmuir*, 2018, **34**(25), 7301–7308.
51 C. B. Uzundal, P. Aydogan Gokturk, S. Suzer and B. Ulgut, Coarse-Grained Electrostatic Model Including Ion-Pairing Equilibrium That Explains DC and AC X-ray Photoelectron Spectroscopy Measurements on Ionic Liquids, *J. Phys. Chem. C*, 2019, **123**(21), 13192–13200.
52 P. Aydogan Gokturk, M. Taner Camci and S. Suzer, Lab-based operando X-ray photoelectron spectroscopy for probing low-volatile liquids and their

interfaces across a variety of electrosystems, *J. Vac. Sci. Technol., A*, 2020, **38**(4), 040805.

53 P. Aydogan Gokturk and S. Suzer, Probing dynamics of non-faradaic processes in ionic liquids at extended time and length scales using XPS with AC modulation, *J. Phys. Chem. C*, 2021, **125**, 9453–9460.

54 H. Zhao, Diffuse-charge-dynamics of ionic liquids in electrochemical systems, *Phys. Rev. E: Stat., Nonlinear, Soft Matter Phys.*, 2011, **84**, 051504.

55 R. F. Stout and A. S. Khair, Moderately nonlinear diffuse-charge dynamics under an ac voltage, *Phys. Rev. E: Stat., Nonlinear, Soft Matter Phys.*, 2015, **92**, 032305.

56 G. Feng, M. Chen, S. Bi, Z. A. H. Goodwin, E. B. Postnikov, N. Brilliantov, M. Urbakh and A. A. Kornyshev, Free and Bound States of Ions in Ionic Liquids, Conductivity, and Underscreening Paradox, *Phys. Rev. X*, 2019, **9**, 021024.

57 M. N. Huang, Y. Jiang, P. Saisanker, G. W. Driver and H. Weingartner, Static Relative Dielectric Permittivities of Ionic Liquids at 25 °C, *J. Chem. Eng. Data*, 2011, **56**, 1494–1499.

58 W.-L. Yuan, X. Yang, L. He, Y. Xue, S. Qin and G.-H. Tao, Viscosity, Conductivity, and Electrochemical Property of Dicyanamide Ionic Liquids, *Front. Chem.*, 2018, **6**, 59.

Faraday Discussions

PAPER

The rise of electrochemical NAPXPS operated in the soft X-ray regime exemplified by the oxygen evolution reaction on IrO_x electrocatalysts

Juan Jesús Velasco Vélez, *[ab] Denis Bernsmeier,[c] Travis E. Jones,[b] Patrick Zeller,[bd] Emilia Carbonio, [d] Cheng-Hao Chuang,[e] Lorenz J. Falling,[bf] Verena Streibel, [b] Rik V. Mom,[bg] Adnan Hammud,[b] Michael Hävecker,[ab] Rosa Arrigo, [h] Eugen Stotz,[b] Thomas Lunkenbein, [b] Axel Knop-Gericke,[ab] Ralph Krähnert [c] and Robert Schlögl[ab]

Received 5th December 2021, Accepted 8th February 2022

DOI: 10.1039/d1fd00114k

Photoelectron spectroscopy offers detailed information about the electronic structure and chemical composition of surfaces, owing to the short distance that the photoelectrons can escape from a dense medium. Unfortunately, photoelectron based spectroscopies are not directly compatible with the liquids required to investigate electrochemical processes, especially in the soft X-ray regime. To overcome this issue, different approaches based on photoelectron spectroscopy have been developed in our group over the last few years. The performance and the degree of information provided by these approaches are compared with those of the well established bulk sensitive spectroscopic approach of total fluorescence yield detection, where the surface information gained from this approach is enhanced using samples with large surface to bulk ratios. The operation of these approaches is exemplified and compared using the oxygen evolution reaction on IrO_x catalysts. We found that all the approaches, if properly applied, provide similar information about surface oxygen speciation. However, using resonant photoemission spectroscopy, we were able to prove that speciation is more involved and complex than previously thought during the oxygen evolution reaction on IrO_x based electrocatalysts. We found that the electrified solid–liquid

[a]*Department of Heterogeneous Reactions, Max Planck Institute for Chemical Energy Conversion, 45470 Mülheim an der Ruhr, Germany. E-mail: velasco@fhi-berlin.mpg.de*

[b]*Department of Inorganic Chemistry, Fritz-Haber-Institut der Max-Planck-Gesellschaft, 14195 Berlin, Germany*

[c]*Department of Chemistry, Chemical Engineering Division, Technical University Berlin, 10623 Berlin, Germany*

[d]*Helmholtz-Zentrum Berlin für Materialien und Energie, BESSY II, 12489 Berlin, Germany*

[e]*Department of Physics, Tamkang University, New Taipei City 25137, Taiwan*

[f]*Materials Sciences Division, Lawrence Berkeley National Laboratory, Berkeley, California 94720, USA*

[g]*Leiden Institute of Chemistry, Leiden University, Einsteinweg 55, 2333 CC Leiden, The Netherlands*

[h]*School of Science, Engineering and Environment, University of Salford, Manchester, M5 4WT, UK*

interface is composed of different oxygen species, where the terminal oxygen atoms on iridium are the active species, yielding the formation of peroxo species and, finally, dioxygen as the reaction product. Thus, the oxygen–oxygen bond formation is dominated by peroxo species formation along the reaction pathway. Furthermore, the methodologies discussed here open up opportunities to investigate electrified solid–liquid interfaces in a multitude of electrochemical processes with unprecedented speciation capabilities, which are not accessible by one-dimensional X-ray spectroscopies.

1. Introduction

In this work, different approaches are used to investigate electrochemical interfaces with photoelectron spectroscopy. These approaches will be exemplified and compared using the oxygen evolution reaction (OER) on IrO_x catalysts as an experimental model. The OER is a bottleneck in electrochemical energy conversion and storage, which are crucial to the solutions to the current global energy challenges.[1] Electrochemical water splitting to produce "green hydrogen" as a storable and clean fuel from renewable sources of energy plays a central role in proposed solutions to global warming, creating new opportunities to replace the use of fossil fuel technologies.[2,3] The electrocatalysts used in OER conditions face two challenges: corrosion and slow kinetics,[4] reducing the process efficiency.[5,6] The catalysts must also remain stable under the strongly oxidizing conditions required for long term industrial applications. Iridium oxide is one of the most promising electrocatalysts for the OER in acidic solutions, because it uniquely combines both of the sought after properties in the same material: low overpotential and high corrosion resistance under harsh acidic conditions and anodic polarization.[7,8] Unfortunately, iridium is among the scarcest in the Earth's crust. This scarcity either demands strategies for its efficient usage, or alternatively electrocatalyst designs based on abundant materials that can mimic the electronic properties of the active sites of iridium. The factors in the electronic structure which make iridium oxide one of the best candidates for the OER are still under debate.[9] A better understanding of the iridium chemistry of the activated catalyst is necessary in order to produce electrodes based on abundant materials that are able to match iridium's electrocatalytic performance. To achieve this goal, it is necessary to describe the electronic properties of IrO_x materials under realistic *operando* conditions and, in particular, with the iridium in contact with liquids.

An in-depth understanding of the atomistic mechanisms underlying different electrochemical processes requires recording large sets of data collected under *operando* conditions to obtain key information on the electrified interface.[10] Desired parameters to be known include the chemical composition at the interface, chemical states of the atoms and their variation as a result of the electrochemical processes, and the structural evolution of the material. Unfortunately, analytical techniques which are able to provide such detailed interfacial information are very limited and hardly compatible with liquids,[11] usually restricting researchers to *ex situ*/post-mortem characterizations. This limitation leads to a loss of important information, as in many cases the intermediates and electrocatalytically active species cannot be detected in a post-process analysis, leaving us blind to key aspects ruling the reaction pathway. X-ray spectroscopic techniques are able to provide the relevant information on the electronic structure in an element specific manner but real electrochemical interfaces are buried and, most of the time, are in the presence of liquid electrolytes. Because of this fact, the electrified interfaces are inaccessible directly to common surface sensitive techniques like photoelectron spectroscopy, requiring new experimental strategies for their operation under these conditions.[12] Recently, different approaches have been developed, enabling the investigation of the electronic structure variation of electrodes under potentiostatic control using near ambient pressure X-ray photoelectron spectroscopy (NAPXPS),[13] from the gas phase up to bulk aqueous electrolytes.[11,12,14–18] Some of the most extensively used approaches entail the use of hard X-rays to allow the photoelectrons to escape through thick electrode membranes[19] (a few nanometers thick) or thin film electrolytes.[20] However, the use of soft X-rays for characterizing electrified interfaces presents significant advantages with respect to hard X-ray techniques, such as higher absorption cross sections, higher surface sensitivity, higher chemical sensitivity, and fine-structure resolution.[15] Unfortunately, performing *in situ/operando* experiments with soft X-rays is technically more challenging than in the hard X-ray regime. In this work, we will show some of the surface sensitive XPS/XAS approaches that have been established in our group in the last years, which overcome the challenge of using soft X-rays for the *in situ/operando* characterization of working electrocatalysts. The performance of these approaches is compared with those of well established bulk sensitive approaches in liquids based on photon-in/photo-out techniques in the fluorescence yield mode[21,22] in order to validate our new approaches based on photoelectron detection. In addition, the unique capabilities of photoelectron based spectroscopy techniques will be discussed in terms of the enhanced speciation, ascribed to the capability to resolve spectra measurements with two different energies simultaneously, binding and excitation energies, using resonant spectroscopy.

In photoelectron spectroscopy, an electron is ejected almost instantaneously using a photon energy higher than the threshold of ionization. However, tuning the photon energy in the vicinity of a core level near the absorption threshold yields a selective enhancement of photoemission from the absorbing atom through the phenomenon of resonance. Thus, the excited state, with a hole in the core level, will rapidly decay by the emission of an electron, as in the Auger decay process, which is shown schematically in Fig. 1. This phenomenon is in the nature of the resonant photoemission spectroscopy (ResPES) operation. Owing to the fact that the excitation energy is close to a core level absorption threshold, the resonant process selects specific chemical elements and is sensitive to particular chemical states, meaning that the information is particularly informative about the electronic structure of the selected atoms. Note that ResPES requires a tunable source of photon energy, as is provided in a synchrotron facility, in order to be able to select the photon excitation energy close to the absorption threshold of the selected core level. This capability simultaneously yields valuable information of the occupied (PDOS) and unoccupied orbital states (XAS). Using ResPES, we find in this work that the oxygen species taking part in the electrochemical oxidation of water on iridium electrodes are more involved and complex that previously stated. It is found that the formation of single coordinated oxygen atoms to iridium (named μ_1-O) yields the formation of peroxo species (named μ_1-OO),

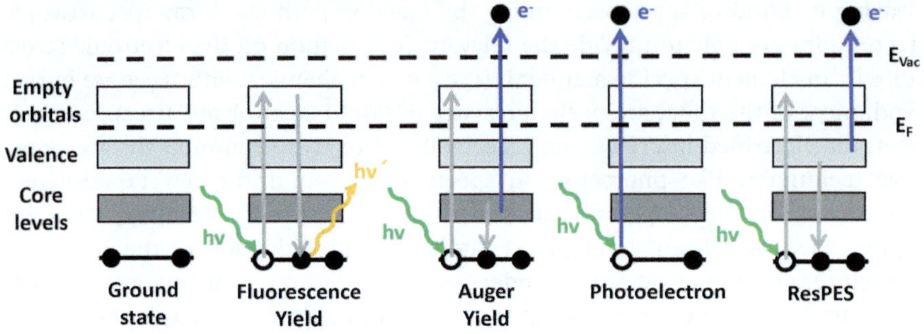

Fig. 1 Relevant electronic transitions for the radiative relaxation fluorescent photon emission, non-radiative Auger electron emission, photoelectron emission and resonant photoemission spectroscopy (ResPES).

which produce dioxygen (O_2) as a product of the oxygen evolution reaction. This result is of prime importance and represents a significant advance in the complete understanding of this complex reaction.

2. Can a bulk sensitive spectroscopy technique provide relevant information on an electrified interface?

First, we investigate the OER on an IrO_x based catalyst using X-ray absorption spectroscopy (XAS), which is an element specific technique that provides information about the electronic structure of the unoccupied orbitals of a target element of interest.[23] XAS is one of most widely used X-ray based spectroscopic techniques to investigate the evolution of the electronic structure of materials, such as energy materials. Note that XAS requires the use of a tunable X-ray source for scanning the energy through the absorption edge of interest. Hence, the measurements are usually performed in a synchrotron light facility. When a sample is illuminated by X-ray radiation of an appropriate energy, the interaction of this radiation with the matter is dominated by photon absorption and electron excitation from an occupied core level to unoccupied energy levels. This process yields the formation of an excited electron and a core hole where the subsequent relaxation process is through photon light emission in total fluorescence yield (TFY) or by the emission of an Auger electron and secondary electrons originating in the cascade processes. The emitted electrons can be collected as the Auger electron yield (AEY) or as the total electron yield (TEY), and also as the partial electron yield (PEY) if the electrons are selected by kinetic energy. A schematic drawing with the different relevant electronic transitions and emission processes is shown in Fig. 1. The AEY requires an electron energy analyzer for its collection, meanwhile the TEY can be collected in different ways, such as collecting the emitted electrons using a current collector electrode (Faraday cup), or grounding the sample and collecting the drain current produced as a consequence of the emission of secondary electrons (TEY-ground). The TFY signal is considered bulk sensitive, owing to the large penetration depth of the photons in the μm range in the soft X-ray regime. Meanwhile, TEY is accepted to be a surface sensitive technique due to the short total escape depth of the photoelectrons (elastic and inelastic).[24,25] While useful for probing catalytic reactions, this surface sensitivity makes the collection of spectra especially challenging in the soft X-ray regime in the presence of liquids, as the photoelectrons and secondary electrons are strongly scattered in dense media.[10]

In order to accomplish *operando* investigations during OER with photoelectron-based techniques, we developed a experimental setup facilitating such measurements. The setup is as follows: the electrochemical cell (EC-cell) body used for this work is made of polyether ether ketone (PEEK), which is chemically inert for most of the liquid electrolytes and electrically insulating. The cell has exchangeable working

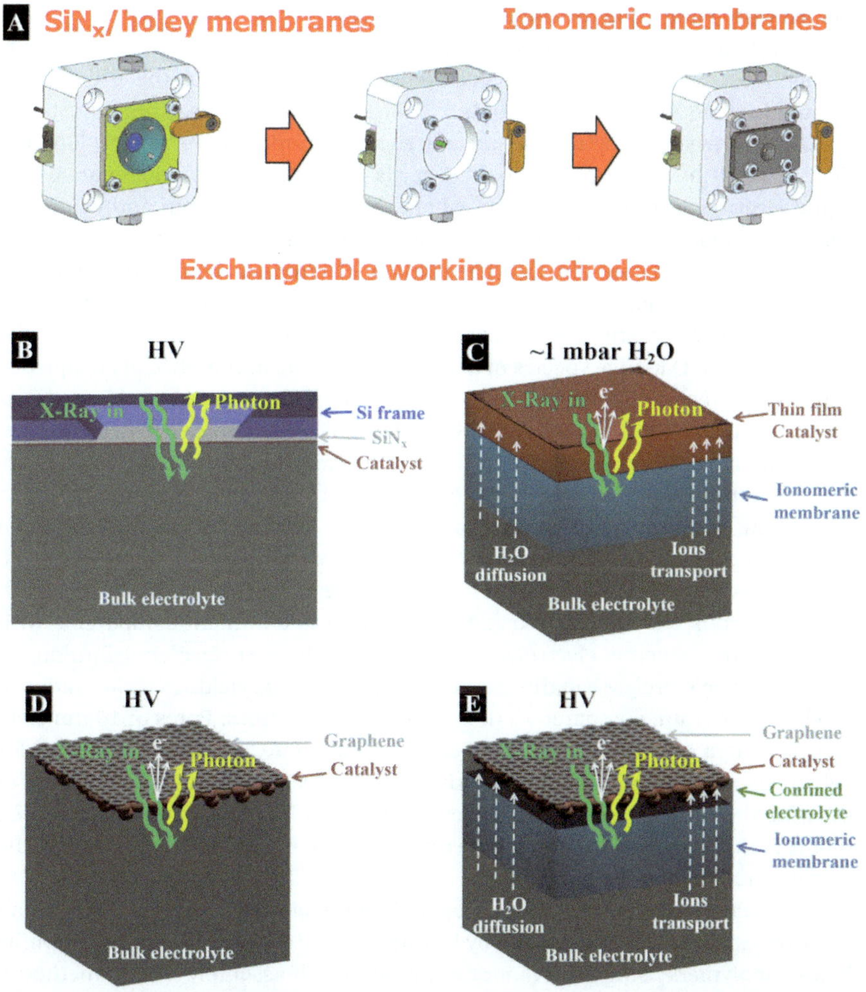

Fig. 2 (A) Drawings of the *operando* EC-cell approach with exchangeable electrodes used in this work. Drawings of the different membranes used for the *operando* electrochemical experiments: (B) catalyst on a SiN_x membrane, (C) thin film electrode on an ionomeric membrane, (D) free standing graphene decorated with the catalyst and (E) capped catalyst on an ionomeric membrane with a graphene membrane.

electrodes (see Fig. 2A), allowing for the use of different types of membranes (as we will discuss later). A Pt wire and a Ag/AgCl electrode (FLEXREF, sourced from WPI Florida, USA) are used as the counter and reference electrodes, respectively. The pulse-free, continuous electrolyte flow is ensured *via* a Perimax16 peristaltic pump from SPETEC. The total fluorescence yield (TFY) signal is acquired using a photodiode (AXUV100 Opto Diode Corp) located in the main vacuum chamber. Note that, in addition, the TEY signal can be collected using a Faraday cup or in the XPS/AEY signal using the hemispherical analyzer. More details about this setup can be found elsewhere.[26] In Fig. 2, the schematic representation of the different membrane assemblies used in this work are shown. Thus, Fig. 2 shows the different approaches based on (B) a catalyst deposited on a SiN_x membrane, (C) a thin film deposited on an ionomeric membrane, (D) a membrane based on free standing graphene decorated with the catalyst and (E) a catalyst on an ionomeric membrane capped with a graphene membrane.

Taking into account the information depth provided by the TEY/AEY and TFY, the simplest way to investigate electrochemical processes in the presence of a liquid is using XAS in the TFY mode.[21] In order to perform measurements based on TFY, in the soft X-ray regime, usually a thin film membrane of SiN_x (100 nm thick) that is quasi-transparent to the X-ray is used. The membrane separates the liquid electrolyte from the vacuum, where the X-ray comes from and where the detector is placed. A thin film is deposited onto this membrane and used as a working electrode (see Fig. 2B). The working electrode is deposited on the SiN_x membrane using many different techniques, such as physical vapour deposition, sputtering, drop casting, spin coating, electrodeposition, *etc.* However, the collection of an O K-edge spectra of a working electrode in presence of an aqueous electrolyte is especially complicated in the TFY mode, since the aqueous electrolyte contribution is dominant due to the bulk sensitivity of this technique. Collecting an O K-edge spectrum of metal oxides in the presence of an aqueous electrolyte therefore requires extremely bulky electrodes, but this may result in a very low surface contribution from the oxide surface in the spectrum. To enhance the surface sensitivity of TFY-XAS, while still having a high material load, we make use of a mesoporous film to increase the surface to bulk ratio. For such a thick, porous electrode, the oxide contribution is dominant/comparable with respect to the aqueous electrolyte, resulting in a high surface contribution. In addition, the electrolyte can diffuse into the porous bulk, yielding higher ratios of products per gram, compared to those of compact samples. Pores of 15 nm yield a surface area of ~140 m^2 of film per m^2 of substrate, compared to 20 m^2 for substrates with 5 nm pores.[27] Note that the surface area is a key parameter to describe the sample performance because it determines the amount of material that it is exposed to the electrolyte per mass, which has a direct influence in the overall catalytic performance.

The mesoporous IrO_x film was prepared on the electrolyte-exposed side of an X-ray transparent SiN_x membrane by iridium acetate and micelles of amphiphilic block-copolymers using an "evaporation-induced self-assembly" (EISA) method.[27] Fig. 3A shows the top-down scanning electron microscopy (SEM) view of such a mesoporous IrO_x-coated SiN_x membrane. The image reveals an average pore size of 20 nm and distance between pores of 27 nm, with a catalyst thin film thickness of 100 nm (as shown in Fig. 3B). Note that for the experiments shown here, three consecutive mesoporous films were deposited onto the SiN_x membrane, yielding

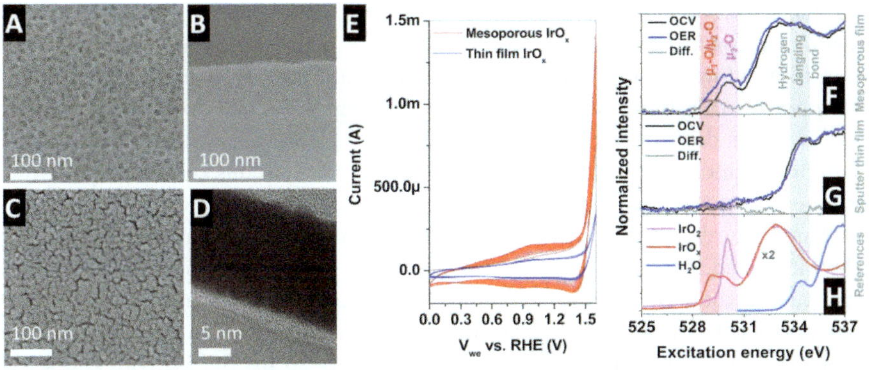

Fig. 3 (A) Top-down SEM view of the IrO_x mesoporous film on the SiN_x membrane. (B) Cross-section SEM image of the IrO_x mesoporous film on the SiN_x membrane. (C) Top-down SEM view of the IrO_{2x} sputtered thin film on the SiN_x membrane. (D) Cross-section SEM image of the IrO_x sputtered thin film on the SiN_x membrane. (E) CV collected in the presence of 100 mM H_2SO_4 (saturated in N_2) with Pt and Ag/AgCl as counter and reference electrodes, respectively, for the mesoporous (red) and sputtered (blue) thin films. Spectra collected under *operando* conditions with XAS-TFY under OCV and OER in 100 mM H_2SO_4 with Pt and Ag/AgCl as the counter and reference electrodes, respectively, for (F) mesoporous IrO_x and (G) sputtered IrO_x thin films. (H) Reference O K-edge spectra.

a ~300 nm thick film. The EISA method yields the formation of well-defined crystalline walls, critical for electrocatalyst stability under harsh acidic conditions in anodic faradaic processes. At the same time, the high surface to bulk area ratio is ensured by the pores, which is quite important for the electrocatalytic performance and material usage optimization. The porous film allows the diffusion of liquids (or gases) within its structure, while the surface to bulk ratio remains high compared to compact thick film electrodes (Fig. 2B shows the detection scheme). A SEM cross-section image of the SiN_x membrane coated with the prepared mesoporous film is shown in Fig. 3B showing the suitability of this method to fabricate mesoporous films onto a relatively fragile thin film SiN_x (100 nm thick) X-ray transparent membrane. The performance of the mesoporous membranes is compared with that of a sputtered thin film deposited onto a SiN_x membrane. Fig. 3C shows the top-down SEM view of a sputtered IrO_x thin film on a SiN_x membrane. The cross section SEM view shows a catalyst thin film thickness of 20 nm, as seen in Fig. 3D.

The electrocatalytic performance of the thin film was investigated by means of cyclic voltammetry (CV) in 100 mM H_2SO_4 (saturated in N_2) with Pt and Ag/AgCl as the counter and reference electrodes, respectively, with a scan rate of 20 mV s^{-1}, as shown in Fig. 3E. The potential conversion to the reversible hydrogen electrode (RHE) was calculated using the Nernstian equation to be $E_{RHE} \approx E_{Ag/AgCl} + 0.254$ V. The cyclic voltammogram (CV) shows two oxidation and two reduction waves.[28,29] In addition, the oxidation of water takes place in the region above 1.4 V *vs.* RHE. Continuous cycling between the open circuit voltage (OCV) and OER potentials (1.854 V *vs.* RHE) in 100 mM H_2SO_4 electrolyte at room temperature results in the development of a larger peak in the oxide region, increasing the peak height with an increasing number of cycles,[30] which is ascribed to the existence of more active highly hydrated iridium (hydroxo) species.[31] These trends are clearly visible in the

CV in Fig. 3E, where it is generally accepted that the active sites are hydrated Ir-oxo-hydroxides formed during the OER.[32–34] Note that the total current in the sputtered thin film is lower due to its lower surface to bulk ratio compared to that of the mesoporous thin film. The molecular level description of the species responsible for the high catalytic activity shown by iridium, however, is still debated[9] and cannot concluded from *ex situ* characterizations but requires the investigation of the electrified interface under working conditions.

Thus, in order to gain deeper knowledge of the mechanism ruling the electrochemical oxidation of water on iridium oxide based catalysts, the electronic structure of the IrO_x mesoporous film was investigated by O K-edge X-ray absorption spectra. The O K-edge provides information ascribed to the electronic transition from the occupied O 1s to the unoccupied O 2p hybridized with the Ir 5d orbitals. The spectrum has two features at *ca.* ~530 eV and ~533 eV ascribed to hybridization with the Ir t_{2g} and e_g orbitals, respectively[35] (see the reference spectra in Fig. 3H). Different O K-edge structures were found depending on the calcination preparation temperature of the mesoporous iridium film EISA template preparation. In the present investigation, we used films prepared at 375 °C, which show higher electrocatalytic activity than those prepared at higher temperatures (500 °C and higher). More experimental details ascribed to the electrode fabrication can be found elsewhere.[36] Using such electrodes, a complete set of measurements as a function of the applied potential was accomplished in the presence of 100 mM H_2SO_4 aqueous electrolyte de-aerated with N_2 (gas). At the same time, the electrolyte was continuously refreshed by a peristaltic pump, minimizing the formation of trapped gas bubbles on the surface of the working electrode.

Fig. 3H (bottom) shows the O K-edge spectra of the reference IrO_2 rutile, amorphous IrO_x and water. The reference spectra are compared with the *operando* spectra collected at open circuit voltage (OCV) and during OER, obtained using a sputtered polycrystalline metallic iridium thin film sample of ~20 nm (Fig. 3G) and a mesoporous film (Fig. 3F). The spectra collected with the sputtered thin film only display one dominant peak at OCV and under OER at around ~534.5 eV, ascribed to hydrogen dangling bonds in the aqueous electrolyte. The sputtered IrO_x film does not show any detectable peak at ~530 eV. In contrast, the O K-edge spectra collected with the mesoporous electrode show a peak at ~530 eV, ascribed to oxygen coordinated to three iridium atoms (named μ_3-O).[37] Upon anodic polarization at OER, a new peak appears at ~528–529 eV, ascribed to the formation of electron deficient active oxygen species with bridging (named μ_2-O) and terminal (named μ_1-O) oxygen atoms.[37–41] Electron-deficient oxygen species have been correlated with the formation of active centers in the electrode catalyst, however its precise atomic structure varies between different iridium (hydr)oxides.[9] Thus, according to these results, a bulk sensitive technique can provide relevant atomistic information of an electrified interface if the sample has a high surface to bulk ratio.

3. Can *operando* approaches based on photoelectron spectroscopy mimic the performance of bulk sensitive spectroscopies?

Bulk sensitive methods, like XAS in the TFY mode, can provide valuable information about the electronic structure of an electrified interface under electrocatalytic reaction conditions using materials with high surface to bulk ratios, like the mesoporous film used in this study. In such systems, the challenge is to compare the information acquired with the bulk sensitive XAS-TFY method to the surface sensitive spectroscopy techniques based on photoelectron spectroscopy, which can reveal information related to the electrified solid–liquid interface directly and more specifically. In this direction, different approaches have been implemented using hard/tender X-ray based methods in recent years, taking advantage of the larger inelastic mean free path (IMFP) of photoelectrons with higher kinetic energies than those generated by soft X-ray energy excitation. Thus, it is possible to investigate electrodes within a thin electrolyte meniscus using the so-called dip & pull[20,42] method. Another technique entails the use of thin Si membranes for separating the liquid from the vacuum side where the analyzer is placed.[19] For using this geometry in the XPS mode, the membrane must be significantly thinner and, again, hard X-rays must be used. However, in the present work, we will focus in the experimental approaches developed to be used in the soft X-ray regime, because in this energy range important elements such as carbon, nitrogen and oxygen have their absorption edge. In addition, the X-ray absorption cross-section is higher in the soft X-ray range, enhancing the signal intensity from the interface between solids and liquids with enhanced spectral resolution (*i.e.* enhanced Auger electron emission compared to the total fluorescence yield in the tender/hard X-ray range). Furthermore, the surface sensitivity of soft X-ray XPS is higher than in the tender/hard regime (subnanometer *vs.* several nanometers).

Note that NAPXPS[13] is operated in the soft X-ray regime at a partial pressure up to a few mbars. At these low partial pressures, it is not possible to allocate aqueous electrolytes in the experimental chamber. Only certain liquid electrolytes with low partial pressures, such as ionic liquids, can be placed directly in the experimental chamber and be used as electrolytes in electrochemical experiments. Taking into account these experimental limitations, the first electrochemical experiments were performed in the gas phase up to partial pressures of a few mbar. Thus, solid oxide electrolytes operating at high temperatures were first used to investigate electrochemical processes under operational conditions.[43,44] Additionally, the use of ionomeric membranes allowed the investigation of electrochemical reactions, *i.e.* OER, at room pressure and low partial pressure of water vapor.[45,46] However, this approach has significant limitations in terms of the membrane hydration (required for high current operations) and chemical potential depending on the partial pressure.

In order to improve the wetting of the ionomeric membrane, it is back-filled with an electrolyte using a peristaltic pump, while the ionomeric membrane separates the liquid electrolyte from the vacuum.[47] This approach allows for an optimal membrane hydration, enabling the transport of ions but not electrons. On top of the ionomeric membrane (Nafion™ in this case), a thin film electrode is deposited. This assembly resembles industrial applications, in acidic media, in which electrochemical reactions are performed using composites of electrodes and ionomeric membranes, which separate the gaseous products from the liquid electrolyte. At the same time, the electrode faces the experimental vacuum chamber, where it is illuminated by the X-ray source and the photoelectrons are collected using the hemispherical analyzer (XPS/AEY/PEY) or with a Faraday cup (TEY). Such a membrane assembly is shown schematically in Fig. 2C. In this configuration, the working electrode is produced by sputtering a thin film directly on the ionomeric membrane. Thus, a ~10 nm Ir thin film is sputtered onto a Nafion™ membrane, which cracks when the membrane swells or shrinks upon wetting or drying. Water can diffuse through these membranes and finally evaporates into the vacuum chamber, leaving the catalyst surface dry. However, in places in which the catalyst is in close contact with the membrane, mass transport can occur.[41] A peristaltic pump refreshes the 100 mM H_2SO_4 electrolyte (saturated in N_2) in a continuous flow behind the ionomeric membrane. The counter (Pt wire) and reference (Ag/AgCl) electrodes are immersed in the bulk liquid.

Fig. 4A shows the Ir 4f XPS and the O K-edge XAS spectra collected under OCV and OER on the surface of the sputtered thin electrode (~10 nm). The *operando* collected O K-edge spectra are compared with those of the reference samples (Fig. 4E). Thus, under reaction conditions, the Ir 4f develops a shoulder at higher binding energies (62.5 eV) associated with the formation of oxidized species, including the active ones.[48] At the same time, the O K-edge shows the formation of

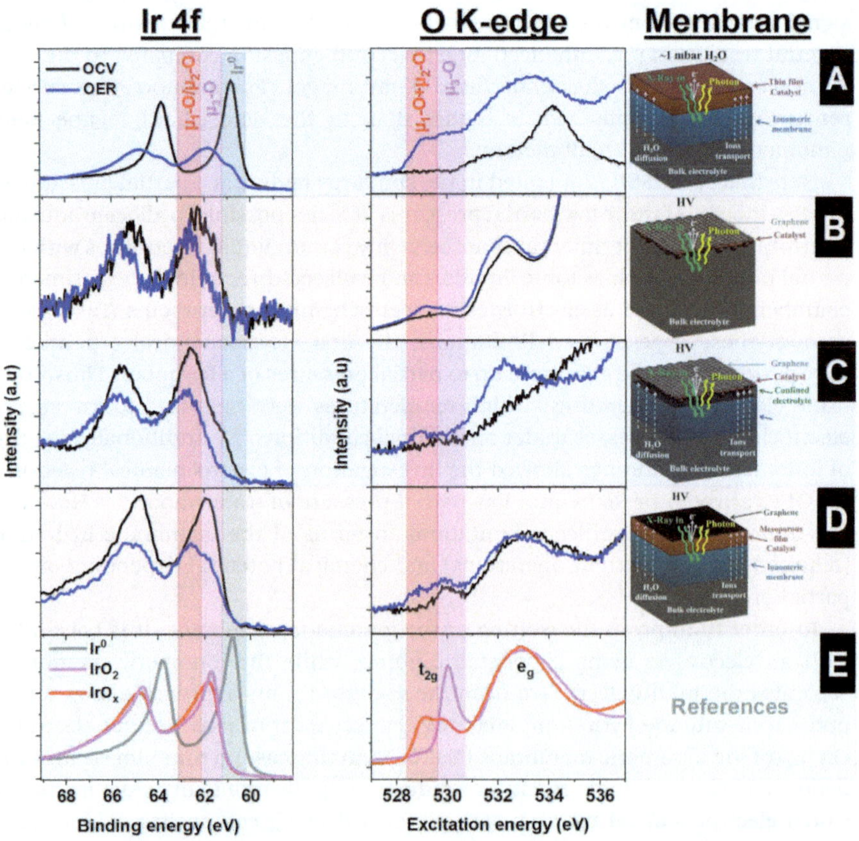

Fig. 4 *Operando* spectra collected using photoelectrons from the O K-edge and Ir 4f with (A) thin film sputtered onto a Nafion™ ionomer membrane, (B) Ir NPs sputtered onto free standing graphene on a holey SiN_x membrane, (C) sputtered Ir NPs "sandwiched" between Nafion™ and graphene, (D) a mesoporous IrO_x thin film "sandwiched" between Nafion™ and graphene, and (E) reference spectra.

a peak at around ~529 eV, ascribed to the formation of oxygen singly bound to iridium (μ_1-O) and doubly coordinated μ_2-O (bridge) species.[37] Therefore, this approach, based on an ionomeric membrane with a sputtered thin film on top, provides detailed information on the active species present under the reaction conditions. However, this setup presents some limitations, since only thin films can be used as working electrodes. In principle, it is difficult to produce a reliable electrode that has good electrical conductivity between the different film patches while being thin enough to provide information of the electrified interface (iridium in close contact with Nafion™). Please note that it is important that the degree of the desiccation cracks formed in the film is a random process, which makes it quite difficult to prepare reproducible samples. Furthermore, with this approach, it is challenging to hydrate the electrode well, which adversely influences the electrochemical performance due to strong mass transport limitations. Metallic iridium present during the OER is therefore ascribed to the side of the electrode facing the vacuum chamber, *i.e.* no contact with either Nafion™ or the liquid electrolyte.

Another approach for the investigation of solid–liquid interfaces with soft X-rays is based on free standing 2D (graphene) membranes.[49–51] With this approach, the X-rays easily penetrate the graphene that, in turn, is also semi-transparent to photoelectrons, *i.e.*, the photoelectrons can escape the confined environment and be detected in the vacuum side where the hemispherical analyzer is placed. To enhance the mechanical stability of larger graphene layers, the graphene is transferred onto a holey SiN_x membrane with an array of micro holes.[50] This approach yields a large free standing graphene area, which is larger than the X-ray beam spot size and increases the signal intensity with respect to a single small graphene covered hole. In addition, this membrane is leak tight to the electrolyte and electrically conductive, allowing for its use as a current collector. A uniform distribution of iridium nanoparticles, covering ~20% of the surface with a size of around ~2 nm, is deposited on top of this membrane by sputtering.[29,48] A schematic representation of this approach is shown in Fig. 2D, where the electrolyte is in close contact with the iridium NPs and graphene. Using this approach, the Ir 4f and the O K-edge spectra were collected, indicating the formation of electron-deficient oxygen species present during the OER, as shown in Fig. 4B. However, the intensity of the peaks observed is lower than in the approach that uses ionomeric membranes coated with a sputtered thin film due to the lower loading of Ir NPs. In addition there is a significant part of the holey SiN_x membrane that is not illuminated by the beam and is, hence, not contributing to the Ir 4f spectra. As an advantage, the graphene membrane uses a similar geometry as the bulk sensitive approaches based on TFY detection with SiN_x membranes, in the sense that the electrochemical experiments are performed with bulk liquid electrolyte. However, this approach suffers from beam damage, requiring graphene of high quality to ensure stability during data acquisition. The illumination of an aqueous electrolyte with an intense X-ray beam yields the formation of radicals from water radiolysis; these radicals damage the graphene layer.[15,52,53] To solve this issue, the stability of the graphene layer can be increased by stacking more layers of graphene on top of each other. However, that would require the use of higher excitation energies (in the tender X-ray regime) or the use of other stable thin membranes, *i.e.* coating the graphene with other materials more stable to the beam effects.[54]

While the free-standing graphene approach is challenged by beam damage with a risk of the electrolyte leaking into the vacuum chamber, using ionomeric membranes coated with a "thick" film is limited by wetting, mass transport, and electrical conductivity. A possible solution is the combination of both approaches based on the transparent graphene and the ionomeric membranes. This is a suitable approach to improve the stability of the free standing graphene (the ionomeric membrane holds the pressure difference) and the evaporation of water (the graphene cover is an evaporation barrier).[26,37,55,56] The underlying idea is to transfer the graphene onto the ionomeric membrane and use it as a photoelectron transparent membrane, current collector and evaporation barrier in the same way as with the free standing graphene approach. There are different techniques to transfer the material of interest to the ionomeric membrane; the easiest consists of depositing NPs of Ir (of around ~1 nm) onto the ionomer membrane by sputtering. After that, a layer or a bi-layer of graphene is transferred on top, leaving the NPs "sandwiched" between the graphene and Nafion™ (Fig. 2E shows a schematic of this approach). By doing this, an electrical contact between the different NPs with the potentiostat is ensured by graphene.

The graphene works as an evaporation barrier, yielding correct membrane wetting and the formation of a confined liquid between the Nafion™ and graphene.[56] In this way, this approach allows the investigation of solid–liquid interfaces, providing similar Ir 4f and O K-edge spectra as the other approaches. However, in the graphene covered samples, the Ir 4f spectra show a much reduced contribution of metallic iridium compared to the sample without the graphene cover, indicating a better electrochemical performance because the metallic iridium is not the active phase, as shown in Fig. 4C. However, this approach has a weakness compared to the "thick" film: the ionomeric membrane itself is more sensitive to beam damage than in the case of the thicker Ir film, requiring that the spectra are collected continuously at fresh spots on the sample. In addition, the charge transport of this approach is limited by the membrane specifications (Nafion™, for example, mostly conducts protons efficiently), but this is not the case for other ions. Thus, this approach can lead to worse charge transport compared to that of free standing graphene, as well as unknown pH of the confined liquid between graphene and the ionomeric membrane, which can lead to an unknown applied potential under certain experimental conditions.

Another method employs the use of mesoporous thin films, which, unlike the sputtered films, allow electrolyte diffusion and mass transport within its volume. This assembly is shown schematically in Fig. 5A. The mesoporous Ir 100 nm film is transferred onto a Nafion™ membrane and capped with a graphene layer to improve the electrical contact and minimize water evaporation. Fig. 5B shows a SEM image of the mesoporous IrO_x film on a Nafion™ membrane, meanwhile Fig. 5C shows the same type of membrane capped with graphene. The layer was prepared by EISA at 375 °C. This configuration has a clear advantage with respect to the use of NPs, as it is less sensitive to beam damage, since a large portion of the incoming radiation of the beam is blocked. Thus, the photon intensity transmitted trough a 100 nm thick iridium layer is around 10% of the original at an excitation energy of 600 eV, according to the calculation of Henke *et al.*[57] This method therefore allows the collection of spectra over long periods of time in the same sample spot. The mesoporous films are transferred using a hot press approach between the Nafion™ membrane and the IrO_x mesoporous film grown

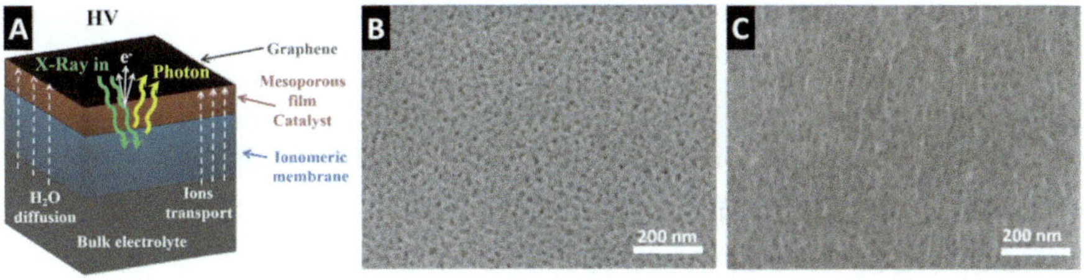

Fig. 5 (A) Drawings of the capped mesoporous film on an ionomeric membrane (Nafion™) with a graphene membrane. SEM images of (B) a Nafion™ membrane coated with a mesoporous IrO_x film and (C) a Nafion™ membrane coated with a mesoporous IrO_x film capped with a graphene layer used in the collection of the *operando* X-ray spectra.

on a silicon wafer.[18] It is also beneficial that this approach can be directly compared with the experiments performed in TFY with the SiN_x X-ray membrane. The fact that the electrolyte diffuses through the pores of the film allows the collection of the spectra with the electrolyte in contact with the solid electrode. Thus, the Ir 4f and the O K-edge spectra resemble the spectra collected with the other approaches, with significantly higher intensities compared to those from the approach based on sputtered NPs, as shown in Fig. 4D. Therefore, these measurements confirm the presence of electron deficient oxygen species during OER. However, not every material is easily prepared as a mesoporous film using the EISA approach, which hinders the application of this approach. In addition, the mesoporous films can suffer from mass transport limitation compared to other electrode geometries.

Taking into account the four different approaches compatible with photoelectron spectroscopy shown in this section, it is clear that all of them provide similar information and offer different advantages and disadvantages. The spectra collected with photoelectrons are similar to those collected in the TFY mode, revealing the formation of electron deficient oxygen (μ_1-O and μ_2-O) during the OER on IrO_x catalysts independently of the experimental approach selected. Consequently, we can conclude that similar information can be extracted using surface sensitive measurements compare to those methods based on TFY with

Table 1 Comparison between the different sample assemblies

Sample assembly	Preparation difficulty	Beam damage	Electrochemical performance	Spectroscopy method
0.1 M H_2SO_4/Ir thin film/SiN_x/HV	Low	Low	Excellent	Photons
0.1 M H_2SO_4/mesoporous Ir/SiN_x/HV	High	Low	Excellent	Photons
0.1 M H_2SO_4/Nafion™/Ir thin film/HV	Low	Medium	Bad	Photons/electrons
0.1 M H_2SO_4/Ir NPs/graphene/holey SiN_x/HV	Extreme	Extreme	Excellent	Photons/electrons
0.1 M H_2SO_4/Nafion™/Ir NPs/graphene/HV	Medium	High	Good	Photons/electrons
0.1 M H_2SO_4/Nafion™/mesoporous Ir/graphene/HV	High	Medium	Good	Photons/electrons

surface enhanced sensitivity. The advantages and disadvantages of each approach discussed here are summarized in Table 1.

4. Can X-ray photoelectron spectroscopy provide additional speciation information not accessible by only XAS?

The well-established techniques based on bulk-sensitive methods, such as TFY, can provide similar information as X-ray photoelectron spectroscopy under certain conditions, as we have shown in this work (*i.e.* if a mesoporous sample is used). If that is the case, one could ask if the development of techniques compatible with photoelectron spectroscopy really deserves the time and effort needed for their implementation. To answer this question and to justify this effort, one can argue that photoelectron spectroscopy is a really surface-sensitive method, meanwhile TFY is not. Against this argument, we show in this work that, after all, in case of the IrO_x system, very similar conclusions are drawn from all approaches. Another advantage of XPS is that it provides information related to other core levels. However, the TFY measurements can easily be complemented by collecting other edges of interest, for example the Ir L-edges. Hence, the additional core level information from XPS is also not a strong argument to justify the development of novel photoelectron spectroscopy approaches. Another argument in favor of photoelectron compatible techniques is that it is possible to detect changes in the potential drop induced by the ions within the electrical double layer, which is an excellent method to distinguish between the voltage and the electrochemical potential applied. XAS methods cannot provide this information. In general, though, the electrochemical performance of the approaches based on SiN_x membranes are significantly better than those used for XPS and, usually, the charge transport is not an issue. Finally, if these three arguments themselves do not justify the development of photoelectron spectroscopy compatible *operando* strategies, what else will justify it?

To answer this question, one should go one step beyond the surface sensitivity of X-ray photoelectron spectroscopy and explore other opportunities that this technique offers for the identification of the active species. ResPES is a technique which simultaneously provides information about occupied and unoccupied states of interest. Similar measurements can be done using resonant inelastic X-ray scattering (RIXS), however this technique requires substantial incident photon flux (inducing significant beam damage) to obtain the substantially scattered photons needed for collecting spectra with enough intensity and energy resolution. Therefore, RIXS cannot compete with ResPES in term of the energy resolution and signal intensity under electrochemical *in situ*/*operando* conditions. Unfortunately, ResPES is hardly compatible with liquids, requiring sophisticated technical approaches for its use under reaction conditions, whereas NAPXPS offers these capabilities. In addition, the long time necessary to collect these measurements (more than one hour) makes it necessary to use samples that are stable under harsh reaction conditions (*i.e.* corrosion/dissolution resistance) and X-ray beam illumination (beam damage). In order to perform ResPES measurements, the best approach is the one based on an ionomeric membrane coated with a mesoporous film capped with a graphene layer. This approach combines good electrochemical performance and long-term stability under harsh OER conditions and X-ray beam illumination. Fig. 5C shows a SEM image of the mesoporous IrO_x film on Nafion™ capped with a graphene membrane. Please note that the size and density of the pores are similar to those in Fig. 3A, which correspond to the mesoporous film on the SiN_x membrane prepared by EISA at 375 °C.

Using this assembly, ResPES measurements are collected for the valence band using the photon energies in the vicinity of the O 1s core level. Note that O 2p and Ir 5d orbitals are hybridized due to the covalent Ir–O bonding. Fig. 6A shows the ResPES collected through the O K-edge under OCV conditions, which is ascribed to the transition of an electron from an occupied O 1s level to the unoccupied O 2p orbitals. The ResPES of the mesoporous IrO_x film[36] collected in the vicinity of the O 1s shows the existence of two dominant resonant peaks, as shown in Fig. 6A. Selected valence band spectra are shown on top of the colour map and the full range AEY spectrum is shown on the right hand side. At OCV, there is a resonant peak at ~530 eV excitation energy (EE) and a 15.6 eV biding energy (BE), highlighted in red, which is ascribed to the existence of μ_3-O and μ_1-OH species. Otherwise, under OER conditions, the ResPES map shows the formation of other strong resonant peaks ascribed to the formation of μ_1-OO peroxo (532.5 eV EE; 14 eV BE),[58] μ_2-O (529.2 eV EE; 12.2 eV BE) and μ_1-O (528.9 eV EE; 8.8 eV BE), as shown in Fig. 6B. The full AEY spectra collected are similar to the O K-edge absorption spectra found under OER, with a peak located at a lower excitation energy. However, in addition to the information extracted from the XAS characterization, the ResPES map shows that the absorption peak below 530 eV is a consequence of at least two different species, as indicated in Fig. 6B by the peaks at (529.2 eV EE; 12.2 eV BE) and (528.9 eV EE; 8.8 eV BE). These results indicate the existence of another important intermediate species during the reaction which is

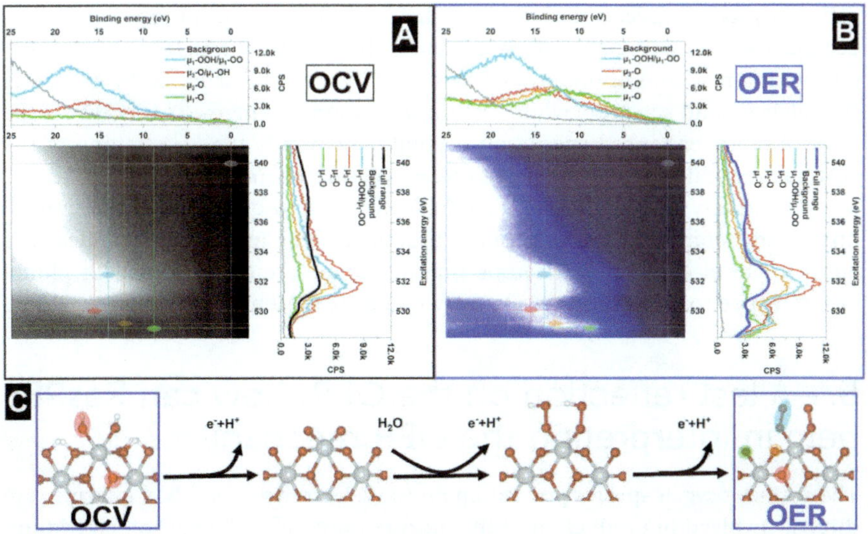

Fig. 6 ResPES collected in the valence band using excitation energy through the O K-edge (A) at OCV and (B) under OER conditions. (C) Representation of the species found at OCV and under reaction conditions and the reaction pathway.

different to the singly coordinated μ_1-O (terminated oxygen) and doubly coordinated μ_2-O (bridge) species. Unlike previous results,[37] the μ_1-O active species can easily be distinguished from the μ_2-O, indicating clearly that these species are key to the OER reaction. In addition, the peak appearing at (532.5 eV EE; 14 eV BE) can be ascribed for the first time to the existence of a peroxo species[58] during the OER conditions. Taking into account the *operando* ResPES measurements, Fig. 6C shows a scheme of the reaction pathway, where under OCV conditions the dominant species are μ_1-OH and μ_3-O. Once the electrode is under OER conditions, the μ_1-OH species undergoes deprotonation to form the active μ_1-O species, which reacts to form μ_1-OOH (hydroperoxo) and μ_1-OO (peroxo) species. The results found in this work are of prime importance and have not been described before in the literature, highlighting the existence of different oxygen species during the OER. These results indicate that the reaction pathway yielding dioxygen is dominated by the formation of peroxo groups. The ResPES results clearly indicate that the speciation is more complex than previously thought, indicating the existence of wide ranging oxygen species under OER conditions.

According to these results, the ResPES characterization of solid–liquid interfaces under reaction conditions is a powerful technique to provide detailed speciation information, more than XAS alone can. The advantage of this technique, with respect to XAS, is that ResPES provides resolved spectra in binding and excitation energies simultaneously. Thus, XAS can provide 2D maps if the spectra are plotted *versus* another parameter, such as applied potential in this work. Otherwise, if the ResPES measurements are plotted *versus* the applied potential, a 3D map can be formed. Thus, ResPES map spectra allow a better speciation of the active species present during the *operando* reactions. Therefore, photoelectron spectroscopy based techniques have a significant advantage with respect to XAS apart from the enhanced surface sensitivity and information of the potential drop, as it is possible to simultaneously collect information on the occupied and unoccupied orbitals, which yields accurate information about the different process/species that undergo the resonance event. Information on the occupied states can be gained by collecting the ResPES through the absorption edge of interest, and information about the PDOS can be acquired by collecting the valence band spectrum while exciting at a resonant energy. This ability of the ResPES yields enhanced speciation capability, which is the cornerstone of the accurate identification of the active species present during the reactions. This is a key capability that should be systematically used in experiments that aim to understand complex electrochemical processes, and justify the last decade's effort to get electrochemical setups running which are compatible with photoelectron spectroscopies in the soft X-ray range, as presented in this work.

5. A last reflection on the OER: how can ResPES help in interpreting the OER mechanism?

The surface oxygen species present on an IrO_x/electrolyte electrified interface are directly involved in production of the dioxygen product.[59,60] Thus, on an iridium (hydr)oxide catalyst, the number of singly- and doubly-coordinated oxygen atoms with iridium (μ_1-O and μ_2-O species, respectively) is larger for defective surfaces and more active during the OER. For structures with higher crystallinity (less defective), the dominant coordination environment is the triple iridium-coordinated oxygen atoms (μ_3-O species), being less active in the OER. Thus, in the presence of an iridium active center and water molecules, the system undergoes the following reaction:

$$Ir^* + H_2O \rightarrow Ir\text{–}OH + e^- + H^+, \tag{1}$$

yielding a protonated μ_1-OH species on the surface electrode surface. Next, when an anodic potential is applied, the surface undergoes an oxidation (deprotonation) as follows:

$$Ir\text{–}OH \rightarrow Ir\text{–}O + e^- + H^+, \tag{2}$$

forming active μ_1-O species at the electrified interface. The μ_1-O (active) or μ_2-O (inactive) species are ascribed to the X-ray absorption peak at ~529 eV in the O K-edge (XAS),[37,41] which becomes more intense with increasing potential. However, spectroscopically distinguishing between μ_1-O and μ_2-O species with XAS is not easy, though ResPES indicates clearly that the peak at around ~529 eV is the consequence of two different species located at (529.2 eV EE; 12.2 eV BE) and (528.9 eV EE; 8.8 eV BE), which are ascribed to μ_2-O (bridge) and μ_1-O (terminal oxygen atoms). Thus, these results indicate that the formation of singly coordinated species (terminal oxygen atoms) is strongly correlated to the catalytic activity shown by iridium (hydr)oxide. These species are stable on the surface of the catalysts and are therefore detectable with X-ray spectroscopy. This part of the reaction which was previously known is shown in the upper part of Fig. 7 ("Orbis Terrarum", mapped), even the complete speciation was not possible with previous methods based only on XAS. Nevertheless, the existence of these two species was confirmed in this study without a doubt using *operando* ResPES.

Furthermore, OER involves the adsorption of water, the transfer of four protons and electrons and the evolution of dioxygen products. This reaction is led by the formation of the O–O bond, where the exact mechanism ruling this bond formation is still under discussion in the literature. Each step of the reaction has several transition states where the oxygen–oxygen bond formation is the key factor dominating the rate step. The formation of the oxygen–oxygen bond is proposed to be produced by two different pathways: (a) through water adsorption on an active site or (b) through the oxygen–oxygen coupling of two neighbouring sites. We consider a simplified model for the absorption of water onto a deprotonated μ_1-O species as:

$$Ir\text{–}O + H_2O \rightarrow Ir\text{–}OOH + e^- + H^+, \tag{3a}$$

where the reaction can take place in three different ways: with proton transfer first, with electron transfer first or simultaneously.[9] If the proton is transferred to a neighboring site, the reaction is purely chemical and needs to be written as:

$$Ir\text{–}O + Ir\text{–}O + H_2O \rightarrow Ir\text{–}OOH + Ir\text{–}O\text{–}H \tag{3b}$$

Independent of the reaction pathway (a or b), this process yields the formation of a hydroperoxo species on the surface. The hydroperoxo is transformed into a peroxo as:

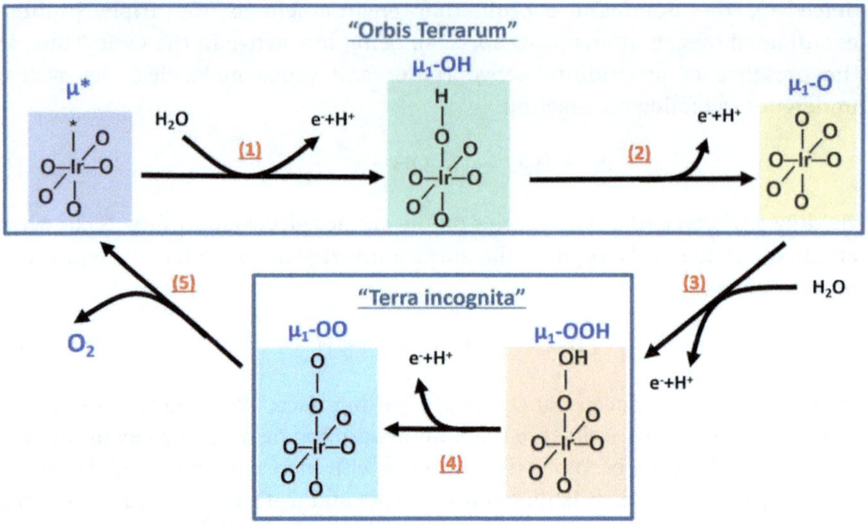

Fig. 7 Schematic representation of the simplified reaction pathway. The upper part shows the intermediates that can be considered to be known according to the *operando* spectroscopy and literature available ("Orbis Terrarum"). The bottom part represents part of the reaction pathway that has not been probed experimentally ("Terra incognita") until the current ResPES experiments presented in this work.

$$\mathrm{IrOOH} \rightarrow \mathrm{Ir\text{–}OO} + \mathrm{e}^- + \mathrm{H}^+, \tag{4}$$

finally yielding dioxygen:

$$\mathrm{Ir\text{–}OO} \rightarrow \mathrm{Ir}^* + \mathrm{O}_2 \tag{5}$$

Note that an alternative pathway that is still considered is the direct coupling of lattice oxygen,[61] which can be written as

$$\mathrm{Ir\text{–}O} + \mathrm{Ir\text{–}O} \rightarrow 2\mathrm{Ir}^* + \mathrm{O}_2 \tag{3c}$$

The reaction pathways involving the formation of the oxygen–oxygen bond are described schematically in Fig. 7 bottom, where the exact reaction pathway is unknown ("Terra incognita", not mapped) in the literature and has been proposed theoretically,[59,60] but not empirically. However, the *operando* ResPES experiments shown in this work indicate that the OER involves the formation of peroxo species, meaning that the speciation is more complex during reaction conditions involving different oxygen species. Thus, the reaction pathway yielding oxygen–oxygen bond formation is dominated by the presence of a peroxo species and does not involve the direct coupling of two oxygen neighbors in the iridium electrode lattice. Therefore, this work represents a mapping of the reaction pathway in the "Terra incognita", going far beyond the current understanding of the OER on iridium (hydr)oxide catalysts. In short, the OER on iridium electrodes is dominated by the formation of active terminal oxygen species (μ_1-O, confirmed in this work), which yield the formation of peroxo species (μ_1-OO, probed for the first time in this work), controlling the reaction pathway to dioxygen (O_2) evolution.

6. Conclusions

Detailed insights into the electronic structure of an electrified interface requires the collection of relevant X-ray spectroscopy data (XPS, NEXAFS and ResPES) under *operando* conditions. Typically, the collection of this information is restricted in the soft X-ray regime to the investigation of *ex situ* samples, especially when using photoelectrons. In order to overcome this limitation and investigate the electronic structure of electrified solid–liquid interfaces under reaction conditions, different approaches compatible with photoelectron spectroscopy have been developed in our group in recent years. The approaches presented here, using the example of iridium oxide in a sulfuric acid aqueous electrolyte, involve a different degree of complexity and their use has different advantages and disadvantages. The operation and the atomistic information provided by each method was compared with the bulk-sensitive approach (NEXAFS photon-in/photo-out). The performance of each approach was exemplified by comparing the spectra collected during the OER on IrO_x electrodes. It was found that, independent of the approach used, a peak at around ~529 eV appears under OER conditions, which is ascribed to electron deficient oxygen species. However, by using ResPES, it was possible to prove clearly that this peak is a consequence of two different species, named μ_2-O (bridge oxygen) and μ_1-O (terminal oxygen). In addition, the formation of μ_1-OO (peroxo) species was found under the reaction conditions, indicating that different oxygen species take part in controlling the reaction pathway. This result is of prime importance and represents a substantial advance in the understanding of the complex OER on IrO_x electrodes. Even if it is accepted that the terminal oxygen atoms are the catalytically active centers, we can conclude in this work that the OER is dominated by the formation of peroxo species and not by the oxygen lattice coupling reaction. Thus, ResPES provides better speciation capabilities as far as the spectral information can be depicted simultaneously depending on the excitation and binding energies. Because of this, ResPES simultaneously provides information related to the occupied and unoccupied states. This unique capability of ResPES opens up the possibility of describing steady state intermediates more accurately. To conclude, one should remark at this point that the methodologies presented here can be implemented to investigate other relevant electrochemical processes, which shows the paramount importance of the photoelectron compatible methods discussed in this work.

Author contributions

Operando mesoporous films: J. J. V. V. and R. K. conceived of the idea and designed the experiments for using mesoporous thin film electrodes on SiN_x and ionomeric membranes for *operando* X-ray spectroscopy electrochemical characterization. D. B. fabricated the mesoporous electrodes. P. Z., E. C., C. H. C. and J. J. V. V. performed the experiments. J. J. V. V. and T. E. J. analyzed the data. *Operando* ionomeric membranes: R. A. conceived of and designed the idea to use ionomeric membranes coated with a thin film for *operando* photoelectron spectroscopy. V. P. and J. J. V. V. performed the experiments. V. P. and T. E. J. analyzed the data. *Operando* free standing graphene: J. J. V. V. conceived of the idea and designed the experiments for using free standing graphene on holey SiN_x

membranes decorated with NPs for *operando* photoelectron spectroscopy. J. J. V. V. and V. P. performed the *operando* experiments. J. J. V. V. and T. E. J. analyzed the data. *Operando* ionomeric membranes capped with graphene: J. J. V. V. conceived of the idea. L. J. F. and R. V. M. developed the experimental approach for transferring PMMA free graphene onto the ionomeric membranes. L. J. F. performed the *operando* experiments. L. J. F. and T. E. J. analyzed the data. J. J. V. V., M. H. and E. S. developed the *operando* X-ray spectroscopy electrochemical cell with exchangeable electrodes. A. H. and T. L. performed the electron microscopy analysis of the different samples. A. K. G. and R. S. led the project. J. J. V. V. wrote the manuscript. All authors discussed the results and commented on the manuscript.

Conflicts of interest

There are no conflicts of interest to declare.

Acknowledgements

We thank DAAD for the financial support through the framework of the Taiwanese–German collaboration (project ID 57218279 and 57392335). C.H.C. acknowledges financial support from projects 106-2112-M-032-003. We thank Helmholtz Zentrum Berlin für Materialien und Energie for allocating beamtime for our experiments within the proposal numbers 192-08521 and 191-08014.

References

1 R. Schlögl, *ChemSusChem*, 2010, **3**(2), 209–222.
2 J. A. Turner, *Science*, 2004, **305**(5686), 972–974.
3 S. Chu and A. Majumdar, *Nature*, 2012, **488**(7411), 294–303.
4 M. G. Walter, E. L. Warren, J. R. McKone, S. W. Boettcher, Q. Mi, E. A. Santori and N. S. Lewis, *Chem. Rev.*, 2010, **110**(11), 6446–6473.
5 K. Gong, F. Du, Z. Xia, M. Durstock and L. Dai, *Science*, 2009, **323**(5915), 760–764.
6 Y. Gorlin and T. F. Jaramillo, *J. Appl. Chem. Sci.*, 2010, **132**(39), 13612–13614.
7 Y. Lee, J. Suntivich, K. J. May, E. E. Perry and Y. Shao-Horn, *J. Phys. Chem. Lett.*, 2012, **3**(3), 399–404.
8 T. C. Wen and C. C. Hu, *J. Electrochem. Soc.*, 1992, **139**(8), 2158.
9 L. J. Falling, J. J. Velasco-Vélez, R. V. Mom, A. Knop-Gericke, R. Schlögl, D. Teschner and T. E. Jones, *Curr. Top. Electrochem.*, 2021, 100842.
10 J. J. Velasco-Velez, C. H. Wu, T. A. Pascal, L. F. Wan, J. Guo, D. Prendergast and M. Salmeron, *Science*, 2014, **346**(6211), 831–834.
11 D. M. Itkis, J. J. Velasco-Velez, A. Knop-Gericke, A. Vyalikh, M. V. Avdeev and L. V. Yashina, *ChemElectroChem*, 2015, **2**(10), 1427–1445.
12 A. Knop-Gericke, V. Pfeifer, J. J. Velasco-Velez, T. Jones, R. Arrigo, M. Hävecker and R. Schlögl, *J. Electron Spectrosc. Relat. Phenom.*, 2017, **221**, 10–17.
13 M. Salmeron and R. Schlögl, *Surf. Sci. Rep.*, 2008, **63**(4), 169–199.
14 J. J. Velasco-Velez, M. Hävecker, A. Knop-Gericke, C. Schwanke, L. Xi, K. M. Lange, J. Xiao, M. F. Tesch, R. Golnak, T. Petit, L. Puskar, U. Schade,

M. Borgwardt, I. Kiyan, R. Seidel and E. F. Aziz, *Synchrotron Radiat. News*, 2017, **30**(2), 14–19.
15 A. Klyushin, R. Arrigo, V. Pfeifer, T. Jones, J. Velasco Velez and A. Knop-Gericke, Catalyst Electronic Surface Structure Under Gas and Liquid Environments, in *Encyclopedia of Interfacial Chemistry*, 2018, pp. 615–631.
16 V. Streibel, M. Hävecker, Y. Yi, J. J. V. Vélez, K. Skorupska, E. Stotz, A. Knop-Gericke, R. Schlögl and R. Arrigo, *Top. Catal.*, 2018, **61**(20), 2064–2084.
17 E. A. Carbonio, J. J. Velasco-Velez, R. Schlögl and A. Knop-Gericke, *J. Electrochem. Soc.*, 2020, **167**(5), 054509.
18 J. J. Velasco-Vélez, L. J. Falling, D. Bernsmeier, M. J. Sear, P. C. Clark, T. S. Chan, E. Stotz, M. Hävecker, R. Krähnert, A. Knop-Gericke, C. H. Chuang, D. E. Starr, m. Favaro and R. V. Mom, *J. Phys. D: Appl. Phys.*, 2021, **54**(12), 124003.
19 T. Masuda, H. Yoshikawa, H. Noguchi, T. Kawasaki, M. Kobata, K. Kobayashi and K. Uosaki, *Appl. Phys. Lett.*, 2013, **103**(11), 111605.
20 S. Axnanda, E. J. Crumlin, B. Mao, S. Rani, R. Chang, P. G. Karlsson, M. O. M. Edwards, M. Lundqvist, R. Moberg, P. Ross, Z. Hussain and Z. Liu, *Sci. Rep.*, 2015, **5**(1), 9788.
21 J. H. Guo, Y. Luo, A. Augustsson, S. Kashtanov, J. E. Rubensson, D. K. Shuh, H. Agren and J. Nordgren, *Phys. Rev. Lett.*, 2003, **91**(15), 157401.
22 P. Jiang, J. L. Chen, F. Borondics, P. A. Glans, M. W. West, C. L. Chang and J. Guo, *Electrochem. Commun.*, 2010, **12**(6), 820–822.
23 J. Stöhr, *NEXAFS spectroscopy*, Springer Science & Business Media, 1992, p. 25.
24 S. L. Schroeder, G. D. Moggridge, E. Chabala, R. M. Ormerod, T. Rayment and R. M. Lambert, *Faraday Discuss.*, 1996, **105**, 317–336.
25 B. H. Frazer, B. Gilbert, B. R. Sonderegger and G. De Stasio, *Surf. Sci.*, 2003, **537**(1–3), 161–167.
26 J. J. Velasco-Velez, R. V. Mom, R. V. Sandoval-Diaz, L. J. Falling, C. H. Chuang, D. Gao, T. E. Jones, Q. Zhu, R. Arrigo, B. Roldan-Cuenya, A. Knop-Gericke, T. Lunkenbein and R. Schlögl, *ACS Energy Lett.*, 2020, **5**(6), 2106–2111.
27 E. Ortel, T. Reier, P. Strasser and R. Kraehnert, *Chem. Mater.*, 2011, **23**(13), 3201–3209.
28 S. Song, H. Zhang, X. Ma, Z. Shao, R. T. Baker and B. Yi, *Int. J. Hydrogen Energy*, 2008, **33**(19), 4955.
29 J. J. Velasco-Vélez, T. E. Jones, V. Streibel, M. Hävecker, C. H. Chuang, L. Frevel, M. Plodinec, A. Centeno, A. Zurutuza, R. Wang, R. Arrigo, R. Mom, S. Hofmann, R. Schlögl and A. Knop-Gericke, *Surf. Sci.*, 2019, **681**, 1–8.
30 S. Gottesfeld and S. Srinivasan, *J. Electroanal. Chem. Interfacial Electrochem.*, 1978, **86**(1), 89–104.
31 L. Ouattara, S. Fierro, O. Frey, M. Koudelka and C. Comninellis, *J. Appl. Electrochem.*, 2009, **39**(8), 1361–1367.
32 E. J. Frazer and R. Woods, *J. Electroanal. Chem. Interfacial Electrochem.*, 1979, **102**(1), 127–130.
33 E. Willinger, C. Massué, R. Schlögl and M. G. Willinger, *J. Appl. Chem. Sci.*, 2017, **139**(34), 12093–12101.
34 C. Massué, V. Pfeifer, X. Huang, J. Noack, A. Tarasov, S. Cap and R. Schlögl, *ChemSusChem*, 2017, **10**(9), 1943–1957.
35 H. M. Tsai, P. D. Babu, C. W. Pao, J. W. Chiou, J. C. Jan, K. P. Krishna Kumar, F. Z. Chien and W. F. Pong, *Appl. Phys. Lett.*, 2007, **90**(4), 042108.

36 R. Sachse, M. Pflüger, J. J. Velasco-Vélez, M. Sahre, J. Radnik, M. Bernicke, D. Bernsmeier, V. D. Hodoroaba, M. Krumrey, P. Strasser, R. Krähnert and A. Hertwig, *ACS Catal.*, 2020, **10**(23), 14210–14223.

37 L. J. Frevel, R. Mom, J. J. Velasco-Vélez, M. Plodinec, A. Knop-Gericke, R. Schlögl and T. E. Jones, *J. Phys. Chem. C*, 2019, **123**(14), 9146–9152.

38 V. Pfeifer, T. E. Jones, J. J. Velasco Vélez, C. Massué, R. Arrigo, D. Teschner, F. Girgsdies, M. Scherzer, M. T. Greiner, J. Allan, M. Hashagen, G. Weinwerg, S. Piccini, M. Hävecker, A. knop-Gericke and R. Schlögl, *Surf. Interface Anal.*, 2016, **48**(5), 261–273.

39 V. Pfeifer, T. E. Jones, S. Wrabetz, C. Massué, J. J. V. Vélez, R. Arrigo, M. Scherzer, S. Piccini, M. Hävecker, A. Knop-Gericke and R. Schlögl, *Chem. Sci.*, 2016, **7**(11), 6791–6795.

40 V. Pfeifer, T. E. Jones, J. J. V. Vélez, C. Massué, M. T. Greiner, R. Arrigo, D. Teschner, F. Girgsdies, M. Scherzer, M. T. Greiner, J. Allan, M. Hashagen, G. Weinwerg, S. Piccini, M. Hävecker, A. Knop-Gericke and R. Schlögl, *Phys. Chem. Chem. Phys.*, 2016, **18**(4), 2292–2296.

41 V. Pfeifer, T. E. Jones, J. J. V. Vélez, R. Arrigo, S. Piccini, M. Hävecker, A. Knop-Gericke and R. Schlögl, *Chem. Sci.*, 2017, **8**(3), 2143–2149.

42 M. Favaro, B. Jeong, P. N. Ross, J. Yano, Z. Hussain, Z. Liu and E. J. Crumlin, *Nat. Commun.*, 2016, **7**(1), 12695.

43 C. J. Zhang, M. E. Grass, A. H. McDaniel, S. C. DeCaluwe, F. El Gabaly, Z. Liu, K. F. McCarty, R. L. Farrow, M. A. Linne, Z. Hussain, G. S. Jackson, H. Bluhm and B. W. Eichhorn, *Nat. Mater.*, 2010, **9**(11), 944–949.

44 A. Toghan, R. Arrigo, A. Knop-Gericke and R. Imbihl, *J. Catal.*, 2012, **296**, 99–109.

45 H. G. Sanchez Casalongue, S. Kaya, V. Viswanathan, D. J. Miller, D. Friebel, H. A. Hansen, J. K. Norskov, A. Nilsson and H. Ogasawara, *Nat. Commun.*, 2013, **4**(1), 1–6.

46 H. G. Sanchez Casalongue, M. L. Ng, S. Kaya, D. Friebel, H. Ogasawara and A. Nilsson, *Angew. Chem., Int. Ed.*, 2014, **53**(28), 7169–7172.

47 R. Arrigo, M. Hävecker, M. E. Schuster, C. Ranjan, E. Stotz, A. Knop-Gericke and R. Schlögl, *Angew. Chem., Int. Ed.*, 2013, **52**(44), 11660–11664.

48 J. J. Velasco-Vélez, E. A. Carbonio, C. H. Chuang, C. J. Hsu, J. F. Lee, R. Arrigo, M. Hävecker, R. Wang, M. plodinec, F. R. Wang, A. Centeno, A. Zurutuza, L. J. Falling, R. V. Mom, S. Hofmann, R. Schlögl, A. Knop-Gericke and T. E. Jones, *J. Appl. Chem. Sci.*, 2021, **143**(32), 12524–12534.

49 A. Kolmakov, D. A. Dikin, L. J. Cote, J. Huang, M. K. Abyaneh, M. Amati, L. Gregoratti, S. Günther and M. Kiskinova, *Nat. Nanotechnol.*, 2011, **6**(10), 651–657.

50 J. J. Velasco-Vélez, V. Pfeifer, M. Hävecker, R. S. Weatherup, R. Arrigo, C.-H. Chuang, E. Stotz, G. Weinberg, M. Salmeron, R. Schlögl and A. Knop-Gericke, *Angew. Chem., Int. Ed.*, 2015, **54**(48), 14554–14558.

51 S. Nemšák, E. Strelcov, H. Guo, B. D. Hoskins, T. Duchoň, D. N. Mueller, A. Yulaev, I. Vlassiouk, A. Tselev, C. M. Schneider and A. Kolmakov, *Top. Catal.*, 2018, **61**(20), 2195–2206.

52 C. Arble, H. Guo, E. Strelcov, B. Hoskins, P. Zeller, M. Amati, L. Gregoratti and A. Kolmakov, *Surf. Sci.*, 2020, **697**, 121608.

53 T. Gupta, E. Strelcov, G. Holland, J. Schumacher, Y. Yang, M. B. Esch, V. Aksyuk, P. Zeller, M. Amati, L. Gregoratti and A. Kolmakov, *ACS Nano*, 2020, **14**(10), 12982–12992.

54 Y. H. Lu, C. Morales, X. Zhao, M. A. Van Spronsen, A. Baskin, D. Prendergast, P. Yang, H. S. Bechtel, E. S. Barnard, D. F. Ogletree, V. Altoe, L. Soriano, A. M. Schwartzberg and M. Salmeron, *Nano Lett.*, 2020, **20**(9), 6364–6371.

55 R. Mom, L. Frevel, J. J. Velasco-Vélez, M. Plodinec, A. Knop-Gericke and R. Schlögl, *J. Appl. Chem. Sci.*, 2019, **141**(16), 6537–6544.

56 L. J. Falling, R. V. Mom, L. E. Sandoval Diaz, S. Nakhaie, E. Stotz, D. Ivanov, M. Hävecker, T. Lunkenbein, A. Knop-Gericke, R. Schlögl and J. J. Velasco-Vélez, *ACS Appl. Mater. Interfaces*, 2020, **12**(33), 37680–37692.

57 B. L. Henke, E. M. Gullikson and J. C. Davis, *At. Data Nucl. Data Tables*, 1993, **54**(2), 181–342.

58 S. Thürmer, I. Unger, P. Slavicek and B. Winter, *J. Phys. Chem. C*, 2013, **117**(43), 22268–22275.

59 I. C. Man, H. Y. Su, F. Calle-Vallejo, H. A. Hansen, J. I. Martínez, N. G. Inoglu, J. Kitchin, T. F. Jaramillo, J. K. Nørskov and J. Rossmeisl, *ChemCatChem*, 2011, **3**(7), 1159–1165.

60 J. A. Gauthier, C. F. Dickens, L. D. Chen, A. D. Doyle and J. K. Nørskov, *J. Phys. Chem. C*, 2017, **121**(21), 11455–11463.

61 O. Kasian, S. Geiger, T. Li, J. P. Grote, K. Schweinar, S. Zhang, C. Scheu, D. Raabe, S. Cherevko, B. Gault and K. J. Mayrhofer, *Energy Environ. Sci.*, 2019, **12**(12), 3548–3555.

Faraday Discussions

ROYAL SOCIETY OF CHEMISTRY

PAPER

Dynamics over a Cu–graphite electrode during the gas-phase CO_2 reduction investigated by APXPS

Rosa Arrigo, *[a,b] Raoul Blume,[c,d] Alexander Ian Large, [b] Juan Jesus Velasco-Vélez, [c,d] Michael Hävecker, [c,d] Axel Knop-Gericke[c,d] and Georg Held [b]

Received 14th December 2021, Accepted 7th February 2022
DOI: 10.1039/d1fd00121c

The electrocatalytic conversion of CO_2 to fuels and chemicals using renewable energy is a key decarbonization technology. From a technological viewpoint, the realization of such process in the gas phase and at room temperature is considered advantageous as it allows one to circumvent the limited CO_2 solubility in liquid electrolytes and CO_2 transport across the electrical double layer. Yet, electrocatalysts' performances reported so far are promising but not satisfactory. To inform the design of new materials, in this study, we apply ambient pressure X-ray photoelectron and absorption spectroscopies coupled with on-line gas detection *via* mass spectrometry to investigate *in situ* performance and interface chemistry of an electrodeposited Cu on graphitic carbon support under conditions of CO_2 reduction. We use the ISISS beamline at the synchrotron facility BESSY II of the HZB and the electrochemical cell based on polymeric electrolyte membrane previously developed. We show that under cathodic potential in which methanol is formed, a fraction of the electrode with a predominantly Cu(I) electronic structure undergoes reduction to metallic Cu. The C speciation is characterized by C–O and sp^3 CH_3 species whereas no atomic C was formed under this condition. We also show the important role of water in the formation of methanol from accumulated surface CH_3 species.

Introduction

The gas phase CO_2 electrocatalytic conversion (GPCEC) has been explored as an alternative to the most commonly investigated liquid phase approach.[1,2] The low-

[a]*University of Salford, School of Science, Engineering and Environment, Cockcroft Building, Greater Manchester M5 4WT, UK. E-mail: r.arrigo@salford.ac.uk*

[b]*Diamond Light Source Ltd, Harwell Science & Innovation Campus, Didcot, Oxfordshire OX11 0DE, UK*

[c]*Department of Inorganic Chemistry, Fritz-Haber-Institut der Max-Planck Gesellschaft, Faradayweg 4-6, 14195 Berlin, Germany*

[d]*Max-Planck-Institut für Chemische Energiekonversion, Stiftstrasse 34-36, 45470 Mülheim an der Ruhr, Germany*

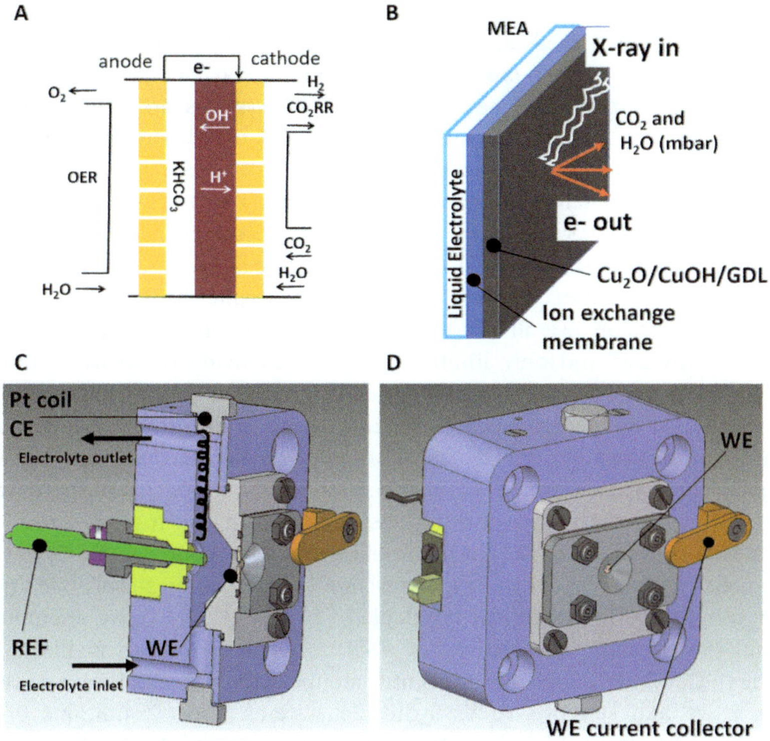

Fig. 1 (A) Schematic representation of the gas phase CO_2 electroreduction; (B) schematic representation of the membrane electrode assembly used in the *in situ* study. (C) Cross sectional view of the three-electrode *in situ* electrochemical cell for ambient pressure XPS; (D) front view of the *in situ* cell showing the mounting of the lid composed of two plates hosting the working electrode (WE).

temperature-GPCEC uses a similar technology to a polymeric electrolyte membrane-based fuel cell in which the electrocatalyst is loaded on a gas-diffusion electrode, generally a graphite cloth. The electrolyte is sandwiched between cathode and anode and serves as supplier of ions; depending on the type of polymeric membrane used, either cations or anions are shuttled through the membrane to the relevant electrode compartment. As a consequence of the interfacial electrode reactions, electrons flow from the anode to the cathode *via* an external circuit: the former is immersed into a liquid electrolyte and catalyses the oxidation of water to molecular oxygen; the latter one is in contact with a humidified stream of carbon dioxide, which is reduced to higher value-added compounds. Fig. 1A and B depict the GPCEC design and the membrane electrode assembly, respectively. In view of a process scale up, GPCEC presents some advantages; amongst these are: (a) easy operation and recycling of unconverted downstream CO_2; (b) enhanced CO_2 confinement and coverage on the electrode surface; (c) reduced overpotential for CO_2 reduction; (d) potentially a suppression of parasitic hydrogen evolution reaction; (e) easier separation of soluble products.[3]

C-supported Pt,[4] Cu,[2] and Fe[5,6] based systems were explored as cathode materials, in some cases showing interesting selectivity towards C_2 (the subscript indicates the number of C atoms in the molecule) and C_{2+} compounds, including alkenes, aldehydes and alcohols. Interestingly, Fe oxyhydroxide systems directly impregnated on a carbon-based gas diffusion layer (GDL) enable C_3 product formation,[6,7] whereas monometallic Cu[2] and Pt[4] systems are characterized by a predominant C_1 productivity. Gutierrez-Guerra *et al.* used Cu and Cu–C sputtered on GDL in the GPCEC, and showed a shift from C_1 chemistry (methanol) for metallic Cu to C_2 chemistry (acetaldehyde) for the Cu–C system.[2] Therein, the higher CO_2 specific electrocatalytic reduction rates and higher selectivity to methanol formation of the Cu system was attributed to the larger particle size and lower CuO/Cu ratio if compared to the Cu–C electrode. Despite the appealing considerations, unsatisfactory overall efficiencies and poor stability are generally being reported for GPCEC and only limited progress was made in terms of material performances, for example, by using bimetallic Cu-based systems and special cell configurations.[8] It is generally recognised that understanding the mechanism of the electrocatalyst activation and deactivation at the molecular level can guide the design of improved electrodes. *Operando* spectroscopy serves this purpose well as demonstrated by the many contributions published in recent years on the electrocatalytic conversion of CO_2 in liquid electrolytes.[9] In contrast, the GPCEC has been poorly analysed by *operando* spectroscopy. Recently, by means of an *in situ* XPS study using the electrochemical cell depicted in Fig. 1C and D, we attempted to explain the peculiar C_3 selectivity of a GDL-supported ferrihydrite-like system during GPCEC.[7] XPS enabled us to identify atomic carbon species forming over the most active and selective Fe oxyhydroxide systems under conditions of CO_2 reduction. This opens the quest whether the formation of interstitial atomic C is required for C_3 chemistry similar to the role of interstitial C which has been observed for nitrogenase enzymes.[10] The involvement of atomic C species in C_3 chemistry has also been discussed in a recent computational study over Cu model catalysts.[11] This motivates further *in situ* studies aimed at exploring the general validity of the mechanism of CO_2 reduction involving the formation of atomic C. Herein we report an ambient pressure X-ray photoelectron spectroscopy (APXPS) and near edge X-ray absorption fine structure (NEXAFS) spectroscopy study on a Cu-based electrode system directly electrodeposited on GDL. By monitoring the gas composition *via* on-line mass spectrometry, we show that these Cu systems are active in the CO_2 reduction to methanol at rather low potentials under the pressure conditions realized in the *in situ* chamber. In contrast to the dynamics observed for C_3-selective FeOOH-based electrocatalysts,[7] the analysis of the chemical composition and the structure of the interface for this Cu system does not reveal the formation of atomic C, but a mechanism involving the reduction of adsorbed C–O species into chemisorbed-CH_3 species already at −1 V *vs.* the Ag/AgCl reference electrode. This is consistent with the poor solubility of C in Cu and corroborates the hypothesis that for realizing C_3 chemistry a metal phase such as Fe allowing higher flexibility to accommodate interstitial C atoms and high mobility of those is required, similarly to the role of C in Fe–V nitrogenase enzymes, the only nitrogenase enzymes capable of C–C coupling.[10]

Results and discussion

The Cu-based electrocatalyst used in this study was obtained by electrodeposition on a graphitic GDL from a $CuSO_4$ solution, followed by reduction in KCl. Herein,

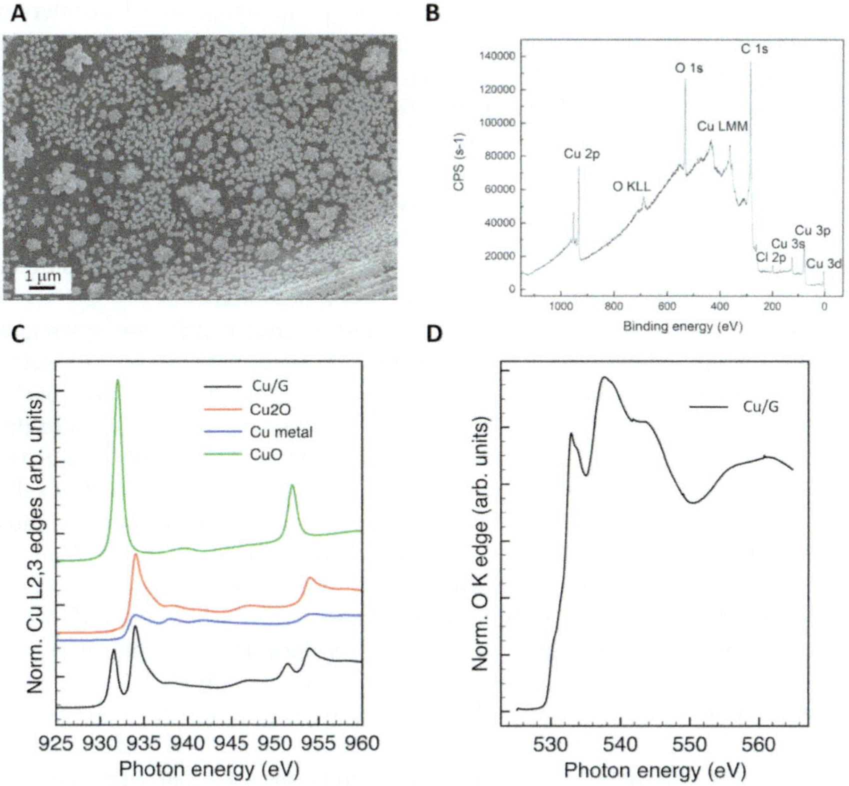

Fig. 2 (A) Scanning electron micrograph of Cu/G in secondary electron mode at 1.6 kV; (B) XPS survey spectrum of Cu/G indicating the presence of Cu, O, C, Cl and small amount of S; (C) Cu $L_{3,2}$ NEXAFS spectrum of Cu/G measured in UHV and spectra for reference CuO, Cu_2O and metallic Cu; (D) O K edge NEXAFS spectrum of Cu/G.

this sample will be referred to as Cu/G. The scanning electron micrograph in Fig. 2A shows the rounded cube-like morphology of the Cu phase in the as prepared electrode, whose chemical composition, as investigated by XPS (Fig. 2B), indicates mainly the presence of Cu, O, Cl. NEXAFS spectra at the Cu $L_{2,3}$ and O K edges measured in total electron yield (TEY) mode are shown in Fig. 2C and D, respectively. Accordingly, Cu(I) is the dominant oxidation state as indicated by the intense Cu L_3 edge absorption edge resonance at 934 eV, which is assigned to the 2p → 4p transition.[9,12] Additionally, a strong resonance related to the 2p → 3d transition at 931 eV is characteristic for Cu(II) species. Consistently, resonances at 530 eV and 532 eV in the O K NEXAFS are assigned to Cu(II) and Cu(I) species, respectively.[13]

Fig. 1B depicts the electrode assembly used for the *in situ* study. We used an anion-exchange membrane, sandwiched between the body of the cell and the GDL-based working electrode (WE). The equilibrium pressure generated by the transport of the aqueous potassium bicarbonate electrolyte from the inner compartment of the cell to the evacuated XPS chamber, through the membrane, is in the range of 0.17–0.2 mbar. Note that the TEY-C K edge NEXAFS spectra indicate that the autogenous gas composition is mainly composed of water with some CO_2 (not shown). Nevertheless, for simplicity, this condition will be referred to as under water enriched conditions. The equilibrium pressure in the APXPS chamber generated by the liquid electrolyte diffusing from the inner compartment of the *in situ* cell is 0.17–0.2 mbar in flow-controlled mode; however the total pressure during the *in situ* experiments is regulated at approximately 0.3–0.36 mbar. *In situ* electroanalytical and spectroscopic measurements were simultaneously performed first under water enriched conditions at a total pressure of 0.3 mbar. Subsequently, a flow of CO_2 is added to the water enriched atmosphere (under equilibrium pressure in flow-controlled mode) up to a total pressure of 0.33–0.36 mbar, before performing another set of measurements.

The cyclic voltammograms (CV) measured *in situ* under the gas-phase composition as indicated are reported in Fig. 3A. Here, redox waves attributed to Cu reduction can be inferred to be present in each case. The reduction of the water partial pressure when CO_2 is added to the reaction mixture leads to the reduction of the current exchanged in the range from −0.75 to −1 V *vs.* Ag/AgCl, indicating that the dominant contribution to the charge exchanged is from the water reduction. This is in strong contrast with the current/voltage trends observed for FeOOH/NC electrodes in the potential range from 0 to −1 V *vs.* Ag/AgCl under similar experimental conditions.[7] In that case, the occurrence of reduction reactions involving CO_2 in the gas-phase was observed as a shift of the onset potential to higher value and a decrease of the current exchanged only at low potential. Such a behaviour is consistent with a carbon dioxide reduction reaction (CO_2RR) and hydrogen evolution reaction (HER) competing for the same sites on the electrode surface, where the CO_2RR occurs with lower overpotential than the HER suffering diffusion limitations under low water partial pressure.

In this study, the *in situ* spectroscopic measurements were performed during a constant potential measurement (chronoamperometry (CA)), in the potential range from open circuit potential (−0.05 V) to −1.7 V *vs.* Ag/AgCl, first under the

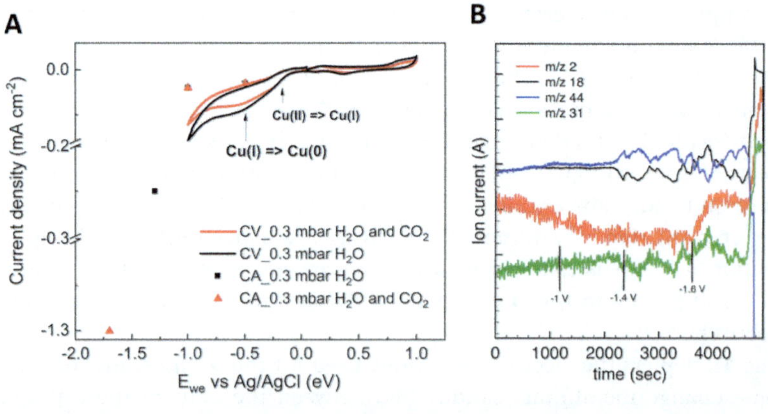

Fig. 3 (A) Cyclic voltammetry (CV) and chronoamperometry (CA) data during the *in situ* APXPS and NEXAFS measurements at the conditions as indicated for Cu/G. The voltage sweep rate is 50 mV s^{-1}. The arrows indicate the potentials for the indicated redox reactions from ref. 9c. (B) Evolution of relevant mass fragments in the mass spectrum during *in situ* characterization at constant potential (CA) as indicated for the Cu/G under CO_2/H_2O mixed atmosphere. The total pressure is 0.3 mbar.

0.3 mbar water enriched atmosphere and then at the same potentials after adding CO_2 to the autogenous atmosphere. The charge exchanged at the WE interface during constant potential measurements is also shown in Fig. 3A. The values obtained under water enriched and water–carbon dioxide mixed atmospheres indicate a dominant contribution of liquid electrolyte related redox processes whereas the effect of the addition of CO_2 is not evident, consistently with the CV measurements. The corresponding mass spectrum recorded during the potential dependent study under water/carbon dioxide mixed atmosphere provides more insights into the electrocatalytic performances of the Cu/G electrode under gas phase CO_2 reduction (Fig. 3B). Particularly, HER is clearly observed when the potential at the WE is set to −1.6 V *vs.* Ag/AgCl (*m*/*z* 2 fragment in red). The variation of the water transfer through the membrane above −1.4 V leads to a continuous change in opposite direction of the water (*m*/*z* 18) and CO_2 (*m*/*z* 44) signals. This also causes the current to fluctuate during the measurements (not shown). The poor control over the water transport through the polymeric membrane is a feature of the cell which was also observed in a previous study,[7,14] and makes the analysis of the CO_2-reduction product related fragments difficult. Nevertheless, we observe that the *m*/*z* 31 follows the evolution of the *m*/*z* 18 attributed to water, rather than the *m*/*z* 44, due to CO_2. The mass fragment *m*/*z* 31 can be safely associated to the formation of methanol, which was already confirmed as a CO_2 reduction product under condition of gas-phase CO_2 reduction for similar electrodes.[15] In contrast, the formation of acetaldehyde, which was another CO_2 reduction product reported in the literature for similar Cu/G systems, is not confirmed in our experiment by the evolution of the *m*/*z* 29 and therefore can be safely ruled out. Interestingly, above 4500 seconds of reaction (Fig. 3B) and at constant potential of −1.6 V *vs.* Ag/AgCl the water pressure suddenly increases causing the CO_2 pressure to decrease. This is accompanied not only by an increase in the HER, but also the formation of methanol. This result shows that both the HER and the CO_2RR are controlled by diffusion under the gas-phase conditions realized here. Moreover, water is needed for the CO_2 reduction to methanol to occur. The higher formation of methanol under low CO_2 partial pressure also indicates that its formation derives from the desorption of surface adsorbed species, *via* hydrolysis and formation of the alcohol functional group.

In the following, we will analyse the surface composition of the electrode under the gas phase compositions and potentials as described above.

Firstly, we analyse the Cu electronic state of the electrode. To this goal, the Cu/G electrode was characterized under non-electrochemical conditions under 0.3 mbar CO_2 at room temperature and at 473 K. The Cu $2p_{3/2}$ XPS and Cu $L_3M_{45}M_{45}$ Auger electron spectroscopy (AES) spectra are reported in Fig. 4A together with the corresponding secondary electrons (SE) scanning electron micrograph (SEM) taken after the experiment. The Cu L_3 edge NEXAFS are reported in Fig. 4B. The freshly prepared Cu/G is characterized by a main Cu $2p_{3/2}$ peak with maximum at a binding energy (BE) of 932.8 eV; a peak at this BE is associated to Cu(I) species.[12,13] The Cu $L_3M_{45}M_{45}$ Auger spectrum of this sample (blue spectrum) shows additional contribution at lower kinetic energy when compared to the Cu_2O reference spectrum. This indicates the presence of Cu(II) species, consistent with the Cu L edge NEXAFS spectrum reported in Fig. 2C. Note that the Cu $2p_{3/2}$ XPS spectrum of the fresh Cu/G differs substantially from the spectrum of a Cu(II)

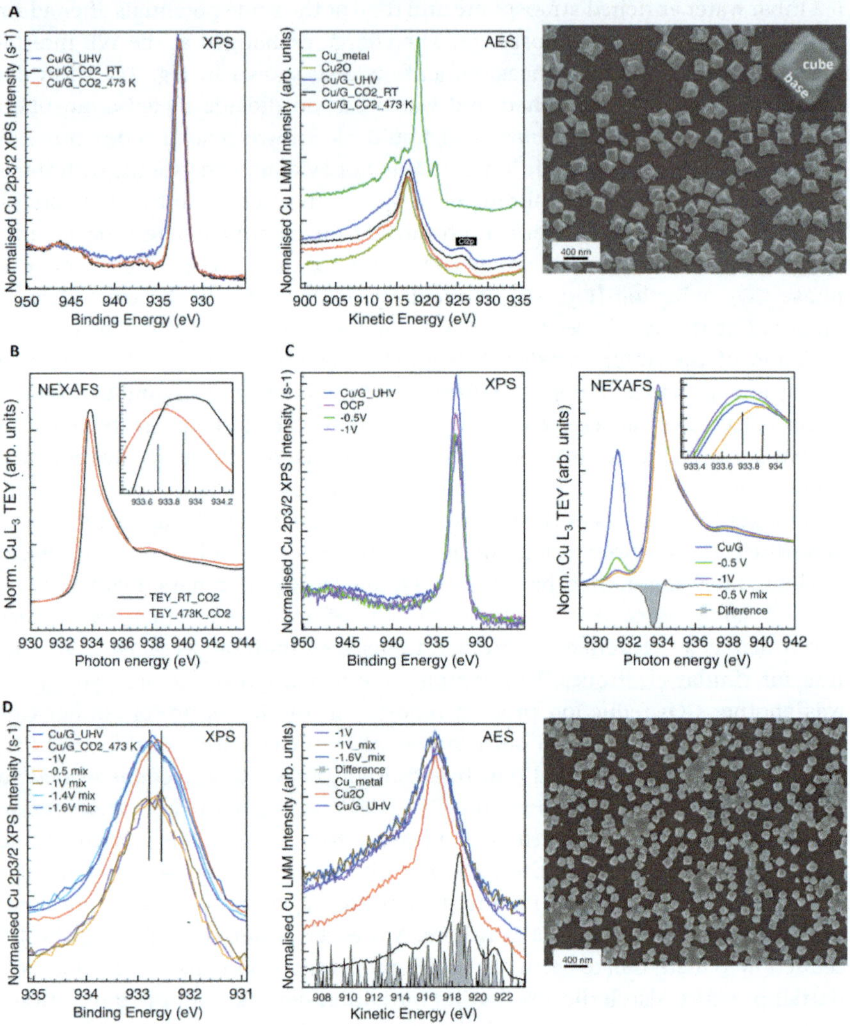

Fig. 4 (A) from left to right, Cu $2p_{3/2}$ XPS spectra measured at KE = 450 eV and Cu LMM AES spectra for Cu/G under non-electrochemical conditions as indicated; SEM micrograph in secondary electron mode at 1.6 kV acceleration voltage for Cu/G after the *in situ* thermal treatment under CO_2 at 473 K. The sample was exposed to air before SEM measurements. (B) Total electron yield (TEY) Cu L_3 NEXAFS spectra measured for Cu/G under non-electrochemical conditions as indicated in the legend. Each NEXAFS spectrum is obtained by merging 6–8 spectra. (C) From left to right, Cu $2p_{3/2}$ XPS spectra measured at KE = 450 eV and TEY Cu L_3 NEXAFS spectra measured for Cu/G under electrochemical conditions as indicated in water enriched atmosphere. Each NEXAFS spectrum is obtained by merging 6–8 spectra. Difference: (spectrum −0.5 V mix) minus (spectrum −1 V). (D) From left to right, Cu $2p_{3/2}$ XPS spectra measured at KE = 450 eV and Cu LMM AES spectra for Cu/G under electrochemical conditions as indicated in CO_2/water mixed atmosphere. The Cu $2p_{3/2}$ spectra in this figure were normalised to the peak maximum to better evaluate peak broadening. Difference: (spectrum −1.6 V mix) minus (spectrum −1 V mix); SEM micrograph in secondary electron mode at 1.6 kV acceleration voltage for Cu/G after the *in situ* electrochemical experiments. In each experiment, the potential is varied from higher (−0.5 V *vs.* Ag/AgCl) to lower (−1.6 V *vs.* Ag/AgCl). The subscript "mix" is used in the figure legends to indicate the experiments carried out under a mixed CO_2/H_2O atmosphere.

O phase, which is characterized by a much more intense satellite structure;[12,13] thus it is quite probable that the Cu(II) species are located at the surface of a predominant Cu_2O phase as a chloride, sulphate or hydroxide species (see survey spectrum in Fig. 2B).The treatment under 0.3 mbar CO_2 at RT produces the reduction of the Cu(II) species (the resonance at 931 eV is no longer present in Fig. 4B). The loss of Cu(II) species is also observed in the Cu $2p_{3/2}$ XPS spectrum (Fig. 4A) as a reduction of the broadening of the main peak on the high binding energy side and a downshift of the main peak of about 0.2 eV. The Cu LMM spectra in Fig. 4A consistently show the disappearance of the low kinetic energy component leaving a spectrum, which resembles more closely the one of the Cu_2O reference. Additionally, the more bulk sensitive TEY Cu L_3 edge NEXAFS spectrum for the Cu/G thermally treated at 473 K under CO_2 in Fig. 4B shows a shift to lower energy of the resonance due to the 2p → 4p transition, when compared to the spectrum at room temperature, and an increase in the post-edge intensity above 937 eV, the latter one indicating the formation of a small amount of metallic Cu. The metallic component is not clearly distinguishable in the XPS and Cu LMM spectra presented in Fig. 4A (spectra in black and red overlap). We postulate that the cubic particles in the SEM image in Fig. 4A are related to the Cu_2O phase, whereas the flat base between the cubes and the support might be the more reduced metallic Cu phase.

The Cu $2p_{3/2}$ XPS and Cu LMM Auger spectra measured during the *in situ* electrocatalytic tests are reported in Fig. 4C and D. In particular, Fig. 4C refers to the first set of measurements performed under a gas atmosphere generated by the transport of the liquid electrolyte from the inner compartment through the membrane and into the XPS chamber. As mentioned before, this is mainly composed of water, but the presence of a low intense resonance at 291 eV in the TEY C K edge NEXAFS (not shown) indicates CO_2 is also present in the gas-phase. With respect to the Cu/G (blue spectrum), under this *in situ* electrochemical conditions, both XPS and NEXAFS spectra are consistent that the cathodic polarization of the surface leads to a gradual reduction of the Cu(II) species to Cu(I) (see spectra from open circuit potential (OCP) to −0.5 V *vs.* Ag/AgCl). At −1 V *vs.* Ag/AgCl, the maximum of the main peak in the Cu $2p_{3/2}$ XPS shifts to even lower BE; in the corresponding NEXAFS spectrum in Fig. 4C, the main resonance is found at the same photon energy as at the preceding potentials but the intensity is higher, whereas the resonance related to the Cu(II) species is lower. Similar to the thermal treatment in Fig. 4B, the more cathodic polarization of the surface leads to an increase of the post-edge intensity above 937 eV in the Cu L edge NEXAFS in Fig. 4C (see spectra in green and purple), consistent with the formation of metallic Cu in a predominant Cu(I) phase. We can safely infer that the small shift of the main XPS peak to lower energy is due to the concomitant Cu(II) → Cu(I) reduction; however the Cu(I)/Cu(0) reduction cannot be excluded, as this reduction also manifests itself in the XPS spectrum as a more pronounced shift of the main peak to lower BE. Now we analyse the impact of the increase of the CO_2 partial pressure by comparing in Fig. 4C the Cu L_3 edge NEXAFS spectrum at −0.5 V *vs.* Ag/AgCl under water (green spectrum) and a water/CO_2 mixture (orange spectrum). Whilst the intensity above the edge is very similar, the main resonance associated with Cu(I) species is shifted to higher energy when CO_2 is added to the reaction mixture (see difference spectrum −0.5 V mix minus −1 V). As it can be seen from the insets in Fig. 4B and C, the addition of CO_2 in the gas phase leads to a consistent 0.15 eV upshift of the main resonance of the Cu L_3 NEXAFS spectra. The origin of this phenomenon should be further investigated. In the Cu LMM spectrum we observe a broadening of the peak at lower kinetic energy (see for example spectra relative to the cathodic polarization at −1 V and −1 V mix *vs.* Ag/AgCl in Fig. 4D). We attribute these spectral changes to the chemisorption of CO_2-related species. Fig. 4D shows the evolution of the Cu $2p_{3/2}$ at different cathodic potential under mixed gases atmosphere. With respect to the initial spectra of Cu/G in UHV, the main Cu $2p_{3/2}$ peak is broader under cathodic polarization and composed of two contributions: one at higher BE as represented by the starting Cu electrode containing also Cu(II) species (blue) and one at lower BE as represented by the thermally treated sample containing also metallic Cu (red spectrum). We also see that the larger peak FWHM when the potential is lowered is due to the increase of the relative abundance of the peak at the lower BE, with the highest value achieved when the potential is −1 V and −1.6 V *vs.* Ag/AgCl in mixed atmosphere. Note that at −1.6 V, the mass spectrum (Fig. 3B) showed HER. Note also that the Cu 2p XPS spectrum at −1.6 V is the only one corresponding to an electrode state over which methanol is being formed and released in the gas-phase. The *in situ* reduction of some of the Cu(I) to metallic Cu is confirmed by the more bulk sensitive Cu $L_3M_{45}M_{45}$ spectra in Fig. 4D. A small component is visible on the high energy side of the spectrum (918.8 eV) relative to the measurements at −1.6 V. The difference spectrum between the spectrum at 1.6 V and −1 V *vs.* Ag/AgCl shows an envelope consistent with a metallic component.

These results show that, although most of the sample remains in a Cu(I) oxidation state, potential-dependent electronic structure dynamics are seen, similar in nature to those reported for the liquid phase reduction of CO_2 over Cu electrode.[9] In analogy to the SEM image recorded for the annealed sample, it is possible to localize the electroactive interface by SEM recorded after these *in situ* measurements (Fig. 4D). Comparing this to the annealed sample, it is possible to infer that the cuboid particles with rounder corners are attributed to the Cu_2O phase, whereas the more shapeless Cu component on the carbon support is the electroactive surface.

Finally, we analyse the C 1s XPS spectra with the goal to identify the C species under methanol electrosynthesis. We are particularly interested in understanding whether C_1 chemistry occurs through the formation of atomic carbon species as seen in the case of FeOOH/NC systems selective to isopropanol.[7]

Therefore, the C 1s XPS spectra were fitted using the model described in ref. 7 and summarized in Table 1 of the Experimental section. Accordingly, 5 main components were considered as shown in Fig. 5A for the Cu/G measured under UHV conditions. The C1 component at a BE of 284 eV describes the electronic structure of C atoms surrounding in-plane point defects in the graphene layers where localized charge accumulation occurs. The C2 component at a BE of 284.3 eV is the sp^2 carbon in graphite. The C3 component at 284.8 eV is assigned to disordered graphite as for example the sp^2/sp^3 moieties and dangling bonds at the periphery of the graphene layers or graphitic domains with large holes. The C1, C2 and C3 components are associated with the carbon support. The C4 component at 285.5 eV is related to sp^3 carbon. This can be due either to an amorphous carbon component of the support or chemisorbed, reduced species related to CO_2 on the electrocatalytic surface. Various forms of C–O(H) species are found at binding energy above 286 eV. Particularly, the C5 component at 288.3 eV

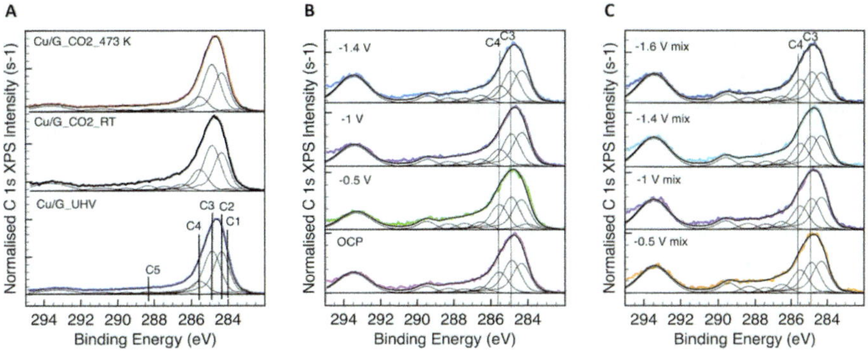

Fig. 5 (A) Fitted C 1s spectra measured for the Cu/G under non-electrochemical conditions as indicated. (B) Fitted C 1s spectra measured for the Cu/G under *in situ* electrochemical conditions as indicated under a water enriched atmosphere. (C) Fitted C 1s spectra measured for the Cu/G under *in situ* electrochemical conditions as indicated under a water/CO_2 mixed atmosphere.

is generally associated with carboxylate species. A comprehensive analysis of the C 1s XPS spectrum of CO_2–H_2O adsorbate species over Cu(111) measured by APXPS was reported earlier,[17,18] including C–O(H) species (BE = ~286.3 eV), formate species (BE = ~287.3 eV), chemisorbed bend-CO_2 species (BE = ~287.9 eV), physisorbed linear CO_2 (BE = ~288.4 eV) and carbonate (–CO_3) species (BE = ~289.4 eV). Price *et al.* found gas-phase components such as CO and CO_2 in the region between 289–292 eV and CH_4 at 284.5 eV.[16] We do not expect a large contribution of gas-phase components under the operating conditions (see Experimental part).

The peak at 293.5 eV is associated with the K $2p_{3/2}$. A small amount of potassium is present in the fresh sample (Fig. 5A) due to the synthesis procedure; a larger amount of K is found during the *in situ* experiments due to the transport of the $KHCO_3$ electrolyte from the inner compartment of the cell to the cathode side under polarization (Fig. 5B and C). The effect of the addition of 0.3 mbar CO_2 at room temperature on the C speciation over the Cu/G under non-electrochemical conditions is evaluated in Fig. 5A as well. It is apparent that the C4 component (sp^3 C) and all the C–O(H) species including the C5 component increase upon addition of CO_2. In an APXPS study of Cu(111) upon exposure of pure CO_2, Favaro *et al.*[17] observed the formation of C–O(H) species above 286 eV, indicating this as a result of the dissociative CO_2 chemisorption. Moreover, Deng *et al.* observed the formation of chemisorbed methoxy groups (BE = 285.2 eV) over polycrystalline Cu as a product of the reaction of water with CO_2 at room temperature.[19] Due to the fact that during this experiment we observe a reduction of Cu(II) to Cu(I), one could assume that the reduction of surface Cu(II)–O(H) species might be the source of water for this reaction. Therefore, we could infer that in our study the C4 component is also due to methoxy species. In Fig. 5A, the increase of the temperature up to 473 K under this gas atmosphere, leads to a reduction of the amount of C–O(H) oxygenated C species (above 286 eV) as well as the C4 species, whereas the C3 component further increases indicating some contribution of chemisorbed sp^3 C species on Cu to this component as well; we postulate that these originate from the deoxygenation of the methoxy species. The reduction of the oxygenated species is accompanied by the

Cu(II) → Cu(I) and Cu(I) → Cu(0) reductions as indicated in the spectral changes in the Cu L_3 edge NEXAFS (Fig. 4B) and the Cu $2p_{3/2}$ (Fig. 4D) as discussed above. We noted that the dissociative chemisorption of CO_2 over Cu_2O is not expected,[19] thus, either the exposure of Cu/G to CO_2 produces already metallic sites on the surface or the reaction is photoinduced. We must clarify that we were not able to identify any beam damage effect on the Cu L edge NEXAFS by time resolved experiments within the time frame of 10 measurements, the spectra being consistently the same under a given reaction condition. Nevertheless, during the *in situ* electrocatalytic study reported in this work, we took particular care to minimize any photoinduced reduction reactions by changing the sample spot at each measurement.

We then discuss the *in situ* data under 0.3 mbar water (water enriched conditions). Notably in Fig. 5B, the C4 and C5 components are more abundant under open circuit potential than under any of the experimental conditions reported in Fig. 5A. These increase further at −0.5 V *vs.* Ag/AgCl, later decreasing again at more negative voltages indicating their consumption. A consideration follows that under water enriched conditions these species are not replenished. In contrast, when the electrode is under CO_2/water mixture (Fig. 5C), these chemisorbed C4 species (probably $-OCH_3$ species) are more abundant at any of the potentials investigated if compared to the water enriched conditions (Fig. 5B) and remain so even at the lowest potential investigated. However, a small consumption of the C–O(H) species above 286 eV (at BE higher than the C4 component) is visible when the potential is lowered from −0.5 V to −1.6 V *vs.* Ag/AgCl whereas the abundance of the C4 sp^3 C component remains constant. It is important to note that the C 1s spectrum at −1.6 V *vs.* Ag/AgCl is measured when methanol is detected in the gas phase. The formation of chemisorbed C–O(H) species at BEs above 286 eV in the C 1s XPS spectrum were described earlier[18] as the first activation step of CO_2 over a metallic Cu surface; the process seems facilitated by subsurface O species which also induce Cu^+ sites. The fact that we see a small reduction of these C–O(H) under methanol formation condition can be a natural consequence of the subsequent transformation of these species into more reduced adsorbates due to the potential. Under this condition, a correlation can be attempted with the formation of C4 species and the increase of the FWHM of the Cu $2p_{3/2}$ peak towards the lower BE side, the latter one describing the Cu state at the electroactive interface on this system. The C4 species ($-CH_3$ and $-OCH_3$) accumulate on the surface and are then desorbed as methanol by the action of water. Of interest, the atomic C species below 284 eV is not found in any of these experiments, in contrast to the FeOOH systems previously reported.[7]

Experimental

Preparation of the Cu/G electrode

The catalysts were synthesized by electrodeposition of Cu on C paper of size 1 cm × 2 cm (Fuelcells). The electrodeposition was carried out using an H-shaped electrochemical cell and a Gamry potentiostat (Interface 1010E) with a three-electrode method. A Pt wire (99.99%) and an Ag/AgCl (saturated KCl) were used as counter and reference electrodes, respectively. During the experiment, only 1 cm × 1 cm area of carbon paper (10 mg) was immersed in the electrolyte with a total charge exchanged corresponding to circa 0.6% in weight of Cu. The depositions were carried under a potential of −0.7 V *vs.* Ag/AgCl for 300 seconds in 5 mM of $CuSO_4$ aqueous electrolyte, prepared using commercial $CuSO_4$ in powder form (Sigma Aldrich, anhydrous powder, 99.99%). For the electrodeposition the electrolyte was continuously saturated with pure N_2 gas. Subsequently, the electrode was transferred to a N_2-saturated, 5 mM solution of aqueous KCl electrolyte prepared from a reagent in powder form (Sigma Aldrich, 99%) and reduced under a potential of −0.7 V *vs.* Ag/AgCl for 300 seconds. During the reduction, N_2 was continuously bubbled into the electrolyte.

In situ electrochemical flow cell

In this study, we use the same three-electrode electrochemical flow cell for *in situ* spectroscopy previously reported in ref. 7, which is adapted to the APXPS end-station of the ISISS beamline at BESSY II/HZB (Fig. 1C and D). In this cell configuration, the compartment housing the counter electrode (CE), a Pt wire or coil, and the reference electrode (RE) a Ag/AgCl (3 M) (DRIREF-2SH, World Precision Instruments, USA) is separated from the evacuated chamber by a membrane electrode assembly (MEA) based on an ion membrane PEM (fumasep FAD-PET-75, Fumatech) and the WE placed on top of the membrane. The WE is facing the evacuated chamber and thus exposed to the X-rays (Fig. 1B). The CE and RE are immersed in a CO_2-saturated 0.1 M $KHCO_3$ solution flowing continuously in the inner compartment of the *in situ* cell. Before use the membrane was cut in small pieces to fit the *in situ* cell, was exchanged in 0.5 M NaCl at 298 K and for 72 h, and then left in 0.1 M $KHCO_3$ solution until use. The WE of this MEA consists of the Cu/G sandwiched between the PEM and the lid of the *in situ* cell. The PEM is kept hydrated by the aqueous electrolyte diffusing through the polymeric electrolyte and then into the evacuated measurement chamber, generating an equilibrium water pressure ranging between 0.17–0.2 mbar. To this atmosphere, a constant flow of 4 mL min^{-1} of CO_2 (99.995%) is added into the chamber using a mass flow controller to achieve a total pressure of 0.30–0.36 mbar. A potentiostat (SP-300, Bio-Logic Science Instruments SAS, France) is used to control the potential of the WE with respect to the RE and to measure the current between CE and WE. Electroanalytical techniques such as cyclic voltammetry and chronoamperometry are used to investigate the current exchanged at the WE interface. Before the spectroscopic measurements, 4 cycles of cyclic voltammetry (50 mV s^{-1}) are measured under 0.3 mbar of the water partial pressure generated by the transport of the electrolyte through the membrane and then after adding a constant flow of CO_2 to this atmosphere. The chronoamperometric experiments were performed at different voltages from higher to lower potentials. A quadrupole mass spectrometer (QMS) is used to monitor the gas composition on-line in the XPS chamber. The m/z 2, 12, 14, 15, 16, 17, 18, 28, 30, 31, 32, 44, were monitored in each *in situ* experiment.

Spectroscopic measurements

The WE is grounded to the electron analyser to prevent shifts of the core levels due to the applied potential. Due to the beam sensitivity of Cu, each spectrum was measured on a fresh spot by moving the manipulator a distance as large as the beam spot size. Moreover, a voltage of 90 V was applied to the aperture of the differential pumping stage of the electron analyser during measurements, resulting in the suppression of most of the gas-phase signal in the XPS spectra.

Table 1 Assignment of C 1s components based on the fitting described in ref. 7

Label	Binding energy (BE)/eV	Assignment	FWHM
C1	284	Defective sp^2 carbon	0.6
C2	284.3	sp^2 carbon	0.5
C3	284.8	Disordered carbon (larger holes in basal graphitic planes, non-saturated bonds, amorphous carbon phase)	0.65
C4	285.5	sp^3 C–H bonds/CO bonds	0.6
C5	288.3	COO bonds	1.1[a]

[a] The same fitting parameter was used for the other peaks at BEs above 286 eV.

Survey XPS measurements were performed to determine the elemental composition of the electrocatalyst. Moreover, high resolution XPS measurements at the core levels Cu 2p, C 1s, O 1s, were performed by applying a suitable excitation energy corresponding to a kinetic energy (KE) of the photo-emitted electrons of 450 eV. The pass energy (E_p) was set to 20 eV. Cu LMM Auger spectra were measured with an excitation energy of 1125 eV and $E_p = 50$ eV. The spectra were collected with a spot size of ≈150 µm × 80 µm. The beamline settings were exit slit (ES): 111 µm and fix focus constant (cff): 2.25.

Auger Electron Yield (AEY) NEXAFS spectra were measured with an analyser setting of $E_p = 50$ eV and electron kinetic energies (KE) of 900 eV, 520 eV, and 270 eV for the Cu L-, O K-, and C K-edges, respectively. The Total Electron Yield (TEY) NEXAFS spectra were recorded using a Faraday cup placed in the XPS chamber in close proximity to the WE. The beamline settings were exit slit (ES): 60 µm and fix focus constant (cff): 2.25.

XPS fitting and NEXAFS data processing

In this work, only the fitting of the C 1s was attempted following the model reported in ref. 7. The binding energy (BE) is referenced to the Fermi edge when possible or to the maximum of the correct value of the C 1s peak was calibrated by measuring the C 1s second order peak at the same excitation energy.

The spectra were fitted following the Levenberg–Marquardt algorithm to minimize the χ^2. Peak shapes were modelled by using Doniach–Sunjic functions convoluted with Gaussian profiles. Background correction was performed using a Shirley background. The assignment of the components in the C 1s XPS spectra is given in Table 1.

The NEXAFS spectra were energy calibrated using features in the refocusing mirror drain current. Moreover, they were intensity normalised to the edge jump after linear background subtraction. A series of spectra were measured (generally 6–8 spectra) and then merged together. The merged spectrum was then reported in the figures.

Conclusions

Herein, we investigated by means of surface sensitive XPS and NEXAFS, the structural dynamics occurring over a C-supported electrodeposited Cu electrode during the gas-phase CO_2 reduction. It was shown that the Cu electrode, whose dominant phase is a Cu(I) oxide phase, is characterized by an increase of the CO_2-related chemisorbed species at potentials above the potential at which CO_2RR to methanol was experimentally proven. The abundance of these species is higher in the case of CO_2/water mixed atmosphere than in the case of water atmosphere. At −1.6 V *vs.* Ag/AgCl, when methanol formation was detected, the Cu electrode partially undergoes a reduction to metallic Cu and at the same time the C–O(H) species are reduced whereas surface sp^3 carbon species such as $-OCH_3$ fragments remain highly abundant. The results presented here indicate the fundamental role of water in realizing the desorption of the CH_3 fragments as methanol. Under the conditions explored herein, this reaction was limited by diffusion due to the slow water transport through the membrane which required a significant overpotential. Nevertheless, due to the peculiar conditions realized in these experiments it was possible to identify CH_3 species accumulated on the surface and exclude the participation of atomic C in the formation of methanol.

Author contributions

R. A. conceived and supervised the study. R. A., A. I. L., J. J. V.-V. and M. H. measured *in situ* data. R. A. and R. B. analysed the spectroscopic *in situ* data. R. A. drafted and finalized the manuscript. A. K.-G. and G. H. gave general advice during the study and for finalizing the manuscript.

Conflicts of interest

There are no conflicts to declare.

Acknowledgements

Measurements were carried out at the ISISS beamline at Helmholtz-Zentrum Berlin für Materialien und Energie. We thank HZB for the allocation of synchrotron radiation beam-time under the proposal ST 18106826. The research leading to the *in situ* spectroscopic results has been supported by the project CALIPSOplus under the Grant Agreement 730872 from the EU Framework Programme for Research and Innovation HORIZON 2020. R. A. acknowledges The UK Catalysis Hub for support provided *via* the membership of the UK Catalysis Hub Consortium and funded by EPSRC (portfolio grants EP/K014706/1, EP/K014668/1, EP/K014854/1, EP/K014714/1, EP/I019693/1). Dr Manfred E. Schuster is acknowledged for providing in-kind SEM measurements.

Notes and references

1 (*a*) S. Pérez-Rodríguez, F. Barreras, E. Pastor and M. J. Lázaro, *Int. J. Hydrogen Energy*, 2016, **41**, 19756–19765; (*b*) G. Centi, S. Perathoner, G. Win and M. Gangeri, *Green Chem.*, 2007, **9**, 671–678.

2 (*a*) N. Gutiérrez-Guerra, L. Moreno-López, J. C. Serrano-Ruiz, J. L. Valverde and A. de Lucas-Consuegra, *Appl. Catal., B*, 2016, **188**, 272–282; (*b*) I. Merino-Garcia, J. Albo, J. Solla-Gullón, V. Montiel and A. Irabien, *J. CO2 Util.*, 2019, **31**, 135–

142; (*c*) X. Dong, G. Li, W. Chen, C. Zhu, Y. Song, N. Sun and W. Wei, *Chin. Sci. Bull.*, 2021, **66**, 816–824, DOI: 10.1360/TB-2020-0512.
3 D. Sebastián, A. Palella, V. Baglio, L. Spadaro, S. Siracusano, P. Negro, F. Niccoli and A. S. Aricò, *Electrochim. Acta*, 2017, **241**, 28–40.
4 K. Hara and T. Sakata, *J. Electrochem. Soc.*, 1997, **144**, 539.
5 C. Genovese, C. Ampelli, S. Perathoner and G. Centi, *Green Chem.*, 2017, **19**, 2406–2415.
6 R. Arrigo, M. E. Schuster, S. Wrabetz, F. Girgsdies, J.-P. Tessonnier, G. Centi, S. Perathoner, D. S. Su and R. Schlögl, *ChemSusChem*, 2012, **5**, 577–586.
7 R. Arrigo, R. Blume, V. Streibel, C. Genovese, A. Roldan, M. E. Schuster, C. Ampelli, S. Perathoner, J. J. Velasco Velez, M. Havecker, A. Knop-Gericke, R. Schlogl and G. Centi, *ACS Catal.*, 2022, **12**(1), 411–430.
8 X. Dong, G. Li, W. Chen, C. Zhu, T. Li, Y. Song, N. Sun and W. Wei, *Mater. Adv.*, 2021, **2**, 241–247.
9 (*a*) J.-J. Velasco-Vélez, T. Jones, D. Gao, E. Carbonio, R. Arrigo, C.-J. Hsu, Y.-C. Huang, C.-L. Dong, J.-M. Chen, J.-F. Lee, P. Strasser, B. Roldan Cuenya, R. Schlögl, A. Knop-Gericke and C.-H. Chuang, *ACS Sustainable Chem. Eng.*, 2019, **7**, 1485–1492; (*b*) J.-J. Velasco-Vélez, R. Mom, L.-E. Sandoval-Diaz, L. J. Falling, C.-H. Chuang, D. Gao, T. Jones, Q. Zhu, R. Arrigo, B. Roldan Cuenya, A. Knop-Gericke, T. Lunkenbein and R. Schlögl, *ACS Energy Lett.*, 2020, **5**, 2106–2111; (*c*) J.-J. Velasco-Vélez, C.-H. Chuang, D. Gao, Q. Zhu, D. Ivanov, H. S. Jeon, R. Arrigo, R. V. Mom, E. Stotz, H. L. Wu, T. E. Jones, B. Roldan Cuenya, A. Knop-Gericke and R. Schlögl, *ACS Catal.*, 2020, **10**, 11510–11518; (*d*) A. Eilert, F. S. Roberts, D. Friebel and A. Nilsson, *J. Phys. Chem. Lett.*, 2016, **7**, 1466–1470.
10 Y. Hu and M. W. Ribbe, *Angew. Chem., Int. Ed.*, 2016, **55**, 8216–8226.
11 H. Peng, M. T. Tang, X. Liu, P. Schlexer Lamoureux, M. Bajdich and F. Abild-Pedersen, *Energy Environ. Sci.*, 2021, **14**, 473–482.
12 P. Jiang, D. Prendergast, F. Borondics, S. Porsgaard, L. Giovanetti, E. Pach, J. Newberg, H. Bluhm, F. Besenbacher and M. Salmeron, *J. Chem. Phys.*, 2013, **138**, 024704.
13 M. T. Greiner, T. E. Jones, B. E. Johnson, T. C. R. Rocha, Z. J. Wang, M. Armbrüster, M. Willinger, A. Knop-Gericke and R. Schlögl, *Phys. Chem. Chem. Phys.*, 2015, **17**, 25073–25089.
14 (*a*) V. Pfeifer, M. Hävecker, J. J. Velasco Vélez, A. Knop-Gericke, R. Schlögl and R. Arrigo, *Top. Catal.*, 2019, **61**, 2064–2084; (*b*) R. Arrigo, M. Hävecker, M. E. Schuster, C. Ranjan, E. Stotz, A. Knop-Gericke and R. Schlögl, *Angew. Chem., Int. Ed.*, 2013, **52**, 11660–11664; (*c*) A. Knop-Gericke, V. Pfeifer, J. J. Velasco-Vélez, T. Jones, R. Arrigo, M. Hävecker and R. Schlögl, *J. Electron Spectrosc. Relat. Phenom.*, 2017, **221**, 10–17.
15 N. Gutiérrez-Guerra, J. A. González, J. C. Serrano-Ruiz, E. López-Fernández, J. L. Valverde and A. de Lucas-Consuegra, *J. Energy Chem.*, 2019, **31**, 46–53.
16 R. Price, T. Eralp-Erden, E. Crumlin, *et al.*, *Top. Catal.*, 2016, **59**, 516–525.
17 M. Favaro, H. Xiao, T. Cheng, W. A. Goddard III, J. Yano and E. J. Crumlin, *Proc. Natl. Acad. Sci. U. S. A.*, 2017, **114**(26), 6706–6711.
18 H. Yang, J. Qian, *et al.*, *Nat. Commun.*, 2019, **10**, 1875.
19 X. Deng, A. Verdaguer, T. Herranz, C. Weis, H. Bluhm and M. Salmeron, *Langmuir*, 2008, **24**, 9474–9478.

Faraday Discussions

ROYAL SOCIETY OF CHEMISTRY

PAPER

Evolution of surface and sub-surface morphology and chemical state of exsolved Ni nanoparticles†

Heath Kersell, *[a] Moritz L. Weber, [bc] Lorenz Falling,[a] Qiyang Lu,‡[a] Christoph Baeumer, [d] Nozomi Shirato,[e] Volker Rose,[ef] Christian Lenser, [c] Felix Gunkel [b] and Slavomír Nemšák *[a]

Received 18th December 2021, Accepted 4th March 2022

DOI: 10.1039/d1fd00123j

Nanoparticle formation by dopant exsolution (migration) from bulk host lattices is a promising approach to generate highly stable nanoparticles with tunable size, shape, and distribution. We investigated Ni dopant migration from strontium titanate (STO) lattices, forming metallic Ni nanoparticles at STO surfaces. *Ex situ* scanning probe measurements confirmed the presence of nanoparticles at the H_2 treated surface. *In situ* ambient pressure X-ray photoelectron spectroscopy (AP-XPS) revealed reduction from Ni^{2+} to Ni^0 as Ni dopants migrated to the surface during heating treatments in H_2. During Ni migration and reduction, the Sr and Ti chemical states were mostly unchanged, indicating the selective reduction of Ni during treatment. At the same time, we used *in situ* ambient pressure grazing incidence X-ray scattering (GIXS) to monitor the particle morphology. As Ni migrated to the surface, it nucleated and grew into compressed spheroidal nanoparticles partially embedded in the STO perovskite surface. These findings provide a detailed picture of the evolution of the nanoparticle surface and subsurface chemical state and morphology as the nanoparticles grow beyond the initial nucleation and growth stages.

[a]*Advanced Light Source, Lawrence Berkeley National Laboratory, Berkeley, California, 94720, USA. E-mail: hkersell@lbl.gov; snemsak@lbl.gov*

[b]*Peter Gruenberg Institute (PGI-7) and JARA-FIT, Forschungszentrum Jülich GmbH, 52425 Jülich, Germany*

[c]*Institute of Energy and Climate Research (IEK-1), Forschungszentrum Jülich GmbH, 52425 Jülich, Germany*

[d]*MESA+ Institute for Nanotechnology, University of Twente, Faculty of Science and Technology, 7500 AE Enschede, The Netherlands*

[e]*Center for Nanoscale Materials, Argonne National Laboratory, Lemont, Illinois, 60439, USA*

[f]*X-ray Science Division, Advanced Photon Source, Argonne National Laboratory, Lemont, Illinois, 60439, USA*

† Electronic supplementary information (ESI) available. See DOI: 10.1039/d1fd00123j

‡ Current address: Key Laboratory of 3D Micro/Nano Fabrication and Characterization of Zhejiang Province, School of Engineering, Westlake University, Hangzhou, Zhejiang, China.

1. Introduction

Nanoparticle catalysts supported on transition metal oxides play a critical role in chemical production and conversion, as well as energy conversion and storage processes.[1–4] Catalytic activity is affected by the nanoparticles' composition, size, structure, stability, and their interaction with the support. Decreasing the nanoparticle size increases the specific surface area and the density of low-coordination sites, often increasing activity. However, processes like nanoparticle coarsening, agglomeration, and coking can decrease activity through structural changes and decreases in available active sites.

In perovskite host lattices, catalytically active dopants can segregate to the oxide surface to form nanoparticles in a process known as 'exsolution'.[5] Recently, exsolution from perovskite hosts during reducing treatments produced metallic nanoparticles with excellent stability and coking resistance, as well as tunable size and distribution.[6–9] Particles exsolved from perovskite hosts exhibit strong metal–oxide interactions, producing nanoparticles partially embedded (socketed) in the oxide surface.[6,8,10] Surface enrichment by such exsolved metallic nanoparticles can enhance the electrocatalytic activity and surface conductivity of electrode materials.[11]

While the aforementioned properties of socketed nanoparticles are promising for a variety of applications, differences in perovskite host structure and composition, and differences in growth or reaction environments can lead to variations in particle size and shape.[12,13] Spherical, cubic, conical, and even core–shell particles have been observed, depending on the reaction environment.[10,13–15] Because particle shape can influence exposed active sites and activity, it is crucial to understand the evolution of the particle morphology during exsolution. To this end, recent investigations have probed the early stages of nanoparticle nucleation and growth to better understand the mechanisms leading to nanoparticle formation.[7,13] Because reaction environments influence the size, shape, and distribution of exsolved nanoparticles, it is crucial to understand their evolution during growth.

In this work, we monitor the chemistry and morphology of Ni nanoparticles as they migrate from $SrTi_{0.9}Nb_{0.05}Ni_{0.05}O_{3-\delta}$ (STNNi) hosts during reducing treatments in H_2 using a recently developed combination of ambient pressure X-ray photoelectron spectroscopy (AP-XPS), and ambient pressure grazing incidence X-ray scattering (AP-GIXS).[16] Ni reduces as it migrates to the surface, forming metallic nanoparticles partially embedded within the perovskite host lattice. Meanwhile, the oxidation state of the Sr and Ti in the host is mostly unchanged during the exsolution process. We used *ex situ* atomic force microscopy (AFM) and synchrotron X-ray scanning tunneling microscopy (SX-STM) to probe the *ex situ* surface structure and Ni distribution. Using this information to understand boundary conditions for the nanoparticle sizes and shapes, we used AP-GIXS to monitor the nanoparticle size and subsurface morphology during growth.

By combining AP-XPS and AP-GIXS in the same measurements, we reveal the evolution of chemistry and morphology during socketed nanoparticle growth by reduction in H_2 gas environments. Because both techniques possess tunable depth sensitivity, this approach is well suited to further studies of the chemistry and morphology of exsolved nanoparticles during exsolution and reactions, and a host of other systems with dynamic structure and chemistry can also benefit from this approach.[17–19]

2. Methods

STNNi thin films of 30 unit cells (11.7 nm) were synthesized by reflection high-energy electron diffraction (RHEED)-controlled pulsed laser deposition (PLD) on (001) Nb-doped and TiO_2-terminated $SrTiO_3$ single-crystalline substrates (Shinkosha Co. Ltd., Yokohama, Japan). The thin films were deposited at T = 650 °C and at an oxygen pressure of $p(O_2) = 0.108$ mbar. A KrF excimer laser was applied for the ablation of the ceramic target material ($\lambda = 248$ nm) using a laser fluence of $F = 1.14$ J cm^{-2} and a repetition rate of $f = 5$ Hz. Here, two different PLD systems were used for the sample growth, which mainly differ in the respective heating systems. While an IR-diode laser ($\lambda = 925$ nm) is applied for heating in one of the systems (Surface GmbH), a resistive heater is applied for the other system (TSST – Twente Solid State Technology). In addition, slight differences in the target-to-substrate distance existed between systems, with $d = 57$ mm for the former and $d = 60$ mm for the latter PLD system. After the thin film growth, the samples were quenched to room temperature. Notably, the cooling rate was lower for the resistive heater, where the thermal history of the sample may cause small differences in the exsolution behavior.

The samples were 10 mm × 10 mm in size and were cut into 5 mm × 5 mm pieces after the thin film growth, which enabled the analysis of four nominally equal samples. For *ex situ* surface imaging, several samples were annealed under continuous 4% H_2/Ar gas flow (t = 5 h, T = 800 °C) in order to induce nickel exsolution. AFM was applied to investigate the surface morphology in the as-prepared state and after thermal reduction of the sample using tapping mode (AFM; Cypher, Oxford Instruments Asylum Research Inc., Santa Barbara, USA), with a tip curvature of approximately 8 nm. In order to obtain reference values for the nanoparticle dimensions, annealing experiments were carried out *ex situ* using one piece (5 × 5 mm) of four nominally equal samples that were cut from 10 × 10 mm^2 thin films that were used for the XTIP and AP-XPS/AP-GIXS investigations. The surface morphology was analyzed using the software package Gwyddion version 2.52, applying a lower threshold of 1 nm for the detection of the nanoparticles. The nanoparticle properties (nanoparticle density, median of the maximum nanoparticle height and the average nanoparticle diameter) were evaluated on the basis of a representative AFM scan with a size of 4 × 4 μm^2.

APXPS and APGIXS measurements were made on the sample spot, at the same conditions, in the same measurement set, using a recently developed ultrahigh vacuum system capable of simultaneous AP-XPS and AP-GIXS measurements. This system is located at undulator beamline 11.0.2.1 of the Advanced Light Source, which provides soft X-ray radiation between 160 and 1800 eV. The base pressure was 1×10^{-9} Torr at the start of the experiments, with water comprising most of the background gas. Photoelectron spectroscopy was conducted using a SPECS Phoibos ambient pressure hemispherical analyzer with 1D delay line electron detection. Binding energies (BEs) of core-level electrons were referenced using known BEs of Ti $2p_{3/2}$ (459.3 eV) and Sr 3d (134.2 eV) for the pristine STO and STNNi states.[20] The sample was mounted on a 4-axis manipulator, with three Cartesian translation axes supplemented by polar angle rotation (enabling control of the X-ray incidence angle). Sample holders were equipped with a Pt button heater, permitting heating temperatures up to ~800 °C. To measure X-ray scattering from the STNNi surfaces, we used a CCD camera (Andor Ikon-L SO), which was isolated from the gas environment by an X-ray transparent 150 nm thick silicon nitride window (Silson Ltd.). Differential pumping was applied behind the X-ray window to maintain vacuum at the CCD chip surface and to prevent water condensation on the CCD chip, which was cooled to −50 °C during operation. The camera was mounted on a 2-axis universal mount, covering ±12° azimuthal and 20° polar scattering angles. More details on this setup can be found in ref. 16. A stainless steel beamstop was applied to block the reflected specular X-ray beam from saturating or even damaging the CCD camera. The beamstop appears as a vertical black rectangle along $\theta_f = 0°$ in all X-ray scattering data shown in this work, and additionally is the source of the circular feature near $\alpha_f = 0°$ in the experimentally obtained scattering patterns shown in the next section.

Analysis of grazing incidence X-ray scattering data requires the iterative simulation of X-scattering patterns emerging from various surface and near-surface geometries. BornAgain, a software package employing the distorted-wave Born approximation for simulating and fitting scattering at grazing incidence, was used to iteratively vary near-surface sample geometries and obtain agreement between the experimental and simulated scattering patterns.[21]

To probe the surface elemental composition of STNNi films after exposure to reducing treatments, we conducted X-ray absorption spectroscopy with an STM tip apex as the XAS detector at the XTIP beamline (4-ID-E) of the Advanced Photon Source.[22–24] The tip-sample junction of an STM was targeted with X-rays, and the X-ray absorption intensity was measured with the STM tip apex as a probe, with incident photons at 847 eV (before the Ni L_3-edge) and 852.2 eV (at the Ni L_3-edge) across a 216 nm × 300 nm region in 12 nm intervals. This approach provides the Ni XAS intensity at intervals smaller than the features expected on the STNNi surfaces discussed in the results. To prevent the detection of photoelectrons emitted from regions far from the tip apex, a coaxial "smart" tip was fabricated with a PtIr center, coated by layers of SiO_2, Ti, and Au.[23] To extract the X-ray induced contributions to the STM tunneling current, the incident X-ray beam was chopped at 650 Hz, and a phase sensitive filter extracted all beam-induced currents at the chopping frequency.[25] Throughout the course of the measurement, the sample was kept under 2×10^{-9} Torr and at room temperature. These measurements are presented in the ESI.†

3. Experimental results

3.1. *In situ* XPS of STNNi during reducing treatments

Fig. 1 shows the Ni 3p and Ti 3s photoemission spectra with 900 eV incident photons for an STNNi sample after progressive reducing treatments in 1 Torr H_2. In the as-prepared state, the Ni 3p feature above 68 eV is characteristic of Ni^{2+}.[7] Heating in 1 Torr of H_2 gas with increasingly higher temperatures and longer times shifts the position of the Ni 3p feature to 66.5 eV, characteristic of a transition to Ni^0. At the same time, the Ni 3p intensity increases notably compared to that of Ti 3s. This indicates that the surface becomes increasingly enriched in Ni^0 as the duration and temperature of the reducing treatments increase. This is in

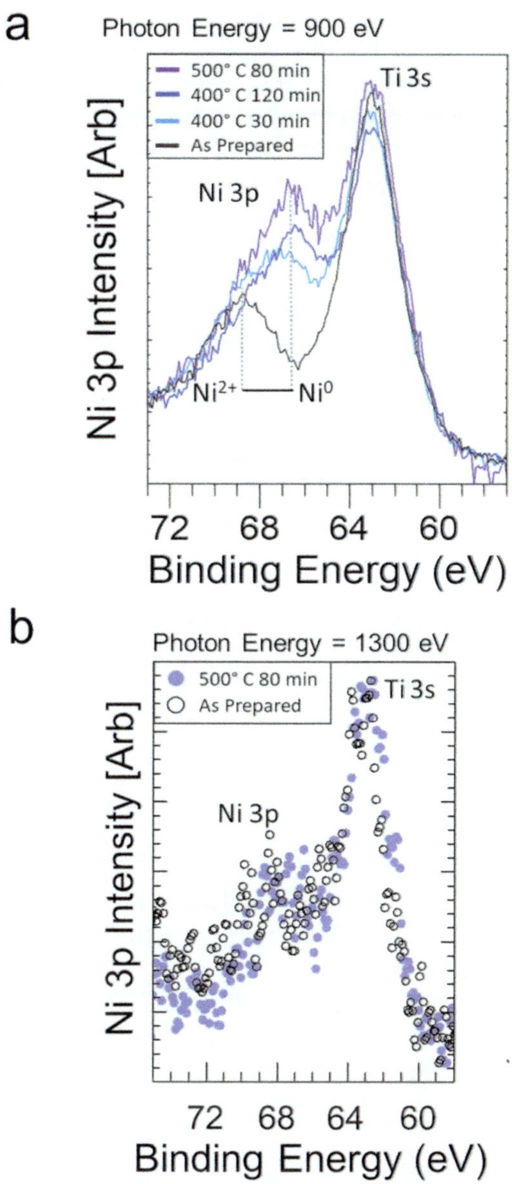

Fig. 1 (a) Ni 3p and Ti 3s regions at 900 eV photon energy are shown on STNNi after reducing in 1 Torr H_2 at increasing annealing times and temperatures. (b) Ni 3p and Ti 3s regions at 1300 eV photon energy measured on as prepared STNNi (black circles) and after annealing in 1 Torr H_2 at 500 °C. The increased photon energy in (b) is accompanied by a decrease in the photon flux, hence the results in (b) are displayed by dots rather than lines.

agreement with the tendency of Ni in STNNi to accumulate and form Ni-rich nanostructures fully embedded beneath the host surface, which then migrate to the surface to form nanoparticles upon reduction.[9]

To probe the sample composition deeper into the subsurface, we also measured the Ni 3p and Ti 3s regions with 1300 eV incident photons (Fig. 1b).

This increased the kinetic energy of the Ni 3p and Ti 3s electrons, thus increasing their inelastic mean free path (IMFP) from ~1.63 nm at 900 eV to ~2.23 nm at 1300 eV, and thus also increasing their probing depth.[26] In this greater depth, the Ni 3p area increased less than in the near-surface region probed in Fig. 1. Hence, while Ni migrates to the surface during reducing treatments, the sample composition was less affected in the sub-surface/bulk.

We also monitored the Sr and Ti composition and oxidation states during Ni reduction and migration. The Sr 3d and Ti 2p intensities both initially decrease (Fig. 2a and b), consistent with Ni enrichment at the STNNi surface. In spite of the reducing treatments and Ni migration, the Sr and Ti oxidation states remain mostly stable during the experiment. More quantitative discussion of the

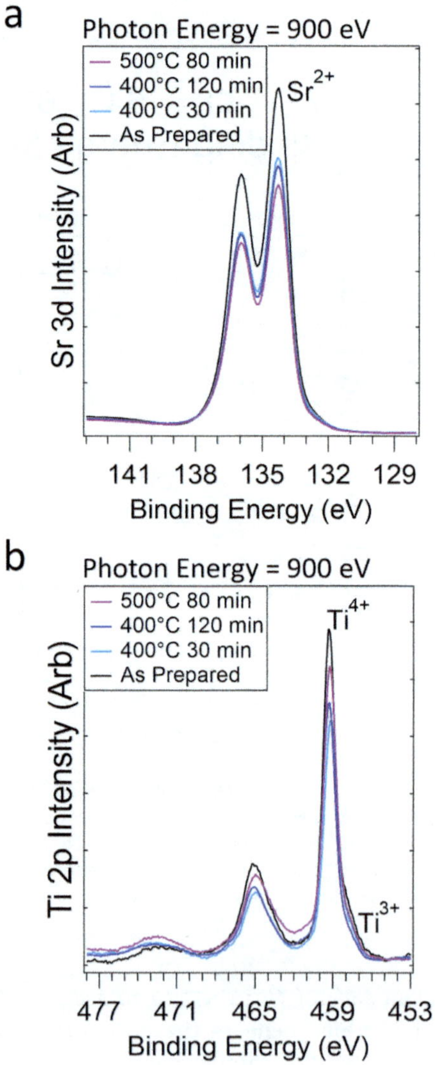

Fig. 2 Sr 3d (a) and Ti 2p (b) regions after increasingly reducing treatments in 1 Torr of H_2 gas. $E_{photon} = 900$ eV.

intensities is provided in the accompanying *Faraday Discussion*. Interestingly, previous investigations revealed the formation of Ni-rich nanostructures fully embedded beneath the STNNi surface.[9] The stability of the Sr and Ti oxidation states could suggest the migration of Ni from these nanostructures, leaving the perovskite structure otherwise intact. However, further investigation is necessary to confirm the subsurface sites from which the Ni migrates.

3.2. *Ex situ* AFM and elemental distribution

To understand the surface structure before and after Ni exsolution, we measured *ex situ* AFM on STNNi films in their as-prepared state, and after reducing at 800 °C for five hours in a 1 : 24 H_2 : Ar mixture. The as-prepared films exhibited a stepped-terrace structure (Fig. 3a) with 0.4 nm step height, equivalent to one $SrTiO_3$ unit cell.[27] After reducing in H_2, the surface was decorated with high contrast structures, averaging approximately 27.8 nm wide and 7.1 nm high (Fig. 3b). Lower contrast features were also present, with a height of 0.4 nm, consistent with the height of one $SrTiO_3$ unit cell (visible in Fig. 3b, inset). These lower contrast features possessed more varied geometries than the higher contrast features, had a typical width of 27.5 nm, and populated the surface more densely than the higher contrast features. The three-dimensional morphology of the brighter features is consistent with nanoparticle formation upon Ni exsolution,[9,12] while the 0.4 nm height of the low contrast features is consistent with the height of one $SrTiO_3$ unit cell. This suggests that the latter features are related to a relaxation or reconstruction of the STNNi host lattice during the reduction and nanoparticle formation. To further support the identification of the high contrast features as Ni nanoparticles, we also conducted X-ray absorption spectroscopy (XAS) using SX-STM to locally probe the surface composition of such a film already exposed to reducing treatment. The results are presented in the ESI,† and agree with the assignment of the high contrast features as exsolved Ni nanoparticles. The 0.4 nm height of the low contrast features observed in AFM suggests a possible

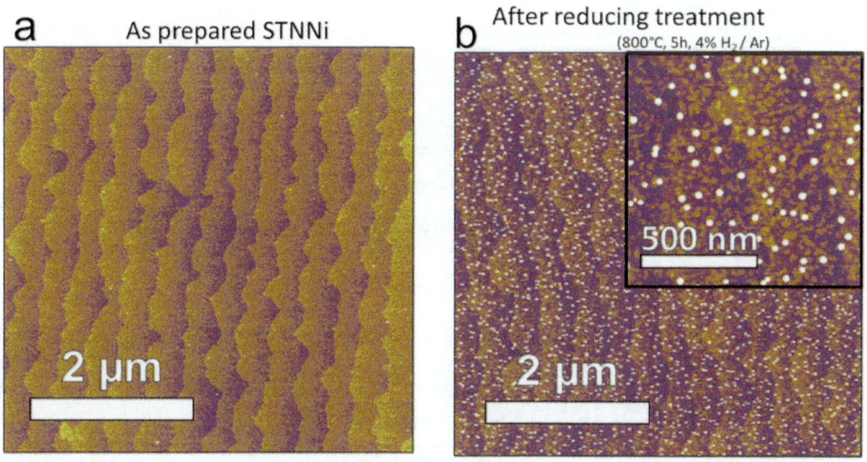

Fig. 3 *Ex situ* AFM imaging of STNNi surfaces in the as prepared state (a) and after reducing treatment in an H_2 : Ar mixture at 800 °C (b).

transformation of the STO host surface upon reduction and nanoparticle formation. Indeed, one potential explanation for the emergence of the lower contrast features is that Ni nanoparticle formation leaves behind subsurface Ni vacancies, generating structural instability in the host lattice.

3.3. GIXS of STNNi during reducing treatments

To understand the nucleation, growth, and evolution of Ni nanoparticles during reducing treatments, we collected GIXS from STNNi surfaces during the *in situ* XPS measurements shown in Fig. 1 and 2. This was done using a recently developed system enabling simultaneous *in situ* APXPS and APGIXS measurements on the same sample region.[16] The X-ray scattering geometry is shown in Fig. 4a, along with an illustration of its orientation with respect to the APXPS aperture of the hemispherical analyzer. This aperture provides a pathway for photoelectrons to travel from the sample to the electron energy analyzer, while limiting gas flow through the aperture.

X-ray scattering patterns (Fig. 4b–e) were collected during the measurements shown in Fig. 1 and 2 on probing regions overlapping with the XPS measurements, under the same conditions. This enabled the direct correlation between chemical, compositional, and structural transformations at each condition during *in situ* experiments. The sample that produced the scattering pattern in

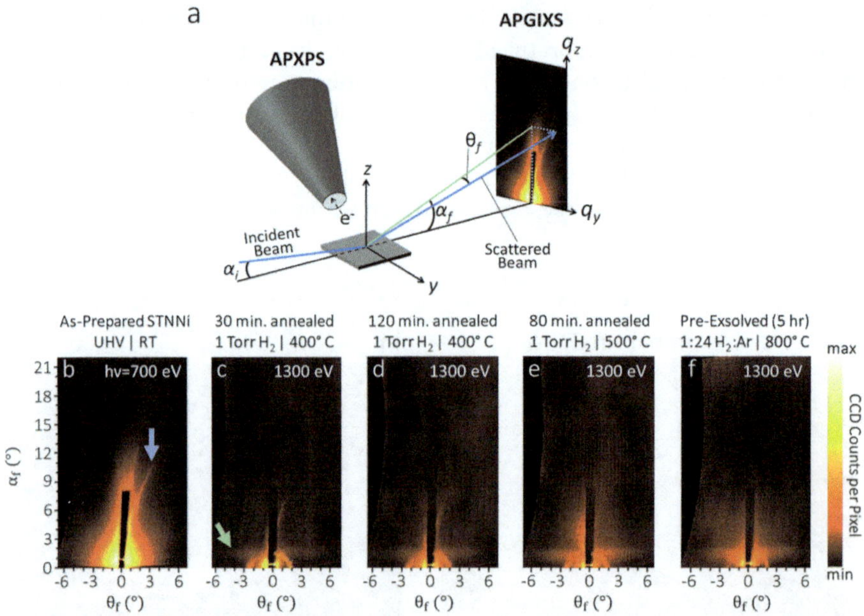

Fig. 4 (a) Combined AP-XPS-GIXS layout. Scattering patterns (b–f) measured from STNNi under progressively reducing conditions. (b) as prepared ('pristine'), (c) after annealing at 400 °C for 30 minutes in 1 Torr H_2 gas (green arrow shows the emergence of the Yoneda line), (d) annealed at 400 °C for 90 more minutes in 1 Torr H_2, (e) annealed at 500 °C for 80 more minutes in 1 Torr H_2, and (f) on an STNNi film pre-annealed *ex situ* at 800 °C in a 1 : 24 H_2 : Ar mixture for five hours and transferred through air to measure GIXS. [Scattering parameters: (a) $h\nu = 700$ eV, $\alpha_i = 1°$; (b–e) $h\nu = 1300$ eV, $\alpha_i = 0.4°$. Color bar is in logarithmic scale covering 4.3 orders of magnitude].

Fig. 4f was reduced *ex situ* by exposure to a 1 : 24 H_2 : Ar mixture at 800 °C for five hours prior to measurement, and was used as a limiting case for the *in situ* measurements. AFM on that sample revealed nanoparticles of 54 nm lateral width (not shown here). Hence, we limited the lateral size of the simulated particles to ~50 nm.

Before beginning the exsolution process by reducing in H_2, measurement of pristine STNNi targeted by 700 eV incident photons yields a clear scattering feature (indicated by an arrow in Fig. 4b) that is asymmetric with respect to the incident beam direction, appearing as a streak to one side of the beamstop. Asymmetric scattering with respect to the scattering plane (which is parallel to the incident beam direction and normal to the sample surface) requires scattering centers oriented asymmetrically with respect to the scattering plane. Faceted surface structures aligned off-axis from the scattering plane are one example of oriented features that can produce this behavior. In fact, our sample was mounted with the [100] crystal direction of the STNNi film rotated by approximately 10° azimuthally from the scattering plane, potentially explaining the presence of this asymmetric streak in the scattering. As we will discuss next, however, this feature is attributed to the STNNi step structure, which can have an orientation instead related to miscut directions during crystal fabrication.

In order to increase our capability to capture the form factor of smaller objects, subsequent measurements were done at a higher incident photon energy of 1300 eV. As a secondary effect, increasing the excitation energy also decreased the photon flux by almost an order of magnitude, slightly affecting the signal/noise ratio of following datasets. Further, at low angles, the polar scattering angle, α_f, varies inversely with the photon energy.[28]

After starting the exsolution process, when the STNNi was heated for increasing times and to higher temperatures in H_2, the asymmetric streak became more diffuse, and a new horizontally elongated diffuse feature emerged near $\alpha_f = 1.5°$ (Fig. 4c–f). This feature is related to the nucleation and growth of Ni nanoparticles partially embedded at the sample surface. We will next interpret these features within qualitative agreement with simulated scattering patterns.

3.4. Qualitative interpretation of Ni nanoparticles' morphology

X-ray scattering simulations were based on a model of partially embedded particles of varying size, shape, and nearest neighbor distance, together with linear/rectangular features mimicking terrace steps. Optimizing the geometry and distribution for each scattering pattern enables us to reveal trends in sample structure transformations emerging under each condition. To limit the simulation time, it is generally necessary to restrict simulations to a feasible set of sample geometries. Hence, based on information from *ex situ* AFM and SX-STM, we restricted the Ni particle geometries to spheres, compressed spheroids, hemiellipsoids, and hemispheres. These were allowed to have various distributions of diameters, heights, and nearest neighbor distances. The largest average particle sizes considered were 50 nm in diameter. The height of the nanoparticles above the STNNi film surface was limited to 8 nm.

Before reduction, the pristine STNNi surface produces an asymmetric streak with respect to the incident beam (blue arrow, Fig. 4b). This streak is consistent with a structure of a several hundred nanometer periodicity, oriented asymmetrically with respect to the incident beam. The only sample feature fitting this description is the STNNi step structure. As such, we treated the STNNi steps by a periodic, partially disordered array of extremely elongated rectangular islands with 0.4 nm height, equal to the size of one $SrTiO_3$ unit cell (Fig. 5a).

Upon nanoparticle formation, the best qualitative agreement with the experiment was obtained for ellipsoidal particles embedded partially within the STNNi surface (Fig. 6a–f). For comparison, Fig. 6g–l show the result for spherical particles, partially embedded in the STNNi surface. In both cases, the average particle height above the surface was restricted to ≤8 nm. At 8 nm diameter, the spheres and the spheroids both produce a diffuse scattering feature near $\alpha_f = 1.5°$, in agreement with the experiment. We note that $\alpha_f = 1.5°$ is in the proximity of the critical angle for STNNi at 1300 eV, which explains the enhanced scattering intensity, otherwise known as the Yoneda line, at this region. The feature is relatively broad, since the exsolved nanoparticles cause increased film roughness compared to the initial atomically flat surface.

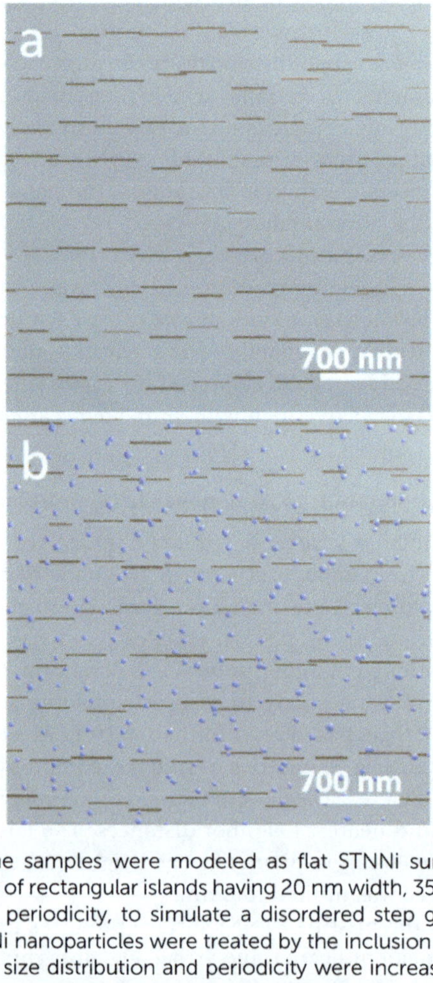

Fig. 5 (a) The pristine samples were modeled as flat STNNi surfaces with a partially ordered square lattice of rectangular islands having 20 nm width, 350 ± 75 nm length, and 400 ± 45 nm lattice periodicity, to simulate a disordered step geometry. (b) Surfaces containing exsolved Ni nanoparticles were treated by the inclusion of a disordered lattice of Ni particles whose size distribution and periodicity were increased with conditions to model the experimental particle geometries and distributions.

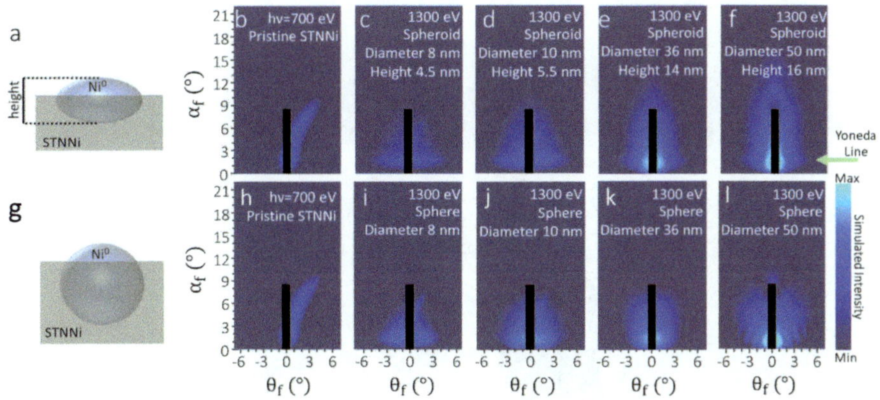

Fig. 6 Simulated scattering patterns for partially embedded spheroidal (a–f) and spherical (g–l) particles: the [100] surface direction of STNNi is rotated 11° azimuthally from the scattering plane. Each sequence begins with scattering from a pristine STNNi surface before nanoparticle formation (b and h), and is followed by scattering patterns from surfaces populated by Ni^0 nanoparticles of increasing diameter and height (b–f and h–l). The color bar is in $\log_{10}$, with arbitrary units crossing 5.5 orders of magnitude.

As the particle diameter increased to 50 nm, the spherical particles produced patterns deviating significantly from the experiment (Fig. 6k and l), with a compression of the Yoneda line and the development of a strong modulation in the form factor along q_y (*i.e.* along θ_f), which are not observed in the experiment. Meanwhile, spheroidal particles produced reasonable agreement across the entire range of particle sizes considered, with q_y modulations absent in the simulations due much lower nanoparticle heights.

The finding that exsolved Ni^0 nanoparticles adopt a significantly compressed spheroidal morphology is somewhat in contrast with previous investigations on smaller exsolved particles, which often appear more symmetric.[8,13] However, differences in the sample growth and exsolution environment significantly affect particle shape.[10,13–15] For example, the STNNi films measured in our experiments are only 12 nm thick, potentially limiting the nanoparticle height, while their lateral size is unrestricted. Moreover, Neagu *et al.* demonstrated the dependence of Ni nanoparticle shape on the concentration of Ni atoms in the perovskite host lattice.[13] It is also possible that a competition between surface and interfacial energies between the vacuum, Ni^0, and the STNNi host also favors the growth of increasingly compressed particles.

3.5. Ni nanoparticle formation quantified from XPS and GIXS experiments

Finally, to understand the correlation between Ni segregation and nanoparticle formation, we quantified both the metallic Ni intensity from the Ni 3p XPS region, and the integrated intensity of the Yoneda line feature, both at progressive stages of the exsolution process. It has been previously established that the Yoneda intensity can be used to assess roughness/inhomogeneities in planar systems, similar to the STNNi films presented here.[29]

The XPS intensity ratio between Ni^0 3p and Ti 3s is plotted in Fig. 7a, and the integrated Yoneda intensities are plotted in Fig. 7b. Ni 3p and Ti 3s intensities

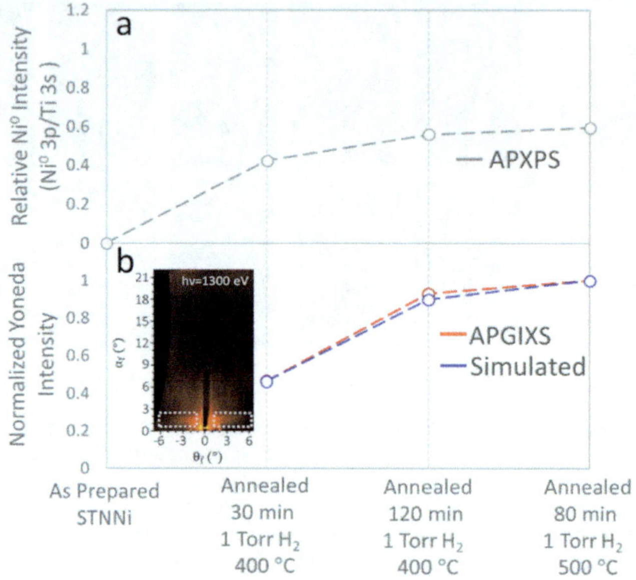

Fig. 7 (a) Relative Ni enrichment determined from Ni 3p intensity at the STNNi surface as a function of the reducing treatment. (b) Experimental (red) and simulated (blue) Yoneda line intensity as a function of treatment. The location of the Yoneda line is indicated by the dotted boxes in the inset scattering pattern. The intensity of the last point in (b) is normalized to 1 for both the APGIXS and simulated scattering patterns.

were determined by fitting the Ni 3p (two Voigt doublets, corresponding to Ni^{2+} and Ni^0) and Ti 3s (single Voigt spectral peak) regions (ESI†). Yoneda intensities were integrated in both the experimental and simulated scattering patterns across the region, illustrated by the inset in Fig. 7b.

No metallic Ni appears in the Ni 3p region of the as-prepared sample. Three further Ni 3p measurements upon annealing in H_2 show the increased intensity of the Ni^0 doublet (Fig. 1a and ESI†). The Ni^0 : Ti ratio first increases to ~0.42, and gradually evolves to ~0.59 (Fig. 7a). The Ni^0 XPS intensity increases by a factor of 1.4 between the first annealing condition (annealed for 30 min at 400 °C in H_2) and the last step of exsolution (80 min at 500 °C in H_2). The ratio of the Ni^0 3p to Ti 3s intensities for each condition was used, along with the experimentally obtained scattering patterns, to determine the particle sizes used in the simulations shown in Fig. 6.

The qualitative agreement between the increasing trends in the Ni^0 XPS signal and Yoneda line intensity confirms the increasing amount and size of the Ni nanoparticles. However, the Yoneda line intensity increases by a factor of ~2 between the first and the last step in the exsolution process. The explanation for these different amounts of signal enhancement (~1.4 in Ni^0 XPS and ~2 in the X-ray scattering Yoneda feature) can be found in the different quantities that XPS and X-ray scattering measure.

Even at a very grazing incidence, the information depth in soft X-ray photoelectron spectroscopy is still limited by the photoelectron inelastic mean free path (IMFP), or more precisely by the electron attenuation length. For the excitation energies used here, the photoelectron IMFP in STO is less than 2 nm, significantly less than either the thickness of the STNNi film or the expected height of the exsolved nanoparticles. On the other hand, the penetration depth of X-rays and the evanescent wave associated with the incident X-ray wave cover the entire thickness of the STNNi film and penetrate through the nanoparticles in their entirety. As a direct consequence of the information depth difference, the XPS signal detecting Ni^0 atoms can only account for the topmost part (*ca.* 2–4 nm) of the nanoparticles, while the X-ray scattering signal originates from the scatterers throughout their entire volume.

For a fully quantitative comparison, corresponding photoelectron spectra (Ti 3s and Ni 3p) were simulated using the SESSA package[30] at each stage of the exsolution. The size and density of nanoparticles in the SESSA models were optimized to yield the same Ni^0 : Ti ratio in the simulated and experimental data. In a self-consistent manner, the same modeled size/density of nanoparticles were then iteratively used to generate scattering simulations in the BornAgain package (Fig. 6b–f). This structural and chemical model of the nanoparticles yields an excellent agreement for both spectroscopic and scattering data in Fig. 7a and b.

Both the Ni^0 XPS and Yoneda line intensities increase less between the last two conditions (120 min at 400 °C and 80 min at 500 °C). It is likely that the exsolution process began depleting the bulk of the STTNi film of nickel, and the nanoparticles' volumes continued slowly increasing. Within the accuracy explained above, the quantitative agreement between the increase in the Ni^0 XPS signal and the Yoneda line intensity also confirm the correlation between the morphology of the exsolved nanoparticles (elongated spheroid like particles) and their chemistry being metallic Ni. This observation of the exsolution process by both X-ray photoelectron spectroscopy and X-ray scattering provides a strong link between morphology and chemistry in the Ni nanoparticle/STNNi system.

4. Conclusions

Using *in situ* APXPS, we found that Ni migrated to the STNNi surface during reductive annealing treatments in 1 Torr of H_2 gas. Ni^{2+} was reduced to Ni^0 upon migration to the surface, while the Sr and Ti oxidation states were mostly unchanged. Increasing the XPS probing depth by increasing the photon energy revealed a less dramatic increase in Ni concentration, suggesting that the near-surface is less Ni enriched than the surface. *Ex situ* scanning probe measurements revealed approximately spherical Ni surface morphologies grown in response to reducing treatments. Throughout the nanoparticle growth process, *in situ* GIXS revealed the formation of spheroidal nanoparticles with a compressed geometry (Fig. 6a).

Because reaction environments influence the size, shape, distribution, and oxidation state of exsolved nanoparticles, it is crucial to understand their structural and chemical evolution *in situ* and under reaction conditions. Previous investigations have provided valuable insights into the atomic scale nucleation and early-stage growth of exsolved nanoparticles. To achieve atomic scale resolution, however, it is often necessary to use methods such as environmental transmission electron microscopy, which place stringent constraints on the sample geometry, such as the use of oriented, electron transparent samples. Here, we were able to simultaneously probe the dynamic Ni atom migration, chemical state evolution, and nanoparticle growth and morphology as the particles grew beyond the initial stages of nucleation and growth, across the region illuminated by incident X-rays, hundreds of microns in width. The X-ray methods used here provide both surface and sub-surface information, which makes them perfect complementary techniques to scanning probes, such as STM or AFM.

Because exsolved nanoparticles were previously shown to have enhanced stability and coking resistance, these findings provide a detailed picture of the behavior of an important emerging class of catalysts. Additionally, this study serves as a proof of concept for monitoring *in situ* structural and chemical evolution in ambient gas and reaction environments by combining APXPS and APGIXS. For example, these methods can also be applied to directly correlate catalyst restructuring and chemical state evolution during reactions. Long standing questions of structure–activity relationships for many catalysts remain unanswered due to challenges in measuring the structure and chemical state of the catalyst *in operando*. Combining X-ray spectroscopies, scattering, and mass spectrometry/gas chromatography opens up possibilities to study how these transformations result from, and at the same time often drive, chemical activity.

Conflicts of interest

There are no conflicts of interest to declare.

Acknowledgements

This work was associated with the development of instrumentation through the Laboratory Directed Research and Development (LDRD) Program at Lawrence Berkeley National Laboratory under a grant titled "Correlation of structural and chemical processes at interfaces under operating conditions using multimodal ambient pressure X-ray photoelectron spectroscopy and surface X-ray scattering." S. N. and H. K. acknowledge this LDRD grant for salary support. Part of this work was performed at beamline 11.0.2 of the Advanced Light Source, a U.S. DOE Office of Science User Facility, which is funded under Contract No. DE-AC02-05CH11231. LJF acknowledges support from the Alexander von Humboldt Foundation, Bonn, Germany. Part of this work was performed at the Advanced Photon Source and the Center for Nanoscale Materials, U. S. Department of Energy Office of Science User Facilities, and supported by the U. S. Department of Energy, Office of Science, under Contract No. DE-AC02-06CH11357. The authors thank D. Rosenmann and Y. Liu for their help with smart tip fabrication at XTIP.

References

1 T. Mallat and A. Baiker, Oxidation of Alcohols with Molecular Oxygen on Solid Catalysts, *Chem. Rev.*, 2004, **104**, 3037–3058.

2 Q. Zhang, E. Uchaker, S. L. Candelaria and G. Cao, Nanomaterials for energy conversion and storage, *Chem. Soc. Rev.*, 2013, **42**, 3127.

3 A. S. Aricò, P. Bruce, B. Scrosati, J.-M. Tarascon and W. van Schalkwijk, Nanostructured materials for advanced energy conversion and storage devices, *Nat. Mater.*, 2005, **4**, 366–377.

4 Y. Nishihata, J. Mizuki, T. Akao, H. Tanaka, M. Uenishi, M. Kimura, T. Okamoto and N. Hamada, Self-regeneration of a Pd-perovskite catalyst for automotive emissions control, *Nature*, 2002, **418**, 164–167.

5 J. H. Kim, J. K. Kim, J. Liu, A. Curcio, J.-S. Jang, I.-D. Kim, F. Ciucci and W. Jung, Nanoparticle Ex-solution for Supported Catalysts: Materials Design, Mechanism and Future Perspectives, *ACS Nano*, 2021, **15**, 81–110.

6 D. Neagu, G. Tsekouras, D. N. Miller, H. Ménard and T. S. Irvine, *In situ* growth of nanoparticles through control of non-stoichiometry, *Nat. Chem.*, 2013, **5**, 916–923.

7 H. Han, J. Park, S. Y. Nam, K. J. Kim, G. M. Choi, S. S. P. Parkin, H. M. Jang and J. T. S. Irvine, Lattice strain-enhanced exsolution of nanoparticles in thin films, *Nat. Commun.*, 2019, **10**, 1471.

8 D. Neagu, T.-S. Oh, D. N. Miller, H. Ménard, S. M. Bukhari, S. R. Gamble, R. J. Gorte, J. M. Vohs and J. T. S. Irvine, Nano-socketed nickel particles with enhanced coking resistance grown in situ by redox exolution, *Nat. Commun.*, 2015, **6**, 8120.

9 M. L. Weber, M. Wilhelm, L. Jin, R. Dittmann, R. Waser, O. Guillon, C. Lenser and F. Gunkel, Exsolution of Embedded Nanoparticles in Defect Engineered Perovskite Layers, *ACS Nano*, 2021, **15**(3), 4546–4560.

10 D. Neagu, E. I. Papaioannou, W. K. Ramli, D. N. Miller, B. J. Murdoch, H. Ménard, A. Umar, A. J. Barlow, P. J. Cumpson, J. T. Irvine and I. S. Metcalfe, Demonstration of chemistry at a point through restructuring and catalytic activation at anchored nanoparticles, *Nat. Commun.*, 2017, **8**, 1855.

11 Y. Li, W. Zhang, Y. Zheng, J. Chen, B. Yu, Y. Chen and M. Liu, Controlling cation segregation in perovskite-based electrodes for high electro-catalytic activity and durability, *Chem. Soc. Rev.*, 2017, **46**, 6345.

12 K. J. Kim, H. Han, T. Defferriere, D. Yoon, S. Na, S. J. Kim, A. M. Dayaghi, J. Son, T.-S. Oh, H. M. Jang and G. M. Choi, Facet-Dependent *in Situ* Growth of Nanoparticles in Epitaxial Thin Films: The Role of Interfacial Energy, *J. Am. Chem. Soc.*, 2019, **141**, 7509–7517.

13 D. Neagu, V. Kyriakou, I. L. Roiban, M. Aouine, C. Tang, A. Caravaca, K. Kousi, I. Schreur-Piet, I. S. Metcalfe, P. Vernoux, M. C. M. van de Sanden and M. N. Tsampas, *In Situ* Observation of Nanoparticle Exsolution from Perovskite Oxides: From Atomic Scale Mechanistic Insight to Nanostructure Tailoring, *ACS Nano*, 2019, **13**(11), 12996–13005.

14 S. Liu, Q. Liu and J.-L. Luo, HighlyStable and Efficient Catalyst with *In Situ* Exsolved Fe-Ni Alloy Nanospheres Socketed on an Oxygen Deficient Perovskite for Direct CO_2 Electrolysis, *ACS Catal.*, 2016, **6**, 6219–6228.

15 W. K. W. Ramli, E. Papaioannou, D. Naegu and I. S. Metcalfe, Exsolution of nickel alloys anchored nanoparticles on perovskite oxides for CO oxidation, *IOP Conf. Ser.: Mater. Sci. Eng.*, 2020, **778**, 012059.

16 H. Kersell, P. Chen, H. Martins, Q. Lu, F. Brausse, B.-H. Liu, M. Blum, S. Roy, B. Rude, A. Kilcoyne, H. Bluhm and S. Nemsák, Simultaneous ambient pressure X-ray photoelectron spectroscopy and grazing incidence X-ray scattering in gas environments, *Rev. Sci. Instrum.*, 2021, **92**, 044102.

17 B. Eren, D. Zherebetskyy, L. L. Patera, C. H. Wu, H. Bluhm, C. Africh, L.-W. Wang, G. A. Somorjai and M. Salmeron, Activation of Cu(111) surface by decomposition into nanoclusters driven by CO adsorption, *Science*, 2016, **351**(6272), 475–478.

18 J. Wang, M. McEntee, W. Tang, M. Neurock, A. P. Baddorf, P. Maksymovych and J. T. Yates, Jr., Formation, Migration, and Reactivity of Au-CO Complexes on Gold Surfaces, *J. Am. Chem. Soc.*, 2016, **138**, 1518–1526.
19 T. W. Hansen, A. T. DeLaRiva, S. R. Challa and A. K. Datye, Sintering of Catalytic Nanoparticles: Particle Migration or Ostwald Ripening, *Acc. Chem. Res.*, 2013, **46**(8), 1720–1730.
20 S. Nemšák, G. Conti, A. X. Gray, G. K. Palsson, C. Conlon, D. Eiteneer, A. Keqi, A. Rattanachata, A. Y. Saw, A. Bostwick, L. Moreschini, E. Rotenberg, V. N. Strocov, M. Kobayashi, T. Schmitt, W. Stolte, S. Ueda, K. Kobayashi, A. Gloskovskii, W. Drube, C. A. Jackson, P. Moetakef, A. Janotti, L. Bjaalie, B. Himmetoglu, C. G. Van de Walle, S. Borek, J. Minar, J. Braun, H. Ebert, L. Plucinski, J. B. Kortright, C. M. Schneider, L. Balents, F. M. F. de Groot, S. Stemmer and C. S. Fadley, Energetic, spatial, and momentum character of the electronic structure at a buried interface: The two-dimensional electron gas between two metal oxides, *Phys. Rev. B*, 2016, **93**, 245103.
21 G. Pospelov, W. Van Herck, J. Burle, J. M. C. Loaiza, C. Durniak, J. M. Fisher, M. Ganeva, D. Yurov and J. Wuttke, BornAgain: software for simulating and fitting grazing-incidence small-angle scattering, *J. Appl. Crystallogr.*, 2020, **53**(1), 262–276.
22 H. Kersell, N. Shirato, M. Cummings, H. Chang, D. Miller, D. Rosenmann, S.-W. Hla and V. Rose, Detecting element specific electrons from a single cobalt nanocluster with synchrotron X-ray scanning tunneling microscopy, *Appl. Phys. Lett.*, 2017, **111**, 103102.
23 N. Shirato, M. Cummings, H. Kersell, Y. Li, B. Stripe, D. Rosenmann, S.-W. Hla and V. Rose, Elemental Fingerprinting of Materials with Sensitivity at the Atomic Limit, *Nano Lett.*, 2014, **14**(11), 6499–6504.
24 V. Rose, N. Shirato, M. Bartlein, A. Deriy, T. Ajayi, D. Rosenmann, S.-W. Hla, M. Fisher and R. Reininger, XTIP - the world's first beamline dedicated to the synchrotron X-ray scanning tunneling microscopy technique, *J. Synchrotron Radiat.*, 2020, **27**(3), 836–843.
25 K. Wang, D. Rosenmann, M. Holt, R. Winarski, S.-W. Hla and V. Rose, An easy-to-implement filter for separating photo-excited signals from topography in scanning tunneling microscopy, *Rev. Sci. Instrum.*, 2013, **84**, 063704.
26 C. J. Powell and A. Jablonski, *NIST Electron Inelastic-Mean-Free-Path Database*, National Institute of Standards and Technology, Gaithersburg, MD, 2010.
27 F. W. Lytle, X-Ray Diffractometry of Low-Temperature Phase Transformations in Strontium Titanate, *J. Appl. Phys.*, 1964, **35**, 2212.
28 P. Müller-Buschbaum, A Basic Introduction to Grazing Incidence Small-Angle X-ray Scattering, in *Applications of Synchrotron Light to Scattering and Diffraction in Materials and Life Sciences*, Springer, Heidelberg, 2009, pp. 61–89.
29 C. Frank, J. Novak, R. Banerjee, A. Gerlach, F. Schreiber, A. Vorobiev and S. Kowarik, Island size evolution and molecular diffusion during growth of organic thin films followed by time-resolved specular and off-specular scattering, *Phys. Rev. B: Condens. Matter Mater. Phys.*, 2014, **90**, 045410.
30 W. Smekal, W. S. M. Werner and C. J. Powell, Simulation of electron spectra for surface analysis (SESSA): a novel software tool for quantitative Auger-electron spectroscopy and X-ray photoelectron spectroscopy, *Surf. Interface Anal.*, 2005, **37**, 1059–1067.

Faraday Discussions

ROYAL SOCIETY OF CHEMISTRY

PAPER

In situ surface analysis of palladium–platinum alloys in methane oxidation conditions†

Alexander I. Large, [*a] Roger A. Bennett, [b] Tugce Eralp-Erden[c] and Georg Held[a]

Received 3rd December 2021, Accepted 1st February 2022

DOI: 10.1039/d1fd00113b

Palladium and palladium–platinum foils were analysed using temperature-programmed near-ambient pressure X-ray photoelectron spectroscopy (TP-NAP-XPS) under methane oxidation conditions. Oxidation of palladium is inhibited by the presence of water, and in oxygen-poor environments. Pt addition further inhibits oxidation of palladium across all reaction conditions, preserving metallic palladium to higher temperatures. Bimetallic foils underwent significant restructuring under reaction conditions, with platinum preferentially migrating to the bulk under select conditions.

1 Introduction

Natural gas engines provide a cleaner alternative to traditional combustion engines, but emissions of unburnt potent greenhouse gas methane need to be abated with exhaust catalysts, which typically contain palladium or other precious metals on metal oxide supports. Palladium has been shown to be more active for methane oxidation than platinum, with a mixture of metallic and oxidised palladium generally required for optimal activity.[1–8] Methane is the second largest contributor to radiative forcing after carbon dioxide.[9] Recent international commitments made at COP26 show that reducing methane emissions is still a key component of climate change science. Countries worldwide are pledging to reduce methane emissions by 30% over the next decade, to achieve warming reduction of 0.2 °C by 2050.[10]

Prior work in our research group focused on alumina-supported palladium and palladium–platinum nanoparticle catalysts.[11,12] For complete oxidation of methane, the addition of platinum to palladium catalysts showed benefits under select reaction conditions. A key issue with palladium catalysts in methane

[a]Diamond Light Source, Harwell Campus, Didcot, UK. E-mail: alex.large@diamond.ac.uk

[b]Department of Chemistry, University of Reading, Reading, UK

[c]Johnson Matthey Technology Centre, Sonning Common, Reading, UK

† Electronic supplementary information (ESI) available. See DOI: 10.1039/d1fd00113b

oxidation is deactivation due to water, which is a core problem due to catalysts both operating in wet conditions, and producing water when active.[13–16] Addition of platinum to palladium to form bimetallic Pd–Pt catalysts has shown activity and stability benefits across a range of oxidation reactions.[17–19] Within methane oxidation, Pd–Pt alloying has often been a route to limit deactivation in wet conditions, and to improve long term catalyst stability.[20–26]

Near-ambient pressure X-ray photoelectron spectroscopy (NAP-XPS) is a quantitative analysis tool for chemical surface states under reactive gas pressures of a few mbar. This enables *operando* studies of catalytic materials, to develop understanding of catalyst activity and deactivation. Over the past decade, it has become a core technique for understanding surface chemistry for heterogeneous catalysts.[27–31]

Analysis of "real" powder catalysts can however cause problems for NAP-XPS, such as low signal due to loadings of a few weight percent, and charging due to insulating supports. This either necessitates longer measurement times, or requires the use of smaller amounts of sample, resulting in less catalytic conversion data. Either of these is an undesirable result, given that many NAP-XPS instruments are at synchrotron facilities, where limited experimental time is available.

In this work, palladium, platinum and palladium–platinum foils were studied under methane oxidation conditions. Palladium and platinum foils have previously been studied for methane oxidation, with moderate activity observed over both.[32] Due to the low metal surface area relative to nanoparticle catalysts, equivalent catalytic activity is not expected. Using metallic foils enabled us to use alloys of the same composition as the nanoparticles studied previously, but with much more intense signals in XPS, allowing for faster high quality measurements. This facilitates the use of temperature programmed NAP-XPS (TP-NAP-XPS). The palladium and platinum core levels are continually scanned in a reactive gas environment over the relevant temperature range. This ensures that the temperature ramp can still be fast enough (3.9 °C min^{-1}) to get close to equivalent real catalyst testing ramps. It also provides sufficient temperature resolution in the data (5 °C per set of spectra) to accurately understand where changes are happening. Typically foils will show worse activity than nanoparticle samples relative to the amount of catalyst used, due to the much lower surface area. It is also important to note the structural differences between supported nanoparticles and metal foils used here. Nanoparticles for these reactions from our previous work typically had diameters of 1–10 nm, and were supported on alumina grains which are around 100 μm.[12] The lack of support material could also limit the extent to the deactivation seen in wet conditions, due to prior studies showing migration of hydroxyls from the support to the metal.[33]

Despite the noted differences, this route still provides understanding of the behaviour of Pd and Pd–Pt systems under reaction conditions in a way that was not achievable using more realistic samples.

2 Experimental

Palladium foil (99.95%, 0.25 mm thickness) was supplied by Advent Research Materials Ltd. The foils studied here have not been characterised crystallographically, and prior experiments have shown the size of surface domains on polycrystalline Pd foils to vary significantly, over the range 10–100 μm, which provides a further variation from nanoparticle systems.[34,35] Microscope images of the Pd foil are shown in Fig. S2 of the ESI.† These show a high degree of roughness and it can therefore be assumed that a variation of Pd sites is present at the surface. Palladium–platinum alloy foils (Pd65–Pt35 and Pd88–Pt12, both 0.30 mm thickness) were produced by Goodfellow Cambridge Ltd, by cold-rolling from alloy ingots. The ratios were selected to match previously tested nanoparticle catalysts of the same compositions.[12] Measurements in multiple positions (with spot size of 0.1 mm × 0.1 mm) on individual samples suggest the composition was uniform across the samples both pre- and post-testing.[35] Samples were cut, with 5 × 3 mm (±1 mm) pieces used for the experiments described below.

All XPS data presented were collected in the Tea Pot chamber of the C branch of the B07 versatile soft X-ray beamline (VerSoX) at Diamond Light Source.[35] For all TP-NAP-XPS experiments, a 400 lines per mm grating was used, with exit slit *y*-gap size of 25 μm. At the photon energy used throughout these experiments, 750 eV, the resolution of the beamline was 0.26 eV. Where depth profiling measurements were made, kinetic energies were kept consistent (at 415, 675 and 1500 eV) at both Pd 3d and Pt 4f. This was achieved by varying the photon energy used between 500 eV and 1835 eV. The IMFP for Pd 3d is 7.28 Å at 415 eV, 9.93 Å at 675 eV and 17.58 Å at 1500 eV. For Pt 4f the IMFP is 9.76 Å at 415 eV, 13.63 Å at 675 eV and 24.79 Å at 1500 eV.[36]

Prior to experiments, the foils were "cleaned" in the Tea Pot chamber in oxygen (2 mbar) to 550 °C to remove surface carbon. The chamber was then evacuated, and the sample cooled to 150 °C. This temperature is low enough to be below the relevant catalytic activity window, but high enough to ensure minimal carbon build up on the surface in vacuum (1×10^{-6} mbar). Sample temperature was measured throughout using a thermocouple mounted on the surface of the sample holder.

Core level XPS spectra (Pd 3d, Pt 4f, C 1s, O 1s, EF) were measured in vacuum at 150 °C to provide clean reference data before exposure to reaction conditions. Reaction mixtures of methane, oxygen and water were dosed at 150 °C, in the ratios and total pressures as shown in Table 1. Whilst gases were dosed, the only pumping on the chamber was *via* the differential pumping system of the analyser, through the analyser cone.

Temperature-programmed near-ambient pressure X-ray photoelectron spectroscopy (TP-NAP-XPS) experiments were performed by repeatedly scanning Pd 3d and O 1s regions (and Pt 4f for bimetallic samples) whilst heating the foil in gas. The heating rate was constant throughout the experiments, at 3.5 °C min^{-1}. This

Table 1 Gas ratios and total pressures used throughout NAP-XPS experiments

Mixture	Label	CH_4	O_2	H_2O	Pressure/mbar
Oxygen only	OO	—	1	—	2.40
Stoichiometric dry	SD	1	2	—	0.33
Stoichiometric wet low	SLW	1	2	2	0.55
Stoichiometric very wet	SVW	1	2	100	5.20
Oxygen rich dry	ORD	1	120	—	2.90
Oxygen rich wet	ORW	1	120	100	5.20

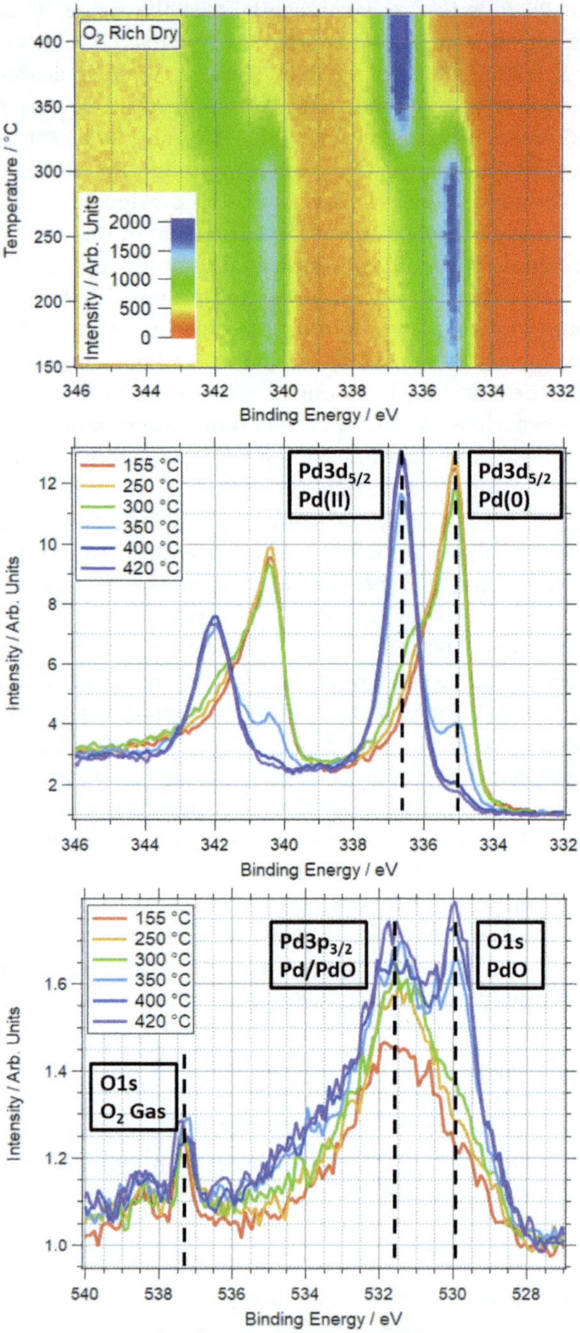

Fig. 1 Top: TP-XPS plot covering the Pd 3d region for Pd foil in ORD conditions. Recorded with 750 eV photon energy. Energy scale corrected based on Fermi edge measured post-ramp. Middle: Pd 3d region XPS spectra taken during TP-AP-XPS ramp in ORD conditions. Each spectrum averaged over 10 °C, centred on indicated temperature. Vertical dashed lines show approximate peak positions of Pd(0) and Pd(II) components of Pd $3d_{5/2}$ peak. Unlabelled peaks on the left side of the spectra are the corresponding Pd $3d_{3/2}$ peaks. These spectra are normalised, hence scaling difference from top plot. Bottom: O 1s region

corresponds to one complete set of spectra every 5 °C. To compensate for the thermal expansion upon heating, the sample manipulator was moved away from the analyser in small steps at the start of each scan. Overall, this gave an approximately consistent background intensity before the Pd $3d_{5/2}$ peak throughout the temperature ramp, to allow for more reliable data analysis. The calibration for the manipulator motion was determined based on the sample motor x position during measurements at various temperatures during the initial cleaning stages for each sample (where the x direction motor moves the sample directly towards/away from the cone of the analyser). For the Pd foil, sample alignment at 150 °C and 500 °C showed a change in x position of −0.12 mm. Assuming a constant expansion rate of the sample over the time of the experiment, this would give a required movement rate of −0.000343 mm $°C^{-1}$. With one set of spectra being recorded every 3.5 °C over the ramp, a step size of −0.0012 mm was used. The first two scans of a temperature-programmed experiment were collected at a fixed temperature (150 °C) to ensure everything was operating correctly before starting the temperature ramp. As a result, the starting x position was increased by double the step size, to ensure the position was correct at the start of the temperature ramp. The final temperature of the ramps varied from 420–590 °C, depending on the conditions. This variation is the result of Pd not always being fully oxidised by the initial maximum temperature of 420 °C. The spectra recorded during the TP-XPS experiments are stacked and presented in an image format, where signal intensity is plotted against both binding energy (x-axis) and temperature (y-axis), as shown in Fig. 1. This visualization method allows the temperatures at which transitions between different chemical states occur to be easily identified. Line profiles performed on this data in both axes provide further information. Line profiles collected over a range of around 10 °C provide relatively low-noise spectra from various stages during the ramp. Vertical line profiles allow changes in intensity at specific peak positions to be identified, showing clearly the temperature range over which states are stable or changing. The set of core level XPS spectra (Pd 3d, Pt 4f, C 1s, O 1s, EF) were measured again in the gas mixture to give the "post-reaction" reference, at the final temperature reached. The reaction gases were then pumped out, and the "regeneration" process was to heat in vacuum to 590 °C (where it was not already at this temperature) to reduce the Pd, before cooling back to 150 °C in vacuum for the following experiment.

All Pd monometallic ramps were performed on a single Pd foil. This was due to the observed reducibility of the foil under high temperature vacuum conditions, meaning that a consistent starting point could be regenerated. All Pd–Pt bimetallic ramps were performed on one sample per Pd : Pt ratio, except for repeat experiments, where noted in the text. For each sample, all TP-XPS experiments were performed on a single sample spot in an attempt to ensure that the starting surface structure was the same for all gas conditions. Crystallographic measurements were not performed, and so the exact phase(s) of Pd and/or Pt measured in each experiment are unknown. This excludes the repeated ramps on the Pd88–

XPS spectra taken during TP-AP-XPS ramp in ORD conditions. Each spectrum averaged over 10 °C at labelled temperature. Vertical dashed lines show approximate peak positions of PdO and O_2 gas components of O 1s peak, and Pd/PdO component of Pd $3p_{3/2}$ peak.

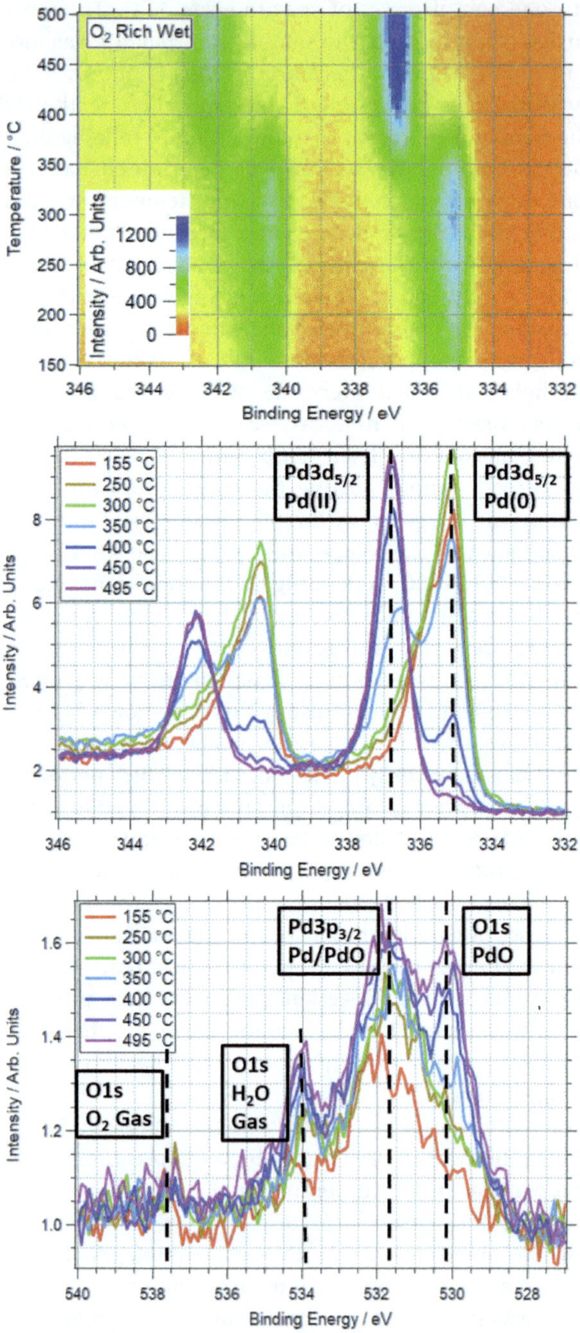

Fig. 2 Top: TP-XPS plot covering the Pd 3d region for Pd foil in ORW conditions. Recorded with 750 eV photon energy. Energy scale corrected based on Fermi edge measured post-ramp. Middle: Pd 3d region XPS spectra taken during TP-AP-XPS ramp in ORW conditions. Each spectrum averaged over 10 °C, centred on indicated temperature. Vertical dashed lines show approximate peak positions of Pd(0) and Pd(II) components of Pd $3d_{5/2}$ peak. Unlabelled peaks on the left side of the spectra are the corresponding Pd $3d_{3/2}$ peaks. These spectra are normalised, hence scaling difference from top plot.

Pt12 foil in stoichiometric dry conditions, which were performed on different pieces of foil.

XPS cross-sections used throughout this experiment are taken or interpreted from the Elettra WebCrossSections website.[37] All presented cross sections are in the units of Mbarn. Ratios are determined based on the normalised peak areas for Pd 3d and Pt 4f, along with the photoionization cross-sections for each line at the used energies. Where cross-section data were not available (*i.e.* for photon energies above 1500 eV) it was determined based on curve-fitting the available data and extrapolating to the relevant higher energies.

3 Results & discussion

3.1 Surface oxidation of Pd foils in reaction conditions

An example of a two dimensional TP-XPS plot is presented in Fig. 1, showing the Pd 3d XPS spectra across a temperature ramp in ORD conditions. The figure also shows individual Pd 3d and O 1s–Pd 3p XPS spectra (averaged over 10 °C) for various key temperatures to demonstrate in more detail the changes in these regions. In Fig. 2 we present the equivalent data set for the oxygen rich wet conditions. This shows the additional presence of the water gas phase peak in the O 1s spectrum (binding energy 534 eV), whilst the Pd states remain essentially unchanged. Changes in relative peak intensity for Pd 3d and O 1s are shown in Fig. S1 in the ESI.† For all other temperature ramps involving monometallic Pd, these data are compiled into a single plot in Fig. 3. This shows the relative intensities at the Pd(0) and Pd(II) Pd $3d_{5/2}$ peak positions (solid and dashed lines at approx. 335.2 and 336.7 eV respectively) for all gas conditions. In all cases, a transition between Pd(0) and Pd(II) is observed during the temperature ramp, as indicated by the relative changes in the peak intensity of the two species. The temperatures of this change increase with the order: oxygen rich dry (ORD) < oxygen only (OO) < stoichiometric dry (SD) < oxygen rich wet (ORW) < stoichiometric wet low (SLW) < stoichiometric very wet (SVW), with cross-over temperatures of approximately 320, 332, 346, 360, 376 and 429 °C respectively.

In stoichiometric reaction conditions (SD, SLW, SVW, methane : oxygen 1 : 2, water varies), an increased water presence is shown to inhibit palladium oxidation. Without water (SD) the transition occurs around 346 °C, whilst with equal amounts of water and oxygen (SLW) this increases to around 376 °C, and further to 429 °C in excess water (SVW, methane : water 1 : 100).

When measured in OO, the Pd(0) to Pd(II) transition is seen to occur around 330 °C. In ORD conditions (with a similar oxygen partial pressure) this is reduced to around 315 °C. The only difference between these gas feeds is a small presence of methane (methane : oxygen 1 : 120). Given the ordinarily reducing nature of methane, the reduction in the oxidation temperature is not expected, as it means that Pd is more likely to oxidise in a more reducing gas mixture. When water is added to this feed (ORW), the transition temperature shifts to around 360 °C, an increase of 45 °C.

Bottom: O 1s region XPS spectra taken during TP-AP-XPS ramp in ORW conditions. Each spectrum averaged over 10 °C at labelled temperature. Vertical dashed lines show approximate peak positions of PdO and O_2/H_2O gas components of O 1s peak, and Pd/PdO component of Pd $3p_{3/2}$ peak.

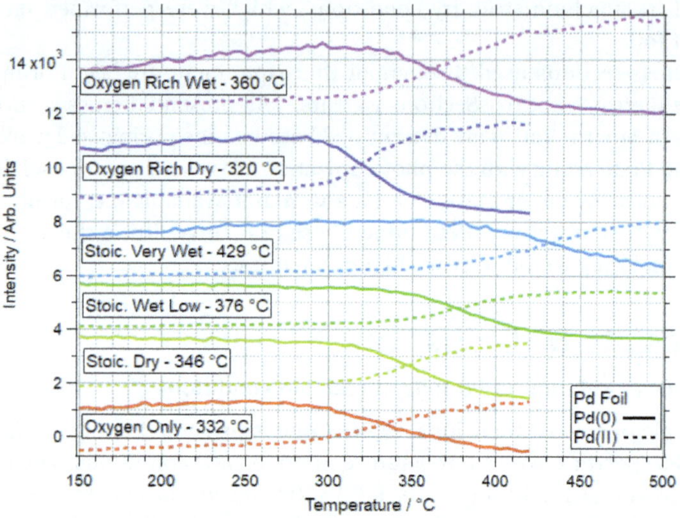

Fig. 3 Intensity plots for Pd(0) and Pd(II) peak positions from Pd 3d XPS measurements in various reaction conditions. Annotated temperatures indicate the point at which the dominant Pd phase changes from reduced to oxidised. Intensity is totalled over a 0.5 eV range, with approximate peak centres of 335.4 and 336.9 eV for Pd(0) and Pd(II) respectively. All recorded with 750 eV photon energy. Within each condition, Pd(0) and Pd(II) intensities are offset by a consistent amount for clarity of presentation.

For equivalent amounts of water and methane, the inhibiting effect is much greater where there is a lower oxygen presence (+80 °C for stoichiometric, +45 °C for oxygen rich). Given that oxygen and water will be competing for surface sites, it is logical that the wet feed with less oxygen results in oxidation only occurring at higher temperatures.

Prior studies have consistently shown water presence to have a detrimental effect on methane oxidation activity over palladium catalysts.[2,38,39] Here we have demonstrated that the presence of water consistently causes Pd to remain reduced at higher temperatures in reaction conditions. Similarly, our data show that higher oxygen partial pressures decrease the Pd oxidation temperature. This explains why these catalysts tend to perform better in more oxygen-rich conditions.

A platinum foil was also studied under SD conditions, matching those used for palladium. For Pt, no change was seen up to the maximum measured temperature of 550 °C. Based on peak positions, Pt remains metallic throughout the ramp, indicating Pt(0) is stable under the dry reaction conditions within the temperature window relevant to methane oxidation. An example is shown in Fig. 4, where the normalised spectra show no change in peak position or intensity between 155–530 °C. This shows the stability of the reduced Pt under a typical reaction condition, as compared to Pd, where the majority of the Pd is oxidised in these conditions below 420 °C.

3.2 Surface oxidation and restructuring of Pd–Pt foils in reaction conditions

Fig. 5 and 6 show the 2-dimensional TP-XPS plots for Pd 3d and Pt 4f, along with line profiles on the Pd(0)/Pd(II) and Pt(0)/Pt(II) peak positions and selected spectra

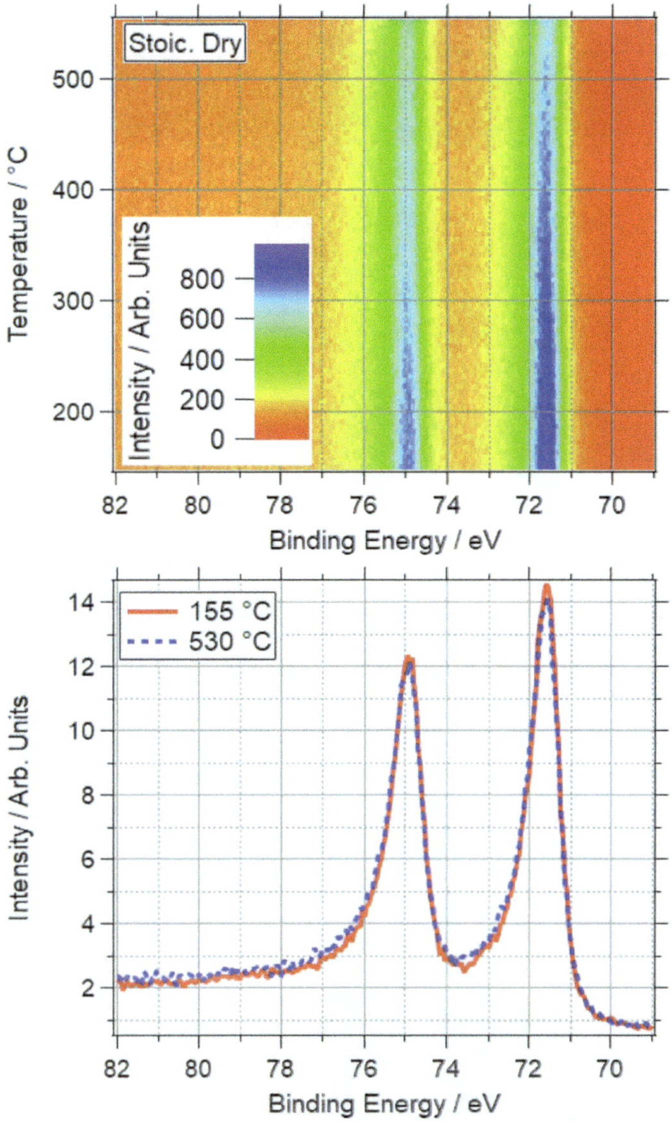

Fig. 4 Top: TP-AP-XPS plot in Pt 4f region during a ramp in stoichiometric dry conditions over platinum foil. Bottom: combined, normalised Pt 4f XPS spectra taken over 10 °C regions at various temperatures.

taken at points throughout this ramp. This particular example shows the spectra for OO conditions for a Pd88–Pt12 foil. All other ramps are summarised in Fig. 7 (Pd) and Fig. 8 (Pt) for Pd88–Pt12 and Fig. 9 (Pd) and Fig. 10 (Pt) for Pd65–Pt35.

For Pd88–Pt12, only 2 conditions led to a majority of Pd(II) on the sample. These conditions were OO and SD. In both cases, the Pt(0) intensity decreases, with the Pt(II) intensity remaining the same. This shows Pt starting to migrate to the bulk, while the remaining surface Pt is partially oxidised. This platinum behaviour and the palladium oxidation begin from around the same temperature,

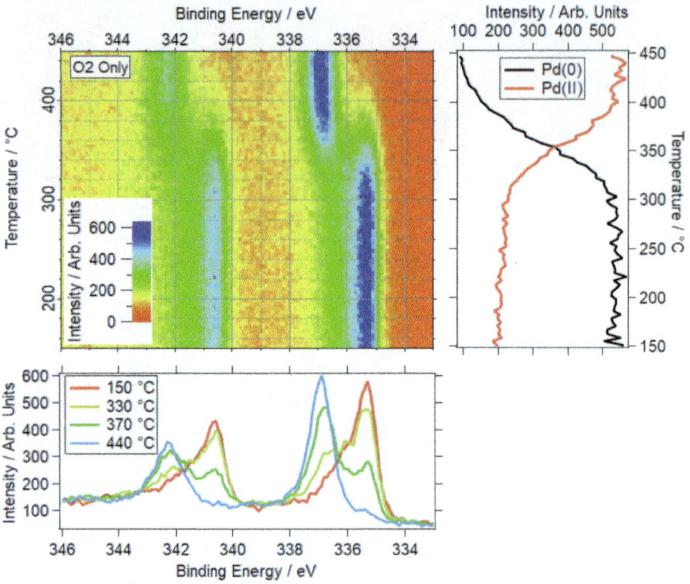

Fig. 5 Various Pd 3d data for Pd88–Pt12 foil in OO. Top left: TP-XPS plot. Top right: Intensity line profiles for Pd(0) and Pd(II) peak positions. Bottom: Pd 3d spectra at select temperatures throughout the ramp, combined over a 10 °C range.

indicating a link between the two. This suggests that the alloy is resistant to Pd oxidation whilst Pt remains well-mixed in the surface region. In both cases, the switch from Pd(0) to Pd(II) occurs at a higher temperature than for Pd foil. In all

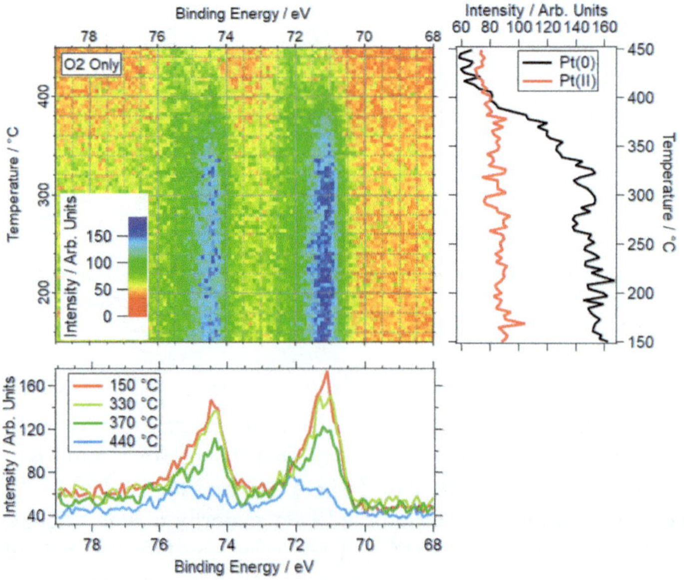

Fig. 6 Various Pt 4f data for Pd88–Pt12 foil in OO. Top left: TP-XPS plot. Top right: intensity line profiles for Pt(0) and Pt(II). Bottom: Pt 4f spectra at select temperatures, combined over a 10 °C range.

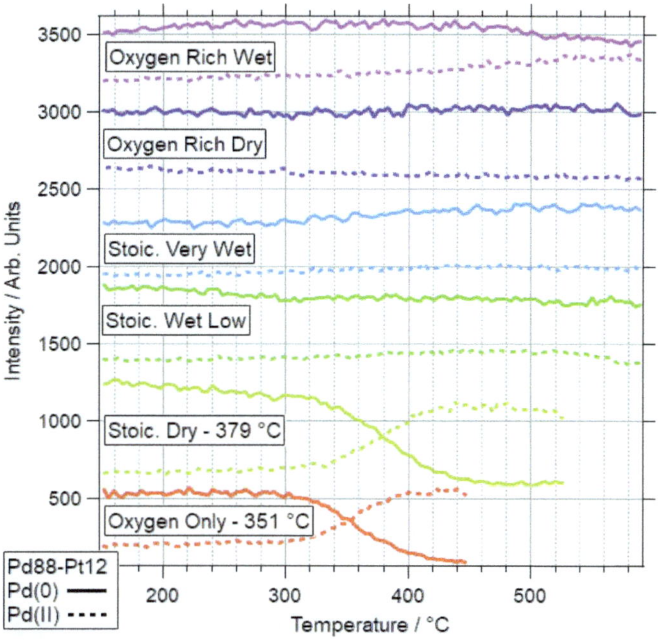

Fig. 7 Pd 3d data for Pd88–Pt12 under various conditions. Data represent line profiles at Pd(0) and Pd(II) peak positions throughout temperature-programmed ramp. Where oxidation is seen to more than 50%, the switchover temperature is quoted.

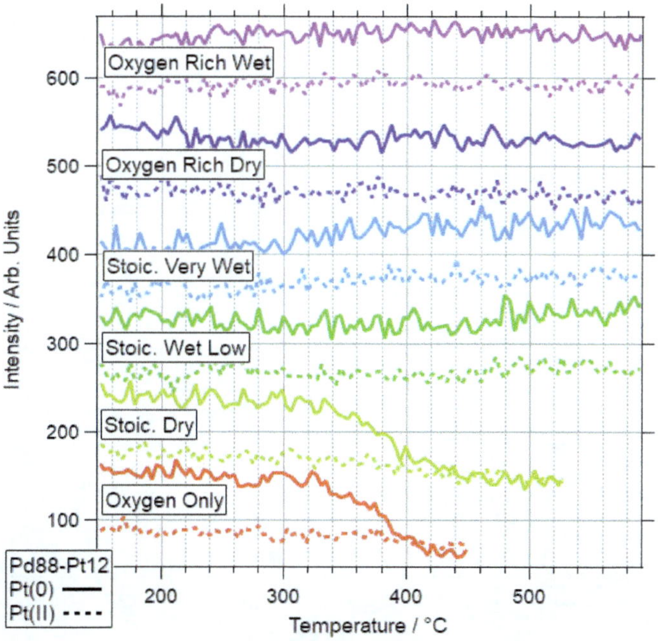

Fig. 8 Pt 4f data for Pd88–Pt12 under various conditions. Data represent line profiles at Pt(0) and Pt(II) peak positions throughout temperature-programmed ramp.

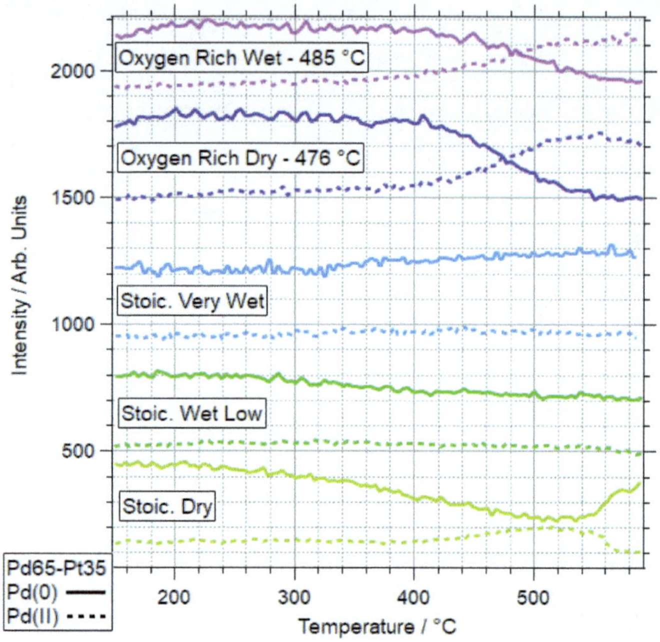

Fig. 9 Pd 3d data for Pd65–Pt35 under various conditions. Data represent line profiles at Pd(0) and Pd(II) peak positions throughout temperature-programmed ramp. Where oxidation is seen to more than 50%, the switchover temperature is quoted.

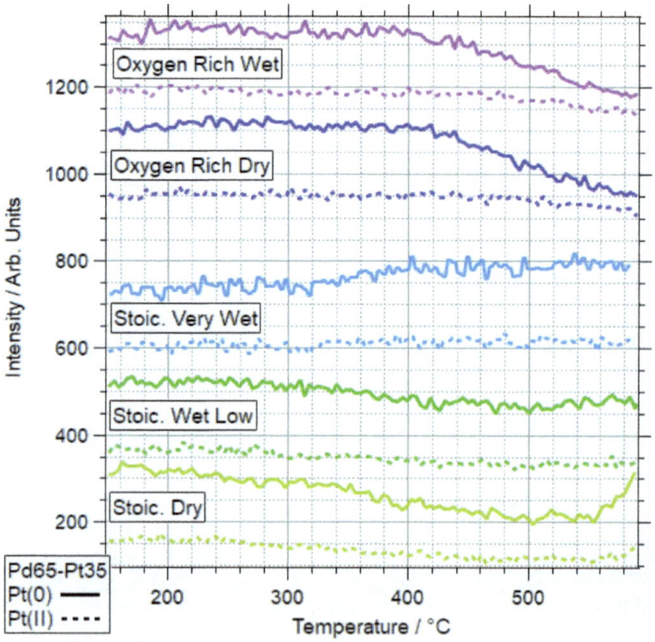

Fig. 10 Pt 4f data for Pd65–Pt35 under various conditions. Data represent line profiles at Pt(0) and Pt(II) peak positions throughout temperature-programmed ramp.

four other conditions, the dominant Pd species by 590 °C was still Pd(0). Only in the case of ORW conditions was significant oxidation seen, although the Pd remained mostly reduced. No change was seen in the Pt signal in this condition, indicating that there is likely to be a point of oxidation at which the migration of platinum occurs. With the experiments presented, the exact point of Pd oxidation at which Pt migration occurs is unclear. Significant changes in Pt(0) are generally seen at slightly higher temperatures than the start of the decrease of Pd(0), indicating the oxidation drives the migration as opposed to the inverse.

The more Pt-rich foil, Pd65–Pt35, shows some similarly varied oxidation behaviours in different conditions. There are still only two conditions which cause Pd to near-fully oxidise, ORD and ORW. These are different to the conditions which lead to oxidation for the more Pd-rich sample, Pd88–Pt12, however, for which neither oxygen rich condition caused full oxidation/restructuring. In both conditions for Pd65–Pt35, the same pattern of behaviour is seen as for the conditions in which Pd88–Pt12 changes, whereby the palladium oxidation is coupled to a reduction in intensity and partial oxidation of platinum. The temperatures at which this change occurs are much higher however, 351 and 379 °C for Pd88–Pt12, compared to 476 and 485 °C for Pd65–Pt35. In the SD gas mixture, the Pd starts to partially oxidise, up to around 550 °C, at which point it promptly reduces back to fully Pd(0) again. This is mirrored in the Pt signal, where the reduced Pt peak decreases in intensity, before sharply increasing above 550 °C. This is another example of the strongly linked behaviours of Pd and Pt in these alloy samples. It is also in direct contrast to the behaviour of Pd88–Pt12 in ORW conditions, where partial Pd oxidation was not coupled to any changes in Pt. Both SLW and SVW conditions saw no change in Pd oxidation or relative intensity over the ramp.

When comparing the Pd65–Pt35 to Pd88–Pt12, there are a number of unusual variations. The only real similarity is the coupled behaviour of Pd (oxidation) and Pt (partial oxidation plus migration). However, the conditions in which they occur, the temperatures at which they are promoted, and the behaviour seen where only partial oxidation of palladium happens all vary with composition. Pd88–Pt12 is more likely to undergo the changes at lower temperature, and does not change significantly in either oxygen rich gas feed (ORD/ORW). Pd65–Pt35 seems to be more oxidation resistant, only oxidising in oxygen rich feeds at very high temperatures, outside of the relevant window for methane oxidation (both 475 °C and above). The greater amount of Pt can be considered as a stabilizer for metallic Pd. Given the greater presence of Pt around Pd in the Pd65–Pt35, it is understandable why this may lead to some of the relative differences seen. It is unclear, based on the information above, why the Pd88–Pt12 is resistant to oxidation in oxygen rich conditions, given that rationally these should be more likely to induce oxidation than the stoichiometric dry condition in which it was seen to oxidise.

With some unanticipated variation in Pd oxidation behaviour in the bimetallic foils, select ramps were repeated for the Pd88–Pt12 sample. First (after a cleaning cycle) was the SD experiment, in which no Pd oxidation was seen up to the maximum temperature of 590 °C – when previously this oxidation had been seen below 400 °C. After this, the sample was exposed to an OO gas feed and another ramp performed. In this case partial oxidation occurred, but again the onset of this was at a higher temperature, with the Pd 3d spectra still showing significant

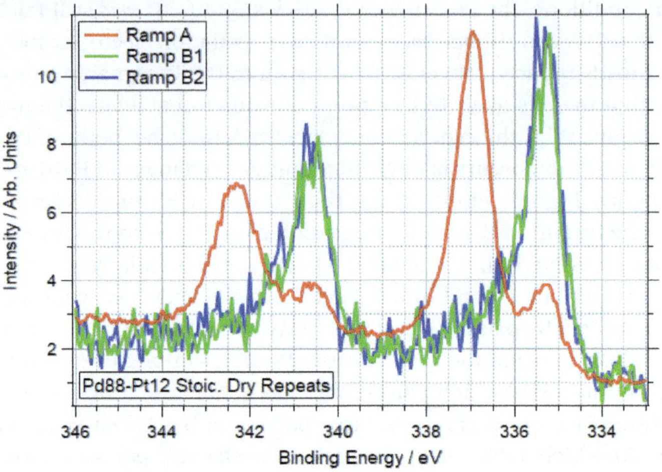

Fig. 11 Pd 3d spectra from 3 separate ramps performed on a Pd88–Pt12 sample in SD gas conditions, at the maximum temperatures used for each (520 °C for A, 590 °C for B1 & B2). Test A is the original, as shown previously. Tests B1 and B2 are performed on a new sample. B1 is post-cleaning only, whilst B2 followed B1, an additional ramp in OO, and additional regeneration stages.

presence of Pd(0) at 590 °C. Following the ramp in oxygen where partial Pd oxidation was seen, and another cleaning cycle, the SD ramp was repeated again. No change was seen from the previous ramp, with palladium remaining reduced throughout. The Pd 3d spectra taken at the end of the ramps in SD conditions are shown in Fig. 11. This variation in behaviour for ramps under identical conditions pointed to a problem with relying on heating in vacuum to regenerate the bimetallic samples to a reliable starting point.

The Pd : Pt ratios were determined after each "regeneration" stage at 150 °C in vacuum and at the highest temperature of the ramp for each gas condition. The

Table 2 Pd : Pt ratio based on XPS data for Pd88–Pt12 and Pd65–Pt35 foils under a range of reaction conditions, along with the % of theoretical value this equates to, and the effective alloy composition at the point of that measurement. Theoretical, based on sample composition, and average, based on the set of tested conditions, included for reference. Measurements made at end temperature of ramp in gas (varies, 450–590 °C depending on gas mixture)

	Pd88–Pt12			Pd65–Pt35		
Gas mixture	Ratio	% of Th.	Eff. ratio	Ratio	% of Th.	Eff. ratio
Theoretical	7.76	100%	88–12	1.97	100%	65–35
Oxygen only	18.81	242%	95–5	N/A	N/A	N/A
Stoichiometric dry	29.50	380%	97–3	3.88	198%	80–20
Stoichiometric wet low	11.50	148%	92–8	3.15	160%	76–24
Stoichiometric very wet	7.47	96%	88–12	3.00	153%	75–25
Oxygen rich dry	7.99	103%	89–11	5.29	269%	84–16
Oxygen rich wet	6.83	88%	87–13	5.13	261%	84–16
Average	13.68	176%	93–7	4.09	208%	80–20

ratios were calculated from the Pd 3d and Pt 4f peak areas weighted with the respective photo-excitation cross-sections at 750 eV photon energy (1.41 and 1.49 Mbarn, respectively). All ratios are summarised in Table 2, which also includes the percentage of the theoretical value that this relates to, and the effective resultant Pd : Pt ratio. These data are presented visually in Fig. 15. Each cleaning stage involved heating to 590 °C in vacuum (where the sample was below this temperature), maintaining for 2 minutes, before cooling back to 150 °C in vacuum.

The analysis of the Pd–Pt foil oxidation data reveals significant alloy restructuring. For the initial experiments on the Pd88–Pt12 foil, the expected Pd : Pt ratio using 750 eV photon energy is approx. 7.75 : 1. Before any reaction ramps had taken place, this ratio for the clean sample was measured as 10.85 : 1, confirming the composition in the near-surface region was more Pd-rich than expected. High resolution spectra taken in OO conditions showed a substantial increase to 18.81 : 1, with this remaining broadly the same at 18.77 : 1 after sample regeneration. Measurements in SD methane oxidation conditions showed the highest Pd : Pt ratio, at 29.50 : 1, which reduced to 22.71 : 1 after regeneration. This represents a significant increase in surface Pd compared to the previous regeneration step. The same sample was then exposed to SLW conditions, where the ratio reduced back to 11.50 : 1, much closer to the clean starting point for the foil. Regeneration resulted in a minor increase in Pd : Pt – unlike both previous phases where it had resulted in a decrease – to 12.64 : 1. Exposure to the SVW condition further decreased the Pd : Pt ratio to 7.47 : 1, the lowest observed ratio for this sample in any conditions. This is within 5% of the expected value for this sample. Given the low oxygen content and high water presence, this result is not surprising. The following regeneration step resulted in an increase up to 10.75 : 1. The ORD experiment followed, with another reduction in Pd : Pt, to 7.99 : 1, which again increased after regeneration, to 10.53 : 1. The final ramp performed on this sample was in ORW conditions, where the Pd : Pt ratio decreased to 6.83 : 1.

The analysis of all data shows a general trend that wet conditions result in a higher relative Pt concentration in the near-surface region than dry conditions, with Pt concentrations increasing with the relative water content of the gas feed. With regards to oxygen levels, a greater Pd : Pt ratio was generally observed in stoichiometric conditions rather than the oxygen rich conditions. Whilst this could imply that the oxygen was effectively pulling Pt from the bulk, the Pt peaks show it to be primarily reduced in these conditions. Regenerating stages after each ramp produced different variations in relative metallic ratios. Initially there was variation, with some increases and decreases each seen. From the SLW experiment onwards, the cleaning stages tended to revert the Pd : Pt to much closer to the expected value.

In only 2 experiments did the majority of the palladium end up oxidised by the end of the ramp, with these being the oxygen only and stoichiometric dry experiments. In both cases, the total Pt signal is seen to decrease, with a significant presence of oxidised Pt by the end of the ramps. The oxygen rich wet ramp also showed partial Pd oxidation, though the majority of it remained reduced. In this experiment, no oxidation of Pt was observed. In most of the cases where no palladium oxidation occurs, no significant change in Pd : Pt is observed during the ramps. At the extreme ratios measured, the variation is effectively between

Pd87–Pt13 and Pd97–Pt3, representing a 10% change in percentage points for both Pd and Pt from the original expected composition.

The same set of experiments were performed on the Pd65–Pt35 foil, with the exception that the oxygen only ramp was omitted. The expected Pd : Pt ratio for this sample, adjusted for cross sections, is 1.97 : 1. Post the initial cleaning, this ratio was 3.41 : 1, much higher than the expected value, similarly to Pd88–Pt12. The initial ramp in ORW conditions caused an increase in Pd : Pt to 5.13 : 1, which is the opposite to the effect caused for the more Pd-rich alloy. Regeneration increased this further to 5.83 : 1. The SD experiment followed, with a decrease to 3.88 : 1, which increased to 4.22 : 1 after regeneration. The ORD ramp increased this to 5.29 : 1, and the following regeneration cycle caused a slightly relative increase in Pt, with a ratio of 4.99 : 1. The SLW experiment saw a significant decrease in ratio to 3.15 : 1, which increased again to 3.43 : 1 with regeneration. The final ramp was in SVW conditions, where the Pd : Pt decreased to 3.00 : 1, the lowest ratio of Pd : Pt for this sample after the initial cleaning stage. This was the lowest value achieved at any stage, but still more than 50% higher than the expected ratio, confirming that after any treatments this alloy was more Pd-rich in the near-surface region. In stoichiometric conditions, the increase of water led to a relative increase in the Pt present in this near-surface area, consistent with the Pd88–Pt12 foil.

As a result of the variations in identical conditions, additional experiments were performed on various Pd–Pt foils to determine the consistency and lasting effects of exposure to reaction conditions for samples of these compositions. This included depth-profiling XPS analysis across various spots on fresh, annealed and used foils, using kinetic energies of 415, 675 and 1500 eV. This variation in energies provides more surface and bulk sensitive measurements, to provide insight into the restructuring of alloys under reaction conditions.

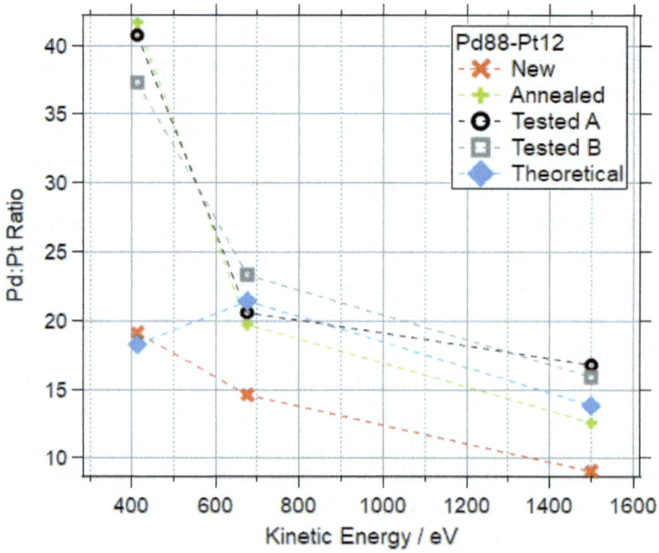

Fig. 12 Pd : Pt ratios for Pd88–Pt12 samples at a range of kinetic energies after various treatments.

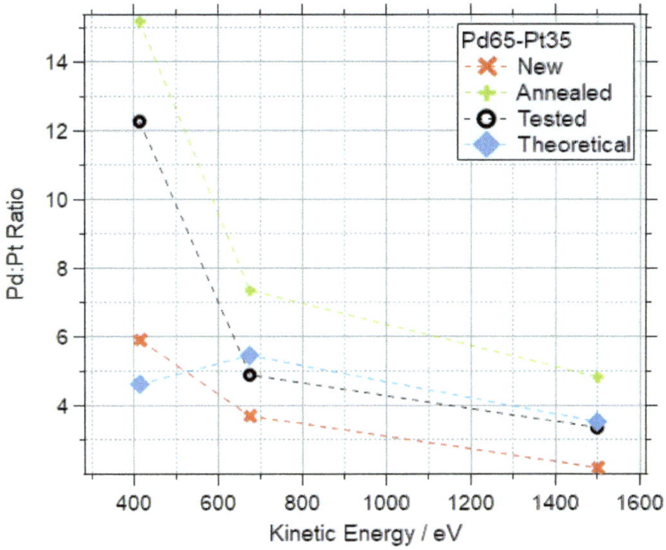

Fig. 13 Pd : Pt ratios for Pd65–Pt35 samples at a range of kinetic energies after various treatments.

Fig. 12 and 13 show the Pd : Pt ratios for Pd88–Pt12 and Pd65–Pt35 foils respectively. For each, data for a "new" (as received, untreated), "cleaned" (exposed to oxygen up to 600 °C) and "tested" (exposed to multiple reaction conditions, up to 600 °C) are shown, along with the theoretical value based on the photoionization cross-sections.

For both compositions, the new foil is much more Pt rich at all measured photon energies, indicating that Pd is disproportionately buried in the bulk of the sample before treatments. For cleaned and tested samples, the more surface-sensitive measurements (415 and 675 eV kinetic energy) show a Pd-enrichment, to varying degrees. For both compositions, annealing and testing under various reaction mixtures caused an increase in Pd : Pt of around 100% in the most surface sensitive measurements (415 eV KE). The bulk changes are less extreme, with all Pd88–Pt12 samples retaining a Pd : Pt ratio between 12 and 18 when measured at 1500 eV kinetic energy. The bulk of the tested Pd65–Pt35 samples remains similar to the theoretical values, whilst annealing causes a Pd enrichment in the bulk.

To study the ratio change at different depths under heating, a Pd65–Pt35 sample was step-annealed in oxygen. Data are shown in Fig. 14. Throughout the ramp, the sample is more Pd-rich than the theoretical value at all measured depths. Between 150–450 °C the Pd : Pt ratio at 415 eV (KE) generally increases. The 675 eV scans show a smaller change, with a slight increase in relative Pt content. The most bulk sensitive spectra showed similarly small, but inconsistent changes. At 550 °C there is a clear increase in Pd : Pt at all measured depths. The low kinetic energy measurement was repeated for this temperature, approximately 1 hour after the initial measurements, and showed a significant increase in Pd : Pt ratio, from 20 : 1 to 71 : 1. The Pd 3d and Pt 4f spectra for the lowest 415 eV kinetic energy points are also shown here. This shows the gradual

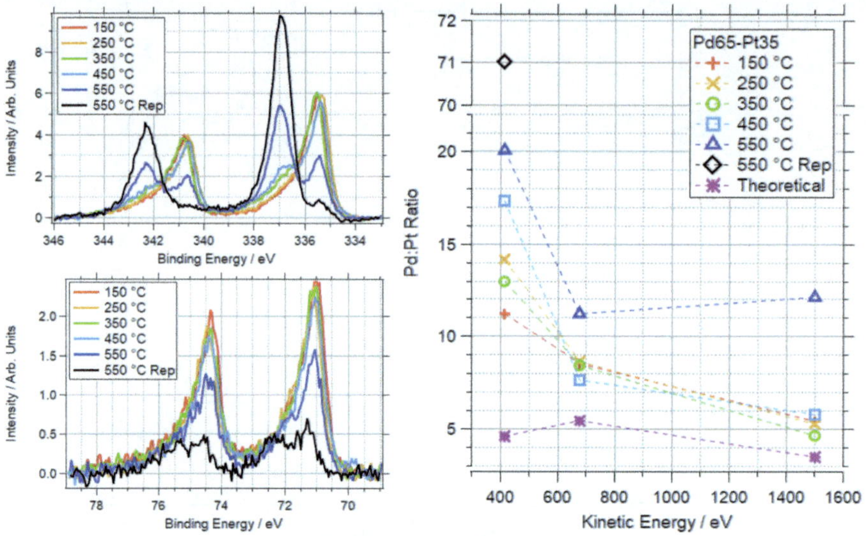

Fig. 14 Plots for Pd65–Pt35 foil in 2 mbar ramp in oxygen. Top left: Pd 3d region XPS data, all recorded at 750 eV photon energy, approx. 415 eV kinetic energy. Bottom left: Pt 4f region XPS data, all recorded at 490 eV photon energy, approx. 415 eV kinetic energy. Right: Pd : Pt ratios, as determined by NAP-XPS. Repeat measurement is approx. 1 hour after initial measurement.

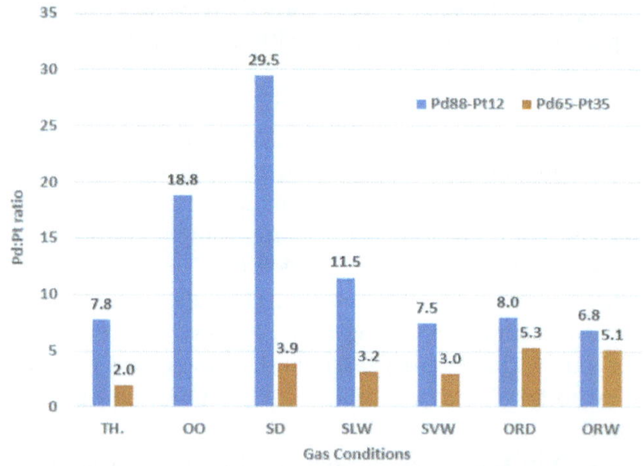

Fig. 15 Pd : Pt ratios based on XPS data for Pd88–Pt12 and Pd65–Pt35 foils under a range of reaction conditions. Theoretical values (TH.) shown for reference.

oxidation of palladium after 350 °C, coupled with the oxidation and decrease in intensity for Pt over the same range.

4 Conclusions

In this work we have demonstrated the effects of water and oxygen on the oxidation of palladium in Pd model catalysts under complete methane oxidation conditions. Palladium consistently underwent full oxidation from Pd(0) to Pd(II) under all tested conditions, with changes happening in the temperature range of 320–430 °C. Increased oxygen caused Pd oxidation at lower temperatures, whilst increased water inhibited the oxidation.

Two bimetallic foils, Pd88–Pt12 and Pd65–Pt35, were also studied to understand previously seen benefits of bimetallic catalysts of these compositions. For these two samples, Pt inhibits Pd oxidation to varying extents. Pd88–Pt12 underwent oxidation of Pd only in pure oxygen, and dry stoichiometric reaction conditions, remaining reduced in all other tested conditions. Contrastingly Pd65–Pt35 only showed Pd oxidation in oxygen rich (dry and wet) conditions.

Under conditions where palladium does oxidise in Pd–Pt foils, platinum partially oxidises and migrates into the bulk of the foil. Depth profiling of bimetallic foils showed that the surface tends to be more Pd enriched, whilst the bulk measurements showed an increased Pt concentration. Annealing and testing in reaction conditions repeatedly caused the surface region to become much more Pd-rich than the average sample composition. Bimetallic foils which had been exposed sequentially to multiple reaction conditions showed relatively small differences in Pd : Pt ratio to those which had only been annealed, as compared to the fresh samples. Combining these results with prior studies which showed Pd–Pt catalysts were more stable long term than Pd equivalents, we suggest that the key to long term bimetallic catalyst stability is the retention of reduced Pd at the catalyst surface. Based on the simultaneous Pd oxidation and Pt migration observed throughout, we suggest that keeping Pd reduced is a result of Pd and Pt being well mixed in the surface region of the catalyst.

Author contributions

A. I. L was responsible for investigation, data analysis and writing. R. A. B. was responsible for investigation. T. E.-E was responsible for conceptualization. G. H. was responsible for conceptualization, investigation and writing.

Conflicts of interest

There are no conflicts to declare.

Acknowledgements

A. I. L. acknowledges support from a studentship from Johnson Matthey and University of Reading. This work was carried out with the support of Diamond Light Source, instrument B07-C (proposals SI27121 and SI24584).

Notes and references

1 K. I. Muto, N. Katada and M. Niwa, *Appl. Catal., A*, 1996, **134**, 203–215.
2 R. Burch, F. J. Urbano and P. K. Loader, *Appl. Catal., A*, 1995, **123**, 173–184.
3 P. Gélin and M. Primet, *Appl. Catal., B*, 2002, **39**, 1–37.
4 M. Monai, T. Montini, R. J. Gorte and P. Fornasiero, *Eur. J. Inorg. Chem.*, 2018, **2018**, 2884–2893.
5 J. G. Firth and H. B. Holland, *Trans. Faraday Soc.*, 1969, **65**, 1121–1127.

6 J. G. McCarty, *Catal. Today*, 1995, **26**, 283–293.
7 Y. Pan, C. Liu and P. Shi, *Appl. Surf. Sci.*, 2008, **254**, 5587–5593.
8 Y. H. C. Chin, M. García-Diéguez and E. Iglesia, *J. Phys. Chem. C*, 2016, **120**, 1446–1460.
9 T. F. Stocker, D. Qin, G.-K. Plattner, M. M. Tignor, S. K. Allen, J. Boschung, A. Nauels, Y. Xia, V. Bex and P. M. Midgley, *IPCC, 2013: Climate Change 2013: the Physical Science Basis. Contribution of Working Group I to the Fifth Assessment Report of the Intergovernmental Panel on Climate Change*, Cambridge University Press, Cambridge, United Kingdom and New York, NY, USA, 2013, pp. 1–1535.
10 U. D. of State, *United States, European Union, and Partners Formally Launch Global Methane Pledge to Keep 1.5C Within Reach*, 2021, https://www.state.gov/united-states-european-union-and-partners-formally-launch-global-methane-pledge-to-keep-1-5c-within-reach/.
11 R. Price, T. Eralp-Erden, E. Crumlin, S. Rani, S. Garcia, R. Smith, L. Deacon, C. Euaruksakul and G. Held, *Top. Catal.*, 2016, **59**, 516–525.
12 A. Large, J. Seymour, W. Quevedo Garzon, K. Roy, F. Venturini, D. C. Grinter, L. Artiglia, E. Brooke, M. B. De Gutierrez, A. Raj, K. R. Lovelock, R. A. Bennett, T. Eralp-Erden and G. Held, *J. Phys. D: Appl. Phys.*, 2021, **54**, 174006.
13 C. F. Cullis, T. G. Nevell and D. L. Trimm, *J. Chem. Soc., Faraday Trans. 1*, 1972, **68**, 1406–1412.
14 R. Burch and F. J. Urbano, *Appl. Catal., A*, 1995, **124**, 121–138.
15 J. C. Van Giezen, F. R. Van Den Berg, J. L. Kleinen, A. J. Van Dillen and J. W. Geus, *Catal. Today*, 1999, **47**, 287–293.
16 F. Zhang, C. Hakanoglu, J. A. Hinojosa and J. F. Weaver, *Surf. Sci.*, 2013, **617**, 249–255.
17 H. Zhang, M. Jin, H. Liu, J. Wang, M. J. Kim, D. Yang, Z. Xie, J. Liu and Y. Xia, *ACS Nano*, 2011, **5**, 8212–8222.
18 Y. Han, J. Kim, S. U. Lee, S. I. Choi and J. W. Hong, *Chem. – Asian J.*, 2020, **15**, 1324–1329.
19 A. P. Wong, E. A. Kyriakidou, T. J. Toops and J. R. Regalbuto, *Catal. Today*, 2016, **267**, 145–156.
20 K. Narui, H. Yata, K. Furuta, A. Nishida, Y. Kohtoku and T. Matsuzaki, *Appl. Catal., A*, 1999, **179**, 165–173.
21 K. Persson, A. Ersson, S. Colussi, A. Trovarelli and S. G. Järås, *Appl. Catal., B*, 2006, **66**, 175–185.
22 A. I. Osman, J. K. Abu-Dahrieh, F. Laffir, T. Curtin, J. M. Thompson and D. W. Rooney, *Appl. Catal., B*, 2016, **187**, 408–418.
23 G. Lapisardi, L. Urfels, P. Gélin, M. Primet, A. Kaddouri, E. Garbowski, S. Toppi and E. Tena, *Catal. Today*, 2006, **117**, 564–568.
24 S. A. Yashnik, Y. A. Chesalov, A. V. Ishchenko, V. V. Kaichev and Z. R. Ismagilov, *Appl. Catal., B*, 2017, **204**, 89–106.
25 R. Abbasi, G. Huang, G. M. Istratescu, L. Wu and R. E. Hayes, *Can. J. Chem. Eng.*, 2015, **93**, 1474–1482.
26 P. Xu, Z. Wu, J. Deng, Y. Liu, S. Xie, G. Guo and H. Dai, *Chin. J. Catal.*, 2017, **38**, 92–105.
27 H. Ali-Löytty, M. W. Louie, M. R. Singh, L. Li, H. G. Sanchez Casalongue, H. Ogasawara, E. J. Crumlin, Z. Liu, A. T. Bell, A. Nilsson and D. Friebel, *J. Phys. Chem. C*, 2016, **120**, 2247–2253.

28 W. C. Chueh, A. H. McDaniel, M. E. Grass, Y. Hao, N. Jabeen, Z. Liu, S. M. Haile, K. F. McCarty, H. Bluhm and F. El Gabaly, *Chem. Mater.*, 2012, **24**, 1876–1882.

29 D. A. Svintsitskiy, T. Y. Kardash, O. A. Stonkus, E. M. Slavinskaya, A. I. Stadnichenko, S. V. Koscheev, A. P. Chupakhin and A. I. Boronin, *J. Phys. Chem. C*, 2013, **117**, 14588–14599.

30 M. F. Lichterman, S. Hu, M. H. Richter, E. J. Crumlin, S. Axnanda, M. Favaro, W. Drisdell, Z. Hussain, T. Mayer, B. S. Brunschwig, N. S. Lewis, Z. Liu and H. J. Lewerenz, *Energy Environ. Sci.*, 2015, **8**, 2409–2416.

31 L. Lukashuk, N. Yigit, R. Rameshan, E. Kolar, D. Teschner, M. Hävecker, A. Knop-Gericke, R. Schlögl, K. Föttinger and G. Rupprechter, *ACS Catal.*, 2018, **8**, 8630–8641.

32 M. Aryafar and F. Zaera, *Catal. Lett.*, 1997, **48**, 173–183.

33 P. Velin, M. Ek, M. Skoglundh, A. Schaefer, A. Raj, D. Thompsett, G. Smedler and P. A. Carlsson, *J. Phys. Chem. C*, 2019, **123**, 25724–25737.

34 D. Vogel, C. Spiel, M. Schmid, M. Stöger-Pollach, R. Schlögl, Y. Suchorski and G. Rupprechter, *J. Phys. Chem. C*, 2013, **117**, 12054–12060.

35 G. Held, F. Venturini, D. C. Grinter, P. Ferrer, R. Arrigo, L. Deacon, W. Q. Garzon, K. Roy, A. Large, C. Stephens, A. Watts, P. Larkin, M. Hand, H. Wang, L. Pratt, J. J. Mudd, T. Richardson, S. Patel, M. Hillman and S. Scott, *J. Synchrotron Radiat.*, 2020, **27**, 1153–1166.

36 S. Tanuma, C. J. Powell and D. R. Penn, *Surf. Interface Anal.*, 1991, **17**, 911–926.

37 *Atomic Calculation of Photoionization Cross-Sections and Asymmetry Parameters*, https://vuo.elettra.eu/services/elements/WebElements.html.

38 A. Raj, *Johnson Matthey Technol. Rev.*, 2016, **60**, 228–235.

39 N. Sadokhina, G. Smedler, U. Nylén, M. Olofsson and L. Olsson, *Appl. Catal., B*, 2017, **200**, 351–360.

Faraday Discussions

ROYAL SOCIETY OF CHEMISTRY

PAPER

Selective hydrogenation of graphene on Ir(111): an X-ray standing wave study

Claus F. P. Kastorp,[a] David A. Duncan,[b] Anders L. Jørgensen,[c] Martha Scheffler,[a] John D. Thrower,[a] Tien-Lin Lee,[b] Liv Hornekær[a] and Richard Balog*[a]

Received 14th December 2021, Accepted 11th January 2022
DOI: 10.1039/d1fd00122a

A combined high resolution X-ray photoelectron spectroscopy and X-ray standing wave study into the adsorption structure of hydrogenated graphene on Ir(111) is presented. By exploiting the unique absorption profiles and significant modulations in signal intensity found within the X-ray standing wave results, we refine the fitting of the C 1s X-ray photoelectron spectra, allowing us to disentangle the contributions from hydrogenation of graphene in different high-symmetry regions of the moiré supercell. We clearly demonstrate that hydrogenation in the FCC regions results in the formation of a graphane-like structure, giving a standalone component that is separated from the component assigned to the similar structure in the HCP regions. The contribution from dimer structures in the ATOP regions is found to be minor or negligible. This is in contrast to the previous findings where a dimer structure was assumed to contribute significantly to the sp^3 part of the C 1s spectra. The corrugation of the remaining pristine parts of the H-graphene is shown to increase with the H coverage, reflecting an increasing number and size of pinning centers of the graphene to the Ir(111) substrate with increasing H exposure.

Introduction

The chemical functionalisation of graphene with elements such as fluorine,[1,2,25] oxygen,[8,12,13,33] and hydrogen[3–6,18,19,23,28] has attracted attention because of the resulting changes in material properties. For hydrogen, these effects include ferromagnetism,[23,31] increased spin–orbit coupling,[26] and the opening of an electronic band gap,[5,14,24,29] which additionally can be controlled by the degree and pattern of hydrogenation.[15,19] Other potential applications of hydrogenated graphene include protective coatings,[21] as well as solid-state hydrogen storage.[27,30]

Graphene on Ir(111) exhibits a moiré pattern which can be closely modelled by ten-by-ten graphene unit cells for each nine-by-nine Ir surface atoms.[11] This supercell contains three high-symmetry regions of interest, namely the HCP, FCC,

[a]*Department of Physics and Astronomy, Aarhus University, Aarhus, Denmark. E-mail: balog@phys.au.dk*
[b]*Diamond Light Source Ltd., Harwell Science and Innovation Campus, Didcot, OX11 0DE, UK*
[c]*The Mads Clausen Institute, SDU NanoSYD, Sønderborg, Denmark*

and ATOP regions. In the HCP and FCC regions, every second C atom is directly above an Ir surface atom and every second above a hollow site.[11] The C and Ir atoms interact more strongly due to the geometry, and the graphene sheet is closer to the Ir(111) surface than in the ATOP regions, 3.29 Å above the substrate for pristine graphene in the FCC regions and 3.27 Å for the HCP regions,[9] based on *ab initio* density functional theory (DFT). In the ATOP region, where the Ir atoms are positioned directly below the centre of the carbon rings,[11] a higher average height with a maximum of 3.62 Å is found.

It is known from earlier works that graphene on Ir(111) can be functionalised to varying levels by exposure to hot (T > 1700 K) atomic hydrogen, at varying substrate temperatures.[19] We have previously shown, by a combination of scanning tunnelling microscopy (STM), X-ray standing waves (XSW) and DFT, that for hydrogenation at high substrate temperatures (HT), *ca.* 600 K, hydrogenation only occurs at the HCP regions[18,19] leading to the formation of graphane-like clusters where the C atoms alternately bind to a hydrogen atom above the graphene and to a surface Ir atom at the graphene–Ir(111) interface.[18] Hydrogenation at intermediate temperatures (IT), *ca.* 550 K, results in the functionalisation of both the HCP and FCC regions, as observed by scanning tunnelling microscopy (STM).[19] It is expected that the functionalised FCC regions resemble the structures observed in HCP regions due to similar registry between the graphene sheet and the Ir surface in these two regions.

Hydrogenation at room temperature (RT), *ca.* 300 K, results in a higher level of functionalisation, as observed in STM and X-ray photoelectron spectroscopy (XPS).[4,5,19] Since, in the ATOP region, the C and Ir atoms are out of registry, graphane-like clusters are not expected to form in the same way as in the HCP and FCC regions. As STM and DFT studies show the formation of H-dimers on both graphite[16,17] and graphene on SiC(0001)[6] after exposure to atomic hydrogen, it has been suggested that dimers might also form in the ATOP regions of graphene on Ir(111).[4] The presence of dimer structures in the ATOP regions has, however, never been imaged directly by STM, as this technique itself is not capable of resolving any detailed structures at high level of functionalisation due to the large-scale changes in electronic structure induced by the hydrogenation.[5,19]

The complex evolution of the structure of hydrogenated graphene is clearly seen in the C 1s photoemission spectra, which become more complex with increasing extent of functionalisation. In earlier XPS works, the C 1s spectrum of fully saturated H-graphene exhibits a minimum of four components labelled C_a, C_b, C_c, and C_d in order of decreasing binding energy.[4] Here, the C_c component was attributed to the pristine sp^2 carbon, and even at the saturation level of functionalisation it has a minimum of *ca.* 30% of the total C 1s intensity. The structure of this remaining pristine graphene remains unclear. Similarly, the same study reveals that the component C_b represents sp^3-hybridised C in graphene-like clusters. This structure was recently determined by XSW in a sample hydrogenated at 600 K.[18] However, the structures of the remaining two components, C_a and C_d, are to date unstudied by XSW, although it has been suggested that the C_a component predominantly represents dimers forming in the ATOP regions and that the C_d component represents sp^2-hybridised C atoms neighbouring sp^3-hybridised C atoms.[4]

In this work, we expand the analysis of these more complicated photoemission spectra using a combination of high resolution XPS and XSW measurements obtained as a function of the substrate temperature during hydrogenation. Our results lead us to question the standard view of the assignment of peaks and provide greater insight into the selective functionalisation of graphene. In particular, in addition to presenting novel XSW results for the IT and RT hydrogenated samples, we also refine the analysis of the HT XSW data[18] in order to fit the data consistently across the temperature dependent data sets.

Experimental procedures

Experimental setup

The XPS and XSW measurements were performed at the permanent end station of the I09 beamline[22] at the Diamond Light Source. The photoelectron spectroscopy data was acquired using a Scienta EW4000 HAXPES analyser mounted in the horizontal plane at a 90° angle from the incident radiation. The light source consists of two separate undulators for providing soft and hard x-rays that irradiate onto overlapping spots on the sample. The soft X-ray light was used to obtain the high resolution XPS data (HR-XPS) and the hard X-ray light to obtain the hard X-ray photoelectron spectroscopy (HAXPES) and XSW data.

Graphene growth and hydrogenation

Graphene was produced on the (111) face of an Ir single crystal *via* the chemical vapour deposition method.[10]

First, the room temperature sample was exposed to a pressure of $p = 2 \times 10^{-7}$ mbar of ethylene (C_2H_4) for two minutes. Then, the sample was quickly heated to 1470 K while maintaining the ethylene pressure, and then cooled to *ca.* 1220 K. At this point, the pressure of ethylene was increased to 4×10^{-6} mbar for four minutes. The flow of ethylene was shut off, and the sample flash annealed to 1470 K before cooling to room temperature.

Prior to the functionalisation with hot atomic hydrogen, the quality of the graphene was examined by XPS.

Hot atomic hydrogen was produced by passing molecular hydrogen gas through a tungsten capillary heated by an electric current ($T > 1700$ K). The sample was heated to the desired temperature before being exposed to the H-beam until saturation coverage was achieved for a given sample temperature (background H pressure of 5×10^{-7} mbar for 45 min).

Each sample was allowed to cool down to room temperature before any photoelectron measurements were performed.

X-ray photoelectron spectroscopy and standing wave measurements

For each hydrogenated sample, several photoemission spectra of the C 1s orbital were acquired: a high-resolution spectrum acquired at a photon energy of 435 eV, as well as a measurement series of 42 HAXPES for the HT measurement and 40 HAXPES each for the IT and RT measurements with photon energies from within a narrow region around the Bragg energy. For Ir(111), this corresponds to 2798 eV at normal incidence. For each photon energy, the reflectivity of the sample was measured as well. Due to the low count rates, the XSW measurement series were repeated on several different spots on the sample, and the spectra acquired at the same photon energy were summed to obtain a single measurement series with a better signal-to-noise ratio.

XSW exploits the standing wave field created when the photon energy is close to the Bragg energy. Interference between the incoming and diffracted light creates a standing wave, whose periodicity is that of the Ir crystal between the (111)-planes, $d^{111} = 2.216$ Å. The phase variation with photon energy provides a characteristic modulation of each individual C 1s component in the spectra. This modulation depends on the average adsorption height for the component.[32]

The resulting absorption curves were fitted using dynamical diffraction theory[7] using two dimensionless structural parameters, both taking values between 0 and 1: the coherent position (P^{111}) and the coherent fraction (F^{111}). The coherent position corresponds to the average height of the atomic species expressed as a fraction of the standing wavelength. A coherent position of 0 or 1 corresponds to being in-plane with the extension of the crystal lattice, and two atoms positioned exactly one lattice periodicity apart will be indistinguishable.[20,32] Given the periodicity of the standing wavefield, the average height h equals $h = d^{111}(n + P^{111})$, where n is an integer. A value of $n = 1$ has been found to be appropriate for pristine graphene,[9] as well as graphane-like clusters in H-graphene on Ir(111),[18] and it will be assumed to apply throughout this article.

The coherent fraction is often interpreted as an order parameter. Flat structures will have high coherent fractions, and thus a decreasing value of the coherent fractions will often indicate an increase in the level of corrugation.[20]

In the HAXPES data, components have significantly larger Gaussian widths, which is primarily an effect of the energy resolution at different photon energies (*ca.* 0.01% of the photon energy), as well as the higher pass energy used for XSW measurements compared to HR-XPS.

Fitting procedures

The analysis of the spectra was performed with KolXPD.[34] All spectra have been fitted using a linear background and a number of peaks to represent individual components. Within a spectrum, all components were forced to have the same Lorentzian width. For each high-resolution spectrum, as few components as possible were used. The acquired number of components and binding energies were then used to fit the XSW spectra. All acquired fitting parameters are tabulated in Table 1.

Correction of the binding energies with respect to the Fermi level is possible for the HR-XPS, but it is not practical for the HAXPES due to the low count rates. Rather, the components in HAXPES and HR-XPS have been assumed to exhibit the same separations in binding energy as well as relative intensities.

When fitting the XSW spectra, all spectra acquired for one sample were described by components that can only vary in intensity across spectra, while the widths and binding energies of each component were kept constant for all photon energies. The process of finding the best possible fitting parameters is highly iterative, and several criteria have to be considered:

Firstly, any set of fitting parameters should provide as high coherent fractions as possible for all components, since higher values are mathematically more likely to represent a single atomic species.[20]

Secondly the relative intensities of the different components were assumed in the off-Bragg spectra (2792 eV) to be similar to those of the HR-XPS.

Table 1 Fitting parameters for each component in the C 1s photoemission spectra, as well as for the relative absorption curves. The Lorentzian width (L. wid.) is the same for all components within a spectrum, as well as for the high-resolution spectra and the corresponding HAXPES. The HT spectra are taken at a different time and with different settings than the IT and RT spectra, and thus the difference in the Lorentzian width is of little concern. The Gaussian width (G. wid.) refers to the high-resolution spectra and is generally wider in the HAXPES. The share refers to the relative intensity of that component calculated as the intensity of each component divided by the sum of the intensities of all components in the spectrum. The spectra taken at 435 eV and 2792 eV on the same sample should provide similar shares for all components. The intensity of the C_d component is too low for structural analysis, and its median height is therefore omitted

	Values from ref. 18	HT	IT	RT
Err/N	3.80	2.25	1.84	1.25
L. wid./eV	0.13	0.10	0.13	0.12
C_a BE/eV			$C_c + 0.73$	$C_c + 0.73$
Share			10%	34%
G. wid./eV			0.60	0.62
F^{111}			0.61(5)	0.55(4)
P^{111}			0.03(2)	0.05(2)
Height/Å			2.28(4)	2.33(4)
C_b BE/eV	$C_c + 0.44$	$C_c + 0.30$	$C_c + 0.30$	$C_c + 0.30$
Share		33%	34%	31%
G. wid./eV		0.70	0.60	0.63
F^{111}	0.41(2)	0.56(6)	0.67(8)	0.65(3)
P^{111}	0.02(3)	0.05(3)	0.10(3)	0.02(1)
Height/Å	2.26(7)	2.33(7)	2.44(7)	2.26(2)
C_c BE/eV	284.18	284.18	284.24	284.19
Share		67%	56%	26%
G. wid./eV		0.26	0.24	0.27
F^{111}	0.48(3)	0.65(3)	0.52(2)	0.48(4)
P^{111}	0.56(1)	0.57(1)	0.56(1)	0.45 (2)
Height/Å	3.46(2)	3.48 (2)	3.46(2)	3.21(4)
C_d BE/eV				$C_c - 0.21$
Share				9%
G. wid./eV				0.46
F^{111}				0.21(9)
P^{111}				0.00(8)

Thirdly, the Lorentzian widths, which arise due to the life-time broadening of the core hole, were set to the value obtained from the fit of the HR-XPS.

Fourthly, the C_a and C_b components in the RT spectra were forced to have the same Gaussian width in the XSW spectra, as they have similar widths in the high-resolution spectrum.

Fifthly, coherent fractions above 0.8 were considered artificially high due to the inherent corrugation within the graphene layer.

Results and discussion

Photoemission results

In Fig. 1 XPS and XSW C 1s photoemission spectra acquired on a sample hydrogenated at 600 K (HT) are displayed. At this temperature, only the HCP regions are expected to functionalise.[18,19] The HR-XPS data obtained at 435 eV

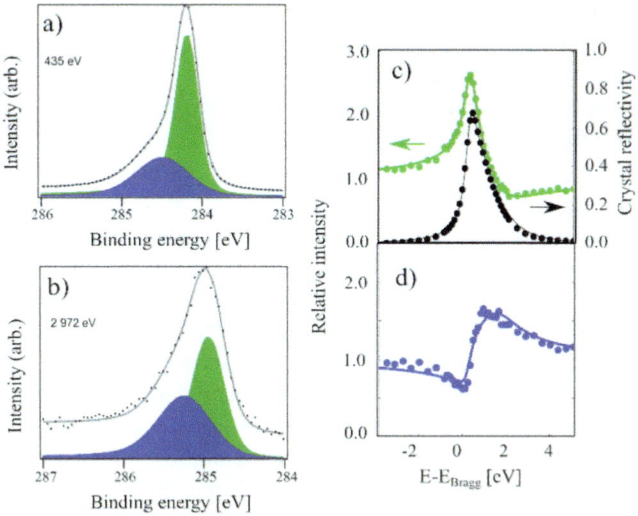

Fig. 1 C 1s photoemission spectra acquired on a sample hydrogenated at 600 K (referred to as HT in the text). (a) The high-resolution spectrum acquired at a photon energy of 435 eV. The C_b component (blue) and C_c component (green) area separated by 0.30 eV in binding energy. (b) The same sample measured at a photon energy of 2792 eV. (c) Relative absorption of the C_c component with best fit to structural parameters (green) and the reflectivity of the crystal (black) as a function of photon energy. Arrows indicate proper axes. (d) Relative absorption of the C_b component with best fit to the structural parameters. This data has been reproduced from ref. 18 with permission from the Royal Society of Chemistry, and it has been revisited and reanalysed for this work. The exact fitting parameters can be found in Table 1.

photon energy and the XSW data obtained at a photon energy of 2972 eV can both be fitted using two components (C_c and C_b). Previously, when analysing the HT spectra,[18] we constrained our fits to the fitting parameters described in ref. 4. However, by using the fitting procedure laid out above and fitting the HT spectra with the two components (C_c and C_b), we obtained the best fit with a binding energy shift between the two components of 0.30 eV, instead of the 0.44 eV used in the fitting of our prior work, as shown in Fig. 1. This resulted in not only a better numerical fit to the HR-XPS (normalised error of the fit decreasing from 3.80 to 2.26), but also, notably, this subtle change in the relative binding energy of the two peaks results in a significant difference in the resulting coherent fractions of both species. Specifically, for C_c/C_b respectively, the coherent fractions increased from 0.41(2)/0.43(5) to 0.65(3)/0.56(7). Nevertheless, the coherent position did not change significantly, 0.56(1)/0.02(3) and 0.57(1)/0.05(3), corresponding to average heights of 3.46(2) Å/2.26(7) Å and 3.48(2) Å/2.33(7) Å, respectively. All structural values are tabulated in Table 1.

The average adsorption height for the sp^2 species is close to the values observed for pristine graphene in other works, 3.38(4) Å in ref. 9 and 3.41(4) Å in ref. 18, although the coherent fraction is significantly higher for the measurements on pristine graphene, 0.74(4) in ref. 9 and 0.74(1) for ref. 18, indicating an increased corrugation in the HT functionalised graphene. The average adsorption height for the C_b component of 2.26(7) Å is in good agreement with the theoretically predicted height for C atoms in sp^3-hybridised graphene-like clusters. As the C_b component comprises 33% of the overall signal, the HCP regions are likely fully hydrogenated.

In Fig. 2 the HR-XPS and XSW measurements for the sample hydrogenated at 550 K (IT) are displayed. The intensity is visibly increased on the high binding energy end of the spectrum compared to that of HT. Utilising the described fitting approach, we fit the IT spectra with three components: C_a, C_b and C_c, as shown in Fig. 2. The binding energy for the C_a component was found at 0.73 eV above that of the C_c component, with a C_a–C_b separation of 370 meV, thus 60 meV smaller than that reported in ref. 4. Fitting of the XSW data resulted in very similar coherent position for C_c as that found for the HT sample, 0.56(1), corresponding to 3.46(2) Å on average, though the coherent fraction decreased significantly to 0.52(2). This indicates a higher degree of corrugation in the remaining sp^2 regions. The coherent fraction of the C_b component in the IT sample increased to 0.67(9) and the coherent position also increased slightly to 0.10(4)/2.44(7) Å. Both values, however, have overlapping confidence intervals with the HT values. The new C_a component has a coherent fraction and coherent position almost identical to those of the C_b component of the HT sample: 0.61(5) and 0.03(2)/2.28(4) Å, respectively.

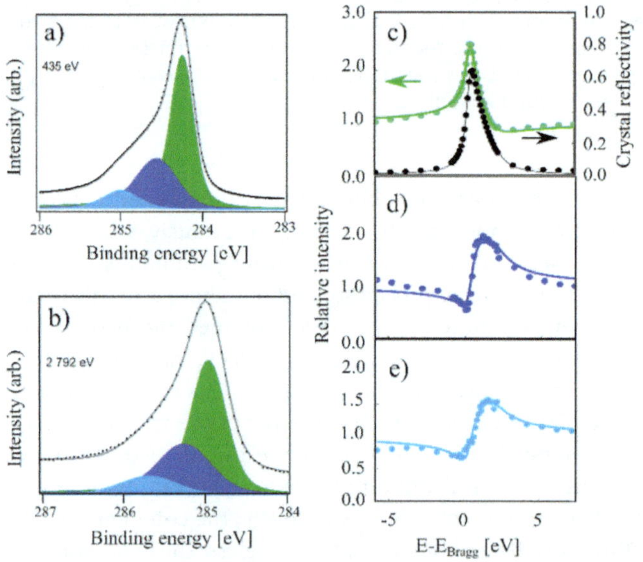

Fig. 2 C 1s photoemission spectra acquired on a sample hydrogenated at 550 K (referred to as IT in the text). (a) The high-resolution spectrum acquired at a photon energy of 435 eV. Along with the C_b (dark blue) and C_c (green) components observed at HT, the C_a component (light blue) is present. (b) The same sample measured at a photon energy of 2792 eV. (c) Relative absorption of the C_c component with best fit to structural parameters (green) and the reflectivity of the crystal (black) as a function of photon energy. Arrows indicate proper axes. (d) Relative absorption of the C_b component with best fit to the structural parameters. (e) Relative absorption of the C_a component with best fit to the structural parameters. In the XSW spectra we see evidence of beam damage in the form of significant hydrogen desorption. This affects the quality of the XSW fits. The exact fitting parameters can be found in Table 1.

We note that the IT spectra shows significant beam damage in the XSW spectra for the C_b species in the form of hydrogen desorption occurring during the measurement. This is exemplified by the poor fit of the XSW absorption profile, where the low photon energy signal is significantly more intense than the high photon energy signal. This naturally influences the quality of the fit. Our data shows that this effect is present but is far weaker in the HT and RT spectra than in the IT spectra.

Finally, in Fig. 3, HR-XPS and XSW spectra for the room temperature (RT) sample are presented. Compared to the IT and HT spectra, this sample exhibits a strong increase of the higher binding energy components, as well as a small shoulder at lower binding energy. To fit this spectrum, we added the fourth component C_d at 0.21 eV below the C_c component (compared to 0.27 eV from ref. 4). However, we note that C_d is very weak, thus the fit is likely not unique. For the RT spectrum we obtain a 34 : 31 : 26 : 9 ratio for the four components: C_a, C_b, C_c and C_d, respectively. As the remaining C_c coverage is comparable to the 30% of C_c remaining at the highest coverage in ref. 4, we expect that this sample is hydrogenated to saturation coverage. Including a fifth component in the fit on the

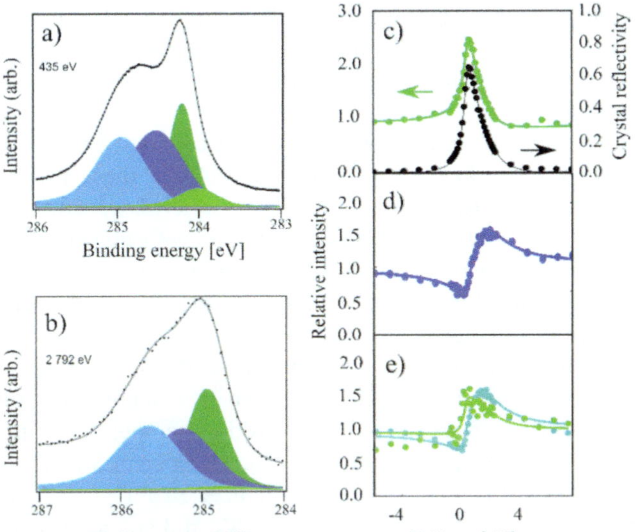

Fig. 3 C 1s photoemission spectra of a sample which was hydrogenated at room temperature (referred to as RT in the text). (a) The high-resolution spectrum acquired at a photon energy of 435 eV. All four components, C_a (light blue), C_b (dark blue), C_c (dark green), and C_d (light green), are present. (b) Photoemission spectrum acquired at a photon energy of 2792 eV using all components, as in (a). The low intensity of the C_d component cannot be properly replicated at high photon energies due to the much lower resolution. (c) Relative absorption of the C_c component with best fit to structural parameters (green) and the reflectivity of the crystal (black) as a function of photon energy. Arrows indicate proper axes. (d) Relative absorption of the C_b component with best fit to the structural parameters. (e) Relative absorption of the C_a component (blue) and C_d component (green) with best fit to the structural parameters. The C_d component produces no discernible signal apart from a replication of the reflectivity. The exact fitting parameters can be found in Table 1.

high binding energy side did not result in a significant improvement to the fit. The fitting of the XSW data shows a largely unchanged coherent fraction for C_c compared to IT at 0.48(4), but a significant decrease in the coherent position down to 0.45(2)/3.21(4) Å. C_a and C_b exhibit coherent fractions of 0.55(3) and 0.65(3), respectively, which are similar to the values obtained in the IT and HT samples, but with a slightly increased coherent position for C_a, 0.05(4)/2.33(9) Å, as well as a slightly decreased coherent position of C_b, 0.02(1)/2.26 Å.

Finally, the intensity of the C_d component is too low to be resolved in the XSW spectrum. The absorption curve for this component in Fig. 3e simply replicates the general reflectivity of the crystal with a coherent position that is indistinguishable from 0 and with a low coherent fraction. Therefore, the resulting values are most likely not representative of the structure of this species.

Functionalised structures

We note that, in the fitting presented here, the relative coverage of the C_b component remains stable across all three samples (*ca.* 30% of the overall signal). As this is the case even in the HT sample, where the HCP regions are functionalized, this strongly indicates that the C_b component exclusively represents sp^3-hybridised C atoms and, thus, graphene-like clusters, as observed in ref. 18 within the HCP regions. This assignment is further supported by the fact that the coherent position and the coherent fraction of the C_b component barely differs as a function of the sample preparation when taking the beam damage in the IT spectra into account.

The strong agreement between the values of the coherent fraction and position for the C_a and C_b components indicate that both of these two components represent the same hydrogenation structure. Therefore, both components represent graphane-like structures. In combination with previous STM and DFT results[19] that correlate lower substrate temperatures with increasing hydrogenation in the FCC regions, this indicates that the C_a component relates to graphane-like structures in FCC regions.

In principle, the C_a component could also arise due to the increasing size of graphene-like clusters in the HCP regions when H is deposited at intermediate temperature. However, based on the deposition temperature and the high observed intensity of the sp^3 component in the HT spectra, we expect that both the HT and IT samples contain fully functionalized HCP regions.[19] Note that for the relative intensities of the three signals (C_a, C_b and C_c), a ratio of 10 : 34 : 56 (respectively) is observed at IT. As the 34% intensity of the overall signal in the C_b component is in good agreement with the value obtained for the HT sample, any additional functionalisation in the FCC regions must contribute to the C_a component. This is consistent with earlier STM measurements performed on both types of samples, which show that the FCC regions are functionalised to a much lower degree than the HCP regions at elevated temperatures.[19] Therefore, hydrogenating only around a third of the FCC carbon atoms (*ca.* 10% of the overall signal) at 550 K is a realistic figure.

This significant difference in reactivity, as well as core level shift between the functionalised C atoms in the FCC and HCP regions, is striking, since the registry between the graphene sheet and the surface layer of Ir atoms is identical. Instead, the difference between the regions stems from the position of Ir atoms in the second and third Ir layer below the surface with respect to the C atom binding to H. In the more reactive HCP sites, the C atom is positioned above an Ir atoms in the third layer from the surface, while the C atoms in the less reactive FCC region are above an Ir atom in the second layer. Such an effect is surprising, but differences in the reactivity of the HCP and FCC regions of graphene in Ir(111) have previously been observed elsewhere.[5,18,19]

By comparing all three samples (HT, IT, and RT), the coherent fraction of the C_c component, representing the pristine parts of the graphene sheet, is seen to gradually decrease with increasing level of functionalisation, though the decrease in coherent fraction between the IT and RT cases is not significant. As the graphane-like clusters in the HCP and FCC regions grow, a larger part of the graphene sheet is pinned to the substrate. The observed lower coherent fraction is, thus, a result of an increasing corrugation in the remaining pristine parts. This is in broad agreement with previous DFT calculations.[18] The coherent position of the C_c component remains constant in the HT and IT spectra compared to the value obtained for pristine graphene,[9,18] but decreases significantly in the RT sample. This signifies that the sp^2 C atoms are being strongly pulled down towards the Ir substrate with the increasing relative coverage of graphane-like regions from IT (41%) to RT (65–74%, the latter number including the C_d component). Note that most of the difference observed between the IT and RT spectra is due to increasing hydrogenation in the FCC regions rather than the ATOP regions.

What remains unclear is the extent of hydrogenation in the ATOP regions and its contribution in the RT sample to the C 1s spectra. The earlier proposed dimer structures in ATOP regions could in principle be assigned to the C_d component, as this component appears only in the RT spectra. It is, however, not considered likely, due to the high thermal stability of the C_d component.[4]

Neither is the C_a component likely to represent such structures. The relative area of the C_a component in the RT spectra is 34%, and it is only marginally more intense than the C_b component at 31%. As the sp^3 C atoms in the FCC region were assigned to the C_a component, realistically at most 10% of the C_a signal ($\sim$3% of all C atoms) may arise from hydrogenation in the ATOP regions. Our analysis reveals that there is no notable change in the structural parameters for the C_a component, compared to the IT sample. However, any major functionalization of ATOP regions is expected to increase the coherent position, as well as reduce the coherent fraction, due to the large spread of such H clusters in the corrugated graphene. Furthermore, there is only an insignificant decrease in the coherent fraction of the C_c component in going from the IT to the RT sample, indicating no significant rehybridisation. Hence, the results presented here do not support the presence of dimer structures in ATOP regions at any significant coverage.

The excess of functionalisation beyond what is expected by the hydrogenation of the FCC and HCP regions alone could be due to the H-clusters expanding to the out of registry regions at the edges of the graphane-like clusters. Such a change would be consistent with the decrease in coherent position of the C_c component from the IT to the RT sample, as the majority of the graphene sheet is now pinned to the substrate. Since the sp^3 C atoms near these edges are not directly above an Ir atom, this would lead to a strong deformation around these atoms. Such regions may potentially contribute to the C_d component.

Conclusions

The XPS and XSW analysis of the C 1s core level is performed for three separate samples of hydrogenated graphene on Ir(111) with varying levels of hydrogenation. At least four components are necessary for a full description of the spectra. These components are labelled C_a, C_b, C_c, and C_d in order of decreasing binding energy. By selective hydrogenation of the moiré parts, the contributions of each part of the supercell to individual components have been identified as follows: the C_b component can be ascribed exclusively to sp^3 in graphane-like carbon in the HCP regions. Hydrogenation of the FCC region results in a graphane-like configuration, as was previously determined for the hydrogenation of the HCP region.[18] The C_a component is primarily related to the sp^3 carbon in these clusters, with a possible minor contribution from ATOP region functionalisation. Our results indicate a surprising significant difference in binding energy, as well as reactivity, for the carbon atoms in the HCP and FCC regions.

The C_c component, representing pristine sp^2-hybridised carbon, was shown to increase its corrugation with increasing functionalisation, and it was found to move towards the Ir substrate at the highest level of hydrogenation. Even at the saturation level of hydrogenation, this component remains at above 25% relative intensity. Finally, the C_d component, which only appears in samples hydrogenated at room temperature, has too low an intensity for a confident analysis using XSW.

Conflicts of interest

There are no conflicts of interest to declare.

Acknowledgements

We acknowledge the Diamond Light Source Ltd. for beamtime at the I09 beamline under the SI16243-1 and SI16243-2 proposals and thank the staff for their assistance during our experiments. The research at Diamond presented in this article has been supported by the project CALIPSOplus under the Grant Agreement 730872 from the EU Framework Programme for Research and Innovation HORIZON 2020. We gratefully acknowledge financial support from VILLUM FONDEN (research grant no. 17547 and grant no. 11744 - Villum Centre of Excellence for Dirac. Materials), Innovation Fund Denmark (NIAGRA, grant no. 5184-00018B), the European Research Council (CoG GRANN, grant no. 648551), and the Danish National Research Foundation through the Center of Excellence "InterCat" (Grant agreement no. DNRF150).

Notes and references

1 N. Akhtar, G. Anemone, D. Farias and B. Holst, Fluorinated graphene provides long lasting ice inhibition in high humidity, *Carbon*, 2019, **141**, 451–456.

2 A. Avsar, J. H. Lee, G. K. W. Koon and B. Özyilmaz, Enhanced spin–orbit coupling in dilute fluorinated graphene, *2D Mater.*, 2015, **2**(4), 044009.

3 J. Balakrishnan, G. K. W. Koon, M. Jaiswal, A. H. Castro Neto and B. Özyilmaz, Colossal enhancement of spin–orbit coupling in weakly hydrogenated graphene, *Nat. Phys.*, 2013, **9**(5), 284–287.

4 R. Balog, M. Andersen, B. Jørgensen, Z. Sljivancanin, B. Hammer, A. Baraldi, R. Larciprete, P. Hofmann, L. Hornekaer and S. Lizzit, Controlling hydrogenation of graphene on Ir (111), *ACS Nano*, 2013, **7**(5), 3823–3832.

5 R. Balog, B. Jørgensen, L. Nilsson, M. Andersen, E. Rienks, M. Bianchi, M. Fanetti, *et al.*, Bandgap opening in graphene induced by patterned hydrogen adsorption, *Nat. Mater.*, 2010, **9**(4), 315–319.

6 R. Balog, B. Jørgensen, J. Wells, E. Lægsgaard, P. Hofmann, F. Besenbacher and L. Hornekær, Atomic hydrogen adsorbate structures on graphene, *J. Am. Chem. Soc.*, 2009, **131**(25), 8744–8745.

7 B. W. Batterman, Effect of dynamical diffraction in X-ray fluorescence scattering, *Phys. Rev.*, 1964, **133**(3A), A759.

8 A. Buchsteiner, A. Lerf and J. Pieper, Water dynamics in graphite oxide investigated with neutron scattering, *J. Phys. Chem. B*, 2006, **110**(45), 22328–22338.

9 C. Busse, P. Lazić, R. Djemour, J. Coraux, T. Gerber, N. Atodiresei, V. Caciuc, *et al.*, Graphene on Ir (111): physisorption with chemical modulation, *Phys. Rev. Lett.*, 2011, **107**(3), 036101.

10 J. Coraux, M. Engler, C. Busse, D. Wall, N. Buckanie, F.-J. Meyer Zu Heringdorf, R. Van Gastel, P. Bene and T. Michely, Growth of graphene on Ir (111), *New J. Phys.*, 2009, **11**(2), 023006.

11 A. T. N'Diaye, J. Coraux, T. N. Plasa, C. Busse and T. Michely, Structure of epitaxial graphene on Ir (111), *New J. Phys.*, 2008, **10**(4), 043033.

12 D. A. Dikin, S. Stankovich, E. J. Zimney, R. D. Piner, G. H. Dommett, G. Evmenenko, S. T. Nguyen and R. S. Ruoff, Preparation and characterization of graphene oxide paper, *Nature*, 2007, **448**(7152), 457–460.

13 D. R. Dreyer, S. Park, C. W. Bielawski and R. S. Ruoff, The chemistry of graphene oxide, *Chem. Soc. Rev.*, 2010, **39**(1), 228–240.

14 D. C. Elias, R. R. Nair, T. M. G. Mohiuddin, S. V. Morozov, P. Blake, M. P. Halsall, A. C. Ferrari, D. W. Boukhvalov, M. I. Katsnelson, A. K. Geim and K. S. Novoselov, Control of graphene's properties by reversible hydrogenation: evidence for graphane, *Science*, 2009, **323**(5914), 610–613.

15 D. V. V. D. Haberer, D. V. Vyalikh, S. Taioli, B. Dora, M. Farjam, J. Fink, D. Marchenko, T. Pichler, K. Ziegler, S. Simonucci and M. S. Dresselhaus, Tunable band gap in hydrogenatedquasi-free-standing graphene, *Nano Lett.*, 2010, **10**(9), 3360–3366.

16 L. Hornekær, E. Rauls, W. Xu, Ž. Šljivančanin, R. Otero, I. Stensgaard, E. Lægsgaard, B. Hammer and F. Besenbacher, Clustering of chemisorbed H (D) atoms on the graphite (0001) surface due to preferential sticking, *Phys. Rev. Lett.*, 2006, **97**(18), 186102.

17 L. Hornekær, Ž. Šljivančanin, W. Xu, R. Otero, E. Rauls, I. Stensgaard, E. Lægsgaard, B. Hammer and F. Besenbacher, Metastable structures and recombination pathways for atomic hydrogen on the graphite (0001) surface, *Phys. Rev. Lett.*, 2006, **96**(15), 156104.

18 A. L. Jørgensen, D. A. Duncan, C. F. P. Kastorp, L. Kyhl, Z. Tang, A. Bruix, M. Andersen, *et al.*, Chemically-resolved determination of hydrogenated graphene–substrate interaction, *Phys. Chem. Chem. Phys.*, 2019, **21**(25), 13462–13466.

19 J. H. Jørgensen, A. G. Cabo, R. Balog, L. Kyhl, M. N. Groves, A. M. Cassidy, A. Bruix, *et al.*, Symmetry-driven band gap engineering in hydrogen functionalized graphene, *ACS Nano*, 2016, **10**(12), 10798–10807.
20 *The X-Ray Standing Wave Technique: Principles and Applications*, ed. A. Kazimirov, World Scientific, Hackensack, NJ, 2013.
21 L. Kyhl, R. Balog, A. Cassidy, J. Jørgensen, A. Grubisic-Čabo, L. Trotochaud, H. Bluhm and L. Hornekær, Enhancing graphene protective coatings by hydrogen-induced chemical bond formation, *ACS Appl. Nano Mater.*, 2018, **1**(9), 4509–4515.
22 T.-L. Lee and D. A. Duncan, A two-color beamline for electron spectroscopies at diamond light source, *Synchrotron Radiat. News*, 2018, **31**(4), 16–22.
23 W. -K. Lee, K. E. Whitener Jr, J. T. Robinson and P. E. Sheehan, Patterning Magnetic Regions in Hydrogenated Graphene *Via* E-Beam Irradiation, *Adv. Mater.*, 2015, **27**(10), 1774–1778.
24 H. McKay, D. J. Wales, S. J. Jenkins, J. A. Verges and P. L. de Andres, Hydrogen on graphene under stress: molecular dissociation and gap opening, *Phys. Rev. B: Condens. Matter Mater. Phys.*, 2010, **81**(7), 075425.
25 R. R. Nair, W. Ren, R. Jalil, I. Riaz, V. G. Kravets, L. Britnell, P. Blake, F. Schedin, A. S. Mayorov, S. Yuan and M. I. Katsnelson, Fluorographene: a two-dimensional counterpart of Teflon, *Small*, 2010, **6**(24), 2877–2884.
26 A. H. C. Neto and F. Guinea, Impurity-induced spin–orbit coupling in graphene, *Phys. Rev. Lett.*, 2009, **103**(2), 026804.
27 L. Schlapbach, and A. Züttel, Hydrogen-storage materials for mobile applications, *Materials for Sustainable Energy: a Collection of Peer-Reviewed Research and Review Articles from Nature Publishing Group*, 2011, pp. 265–270.
28 D. Smith, R. T. Howie, I. F. Crowe, C. L. Simionescu, C. Muryn, V. Vishnyakov, K. S. Novoselov, Y. J. Kim, M. P. Halsall, E. Gregoryanz and J. E. Proctor, Hydrogenation of graphene by reaction at high pressure and high temperature, *ACS Nano*, 2015, **9**(8), 8279–8283.
29 J. O. Sofo, A. S. Chaudhari and G. D. Barber, Graphane: A two-dimensional hydrocarbon, *Phys. Rev. B: Condens. Matter Mater. Phys.*, 2007, **75**(15), 153401.
30 V. Tozzini and V. Pellegrini, Prospects for hydrogen storage in graphene, *Phys. Chem. Chem. Phys.*, 2013, **15**(1), 80–89.
31 M. V. Ulybyshev and M. I. Katsnelson, Magnetism and interaction-induced gap opening in graphene with vacancies or hydrogen adatoms: Quantum Monte Carlo study, *Phys. Rev. Lett.*, 2015, **114**(24), 246801.
32 D. P. Woodruff, Adsorbate structure determination using photoelectron diffraction: Methods and applications, *Surf. Sci. Rep.*, 2007, **62**(1), 1–38.
33 Z. Xu, H. Sun, X. Zhao and C. Gao, Ultrastrong fibers assembled from giant graphene oxide sheets, *Adv. Mater.*, 2013, **25**(2), 188–193.
34 *KolXPD, Version 1.8.0 (Build 65); Software for Spectroscopy Data Measurement and Processing*, https://www.kolibrik.net/kolxpd/, accessed Dec. 2021.

Faraday Discussions

ROYAL SOCIETY OF CHEMISTRY

PAPER

Identifying chemical and physical changes in wide-gap semiconductors using real-time and near ambient-pressure XPS

Simon Astley,[a] Di Hu,[a] Kerry Hazeldine, [a] Johnathan Ash,[a] Rachel E. Cross, [a] Simon Cooil,[ab] Martin W. Allen, [c] James Evans,[d] Kelvin James,[d] Federica Venturini,[e] David C. Grinter,[e] Pilar Ferrer, [e] Rosa Arrigo,[e] Georg Held, [e] Gruffudd T. Williams[f] and D. Andrew Evans *[a]

Received 13th December 2021, Accepted 14th March 2022

DOI: 10.1039/d1fd00119a

Photoelectron spectroscopy is a powerful characterisation tool for semiconductor surfaces and interfaces, providing in principle a correlation between the electronic band structure and surface chemistry along with quantitative parameters such as the electron affinity, interface potential, band bending and band offsets. However, measurements are often limited to ultrahigh vacuum and only the top few atomic layers are probed. The technique is seldom applied as an *in situ* probe of surface processing; information is usually provided before and after processing in a separate environment, leading to a reduction in reproducibility. Advances in instrumentation, in particular electron detection has enabled these limitations to be addressed, for example allowing measurement at near-ambient pressures and the *in situ*, real-time monitoring of surface processing and interface formation. A further limitation is the influence of the measurement method through irreversible chemical effects such as radiation damage during X-ray exposure and reversible physical effects such as the charging of low conductivity materials. For wide-gap semiconductors such as oxides and carbon-based materials, these effects can be compounded and severe. Here we show how real-time and near-ambient pressure photoelectron spectroscopy can be applied to identify and quantify these effects, using a gold alloy, gallium oxide and semiconducting diamond as examples. A small binding energy change due to thermal expansion is followed in real-time for the alloy while the two semiconductors show larger temperature-induced changes in binding energy that, although superficially similar, are identified as having

[a]*Department of Physics, Aberystwyth University, Aberystwyth SY23 3BZ, UK. E-mail: a.evans@aber.ac.uk*

[b]*Centre for Materials Science and Nanotechnology, University of Oslo, Oslo, 0318, Norway*

[c]*Department of Electrical and Computer Engineering, University of Canterbury, Christchurch 8014, New Zealand*

[d]*Diamond Centre Wales Ltd, Talbot Green, RCT, CF72 9FG, UK*

[e]*Diamond Light Source, Harwell Science and Innovation Campus, Didcot, OX11 0DE, UK*

[f]*Element Six Ltd., Global Innovation Centre, Harwell, OX11 0QR, UK*

different and multiple origins, related to surface oxygen bonding, surface band-bending and a room-temperature surface photovoltage. The latter affects the p-type diamond at temperatures up to 400 °C when exposed to X-ray, UV and synchrotron radiation and under UHV and 1 mbar of O_2. Real-time monitoring and near-ambient pressure measurement with different excitation sources has been used to identify the mechanisms behind the observed changes in spectral parameters that are different for each of the three materials. Corrected binding energy values aid the completion of the energy band diagrams for these wide-gap semiconductors and provide protocols for surface processing to engineer key surface and interface parameters.

1 Introduction

X-ray photoelectron spectroscopy (XPS) is widely used for the quantitative analysis of solid surfaces, adsorbates and interfaces.[1] A simple energy transfer picture for the photoemission process is illustrated in Fig. 1 for a metal (a gold-based alloy), an n-type semiconductor (β-Ga_2O_3) and a p-type wide-gap semiconductor (diamond). This assumes that the single-particle energy levels are unchanged on photoexcitation and ignores final state effects and electron energy losses. In this simple picture, the measured kinetic energy of photoelectrons emitted from the surface of a metal excited by an X-ray of frequency ν is usually assumed to be given by eqn (1).

$$E_K = h\nu - E_B - \phi_A \quad (1)$$

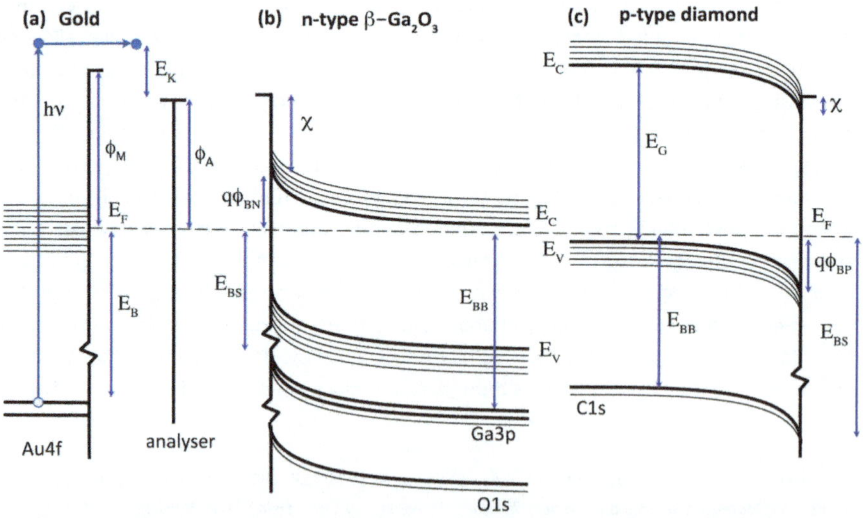

Fig. 1 Energy band diagrams for (a) a metal (gold), (b) an n-type semiconductor (β-Ga_2O_3) and (c) a p-type semiconductor (diamond). The emission process is illustrated for the excitation, emission and collection of a photoelectron from a metal core state in (a). The influence of the potential barriers, ϕ_{BN} and ϕ_{BP}, on the core level binding energies relative to the Fermi level, E_F, are shown in (b) and (c).

This equation is often further simplified by omitting the work function of the electron energy analyser ϕ_A as this is taken to be constant for a given spectrometer. Measurement of the kinetic energy of photoelectrons emitted following excitation by photons of a fixed energy $h\nu$ is then assumed to directly yield the binding energy of the core electron, E_B, that is usually defined with reference to the Fermi level that can be directly measured for the metal and inferred from the photoelectron core level spectra for a clean, elemental surface. For a metal therefore, eqn (1) is usually a reasonable approximation, especially using fixed laboratory sources such as Mg and Al Kα X-rays coupled to calibrated, energy-dispersing analysers with appropriate electron lens parameters and a fixed experimental geometry. This combination can provide reliable energy values with an overall energy resolution, determined by the lifetime and instrumental broadening, that is sufficient to identify the chemical environment of elements on solid surfaces.[2,3] With these assumptions, the binding energy can be used quantitatively to infer the surface chemistry of metal surfaces, for example identifying alloy composition,[4] oxidation[5] and adsorbates.[6] The reproducibility of clean, inert metal surfaces has led to their extensive use for the study of surface reactions and for thin film growth. Advances in instrumentation, in particular the efficiency of electron detectors, is now enabling real-time and *in-operando* measurements of these surface processes.

For semiconductors, photoelectron spectroscopy can provide direct measurement of key parameters such as the valence and conduction band edges (E_V, E_C), electron affinity (χ) and interface potential (ϕ_B) in order to determine the energy band diagrams shown in Fig. 1b and c that are widely used to design and model devices based on semiconductor and metal interfaces.[7] In this case, the core level binding energy is not uniquely defined for a given semiconductor since the energy difference between the Fermi level and the band edges depends on the bulk doping level and on the band bending, defined by the potential barrier ϕ_{BN},ϕ_{BP}, that results from charges in the depletion region and at the surface.[8] This is analogous to the potential energy barrier at a metal–semiconductor or p–n junction and for semiconductor surfaces, correlations can be determined between the energy band diagram shown in Fig. 1 and the surface chemistry, inferred from the line-shape analysis of photoelectron core level spectra.[9]

For many wide gap semiconductors, doping is a particular challenge and the acceptors or donors may not be fully ionised at room temperature. This results in a high resistivity that can lead to a photovoltage or photocurrent induced charging.[10] The Fermi level at the surface does not then align with the analyser, with the charging offset, E_q, affecting the measured kinetic energy as shown in eqn (2).

$$E_K = h\nu - E_{BS} - \phi_A \pm E_q \tag{2}$$

In a surface-sensitive XPS measurement, where the electron escape depth is small in comparison with the depletion width, the measured core level binding energy is that at the surface, E_{BS}, rather than in the bulk, E_{BB}. These are related by eqn (3).

$$E_{BS} = E_{BB} \pm (q\phi_B - \zeta) \tag{3}$$

For the n-type and p-type semiconductors in Fig. 1b and c, respectively,

$$E_{BS} = E_{BB} - (q\phi_{BN} - E_C + E_F) \tag{4}$$

$$E_{BS} = E_{BB} + (q\phi_{BP} - E_F + E_V) \tag{5}$$

The kinetic energy is then given by eqn (6),

$$E_K = h\nu - E_{BB} \pm (q\phi_B - \zeta) - \phi_A \pm E_q \tag{6}$$

The quantity, ζ, is determined by the acceptor (donor) binding energy and is often ignored for moderately-doped semiconductors as its value is ideally less than the thermal energy, kT.

However, for wide-gap semiconductors such as diamond, these binding energies are often larger than kT (the B acceptor is around 15 $\times$ kT and the P donor is around 25 $\times$ kT at room temperature). This makes doping difficult and increases the room temperature resistivity, introducing a further energy term, E_q, due to surface charging and photovoltage generation during exposure to X-rays.[11]

For semiconductors, therefore, the core level binding energy depends on a number of factors that must be identified and quantified and these cannot always be uniquely defined in a single XPS measurement. Errors can accumulate within the many terms and any spatial inhomogeneity in the chemical and electronic environment within the measurement volume can also change the lineshape through increased broadening or additional components.[12]

To complete the electron energy landscape for a semiconductor interface, measurement of the valence band maximum, E_V, is needed to determine the band-bending and potential barrier, ϕ_B. For soft X-ray excitation, core and valence photoelectron spectra can be measured at the same photon energy;[13] for laboratory X-ray sources such as Mg and Al Kα, the usual approach is to record the valence band spectra using UV excitation such as He I. These measurements can also provide the metal work function, ϕ_M, and the semiconductor electron affinity, χ.[14] The conduction band maximum, E_C, requires a value for the energy band gap, E_G, for example from optical measurements, and ζ is calculated from the doping concentration and acceptor (donor) binding energy.[15]

The surface band bending for both Ga_2O_3 and diamond has been widely studied and has been reported to be sensitive to the surface composition. In both cases, the Fermi level at the bare surface is pinned within a range that is consistent with the barrier heights measured for metal contacts and that are clustered around the charge neutrality level that lies in the upper half of the energy band gap for Ga_2O_3 and the lower half of the energy band gap for diamond. This results in similar potential barriers of 1–2 V for both semiconductors that have similar room-temperature band gaps of 4.7 eV (Ga_2O_3) and 5.6 eV (diamond), as shown in Fig. 1.

XPS studies of semiconductor surfaces can therefore in principle provide a correlation between the energy band profile and changes in the surface chemical environment induced, for example, by adsorption, heating and irradiation.[16–18] However, additional measurements are required to identify the influence of sample inhomogeneity and the influence of the measurement method such as final state effects, radiation damage and surface charging.[19,20]

Here we show how different mechanisms are responsible for the observed core level binding energy changes for a metal alloy and two wide-gap semiconductors when the temperature is cycled between room temperature and 600 °C. Using real-time and near-ambient pressure XPS, where temperature and pressure are the only variables, the dominant physical and chemical effects are identified and corrected.

2 Experimental

2.1 Materials

A gold–silver–copper alloy was prepared by the Diamond Centre Wales (DCW Ltd) using injection mold casting with a predicted composition of 75% gold, 15% silver and 10% copper.[21] The sample was then ultrasonically cleaned using acetone, isopropyl alcohol and methanol before mounting in the spectrometer.

The β-Ga_2O_3 single crystal used in this study was a Sn-doped ($\bar{2}01$) single crystal wafer grown using the edge-defined film-fed growth method,[22] (Tamura Corporation (Japan)). The carrier concentration was measured to be 4.46×10^{18} cm^{-3}, corresponding to a bulk resistivity of 0.020 Ω cm. The band gap of the ($\bar{2}01$) β-Ga_2O_3 wafer was measured optically to be 4.71 ± 0.02 eV; Ti/Au contacts were deposited by e-beam evaporation on part of the front and back sides of each sample. The β-Ga_2O_3 sample ($5 \times 5 \times 0.65$ mm^3) investigated here was immersed in 1 M of NaOH at 60 °C for 30 minutes, before drying in N_2 gas and mounting into the spectrometer.

The p-type diamond (001) sample was a CVD-grown, B-doped single crystal, with a boron concentration of around 10^{15} cm^{-3} (Element Six Ltd.).[23] The diamond ($6 \times 8 \times 1$ mm^3) was scaife-polished and oxidised using 10 mL of sulphuric acid and 0.5 g potassium nitrate, followed by cleaning in water and organic solvents and drying in N_2 gas before mounting in the spectrometer.

2.2 Methods

Ultrahigh vacuum measurements were recorded using a Mg Kα X-ray excitation source, with photoelectrons continuously energy analysed and detected using a Phoibos 100 analyser with a 2d CCD detector. In the spectrometer, UPS measurements using a SPECS UVS 300 He I/He II UV source with photoelectrons are possible with the same analyser, in the same sample position. Real-time monitoring of core level photoelectron spectra during temperature cycling was enabled by operating the multichannel detector in snapshot mode.[24] Programmed annealing cycles were set up by varying the power of a graphite–boron nitride heater and monitoring the temperature, pressure and photoelectron spectra. The sample heating is indirect with the metal directly connected to an external earth to avoid the influence of the bias applied to the heater. A typical cycle between room temperature and 600 °C for the metal alloy required a total time of 5 hours to sequentially record Au4f, Ag3d, and Cu2p snapshot spectra every 20 s, each with a 6 s dwell time. The annealing cycles for the semiconductor samples over the same temperature range involved the collection of Ga2p, Ga3p and O1s core level snapshot spectra for the β-Ga_2O_3 and the C1s and O1s core level snapshot spectra for the diamond. Near-ambient pressure XPS was carried out at beamline B07-C at Diamond Light Source (DLS) with photoelectrons excited by photons of energy 400 eV to 840 eV and collected in a Phoibos 150 NAP analyser, equipped with a delay line detector operating in scanning (fixed transmission) mode, at pressures from UHV to 10 mbar of oxygen. O1s and C1s core level spectra for the diamond at near-ambient pressures were recorded with higher surface sensitivity at selected temperatures between room temperature and 400 °C at an oxygen pressure of 1 mbar.[25]

3 Results and discussion

3.1 Gold alloy

Inert transition metals such as gold, copper and silver provide useful reference samples for XPS measurements. A clean surface can be prepared *in situ* that is representative of the bulk and the photoelectron spectra measured with typical X-ray excitation have narrow emission peaks covering the full range of kinetic energy. This enables the calibration of the electron analyser to provide reliable reference energy scales for the measurement of binding energy shifts in metal alloys and in non-metals such as semiconductors. A reference metal sample is also useful to identify non-equilibrium charging effects in high resistance materials.

Detailed core level emission spectra were recorded with high energy and temporal resolution using multichannel electron detection as shown in Fig. 2a. The Au4f doublet in a Au–Ag–Cu alloy was measured in snapshot mode during a programmed temperature cycle between room temperature (20 °C) and 600 °C. Temperature series for metal alloys are useful to directly monitor effects such as surface desorption and changes in alloy composition. Within each series, the metal core level spectra were fitted using asymmetric pseudo-Voigt functions with fixed Gaussian and Lorentzian mixing ratio and width of 0.7 and 0.4 eV respectively. The variation of peak position with temperature extracted from the fitting sequence for a typical series is shown in Fig. 2b.

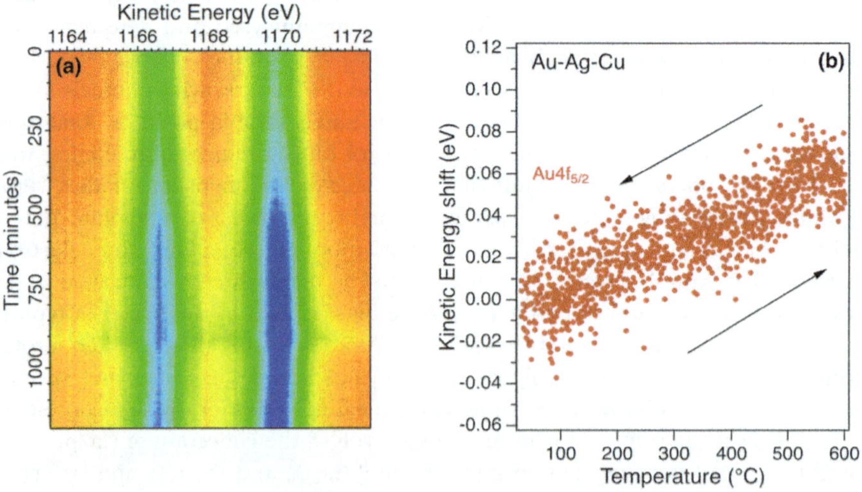

Fig. 2 (a) Real-time photoelectron spectroscopy for the 4f core level in a Au–Ag–Cu alloy recorded in snapshot mode during an annealing cycle from room temperature to 600 °C. (b) Curve fitting reveals a reversible kinetic energy shift of +60 meV of the $Au4f_{5/2}$ photoelectrons as the alloy is heated.

There is an initial increase in peak intensity due to the desorption of surface species with no re-adsorption during cooling. The lineshape is unchanged other than a small increase in broadening which suggests that the surface gold environment is not significantly changed. However, there is a reversible shift in kinetic energy for the fitted Au4$f_{5/2}$ component as shown in Fig. 2b. This change in energy is observed for each core level and is interpreted as due to the thermal expansion of the alloy. Metals such as gold are expected to have an increase in binding energy when heated, while alkali metals have a decrease in binding energy.[26–29] The gold alloy in this case has a rate of change of 0.1 meV K^{-1} that compares with measurements for elemental gold and suggests a uniform expansion without a significant structural phase change. Relative changes in the peak intensities for the Au, Ag and Cu reflect the alloy composition to reveal stoichiometric changes at the surface as the temperature is increased. The energy shift is reliably measured in real-time with a fixed geometry by minimising experimental uncertainty due to changes in sample position. Calibration of the energy scale using pure metals reveal that the initial binding energy of the Au4f core levels for this alloy differ by less than 0.1 eV from the reference gold spectra due to a charge transfer effect in the alloy.[30] Changes in surface alloy composition with temperature has relevance for soldering application in sectors such as jewellery and dentistry, where the joints become fail points.[31,32] These studies enable metal migration of elements such Cu and Ag to be followed in real time, to better mitigate risks.[33]

3.2 n-type β-Ga_2O_3

The temperature sequence for the Ga2$p_{3/2}$ core level photoelectron spectrum for β-Ga_2O_3 is shown in Fig. 3a. Also recorded in the sequence were the Ga3$p_{3/2}$ and

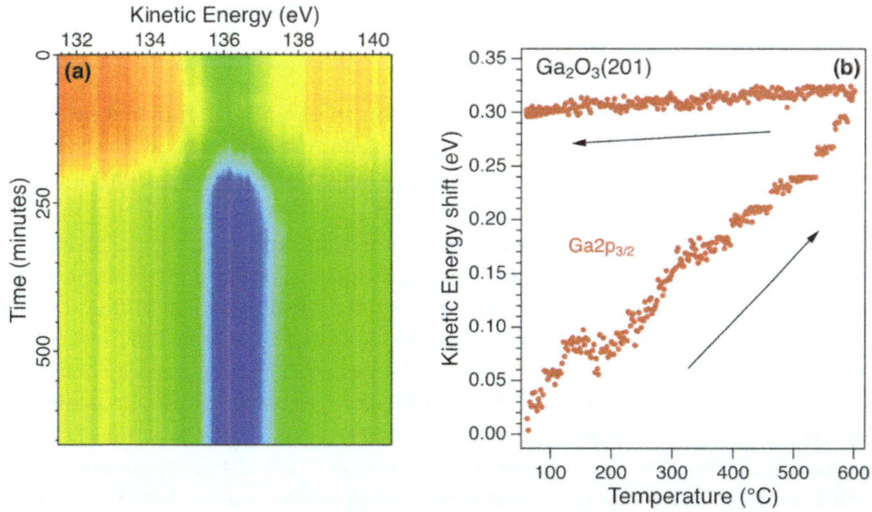

Fig. 3 (a) Real-time photoelectron spectroscopy for an n-type β-Ga_2O_3 single crystal during an annealing cycle from room temperature to 600 °C. The Ga2$p_{3/2}$ core level increases in intensity up to 600 °C and shows a significant, irreversible energy shift of 300 meV to higher kinetic energy (b).

O1s photoelectron emission spectra. The initial increase in intensity is due to the desorption of weakly-adsorbed species and is reflected in an increase in the corresponding O1s photoelectron emission and a decrease in the C1s photoelectron emission. The Ga and O core level emission spectra for the initial surface were fitted with two components and no additional components were required to fit the sequence shown in Fig. 3a. The components are broad such that identification of all surface species requires a higher surface sensitivity than available using these laboratory sources. For example, it has been shown using synchrotron radiation excitation that different adsorption sites are possible for surface OH species and that these are sensitive to the surface treatment.[34]

The peak position for the $Ga2p_{3/2}$ core level, extracted from the real-time sequence is shown in Fig. 3b. There is a significant increase in kinetic energy (decrease in binding energy) as the temperature is increased to 600 °C that is not reversed when the temperature is decreased to 20 °C. This irreversible shift is interpreted as a change in the potential barrier due to a change in surface states and Fermi level pinning. There is no evidence of surface charging for this semiconductor due to its low potential barrier, high doping concentration and low resistivity. During heating to 600 °C, the relative Ga : O ratio is increased as the surface OH species are desorbed irreversibly in vacuum. The O1s core level spectrum can be fitted with two components, a larger bulk peak corresponding to oxygen in Ga_2O_3 and a smaller low binding energy component due to surface OH groups. The latter component is reduced on heating as the OH is desorbed with the total area of the O1s reducing by 3%.[35] There is a corresponding change in the Ga core level lineshape as the Ga–OH component is removed on heating. The binding energies of the core levels in this n-type semiconductor are therefore correlated with the surface composition which affects the surface electronic states and hence potential barrier.[36] Knowledge of the thermal stability of semiconducting oxides is essential for device applications where the surface termination and band profile change with temperature.[37]

3.3 p-type diamond

The room temperature diamond photoelectron spectrum is dominated by emission from the C1s core level with a small emission from the O1s core level corresponding to a monolayer coverage (1 mL). The initial annealing of this surface in ultrahigh vacuum (UHV) to 400 °C desorbs residual surface contaminants, resulting in a small increase in core level emission and this is evident in the sequence of C1s snapshot spectra for the oxidised diamond (001) surface in Fig. 4a. This surface is stable when the maximum annealing temperature is below 400 °C, with both C1s and O1s core level emission spectra showing a reversible change in kinetic energy without change in lineshape. This reversible change is sensitive to the X-ray flux and is interpreted as a room temperature photovoltage, as reported for a H-terminated diamond (111) crystal with a similar boron concentration measured in the same environment.[38] For the (111) diamond, the observed shift and its saturation at higher temperature was modelled using the diamond resistivity as the main temperature-dependent variable. At room temperature, the photoexcitation process generates electrons and holes in the depletion region that are separated, with electrons accumulating at the surface when the resistance to earth in the diamond is high. Under these conditions, the

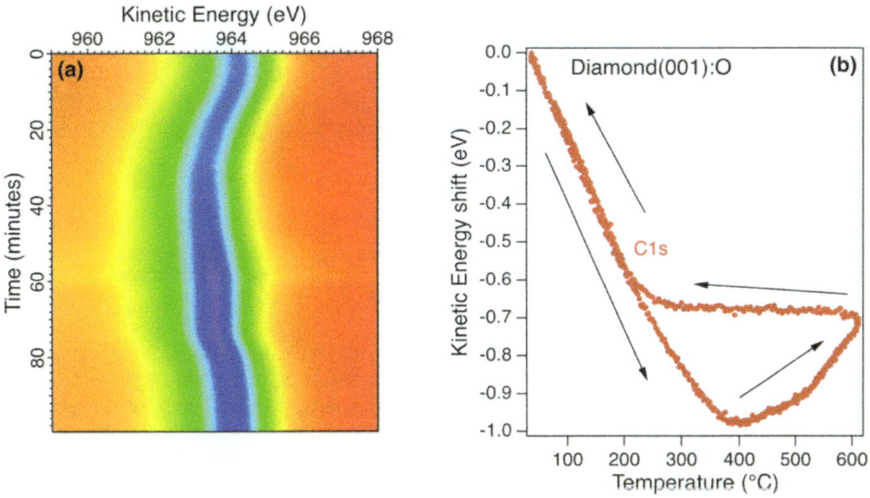

Fig. 4 (a) Real-time photoelectron spectroscopy for a p-type diamond single crystal during an annealing cycle from room temperature to 600 °C. (b) The kinetic energy of the dominant component of the C1s core level emission spectrum exhibits a reversible decrease of 960 meV on heating to 400 °C due to the removal of a surface photovoltage. Between 400 °C and 600 °C the energy shift is reversed and is unchanged on cooling to 200 °C. A surface photovoltage returns on cooling to room temperature.

surface is at a negative potential relative to the electron analyser and this leads to an increase in the measured electron binding energy and a decrease in the kinetic energy. As the temperature is increased, the diamond resistance and the depletion width decrease and the surface photovoltage reduces to zero. At room temperature, the bulk diamond peak for the p-type diamond (001) (Fig. 4a) was found to have an apparent binding energy similar to graphite rather than its true value of 285.4 eV that is only revealed at the higher temperature.[38,39] This surface photovoltage was also observed for this diamond when irradiated with He I and synchrotron radiation at room temperature and was found to persist when measured at near-ambient pressure of oxygen. Ultraviolet Photoelectron Spectroscopy (UPS) measurements for this moderately-B-doped diamond at the higher temperature yield a barrier height that is close to that predicted by the charge neutrality level but is dependent on the surface preparation.[40]

Above 400 °C there is a loss of surface oxygen and a reversal of the peak shift as shown in Fig. 4b for the sequence of C1s snapshot spectra recorded during a temperature cycle between room temperature and 600 °C. During the cooling phase, the peak position remained constant between 600 °C and 200 °C, returning to an energy close to the starting value at room temperature. The irreversible shift at higher temperatures is similar to that observed for the Ga oxide, corresponding to a decrease in the potential barrier for the p-type diamond as the surface oxygen is desorbed. Below 200 °C therefore, the surface photovoltage dominates and for this moderately-doped diamond, the true potential barrier is only correctly measured at the higher temperature when the resistance is sufficiently low to ensure equilibrium.

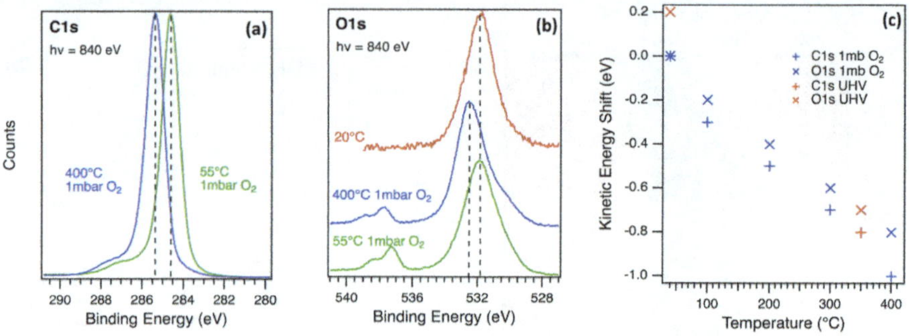

Fig. 5 Near-ambient pressure photoelectron emission spectra for the C1s (a) and O1s (b) core level in a p-type diamond (001) single crystal measured at 55 °C and 400 °C in 1 mbar of O_2 and at 20 °C in UHV. The incident photon energy for the synchrotron radiation excitation source was 840 eV in both cases. The kinetic energy of all peaks (c) changes with temperature during annealing in UHV and in 1 mbar of O_2 with a reversible shift as observed for higher energy X-ray excitation in UHV (Fig. 4).

To separate the influence of surface photovoltage from chemical changes, the annealing of the oxygen-terminated diamond (001) surface was repeated under 1 mbar of molecular oxygen as shown in Fig. 5. The C1s spectrum (Fig. 5a) is dominated by a main peak at a binding energy of 284.6 eV at 55 °C corresponding to C–C bonding in the bulk diamond, with a broad, multi-component shoulder at higher binding energy associated with C atoms in different C–O bonding environments. Non-diamond carbon contamination was minimal on this surface as there is a negligible component at low binding energy. The thermal annealing cycle resulted in small line shape changes but the main observation in Fig. 5a is a rigid energy shift that affects all peaks and components as observed in UHV. The O1s spectrum (Fig. 5b) is dominated by an ether (C–O–C) component at a binding energy of 531.9 eV at 55 °C, with lower intensity components due to other species such as ketones at lower binding energies.[41] The doublet at 537.3 eV and 538.4 eV corresponds to photoelectron emission from molecular oxygen in the gas phase above the surface.[42,43] The O1s emission peak for the oxidised diamond surface at the lower temperatures (Fig. 5b) has a similar lineshape when measured at UHV and in 1 mbar of oxygen confirming that the surface remained oxidised during the 400 °C annealing cycle in oxygen. At 400 °C, the substrate and gas phase peaks are shifted to higher binding energy, with a similar shift for the C1s core level. This energy shift in both core levels is dependent on temperature as shown in Fig. 5c and seems to be insensitive to the pressure. The reversible energy shift in the photoelectron spectra is similar to that observed in UHV and in this case can only be associated with physical effects. This surface processing method of annealing in oxygen has been shown to provide the optimal surface for diamond quantum technology applications.[44]

Correction for the surface photovoltage effect in diamond and confirmation of equilibrium measurement of β-Ga_2O_3 enables a direct comparison of the energy band diagrams for the two wide-gap semiconductors as shown in Fig. 6. In both cases, there are several possible adsorption sites and oxygen-containing species which have different photoelectron binding energies, making their precise identification difficult, especially in laboratory XPS measurements. The density of

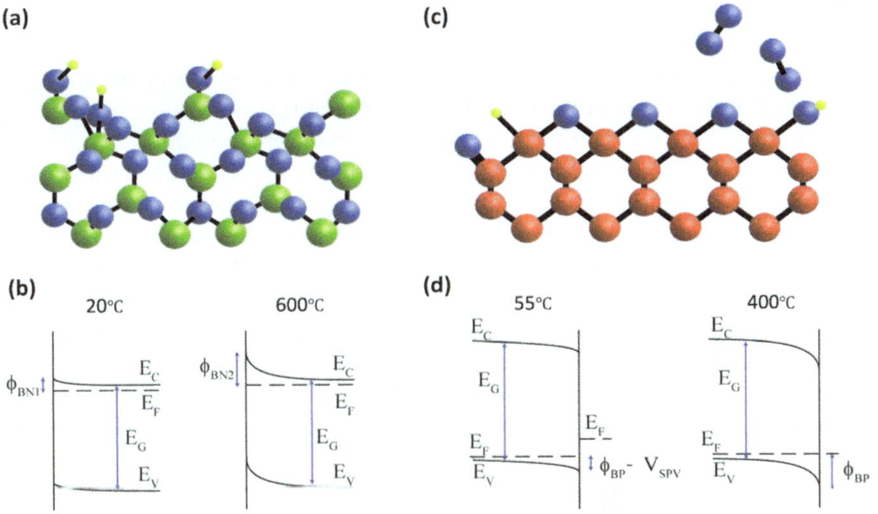

Fig. 6 (a) Schematic diagram of the β-Ga_2O_3 ($\bar{2}01$) surface with calculated atomic coordinates from Anvari *et al.*[35] The surface is terminated by OH groups that are desorbed during heating from room temperature to 600 °C. This affects the band bending as shown in (b). (c) The oxygen-terminated diamond (001) surface with calculated atomic coordinates from ref. 44 and 45 is preserved by heating to 400 °C in 1 mbar O_2. The apparent change in core level binding energy at high temperature is due to the removal of a surface photovoltage with an unchanged potential barrier as shown in (d).

electronic states that determine the surface Fermi level pinning is below the sensitivity of the measurements and cannot be directly measured in general. However, the parallel monitoring of photoelectron spectra during temperature annealing can reveal these correlations indirectly as demonstrated for these two wide-gap semiconductors. In Fig. 6a, only one possible OH site, bonded to surface Ga atoms, is shown to illustrate the desorption that is observed during heating to 600 °C. During this processing, the potential barrier for this n-type semiconductor increases from ϕ_{BN1} to ϕ_{BN2} as shown in Fig. 6b. For β-Ga_2O_3, therefore, the electronic properties are strongly dependent on adsorbed OH species, with the upward band bending increased irreversibly for the NaOH-treated surface when OH species are desorbed. For p-type diamond, the oxidised (001) surface is dominated by bridge C–O–C bonding, although other bonding configurations such as C–OH, C–H and C=O are usually present. This is shown schematically in Fig. 6c, with oxygen desorption above 400 °C prevented by measurement at a pressure of 1 mbar of molecular oxygen. Under these conditions, the core level binding energy is reversible, and the band bending is determined by a surface photovoltage, V_{SPV}, and the potential barrier ϕ_{BN} is unchanged as shown in Fig. 6d. The true barrier height is close to that measured for most metal contacts[17] that suggests a Fermi level pinning close to the charge neutrality level. Above 400 °C, surface oxygen is desorbed in vacuum and this reduces the downward band bending on this p-type diamond surface. The potential barriers for both n-type β-Ga_2O_3 and p-type diamond are thus sensitive to surface oxygen composition.

4 Conclusions

Real-time measurements reveal shifts in measured kinetic energy for all three samples, but the origin in each case is different. For the metal alloy, the binding energy shift of 60 meV is reversible and is due to thermal expansion. For the n-type β-Ga_2O_3, the shift of 300 meV to lower binding energy is due to a change in surface state pinning of the Fermi level as surface OH species are desorbed. For the p-type diamond, there is a reversible shift at temperatures below 400 °C due to the removal of a surface photovoltage and an irreversible shift above 400 °C, due to a change in surface Fermi level pinning position as the surface oxygen is desorbed. When the oxygen is replenished by heating in an oxygen environment, the shift is fully reversible. Corrected binding energy values aid the completion of the energy band diagrams for these wide-gap semiconductors and provide protocols for surface processing to engineer key surface and interface parameters.

Author contributions

SA, DH, KH, JA, RC, SC, MA, GW and AE contributed to the research and investigation using experimental measurements at Aberystwyth. SA, KH, JA, RC, SC, FV, GH, DG, PF, RA and AE contributed to the research and investigation using synchrotron radiation at the Diamond Light Source. JE and KJ provided and prepared the metal alloy materials. All authors contributed to the analysis and interpretation of measurements and to the final manuscript.

Conflicts of interest

There are no conflicts to declare.

Acknowledgements

The authors acknowledge the financial support of the EPSRC CDT in Diamond Science and Technology, the Diamond Light Source (instrument B07-C/VerSoX; proposal SI18182), the European Regional Development Fund through the Solar Photovoltaic Academic Research Consortium (SPARC II) and the Centre for Photonics Expertise (CPE) operated by the Welsh Government. We thank R. F. Martinez-Gazoni and R. J. Reeves of the School of Physical and Chemical Sciences, University of Canterbury, New Zealand, for their help in the preparation and characterisation of the β-Ga_2O_3 samples.

References

1 M. P. Seah, I. S. Gilmore and S. J. Spencer, *J. Electron Spectrosc. Relat. Phenom.*, 2001, **120**, 93–111.
2 J. C. Fuggle and S. F. Alvardo, *Phys. Rev. A: At., Mol., Opt. Phys.*, 1980, **22**, 1615–1624.
3 M. Descostes, F. Mercier, N. Thromat, C. Beaucaire and M. Gautier-Soyer, *Appl. Surf. Sci.*, 2000, **165**, 288–302.
4 K. J. Kim, D. W. Moon, C. J. Park, D. Simons, G. Gillen, H. Jin and H. J. Kang, *Surf. Interface Anal.*, 2007, **39**, 665–673.

5 J. M. Cerrato, M. F. Hochella, W. R. Knocke, A. M. Dietrich and T. F. Cromer, *Environ. Sci. Technol.*, 2010, **44**, 5881–5886.
6 J. Yan, A. D. Carl, A. R. Maag, J. C. Macdonald, P. Müller, R. L. Grimm and S. C. Burdette, *Dalton Trans.*, 2019, **48**, 4520–4529.
7 W. Mönch, *Rep. Prog. Phys.*, 1990, **53**, 221–278.
8 T. Veal, P. Jefferson, L. Piper, C. F. McConville, T. Joyce, P. Chalker, L. Considine, H. Lu and W. Schaff, *Appl. Phys. Lett.*, 2006, **89**, 202110.
9 D. Caliskan, H. Sezen, E. Ozbay and S. Suzer, *Sci. Rep.*, 2015, **5**, 14091.
10 D. R. Baer, K. Artyushkova, H. Cohen, C. D. Easton, M. Engelhard, T. R. Gengenbach, G. Greczynski, P. Mack, D. J. Morgan and A. Roberts, *J. Vac. Sci. Technol., A*, 2020, **38**, 031204.
11 J. Y. Tsao, *et al.*, *Adv. Electron. Mater.*, 2018, **4**, 1600501.
12 C. J. Powell and M. P. Seah, *J. Vac. Sci. Technol., A*, 1990, **8**, 735–763.
13 G. A. Rooke, *J. Res. Natl. Bur. Stand., Sect. A*, 1970, **74**, 273–279.
14 L. Diederich, O. M. Küttel, P. Aebi and L. Schlapbach, *Surf. Sci.*, 1998, **418**, 219–239.
15 S. Rafique, L. Han, S. Mou and H. Zhao, *Opt. Mater. Express*, 2017, **7**, 3561–3570.
16 S. Santucci, C. Cantalini, M. Crivellari, L. Lozzi, L. Ottaviano and M. Passacantando, *J. Vac. Sci. Technol., A*, 2000, **18**, 1077.
17 D. A. Evans, H. J. Steiner, S. Evans, R. Middleton, T. S. Jones, S. Park, T. U. Kampen, D. R. T. Zahn, G. Cabailh and I. T. McGovern, *J. Phys.: Condens. Matter*, 2003, **15**, S2729.
18 J. Wu, B. Wang, F. Yi, S. Deng, N. Xu and J. Chen, *RSC Adv.*, 2018, **8**, 752–760.
19 F. S. Roberts, S. L. Anderson, A. C. Reber and S. N. Khanna, *J. Phys. Chem. C*, 2015, **119**, 6033–6046.
20 M. Bär, J.-P. Theisen, R. G. Wilks, F. Erfurth, R. Félix, D. Gerlach, S. Haas, L. Tati Bismaths, F. Reinert, F. Kronast, T. P. Niesen, J. Palm, S. Visbeck and L. Weinhardt, *J. Appl. Phys.*, 2013, **113**, 193709.
21 B. Henriques, P. Pinto, J. Souza, J. C. Teixeria, D. Soares and F. S. Silva, *Gold Bull.*, 2013, **46**, 117–225.
22 A. Kuramata, K. Koshi, S. Watanabe, Y. Yamaoka, T. Masui and S. Yamakoshi, *Jpn. J. Appl. Phys.*, 2016, **55**, 1202A2.
23 T. Kwak, J. Lee, G. Yoo, H. Shin, U. Choi, B. So, S. Kim and O. Nam, *Phys. Status Solidi A*, 2020, **217**, 1900973.
24 T. R. Chase, D. A. Evans, D. P. Langstaff and A. Bushell, *Nucl. Instrum. Methods Phys. Res., Sect. B*, 2005, **238**, 219–223.
25 G. Held, F. Venturini, D. Grinter, P. Ferrer, R. Arrigo, L. Deacon, W. Quevedo Garzon, K. Roy, A. Large, C. Stephens, A. Watts, P. Larkin, M. Hand, H. Wang, L. Pratt, J. J. Mudd, T. Richardson, S. Patel, M. Hillman and S. Scott, *J. Synchrotron Radiat.*, 2020, **27**, 1153–1166.
26 D. M. Riffe, G. K. Wertheim, D. N. Buchanan and P. H. Citrin, *Phys. Rev. B: Condens. Matter Mater. Phys.*, 1992, **45**, 6216–6225.
27 D. M. Riffe, W. Hale, B. Kim and J. L. Erskine, *Phys. Rev. B: Condens. Matter Mater. Phys.*, 1996, **54**, 17118–17122.
28 T. Durakiweicz, J. A. J. Arko, J. J. Joyce, D. J. P. Moore and A. S. Halas, *Surf. Sci.*, 2001, **478**, 72–82.
29 K. Shen, G. Li and W. Wang, *Trans. Nonferrous Met. Soc. China*, 2016, **26**, 2900–2909.

30 D. Wang, X. Cui, Q. Xiao, Y. Hu, Z. Wang, Y. M. Yiu and T. K. Sham, *AIP Adv.*, 2018, **8**, 065210.

31 I. Watanabe, E. Watanabe, M. Atsuta and T. Okabe, *J. Prosthet. Dent.*, 1997, **78**, 260–266.

32 R. Süss, E. van der Lingen, L. Glaner and M. du Toit, *Gold Bull.*, 2004, **37**, 196–207.

33 F. W. Jones and C. Sykes, *Proc. R. Soc. London, Ser. A*, 1938, **166**, 376–390.

34 R. M. Gazoni, L. Caroll, J. I. Scott, S. Astley, D. A. Evans, A. J. Downard, R. J. Reeves and M. W. Allen, *Phys. Rev. B*, 2020, **102**, 035304.

35 R. Anvari, D. Spagnoli, G. Parish and B. Nener, *Chem.–Eur. J.*, 2018, **24**, 7445–7455.

36 J. E. N. Swallow, J. B. Varley, L. A. H. Jones, J. T. Gibbon, L. F. J. Piper, V. R. Dhanak and T. D. Veal, *APL Mater.*, 2019, **7**, 022528.

37 R. Heinhold, S. P. Cooil, D. A. Evans and M. W. Allen, *J. Phys. Chem. C*, 2014, **118**, 24575–24582.

38 G. T. Williams, S. P. Cooil, O. R. Roberts, S. Evans, D. P. Langstaff and D. A. Evans, *Appl. Phys. Lett.*, 2014, **105**, 061602.

39 S. J. Schmeig and D. N. Belton, *Surf. Sci. Spectra*, 1992, **1**, 333.

40 S. J. Sque, R. Jones and P. R. Briddon, *Phys. Rev. B: Condens. Matter Mater. Phys.*, 2006, **73**, 085313.

41 F. Maier, J. Ristein and L. Ley, *Phys. Rev. B: Condens. Matter Mater. Phys.*, 2001, **64**, 165411.

42 J. Schnadt, J. Knudsen, J. N. Andersen, H. Siegbahn, A. Pietzsch, F. Hennies, N. Johansson, N. Mårtensson, G. Öhrwall, S. Bahr, S. Mähld and O. Schaff, *J. Synchrotron Radiat.*, 2012, **19**, 701–704.

43 P. E. Pehrsson and T. W. Mercher, *Surf. Sci.*, 2000, **460**, 59–66.

44 S. Sangtawesin, B. L. Dwyer, S. Srinivasan, J. J. Allred, L. V. H. Rodgers, K. De Greve, A. Stacey, N. Dontschuk, K. M. O'Donnell, D. Hu, D. A. Evans, C. Jaye, D. A. Fischer, M. L. Markham, D. J. Twitchen, H. Park, M. D. Lukin and N. P. de Leon, *Phys. Rev. X*, 2019, **9**, 03105.

45 D. Petrini and K. Larsson, *J. Phys. Chem.*, 2007, **111**, 795–801.

Faraday Discussions

ROYAL SOCIETY OF CHEMISTRY

PAPER

Reaction surfaces and interfaces of metal sulfides: cryo-XPS meets HAXPES and DFT†

Yuri Mikhlin, * Vladimir Nasluzov, Yevgeny Tomashevich, Sergey Vorobyev, Alexander Romanchenko and Maxim Likhatski

Received 29th November 2021, Accepted 6th January 2022
DOI: 10.1039/d1fd00104c

We report here the application of the low-temperature X-ray photoelectron spectroscopy (cryo-XPS) of fast-frozen dispersions as a quasi *in situ* technique for a case study of metal sulfides reacted in acidic aqueous solutions under non-oxidizing and moderate oxidizing conditions. The sulfide surfaces are known to tend to be depleted in metals, producing essentially sulfur-enriched surfaces and extended underlayers on Fe- and Cu-bearing sulfides, which have previously been examined using depth-sensitive HAXPES and cryo-XPS. The current study is focused on zinc and lead sulfides (natural sphalerite and galena), for whom both the experiment and theoretical DFT simulations suggest a low stability of sulfur-excessive structures. Cryo-XPS revealed the complicated behavior of the minerals under non-oxidative etching conditions, in particular, a notable concentration of polysulfide for PbS in dilute perchloric acid and a very minor one in hydrochloric acid. Oxidative etching with Fe^{3+} cations produced polysulfide anions and then elemental sulfur, which both volatized in the ultra-high vacuum at room temperature; the species can, nonetheless, be distinguished by considering the binding energies, electrostatic charging and evaporation rates. The cryo-XPS also detected interfacial products, *e.g.*, ferrous chloride. DFT found that S-excessive centers are unstable in the undersurface regions of both materials, but are less unfavorable for ZnS surfaces, in agreement with the experimental data. It was concluded that cryo-XPS allows us to greatly reduce distortions of the interface composition in comparison with conventional techniques.

1. Introduction

The main limitation of X-ray photoelectron spectroscopy as a method for the examination of interfacial structures and processes is the necessity of an ultra-high vacuum environment, in which the interface decomposes and solid surfaces undergo substantial modification. At present, near-ambient pressure

Institute of Chemistry and Chemical Technology, Krasnoyarsk Science Center of the Siberian Branch of the Russian Academy of Sciences, Akademgorodok, 50/24, Krasnoyarsk, 660036, Russia. E-mail: yumikh@icct.ru
† This paper was not presented at the meeting.

(NAP) XPS[1–5] has attracted much attention as a promising tool for the examination of solid/gas and solid/liquid interfaces. However, the techniques require complex installation, often including synchrotron radiation facilities. Cryogenic, or low-temperature XPS (cryo-XPS) has emerged as an alternative method for studying the processes at the boundary with aqueous media. As proposed by Shchukarev,[6–9] particulate materials of interest are separated from aqueous phases by centrifugation and the wet pastes are placed and rapidly frozen on a sample holder which is pre-cooled to a liquid nitrogen temperature (LN2) in the entry lock chamber before the evacuation. In contrast to a "freeze-drying" procedure[10] that means, in fact, the cooling of desiccated samples, the fast-freezing allows both volatile surface species and a thin, 1–2 nm, film of the aqueous phase containing near-surface solutes, adsorbates and adhered solid species to be kept. Cryo-XPS has been successfully used for the investigation of many objects, for example, electrical double layers,[8,11] the products of sorption and chemical reactions, including at metal sulfides,[12–15] living cells and viruses,[16] and so forth. However, a number of methodological questions remain unclear, requiring further investigation with a wide range of objects.

Metal sulfides are abundant minerals, which are the main source of basic and precious metals and play crucial roles in many natural processes.[17] On the other hand, metal chalcogenides are indispensable functional materials often utilized as thin films and nanoscale particles, whose surface characteristics are critically important in many fields of materials science, including sensors, photovoltaics, biomedicine, energy conversion and storage, and others (for example, in ref. 18 and references therein). Metal sulfides undergo significant corrosion upon contact with the atmosphere and aqueous media, which has been studied using XPS methods[12–15,17–27] but is still far from being completely understood. An important peculiarity of metal sulfide reactivity is the preferential release of metal cations into aqueous solution or metal (hydr)oxide or/and another phase, leaving the surface and under-surface layer depleted in metal and enriched in sulfur. These phenomena are detectable mainly due to S–S bonding and the formation of di- and polysulfide anions.[19–33] The nature of the metal-deficient structures, which are disordered, non-equilibrium and fairly thick in many cases, and the mechanisms behind their formation and decomposition yielding elemental sulfur as the main S-bearing product, remains poorly understood.

X-ray absorption spectroscopy (XANES) and some other techniques with probing depths larger than that of conventional XPS have revealed that the non-stoichiometric regions can be essentially, up to dozens and hundreds of nm thick.[25–27] We recently established, by applying synchrotron-based high energy

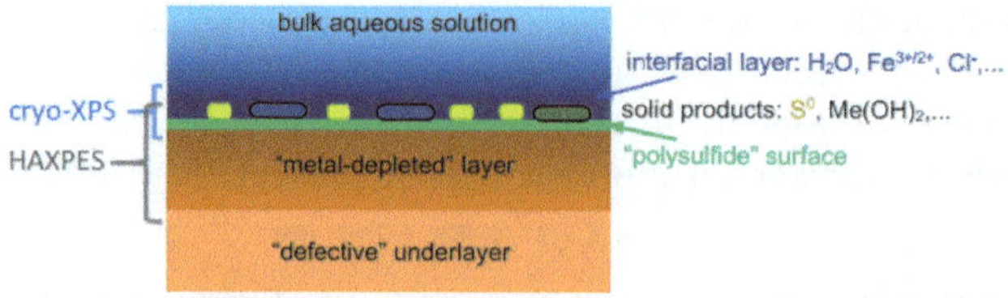

Fig. 1 Scheme illustrating a stratified structure of the reaction interfaces at metal sulfides and arbitrary probing depths for their analysis with HAXPES and cryo-XPS.

photoemission spectroscopy (HAXPES) together with XANES and DFT+U calculations, that the metal-deficient sulfide regions formed for the iron sulfides FeS_2 and Fe_9S_{10} (ref. 29) and chalcopyrite, $CuFeS_2$,[30] in air or aqueous solutions are stratified into at least three layers (Fig. 1). The outermost surface layers with the highest excess of S contain polysulfide species; the underlying regions, which are notably depleted in metals with mono- and disulfide anions (and probably trisulfide species for pyrrhotite $Fe_{1-x}S$, to be published), varied in thickness from few nm to several tens of nm. The nearly-stoichiometric "defective" regions extend down to hundreds of nm in depth and manifest themselves through distortions of the XANES spectra and physical characteristics (conductivity, *etc.*). It is likely that the buried layers remain almost intact over the UHV experiments but the outer "polysulfide" layers, interfaces and reaction products can be substantially modified or just disappear, like elemental sulfur and, partially, polysulfide species, and must be studied by applying cryo-XPS.[31,32] Aside from the elusive sulfur species, cryo-XPS was able to find out[32] high concentrations of Fe atoms with unusual oxidation and spin states of Fe atoms, *e.g.*, high-spin Fe^{2+}–S and Fe^{3+}–S at the pyrite–water interface instead of singlet Fe^{2+} in the solid bulk, which are missed in the room temperature spectra.

The metal depletion and stratification phenomena are especially pronounced for iron and copper sulfides,[20–33] whereas "metal-depleted" strata can be absent or very thin and undetectable with HAXPES, depending on the metal sulfide nature and the reaction conditions. This appears to be the case for pyrite, FeS_2,[29] and also for sphalerite, ZnS, and galena, PbS, which are the main minerals of these metals and are important semiconductors (with band gaps of about 3.6 eV and 0.4 eV, respectively) for optoelectronics and other applications. While elemental sulfur has been not been found in early XPS studies,[19] Buckley and Woods[20–22] have detected sulfur and polysulfide species in dried and frozen mineral samples which were preliminary oxidized in air and aqueous solutions; they introduced the terms "metal-deficient" and "polysulfide" surfaces for the samples with an excess of S but minor S–S bonding and with significant one, accordingly. Although no HAXPES studies uncovered buried "metal-depleted" strata on PbS and ZnS, the changes of semiconducting properties revealed in a number of studies suggest the existence of extended defective layers at the materials.

Here, we report on the cryo-XPS examination of fast-frozen particulate sphalerite and galena treated in different diluted non-oxidizing acids and oxidized in aqueous ferric iron solutions. The reaction conditions, *i.e.* moderate concentrations, time and temperature, were selected in order to register both the products and underlying sulfide surfaces (Fig. 1). The experimental results for sphalerite and galena are compared with DFT simulations of metal-deficient quasi-bulk and surface structures, and with previous cryo-XPS, HAXPES and DFT results on $CuFeS_2$, FeS_2 and $Fe_{1-x}S$. The findings are discussed in terms of the real composition of the reaction interfaces and prospects of the cryo-XPS technique.

2. Materials and methods

2.1. Materials and preparation

Natural sphalerite, with the chemical composition (wt%) S 32.1%, Zn 65.9%, Fe 2.09%, Pb 0.04%, Cu 0.05%, and lead sulfide, with the total concentration of impurities less than 0.2%, were both from the Zhairem ore deposit, Kazakhstan.

The selected mineral samples contained no inclusions of foreign phases according to the mineralogical and XRD analyses. The samples were ground in a jasper mortar to a particle size of less than 75 μm immediately before the experiments. Analytical grade chemicals and deionized water were used to prepare the solutions. About 0.5 g of the metal sulfide and 20 mL of 0.1 M HCl, 0.1 $HClO_4$, 0.05 M $FeCl_3$ + 0.05 M HCl, or similar solutions were mixed and agitated in a thermostated glass at 25 °C or 50 ± 0.5 °C for a predetermined time. The dispersion was loaded into a polypropylene beaker and centrifuged at 5000 rpm (10 min); the supernatant was discarded.

2.2. XPS measurements

The cryo-XPS experiments were performed using a SPECS photoelectron spectrometer (SPECS Surface Nano Analysis GmbH) equipped with a PHOIBOS 150 MCD9 energy analyzer; the entry lock chamber was modified in-house to allow freezing of the samples to LN2 temperatures. The wet paste obtained by centrifugation was promptly placed on a Ti sample holder kept at −150 °C and frozen in the spectrometer air-lock under dry nitrogen gas for 60 s, and then evacuated to a vacuum of 10^{-7} mbar. The sample was transferred to the pre-cooled manipulator in the analytical chamber with the vacuum and the sample temperature maintained at 10^{-9} mbar and about −155 °C, respectively. After the cryo-XPS experiments, the specimens were warmed in the spectrometer vacuum up to room temperature (RT) overnight and reexamined. The spectra were also collected at RT from the samples, which were reacted under the same conditions and dried in air or vacuum after centrifugation. Photoelectron spectra were acquired at electron emission normal to the sample surface using a monochromated Al Kα source (1486.7 eV). The binding energy (BE) scale was referenced to the C 1s peak at 285.0 eV of an adventitious contamination; an electron flood gun FG-20 (SPECS) was typically applied to eliminate electrostatic charging. The high-resolution spectra were fitted with Gaussian–Lorentzian peak profiles after subtraction of the Shirley-type background. Spin–orbit splitting and an intensity ratio for S $2p_{3/2,1/2}$ doublet were 1.19 eV and 2 : 1, respectively. CasaXPS software package was utilized for spectra processing.

2.3. DFT computational details

The DFT calculations were performed using the GGA+*U* scheme[34] with a PW91 parameterization of the exchange–correlation functional[35] and the Hubbard-type on-site correction parameter *U* (for ZnS), as implemented in the Vienna *ab initio* simulation package (VASP).[36] The *U* value was set to be equal to 7 eV for the Zn 3d states, as justified elsewhere.[37] Spin unrestricted eigenfunctions of the projector augmented wave method,[38] with explicit treatment of the $3s^23p^4$ electrons for S atoms, $4s^23d^{10}$ electrons for Zn atoms and $6s^26p^2$ electrons for Pb atoms, were generated using plane waves restricted by an energy cut off of 300 eV. Numerical integration in the reciprocal space was carried out using a 5 × 5 × 5 Monkhorst–Pack *k*-points grid.[39] Translation unit cells for defects were constructed from cubic symmetry unit cells of the bulk of PbS and ZnS. To model defects of the PbS(001) surface, we used a 64-atom translation cell as a four-layer slab with a vacuum gap of 9.1 Å. The related system without a vacuum gap was used to calculate quasi-bulk defects. To model defects of the ZnS(110), a surface 32-atom translation cell (four-layer slab) with a vacuum gap of 7.6 Å was utilized, and the same construction without a gap was used for modelling quasi-bulk defects. The formation energy E_f of defects with x cationic vacancies corresponds to the energy of the arbitrary reactions

$$Pb_{32}S_{32} + x(H_2O^{sol} + 0.5O_2) \rightarrow Pb_{32-x}S_{32} + xPb(OH)_2{}^{sol},$$

$$Zn_{16}S_{16} + x(H_2O^{sol} + 0.5O_2) \rightarrow Zn_{16-x}S_{16} + xZn(OH)_2{}^{sol},$$

where $x = 1$–3, and the index "sol" indicates that the total molecular energies of H_2O and $M(OH)_2$ included solvent effects.[40] To compare the different structures, the formation energies E_f were normalized per one removed metal atom.

3. Results and discussion

3.1. Cryo-XPS of the PbS surfaces

Fig. 2 shows the photoelectron spectra of lead sulfide reacted in 0.1 M hydrochloric acid, which was fast-frozen and measured at LN2 temperature and then thawed overnight in the UHV. The atomic ratio of Pb/S is close to 1 with an insignificant, within the error margin, excess of Pb. The signal of oxygen is small (the ratio of Pb/O > 2), suggesting the hydrophobic character of the surface with a low content of hydrophilic products and a minor amount of adjacent water. The concentration of Cl is small (Pb/Cl = 4.5), with the high-resolution Cl 2p spectrum fitted well using one doublet (BE of Cl $2p_{3/2}$ peak of 198.2 eV) attributable to chloride anions. The Pb 4f spectrum can be fitted with a major component of lead sulfide with the Pf $4f_{7/2}$ peak at 137.8 eV, and a signal of about 11% of total Pb at 138.8 eV. It follows from these intensities and BEs that the minor Pb species is mainly lead chloride, $PbCl_2$, with some contributions from lead oxide, as the O 1s bands contain a signal from O^{2-} ions at 529.5 eV, and hydroxide. Interpretation of the S 2p spectra is complicated due to its overlapping with a secondary structure of the strong Pb 4f band, while the S 2s lines are wide and less informative. The comparison of a number of specimens, as well as the literature data,[14,20] suggest that the feature at 165–166 eV arises from the Pb 4f spectra, that is arbitrary fitted with a S 2p doublet, shown in grey colour (Fig. 2–4) for convenience. A small doublet at 163 eV (about 6% of total sulfur) may originate from S–S species, although its exact origin and parameters are uncertain.

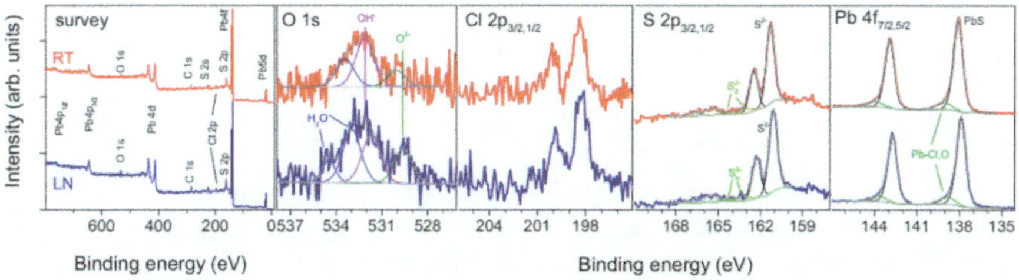

Fig. 2 X-ray photoelectron spectra acquired from PbS reacted in 0.1 M HCl for 30 min at 25 °C, fast-frozen after centrifugation and measured at LN2 temperature and the same samples thawed overnight in the vacuum and measured at room temperature (RT).

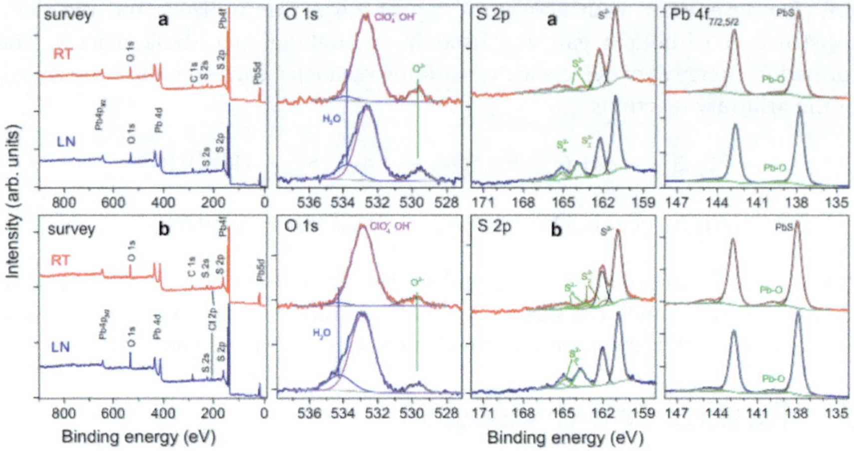

Fig. 3 X-ray photoelectron spectra acquired from PbS reacted in (a) 0.1 M $HClO_4$ and (b) 0.1 M $HClO_4$ + 1 mM Fe^{3+} for 30 min at 25 °C, fast-frozen after centrifugation and measured at LN2 temperature, and the same samples thawed overnight in the vacuum and measured at room temperature (RT).

The spectra of the sample warmed up to room temperature and the ones dried in air or vacuum without freezing collected at RT (not shown in figures) only slightly differ. The main differences are observed in the O 1s spectra, where the contributions of water at 534.6 and 533.5 eV notably declined. Therefore, the reaction of PbS in hydrochloric acid solutions causes minimal alterations of the solid sulfide surface and interfacial layer, which are detected both with cryo- and routine RT-XPS. Similar behaviour was also observed for galena reacted in 0.1 M acetic acid (not shown in the figures), including after a prolonged oxidation in ambient air before etching.

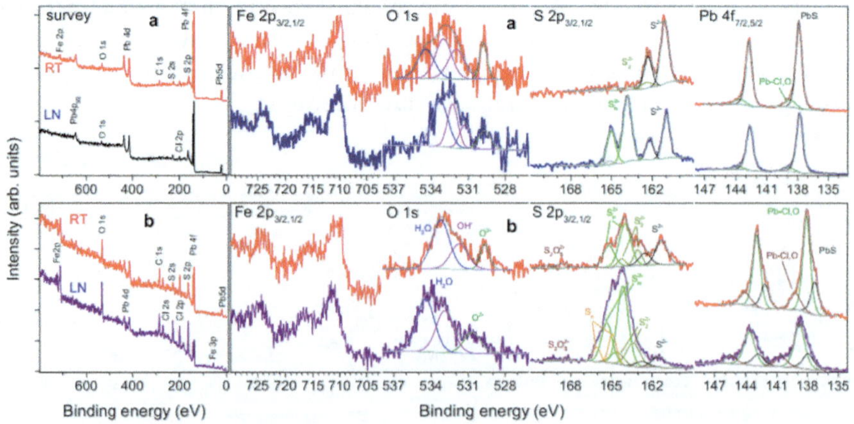

Fig. 4 X-ray photoelectron spectra of PbS reacted in 0.05 M HCl + 0.05 M $FeCl_3$ (a) for 1 min and (b) for 10 min at 50 °C, fast-frozen after centrifugation and measured at LN2 temperature, and the same samples thawed overnight in the vacuum and measured at room temperature (RT).

The effect of perchloric acid etching is different (Fig. 3). Although the Pb/S ratio stays close to 1 too, the S 2p spectra show a notable, almost 20 rel.%, component at 163.4 eV that should be assigned to polysulfide rather than elemental sulfur. The spectra also show a larger oxygen signal, mainly from perchlorate anions (532.4 eV). The content of perchlorate (Cl $2p_{3/2}$ at 207.8 eV) approaches 5% and partially correlates with a component of oxidized Pb, which also encompasses some hydr(oxide) species, and is lower than in HCl solution. A prolonged acidic treatment or addition of small quantity of Fe^{3+} cations (spectra b) somewhat enhances the S 2p doublet at 163.4 eV. Upon warming up in the UHV, as well as for standard preparation of the specimens without freezing, the signal at 163.4 eV decreases by several times or almost disappears; other spectra change insignificantly, except for the loss of water.

The results highlight a difference in the reaction mechanisms of the acidic dissolution of PbS in hydrochloric and perchloric acids, which were previously observed by various reliefs of the etched surfaces using scanning probe microscopy and the Fermi level positions in the band gap as detected by tunneling spectroscopy.[41] It was proposed that these acids produce different defects at the PbS surface and in the undersurface layer. The current findings suggest that Pb cations leave the surface and interfacial layer more easily because of the higher solubilisation in perchlorate solutions than in chloride media.

Fig. 4 presents the typical spectra of PbS oxidized in ferric chloride solutions in the early and advanced reaction stages. Even short, 1 min contact with 50 mM $FeCl_3$ solution (spectra a in the upper panels) results in a large excess of sulfur (Pb/S = 0.12) along with high amounts of Fe and Cl (with ratios of Pb/Fe and Pb/Cl of about 1.2 and 2.2, respectively), and a rather low concentration of oxygen. The S 2p spectral region is dominated by a doublet at the BE of 163.4 eV, with the contribution of sulfide anions (PbS) being as low as 10 rel.%. Furthermore, there is a very small doublet at ~168 eV that is due to sulfite or thiosulfate as minor by-products of sulfide oxidation. About 20% of lead occurs as Pb chloride, oxide and hydroxide, and the rest is PbS. The main part of chloride is bonded to Fe, since the Fe 2p spectra are similar to that of $FeCl_2$ with the characteristic satellite near 715 eV,[32,40] *i.e.* this is the reduced reactant kept at the interface. We refrain, however, from detailed fitting of the Fe 2p spectra composed of multiplet sets of lines[42] from several chemical forms of Fe. The sample warmed up to RT in the spectrometer vacuum lost the major share or almost all excessive S with the BE of about 163.5 eV (the case shown in Fig. 4a), while some disulfide species (BE of 162.5 eV) became observable. Smaller changes in the spectra of other elements suggest, in particular, the partial hydrolysis of Fe and Pb chlorides.

The galena oxidized under more severe conditions exhibits not only higher excesses of sulfur but also large concentrations of Fe, Cl and O (Fig. 4b). The S 2p band can be fitted using doublets at 164.3 eV, 163.6 eV, 162.7 eV and 161 eV, which can be assigned to elemental sulfur, poly-, di- and monosulfide species, along with very minor S–O species (~168 eV). However, the fitting results are not unequivocal because of non-uniform electrostatic charging that is seen in the Pb 4f spectrum too. The charging can be explained by the presence of the insulating phase of elemental sulfur, entrapping large quantities of the products. After thawing the sample to RT over 16 h, the Pb/S ratio increases by several times and "elemental sulfur" (the component with the BE larger than 164 eV) disappears, but the charging still persists due to the entrapped nonvolatile products.

The findings can be rationalized as follows. Excessive sulfur produced by the oxidation first forms the polysulfide-like species S_n^{2-} (n may be $\gg 2$), which are adsorbed on the reacted surfaces and/or covalently bonded with a few lattice Pb^{2+} cations *via* some S atoms ("disulfide"). The signatures of the elemental sulfur phase are the S $2p_{3/2}$ peak at the BE at 164 eV and higher, and inhomogeneous electrostatic charging of the specimens. The last effect is due to both dielectric sulfur and reaction products caught by the sulfur (Fig. 1). Elemental sulfur is more volatile than polysulfide attached to the sulfide surface but its quantities are often much larger, so their evaporation could be uncompleted during the XPS experiment. Furthermore, hydrophobic elemental sulfur tends to form protrusions and similar entities[43] rather than being uniformly distributed through aqueous interfaces, further slowing down the volatilization. It is also evident from comparison with previous studies[29–32] that the evaporation of polysulfide species is retarded on metal sulfides with underlying essentially metal-deficient layers with S–S bonding, which are detected with HAXPES and/or XANES, that is, at chalcopyrite[30,31] and pyrrhotite.[29,32]

3.2. Cryo-XPS of sphalerite

The spectra of sphalerite etched in 0.1 M HCl under non-oxidative conditions for 2 h are shown in Fig. 5. Akin to galena, the metal/sulfur ratios stay close to 1, also taking into consideration impurities of Fe and Cu (sphalerite surfaces are often enriched in copper relative to the volume). The interface also comprises considerable oxygen and chloride ions with the atomic Zn/O and Zn/Cl ratios being in the order of 1. The Zn $2p_{3/2}$ maximum can be fitted with the main peak of ZnS and small ones at higher BEs from Zn hydroxide and chloride.[15,22,44] However, the chloride ions are largely bonded in ferrous chloride, as it follows from the Fe 2p spectrum. The S 2p spectrum contains a clear contribution of disulfide anions (10–12 rel.%), which is kept after warming the sample up to RT in the vacuum but

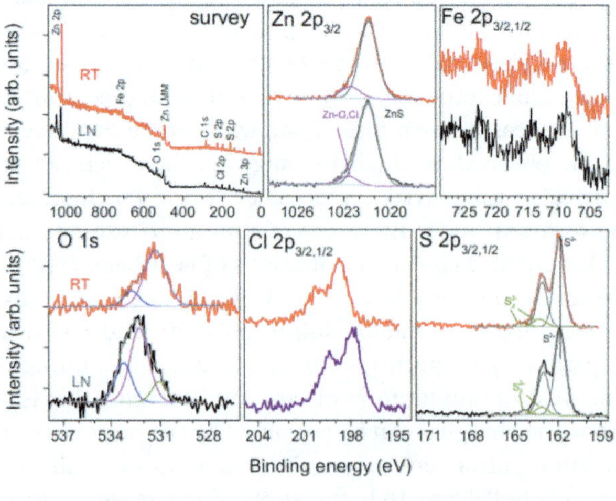

Fig. 5 X-ray photoelectron spectra of ZnS reacted in 0.1 M HCl at 50 °C for 2 h, fast-frozen after centrifugation and measured at LN2 temperature, and this sample thawed overnight in the vacuum and measured at room temperature (RT).

almost missed in the conventional procedures of air or vacuum drying. The sample fast-frozen and thawed in the vacuum shows slightly modified spectra and reduced concentrations of oxygen and chlorine that can be rationalized in terms of the hydrolysis and evaporation of HCl in the vacuum, which is more pronounced than that for PbS. Interestingly, the concentrations of disulfide species and interfacial reaction products decrease after a longer acidic etching (*e.g.*, for 4 h). This correlates with the slowing of the dissolution rate reported previously, probably owing to the changed properties of the buried "defective" region.

Fig. 6 shows the XPS spectra of fast-frozen and then warmed up sphalerite that was preliminarily etched in oxidizing 0.05 M $FeCl_3$ solution. The reaction creates excessive sulfur (Zn/S $\sim$ 0.5), exhibiting roughly equal doublets at 161.7 eV, from monosulfide, and 163.6 eV, probably from polysulfide, together with a small disulfide component at 162.4 eV (about 6 rel.%). The asymmetric Zn 2p band can be fitted with peaks of oxidized and probably solubilized Zn chloride and hydroxide species at 1022.8 eV (this agrees with the kinetic energy of the Auger Zn L_3MM peak of 990.4 eV)[44] and a low ZnS contribution.

This rationalization concurs with rather high concentrations of Cl and O (Zn/Cl $\sim$ 1 and Zn/O $\sim$ 0.4); in difference to PbS etched under similar conditions, no signal of O^{2-} and a moderate content of Fe (Zn/Fe $\sim$ 3) were observed. So, the reaction interface is dominated by dissolved or/and hydrophilic surface Zn species above "polysulfide" surface. After the warming, the high-energy S 2p component drastically decreases but still occurs, and the main part of water evaporates, while chloride ions stay and form ferrous chloride. Some accumulation of Cu on the surface takes place, as also observed by other researchers for mineral samples.[20–22] In general, the oxidative leaching of sphalerite is slower than that of galena, and yields polysulfide species rather than elemental sulfur under the same reaction conditions (Fig. 4 and 6). The oxidation of sphalerite using higher concentrations of ferric iron, enhanced temperature, or a longer

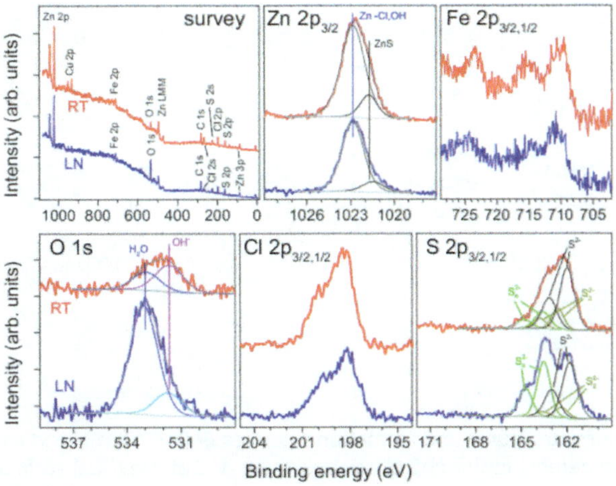

Fig. 6 X-ray photoelectron spectra of ZnS reacted in 0.05 M HCl + 0.05 M $FeCl_3$ (a) at 50 °C for 10 min, fast-frozen after centrifugation and measured at LN2 temperature, and that thawed overnight in the vacuum and measured at room temperature (RT).

reaction time produces elemental sulfur too; the cryo-XPS spectra show essential inhomogeneous electrostatic charging, making the analysis of the samples containing the wide-gap ZnS very difficult.

3.3. DFT simulation

Fig. 7 shows several examples of defective structures involving Pb vacancies and S–S bonding in quasi-bulk PbS and on the PbS (100) surface. The formation energy for one Pb vacancy in the bulk is positive and increases for the formation of divacancies or associated S–S bonds (compare the structures 3, 4, 6, 7). The energy somewhat decreases after the removal of two and three cations and the formation of S_2 and S_3 anionic groups (configurations 1 and 2), but remains fairly large, indicating that the existence of the metal-depleted undersurface structures is unfavorable. Pb vacancies on the PbS surface are unstable too, and their formation energies slightly decrease due to S–S bonding. Meanwhile, the analogous values calculated, for instance, for the S-excessive surface layers of $CuFeS_2$ (ref. 30 and 31) and CuS[33] were negative. This means that both surface and undersurface Pb-deficient structures at PbS are energetically unfavorable.

The DFT+U simulation of metal-deficient ZnS (Fig. 8) shows that the deletion of one Zn atom creates quasi-bulk structures with formation energies larger than 2 eV. The removal of additional Zn cations and S–S bonding require slightly lower energies, comparable with or larger than those for PbS. At the same time, the extraction of surface Zn atoms followed by S–S bonding is more advantageous than for PbS but the formation energies are still positive. It is interesting that the formation energies are lower by several tenths of eV for Cu-doped ZnS (not shown in the figures). This may be a reason, in particular, for the accumulation of Cu and some other impurities at oxidized mineral samples and, on the other hand, larger stability of the metal-deficient and polysulfide entities,[15,20–22] and it may explain some discrepancies between the current and the previous work.

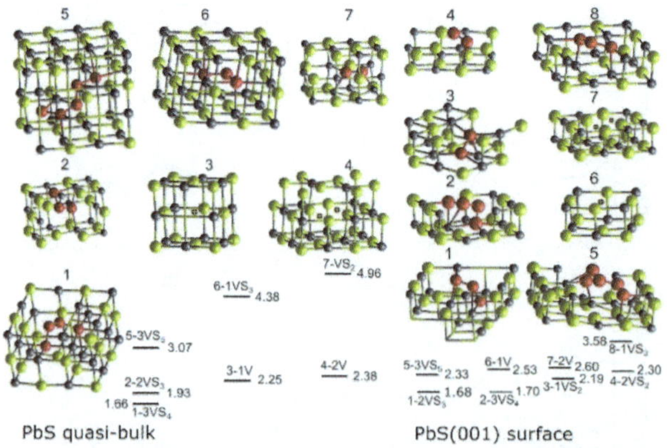

Fig. 7 Equilibrium structures and formation energies E_f (eV) normalized to one extracted Pb atom for the defects in the PbS quasi-bulk layer and at the (001) surface produced by the removal of Pb atoms from $Pb_{32}S_{32}$ unit cells. For example, 6-1VS_3 means structure number 6 with one Pb vacancy and S_3 group, having the formation energy of 4.38 eV. Black spheres are Pb atoms and yellow are S, red spheres represent atoms of S_2 and S_3 groups, and black ones with crosses indicate the positions of the Pb vacancies.

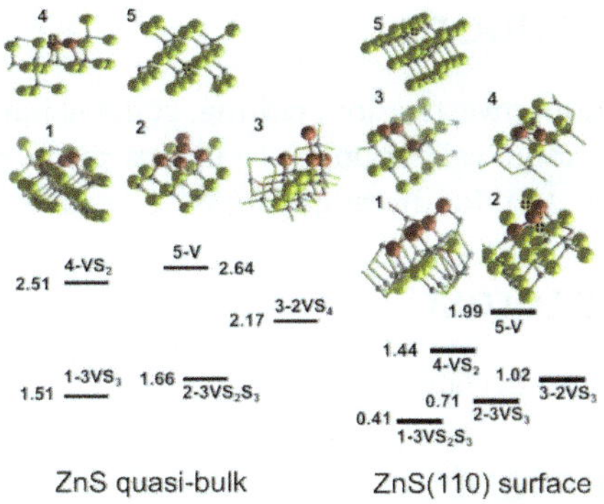

Fig. 8 Equilibrium structures and formation energies E_f (eV) normalized to one extracted cation for the quasi-bulk ZnS layer and the (110) surface produced by the removal of one, two or three Zn atoms from $Zn_{16}S_{16}$ unit cells. The designations are analogous to those in Fig. 7.

The DFT results indicate that, akin to PbS, the metal-depleted under-surface regions and, to a lesser extent, polysulfide surfaces at the reacted ZnS are not energetically stable. The findings can explain why no essentially depleted near-surface layers form at PbS and ZnS and, although the polysulfide species arise at the reacted surfaces, they vanish much more easily than from the surfaces of Cu and Fe sulfides.[30–32] Moreover, we cannot fully exclude that some sulfur was lost even in the cryo-XPS experiment, particularly for PbS.

4. Conclusions

Cryo-XPS study of fast-frozen dispersions of natural sphalerite ZnS and galena PbS in aqueous media has found several phenomena, which are commonly missed using conventional XPS, in addition to the well-known volatilization of elemental sulfur and elusive aqueous species. The low-temperature spectra of PbS etched in non-oxidizing 0.1 M HCl solutions show very low, if any, concentrations of S–S bonds, but the polysulfide species arise in 0.1 M $HClO_4$ solution. The polysulfides are unstable and almost completely disappear after the warming up to room temperature in the spectrometer vacuum. In the case of ZnS, the content of excessive sulfur was modest and its decreased upon prolonging the reaction time. These facts suggest that different defective centers emerged and different reaction mechanisms are involved in various solutions or even in the same system at different reaction stages. Oxidative etching with aqueous Fe^{3+} cations produces polysulfide anions and then elemental sulfur, which both volatize upon thawing the samples in the ultra-high vacuum. These S species can be discriminated *via* comparing their binding energies, electrostatic charging and the evaporation rates. Cryo-XPS can also specify the interfacial products, such as lead, zinc and iron chlorides and hydr(oxides). DFT simulation found that the metal-depleted bulk structures at PbS and ZnS are energetically unfavorable, in contrast to the structures detected using HAXPES at Fe- and Cu-bearing sulfides. The S–S species are somewhat more stable at the ZnS (110) surface and their formation energy decreases for Cu-doped sphalerite, in agreement with the experimental evidence.

Author contributions

Y. M.: conceptualization, investigation, writing, supervision; V. N.: methodology, investigation, visualization, validation; Y. T.: investigation, formal analysis; S. V.: investigation, validation; A. R.: investigation; M. L.: validation, visualization.

Conflicts of interest

There are no conflicts to declare.

Acknowledgements

This research work was supported by the Ministry of Science and Higher Education, project 0287-2021-0014 for the Institute of Chemistry and Chemical Technology of the Siberian Branch of RAS. The facilities of the Krasnoyarsk Regional Research Equipment Center of the SB RAS were employed in the work.

References

1 M. Salmeron and R. Schlögl, *Surf. Sci. Rep.*, 2008, **63**, 169.
2 C. Arble, M. Jia and J. T. Newberg, *Surf. Sci. Rep.*, 2018, **73**, 37.
3 D. I. Patel, T. Roychowdhury, V. Jain, D. Shah, T. G. Avval, S. Chatterjee, S. Bahr, P. Dietrich, M. Meyer, A. Thißen and M. R. Linford, *Surf. Sci. Spectra*, 2019, **26**, 016801.
4 D. E. Starr, M. Favaro, F. F. Abdi, H. Bluhm, E. J. Crumlin and R. van de Krol, *J. Electron Spectrosc. Relat. Phenom.*, 2017, **221**, 106.
5 R. Endo, D. Watanabe, M. Shimomura and T. Masuda, *Appl. Phys. Lett.*, 2019, **114**, 173702.
6 A. Shchukarev and S. Sjöberg, *Surf. Sci.*, 2005, **584**, 106.
7 A. Shchukarev, *Adv. Colloid Interface Sci.*, 2006, **122**, 149.
8 A. Shchukarev, *J. Electron Spectrosc. Relat. Phenom.*, 2010, **176**, 13.
9 A. Shchukarev and M. Ramstedt, *Surf. Interface Anal.*, 2017, **49**, 349.
10 R. S. C. Smart, *Miner. Eng.*, 1991, **4**, 891; I. Kartio, K. Laajalehto, E. Suoninen, S. Karthe and R. Szargan, *Surf. Interface Anal.*, 1992, **18**, 807; P. Nowak, K. Laajalehto and I. Kartio, *Colloids Surf., A*, 2000, **161**, 447.
11 P. A. Kozin, A. Shchukarev and J.-F. Boily, *Langmuir*, 2013, **29**, 12129; J. Škvarla, M. Kaňuchová, A. Shchukarev, A. Girová and I. Brezáni, *Colloids Surf., A*, 2020, **586**, 124234.
12 M. Wan, A. Shchukarev, R. Lohmayer, B. Planer-Friedrich and S. Peiffer, *Environ. Sci. Technol.*, 2014, **48**, 5076.
13 M. Khoshkhoo, M. Dopson, A. Shchukarev and A. Sandström, *Hydrometallurgy*, 2014, **149**, 220.
14 Y. Mikhlin, A. Karacharov, Y. Tomashevich and A. Shchukarev, *J. Electron Spectrosc. Relat. Phenom.*, 2016, **206**, 65.
15 Y. Mikhlin, A. Karacharov, Y. Tomashevich and A. Shchukarev, *Vacuum*, 2016, **125**, 98.
16 A. Shchukarev, Z. Gojkovic, C. Funk and M. Ramstedt, *Appl. Surf. Sci.*, 2020, **526**, 146538; A. Shchukarev, E. Backman, S. Watts, S. Salentinig, C. F. Urban and M. Ramstedt, *Front. Chem.*, 2021, **9**, 666853.

17 K. M. Rosso and D. J. Vaughan, *Rev. Mineral. Geochem.*, 2006, **61**, 505; K. M. Rosso and D. J. Vaughan, *Rev. Mineral. Geochem.*, 2006, **61**, 557.
18 M. G. Kanatzidis, *Inorg. Chem.*, 2017, **56**, 3158; B. Yan, X. Li, W. Xiao, J. Hu, L. Zhang and X. Yang, *J. Mater. Chem. A*, 2020, **8**, 17848; Y. Zhang, Y. Zhang, H. Zhang, L. Bai, L. Hao, T. Ma and H. Huang, *Coord. Chem. Rev.*, 2021, **448**, 214147.
19 S. Evans and E. Raftery, *J. Chem. Soc., Faraday Trans. 1*, 1982, **78**, 3545.
20 A. N. Buckley and R. Woods, *Appl. Surf. Sci.*, 1984, **17**, 401.
21 A. N. Buckley and R. Woods, *Aust. J. Chem.*, 1984, **37**, 2403; A. N. Buckley and R. Woods, *Appl. Surf. Sci.*, 1985, **20**, 472; A. N. Buckley and R. Woods, *Appl. Surf. Sci.*, 1985, **22**, 280.
22 A. N. Buckley, R. Woods and H. J. Wouterlood, *Int. J. Miner. Process.*, 1989, **26**, 29; A. N. Buckley, H. J. Wouterlood and R. Woods, *Hydrometallurgy*, 1989, **22**, 39.
23 R. S. C. Smart, W. M. Skinner and A. R. Gerson, *Surf. Interface Anal.*, 1999, **28**, 101.
24 A. R. Gerson and T. Bredow, *Surf. Interface Anal.*, 2000, **29**, 145.
25 Y. Mikhlin, V. Varnek, I. Asanov, Y. Tomashevich, A. Okotrub, A. Livshits, G. Selyutin and G. Pashkov, *Phys. Chem. Chem. Phys.*, 2000, **2**, 4393.
26 S. L. Harmer, J. E. Thomas, D. Fornasiero and A. R. Gerson, *Geochim. Cosmochim. Acta*, 2006, **70**, 4392.
27 Y. Mikhlin, *Appl. Sci.*, 2020, **10**, 5138.
28 M. Fantauzzi, B. Elsener, D. Atzei, A. Rigoldi and A. Rossi, *RSC Adv.*, 2015, **5**, 75953.
29 Y. Mikhlin, Y. Tomashevich, S. Vorobyev, S. Saikova, A. Romanchenko and R. Félix, *Appl. Surf. Sci.*, 2016, **387**, 796.
30 Y. Mikhlin, V. Nasluzov, A. Romanchenko, Y. Tomashevich, A. Shor and R. Félix, *Phys. Chem. Chem. Phys.*, 2017, **19**, 2749.
31 V. Nasluzov, A. Shor, A. Romanchenko, Y. Tomashevich and Y. Mikhlin, *J. Phys. Chem. C*, 2019, **123**, 21031.
32 Y. Mikhlin, A. Romanchenko and Y. Tomashevich, *Appl. Surf. Sci.*, 2021, **549**, 149261.
33 Y. Mikhlin, V. Nasluzov, A. Ivaneeva, S. Vorobyev, M. Likhatski, A. Romanchenko, A. Krylov, S. Zharkov and D. M. Meira, *Mater. Chem. Phys.*, 2020, **255**, 123600.
34 J. P. Perdew, J. A. Chevary, S. H. Vosko, K. A. Jackson, M. R. Pederson, D. J. Singh and C. Fiolhais, *Phys. Rev. B: Condens. Matter Mater. Phys.*, 1992, **46**, 6671; J. P. Perdew, J. A. Chevary, S. H. Vosko, K. A. Jackson, M. R. Pederson, D. J. Singh and C. Fiolhais, *Phys. Rev. B: Condens. Matter Mater. Phys.*, 1993, **48**, 4978.
35 P. E. Blöchl, *Phys. Rev. B: Condens. Matter Mater. Phys.*, 1994, **50**, 17953.
36 G. Kresse and J. Hafner, *Phys. Rev. B: Condens. Matter Mater. Phys.*, 1993, **47**, 558; G. Kresse and J. Furthmüller, *Phys. Rev. B: Condens. Matter Mater. Phys.*, 1996, **54**, 11169.
37 E. Şaşıoğlu, C. Friedrich and S. Blügel, *Phys. Rev. B: Condens. Matter Mater. Phys.*, 2011, **83**, 121101.
38 G. Kresse and D. Joubert, *Phys. Rev. B: Condens. Matter Mater. Phys.*, 1999, **59**, 1758.
39 H. J. Monkhorst and J. D. Pack, *Phys. Rev.B: Solid State*, 1976, **13**, 5188.

40 A. Klamt and G. Schüürmann, *J. Chem. Soc., Perkin Trans. 2*, 1993, 799.
41 Y. Mikhlin, A. Romanchenko and A. Shagaev, *Appl. Surf. Sci.*, 2006, **252**, 5645.
42 A. P. Grosvenor, B. A. Kobe, M. C. Biesinger and N. S. McIntyre, *Surf. Interface Anal.*, 2004, **36**, 1564; M. C. Biesinger, B. P. Payne, A. P. Grosvenor, L. W. M. Lau, A. R. Gerson and R. S. C. Smart, *Appl. Surf. Sci.*, 2011, **257**, 2717.
43 Y. Li, G. Qian, P. L. Brown and A. R. Gerson, *Geochim. Cosmochim. Acta*, 2017, **212**, 33.
44 D. Barreca, A. Gasparotto, C. Maragno, E. Tondello and T. R. Spalding, *Surf. Sci. Spectra*, 2002, **9**, 54.

Faraday Discussions

DISCUSSIONS

In situ methods: discoveries and challenges: general discussion

Rosa Arrigo, Damien Aureau, Prajna Bhatt, Mark A. Buckingham, James J. C. Counter, Giulio D'Acunto, Philip R. Davies, D. Andrew Evans, Wendy R. Flavell, Joshua S. Gibson, Shaoliang Guan, Georg Held, Mark Isaacs, J. Matthias Kahk, Claus F. P. Kastorp, Heath Kersell, Alenka Krizan, Alexander I. Large, Robert Lindsay, Johannes Lischner, Patrick Lömker, David Morgan, Slavomír Nemšák, Anders Nilsson, David Payne, Benjamen P. Reed, Olivier Renault, Günther Rupprechter, Alexander G. Shard, Mzamo Shozi, Mathieu G. Silly, William S. J. Skinner, Francine Solal, Kelsey A. Stoerzinger, Sefik Suzer, Juan Jesús Velasco Vélez, Marc Walker and Robert S. Weatherup

DOI: 10.1039/d2fd90025d

Sefik Suzer opened discussion of the introductory lecture by Wendy Flavell: What are your thoughts about cryo-XPS? Is environmental XPS similar to cryo-XPS? A lot of biologists ask when cryo-XPS will be widely available. In your opinion, when do you think that the instrument makers will consider developing it as an easy-to-use attachment to commercial and/or home-made instruments?

Wendy Flavell answered: Possibly, as it seems that it is beginning to be used as a "go-to-tool" by biological scientists. However, as studies of the hydrated state under near-ambient pressure are now becoming possible, I wonder how useful this will be in the long-term. I think there is still a discussion to be had about whether or not the process of freezing changes the XPS. For example, in studies of *Escherichia coli*,[1] NAP-XPS using the SPECS EnviroESCA has been used to show that freeze drying (which has been commonly used to study biological systems by XPS[2]) alters the composition of carbon species at the surface, compared with the hydrated state.[1] In later studies of the bacterium *Pseudomonas fluorescens*, it has been shown that this difference can be minimised by using fast freezing in cryo-XPS.[3]

1 M. Kjærvik, K. Schwibbert, P. Dietrich, A. Thissen and W. E. S. Unger, *Surf. Interface Anal.*, 2018, **50**, 996–1000.
2 P. G. Rouxhet and M. J. Genet, *Surf. Interface Anal.*, 2011, **43**, 1453–1470.
3 M. Kjærvik, M. Ramstedt, K. Schwibbert, P. M. Dietrich and W. E. S. Unger, *Front. Chem.*, 2021, **9**, DOI: 10.3389/fchem.2021.666161.

Anders Nilsson remarked: The phrase ambient pressure XPS (APXPS) – I think it's misunderstood. At the ALS in Berkeley we built beamline 11.2 for molecular environmental science and APXPS was one of the unique instruments. We were funded for an NSF molecular environmental science institute between Stanford and Berkeley with the goal to look at interfaces. The partial pressure of water at room temperature is 20 Torr and with a humidity of 10–20% simulating California climate a pressure of 2–40 Torr would represent ambient conditions. It was in this context that the words ambient pressure were used. The intention was not to indicate atmospheric pressure.

Philip Davies asked: You discuss many aspects of the way photoelectron spectroscopy will develop in the future but what about improvements in its lateral resolution?

Wendy Flavell responded: I decided to focus on three key themes of this meeting – advances in *in situ* and time-resolved measurements, and in extending the sampling depth of photoelectron spectroscopy, but there is undoubtedly a strong drive to combine these with improved lateral (and in some cases, spin) resolution. I think to some extent this is driven by device applications, because we need to study so many laterally inhomogeneous samples. Scanning photoelectron microscopy (SPEM) is implemented with NAP-XPS for example at the ESCA Microscopy beamline at Elettra (Trieste), offering sub-micron resolution.[1,2] The use of a 3D delay-line detector with a lateral resolution lens at the SpAnTeX beamline at BESSY II (Berlin)[3] has opened up the possibility of combined time- and lateral resolution (better than 30 microns) in NAP-XPS at ≥20 mbar with tender X-rays. Developments in micro- and nano-angle-resolved photoemission have been recently reviewed by King *et al.*[4] Although developed primarily for their valence band mapping capability, the full-field imaging momentum microscope[5] offer great possibilities for microspectroscopy. In the hard X-ray regime, this can be used to resolve lateral structures on the sub-μm scale. A hard X-ray photoelectron emission microscope (HAXPEEM) with sub-micron-scale lateral resolution for structures buried more than 10 nm below the surface is now operated at beamline P22 at PETRA-III (DESY, Hamburg).[6]

1 M. Amati, V. Bonanni, L. Braglia, F. Genuzio, L. Gregoratti, M. Kiskinova, A. Kolmakov, A. Locatelli, E. Magnano, A. A. Matruglio, T. O. Menteş, S. Nappini, P. Torelli and P. Zeller, *J. Electron Spectrosc. Relat. Phenom.*, 2019, 146902.
2 M. Amati, L. Gregoratti, P. Zeller, M. Greiner, M. Scardamaglia, B. Junker, T. Ruß, U. Weimar, N. Barsan, M. Favaro, A. Alharbi, I. J. T. Jensen, A. Ali and B. D. Belle, *J. Phys D: Appl. Phys.*, 2021, **54**, 204004.
3 M. Favaro, P. C. J. Clark, M. J. Sear, M. Johansson, S. Maehl, R. van de Krol and D. E. Starr, *Surf. Sci.*, 2021, **713**, 121903.
4 P. D. C. King, S. Picozzi, R. G. Egdell and G. Panaccione, *Chem. Rev.*, 2021, **121**, 2816–2856.
5 D. Kutnyakhov, R. P. Xian, M. Dendzik, M. Heber, F. Pressacco, S. Y. Agustsson, L. Wenthaus, H. Meyer, S. Gieschen, G. Mercurio, A. Benz, K. Bühlman, S. Däster, R. Gort, D. Curcio, K. Volckaert, M. Bianchi, C. Sanders, J. A. Miwa, S. Ulstrup, A. Oelsner, C. Tusche, Y.-J. Chen, D. Vasilyev, K. Medjanik, G. Brenner, S. Dziarzhytski, H. Redlin, B. Manschwetus, S. Dong, J. Hauer, L. Rettig, F. Diekmann, K. Rossnagel, J. Demsar, H.-J. Elmers, P. Hofmann, R. Ernstorfer, G. Schönhense, Y. Acremann and W. Wurth, *Rev. Sci. Instrum.*, 2020, **91**, 013109.
6 C. Schlueter, A. Gloskovskii, K. Ederer, I. Schostak, S. Piec, I. Sarkar, Yu. Matveyev, P. Lömker, M. Sing, R. Claessen, C. Wiemann, C. M. Schneider, K. Medjanik, G. Schönhense, P. Amann, A. Nilsson and W. Drube, *AIP Conf. Proc.*, 2019, **2054**, 040010.

Philip Davies remarked: I was very interested in your comment on the possible presence of hydroxyl radicals during NAP experiments in the presence of water. As you know hydroxyl radicals have an important role in processes such as photocatalysis[1] and the ability to study reactions involving such species directly with XPS could be very useful.

1 Y. Nosaka and A. Nosaka, Understanding hydroxyl radical (˙OH) generation processes in photocatalysis, *ACS Energy Lett.*, 2016, **1**(2), 356–359, DOI: 10.1021/acsenergylett.6b00174.

Wendy Flavell responded: Yes, it's exciting that we can start to gain some evidence from NAP-XPS about the role of ˙OH in water-based systems. The example I mentioned is the first NAP-XPS study of radiation damage in fully hydrated double-strand DNA.[1] XPS in vacuum, nitrogen and water was used to deconvolute direct DNA damage, caused by photons and low energy electrons, from the indirect effects caused by ionisation (radiolysis) of water molecules to form reactive oxygen species such as ˙OH and H_2O^+. The latter initiate indirect damage processes which lead to deoxyribose and nucleobase modification and base release. In the presence of water, the authors show that these processes predominate over the strand breakage which is the dominant type of damage in dry DNA, although damage of all types increases when water is introduced. That said, I should add that the signals due to the ˙OH radicals themselves are not identified, but spectral changes due to damage known to be caused by ˙OH radicals (for example to C=C and C=N bonds) are observed.[1]

1 M. B. Hahn, P. M. Dietrich and J. Radnik, *Commun. Chem.*, 2021, **4**, 1–8.

Olivier Renault said: Thank you very much for the inspiring presentation about the huge progress made in the field and the empty space to be filled in the future. Maybe as a complement one would need to keep in mind two aspects: first, the necessity of improved imaging capabilities in photoelectron spectroscopy to tackle the challenge of laterally-heterogeneous samples (as performed especially by small-probe scanning and by PEEM). Secondly, and pretty much connected with the first aspect but also with the field of time-resolution, there is a more and more pressing need for proper data science both because of low signal levels that may be encountered in data acquisition, and because of "data deluge" where the amount of collected spectra during a particular experiment cannot be handled by conventional means.

Wendy Flavell responded: I completely agree with your first point about improved imaging capability, as outlined in my response to Philip Davies. You raise a very important point about the data avalanche, too, and it is certainly already the case that in some areas of FEL science, more data is accumulated than can be reasonably analysed. For example if we take a time-resolved photoemission experiment using time-of-flight momentum microscopy, like the one I showed at FLASH (Hamburg),[1] then, even without parallel spin detection, for each photoemitted electron, the number of parameters to be recorded is four (the 2D momentum coordinates, k_x and k_y, the time of flight (yielding the binding energy) and the pump–probe delay). Added to that each probe pulse and its matching pump pulse is separately characterised and correlated with the corresponding photoemitted electron –and all at high repetition rate. Efficient processing of the photoelectron event streams at a rate of up to tens of megabytes per second is needed. Researchers at FLASH have developed an open-source distributed workflow approach to allow user interaction with the data.[2] I agree that a much more "data-science-oriented" approach will be needed as we move forward.

1 D. Kutnyakhov, R. P. Xian, M. Dendzik, M. Heber, F. Pressacco, S. Y. Agustsson, L. Wenthaus, H. Meyer, S. Gieschen, G. Mercurio, A. Benz, K. Bühlman, S. Däster, R. Gort, D. Curcio, K. Volckaert, M. Bianchi, C. Sanders, J. A. Miwa, S. Ulstrup, A. Oelsner, C. Tusche, Y.-J. Chen, D. Vasilyev, K. Medjanik, G. Brenner, S. Dziarzhytski, H. Redlin, B. Manschwetus, S. Dong, J. Hauer, L. Rettig, F. Diekmann, K. Rossnagel, J. Demsar, H.-J. Elmers, Ph. Hofmann, R. Ernstorfer, G. Schönhense, Y. Acremann and W. Wurth, *Rev. Sci. Instrum.*, 2020, **91**, 013109.
2 R. P. Xian, Y. Acremann, S. Y. Agustsson, M. Dendzik, K. Bühlmann, D. Curcio, D. Kutnyakhov, F. Pressacco, M. Heber, S. Dong, T. Pincelli, J. Demsar, W. Wurth, P. Hofmann, M. Wolf, M. Scheidgen, L. Rettig and R. Ernstorfer, *Sci. Data*, 2020, 7, 442.

Robert Lindsay opened discussion of the paper by Kelsey Stoerzinger: Could you describe what evidence you have for iron cations being present at the surface of your amorphous oxide films?

Kelsey Stoerzinger replied: We observe Fe reduction at elevated temperatures (Fig. 6 in the paper; https://doi.org/10.1039/D1FD00109D) and that the amount of methoxy on the surface increases with greater Fe content (Fig. 2 in the paper).

Georg Held enquired: The method of discriminating and quantifying basic and acidic active sites through reaction products (CO_3 and OCH_3, respectively) is very interesting and useful. I wonder, however, how quantitative the method really is. If the dissociation products desorb from the surface or undergo further reactions the number of sites would be underestimated; if the products diffuse away from the active sites after the reaction they would be overestimated. Could you please comment on these potential problems.

Also, is there an alternative spectroscopic way of quantifying these sites, *e.g.* through resonant photoemission spectroscopy, as in the paper by Velasco Vélez *et al.* (https://doi.org/10.1039/d1fd00114k)?

Kelsey Stoerzinger answered: It is a great point that at elevated temperatures, we do see conversion of methanol (Fig. S11 in the supplementary information of the paper; https://doi.org/10.1039/d1fd00109d) and thus dissociation products are indeed desorbing from the surface, and might lead to an underestimation of the number of sites (indeed, the intensity of these adsorbates also decreases with increasing temperature). At room temperature, however, we don't observe any products by mass spectrometry, suggesting desorption of products is negligible. Resonant photoemission would be a great way to further assess the nature of and quantify active sites.

Günther Rupprechter asked: Concerning the observed products: A couple of years ago, we studied methanol decomposition (and oxidation)[1] on Pd surfaces by XPS and PM-IRAS and found that carbon deposits that build up during the process changed the product distribution from CO/CO_2 to formaldehyde (as formaldehyde decomposition seems hampered on a carbon-modified surface with hollow sites blocked by C). Did you see a selectivity change on your oxide films as well or are CH_x just by-products? For adventitious C, isn't the amount too much? Maybe there is dissociation of the produced CO, as for CO on Co_3O_4 using NAP-XPS we also observed C species.[2]

1 M. Borasio, O. Rodríguez de la Fuente, G. Rupprechter and H.-J. Freund, *J. Phys. Chem. B*, 2005, **109**, 17791, DOI: 10.1021/jp053855c.
2 L. Lukashuk, N. Yigit, R. Rameshan, E. Kolar, D. Teschner, M. Hävecker, A. Knop-Gericke, R. Schlögl, K. Föttinger and G. Rupprechter, *ACS Catal.*, 2018, **8**, 8630, DOI: 10.1021/acscatal.8b01237.

Kelsey Stoerzinger replied: That's very interesting. We didn't see any formaldehyde by mass spectrometry, but it's possible the amount was too low to observe. We primarily observe CO_2 (and water) with small amounts of dimethyl ether from the Al–Mn film. It is possible that some of the adventitious carbon could come from CO dissociation; it would be interesting to look at these films in controlled atmospheres of potential products (*e.g.* CO_2, CO…) to better assess this.

Shaoliang Guan queried: Very rich data information is included in the supplementary information of the paper (https://doi.org/10.1039/d1fd00109d). In the caption of Fig. S5, it states that 'The visibility of Ir 4f is more for 690 eV compared to 490 eV and 350 eV…'. But to me, the spectra of 690 and 490 eV do look similar, only the one for 350 eV is much smaller. Could you please clarify the statement? And why it's much bigger above 400 eV?

Kelsey Stoerzinger answered: To be precise in capturing the distribution of the buried Ir layer we'd have to model the photoelectron intensity considering the relative cross sections of each element, photon flux at each energy from the beamline, and the attenuation of photoelectrons with depth and sample composition. Regardless, the trends in Ir/Al ratios indicate that Ir is buried underneath the $\sim$5 nm thick amorphous Al–Fe oxide, and the minimal intensity at 350 eV suggests it is not present on the surface of the oxide.

Philip Davies enquired: In Fig. 1(b) of the paper (https://doi.org/10.1039/d1fd00109d) the C(1s) signal shows a reasonable concentration of carbonate, you mention it's typically 0.3–0.4 of the total carbon species. That should give a corresponding strong signal in the O(1s) spectra, at what binding energy is that peak occurring? Knowing the carbon : oxygen ratio for the other species at the surface have you tried to quantify the concentration of OH at the surface?

Kelsey Stoerzinger responded: We expect the CO_3 feature in the O 1s is buried in the large, featureless component here labeled OH + OCH_3. Because the amount of CO_3 is similar across all surfaces, this gives a roughly constant offset in the intensity of residual OH + OCH_3 if fit explicitly. Using the C : O RSF to assign specific intensities in the O 1s to OCH_3 and CO_3, we generally get a residual OH O 1s intensity comparable to that of OCH_3, however this is dependent on the specific conditions and the parameters chosen for fitting.

Anders Nilsson asked: We always have to be concerned with beam-induced effects when conducting XPS measurements of adsorbates on surface.

Methanol is a relatively fragile molecule. Have you conducted any test to see if the results are affected by beam-induced effects?

Kelsey Stoerzinger replied: These experiments were performed on a bending magnet beamline, but of course beam-induced effects can be an issue here as well. We have checked for such effects by (i) checking whether the spectra remain consistent over time, and (ii) moving to a new spot that has not yet seen the beam and checking for spectral consistency. On the timescale of measurements, we didn't observe such effects. It is however still possible that the beam increases the tendency for methanol dissociation on the surface, influencing the absolute value of *e.g.* OCH_3 intensity measured here. Even in this case, however, our comparisons across samples in this case remain robust.

Anders Nilsson enquired: To clearly interpret the shoulder towards higher binding energy in the O 1s spectra depends on the exact line shape of the oxide peak. Do you have the clean oxide spectrum in order to determine the exact line shape and can you conduct a subtraction to isolate the shoulder peak?

Kelsey Stoerzinger answered: We do have oxide spectra on a cleaned surface in the absence of *e.g.* water and methanol, however a shoulder in the O 1s spectrum remains. We have fit this shoulder as a strongly bound OH feature due to a previous study of these materials across a wide range of relative humidities,[1] where the intensity of this feature increases with relative humidity.

1 K. A. Stoerzinger, L. J. Enman, E. A. Cochran, J. T. Diulus, R. T. Frederick, K. Artyushkova, E. J. Crumlin, G. S. Herman and S. W. Boettcher, *J. Phys. Chem. C*, 2019, **123**, 27048, DOI: 10.1021/acs.jpcc.9b08227.

Anders Nilsson questioned: It is very interesting that you find the methoxy binding energy is above 286 eV. Why do you think the binding energy is shifted relative to methoxy on metals which is usually around 285.5 eV?

Kelsey Stoerzinger replied: This is a great question that would be interesting to study further by comparison on *e.g.* Cu in its metallic form, the amorphous oxide equivalent (containing Cu^{2+} (ref. 1)), and a crystalline oxide to disentangle the effects of oxidation state and local coordination. This may be related to local charge density on the methoxy group, which might be modeled by DFT or reflected in the vibrational frequency of adsorbates as well; an interesting opportunity for future work.

1 L. J. Enman, M. G. Kast, E. A. Cochran, E. Pledger, M. B. Stevens and S. W. Boettcher, *J. Phys. Chem. C*, 2018, **122**, 13691, DOI: 10.1021/acs.jpcc.8b00239.

J. Matthias Kahk asked: In Fig. 1b of the paper (https://doi.org/10.1039/d1fd00109d), peaks are assigned to adsorbed methoxy and carbonate species. I was wondering, to what extent the detailed atomic structures of these adsorbates on these complex surfaces are understood? It would be desirable to verify the peak assignments by comparing the experimentally determined binding energies to calculated binding energies for the same species, however, core electron binding energies can only be performed if a structural model is available.

Kelsey Stoerzinger responded: It would be very interesting to perform such calculations. We can conclude that methoxy binds to the transition metal site based on its saturation-behavior at low Fe content (Fig. 2 of the paper; https://doi.org/10.1039/d1fd00109d), presumably through an Fe–O bond. For carbonate we unfortunately cannot draw detailed atomic structures solely based on XPS binding energies, but coupling with vibrational spectroscopy would yield additional insight into the mode of adsorption.

Damien Aureau enquired: Nice comparison between different oxide films is given. As the surface preparation must be really important for these chemical reactions at surfaces, was the influence of different parameters involved in the surface preparation studied on one particular oxide? How critical is the recipe used to treat the samples before performing AP-XPS studies?

Kelsey Stoerzinger answered: Our surface preparation method was guided by our experience working with other oxides and bound by the conditions for crystallization of these samples (450 °C in air). We avoid sputtering as this is known to change oxide composition, and instead heat under modest (*e.g.* 150 mTorr here) O_2 pressures to oxidize carbon present from *ex situ* growth. We have not tested other surface preparation methods for these samples, but it would be interesting to do so.

Robert Weatherup queried: Are you able to obtain any additional information from your measurements of gas phase species? For example, have you tried to correlate changes in work function with changes in surface chemistry?

Kelsey Stoerzinger replied: We indeed see changes in the gas phase binding energy (*e.g.* for Al_4Fe_6 in 50 mTorr CH_3OH : 100 mTorr O_2, the O_2 gas phase peak shifts to 0.4 eV higher binding energy relative to the bulk upon increasing the temperature from 25 °C to 275 °C). This could indicate a decrease in sample work function with increasing temperature and the removal of surface adsorbates *via* conversion, but also changes in the surface dipole through which bulk oxygen photoelectrons must escape.

Claus F. P. Kastorp asked: On the note of active sites: do you know what the surface looks like structurally, *e.g.* if it is smooth or porous? Do you believe it is suitable for scanning tunneling microscopy and if so, if it could help you determine the active sites?

Kelsey Stoerzinger answered: From AFM of the as-grown films, they are flat with an RMS roughness ~0.15 nm.[1] As we don't see changes in the oxidation state or intensity of the buried Ir layer used to improve conductivity, we don't see any evidence for porosity under the conditions considered here. With contact to the Ir layer it is likely the films would be amenable for STM, which would be a very interesting measurement to image the active surface.

1 M. G. Kast, E. A. Cochran, L. J. Enman, G. Mitchson, J. Ditto, C. Siefe, P. N. Plassmeyer, A. L. Greenaway, D. C. Johnson, C. J. Page and S. W. Boettcher, *J. Am. Chem. Soc.*, 2016, **138**, 16800, DOI: 10.1021/jacs.6b11084.

Johannes Lischner questioned: When carrying out the peak fitting, it appears as if you do not ensure consistency between the spectra of the different elements. For example, a CO_3 peak is observed in the carbon spectrum, but no corresponding peak is found in the oxygen spectrum. Can you explain this?

Kelsey Stoerzinger replied: Because we have a broad shoulder in the O 1s (in some cases also overlapping with gas phase methanol and potential oxidation products), we have not deconvolved this to isolated CO_3, OCH_3, and OH species. Using the C : O relative sensitivity factor (RSF) to assign specific intensities in the O 1s to OCH_3 and CO_3, we generally get a residual OH O 1s intensity comparable to that of OCH_3, however this is dependent on the specific conditions and the parameters chosen for fitting.

Robert Lindsay asked: In Fig. 1 of the paper (https://doi.org/10.1039/d1fd00109d), only one peak is assigned to adventitious carbon. In other studies, the adventitious carbon layer typically gives rise to several peaks. Could you explain your reason for only assigning one C 1s peak to adventitious carbon? In addition, there are also typically peaks associated with this adventitious carbon layer present in the O 1s spectrum. Can you explain why you have not considered such features in your analysis?

Kelsey Stoerzinger answered: We have assumed a simplified picture of adventitious carbon as primarily arising from C–C and CH_x species and fit it with a singular, relatively broad feature. We unfortunately are not able to isolate other contributions in our experiments.

Joshua S. Gibson remarked: There has been a lot of discussion regarding justification of your peak assignments in the oxygen 1s region, and I was wondering about the suitability of utilising a reference spectrum, given their extensive use in X-ray absorption spectroscopy. Could you have synthesised a hydroxide terminated surface and used this to ascertain the fitting parameters of the hydroxide peak, and do you feel that this is a valuable tool for XPS and should be undertaken more extensively?

Kelsey Stoerzinger responded: This suggestion is a great one for the community. We have previously studied these surfaces in O_2 and a range of relative humidities to obtain such hydroxylated reference spectra.[1]

1 K. A. Stoerzinger, L. J. Enman, E. A. Cochran, J. T. Diulus, R. T. Frederick, K. Artyushkova, E. J. Crumlin, G. S. Herman and S. W. Boettcher, *J. Phys. Chem. C*, 2019, **123**, 27048, DOI: 10.1021/acs.jpcc.9b08227.

Andrew Evans opened discussion of the paper by Giulio D'Acunto: For III–V semiconductors, the group III oxide (*e.g.* In or Ga) is difficult to remove once formed. Is this oxide fully removed by the atomic layer deposition (ALD) process of growing the rare earth oxide layer?

Giulio D'Acunto replied: We found that on InAs the InO_x is easier to remove with respect to the GaO_x in GaAs or GaSb, during the self-cleaning process. However, exposing the surfaces to hydrogen plasma led to the complete removal of the unwanted oxides.[1]

A small note: we performed several experiments using TDMAHf and water to grow HfO_2 ALD (not a rare earth oxide, but a transition metal oxide). This process never removed all the InO_x from the InAs, but leaves behind a monolayer of InO_x which is necessary for the interface between the InAs and the HfO_2.

1 Y.-P. Liu, S. Yngman, A. Troian, G. D'Acunto, A. Jönsson, J. Svensson, A. Mikkelsen, L.-E. Wernersson and R. Timm, *Appl. Surf. Sci.*, 2022, **593**, 153336, DOI: 10.1016/j.apsusc.2022.153336.

Günther Rupprechter enquired: At the end, you showed a short view on "*in situ* and *operando*" XPS measurements of ALD, I guess. What does this refer to? In catalysis, for *operando* one would combine NAP-XPS with *e.g.* MS product analysis (*e.g.* ref. 1). Is the ALD analogue to detect by-products of the ALD process by MS? Or is the excess of precursors too large anyway? If MS does not provide information, why call it *operando*?

1 V. Pramhaas, M. Roiaz, N. Bosio, M. Corva, C. Rameshan, E. Vesselli, H. Grönbeck and G. Rupprechter, *ACS Catal.*, 2021, **11**, 208, DOI: 10.1021/acscatal.0c03974.

Giulio D'Acunto answered: "Originally, the term *operando* refers to characterisation of a working catalyst under realistic pressure conditions (*in situ* in catalysis lingo) and the simultaneous acquisition of reaction data. The term *in situ*, is, however, used differently in the ALD community: it denotes that the sample is transferred from the ALD reactor into a characterisation instrument that is directly connected to the reactor. Hence, exposure to air is avoided, but the characterisation does not take place at the same time as the ALD film growth. We therefore employ the term *operando* to clearly distinguish the characterization during growth and at realistic pressure conditions from such *in situ* characterisation." This is an extract of my PhD thesis.[1]

Regarding the mass spectrometer, at the HIPPIE and SPECIES beamlines, as well as the TEMPO beamline, we tried more than once to collect valuable mass spectra; however, not enough by-products are formed during the ALD process to be detected, and the MS is right now located at a distance of around 1 m away from the ALD cell, therefore we believe that some of the by-products react in the long pipe. We recently changed the position of the MS at the SPECIES beamline to the first differential pumping stage. We hope that this will help it to be more sensitive.

1 G. D'Acunto, *Reaction Mechanisms and Dynamics in the Early Stage of High-κ Oxide Atomic Layer Deposition: Investigations by* In Situ *and* Operando *X-ray Photoemission Spectroscopy*, PhD thesis, Lund, 2022, https://portal.research.lu.se/en/publications/reaction-mechanisms-and-dynamics-in-the-early-stage-of-high-%CE%BA-oxi.

Mathieu Silly communicated: You have shown that after the first half a cycle, the thickness of the HfO_x layer corresponds at most to one monolayer. I would like to know if you have performed more than half a cycle to check that the thickness of the HfO_x is limited to one monolayer or if the oxide of InAs provides a reservoir for oxidizing more than one monolayer of HfO_x.

Giulio D'Acunto communicated in reply: Thanks for your question. We performed more than a half-cycle, and we used water as a precursor. The results are submitted and are currently under review (see also ref. 1). We found that the water mainly removes all the surface ligands (DMA^- and MMI) and leaves the surface covered with hydroxyl (–OH) groups. The interface remains sharp since no trace of As–O_x nor an increase in In–O_x species was observed.

1 G. D'Acunto, *Reaction Mechanisms and Dynamics in the Early Stage of High-k Oxide Atomic Layer Deposition: Investigations by In Situ and Operando X-ray Photoemission Spectroscopy*, PhD thesis, Lund University, 2022, https://portal.research.lu.se/en/publications/reactionmechanisms-and-dynamics-in-the-early-stage-of-high-%CE%BA-oxi.

Alexander Shard queried: Can you provide more details about the method used to calculate oxide thickness?

Giulio D'Acunto responded: The thickness of the oxide was evaluated using two different approaches. The first one consists of approximating the overlayer (oxide) as a uniform layer instead of considering the crystalline structure. It is based on the Beer–Lambert law, modified for two layers (InAs + InAsOxide) and three layers (InAs + InO_x + HfO_2), before and after the first half cycle.[1]

The second one, instead, was used only to calculate the thickness of the oxide layer before the ALD. It is based on the assumption of knowing the crystalline structure of the oxide. It is a self-consistent converging method that takes as an input the experimental ratio between the oxide components and the bulk ones. As a first step, the intensity of the bulk is estimated by considering 200 layers of InAs where each layer gives a contribution of $e^{(-d/\lambda)}$ where d is the distance between each plane of InAs and λ the Inelastic mean free path (IMFP). Then, in the same way, the intensity of the thermal oxide is estimated. For every iteration, the theoretical ratio between the intensity of the oxide and the bulk is compared with the input values, and if they match within an error of 5%, the code stops.

I hope this answer your question.

1 G. D'Acunto, A. Troian, E. Kokkonen, F. Rehman, Y.-P. Liu, S. Yngman, Z. Yong, S. R. McKibbin, T. Gallo, E. Lind, J. Schnadt and R. Timm, *ACS Appl. Electron. Mater.*, 2020, **2**, 3915–3922, DOI: 10.1021/acsaelm.0c00775.

Alexander Shard asked: How did you determine the electron effective attenuation length to calculate oxide thickness?

Giulio D'Acunto answered: We used the formula provided by S. Tanuma, C. Powell and D. Penn,[1] and then corrected for elastic scattering. The attenuation length is typically slightly shorter, around 90% of the IMFP. More detail can be found in the supplementary information of ref. 2 and on page 28 of my PhD thesis.[3]

1 S. Tanuma, C. J. Powell and D. R. Penn, *Surf. Interface Anal.*, 2003, **35**, 268–275, DOI: 10.1002/sia.1526.

2 G. D'Acunto, A. Troian, E. Kokkonen, F. Rehman, Y.-P. Liu, S. Yngman, Z. Yong, S. R. McKibbin, T. Gallo, E. Lind, J. Schnadt and R. Timm, *ACS Appl. Electron. Mater.*, 2020, **2**, 3915–3922, DOI: 10.1021/acsaelm.0c00775.
3 G. D'Acunto, *Reaction Mechanisms and Dynamics in the Early Stage of High-k Oxide Atomic Layer Deposition: Investigations by In Situ and Operando X-ray Photoemission Spectroscopy*, PhD thesis, Lund University, 2022, https://portal.research.lu.se/en/publications/reactionmechanisms-and-dynamics-in-the-early-stage-of-high-%CE%BA-oxi.

Alexander Shard enquired: Did you observe any photoelectron diffraction effects from the single crystal substrate?

Giulio D'Acunto responded: Thanks for the question. The acceptance angle of the electron analyser at the SPECIES beamline at MAX IV Laboratory is around 30° (solid angle). We, therefore, assume that we are integrating over a too large angular area to see any diffraction effects. Moreover, from another unpublished experiment performed at the TEMPO beamline, SOLEIL (France) (with similar angular acceptance), we tried to move the emission angle more than once, and no diffraction effects were observed.

Philip Davies asked: Your paper (https://doi.org/10.1039/d1fd00116g) states that the rate of HfO_2 formation and the removal of the oxide layer are the same. However, for the samples, E, F, H and G, the initial oxide thickness is greater than one monolayer but only a single monolayer of HfO_2 is formed. Can you account for the discrepancy in rates in these cases?

Giulio D'Acunto answered: This is an excellent question, and it is part of the open question of my PhD. One hypothesis is to consider the diffusion of As–O species through the high-κ layer,[1] as suggested by other groups. However, we rule out this hypothesis since depth profile analysis clearly shows no As–O_x and shows that the In–O_x is at the interface. Perhaps it can be due to the nature of the oxide since samples E, F, G and H are InAs with a thick native oxide, in contrast to the others, which are prepared with a uniform thin layer of thermal oxide. The SPECIES beamline, at MAX IV laboratory, since summer 2021 (after the experiment that produced the paper under discussion https://doi.org/10.1039/d1fd00116g), moved the QMS from the outlet of the APcell, to the first differential stage of the electron energy analyzer. Perhaps this will give us more information about the desorbed species during the first half cycle, and it will allow us to find a proper scientific answer to your excellent question.

1 C.-H. Chang, Y.-K. Chiou, Y.-C. Chang, K.-Y. Lee, T.-D. Lin, T.-B. Wu, M. Hong and J. Kwo, Interfacial self-cleaning in atomic layer deposition of HfO_2 gate dielectric on $In_{0.15}Ga_{0.85}As$, *Appl. Phys. Lett.*, 2006, **89**, 242911.

Philip Davies questioned: The O(1s) peak in Fig. 3 of the paper (https://doi.org/10.1039/d1fd00116g) is very symmetric and you assign this to a chemisorbed oxygen but since it was created by reaction with water have you ruled out the possibility that it's actually a hydroxide species?

Giulio D'Acunto responded: This is a really nice question. The possibility of having a hydroxide species was ruled out for two reasons:

(i) The binding energy (BE) position of the O 1s lineshape is the same as of the O 1s of the samples A, B, C and D, prepared only with O_2.

(ii) On another paper, just submitted (but see also ref. 1), we performed also the second half cycle with water, and we clearly see, after the deposition, a new component in the O 1s line that we associated to the hydroxyl group (–OH).

1 G. D'Acunto, *Reaction Mechanisms and Dynamics in the Early Stage of High-k Oxide Atomic Layer Deposition: Investigations by In Situ and Operando X-ray Photoemission Spectroscopy*, PhD thesis, Lund University, 2022, https://portal.research.lu.se/en/publications/reactionmechanisms-and-dynamics-in-the-early-stage-of-high-%CE%BA-oxi.

Mark Isaacs asked: You observe full HfO_2 monolayer coverage from a 50% oxide coverage, and you can see a changing Hf spectrum correlating with the number of TDMA ligands remaining on the Hf centres. Do you see just a shift in the Hf binding energy or a broadening FWHM? If you can quantify the N : Hf : O ratio, could you determine the average number of oxygen bonds to Hf and determine the bridging/terminal environment ratio. Could this give you a method by which to quantify the homogeneity of the surface for submonolayer coverage?

Giulio D'Acunto answered: Thanks for your question. Probably there was a misunderstanding about your first sentence. We did not see a full monolayer of HfO_x from a 50% oxide coverage. From the data presented in Fig. 2(d) in the paper (https://doi.org/10.1039/d1fd00116g), we see a shift in BE of the Hf 4f line as a function of the initial thickness of the InAs oxide layer, but not an increase or decrease in its width (please see the third paragraph of the "Thermal and native oxides of InAs, during and after TDMAHf deposition" section of the paper). Regarding the second part of your question, I can say that in principle, yes, one can quantify the average number of N and O per Hf atom. We did not perform such analysis in this manuscript since it required several unknown parameters to normalize the XP core-level spectra with accurate precision and then determine the composition of the chemisorbed TDMAHf. We did something similar on the HfO_2 ALD on SiO_2, which is currently under revision at *Chem. Mater.*, where we obtained the average number of retained ligands per Hf ion (see also ref. 1).

1 G. D'Acunto, *Reaction Mechanisms and Dynamics in the Early Stage of High-k Oxide Atomic Layer Deposition: Investigations by In Situ and Operando X-ray Photoemission Spectroscopy*, PhD thesis, Lund University, 2022, https://portal.research.lu.se/en/publications/reactionmechanisms-and-dynamics-in-the-early-stage-of-high-%CE%BA-oxi.

David Morgan opened discussion of the paper by Sefik Suzer: You are using a rather simple ionic liquid in your work, do you feel that other ionic liquids, with more bulky ligands for example, will show similar behavior or degree of binding energy shift to that shown in your paper (https://doi.org/10.1039/d1fd00102g)?

Sefik Suzer replied: This is a very important question, and we are trying to extend our work to include more bulky and asymmetric (anion/cation size-wise) ions. There are two important issues related to the size of the ions: (i) their diffusion constants will be different. (ii) Their coverage of the metal electrode surfaces will be different. Both issues must be taken into account and dealt with properly.

David Morgan enquired: I note that the ionic liquid you used was commercially sourced and as evidenced by the survey spectrum shown rather pure. My question however is that many ionic liquids are prone to silicone contamination, especially if prepared in-house. Do you feel therefore that the presence of any insulating material, such as a silicone, will have a significant effect on the phenomena you observe in your paper?

Sefik Suzer responded: Indeed, Si (I believe it is an organo-silicon compound and not charged) is always present as a very small impurity (much less than one atomic percent), in most of the sample surfaces we have worked with. However, I do not believe it has any significant effect on the phenomena we observe. On the other hand, other unwanted and unavoidable ionic (anionic/cationic) species, such as F^- or Cl^-, definitely and significantly affect our measurements.

Benjamen P. Reed asked: Due to their ionic conductivity, ionic liquids tend to exhibit surface charging which develops over extended periods of time when exposed to X-rays or low energy electron sources for charge compensation. The magnitude and rate of the charging depends on the ionic liquid being measured, and generally correlates with the coordination strength of the anion.[1] Did you observe any X-ray induced shifts in your work and would you expect it to have a significant effect on the measured ion dynamics, especially with depth or thickness of the ionic liquid layer? For example, over the gold contacts where you say the ionic liquid layer is thinner.

1 I. J. Villar-Garcia, E. F. Smith, A. W. Taylor, F. Qiu, K. R. J. Lovelock, R. G. Jones and P. Licence, *Phys. Chem. Chem. Phys.*, 2011, **13**, 2797, DOI: 10.1039/C0CP01587C.

Sefik Suzer answered: We have also observed shifts in the measured BE as a function of exposure to X-rays and also to the bias applied over long periods of time like hours. This usually manifests as an overall shift in the peaks. However, such an effect is much less pronounced while AC-biasing, since the polarity of the applied bias is constantly and rapidly alternated. Moreover, when assessing the potential drop, which we are relating to screening dynamics, at different points of the device, we use the BE difference between the twinned peaks, which mostly eliminates the influence of the accumulated surface charge.

Benjamen P. Reed enquired: I'm interested to know if this AC biasing technique could be used as a method to separate out differentially charged chemical environments when measuring inhomogeneous materials? So, for example, could a particular chemical environment have a different energy shift in positive and negative bias, or develops toward an equilibrium charge state at a different rate compared to some other chemical environment? An example where this could be useful would be the complex solid–electrolyte interface formed on cycled Li-ion battery materials.

Sefik Suzer replied: This is an excellent question and one of our Holy-Grails we are still seeking for. The closest related experiment we have carried out was investigating the Au-nanoparticles created within the ionic liquid medium *via* electro-corrosion. We found that the time-dependent behavior of the Au 4f peaks was surprisingly similar to that of the F 1s of the IL. This finding was already published in ref. 1. We are confident that AC-biasing would give valuable information for sufficiently different chemical moieties within inhomogeneous materials, such as the solid-electrolytes in battery systems, as you also kindly advocate.

1 M. T. Camci, B. Ulgut, C. Kocabas and S. Suzer, *ACS Omega*, 2017, **2**, 478–486.

Alenka Krizan asked: The investigated system (coplanar Au electrodes) is symmetrical. Still, in static conditions the potential gradient (shift in binding energy) observed for the ionic liquid on top of the working electrode is larger than the potential gradient within the ionic liquid covering the counter electrode. This trend is observed for both biases, −3 and +3 V. What is your explanation for this asymmetry? *I.e.* why is ionic liquid on top of the counter electrode behaving differently from the ionic liquid on top of the working electrode?

Sefik Suzer responded: I am not quite sure, if and how and where in our paper (https://doi.org/10.1039/d1fd00102g) that information is relayed. But definitely not. The potential gradients measured through the shifts in the BE of F 1s in both electrodes are similar, to the best of our knowledge. Fig. 4(d) of the paper clearly presents no asymmetry with respect to electrodes. However, one must remember that we extract that information from the BE difference between the twinned peaks, which are larger on the electrified side, and gets progressively smaller as we approach the other side, *i.e.* the grounded electrode. Moreover, in order to check if such an asymmetry exists, we make a set of extra measurements on the same device, after reversing the bias on the electrodes through connecting the biasing from the other side (*i.e.* changing the cathode, to be anode and *vice versa*.) This procedure also ensures that there are no significant geometrical differences between the two liquid/solid interfaces, which means the capacitances of the two sides are similar.

Mark A. Buckingham enquired: I really enjoyed reading about your work using XPS and impedance together, I was wondering if there was a way to tangibly link the two? Could XPS be used to inform about the kinetic *vs.* mass transport limitations in the system which impedance typically measures.

Sefik Suzer replied: Thanks very much for your kind assessment of our work. This is exactly our point of calling it Lo-XPIS. However, there is a basic difference since we use square-waves, and electrochemists use sine-waves. We are presently working on issues related with extracting kinetic and mass transport information using our data, but honestly have not been very successful so far.

Robert Lindsay questioned: You concentrated on acquiring data from the F 1s core level, which is only present in the anion. Did you also look at the N 1s signal to compare cation and anion. If so, what were your findings?

Sefik Suzer answered: This is an excellent question, since the N 1s peak(s) do have a richer information content, but are more difficult to utilize, on account of the lower photoemission cross-section. Nevertheless, we have made similar measurements using C 1s, N 1s, O 1s and S 2p peaks, giving more-or-less the same time-dependent shifts.

Robert Weatherup enquired: The simulations consider bulk diffusivity of the ionic liquid, but the XPS measurements are probing the surface. Do you expect differences between the surface and bulk diffusivity to be affecting the alignment between the simulations and experiments.

Sefik Suzer replied: Very good point. Indeed, yes, we do expect the bulk and surface diffusivities to be different. At this point, we do not have any bright ideas to design experiments in order to distinguish between them. However, we already know that the thickness (or the thinness) of the liquid film affects significantly the time constant of the device, which somehow must be related to this difference.

James J. C. Counter asked: At the end of your presentation you make reference to possible application of this technique to investigating solid electrolytes. Have you considered using this technique to investigate solid state batteries? If so do you think it would provide information about the solid–electrolyte interfaces formed?

Sefik Suzer responded: We have not tried anything related with solid electrolytes, yet. However, I do believe that it is possible to use our methodology and extract similar but somewhat different physicochemical information using XPS. I expect the difference(s) will mainly be due to the diffusion time-constants, solid electrolytes being even slower! On the other hand, there is another advantage in working with liquid surfaces. Since the surface of the liquids are continuously replenished, X-ray damage is not very significant, and one can work with the same sample (liquid surfaces) for hours to days.

Philip Davies enquired: To model your data successfully you have to introduce an inverse relationship between the time constant and the ionic liquid film thickness. Can you explain this relationship?

Sefik Suzer answered: Our model reveals that as the IL film gets thinner, the bulk conductivity decreases and the resistance gets larger. Assuming the capacitance(s) of the device do not change significantly, the increase in the resistance would lead to a larger *RC*-time constant.

Mathieu Silly communicated: The numerical simulations of charge dynamics of a parallel plate capacitor show that the electric response is related to ion distribution near both electrodes. As a chemical sensitive technic, is it possible to evidence the charge distribution by XPS and correlate it to the electric response to confirm the simulation?

Sefik Suzer communicated in reply: The only electrical response we were able to record, while recording XPS data under bias, was the overall current passing through the device, but only for long time windows (>10 ms). We do get some correlation with the XPS data, but so far we were not able extract any new information from it.

Alenka Krizan communicated: The simulation model presented in your paper (https://doi.org/10.1039/d1fd00102g) assumes that the electrostatic potential profile on the surface of the ionic liquid is the same as on the Au electrode surface. What are the grounds for assuming there are no potential gradients within the ionic liquid? Ionic liquids are poorer conductors than metals which means they are more likely to develop potential gradients.

Sefik Suzer communicated in reply: To my knowledge, the model does not assume any potential gradient anywhere from the beginning. It only calculates the screening of the initial electrical field imposed at time = 0, as a function of time due to the ion migration. In fact, if one looks at Fig. 6(c), 7(d) and (e) and 8(b) and (c) of the paper (https://doi.org/10.1039/d1fd00102g), one can easily see that the voltage changes as a function of time are different on the metal electrodes, when compared to those in between the electrodes (*i.e.* the IL-only medium). One must also remember that the IL also covers the electrodes.

Anders Nilsson opened discussion of the paper by Juan Jesús Velasco Vélez: Electrochemical measurements with XPS is an enormous challenge. In the past we have conducted *in situ* XPS of IrO_2 nanoparticles[1] and *in situ* L-edge XAS of monolayer Ir on a surface of RuO_x (ref. 2) that shows sign of Ir(v) as the active state. Furthermore, with your proposed μ_3 site we would expect an additional shift towards higher Ir oxidation state with a replacement of Ir–OH by Ir–O bonds on the surface. Should you not interpret your Ir 4f spectra under the oxygen evolution reaction (OER) to contain Ir(v) species?

1 H. G. Sanchez Casalongue, M. L. Ng, S. Kaya, D. Friebel, H. Ogasawara and A. Nilsson, *Angew. Chem., Int. Ed.*, 2014, **53**, 7169.
2 A. F. Pedersen, M. Escudero-Escribano, B. Sebok, A. Bodin, E. Paoli, R. Frydendal, D. Friebel, I. E. L. Stephens, J. Rossmeisl, I. Chorkendorff and A. Nilsson, *J. Phys. Chem. B*, 2018, **122**, 878.

Juan Jesús Velasco Vélez answered: *In situ* measurements based in Ir 4f are not enough to describe accurately the oxidation state of IrO_x during the OER. It is because the Ir(v) is shifted with respect to the position of the Ir(IV) peak, but on the other hand, Ir(III) has a gap which also induces a shift to higher BEs with respect to the Ir(IV). This conundrum can be solved using the Ir L-edge where it is expected that Ir(III) has a lower whiteline intensity compared to Ir(IV) meanwhile Ir(v) has a higher whiteline intensity with respect to Ir(IV). Our *in situ* Ir L_3-edge measurements indicated a higher oxidation state than IV during the OER conditions. However, it is difficult to establish the absolute oxidation state of Ir during the OER as it is complicated by the Ir L_3-edge bulk sensitive measurements because only the surface contributes to the changes in the intensity. These results are explained in more detail in our recent publication.[1]

Nevertheless, it is important to indicate at this point that the most important result is the existence of μ_1-O species which in the long term control the catalytic activity of IrO_x during the OER.

1 J.-J. Velasco-Vélez, *et al.*, *J. Am. Chem. Soc.*, 2021, **143**, 12524.

Anders Nilsson asked: When we are considering resonant photoemission, it is important to establish that this involves a participant Auger decay resulting in a one-hole final state in interference with the photoemission channel. Is this established concerning the O K-edge resonance and how can the peaks be interpreted to be associated with OOH species?

Juan Jesús Velasco Vélez replied: Thanks for your question, it is really relevant and a cornerstone in the complete ResPES spectra interpretation which is not trivial. Please note first that the HOO^- are not stable for a long time as far as all our molecular dynamics calculations indicate a quick deprotonation process for these species. Meanwhile, the OO^- are stable for enough time to be detected. Thus we believe that these species are ascribed most likely to a peroxo rather than to a hydroperoxo. We found that the resonant photoemission at lower excitation energy (from 528 eV to 530 eV) is ascribed to the two-holes final state as a spectator Auger decay occurring right at the threshold photoemission process at a constant kinetic energy (KE) (note that the normal Auger photoemission increases linearly with the excitation energy). For all the one-hole final states (participator Auger) simulated in our calculations, the resonant process take part at low BE (close to the Fermi edge (FE)). Thus we are sure that the μ_1-O, μ_2-O and μ_3-O resonant peaks are due to spectator Auger decay (two electron–hole process).

Regarding the HOO^- and OO^-, our calculations show that the excitation energy is correct (in agreement with the O K-edge measurements presented in the literature). The HOO^- and OO^- have a participator character (*i.e.* appearing at low BEs close to the Fermi edge (FE) in our calculations). Thus, the participator photoemission character is not fitting well with the high BE found in our measurements indicating most likely a spectator Auger decay process for the peroxo species. The reason behind the discrepancy between the measurements and calculations is not clear at the moment but can be closely related to the difficulty to describe the right final state ascribed to the resonant process in our calculations. We are still working on this issue to try to understand completely these measurements. In parallel we will try to do some additional measurements *in situ* adding peroxo in our electrolyte and collecting the ResPES map to get a proper peroxo species reference ("fingerprint") if this approach work properly.

Anders Nilsson enquired: When it comes to the proposed interpretation of OO species in the proposed resonant photoemission process would we expect a very low concentration of such species that puts into question if it can be seen in your data? Theoretical calculations[1] show that Ir–O is the reaction sink and therefore as the reaction proceeds over the barrier we expect OO to be a transient intermediate existing for a very short time. Can you elaborate why you would see such a species?

1 J. Rossmeisl, Z.-W. Qu, H. Zhu, G.-J. Kroes and J. K. Nørskov, *J. Electroanal. Chem.*, 2007, **607**, 83.

Juan Jesús Velasco Vélez answered: The mass transport is an issue with the mesoporous material playing a role in two directions: allowing water molecules to reach the surface of the IrO_x catalyst and letting the products escape. Therefore, we believe that this side effect helps detection of the intermediate and products.

Kelsey Stoerzinger asked: What voltage are you considering for spectra labeled as "OER" conditions? The motivation for asking is twofold: (i) you might accumulate and/or remove adsorbates differently as a function of applied potential (on the RHE scale), which could give you additional information and understanding of the systems or comparison across cell geometries considered, and (ii) depending on the current density sustained at this applied potential and the mass transport in the system, you could develop (micro)bubbles of O_2 on the surface, which might have very different spectral features compared to the solid/liquid or solid/polymer interface.

Juan Jesús Velasco Vélez responded: The applied potential during OER is very well beyond the onset potential and it is 1.6 V *vs.* Ag/AgCl, which is the reference electrode used in our setups. Of course, we investigated the changes in the electronic structure at different applied potentials. Simultaneously the O_2 product evolving from the reaction was detected with online mass spectrometry. Thus the amount of O_2 produced is directly correlated to the process current and applied potential. The species that accumulates at higher applied potential is the μ_1-O, indicating a clear correlation with these species and the electrocatalytic activity. This information will be provided in another manuscript that we are working on, which will include information of other absorption edges to provide a full view of the covalent bond of Ir and O depending on the applied potential.

Yes, we can detect the O_2 formation in total electron yield (TEY) (collected with a "Faraday cup"), however the spectra shown in this work are collected in Auger electron yield (AEY) with 90 V applied in the nozzle, which shifts and broadens the non-grounded species like the gas phase (basically the applied potential in the nozzle suppresses all the gas phase contributions in the XPS and AEY spectra). Because of this fact, it is not possible to see the O_2 product contribution or the H_2O gas phase in the spectra. Notice that the gas phase contribution is a problem for a right spectra interpretation. Otherwise, the bubbles are efficiently eliminated from the surface because we do measurements in a continuous electrolyte flow mode. In addition the membrane is mounted perpendicular to the ground which facilitates the bubbles to be eliminated from the surface. In the case of the ionomeric membranes the gas can be pumped out quite efficiently in the HV side and be detected with the QMS.

Kelsey Stoerzinger queried: Do you have any bubble accumulation on your surface under OER conditions?

Juan Jesús Velasco Vélez answered: As described in the previous answer, the cells were designed in such a way that the bubbles can be removed from the surface efficiently due to:

(i) The cell is operated in continuous flow mode, which helps the bubbles to be removed by the electrolyte flow.

(ii) The cell is mounted perpendicular to the floor, which facilitates the bubbles to leave the electrode surface.

(iii) The bubbles generated in the ionomeric membranes are efficiently pumped down in the HV side. This assembly resembles industrial applications, in acidic media, in which electrochemical reactions are performed using composites of electrodes and ionomeric membranes, which separate the gaseous products from the liquid electrolyte.

Robert Weatherup said: Most of the sample assemblies are covered with graphene, or other sources of carbon contamination. Oxidation of graphene is expected at the potentials where OER occurs.[1] This could produce features in the O K-edge [*e.g.* graphene oxide – see ref. 2] at similar energies to those attributed to μ_3-O. Can you exclude this in your interpretation of the O K-edge?

1 S. Khatun, S. R. Cohen, S. S. Peled, I. Rosenhek-Goldian, R. S. Weatherup and B. Eren, *Adv. Mater. Interfaces*, 2021, **8**, 2100662, DOI: 10.1002/admi.202100662.
2 A. Dey, P. Ghosh, J. Bowen, N. St. J. Braithwaite and S. Krishnamurthy, *Phys. Chem. Chem. Phys.*, 2020, **22**, 7685–7698, DOI: 10.1039/C9CP06174F.

Juan Jesús Velasco Vélez responded: Usually the intensity of the μ_3-O tends to decays when the peaks ascribed to μ_2-O and μ_1-O decay thus even though we cannot exclude graphene oxidation for certain, it is not the dominant feature in the spectra collected.

Philip Davies questioned: Can the amount of peroxo that you suggest is present at the surface be quantified from your data?

Juan Jesús Velasco Vélez replied: Thanks for your question. It is really relevant to correlate the catalytic activity to the presence of the peroxo and μ_1-O species. The amount of species can be correlated qualitatively but not quantitatively, *i.e.* it is possible to see how the intensity of the different peaks increases or decreases with respect to each other. However, it is not easy/possible to give an absolute number.

Robert Weatherup asked: Table 1 in the paper (https://doi.org/10.1039/d1fd00114k) indicates different extents of beam damage from the different sample assemblies. Could you elaborate on whether this refers to different extents of radiolysis for the different assemblies, or if it indicates certain assemblies are more vulnerable to the impact of radiolysis (*e.g.* membrane leaks)?

Juan Jesús Velasco Vélez answered: The beam damage affects the graphene, SiN_x and ionomeric membranes as well as the catalyst. Thus in the presence of aqueous electrolytes the radiolysis process enhances the beam damage effects. The table aims to indicate in an easy way what assembly is more stable in term of beam damage in general taking into account the possibility that the different approaches have for illuminating fresh spots.

David Payne enquired: How many membranes have you broken during the course of this work?

Juan Jesús Velasco Vélez responded: We used three types of membranes during the experiments reported in this work:

(i) SiN_x: It took some time to find the right process to grow the porous IrO_x onto the SiN_x membrane. Basically, if the membrane is cooled down quick (after the deposition of the IrO_x mesoporous layer) it will not be able (most of the time) to keep the difference in pressure between the liquid phase and HV in the experimental chamber. When the membrane was cooled down slowly, most of the membranes were stable and none of them break during the pumping down or during the spectra acquisition. Then we can assure that 99% of the SiN_x membranes coated with the mesoporous IrO_x or sputtered IrO_x are stable if they are prepared properly. Thus the problem of stability with these membranes can be ascribed to thermal stress in the SIN_x due to a quick cooling down process.

(ii) Ionomeric membranes: none of the ionomeric membranes break during the experiments reported in this work. It happens to other users in our beamline that during the sample alignment they punch the membrane with the analyzer nozzle and the membrane breaks, but we can count this as an incident and not as a problem of the membrane stability itself. Nevertheless, not all the ionomeric membranes are compatible with HV experiments; the stability of any new type of ionomeric membrane not used before in HV conditions and under the reaction conditions should be tested before its use in the *in situ*/*operando* experiments.

(iii) Free-standing graphene on SiN_x holey structure: the stability of this membrane depends on many factors. First of all, we only achieved mechanical stability to keep the difference in pressure between the liquid phase and HV with holes of 1 μm in diameter or smaller and with two layers of graphene. Every single membrane was checked with SEM after the graphene transfer onto it. Only the membranes that showed all the surface coated with graphene (~99% of the holey structure coated with two layers of graphene) were used. Those membranes always keep mechanically the difference in pressure between HV and the aqueous electrolyte. These membranes were also stable at different potentials in most of the cases. However, when the membranes are illuminated by the X-ray beam around 80% don't survive for a long time presenting different stability times. Some of them are stable for only a few minutes, and others never break. From our point of view the problems are the number of defects in the graphene layers and the radiolysis process that tends to etch away the non-perfect graphene layer. For example in gas phase experiments, we didn't observe as big a problem with the stability of the graphene membrane as in the case of the liquid phase (aqueous electrolyte), which points to a problem with the radiolysis. Note also that the experiments were done in a bend magnet beamline in the work reported here, in ondulator beamlines the beam damage should be more significant. Otherwise, we aimed to do experiments with lab-based NAPXPS sources (Al or Mg monochromator), where the photon flux is significantly lower, but unfortunately we never accomplished these experiments completely (may be a way to explore to achieve higher stability during spectra acquisition). In addition, for enhancing the stability of these membranes, the transfer process has to be improved to achieve better graphene quality or alternatively increase the number of graphene layers transferred.

Slavomír Nemšák asked: One comment on the presence of Auger lines in the resonant photoelectron spectra that another participant asked. Auger lines can be usually identified relatively easily by measuring complete photon energy ($h\nu$) *vs.* binding energy (BE) valence band scans over a broader region as continuous higher intensity diagonal lines (linear BE *vs.* $h\nu$ dependence). Unfortunately, as presented in the manuscript (https://doi.org/10.1039/d1fd00114k), the color scales of the resonant spectral maps are saturated, which prevents identifying

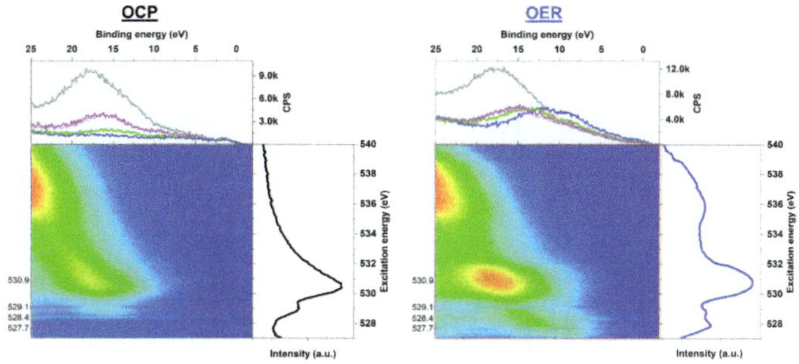

Fig. 1 Resonant spectral maps with a non-saurated color scale.

Augers and other artifacts (core-levels excited with higher order light, for example). It would be very useful for readers if the data could be presented with different color settings (logarithmic, or other non-linear, as needed). If these features become visible, it would be great to discuss them in some detail.

Juan Jesús Velasco Vélez replied: Thanks for your comment. Yes I agree, in Fig. 1 here you can find the unsaturated image with a different color scale. The idea to present the color map saturates was to show clearly the species of interest (*i.e.* low excitation energy) formed during the OER.

Mark A. Buckingham opened discussion of the paper by Rosa Arrigo: Congratulations on this wonderful work, I was wondering if you optimised the deposition at all? I would expect the single quite negative potential you apply to form Cu(0), which looks like it is oxidised with the thermal treatment as shown in your SEM (Fig. 4 in the paper; https://doi.org/10.1039/d1fd00121c) in the presence of CO_2. Is there any dendrite formation in your electrodeposition procedure?

Rosa Arrigo answered: Thank you very much for your feedback. In fact, we have not systematically explored the electrodeposition of Cu but we have adapted an electrodeposition method from the literature. Starting from the aqueous solution of $CuSO_4$, we performed a cyclic voltammetry investigation to identify the potential at which the reduction of the Cu species occurs. The potential of the electrodeposition (−0.7 V *vs.* Ag/AgCl) was chosen at the maximum of the Cu(I) → Cu^0 redox wave. At lower potentials (more negative) we expected a higher contribution of the hydrogen evolution reaction (HER) which could disrupt the electrodeposition. The procedure used in this contribution led to Cu_2O particles with a cauliflower-like morphology, but no dendrites were observed. It may well be that initially we had an electrodeposited metallic Cu and the second step in KCl led to the formation of Cu_2O (ref. 1) but this is something to be verified.

We also explored more negative potentials (−1.1 V *vs.* Ag/AgCl) for the electrodeposition starting from $CuSO_4$ aqueous solution and we observed a higher abundance of the metallic component. See the Cu L-edges NEXAFS spectrum in Fig. 2 here. However from the spectroscopic investigations the deposition was much less efficient in terms of metal deposited. We do not have information about the morphology for this sample.

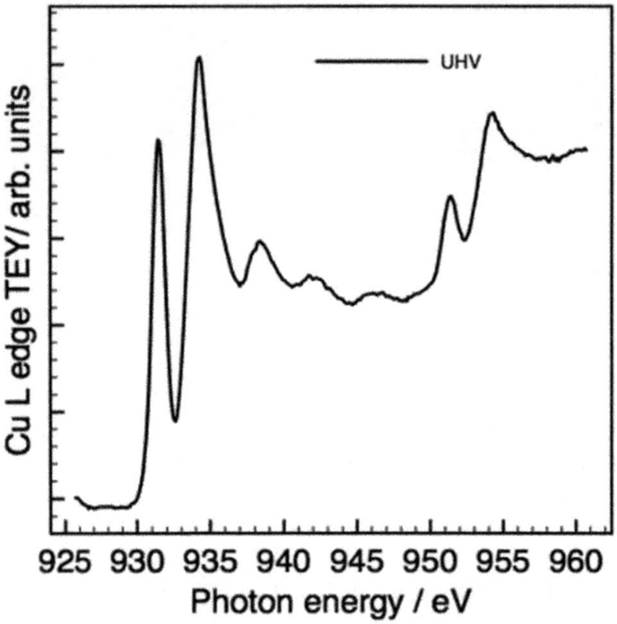

Fig. 2 Cu L-edges NEXAFS spectrum.

1 Y. C. Sun, C. Y. Sun, Z. X. Chen, P. Wang, H. T. Wang, M. Z. Yao, S. Wu and P. Xu, *Inorg. Chem. Front.*, 2021, **8**, 1449–1454, DOI: 10.1039/D0QI01367F.

Mark A. Buckingham said: There is some interesting work recently of *in situ* TEM Cu nanomaterial growth from the group of Roldan-Cuyena[1] in which they form Cu_2O nanocubes but with a very highly resistive cell. In your system it looks like the cubes grow out of the plate, can you comment on the growth? I was also wondering if you see any corrosion in your system in the gas phase reaction? It is common in the typical liquid system.

1 R. M. Arán-Ais, R. Rizo, P. Grosse, G. Algara-Siller, K. Dembélé, M. Plodinec, T. Lunkenbein, S. W. Chee and B. R. Cuenya, *Nat. Commun.*, 2020, **11**, 3489.

Rosa Arrigo responded: It is difficult to compare the *in situ* data from the *Nat. Commun.* paper you are suggesting with our system, due to the fact that the cells are very different. Our electrodeposited Cu_2O are not perfect cubes initially but their surface is rather roughened. Whilst the main phase is Cu_2O, there is a large amount of Cu(II), probably bound to OH^- or Cl^-, which are possibly located at the surface of the particles. Rationalizing the phenomenology of the particles' morphology in terms of growth mechanism is not so trivial and a dedicated study would be needed. Changes in morphology were observed for Cu electrodeposition from different concentrations of the Cu precursor[1] which influences also the pH of the solution. I could suggest that the chemical interaction between the Cu(II) species and the C support and the availability of the Cu(II) species influences the nucleation and growth of the nuclei and thus the particles' morphology and size. The pH is also an issue in the sense that the lower the pH (higher concentration of

H^+) the higher the potential for HER, which occurs simultaneously to the Cu electrodeposition. Bubble formation on the particles could lead to diffusion limited supply of Cu(II) species. An extreme manifestation of this is the dendrite formation, which has been also observed and is known to be a consequence of diffusion-limited electrodeposition (diffusion limited supply of Cu(II) species). You mention that in our system it looks like the cubes grow out of the plate, thus I am assuming you are referring to the sample annealed in UHV at 473 K. Intuitively I would explain this as the results of the desorption of surface chlorine species and the reduction of the surface Cu(II). At 473 K enough energy is supplied for the atoms to diffuse and form a more crystalline Cu_2O phase. It also appears that the interaction of the Cu atoms at the interface with the C support is sufficiently strong to generate this plate-like morphology at the base of the nanostructures. We postulate that Cu is in a reduced state at the plate-like region, however this needs further analysis.

In terms of corrosion, we do not have experimental evidence that this is happening (normally we would expect Cu(II) species increasing) under gas phase CO_2RR conditions, but we observed it on the same Cu/G system when using another cell approach with stagnant liquid electrolyte (paper in preparation).

1 Y. C. Sun, C. Y. Sun, Z. X. Chen, P. Wang, H. T. Wang, M. Z. Yao, S. Wu and P. Xu, *Inorg. Chem. Front.*, 2021, **8**, 1449–1454, DOI: 10.1039/D0QI01367F.

Mark A. Buckingham asked: I was also wondering if you tried to use any other electrolytic conditions? Any kind of further additional electrolyte in the solution?

Rosa Arrigo answered: To electrodeposit Cu, we used in this work a procedure adapted from ref. 1. The method suited our purpose which was to produce a nanostructured metallic Cu film on a C support, study the performance of this in CO_2RR and investigate the dynamics of this system *in situ*. We have not yet explored systematically the electrodeposition of Cu on C, though, but we plan to do it to a certain extent. Many factors influence the efficiency of the electrodeposition as well as the nanostructure and morphology of the resulting electrodeposited Cu, including the nature of the metal precursor, concentration, pH and the substrate. See for example ref. 2. Certainly controlling the electrode structural characteristics is an important aspect, which can be addressed by *in situ* methods. I think there is much to be done in terms of evaluating the "pre-history" of the electrode on the catalytic performances, and this includes the electrodeposition procedure, although I expect that for longer CO_2RR time of reaction, these effects should be levelled up.

1 S. Zhu, B. Jiang, W.-B. Cai and M. Shao, *J. Am. Chem. Soc.*, 2017, **139**(44), 15664–15667, DOI: 10.1021/jacs.7b10462.
2 Y. C. Sun, C. Y. Sun, Z. X. Chen, P. Wang, H. T. Wang, M. Z. Yao, S. Wu and P. Xu, *Inorg. Chem. Front.*, 2021, **8**, 1449–1454, DOI: 10.1039/D0QI01367F.

Anders Nilsson remarked: It is most interesting that you can detect the formation of methanol in the electrochemical reduction of CO_2 on Cu oxide in your XPS study. We would expect that the Cu oxide will not be stable under harsh reducing conditions[1] and eventually Cu metal should appear with some subsurface oxygen. If you waited long enough for this occur would you therefore expect that the product distribution would change to ethylene, methane and other C_{2+} species?

1 H.-Y. Wang, M. Soldemo, D. Degerman, P. Lömker, C. Schlueter, A. Nilsson and P. Amann, *Angew. Chem., Int. Ed.*, 2022, **61**, e202111021.

Rosa Arrigo responded: The paper indicated shows clearly that an apparently Cu metallic phase might contain indeed O species. Indeed there is convincing evidence that subsurface oxygen plays a role in activating CO_2 in the liquid phase electro-catalytic conversion, in which conditions Cu was found to be metallic by means of various Cu core level spectroscopy and grazing incident γ-ray diffraction. However the gas/electrode interface in the gas phase electro-catalytic approaches has been less investigated. Indeed we would expect metallic Cu formation at the negative potentials of CO_2RR and to a certain extent we do see this happening but the experimental environment is such that the electrolyte diffusion is limiting the reaction. Only when the water partial pressure increases is methanol formation observed and the reduction of Cu to metal begins. Therefore, I think that your suggestion is plausible that the predominant Cu_2O is selective to methanol as reported in several contributions. I expect the formation of methane at longer reaction time, similar to the performance of this very system in the liquid phase CO_2RR.

Anders Nilsson asked: To detect adsorbed carbon species on Cu during CO_2 electrochemical reduction is an enormous challenge since there can be a dominating signal from the membrane or the carbon support. I wonder how well can you separate the adsorbed reduced species from electrochemical modifications of the membrane?

Rosa Arrigo replied: Indeed, distinguishing the adsorbed C species from the C substrate or the polymeric membrane is a big challenge. And this is probably one of the reasons why most of the *in situ* soft X-ray studies focus on the metal dynamics and methodologies were optimized to allow these to be captured under conditions as close as possible to relevant conditions. See for example one of our own works.[1] However, it appears to me evident that the knowledge gained so far on this reaction is limited and more analysis is required, focusing on enhancing the surface sensitivity of the analysis (a huge challenge) as well as on looking at the light elements involved. Already previous works, including one of yours,[2] evidenced the importance of the oxygen species in CO_2RR. Given the importance of this topic, in fact, this contribution aims also to show to which extent it is possible to discriminatively assess the C speciation. This goal requires first of all an accurate calibration of the BE of the peaks. We approached this task by using both the FE calibration as well as the C 1s peak measured at the same excitation energy, for which the actual value was determined by measuring the second order C 1s peak. We also generated several reference experiments in non-electrochemical conditions (see for example Cu 2p XPS in Fig. 4A and C 1s XPS in Fig. 5A in the paper (https://doi.org/10.1039/d1fd00121c), whereas the O 1s XPS is not reported but used to check the validity of the calibration), which confirmed the validity of the calibration particularly by considering the BE of the Cu_2O reference at the Cu 2p and O 1s. Secondly, one needs an accurate C 1s fitting. A robust model to fit the C 1s XP spectrum of C materials was previously reported[3] and applied to a similar study (ref. 7 in the paper).[4] We use here the same model to explore whether similar dynamics to a previous study on FeOOH/C could be observed.

Our approach is to identify changes in the C species upon polarization; thus, it is about looking at the differences between spectra measured under controlled conditions.

We noted that under any of the electrochemical conditions the C 1s XPS peak shape as well as the broadening are not changing very much, especially at the lower BE side of the peak. This might well be the case if the components due to the C support (C_1, C_2 and C_3 species in the spectrum) do not change significantly under cathodic polarization. This hypothesis is also supported by the fact that the peak intensity in Fig. 5 of the paper, which was not further normalized after background subtraction, does not change significantly.

Furthermore, the comparison between the C 1s XP spectra under water enriched atmosphere in Fig. 5B and the C 1s spectra in CO_2/water mixture in Fig. 5C of the paper enables us to pin down the contribution of the CO_2-related adsorbed species accumulated or depleted upon polarization. Under the conditions in Fig. 5C, we expect to identify the signature of the CO_2-related adsorbed species. An interesting fact that shows the validity of the approach is that under water conditions, these species are consumed (not replenished) at CO_2RR potentials, in contrast to the mixed CO_2/water conditions.

So, it is safe to say that it is possible to discriminate between the adsorbed species and the C support with an appropriate experimental design.

1 J.-J. Velasco-Vélez, *et al.*, *ACS Sustainable Chem. Eng.*, 2019, 7, 1485–1492.
2 F. Cavalca, *et al.*, *J. Phys. Chem. C*, 2017, **121**, 25003–25009.
3 R. Blume, *et al.*, *ChemCatChem*, 2015, 7, 2871–2882.
4 R. Arrigo, *et al.*, *ACS Catal.*, 2022, **12**, 411–430.

Claus F. P. Kastorp enquired: In the C 1s spectra in Fig. 5 in the paper (https://doi.org/10.1039/d1fd00121c), the measurements done in a water atmosphere show highly functionalised carbon that is otherwise associated with reduced carbon dioxide. What is the origin of this carbon?

Rosa Arrigo responded: Fig. 5B in the paper (https://doi.org/10.1039/d1fd00121c) reports the C 1s XPS during electrocatalysis under 0.3 mbar water enriched conditions. It was explained in the paper that this condition refers to the gas composition generated by the transport of the electrolyte ($KHCO_3$) from the inner compartment of the cell, through the polymeric membrane and into the XPS chamber. Therefore it is a water enriched condition as indicated by the on line mass spectrum but it does contain CO_2. This can be seen also in the C K-edge NEXAFS spectra, which were not included in the original manuscript but are now shown here (Fig. 3). There not only the resonance around 290.5 eV is seen but also the K L-edges resonances indicating the transport of K as well. Nevertheless for simplicity we have indicated this condition as water condition to distinguish from the "mix" condition in which CO_2 is added into the XPS chamber, leading to a significant increase in the CO_2 partial pressure (spectrum in blue in Fig. 3 here). Therefore, we do expect species associated to reduced carbon dioxide under water enriched conditions.

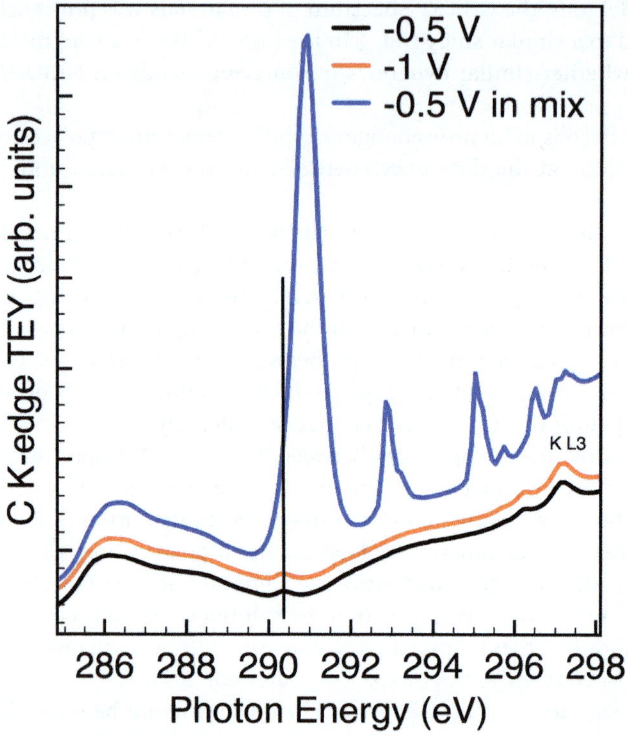

Fig. 3 C K-edge NEXAFS spectra.

Of interest, these species decrease at more negative voltages indicating their consumption, meaning that under water enriched conditions these species are not replenished.

Günther Rupprechter questioned: As shown by the SEM images, your Cu nanoparticles seem to be mostly 100-terminated cubes and related shapes. Maybe this is beneficial for creating a good (smooth?) interface to carbon, but should nanoparticles with rougher surfaces (311, 110) not be more active toward CO_2? HRTEM should reveal whether they have stepped surfaces. You also mentioned that CO is produced. Please note that the CO will lead to Cu atom mobility and roughening (as shown for Cu(100) by PM-IRAS, XPS, TPD and DFT).[1]

1 M. Roiaz, L. Falivene, C. Rameshan, L. Cavallo, S. M. Kozlov and G. Rupprechter, *J. Phys. Chem. C*, 2019, **123**, 8112, DOI: 10.1021/acs.jpcc.8b07668.

Rosa Arrigo replied: The SEM figure of the as prepared sample is reported in the manuscript in Fig. 2A (https://doi.org/10.1039/d1fd00121c). Shown here (Fig. 4) is also another image at higher resolution. Based on this data, I would actually suggest that the surface of the particles (brighter area in the image), which are rather thin along the direction normal to the C support surface (darker region in the image), is rather rough. Under the *in situ* experiments presented in this work, we are able to see only methanol formation in the mass spectrum. CO formation is not confirmed in the *in situ* study, although on a lab

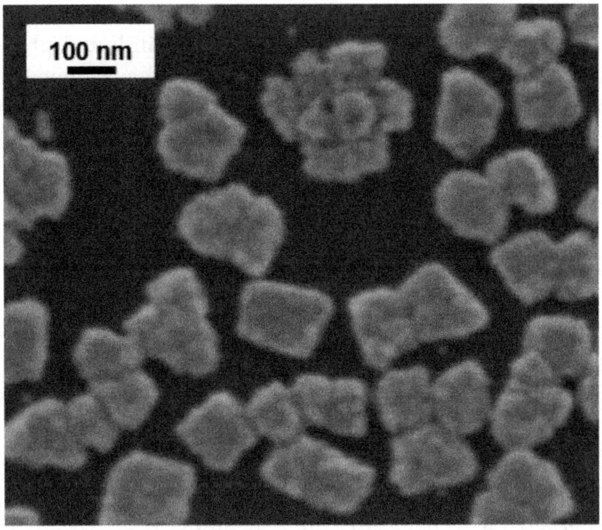

Fig. 4 Scanning electron micrograph of Cu/G in secondary electron mode at 1.6 kV.

bench set up, this sample forms CO. Regardless, we should draw a structure function correlation considering the reactivity data obtained *in situ*. The SEM image in Fig. 4D in the paper refers to the samples after the *in situ* measurements. The morphological changes are consistent with a reordering of the particles and formation of less roughened surfaces. Of course this is an *ex situ* analysis and therefore the dynamics due to that fact that the sample was left to equilibrate under open circuit potential (OCP) and then removed from the electrochemical conditions and exposed to air must be taken into account. Nevertheless, we feel that the surface reordering is real and related to the dynamics under the very low potential for CO_2RR as normally we see the opposite behaviour going to higher potentials (OCP). In general HRTEM analysis on these systems is very challenging due to the thickness of the particles on the one hand; on the other hand, only *in situ* study will allow a careful consideration of the crystallographic details at the surface during CO_2RR. In regard to the behaviour of (100) surfaces under CO_2RR, there seems to be contrasting evidence. See for example:[1]

1 S. H. Lee, J. C. Lin, M. Farmand, A. T. Landers, J. T. Feaster, J. E. Avilés Acosta, J. W. Beeman, Y. Ye, J. Yano, A. Mehta, R. C. Davis, T. F. Jaramillo, C. Hahn, W. S. Drisdell, Oxidation State and Surface Reconstruction of Cu under CO_2 Reduction Conditions from In Situ X-ray Characterization, *J. Am. Chem. Soc*, 2021, **143**, 588–592.

Philip Davies asked: In the C(1s) spectra taken under electrochemical conditions in Fig. 5 in your paper (https://doi.org/10.1039/d1fd00121c), I notice that there is always potassium present at the same time as the high binding energy features. Can you rule out the possibility that these features are due to reaction with the potassium?

Rosa Arrigo responded: This is a very interesting point. We have evidence from XAFS of the formation of $CuCO_3$ on the same system in liquid phase CO_2RR. *Ex situ* microscopy and EDX analysis on the sample after the *operando* study indicate high concentration of potassium in the area where well-defined crystals are found which we attributed to a mixture of Cu and K carbonates. Therefore it might well be that in the case of gas phase CO_2RR the carbonate species seen in the C 1s spectrum are actually due to K_2CO_3.

J. Matthias Kahk enquired: The surface species present on the copper/graphite electrode are discussed with the aid of C 1s XPS in Fig. 5 and Table 1 of the paper (https://doi.org/10.1039/d1fd00121c). A few of the proposed adsorbates are oxygenated species – in addition, copper oxides may be present on the surface, as indicated in Fig. 2 of the paper. I was wondering whether you also measured the O 1s spectra of the catalyst under the same conditions as the C 1s? Copper oxides and molecular adsorbates containing O should be well separated in the O 1s, and it may also be possible to obtain information about the nature of the molecular adsorbates by considering the C 1s and O 1s spectra together.

Rosa Arrigo answered: I agree that it is certainly sensible to see whether the O 1s spectra would provide a picture of the adsorbed species, consistent with the C 1s spectra. Indeed, as you suggest the copper oxides and molecular species are mostly distinguishable in the O 1s spectrum, however the region around 531 eV could still be due to both species. We have not yet attempted a fitting of the O 1s for this contribution but we did it in a previous contribution in a similar study but on a FeOOH/C system (ref. 7 in the manuscript).[1] We found a complex situation with many components above 532 eV, which we generically assigned to adsorbed C–O species. In order to precisely assign the chemical nature of these components more work is needed, involving probably a combined experimental and theoretical approach.

1 R. Arrigo, *et al.*, *ACS Catal.*, 2022, **12**, 411–430.

J. Matthias Kahk asked: In Table 1 of the paper (https://doi.org/10.1039/d1fd00121c), it is indicated that component C4 at 285.5 eV is assigned to both sp^3 carbon (C in the oxidation state zero), and carbon with a single bond to oxygen (*e.g.* a methoxy group or similar). However, one would typically expect the core electron binding energy for C next to O to be higher by about 1 to 1.5 eV than C0. Please comment on this assignment.

Rosa Arrigo responded: This is an interesting point. In regards to the oxidation state, the species which one can assume to be in the oxidation state zero is the C2 component at a BE of 284.3 eV, which is the sp^2 carbon in graphite. And this is consistent with your statement that the C–O bond is found upshifted by about 1–1.5 eV. However one should also consider that final-state polarization might contribute to determining the chemical shift (see the seminal work in ref. 1 and 2. This is specially the case for molecular or oligomeric C species where the screening of the core-hole by the surrounding atoms is less effective.

Blume *et al.*[3] have developed a solid model on the fitting of the C 1s spectrum of carbon materials, based on a large experimental data set combining several C materials and treatments. Accordingly, the C4 component at 285.5 eV is assigned to amorphous sp^3 C. This should contain a large portion of aliphatic carbon species thus containing $(-CH_2-)_n$, which are not to be considered in the oxidation state zero. Moreover, we found that this component increases upon CO_2 chemisorption and under electrochemical conditions. A component at the same BE was found on polycrystalline Cu as a product of the reaction of water with CO_2 at room temperature.

Thus it is clear that the C4 peak is possible due to different species.

1 D. A. Shirley, *J. Electron Spectrosc. Relat. Phenom.*, 1974, **5**(1), 135–148.
2 R. L. Martin and D. A. Shirley, Relation of core-level binding energy shifts to proton affinity and Lewis basicity, *J. Am. Chem. Soc.*, 1974, **96**(17), 5299–5304.
3 R. Blume, *et al.*, *ChemCatChem*, 2015, **7**, 2871–2882.
4 X. Deng, *et al.*, *Langmuir*, 2008, **24**, 9474–9478.

Mzamo Shozi communicated: Very interesting work. Have you looked at other carbon-based supports such as carbon nanotubes? Can they also be easily studied using APXPS and NEXAFS for this type of application, *i.e.* CO_2 reduction?

Rosa Arrigo communicated in reply: We use graphitic carbon paper because this is easily mounted in the *in situ* electrochemical cell and it has sufficient porosity to allow the liquid electrolyte to diffuse and wet the electrode surface. We tried earlier to deposit directly on the polymeric membrane an ink of an activated carbon supported electrocatalyst with liquid Nafion, however the results were not particularly encouraging to pursue this further, probably due to the blocking of the liquid electrolyte and the electrical conductivity of the film. I expect carbon nanotubes present a similar issue. A solution could be to deposit the ink onto the gas diffusion layer but this adds another element of complexity to the spectroscopic analysis.

David Payne asked: In Fig. 4a of the paper (https://doi.org/10.1039/d1fd00121c) the satellite peaks at high binding energy could be due to excitations in the gas phase induced by outgoing photoelectrons. Have you considered this?

Rosa Arrigo answered: This is an interesting point. The separation between the main core line of the Cu2p and the higher BE features is *ca.* 10.2 eV, consistent with observations reported by A. Regoutz *et al.*[1]. The explanation provided therein could be plausible. In fact we do not expect Cu(II)O at 473 K, whereas the features are still present in Fig. 4a in the paper. I am aware that similar features were found under H_2 for Pd (ref. 2) and Ni, due to inelastic scattering of the photoelectrons on the hydrogen molecules. Most interesting is that the peak intensity is pressure dependent. This means that ideally, in reactions in which CO_2 is a product, such a feature could be used to estimate the productivity. We have not been looking into this, but I agree that this is an interesting aspect to study further and extend to other molecules of interest.

1 A. Regoutz, G. Kerherve, I. Villar-Garcia, C. K. Williams and D. J. Payne, *Surf. Sci.*, 2018, **677**, 121–127.
2 D. Teschner, A. Pestryakov, E. Kleimenov, M. Hävecker, H. Bluhm, H. Sauer, A. Knop-Gericke and R. Schlögl, *J. Catal.*, 2005, **230**, 186–194.

Mark Isaacs opened discussion of the paper by Heath Kersell: Are there any suitable peaks in this system to look at the changing inelastic background scatter as the Ni NPs evolve in the system? Would such observations assist with the modelling process of the scattering data?

Heath Kersell replied: While the Ni 3p region can provide information about the changing inelastic background, its position next to the Ti 3s complicates the interpretation of the inelastic background in that region. In principle, the Ni 2p would be a good choice for this, although we need to consider a few constraints. For larger exsolved particles, the depth dimension by far exceeds the information depth of our experiments. This calls for higher excitation energies, leading to more common analyses of inelastic backgrounds found in HAXPES (Hard X-ray Photoelectron Spectroscopy). We should also note that when analyzing the inelastic background, it is important to have a well-calibrated analyzer transmission function. Given these considerations, this approach may prove useful in the very early stages of exsolution, when the particles' depth dimensions remain relatively small. Integrating the analysis of inelastic backgrounds and scattering may provide a powerful tool for analysis of nanostructures, and we do look forward to employing it in the future.

Kelsey Stoerzinger asked: The scattering here was done post-H_2 exposure, in UHV. How high a pressure can we use for scattering, and how does this depend on the nature of the gas molecules and incident photon energy/photoelectron energy?

Heath Kersell answered: The X-ray intensity is increasingly attenuated (together with changes in photon momentum) with increases in the path length through the gas and with increasing gas pressure. The system we used for this investigation is designed with a sample-to-detector distance of approximately 85 cm to facilitate coherent scattering measurements. If we think of "typical" gases in this system, the photon intensity reaching the detector at 1 Torr for 530 eV photons (at the O K-edge resonance) is approximately 0.23, 0.13, and 0.05 times the incident intensity for H_2O, CO, and O_2, respectively. For 730 eV photons, the signal instead becomes 0.5, 0.4, and 0.25 times the incident intensity, and for 1000 eV photons, it becomes approximately 0.7, 0.62, and 0.5. So measuring off resonance will typically be feasible even at 1 Torr. The attenuation is much less significant for the aforementioned gases at lower pressures, even at 500 mTorr, making measurements at resonance generally feasible as well. All values were obtained from the Center for X-ray Optics (CXRO) Henke tables.[1]

For the case in our experiment, H_2 gas presents almost negligible X-ray intensity attenuation for photons beyond 200 eV. Instead, in these measurements the ability to measure during gas exposure was limited more by the ability to continuously pump hydrogen gas through the analyzer in the Torr regime.

1 https://henke.lbl.gov/optical_constants/.

Georg Held enquired: I assume, there is potential for AP-XPS-GIXS to be used in resonance. In the present case, *e.g.* using photon energies of the Ni L_2-edge one might be able to discriminate the diffraction pattern of the Ni nanoparticles from features of the substrate and even achieve chemical contrast (in terms of Ni oxidation state) within the nanoparticles. Has this been tried? Also, could you comment on potential benefits of resonant soft X-ray scattering in this case.

Heath Kersell responded: Measuring X-ray scattering in resonance is certainly a useful approach.[1] By comparing the scattering intensities on- and off-resonance, it is possible to gain information about the distribution of species contributing to the observed structures, including different chemical states. These experiments are possible at our system although we have not performed them in this case due to time constraints. Resonant soft X-ray scattering presents several potential benefits. Perhaps highly relevant to users of combined AP-XPS-GIXS experiments is that the soft X-ray regime contains the absorption edges of C, N, O, and many other elements of interest to interested communities. More generally, this approach can provide a strong supplement to the chemical and morphological insights gained from combining XPS and X-ray scattering in gas phase environments and we are planning on including such experiments in the near future.

1 J. Fink, *et al.*, *Rep. Prog. Phys.*, 2013, **76**, 056502.

Claus F. P. Kastorp asked: Mostly, the measurements of Ti 2p and Sr 3d show a clear trend as function of the hydrogen treatment. However, in the Ti 2p-spectra, there is an undesignated peak at *ca.* 472 eV which shows a completely different behaviour. What is the origin of this component?

Similarly, in the Ti^{4+}-component, the relative intensities of the two measurements at 400 °C for 120 min and 500 °C for 80 minutes seem inverted. Is this real, or merely a quirk of photoemission experiments?

Heath Kersell replied: The component shown at ~472 eV is part of a doublet of charge transfer satellite structures in the Ti 2p region,[1] whose relative intensities increased as a function of treatment, as shown in Fig. 5 here. We should note that the relevant figure in the paper (https://doi.org/10.1039/d1fd00123j) will be updated as we noticed that the figure in the pre-print contained an error. Nevertheless, the total intensity trend is the same as indicated in the question, where the Ti 2p intensity initially decreases, and then increases in the final condition. Regarding this increase of the Ti intensity in the final condition, we also fitted the Sr 3d, Ti 2p, Ni 3p, and Ti 3s regions, and performed an analysis of the relative intensities as a function of condition. While the Sr and Ti intensities decrease overall relative to the Ni 3p, the Ti 2p intensity increases relative to the Sr 3d, especially for the final condition measured. This may be related to a reconstruction of the STNNi surface observed in parallel with the nanoparticle exsolution. Also relating to the relative increase in Ti intensity relative to Sr, STO surfaces have been shown previously to become Ti enriched under reducing conditions, potentially explaining this behavior.[2]

1 J. C. Woicik, *et al.*, *Phys. Rev. B: Condens. Matter Mater. Phys.*, 2020, **101**, 245119.
2 D. Wrana, *et al.*, *Appl. Surf. Sci.*, 2018, **432**, 46.

Wendy Flavell remarked: In the context of the Ti 2p peaks in Fig. 2b of the paper (https://doi.org/10.1039/d1fd00123j), the apparent switch in relative intensity in the Ti $2p_{1/2}$ features on heating (which will change the amount of

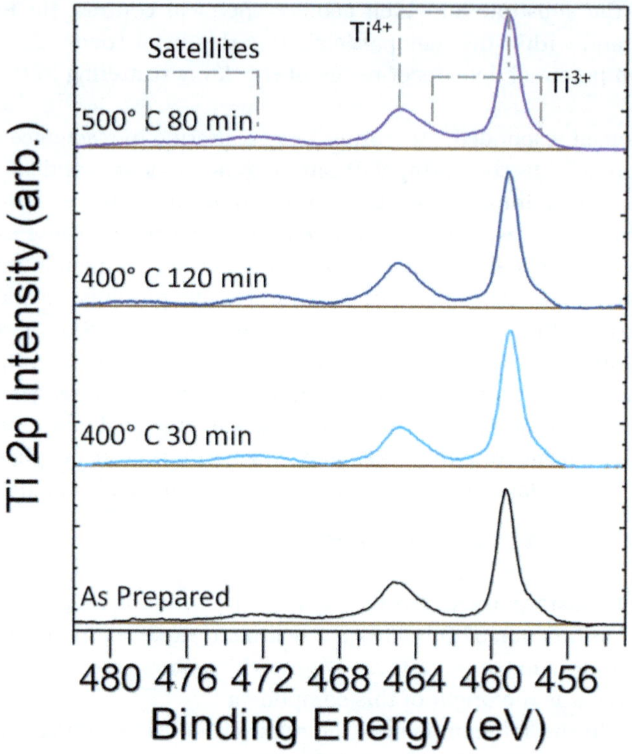

Fig. 5 Ti 2p intensity as a function of reducing treatment. The difference between the maximum and minimum intensities has been set equal for all conditions in order to show relative changes in the Ti 2p composition, and a Shirley background has been subtracted from each spectrum in order to quantitatively show the relative satellite intensities.

Ti(III)) might not be anomalous. I seem to recall there are other features under this peak shape – see *e.g.* ref. 1 for discussion of the charge-transfer satellites (which I think may contribute to the peak at 471 eV), and ref. 2 for discussion of the anomalous broadening of the $2p_{1/2}$ multiplet.

1 J. C. Woicik, *et al.*, *Phys. Rev. B: Condens. Matter Mater. Phys.*, 2020, **101**, 245119.
2 P. S. Bagus, *et al.*, *J. Phys. Chem. C*, 2019, **123**, 7705.

Heath Kersell responded: This is an excellent insight. Indeed, upon fitting it becomes clear there are multiple contributions to these peaks which can change the apparent broadening. With regard to the charge-transfer satellites, we have also presented here the full Ti 2p region (Fig. 5) in response to the previous question regarding trends in the Ti 2p intensities, showing that the charge transfer satellite intensity increased somewhat as a function of reducing treatments and nanoparticle formation.

Patrick Lömker commented: The feature at 471 eV in Fig. 2b of the paper (https://doi.org/10.1039/d1fd00123j) that you are observing could be related to the hydrogen gas excitation (EELS) 1 Π ungerade state (sometimes just referred to as the 2p state) that is described in ref. 1.

The variation in intensity of this feature could in part be explained by different sample-to-analyzer distances as this varies the amount of scattering in the gas phase and thus the height of this peak.

1 K.-P. Huber and G. Hertzberg, *Constants of Diatomic Molecules*, ed. K.-P. Huber and G. Hertzberg, Van Nostrand Reinhold Publishing, New York, 1979, vol. 4, DOI: 10.1016/0022-2860(80)80150-5.

Heath Kersell responded: This is a good insight, as that feature is approximately in the correct energy range for photoelectron energy loss through hydrogen. We note in the text that all measurements are performed after reducing treatments, but we should be sure to clarify here that the measurements were performed in vacuum between treatments. This was necessary because hydrogen gas becomes increasingly difficult to pump over time as it builds up in the pumping lines. So rather than being due to energy loss from excitation of the hydrogen gas, this feature is part of a satellite doublet, which we discussed further in the earlier response to the question regarding trends in the Ti 2p region and its satellites. There, we also show the Ti 2p across a broader energy range. In the text, we focused on the Ti 2p features near 459 eV to demonstrate that the peaks at BEs characteristic of the 4+ and 3+ states are reasonably stable relative to one another.

Philip Davies queried: The strontium titanate films are ~12 nm thick, the nanoparticles are ~8 nm in height and maybe as much as 50 nm in diameter, but the nickel is only present at approximately 1 atom%. Have you calculated whether there is enough nickel in the films to create these large nanoparticles? Has all the nickel been extracted from the films?

Heath Kersell replied: This is an important question as the amount of Ni may be a limiting factor in the relative size and distribution of the particles. The iterative quantitative analysis presented in section 3.5 of the manuscript (https://doi.org/10.1039/d1fd00123j) (where the XPS and scattering intensities constrain the particle periodicity and size) limits the calculated amount of Ni to approximately the amount of Ni present at the surface in each condition. We also did an approximate calculation of the amount of Ni relative to the entire film composition, and found the amount of Ni is less than or equal to ~1% for all conditions studied here. Interestingly, the Ni intensity did increase more slowly as the samples were treated by increasingly longer and higher temperature reduction. This can suggest that the available Ni begins to deplete. As a final note, additional *ex situ* AFM measurements demonstrate that as Ni exsolves in high temperature reducing environments, it tends to coalesce into larger particles, which are fewer in number at the surface. This may happen at the same time as any potential depletion of the available Ni.

William S. J. Skinner asked: In Fig. 1a of the paper (https://doi.org/10.1039/d1fd00123j), where we see the evolution of the Ni 3p core level after heating at 400 °C for 30 minutes and then 120 minutes, I was wondering what you anticipate we would observe if you heated for longer durations? Do you expect that the peak intensity would continue to increase to the extent of that after heating to 500 °C, or that the extent to which it increases will be limited by the temperature?

Heath Kersell responded: As a function of treatment, the Ni 3p intensity will be affected by both the amount of Ni exsolved to the surface, and the geometry of the exsolved Ni nanoparticles. As the surface becomes increasingly Ni-enriched, the Ni 3p intensity increases. However, we also know from *ex situ* AFM measurements that the number of Ni nanoparticles decreases as their size increases, suggesting possible agglomeration under high-temperature exsolution conditions. If the available Ni (potentially as a function of temperature) is depleted, then we expect that the Ni 3p intensity may stop increasing. Further, if the Ni available for exsolution is depleted, and the particles then agglomerate under high temperature reducing conditions, it is possible that some of the Ni signal becomes attenuated as the particles grow in the direction of the surface-normal. So although we did not observe it in this case, there may be a limiting case where the Ni 3p intensity would decrease somewhat after depletion of the available subsurface Ni. The aforementioned behavior may occur as the limiting case at both temperatures, wherein the relative Ni 3p intensity reaches a maximum and then potentially decreases. However, further measurements are needed to determine whether the temperature limits the final amount of exsolved Ni after sufficiently long reducing treatments.

David Payne questioned: How confident are you that you do not also co-exsolve niobium?

Heath Kersell replied: Previous published work on similar samples has come to similar conclusions, observing no indication of Nb exsolution in strontium titanate doped with Ni and Nb. We also performed a set of verification measurements where we measured the Nb 3d region. The relative Nb 3d intensity increased on the order of ~10% compared with the Sr 3d, albeit while heating only to 400 °C. Meanwhile, the relative Ni 3p intensity increases significantly more, and shows a binding energy shift characteristic of Ni reduction from the 2+ state to metallic Ni, pointing to Ni exsolution as the dominant process.

Robert Lindsay requested: Could you describe your procedure to normalise the spectra in Fig. 1 and 2 of the paper (https://doi.org/10.1039/d1fd00123j)?

Heath Kersell responded: In Fig. 1 and 2 of the paper (https://doi.org/10.1039/d1fd00123j), the spectra were normalized by setting the lower binding energy (BE) background in all spectra for a given core level to approximately the same intensity, and scaling each spectrum by the same factor as its lower BE background. This procedure was chosen to demonstrate the approximate stability of the oxidation states, and to give a qualitative picture of spectral shape and relative intensity changes. To ensure a quantitatively accurate description of the relative intensities of the Ni and Ti, we also included fits of the Ni 3p and Ti 3s regions in the supplementary information of the paper. These fits were then used in the quantitative analysis provided with Fig. 7 of the main manuscript. We also performed fits of the Ti 2p, Sr 3d, and Ni 3p regions, and compared their relative intensities as a function of condition. Although we did not present all these trends in the manuscript, they agree with the discussion in the text. We did not attempt to perform an absolute quantitative analysis due to the unknown analyzer transmission function.

Robert Lindsay asked: To gain further understanding of surface evolution as a function of treatment, I was wondering if you have determined the ratios of various core level peaks? For example, you could compare the ratio of Sr 3d and Ti 2p signals in Fig. 2 of the paper (https://doi.org/10.1039/d1fd00123j) to gain further insight into relative surface composition.

Heath Kersell replied: Yes. In addition to the comparison of the Ni 3p to Ti 3s described in the text, we have performed fitting on the Sr 3d and Ti 2p core levels. Although the Sr and Ti intensities both decrease overall relative to the Ni, there is interestingly an increase in the Ti intensity relative to Sr, especially under the final condition. The relative intensities from this analysis as a function of condition are plotted here (Fig. 6). While we did not attempt a full quantitative analysis due to a strongly kinetic energy dependent analyzer transmission function, we clearly see that in the last condition, the Ti 2p intensity increases significantly compared with the Sr 3d. One possible explanation for the relative increase in Ti intensity relative to Sr is that STO surfaces have been shown previously to become Ti enriched under reducing conditions.[1]

1 D. Wrana, *et al.*, *Appl. Surf. Sci.*, 2018, **432**, 46.

Damien Aureau opened discussion of the paper by Alexander I. Large: In Fig. 1 of the paper (https://doi.org/10.1039/d1fd00113b), an impressive and drastic change is observed around 350 °C with the experiment with the heating rate of 3.5 °C min^{-1}. Is it possible to get an idea of the evolution of this particular temperature with the heating rate, mainly for lower rates. Would this change be strongly related or not with the heating recipe?

Alexander I. Large responded: Heating with lower rates, especially over this switchover range, is something we are keen to do, but have not as yet tested. The

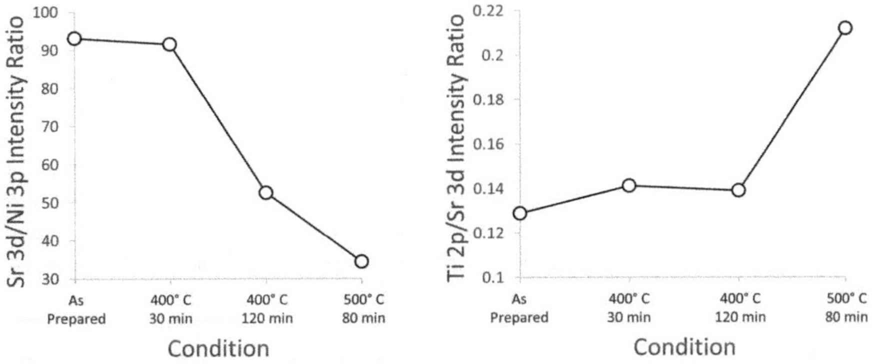

Fig. 6 Relative intensities of Sr 3d, Ni 3p and Ti 2p regions as a function of condition.

rate we are using throughout these experiments, 3.5 °C min^{-1}, is already lower than our comparable heating rates for real catalysts in the same conditions, which is 15 °C min^{-1}. We expect that some variation would be seen in the oxidation behaviour on either increasing to a "real" ramp rate (although this is somewhat impractical whilst retaining good data quality) or by decreasing to a slower ramp rate. 3.5 °C min^{-1} was chosen as a rate as it facilitates recording approximately 1 set of core level spectra every 5 °C, giving a reasonable combination of temperature resolution, ramp rate, data quality and overall experimental time.

Anders Nilsson asked: You have observed the formation of oxide during the oxidation of methane with XPS. I wonder if you have analyzed the O 1s spectra carefully before the strong oxide peak appears to observe if any chemisorbed O can be observed being essential for the methane oxidation? Something similar to what was seen in the case of electrochemical selective propene oxidation on Pd.[1]

1 S. Koroidov, *et al.*, *Catal. Sci. Technol.*, 2021, **11**, 3347, DOI: 10.1039/d0cy02134b.

Alexander I. Large answered: Based on the oxygen spectra we have recorded we have not seen evidence for chemisorbed oxygen species. Some examples of oxygen TP-XPS spectra are shown in Fig. 1 and 2 of the paper (https://doi.org/10.1039/d1fd00113b), and show the quality of the O 1s spectra recorded during the changes. Higher resolution O 1s spectra are only recorded before the oxidation reaction (in vacuum) and after (when typically fully oxidised) and would, therefore, not necessarily give us evidence for this. In order to determine the presence of chemisorbed oxygen during the ramps, additional O 1s TP-XPS would have to be added, which was omitted in favour of higher temperature resolution of the metal states.

Günther Rupprechter questioned: You described Pt migration to the bulk. Based on the driving force for the surface composition changes, is it not rather Pd surface segregation induced by Pd-oxidation? A similar effect can be observed for Si segregation in Pd foil induced by SiO_2 formation on the surface.[1]

1 D. Vogel, Z. Budinska, C. Spiel, R. Schlögl, Y. Suchorski and G. Rupprechter, *Catal. Lett.*, 2013, **143**, 235, DOI: 10.1007/s10562-012-0955-5.

Alexander I. Large replied: It is certainly plausible that the mechanism for the relative changes in surface composition are reliant on Pd surface segregation rather than Pt anti-segregation into the bulk. Based on our measurements we do not have a clear way to distinguish between these two routes, as both effectively produce the same layered composition and resulting spectra.

Rosa Arrigo asked: You have analysed the dynamics of the Pd species mainly in terms of metallic Pd and Pd^{2+}. Have you attempted any fitting of the Pd 3d XPS spectrum to see whether there are other species? For example the two-dimensional Pd_5O_4 surface oxide phase, which was seen under methane oxidation conditions over single crystals?

Alexander I. Large responded: Fitting has been performed on high-resolution spectra taken after the conclusion of the gas ramps, with the samples still exposed to the reaction conditions at the finishing temperature (varied between 420 and 590 °C). In order to achieve good fits we did not find a need for additional peaks, such as Pd^{4+} or other similar higher oxidation states. Pd_5O_4 is not a phase that we had initially considered, but based on the peak positions and the fits which were achieved, it does not seem like this – or any other – Pd state has any significant presence.

Rosa Arrigo enquired: Moving the manipulator back and forth to maintain the intensity approximately consistent sounds difficult; how about normalising for the total intensity of the other measured core levels?

Alexander I. Large answered: Thankfully this experimental procedure is not so difficult. With the control system available on B07-C at Diamond Light Source, we can send a command which essentially tells the analyser to scan a set of core levels as a function of sample manipulator position, over a defined range, with a step size we have determined based on the observed sample/manipulator expansion from previous experiments. This command also includes a temperature measurement, however the temperature control is done separately for simplicity. The procedure is broadly the same for all gas conditions and samples. We had considered other approaches, including the one suggested in the question, however in most cases the thermal expansion of the manipulator is too big and essentially eliminates all photoelectron intensity. The consistency observed with the method used meant that extra normalisation was not necessary.

Philip Davies asked: Your data shows that the Pd88–Pt12 alloy oxidises more rapidly in an O_2/CH_4 mixture than pure O_2. As discussed by Roberts and co-workers,[1,2] this suggests an interaction between the molecular oxygen and the methane at the surface before oxygen dissociation. Have you investigated this by for example looking at the effect of the adsorption temperature?

1 A. F. Carley, S. Yan and M. W. Roberts, Role of oxygen transients in the chemistry of dioxygen at atomically clean metal-surfaces: the Zn(0001)–dioxygen–ammonia system, *J. Chem. Soc., Faraday Trans.*, 1990, **86**(15), 2701–2710, DOI: 10.1039/ft9908602701.
2 A. F. Carley, P. R. Davies and M. W. Roberts, The development of a new concept: the role of oxygen transients, defect and precursor states in surface reactions, *Catal. Lett.*, 2002, **80**(1–2), 25–34.

Alexander I. Large replied: I believe this question is referring to the Pd foil rather than the Pd88–Pt12 alloy, as this is the only case where oxidation of Pd is seen at a lower temperature in an O_2/CH_4 mixture than in just O_2. The references do indeed suggest an interaction between the molecular oxygen and the methane, but we have not investigated this further. This behaviour was only seen over the Pd foil and not over the Pd88–Pt12 foil (Pd65–Pt35 was not tested in oxygen only).

Joshua S. Gibson said: Looking at the O 1s spectra in Fig. 1 and 2 of the paper (https://doi.org/10.1039/d1fd00113b), there appears to be a significant decrease in the intensity of the O_2 gas peak, upon the addition of water. Additionally from the ratios shown in Table 1 of the paper we might expect that the H_2O gas and O_2 gas peaks in Fig. 2 of the paper would be approximately equal. Do you have any thoughts on what has caused these changes, or could they be explained by gas phase reactions?

Alexander I. Large responded: With regards to gas phase reactions changing the composition, I would suggest that this would only be a minor contribution to the observed variation in peak sizes. Based on the residual gas analysis in the first differential pumping stage of the analyser, the pressures of water and oxygen are fairly stable throughout the ramps, and are roughly comparable, given the different pumping rates resultant from the significantly different molecular masses of the two molecules (18 and 32 respectively).

Kelsey Stoerzinger enquired: I'd imagine that under many of these conditions you are indeed oxidizing methane (although the conversion, or amount of reactants converted to products, might be low and show only a small signal in *e.g.* a mass spec, the concentration near the catalyst surface might be appreciable). Have you looked at surfaces in contact with any partial and complete oxidation products, which might help understand changes that happen at elevated temperature?

Alexander I. Large answered: At this stage we have not performed any NAP-XPS experiments with a gas feed exclusively comprised of partial or complete methane oxidation products. Though no clear signal of reaction products are seen in either the mass spectra, or the high resolution XPS measurements, it is certainly possible that there is a presence of products and this has an effect on the observed surface states. It would be interesting to perform measurements at elevated temperatures in the presence of CO_2/H_2O and/or CO/H_2 (complete and partial oxidation products respectively) and measure the surface behaviours of Pd and Pt, to see if this could be correlated to any of the changes seen in this work.

Kelsey Stoerzinger asked: In the Pd 3d profile in Fig. 9 of the paper (https://doi.org/10.1039/d1fd00113b) for Pd65–Pt35, under stoichiometric dry conditions, Pd starts to oxidize (~400 °C) and then is again reduced (~560 °C). This doesn't appear to directly correlate with the behavior in the Pt 4f profile in Fig. 10 of the paper; do you have any thoughts on what is happening here?

Alexander I. Large responded: In general we observe a correlation between Pd oxidation and Pt migration, which typically occurs at the same temperatures for a given sample and condition. For the Pd65–Pt35 foil in stoichiometric dry conditions (as shown in Fig. 9 and 10 in the paper) this behaviour is seen, initially starting above 300 °C. Above 500 °C both of these transitions stop, and the degree of Pd oxidation and the total Pt intensity stabilises. Above 550 °C the initial changes are reversed, *i.e.* Pd is reduced whilst Pt increases in quantity (interpreted as a migration towards the surface). This is the same correlation which has been seen before, just in the reverse direction.

Prajna Bhatt opened discussion of the paper by Claus F. P. Kastorp: The paper (https://doi.org/10.1039/d1fd00122a) mentions removal of a linear background, and I wanted to know the motivation behind this. Is there something to be gained – or rather not lost – when using a linear background removal?

Claus F. P. Kastorp replied: Initially, a Shirley type background was used, and the data was presented as such in my qualifying exam some time ago. Here, it was pointed out that considering we have used synchrotron light, the C 1s core level should have a linear background.

For the individual high resolution spectra, there is no large effect other than a small change in the relative intensities of the components. For the standing wave analysis, the only significant change is in the C_a-component, which provided less uncertain and more robust results with the linear background.

Alexander Shard questioned: The X-ray standing waves (XSW) data for the (111) reflection normal to the surface is shown. Did you try to acquire off-normal reflections to provide information on site occupancy?

Claus F. P. Kastorp answered: In the interest of time, only the direction normal to the surface was measured. Should other reflections be exploited, I would speculate that very little extra information could be obtained from the pristine graphene due to the lattice mismatch between graphene and iridium. This results in a gradual displacement of the carbon atoms, which in turn would smear the signal.

Considering the occupancy of functionalised carbon on the scale of the moiré supercell (*i.e.* HCP, FCC, and ATOP regions) this information could presumably be extracted for each sample, if it had not already been investigated with scanning tunneling microscopy.[1]

On the atomic scale, I would expect that the functionalised clusters themselves are ordered to such a degree that very little additional information could be gathered.

1 J. H. Jørgensen, *et al.*, Symmetry-driven band gap engineering in hydrogen functionalized graphene, *ACS Nano*, 2016, **10**(12), 10798–10807.

Francine Solal requested: Could you explain why the coherent fraction which is one parameter of the simulation is limited to 0.8 and the relationship to surface corrugation?

Claus F. P. Kastorp replied: A coherent fraction of 1 would correspond to a completely flat plane of atoms, and higher corrugation would cause lower values. Effects like thermal vibration will further lower the coherent fraction.

In spectra with large overlaps such as these, there is a danger of overfitting one component and achieving artificially high coherent fractions. To have a guideline for identifying realistic values, I calculated the coherent fraction of carbon atoms positioned at two alternating heights corresponding to a single layer of diamond. On Ir(111), this produced a value of 0.77, and therefore I would not imagine that the sp^3-hybridised species would have a much higher value than that. Thus the limit of 0.8.

Anders Nilsson said: It is an interesting approach to combine photoelectron spectroscopy and standing wave measurements. For understanding the graphene interaction with the iridium substrate the question is if there is any part of the carbon spectrum that can be associated with a direct C–Ir bond after hydrogenation?

Claus F. P. Kastorp responded: Intuitively, since graphene on Ir(111) is normally p-doped, while hydrogenation generally n-dopes the graphene sheet, one could expect alternate energy contributions, and it was previously thought that the origin of the separate C_a and C_b-components were due to a difference in C 1s binding energy between carbon atoms bound to either hydrogen or the iridium surface.

However, since the highly ordered samples (referred to as HT in the main article) can be described accurately with only one additional component for the rehybridised carbon, there is no evidence of a separate C–Ir-contribution in our data.

Wendy Flavell asked: I wonder if you could comment on the precision of the location of C_a and C_b in your XSW measurement? You commented that you can't easily tell the difference, but I think it's really impressive that you can make this measurement at all given the very low X-ray scattering of the carbon compared with Ir.

Claus F. P. Kastorp replied: The binding energy shifts between the components are indeed small, and therefore the high resolution spectra have been crucial in resolving them in XSW, where the overlap is significant and the signal-to-noise ratio is less than ideal. For these reasons, long measurement times have been necessary for the XSW. In terms of binding energies, the shifts used here provide fits with low error as well as high coherent fractions in XSW. The relatively high coherent fractions are also an indication that each component only represents a single chemical species and therefore that the fitting parameters describe the system well.

Marc Walker enquired: Component C_c in the C 1s spectra is ascribed to sp^2-hybridised carbon. Did you include the characteristic asymmetry in the fitting of this peak? It is difficult to tell from the figures.

Claus F. P. Kastorp answered: The fits are indeed done with an asymmetric Doniach–Sunjic peak shape. We have not been investigating the asymmetry systematically, and therefore it has been left out of the analysis.

Georg Held queried: The modulations of the standing wave in XSW are smaller than the diameter of the core-ionized atom. Does the photoionization cross-section depend on the shape of the standing wave or is it the same as in a "normal" XPS experiment, where the exciting photon can be considered as a plane wave?

Claus F. P. Kastorp replied: The model used to fit the standing wave curves only includes the modulation of the X-ray intensity while leaving the cross-section constant, and it seems to be a good job. It is therefore assumed that the cross-section for the photoelectric effect is not affected.

Since the potential experienced by the electron is dominated by the nucleus, I would only expect an effect at very large intensities.

Patrick Lömker said: In Fig. 1a and b of the paper (https://doi.org/10.1039/d1fd00122a) the C_b feature (blue) appears quite broad. This is often the case for spectra where Voigt peaks are fit to a spectrum where a Doniach–Sunjic feature would fit better. Also this feature usually indicates conductivity. As part of the research question (in general about graphenes) is about graphene conductivity, as you also showed in your local presentation, fitting this with an asymmetric peak shape could aid the interpretation of the data.

Claus F. P. Kastorp responded: We have indeed used Doniach–Sunjic type components. If this information has not made it to the final version of the article (https://doi.org/10.1039/d1fd00122a), I take responsibility on behalf of the authors.

The asymmetries have not been investigated systematically, partly because of the large overlap between components, which has made it necessary to either fix the asymmetries across samples or only allow them to change within narrow windows. This is to prevent the fitting algorithm from changing the asymmetry of one component in order to slightly modify the shape of another and therefore making sure that the spectra are comparable across the different samples.

In regards to conductivity and metallicity, there could very well be an effect, even though I have not been investigating it. In a recent publication by David Morgan,[1] it is outlined that the situation can be non-trivial in carbon materials, *e.g.* increased asymmetry in hydrogenated nanoribbons.

If I were asked to speculate, I would guess that some metallicity would remain no matter the conductivity of the graphene due to the interaction with the iridium substrate.

1 D. J. Morgan, *C*, 2021, 7, 51, DOI: 10.3390/c7030051.

David Morgan enquired: As shown in a recent publication,[1] the degree of hydrogenation can influence the asymmetry and examples are shown therein for HOPG and nano-onions for example. I note that in your carbon spectra there is little asymmetry in the main sp^2 peak, do you really expect no asymmetry or for this not to change on hydrogenation?

1 D. J. Morgan, *C*, 2021, 7, 51, DOI: 10.3390/c7030051.

Claus F. P. Kastorp answered: Thank you for the reference, of which I was previously unaware. In this analysis, the asymmetry has been capped at 0.07, which is the value for the sp^2-hybridized carbon as well in the high resolution spectra.

Due to the large overlap between the different components, the asymmetries have been fixed or only allowed to change within a narrow window in order to make the results for the different samples comparable. Even without this assumption, I doubt that much meaningful information could be extracted for the same reason: the components are close to each other in binding energy, and therefore the algorithm may change the asymmetry of one component to slightly modify the shape of another.

Mathieu Silly asked: Do the beamline and experimental setup allow to perform ARPES measurement and have you tried to measure the band structure of the graphene as a function of the hydrogenation rate. As you mentioned during your presentation, by measuring the linear dispersion, it is possible to determine the doping of the graphene layer and to evidence the band gap opening corresponding to graphane species. This measurement would permit to correlate the chemical behavior determined by XPS with the electronic properties of the hydrogenated graphene.

Claus F. P. Kastorp responded: It is indeed possible to do ARPES either at this end station or at the K09 branch of the same beamline. In this case, it was not prioritised, since it has been studied in other works.

For highly ordered samples, the geometry of the functionalisation has been observed to affect the band gap,[1] and therefore the correlation you propose could presumably be done. However, both XSW and ARPES are time-consuming techniques.

1 J. H. Jørgensen, *et al.*, Symmetry-driven band gap engineering in hydrogen functionalized graphene, *ACS Nano*, 2016, **10**(12), 10798–10807.

Kelsey Stoerzinger enquired: You mention (unavoidable) beam-induced hydrogen desorption. Could you mitigate this by, for example, cooling?

Claus F. P. Kastorp replied: Given the high photon energies used in the XSW experiment compared to the desorption energies and the work function of the system, I do not think that cooling would protect significantly against beam damage. That being said, there might be other reasons to cool the sample, such as minimising thermal vibration, and thus being able to determine the structural parameters with greater precision.

On the other hand, cooling to very low temperatures takes time and allows more types of atoms and molecules from the background gas to adsorb on the sample, effectively degrading it.

Kelsey Stoerzinger said: A follow-up regarding potential cooling of the sample to avoid beam-damage, which might increase contamination of the surface. You mentioned these measurements take a long time and are performed overnight. Do you have any means to assess how clean your surface remains over this time frame, or how to determine how much contamination from the chamber might form and how this impacts your spectra over time?

Claus F. P. Kastorp responded: Before and after each XSW spectrum, a C 1s spectrum is acquired at a photon energy well below the Bragg condition to make this very analysis possible. I do not recall such significant changes except for one measurement where a problem with the acquisition software would sometimes cause rigid 50 meV shifts. The issue mentioned in this work is predominantly damage done by the beam while measuring a single spot with standing waves rather than large scale changes over time, which is the reason a measurement series consists of several fast measurements carried out at different spots on the sample.

Since these long measurement times are necessary, the beamline staff go to great lengths to ensure the cleanliness of the XPS chamber.

Rosa Arrigo queried: In Fig. 1 of the paper (https://doi.org/10.1039/d1fd00122a), the fitting model used includes a coherent and a non-coherent component. However at the different excitation energies used (more surface and more bulk sensitive), the ratios between these components are different. Could you please explain why it should be the case?

Claus F. P. Kastorp replied: The ratios between the peak intensities are different, while the ratios between the areas are not, although I agree that it can be difficult to see in the figure.

In both cases, the components are wider in the hard X-ray spectrum than in the soft X-ray spectrum for various reasons. Initially, the sp^2-component is much narrower than the sp^3-component, and therefore the broadening relative to the original width is larger for the sp^2-component. This in turn lowers the peak intensity for any given peak area, and therefore gives the impression that the ratio between the components has changed.

Rosa Arrigo enquired: Can you exclude that molecular H intercalates between iridium and graphene? Thus the peak shifts due to this phenomenon?

Claus F. P. Kastorp responded: This very question was our original motivation for performing these experiments: when graphene on Ir(111) is hydrogenated, does it bind directly to the Ir(111) surface, or does hydrogen intercalate and bind to the graphene sheet from below.

By combining standing wave XPS for the lowest degree of functionalisation (the sample referred to as HT in the main article) and DFT, we were able to show that binding to the substrate matches well to the data, while binding to additional hydrogen from below does not.[1]

In short: while some intercalation can maybe not be ruled out completely, large scale intercalation simply does not fit the data.

1 A. L. Jørgensen, *et al.* Chemically-resolved determination of hydrogenated graphene–substrate interaction, *Phys. Chem. Chem. Phys.*, 2019, **21**(25), 13462–13466, DOI: 10.1039/C9CP02059D.

Rosa Arrigo asked: Would you be able to see point defects by applying the coherent and incoherent fitting model. For example an iridium atom in a vacancy? Although vacancies are in-plane, the Ir atom should be out of the graphene plane. Which of the two peaks does this contribute to?

Claus F. P. Kastorp replied: We have not investigated defects due to the high concentrations required. In the RT-sample, the C_d component constitutes 9% of the total intensity, and yet we are unable to resolve its standing wave signal. Some of this can be mitigated by longer measurement times, but even at concentrations in the vicinity of 5%, the amount of defects would be so high that the experiment is not necessarily meaningful.

As to the binding energy contribution, I have no qualified guess at the present time.

Robert Lindsay enquired: In both Fig. 1(a) and (b) of the paper (https://doi.org/10.1039/d1fd00122a), the two fitted components look to have significantly different widths. Could you explain the physical reason for this variation?

Claus F. P. Kastorp answered: The main difference between the two spectra is the photon energy. Even at a line width of 1/10 000, the photon energy as provided by the beamline, this effect becomes significant for the high energy spectra. Furthermore, these measurements are faster to limit the exposure time to any one spot on the sample, which results in significant broadening.

Finally, several spectra from various spots on the sample have been added to increase the signal, which have not been corrected to the Fermi level, and which could have slightly different characteristics due to inhomogeneities.

Damien Aureau opened discussion of the paper by Andrew Evans: In the Ag–Au–Cu alloy, the shifts observed on Au 4f are presented. What about the other peaks related to the alloy? Changes are mentioned but not totally described. Are they similar? Does quantification reveal stoichiometry changes? Are they partially reversible with temperature?

Andrew Evans replied: In the real-time measurement of the Au(75)–Ag(15)–Cu(10) alloy, the core levels were measured sequentially to provide around 1200 spectra during temperature cycling between room temperature and 600 °C. The Cu 2p core level was below the detection limit up to 300 °C; all metal core levels showed an increase in intensity during the initial heating stage due to the desorption of surface species containing C and O. The Au 4f and Ag 3d core levels showed a similar overall energy shift of around 60 meV. There are stoichiometry shifts, in particular an increase in all core levels with temperature as surface contaminants (C, O) are removed and there is a relative increase in the Cu at the surface at the higher temperatures.

Sefik Suzer said: My first comment is: in the last paragraph of section 3.1 of the pre-print of your manuscript, you had given the shift in the Au 4f peak to be 60 meV K^{-1}, which must be a mistake, since the overall shift is about 60 meV over a range of *ca.* 500 K. Secondly, you are attributing this thermal shift to the lattice expansion of the alloy. However, the total thermal energy is approaching $\sim 2kT$, which corresponds to about 50 meV. Therefore, I am asking whether or not some of this shift might be due to this additional thermal energy the ejected photoelectron might have acquired, before being ejected, as in the hot bands often observed in UV and IR spectroscopies.

Andrew Evans responded: The shift in binding energy for the Au–Cu–Ag alloy is 0.1 meV K^{-1} and this error will be corrected in the manuscript (https://doi.org/10.1039/d1fd00119a). The energy shift in the metal core levels measured for the alloy during heating and cooling is believed to be due to thermal expansion as described, for example in ref. 26–29 of the manuscript (ref. 1–4 here), where the shift is different for different metals and is negative for alkali metals. We have measured other metals and alloys in the same environment and observe a difference in the rate of change of binding energy with temperature, including samples where the slope changes or reverses direction during heating.

1 D. M. Riffe, G. K. Wertheim, D. N. Buchanan and P. H. Citrin, *Phys. Rev. B: Condens. Matter Mater. Phys.*, 1992, **45**, 6216–6225.
2 D. M. Riffe, W. Hale, B. Kim and J. L. Erskine, *Phys. Rev. B: Condens. Matter Mater. Phys.*, 1996, **54**, 17118–17122.
3 T. Durakiweicz, J. A. J. Arko, J. J. Joyce, D. J. P. Moore and A. S. Halas, *Surf. Sci.*, 2001, **478**, 72–82.
4 K. Shen, G. Li and W. Wang, *Trans. Nonferrous Met. Soc. China*, 2016, **26**, 2900–2909.

Sefik Suzer enquired: While recording the spectra of the Ga-oxide, have you observed any effects of X-ray flux? If yes, how can you distinguish between these and other effects, such as, thermal, *etc.*?

Andrew Evans answered: The shift in binding energy measured for the gallium oxide is believed to be due solely to changes in binding energy due to a change in surface band-bending. The surface photovoltage effect is dependent on the X-ray flux such that the binding energy change observed for the diamond is sensitive to the power of the radiation source. Such a change is not observed for the gallium oxide and similar shifts and line-shapes are observed when the surfaces are measured with a laboratory X-ray source and with synchrotron radiation (ref. 34 in the manuscript).[1] The gallium oxide are doped n-type with ohmic contacts fabricated on the reverse and on part of the surface to further reduce surface charging effects.

1 R. M. Gazoni, L. Caroll, J. I. Scott, S. Astley, D. A. Evans, A. J. Downard, R. J. Reeves and M. W. Allen, *Phys. Rev. B*, 2020, **102**, 035304.

Sefik Suzer asked: In Fig. 4 in the paper (https://doi.org/10.1039/d1fd00119a) regarding the spectrum of diamond, in addition to intriguing binding energy shifts, there seem to be intensity fluctuations as well. What is the origin, removal of oxygen or other surface impurities?

Andrew Evans responded: The as-loaded diamond surface shows an increase in core level intensity and a decrease in non-diamond carbon species during an initial annealing cycle between room temperature and 300 °C in addition to a reversible shift in the bulk diamond carbon peak. Subsequent annealing cycles to 300 °C reproduce the binding energy shift but there is little further change in C 1s and O 1s core level line shape or relative intensity, indicating a chemically stable, O-terminated surface. The binding energy shift is therefore temperature-dependent with the binding energy and the core-level spectral line shape unchanged when measured at a fixed temperature within this temperature range. Above 400 °C, oxygen is desorbed, leading to a reversal of the binding energy shift as shown in Fig. 4 of the paper that is due to a chemically-induced change in the band-bending.

Alexander Shard questioned: Do you have any information on the stability of the spectrometer work function in ambient pressure XPS instruments?

Andrew Evans answered: The work function of the analyser is always a concern when the environment can potentially lead to internal changes during for example, real-time monitoring of surface processing and near-ambient pressure measurement. This requires regular calibration of the binding energy to a kinetic energy conversion algorithm. For the real-time measurements presented in this paper, the focus was on the change in kinetic (binding) energy with temperature rather than the absolute values.

Georg Held remarked: Comment on the question raised about the work function of the analyser in AP-XPS experiments: this is generally not known exactly as environment gases adsorbing on the analyser surfaces will alter the work function of both sample and analyser. In synchrotron experiments it is general practice to measure a reference spectrum with known binding energy (*e.g.* Fermi edge, gas-phase core levels) to correct any energy shifts due to work-function changes or mis-calibration of the monochromator.

Benjamen P. Reed asked: My understanding of the data in Fig. 4 of the paper (https://doi.org/10.1039/d1fd00119a) is that surface photovoltage in the case of the p-type (001) diamond is reversible up to 400 °C, and that the desorption of the surface oxygen above 400 °C causes the surface photovoltage to reverse and reach an equilibrium value at ∼−0.7 eV at temperatures above 200 °C. Assuming that re-adsorption of oxygen onto the surface is mitigated during cooling, would you observe the surface photovoltage only decrease to −0.7 eV and plateau on subsequent heating cycles? On the other hand, if one were to dose the chamber with oxygen during the initial cooling, and then heat the diamond back up to 600 °C, would you expect to recover the original energy–temperature profile seen in Fig. 4b? Finally, have you attempted to calculate the activation energy of desorption for the different oxygen species on the surface of the (001) diamond using this data?

Andrew Evans answered: During annealing cycles between room temperature and temperatures below 400 °C, the binding energy shift for the C 1s core level in the p-type diamond is fully reversible as the surface photovoltage is removed during heating and returns on cooling. Above 400 °C, the shift reverses and the final, high temperature binding energy, representing the true value is reduced. On subsequent annealing in vacuum, the reduced binding energy shift persists, for example reaching a maximum of 0.7 eV following the cycle shown in Fig. 4 of the manuscript (https://doi.org/10.1039/d1fd00119a). Above 600 °C, the maximum value is further reduced until all the surface oxygen is adsorbed. In the NAP experiment described in Fig. 5 in the manuscript, the oxygen coverage of the surface at 400 °C is maintained by the oxygen gas to preserve the high temperature C 1s binding energy that represents the maximum band bending for this p-type diamond. Real-time oxygen desorption studies enable the estimation of activation energies by monitoring changes in photoemission peak areas with time at fixed temperatures.

Benjamen P. Reed queried: In Fig. 2a of your paper (https://doi.org/10.1039/d1fd00119a), you state that during the annealing cycle from room temperature to 600 °C, the lineshape of the Au 4f core level does not change other than a small increase in broadening. However, at approximately 800 s into the annealing cycle, it does appear that the Au 4f core level narrows slightly. Could you speculate as to what is happening here? Perhaps there are defects in the metallic matrix that are being annealed out at the temperature corresponding to 800 s in the temperature programme?

Andrew Evans responded: At 800 s, the temperature is at a maximum and marks the transition from the heating cycle to the cooling cycle. Although the Au 4f peak shows an overall shift in energy and an increase in broadening during the heating cycle, there are changes in the energy position, intensity and peak width during the annealing cycles, in particular in relative values of the Au, Cu and Ag parameters, that suggest changes in the surface stoichiometry of the alloy with temperature. There are also changes in the photoelectron background that can affect the peak fitting. Near the maximum temperature at 800 s, the Au 4f peak broadens significantly and increases in intensity before narrowing at the start of the cooling phase. However, during cooling, the peak width is no narrower than that measured at the same temperature during the heating cycle. Further experiments on alloys of different composition are underway to further study these changes at the higher temperatures.

Giulio D'Acunto enquired: Does the rate of annealing (K s^{-1}) have any relevant role in the shifts of the core level? Or are the shifts independent of the annealing rate?

Andrew Evans replied: The rate of annealing is determined by the programmed power profile for the indirect ceramic heater, with the temperature measured by a thermocouple placed close to the sample position. The apparent shift in binding energy can be sensitive to the annealing rate if there is a difference between the surface temperature and the measured temperature. To reduce this error, the surface temperature for the diamond was calibrated by recording a real-time sequence of Raman spectra at the same sample position to define the optimum power delivery profile. With this calibration, the observed shifts are not sensitive to the rate of annealing when there are no changes in surface chemistry.

Alenka Krizan queried: It looks like the slope of the Au peak as a function of temperature (Fig. 2(b) in the paper; https://doi.org/10.1039/d1fd00119a) changes a little bit around 300 °C. Do you have any explanation for this or do you think this is only an apparent change of the slope due to the experimental error?

Andrew Evans responded: The Au 4f and Ag 3d core level spectra, recorded sequentially in real time during heating, showed a similar overall shift, but the profile was not identical. For the Au 4f core level, there was a small change at around 300 °C that suggests a change at the surface of the alloy at a temperature where the Cu 2p becomes detectable. However, the data statistics need improvement to precisely determine variations in the rate of change of binding energy and other fitting parameters for each of the metal core levels. This is currently being addressed in further experiments that include other alloys and pure metals.

Philip Davies asked: In Fig. 5c of the paper (https://doi.org/10.1039/d1fd00119a), why is the shift in the C(1s) peak always greater than that of the O(1s) at any given temperature?

Andrew Evans replied: The C 1s and O 1s spectra were recorded at a fixed synchrotron photon energy, such that the electrons were emitted with different kinetic energy. They also originated from different parts of the sample since the oxygen was located only in the monolayer termination of this single-crystal diamond. The C 1s spectra are thus the more reliable measure of changes in the Fermi level relative to the band edges due to charging and band-bending since the energy shift was determined from the bulk C 1s component that is intense and does not change in shape. In contrast, the O 1s spectra provide a less reliable measure of the energy shift as it is more sensitive to the chemical environment at the surface, is less intense and requires two or more fitting components.

Philip Davies enquired: Under the X-ray beam, the diamond surfaces you have studied develop a significant negative charge. Do you think that negative charge could be used to drive some surface chemistry?

Andrew Evans answered: The room temperature photovoltage effect that occurs when a moderately B-doped, p-type diamond is irradiated with above-bandgap radiation generates a negative charge on the surface, and this is enhanced when the excitation source provides X-rays with a penetration depth that is comparable to the depletion width in the diamond. The surface charge correlates with the X-ray flux. The work function of the diamond surface can be controlled by surface processing, for example, the H-terminated diamond surface has a negative electron affinity. This surface treatment applied to a moderately B-doped diamond would efficiently emit low energy electrons from the surface, providing a method of controlling the surface charge and the emitted electrons.

Conflicts of interest

There are no conflicts of interest to declare.

Faraday Discussions

ROYAL SOCIETY OF CHEMISTRY

PAPER

Gently does it!: *in situ* preparation of alkali metal–solid electrolyte interfaces for photoelectron spectroscopy†

Joshua S. Gibson, [ab] Sudarshan Narayanan, [ab] Jack E. N. Swallow,[ab] Pardeep Kumar-Thakur, [c] Mauro Pasta, [ab] Tien-Lin Lee[c] and Robert S. Weatherup *[ab]

Received 7th December 2021, Accepted 26th January 2022

DOI: 10.1039/d1fd00118c

The key charge transfer processes in electrochemical energy storage devices occur at electrode–electrolyte interfaces, which are typically buried, making it challenging to access their interfacial chemistry. In the case of Li-ion batteries, metallic Li electrodes hold promise for increasing energy and power densities and, when used in conjunction with solid electrolytes, the adverse safety implications associated with dendrite formation in organic liquid electrolytes can potentially be overcome. To better understand the stability of solid electrolytes when in contact with alkali metals and the reactions that occur, here we consider the deposition of thin (~10 nm) alkali metal films onto solid electrolyte surfaces, where the metal is thin enough that X-ray photoelectron spectroscopy can probe the buried electrode–electrolyte interface. We highlight the importance of *in situ* alkali metal deposition by assessing the contaminant species that are present after glovebox handling and the use of 'inert' transfer devices. Consequently, we compare and contrast three available methods for *in situ* alkali-metal deposition; Li sputter deposition, Li evaporation, and Li plating induced by e^- flood-gun irradiation. Studies on both a sulphide solid electrolyte (Li_6PS_5Cl), and a single-layer graphene probe surface reveal that the more energetic Li deposition methods, such as sputtering, can induce surface damage and interfacial mixing that are not seen with thermal evaporation. This indicates that the appropriate selection of the Li deposition method for *in situ* studies is required to observe representative behaviour, and the results of previous studies involving energetic deposition may warrant further evaluation.

[a]*Department of Materials, University of Oxford, Parks Road, Oxford, OX1 3PH, UK. E-mail: robert.weatherup@materials.ox.ac.uk*

[b]*The Faraday Institution, Quad One, Harwell Science and Innovation Campus, Didcot, OX11 0RA, UK*

[c]*Diamond Light Source, Harwell Science and Innovation Campus, Didcot, Oxfordshire, OX11 0DE, UK*

† Electronic supplementary information (ESI) available. See DOI: 10.1039/d1fd00118c

Introduction

Batteries with improved energy densities, power densities, and cycling lifetimes are needed to increase the range and fast-charging capabilities of electric vehicles, and lower their lifetime cost.[1,2] Alkali metal anodes offer very high theoretical specific capacities (3860 mA h g^{-1} for Li),[1,3–5] however the propensity for filament/dendrite formation during cyclic metal stripping/plating renders them unsafe when used with the flammable liquid electrolytes found in conventional Li-ion batteries.[6,7] The use of a solid electrolyte (SE) as the ion (*e.g.* Li^+, Na^+, Mg^{2+} *etc.*) conducting medium avoids flammable organic electrolytes and can mechanically suppress dendrite formation, making the commercial realisation of solid-state batteries (SSBs) a promising prospect.[1,8] Amongst the alkali-metal SSB systems, Li metal anodes have the highest theoretical capacity and lowest electrode potential,[5] and the improvement in energy density this offers has made them the focus of many SSB studies to date.[1–3,9] Given the highly reducing character of Li metal, its reaction with the battery electrolyte to form a stable interfacial layer, or a solid electrolyte interphase (SEI), is of vital importance to the extended battery cycling performance. Many studies have thus focussed on attempting to understand the nature of the reactions that proceed, to inform future battery development.[6,10–15] The exact nature of the interaction layers formed, and therefore the reaction mechanisms, are dependent upon the electrolyte used. For example, LLZO ($Li_7La_3Zr_2O_{12}$) garnets have been suggested to form oxygen deficient layers,[4,16] sulphide glasses to form lithium sulphides and phosphides,[17,18] while lithium phosphorus oxynitride (LiPON) can react to form phosphides and nitrides.[19,20] Argyrodite-type SEs, in particular the chloride variety Li_6PS_5Cl, have attracted much attention given their potentially high Li-ion conductivities (10^{-3} to 10^{-2} S cm^{-1} [21]), and relative ease of preparation, *e.g.* through the cold-pressing of commercially available powders. Despite high Li-ion conductivities and successful cycling having been experimentally observed,[22,23] degradation of the SE when in contact with Li metal leads to poor long-term performance.[24,25] Elucidating the chemical processes occurring at such interfaces and the nature of the SEI formed remains a significant challenge, with the SEI referred to as the "most important and least understood" element within Li-ion batteries.[26]

SSB performance, and the formation of the associated SEI, requires good physical contact between Li and the SE. Methods previously employed to achieve this include pressing Li metal into the solid electrolyte surface[27] and dropping it on as a molten liquid.[28] To obtain chemically specific information about these buried interfaces with surface sensitive techniques, such as X-ray photoelectron spectroscopy (XPS),[18,20,29] post-mortem studies are usually performed, where the Li electrode must subsequently be peeled off. However, this risks mechanical damage to the interfacial layers and can allow undesired side reactions to occur by exposure to trace contaminants within glovebox atmospheres.[27,30,31] A recent systematic study suggested that Li metal from different suppliers may have different passivation layers, also affecting the interfacial chemistry.[32] Alternatively, thin Li films can be deposited onto the solid electrolyte surface through evaporation,[20,33,34] sputtering[11,35] or even plating, driven by electron beam irradiation[10,34] to ensure good physical contact. For characterisation, these can then either be controllably etched away[14] or, if thin-enough, non-destructive probing directly through them may be possible.[10,35] Given the high reactivity of the species involved and variety in Li preparation methods available, it is important to establish what impact these deposition processes may have on the reaction species observed.

Thin films can be deposited *in situ*, or in external deposition systems, prior to surface analysis. External deposition systems generally allow greater convenience, without the necessity for compatibility with the analysis system or the challenges of manipulation under vacuum conditions (at least for glovebox-based systems). The main disadvantage is the fact that even in a glovebox there is the possibility that surfaces react with contaminant species. A glovebox operating at 1 bar with sensors reading <0.1 ppm for water or oxygen could still present a partial pressure up to 10^{-4} mbar, and other contaminants such as CO_2 are often not controlled for. Transfer from the glovebox to the ultra-high vacuum (UHV) conditions typical of surface analysis systems is then required, often with an O-ring sealed transfer vessel, which over time will permit some leakage of air, leading to unwanted surface reactions. Such issues were investigated recently when looking at the surface passivation of Li metal,[32] concluding that the storage and processing history of the Li metal surface is very important for the interpretation of the surface analysis.

A number of approaches have been reported for *in situ* Li deposition in commercial UHV surface analysis systems. Sputtering of a Li metal chip with an Ar^+ beam, leading to deposition of the sputtered material onto an adjacent material surface[11,35] can be achieved relatively easily with a simple modification to the sample holder and using an existing sputter gun used for surface cleaning or depth-profiling. There is, however, the necessity for inert transfer, and consequent surface contamination of the Li metal target that will be sputtered onto the surface. A low energy e^- beam (the e^- charge neutraliser or flood-gun) has also been used to drive surface Li plating, fed from a reservoir of Li metal in contact with another region of the SE.[10,34] However, it has been observed for carbon nanotube and graphene surfaces that the irradiation of sample surfaces with an electron beam leads to increased deposition of amorphous carbon,[36,37] which could affect the observed surface chemistry. *In situ* evaporation often requires more extensive modification of the analysis system. In the case of alkali metals, an adjacent deposition chamber may be preferable to prevent contamination of the analysis chamber with these reasonably high vapour pressure metals[38] during evaporation. The benefit of such a system, however, is that transfer of the sample from the deposition chamber to the analysis chamber occurs under UHV ($\approx 10^{-8}$ mbar), without significant exposure to contaminants.

In this work, we use lab- and synchrotron-based XPS to compare the interfacial chemistry of Li deposited onto Li_6PS_5Cl by thermal evaporation, sputtering, and e^- flood-gun driven Li plating. This reveals that *in situ* deposition is critical to avoiding the conversion of the Li film to hydroxide and carbonate species by reaction with environmental contaminants present in gloveboxes and 'inert' transfer devices. We discuss the different methods for Li deposition and assess the benefits and disadvantages that each may have on the interfacial chemistry seen during SEI formation. In particular, we use Raman spectroscopy to reveal the surface damage induced by each deposition method. This reveals that the use of more energetic deposition methods, such as sputtering, can lead to significant surface damage and interfacial mixing that is not seen with thermal evaporation, and is not expected to be representative of the SEI formed through the mechanical contacting of Li to the SE. These findings will inform improved experimental procedures for alkali metal deposition and handling when performing surface analysis, contributing to an enhanced understanding of the interfacial stability of solid electrolytes and alkali metals. Furthermore, these results indicate that previous studies of SE systems where Li deposition by sputtering or charge neutralisation using Ar^+ ions may require revisiting.

Experimental

Argyrodite (Li_6PS_5Cl) solid electrolyte pellets were prepared by cold pressing commercial powders (~10 μm grain size, purchased from Ampcera™, USA) into 8 mm disks of 700–750 μm thickness by the application of 500 MPa of pressure using a hydraulic press. The handling of Li_6PS_5Cl powders and pellets was performed in an Ar-filled glovebox with H_2O and O_2 contamination below the detectable limits (both <0.1 ppm, partial pressure $<1 \times 10^{-4}$ mbar). Pellets were used as pressed, with no further surface treatment. Single-layer graphene (SLG), used as a model surface for evaluating the surface damage related to different lithium deposition methods, was grown on copper foil substrates by chemical vapour deposition,[39] and transferred to 300 nm SiO_2 on Si(100) substrates (Pi-Kem Ltd) using the well-established wet transfer method.[40,41] In brief, the SLG on copper was supported by polymethylmethacrylate (PMMA), after which the sample was floated on top of a sodium persulphate etchant solution (0.21 M aqueous $Na_2S_2O_8$, Acros Organics >98%) for a minimum of 5 hours. The etched samples were transferred to distilled water to rinse away residual etchant, and then the PMMA coated SLG was picked up onto SiO_2(300 nm)/Si substrates and dried. The PMMA support was then removed by soaking in acetone (>99%, Sigma-Aldrich) for 1 hour, followed by rinsing in isopropyl alcohol (>99.5%, Sigma-Aldrich).

For *in situ* Li deposition, either by sputtering or driven by the e^- flood-gun, bare Li_6PS_5Cl pellets were loaded into an ULVAC Phi Versaprobe III XPS system from glovebox environments using an inert transfer device (ULVAC Phi GmbH). *In situ* sputter deposition of Li has already been described in detail elsewhere.[11,35] Briefly, metallic Li foil (750 μm thick, purchased from Sigma-Aldrich®), was mechanically cleaned by scraping in a glovebox, affixed vertically to an L-shaped *in situ* metal deposition stage (ULVAC Phi GmbH) and inertly transferred to the XPS system. The Li is then bombarded with an Ar^+ ion beam (4 keV, 3×3 mm^2 raster, 2.8 μA beam current or 2 keV, 3×3 mm^2 raster, 1 μA beam current), to sputter Li onto samples placed horizontally beneath the Li foil (see the schematic in Fig. 1). The Li deposition rate is determined based on the attenuation of Cu 3p peaks during calibration experiments on bare Cu foil samples, assuming a continuous layer and using the Beer–Lambert law with calculated values for the inelastic mean free path.[42,43] The attenuation of the Cu 3s and Cu 2p peaks corroborated the deposition rate obtained using the Cu 3p peaks. Li-deposition using a "virtual electrode" created by an e^- flood-gun has been previously demonstrated[10,34] and involves irradiating the surface of the Li_6PS_5Cl pellet with low-energy e^-, such as those generated by an e^- flood-gun within the XPS system. The e^--flooded electrolyte surface draws Li^+ ions from a Li metal reservoir on the underside of the pellet, leading to plating of Li metal on the irradiated surface.

During *in situ* Li deposition, surfaces are studied by X-ray photoelectron spectroscopy (XPS) using a water-cooled monochromatic Al K_α source ($h\nu = 1486.6$ eV, 15 kV anode voltage, 25 W beam power). Charge compensation with the in-built e^- flood-gun or low-energy Ar^+ ion source was avoided for all samples, except the Li sputtered Li_6PS_5Cl surface, or cases where the use of the charge neutraliser was being studied (such as the "virtual electrode" measurements where only the e^- flood-gun is used, or the evaluation of SLG damage). In these cases the electron neutraliser was set to 1.35 V bias to accelerate thelow energy electrons, while the argon ion neutraliser used a beam voltage of 110 V and a−100 V float to provide ions with energies of approximately 10 eV.[71] Core level spectra were collected with a pass energy of 55 eV and a step size of 0.05 eV/step (giving a Ag $3d_{5/2}$ peak with a fwhm of 0.69 eV). The analysis chamber pressure was maintained below 5×10^{-9} mbar during measurement. Data analysis was performed using the CasaXPS software package.[44]

For *in situ* Li evaporation, a purpose built UHV chamber was attached to the I09 end station at the Diamond Light Source synchrotron facility (Didcot, UK), where the bare Li_6PS_5Cl pellets could be loaded from a glovebox environment using an inert transfer device sealed under Ar. The inert transfer device was constructed entirely from UHV compatible components and sealed with a gate-valve with a Viton® seal. Li was deposited by thermal evaporation from a boron nitride crucible heated by a tantalum wire basket. The deposition rate was calibrated for the chamber geometry to be 4 nm min^{-1}, with a chamber pressure below 1×10^{-6} mbar during evaporation. The samples could then be transferred to the analysis chamber whilst maintaining a base pressure below 5×10^{-10} mbar. Once in the analysis chamber, the surfaces were probed with hard X-ray photoelectron spectroscopy (HAXPES), using photon energies of 2.2 keV and 6.6 keV, and soft X-ray photoelectron spectroscopy (SOXPES), such that the kinetic energy of core level photoelectrons was ~315 eV (*i.e.* photon energies of 845 eV, 600 eV, 510 eV, 475 eV, 450 eV and 370 eV for O 1s, C 1s, Cl 2p, S 2p, P 2p and Li 1s, respectively). The use of different photon energies provides depth-resolved data for the studied surfaces with photoelectrons having inelastic mean free paths through Li metal of approximately 1.1 nm, 6.0–7.5 nm and 18.5–19.8 nm for the SOXPES, 2.2 keV and 6.6 keV photons, respectively.[43] To minimise beam damage (as monitored using the S 2p core level) the undulator was detuned to provide an X-ray intensity of $\sim I_0/10$ for HAXPES and $\sim I_0/100$ for SOXPES, where I_0 is the full beam intensity. The defocussed X-ray beam gave dimensions of ~300 (vertical) × 300 (horizontal) µm and impinged upon the sample surface at an angle of 15° to enhance the photoelectron signal and yield a ~300 × 1200 µm illuminated area. The energy distribution curves (EDCs) of the photoelectrons leaving the surface were measured with a concentric hemispherical analyser (VG Scienta EW4000 10 keV, lens acceptance angle ±28°). The hemispherical analyser was operated with a pass energy of 200 eV for HAXPES and 100 eV for SOXPES.

Raman spectra of the SLG samples were collected using a Renishaw inVia confocal Raman microscope, with a 532 nm wavelength laser having a maximum power of 50 mW. Following Li deposition samples were transferred to air to allow Raman spectra to be collected. Raman data was collected over the range 1000 to 3000 cm^{-1} using 1% of the maximum power to avoid laser-induced damage to the SLG. Data was recorded using a 1800 mm^{-1} grating, with 30 acquisitions per scan and a 10 s exposure time. Data was analysed and fitted using Lorentzian line shapes in the Igor Pro software package.[45]

Results and discussion

Fig. 1 shows XPS core level spectra for the C 1s, O 1s, S 2p and Li 1s regions for different stages of Li deposition on Li_6PS_5Cl solid electrolytes, with the peak area of each region normalised to unity (except S 2p (iii), due to its lack of discernible peaks). The top row of spectra, Fig. 1(i), show measurements of the as-loaded bare Li_6PS_5Cl. The O 1s region contains a broad peak at ~531.6 eV, consistent with the presence of Li_2CO_3 and LiOH.[10] The C 1s region shows a contribution at ~288.6 eV, again attributable to Li_2CO_3,[4,46] as well as adventitious C at ~284.8 eV. Changes in the relative intensities of these components are apparent with increasing photon energy, which corresponds to increased information depth, as faster photoelectrons can escape from deeper below the surface. Between SOXPES and 2.2 keV HAXPES, the ratio of Li_2CO_3 to adventitious C peaks in the C 1s region is seen to increase, suggesting that the top surface is covered in adventitious C (*e.g.* hydrocarbons), with a carbonate layer beneath it. For the 6.6 keV HAXPES, the signal to noise ratio decreases significantly due to both the lower photoionisation cross section at higher energies and the increased probing depth of the faster photoelectrons, resulting in the peak attributable to Li_2CO_3 no longer being well-resolved. These variations in peak intensity with depth are thus consistent with a Li_2CO_3 layer on the outermost surface of the pellet that is covered with adventitious C.

The S 2p core level of the as-loaded Li_6PS_5Cl shows a strong doublet with a S $2p_{3/2}$ component at ~161.4 eV, attributable to Li_6PS_5Cl, whilst a much weaker

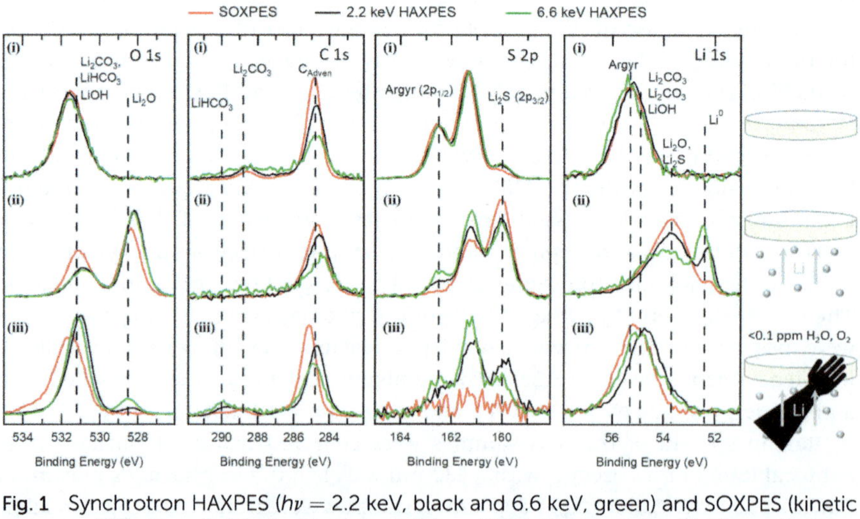

Fig. 1 Synchrotron HAXPES ($h\nu = 2.2$ keV, black and 6.6 keV, green) and SOXPES (kinetic energy = 315 eV, red) data for (i) as-loaded bare Li_6PS_5Cl argyrodite, (ii) 20 nm Li evaporated on Li_6PS_5Cl *in situ* and (iii) 20 nm of Li on Li_6PS_5Cl after having been transferred in an inert Ar atmosphere. Data has been normalised so that the total peak area is the same, with the exception of the S 2p data in (iii), and the binding energy is calibrated such that the Cl $2p_{3/2}$ sits at 198.5 eV (see ESI, Fig. S1†).

doublet with S $2p_{3/2}$ at ~160.1 eV is assigned to S^{2-} or Li_2S.[12] The S 2p shows a slightly lower contribution of Li_2S relative to Li_6PS_5Cl in the SOXPES measurements, which indicates some depletion of S, perhaps related to the formation of Li_2CO_3 at the surface. Nevertheless the fact that the Li_6PS_5Cl doublet is well-resolved even in the SOXPES data indicates that the carbonate layer is relatively thin, with a uniform carbonate layer of >3 nm thickness expected to fully attenuate any Li_6PS_5Cl signal at this energy.[42] The Li 1s core level shows only a single peak for the bare pellet at ~55.4 eV, which is consistent with Li_6PS_5Cl, however likely also includes contributions from Li_2CO_3, $LiHCO_3$ and LiOH, which are expected to have overlapping peaks at ~55 eV for the charge referencing used here.[4,46] The Cl 2p and the P 2p spectra (Fig. S1†) do not reveal any significant variations in peak shape with photon energy. These results thus confirm the presence of surface-contaminants on the Li_6PS_5Cl, particularly Li_2CO_3 and adventitious C, which likely arise from glovebox storage and processing. Since the powders are stored in a glovebox prior to room-temperature pellet pressing, it is likely that each of the micron-sized grains is coated with these contaminants. Although the presence of Li_2CO_3 has not been rigorously assigned in Li_6PS_5Cl previously, we note that it has been observed on many other lithium containing solid electrolytes and has been implicated in increasing interfacial resistance (*e.g.* LLZO garnets,[4] (LiPON)[20] and Li_2S–P_2S_5 sulphides[47]).

Cl 2p was chosen as an internal reference for charge correction (Cl $2p_{3/2}$ = 198.5 eV) as prior reports have indicated that Cl 2p peak distortion is not observed with Li metal deposition.[12] Recoil effects in photoelectron emission, where the outgoing photoelectron transfers momentum to the photoemitting nucleus, has been shown to be well approximated by the use of the momentum equation $\Delta E = E_{KE}m/M$ where ΔE is the recoil shift, E_{KE} is the photoelectron kinetic energy, m is the photoelectron mass and M the nuclear mass. Recoil effects become significant for high photoelectron kinetic energies, or for lighter elements, causing an increase in the experimentally observed binding energy.[48,49] It follows that for the photoelectron energies produced in this work, the Cl 2p is expected to experience recoil shifts of 5 meV, 31 meV and 99 meV, whilst the Li 1s would experience shifts of 25 meV, 168 meV and 513 meV for the SOXPES, 2.2 keV and 6.6 keV photon energies, respectively. Recoil effects are not significant enough (below 0.1 eV) to affect most of the photoelectron peaks nor the choice of Cl 2p for alignment. They will, however, cause significant BE increases in the Li 1s, C 1s and O 1s of ~0.5 eV, 0.3 eV and 0.2 eV for the 6.6 keV (green) data, respectively.

The spectra in the second row, Fig. 1(ii), were acquired from an Li_6PS_5Cl pellet following *in situ* evaporation of ~20 nm of Li onto the surface. The O 1s region shows an oxygen peak at ~528.4 eV, assigned to Li_2O from the reaction of the deposited Li with trace oxygen,[10,11] in addition to the carbonate/hydroxide peak at ~531.2 eV that is already present in the as-loaded bare pellet. The C 1s data show a single peak due to adsorbed hydrocarbons at ~284.8 eV, with the lack of a significant contribution from any carbonate species suggesting that the high BE oxygen peak is primarily attributable to lithium hydroxide, presumably due to reaction with background water. The enhanced hydroxide to oxide peak ratio in the SOXPES data compared to the 2.2 keV and 6.6 keV HAXPES data indicates that the hydroxide species are at the outermost surface of the deposited Li, while the formation of Li_2O by reaction with residual oxygen in the UHV system is more uniform throughout. The S 2p spectra show a change in the ratio of Li_6PS_5Cl to

Li_2S, compared to the as-loaded bare pellet, with Li_2S now being dominant following Li evaporation. The Li 1s spectra show a large component at ~53.7 eV, indicative of interfacial products, such as Li_2S, Li_2P and Li_2O,[10] A metallic Li component is observed at ~52.2 eV, once recoil effects have been accounted for, and is more prominent at greater probing depths (higher photon energies).

The deposition of Li significantly attenuates the Li_6PS_5Cl species observed in the Cl 2p, S 2p, P 2p and Li 1s core levels, as well as the Li_2CO_3 and adventitious C seen at the surface of the bare pellet in the C 1s region. The depth resolution available from the different photon energies shows that the topmost surface layers (SOXPES data, red) appear to contain a greater proportion of Li_2S, following from the reaction of mobile S^{2-} species with metallic Li.[12] With increasing probing depth, the ratio of Li_6PS_5Cl to Li_2S can be seen to increase, starting from complete attenuation of the Li_6PS_5Cl component in the SOXPES data, and increasing when moving from 2.2 keV to 6.6 keV photons (see Fig. 1(ii), S 2p). This behaviour is consistent with a continuous film deposition, suggesting that the Li metal has adequately wet the pellet surface.

The third row of Fig. 1(iii) shows EDCs collected from a surface with 20 nm Li evaporated onto Li_6PS_5Cl, transferred inertly *via* a vacuum suitcase. Samples were sealed under an Ar glovebox environment and spent <6 hours in this environment prior to transfer into UHV for measurement. The O 1s spectra show a broad peak at ~531.3 eV, attributable to Li_2CO_3, $LiHCO_3$ and/or LiOH.[10] The C 1s data show the presence of the adsorbed hydrocarbons at ~284.8 eV, as well as two higher binding energy peaks at ~290.3 eV and ~288.9 eV assigned to $LiHCO_3$ and Li_2CO_3, respectively.[4] With increasing probing depth (increasing photon energy) the high binding energy peak in the O 1s can be observed to decrease in binding energy from ~531.6 eV to 531.2 eV, which is not attributable to recoil effects, and a peak corresponding to Li_2O at 528.4 eV is seen to appear and grow in intensity. This is consistent with the outermost surface being composed predominantly of the higher binding energy carbonate species[34] on top of LiOH and Li_2O. These results support the idea of gaseous CO_2, presumably present in trace amounts in the "inert" glovebox atmosphere, reacting with the surface LiOH that is present following UHV deposition.[14] Additionally, the C 1s shows adventitious hydrocarbon adsorption on the outer surface and an increasing ratio of $LiHCO_3$ to Li_2CO_3 when the probing depth increases. The reaction of CO_2 and LiOH to form Li_2CO_3 releases water;[14] we thus tentatively suggest that at the buried LiOH/Li_2CO_3 interface this results in $LiHCO_3$ formation, whilst closer to the surface we hypothesise that H_2O diffuses out into the low water content "glovebox" atmosphere used for inert transfer, leading to more Li_2CO_3 towards the outer surface.

The S 2p spectra show complete attenuation of the Li_2S and Li_6PS_5Cl species in the SOXPES data, indicating that the carbonate passivation layer has a thickness of >3 nm, as discussed earlier. The Li 1s data show a high binding energy peak at ~54.9 eV, consistent with the presence of Li_2CO_3, $LiHCO_3$ and LiOH. The Li 1s HAXPES data show a shift towards the lower BE side, when we consider the respective ~0.15 eV and ~0.5 eV recoil shifts for 2.2 and 6.6 keV data, indicating the presence of other reacted Li species (*e.g.* Li_2O and Li_2S seen in Fig. 1(ii)) beneath the carbonate layer. However, there is a lack of any metallic Li peak for all probing depths. The 2.2 keV S 2p HAXPES shows the emergence of Li_2S at 160.0 eV, closest to the carbonate layer, with pristine Li_6PS_5Cl observed beneath this (162.4 eV), as shown by the increased Li_6PS_5Cl to Li_2S ratio at greater probing depths. The absence of Li_2S at the outermost surface could be due to either the fact that the Li_2CO_3 is more thermodynamically stable[50] or that Li_2S hydrolyses to $H_2S(g)$, highlighting the importance of ensuring CO_2-free and H_2O-free processing of materials used in alkali-metal batteries. We note that the observed behaviour may not be solely attributable to glovebox exposure but may also arise from contaminants present during 'inert' transfer. However, the impact of trace glovebox contaminants is further highlighted when storing thin (<50 nm) Li films deposited onto Li_6PS_5Cl and Cu foils, which are initially visible to the naked eye as reflective, silvery coatings, but within the course of a few hours disappear as they form less opaque products. On a shorter timescale (<10 min), samples transferred directly from the evaporation chamber to the lab-based XPS in a Viton® sealed vessel also failed to exhibit any metallic Li (see Fig. S3†), indicating the formation of carbonate layers that exceed 10 nm in thickness (based on an inelastic mean free path of 3.5 nm in Li_2CO_3).

The data presented in Fig. 1 highlight the importance of *in situ* deposition with UHV transfer for studying Li metal interfaces with the Li_6PS_5Cl surface, showing that even performing an 'inert' transfer can radically alter the species observed at the surface. We expect this to be equally applicable to other alkali metals, solid electrolytes and many other electrode materials. Although Li is the most reducing of the alkali metals (−3.04 V *vs.* the standard hydrogen electrode), it is the least chemically reactive[51,52] and, thus, *in situ* deposition is expected to be similarly important, if not more so, for studies of electrode interfaces with metallic Na [53] and K.[54]

In Fig. 2, we compare several methods for the *in situ* deposition of alkali metals on the Li_6PS_5Cl electrolyte: evaporation,[34] Ar^+ sputtering of Li metal[11,34,35] and e^- flood-gun irradiation.[10,34] The Li_6PS_5Cl prior to any Li metal deposition (grey) shows the expected peaks in the Cl 2p, S 2p, P 2p, and Li 1s regions, along with common contaminants in the C 1s and O 1s regions, as discussed earlier. Comparison of the data for approximately 10 nm thick Li layers formed by e^- irradiation (blue), Li sputtering (red) and evaporation (green) confirms the presence of metallic Li in all cases, as indicated by a shoulder at 52.2 eV in Fig. 2(f). We note that for the evaporated data, sample transfer involved passing through the sample introduction chamber, which is sealed by a fast access port with a polymer seal. Consequently, the Li surface is briefly exposed to higher pressures (still $<5 \times 10^{-7}$ mbar) which may contribute to slightly increased oxygen contamination. The Cl 2p core level spectra all show the same main features attributable to the Li_6PS_5Cl, but with reduced signal to noise ratio, consistent with attenuation by the Li overlayers. The S 2p core level spectra show that Li_2S is the dominant S environment following Li deposition in all cases. Peak fitting finds that the attenuation of the S 2p and Cl 2p components related to Li_6PS_5Cl are similar for the evaporated Li surface (both with intensities of around 6% of the bare surface), indicating that the deposition method forms a continuous Li layer. In the case of the Li sputtered and e^- flood-gun irradiated surfaces, following Li deposition the S 2p intensity related to Li_6PS_5Cl is reduced to a greater extent than the corresponding Cl 2p peak (5% *vs.* 14% and 8% *vs.* 26% of bare intensity, respectively). Although this may indicate a less continuous Li layer, the large differences more likely correspond to a loss of surface sulphur species, or an increase in surface chlorine concentration during these deposition methods.

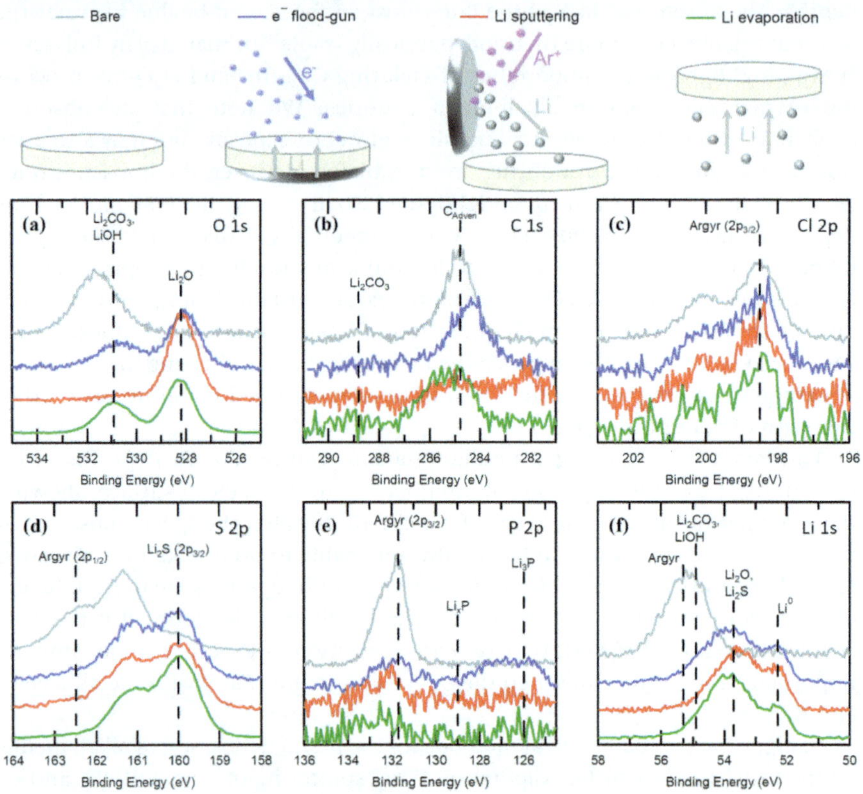

Fig. 2 Photoelectron spectra collected from surfaces following the *in situ* preparation of Li metal surfaces on Li_6PS_5Cl argyrodite (grey) in a lab-based XPS through e⁻ flood-gun irradiation (blue), Li metal sputtering (red) and thermal evaporation (green). Core level spectra are shown for (a) O 1s, (b) C 1s, (c) Cl 2p, (d) S 2p, (e) P 2p and (f) Li 1s. Data has been normalised to have equal peak areas and offset for clarity, with the binding energies calibrated such that the Cl $2p_{3/2}$ sits at 198.5 eV.

Significantly, clear differences are observed in the O and C spectra (Fig. 2(a) and (b), respectively) depending on the Li deposition technique used. Whilst all spectra show some Li_2O associated with the reaction between deposited Li and residual oxygen following Li metal sputtering, peaks related to Li_2CO_3, LiOH and adventitious C are no longer observed. For the other deposition methods, these peaks are still present, albeit with reduced signal to noise ratios following attenuation by the Li overlayer. These results therefore suggest that Li metal sputtering is causing surface adsorbed species to be removed by bombardment with the energetic Li, with similar spectral changes observed following Ar^+ sputtering of both Li_6PS_5Cl and Li surfaces (see ESI, Fig. S2†). Differences between deposition methods are also observed in the P 2p core level, Fig. 2(e), where there appear to be different interfacial species dependent on the deposition method. The electron beam growth of Li shows the presence of Li_3P at ~126 eV, as seen in other work,[12,20] whilst evaporation shows some evidence of intermediate species Li_xP, where the phosphorous is not fully reduced (see ESI, Fig. S1†). For Li metal sputtered surfaces, none of these interfacial species are apparent for the 10 nm thick Li film, in contrast to the findings of a previous study for 14 nm Li films,[12] although we note that sample cooling to −85 °C was employed in that work to suppress the removal of volatile species, whereas deposition herein was performed at ~20 °C. It is observed that for thinner sputtered Li films (generated by shorter sputtering times), Li_xP and Li_3P species are initially observed to form, but then decrease in intensity relative to the Li_6PS_5Cl P 2p peak with increasing sputtering time. Given the sputter-induced removal of O and C containing species during Li sputtering, we attribute the loss of generated phosphorous species to a similar process. We further note that bombardment of the surface layers by energetic Li during sputtering is also likely to cause a degree of interfacial mixing, affecting the observed reaction species and surface layering.

Previous work comparing Li metal deposition methods on $Li_7La_3Zr_2O_{12}$ (LLZO) garnets concluded that the deposition method chosen could induce different chemical reactions at the Li metal–LLZO interface.[34] In the case of the magnetron sputtering of Li, it has been argued that energetic Ar^+ ions can give the Li metal atoms additional energy comparable to that of an electrochemical overpotential, thus enabling kinetic barriers for the formation of an oxygen deficient interphase (ODI) at Li–LLZO interfaces to be overcome. Li deposition by e-beam evaporation was observed to not form an ODI, until the Li layer is bombarded with Ar^+ ions. "Virtual electrode" e^- flood-gun deposition was also observed to form an ODI. We suggest that these differences can be described in terms of the near-surface damage induced by the different Li metal deposition methods, rather than the kinetic stability of the reaction species. The observed results in both ref. 34 and the present study are consistent with sputtering Li metal onto a surface being more energetic and, consequently, more damaging than Li deposition by evaporation. Studies have found that the energy of sputtered atoms following impact by an incident Ar^+ beam can be up to 50 eV, with the most likely energy being ~5 eV,[55,56] whilst for the thermal evaporation of Li at the pressures used herein, this is <0.1 eV, according to the ideal gas law, which is consistent with the experimental observations.[57,58] Electrical impedance spectroscopy studies have shown that typical activation energies for lithium plating on different electrode materials (including graphite, lithium nickel cobalt aluminate and lithium iron phosphate) can vary from ~0.50 eV to 0.75 eV, which is an order of magnitude lower than the energy imparted by atoms during lithium sputtering.[59–61] For e-beam evaporation with higher evaporation rates, the vapour becomes "less ideal", leading to higher energy Li deposition, but this is still expected to be well below that of sputtering.

To further investigate the differences between these Li deposition techniques and the energetics involved, SLG grown by chemical vapour deposition (CVD) was transferred to SiO_2(300 nm)/Si substrates and used as a model surface probe. The use of SLG allows for a sensitive and quantitative measure of the defect density generated by each of the lithium deposition methods, through the use of Raman spectroscopy, on a surface whose atomic structure is not expected to be disrupted simply by contact with Li metal,[62] providing a probe of the physical damage induced by the deposition process in isolation. The Raman spectrum of the as-transferred SLG is shown in Fig. 3 (black line) with D (~1348 cm^{-1}), G (~1585 cm^{-1}) and 2D (~2685 cm^{-1}) peaks apparent. The measured spectrum is characteristic of SLG produced by CVD, with a 2D/G ratio of ~6.1 ± 1.3 and a 2D peak that is well fitted by a single Lorentzian with a fwhm of ~28 ± 4 cm^{-1}. The D/

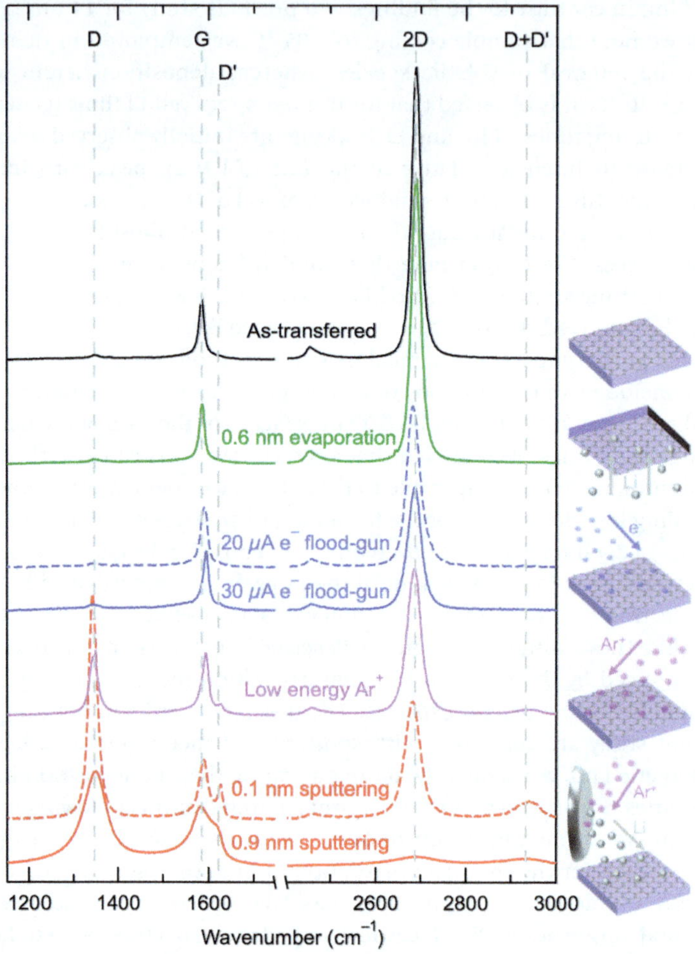

Fig. 3 Raman spectra of SLG on SiO_2(300 nm)/Si for different Li deposition methods. Data are shown for as-transferred SLG (black), 0.6 nm evaporated lithium (green), after irradiation with the e^- flood-gun (blue), after irradiation with the low energy Ar^+ (pink) and after sputtering with Li metal (red). All Raman spectra were collected in air, with spectra normalised to the height of the G peak.

G peak ratio provides a sensitive measure of defect density[63–65] that can be used to compare the extent of physical damage induced by the different deposition processes, and it is ~0.1 for the as-transferred SLG, consistent with a reasonably low defect density. The similarity of the green (0.6 nm Li evaporation) and black data in Fig. 3 shows that the thermal evaporation of Li does not appear to induce graphene defects, with Table 1 showing that the area ratios for D : G are very similar for both surfaces (~0.1).

Electron irradiated SLG surfaces, with conditions that mimic the "virtual electrode" growth of Li used in Fig. 2 (4.5 hours of irradiation) for different e^- flood-gun currents (20 μA, dashed blue, 30 μA solid blue), exhibit 2- to 3-fold increases in the D : G ratio compared to that of the as-transferred SLG (Table 1).

Table 1 Parameters extracted from fitting Lorentzian line shapes to the Raman data for SLG at different stages of Li deposition, shown in Fig. 3. Peak positions and widths were consistent with literature data. The area ratios of D to G peaks were calculated as an indication of the extent of damage induced by the deposition method. Averages were calculated from measurements of at least 3 different sample positions

	$(\text{fwhm G})_{ave}$	$(\text{fwhm 2D})_{ave}$	$(2D/G)_{ave}$	$(D/G)_{ave}$
As-transferred	16.6 ± 1.4	27.8 ± 3.5	6.01 ± 1.31	0.096 ± 0.033
0.6 nm Li evaporation	15.6 ± 1.4	28.7 ± 3.6	5.51 ± 1.71	0.103 ± 0.021
20 mA e^- flood-gun	18.5 ± 1.4	32.2 ± 0.5	4.43 ± 0.17	0.200 ± 0.045
30 mA e^- flood-gun	18.8 ± 0.7	31.7 ± 0.8	4.12 ± 0.67	0.288 ± 0.132
Low energy Ar^+	19.6 ± 0.6	30.3 ± 2.6	4.44 ± 0.25	1.752 ± 0.642
0.1 nm sputtering	24.8 ± 0.5	37.8 ± 0.3	3.17 ± 0.03	3.431 ± 0.026
0.9 nm sputtering	75.2 ± 11	186 ± 20	0.74 ± 0.43	2.644 ± 0.809

High energy electrons (>100 keV) in transmission and secondary electron microscopy are known to create defects in SLG through "knock-on" processes, where an electron knocks out a carbon atom from the graphene lattice.[66] However the energies used herein are well below the "knock-on" threshold.[67] Other work has suggested that lower energy electrons can cause damage *via* radiolysis or localised heating.[66] It is also observed that the low energies used in e^- flood-guns (<20 eV) can cause damage on organic substrate surfaces through C–C and C–H bond scission, especially in high fluences.[68,69] Additionally the low energy electron beam may induce amorphous carbon deposition through the radiolysis of residual hydrocarbons, as is routinely observed for higher energy electron beams,[37,70] which has been implicated in increases in the D peak intensity for SLG surfaces.[36] We therefore suggest that flood-gun irradiation may induce reactions involving adsorbed species on the SLG (*e.g.* polymer residues) or residual gas in the measurement chamber that leads to an increased defect concentration, either through attack of the SLG or deposition of defective carbon.

SLG on SiO_2(300 nm)/Si was also irradiated with low energy Ar^+ ions for 38 minutes, as Ar^+ ion neutralisers are increasingly commonly in surface analysis instruments and often used in conjunction with e^- flood-gun neutralisers for charge compensation.[71] The effect of the irradiation of the graphene surface with the low energy Ar^+ ions is apparent, with a clear change in the measured Raman signal (pink line, Fig. 3), now with a much more significant D peak, as well as the emergence of a D′ peak at $\sim$1620 cm^{-1} and a D + D′ peak at $\sim$2933 cm^{-1}, which indicate the formation of highly defective SLG.[67] These Ar^+ ions have a much larger mass than the electrons (73 000 times), which facilitates a more effective momentum transfer to the carbon atoms within the SLG, despite the low ion energies ($\sim$10 eV herein), causing significant damage. Indeed SLG damage following irradiation by Ar^+ ions with energies as low as 5 eV has been observed.[64] It is calculated that to cause knock-on damage, an Ar^+ ion needs an energy of >31 eV, however Stone–Wales defects can be generated with Ar^+ ions of 5 to 10 eV.[36,64] Our observations are consistent with previous reports of surface damage by the Ar^+ ion and e^- flood-gun[71] that raises concerns around some of the observations and interpretation of data when these have been used.[11,12,35]

Li metal deposition resulting from Ar^+ sputtering of a Li metal target, can also be observed to damage the SLG probe surface (red data, Fig. 3). In both cases, the

Li deposition took place over 38 minutes (the same as the Ar^+ ion irradiation), with the 0.1 nm sputtered Li data (dashed red) measured from a position on the surface that is further away from the Li metal target, whilst the 0.9 nm sputtered Li data (solid red) was closer. It is clearly seen that significant damage to the SLG has occurred, with the D peak dominating the signal and, for the 0.1 nm sputtered Li, a well-defined D′ peak is apparent. For the 0.9 nm sputtered Li surface, the extent of the damage is so great that the peaks have broadened significantly and the 2D peak has also reduced in size, indicating that the SLG has been heavily amorphized. This conclusion is supported by the observation that the D : G ratio for the 0.9 nm Li sputtered surface is smaller than that for the 0.1 nm sputtered Li case, even though a much higher defect density is expected, as established through prior observations of the amorphization trajectory of SLG using Raman spectroscopy.[72] It is clear from our findings with Li evaporation that the presence of Li itself does not damage the SLG surface. However, when Li sputtering is performed, the more energetic Li atoms appear to cause significant physical damage to the SLG.

We note that our results and interpretation thereof, *i.e.* sputter deposition of Li causes damage near surface layers of the substrate (whether it is solid electrolyte, SLG, or likely many other materials), are not consistent with the explanations provided by Connell *et al.*[34] In their study, the differences observed between LLZO–Li interfaces produced by Li metal sputtering and e-beam evaporation are attributed to kinetic stability, requiring an activation energy to be overcome for an oxygen deficient interphase to be produced. Changes in the oxidation state of Zr were observed for surfaces where Li was sputtered on top, but not those where Li was evaporated. Furthermore, it was concluded that the same changes seen for the Li sputtered surface could be observed in the e-beam evaporated surface, if the evaporated Li film was subsequently bombarded with Ar^+ ions. They argue that the Ar^+ bombardment provides the Li atoms with sufficient kinetic energy to overcome the chemical activation energy barrier. We instead suggest that Li deposition by sputtering or Ar^+ ion bombardment after Li evaporation transfers significant kinetic energy not just to the Li atoms, but also the other species close to the surface, hence promoting surface damage. Despite their conclusions that the lack of Zr oxidation for an Ar^+ sputtered surface in the absence of Li means that changes are not solely due to sputter damage, in our view it is highly likely that sputter damage, in conjunction with the highly reducing Li environment, leads to the observed oxidation state changes in the Zr. We argue that their findings point to evidence of a physical damage process due to momentum transfer from the Li metal during sputtering, rather than the resulting sputtered Li simply having an increased kinetic energy to overcome a chemical activation energy barrier, as they propose. Consequently, the interpretation of data regarding the formation of interfacial species where either sputter deposition or activation by Ar^+ bombardment has been employed must be undertaken with care to ensure that the observed effects are not significantly influenced by the energy transfer to the substrate species, potentially causing sputter damage and surface mixing.

From the above comparison of possible Li deposition methodologies, Li evaporation appears to be the most suitable for forming a well-defined interface resembling those produced by pressing Li in contact with a solid electrolyte, and subsequent interfacial studies in the absence of induced damage effects. Using

SLG as a probe surface has allowed the confirmation that Li metal sputtering causes damage through a physical process, rather than just overcoming the kinetic barrier of a chemical process. In light of this, alongside the observation of the flood-gun damage of polymers observed previously,[68,69] we suggest that for *in situ* studies of interfacial reactions at the metal–solid electrolyte interfaces, the preparation of alkali metal layers through thermal or e-beam evaporation is beneficial in avoiding physical damage to the interfacial layers, as well as avoiding the contamination associated with glovebox-based deposition.

Conclusions

Depth resolved SOXPES and HAXPES has confirmed that Li_2CO_3 species are present at the surface of Li_6PS_5Cl argyrodite, on the outermost surface of the powder grains after glovebox storage and processing, as seen for many other Li solid electrolytes. Following the deposition of metallic Li *via in situ* thermal evaporation, reaction of the Li metal with trace gases leads to some level of Li_2O and LiOH formation. This effect is exacerbated if using a polymer-sealed 'inert' transfer device of the type widely used in battery-research, as residual H_2O and CO_2 result in significant formation of surface carbonates and bicarbonates. Surface carbonates are observed to form preferentially to lithium phosphides and lithium sulphides when Li is deposited on Li_6PS_5Cl, highlighting the importance of keeping battery interfaces free from CO_2. These findings indicate that *in situ* surface preparation is key to understanding the ongoing chemistry at these reactive interfaces.

The comparison of different *in situ* Li metal deposition methods has highlighted that this can affect the observed interfacial species. Following Li metal sputtering, the peaks due to Li_2CO_3, LiOH and adventitious C cease to be observed, whilst for Li evaporation and e^- flood-gun growth, these environments are simply attenuated. These results are consistent with the Li metal being deposited by sputtering causing the surface adsorbed species to be removed by the energetic incident Li. Additionally differences are observed in the P 2p photoelectron peak, with the e^- flood-gun growth exhibiting Li_3P, whilst incomplete phosphorus reduction can be observed for thermal evaporation. Sputtering of Li initially forms reduced phosphorous species (Li_xP, Li_3P), but increased Li sputter deposition then leads to their eventual removal. This interpretation is supported by studies using SLG as a probe surface. Thermal evaporation of Li metal does not appear to induce defects in a SLG surface, as measured by the D/G ratio compared to that of as-transferred CVD graphene. Long term irradiation (over 4 hours) with an e^- flood-gun is seen to induce at least a two-fold increase in the D/G ratio, indicative of slight damage to the SLG or the deposition of defective carbon, whilst Li metal sputtering causes the carbon to amorphize almost completely after depositing as little as 1 nm of Li. Additionally, irradiation by a low energy Ar^+ ion neutraliser was seen to cause significant damage to SLG, suggesting that care must be taken when choosing charge compensation methods for delicate and reactive surfaces. Caution is therefore needed when interpreting existing data that has been collected utilising low energy Ar^+ charge compensation, or where *in situ* Li sputtering has been performed.

Our findings are consistent with the idea that Li metal sputtering causes physical damage, in contrast to previous work,[34] which suggested that it simply overcomes kinetic barriers associated with the formation of oxygen deficient interphases at the Li metal–LLZO interface. Following our systematic study of Li deposition methods, we recommend that when studying energy materials interfaces involving alkali metals, future studies should consider deposition methods that are "*in situ*" and non-damaging, such as thermal evaporation.

Author contributions

RSW, MP and T-LL were responsible for funding acquisition and supervision of the work. JSG and RSW were responsible for the conceptualisation of the work. JSG, SN and JENS collected the experimental data, JSG and SN analysed the data and JSG wrote the original draft. PKT and T-LL assisted in sample preparation and data acquisition at Diamond Light Source. All authors contributed to preparation of the manuscript.

Conflicts of interest

There are no conflicts to declare.

Acknowledgements

The authors would like to acknowledge support from the Faraday Institution through Characterisation (FIRG011, FIRG020), SOLBAT (FIRG026) and LiSTAR (FIRG014) projects, as well as the Henry Royce Institute (through the UK Engineering and Physical Science Research Council grant EP/R010145/1) for capital equipment. Additionally we would like to acknowledge Diamond Light Source for synchrotron beam time (SI 25807-2, SI 25807-3).

References

1 M. Pasta, D. Armstrong, Z. L. Brown, J. Bu, M. R. Castell, P. Chen, A. Cocks, S. A. Corr, E. J. Cussen, E. Darnbrough, V. Deshpande, C. Doerrer, M. S. Dyer, H. El-Shinawi, N. Fleck, P. Grant, G. L. Gregory, C. Grovenor, L. J. Hardwick, J. T. S. Irvine, H. J. Lee, G. Li, E. Liberti, I. McClelland, C. Monroe, P. D. Nellist, P. R. Shearing, E. Shoko, W. Song, D. S. Jolly, C. I. Thomas, S. J. Turrell, M. Vestli, C. K. Williams, Y. Zhou and P. G. Bruce, 2020 roadmap on solid-state batteries, *Journal of Physics: Energy*, 2020, **2**(3), 032008.

2 S. Randau, D. A. Weber, O. Kötz, R. Koerver, P. Braun, A. Weber, E. Ivers-Tiffée, T. Adermann, J. Kulisch, W. G. Zeier, F. H. Richter and J. Janek, Benchmarking the performance of all-solid-state lithium batteries, *Nat. Energy*, 2020, **5**(3), 259–270.

3 W. Xu, J. Wang, F. Ding, X. Chen, E. Nasybulin, Y. Zhang and J.-G. Zhang, Lithium metal anodes for rechargeable batteries, *Energy Environ. Sci.*, 2014, **7**(2), 513–537.

4 R. H. Brugge, F. M. Pesci, A. Cavallaro, C. Sole, M. A. Isaacs, G. Kerherve, R. S. Weatherup and A. Aguadero, The origin of chemical inhomogeneity in garnet electrolytes and its impact on the electrochemical performance, *J. Mater. Chem. A*, 2020, **8**(28), 14265–14276.

5 H. Wang, D. Yu, C. Kuang, L. Cheng, W. Li, X. Feng, Z. Zhang, X. Zhang and Y. Zhang, Alkali Metal Anodes for Rechargeable Batteries, *Chem*, 2019, **5**(2), 313–338.
6 J. Liu, X. Gao, G. O. Hartley, G. J. Rees, C. Gong, F. H. Richter, J. Janek, Y. Xia, A. W. Robertson, L. R. Johnson and P. G. Bruce, The Interface between $Li_{6.5}La_3Zr_{1.5}Ta_{0.5}O_{12}$ and Liquid Electrolyte, *Joule*, 2020, **4**(1), 101–108.
7 M. Armand and J. M. Tarascon, Building better batteries, *Nature*, 2008, **451**(7179), 652–657.
8 Y. Zhan, W. Zhang, B. Lei, H. Liu and W. Li, Recent Development of Mg Ion Solid Electrolyte, *Front. Chem.*, 2020, **8**(125), 1–7.
9 J. B. Goodenough and Y. Kim, Challenges for Rechargeable Li Batteries, *Chem. Mater.*, 2010, **22**(3), 587–603.
10 K. N. Wood, K. X. Steirer, S. E. Hafner, C. M. Ban, S. Santhanagopalan, S. H. Lee and G. Teeter, *Operando* X-ray photoelectron spectroscopy of solid electrolyte interphase formation and evolution in $Li_2S–P_2S_5$ solid-state electrolytes, *Nat. Commun.*, 2018, **9**, 10.
11 S. Wenzel, D. A. Weber, T. Leichtweiss, M. R. Busche, J. Sann and J. Janek, Interphase formation and degradation of charge transfer kinetics between a lithium metal anode and highly crystalline Li_7P_3S*11* solid electrolyte, *Solid State Ionics*, 2016, **286**, 24–33.
12 S. Wenzel, S. J. Sedlmaier, C. Dietrich, W. G. Zeier and J. Janek, Interfacial reactivity and interphase growth of argyrodite solid electrolytes at lithium metal electrodes, *Solid State Ionics*, 2018, **318**, 102–112.
13 A. Schwobel, R. Hausbrand and W. Jaegermann, Interface reactions between LiPON and lithium studied by *in situ* X-ray photoemission, *Solid State Ionics*, 2015, **273**, 51–54.
14 A. Sharafi, S. H. Yu, M. Naguib, M. Lee, C. Ma, H. M. Meyer, J. Nanda, M. F. Chi, D. J. Siegel and J. Sakamoto, Impact of air exposure and surface chemistry on $Li–Li_7La_3Zr_2O_{12}$ interfacial resistance, *J. Mater. Chem. A*, 2017, **5**(26), 13475–13487.
15 S. Narayanan, J. S. Gibson, J. Aspinall, R. S. Weatherup and M. Pasta, *In Situ* and *Operando* Characterisation of Li Metal–Solid Electrolyte Interfaces, *Solid State Mater. Sci.*, 2022, **26**(2), 100978.
16 A. Sharafi, H. M. Meyer, J. Nanda, J. Wolfenstine and J. Sakamoto, Characterizing the $Li–Li_7La_3Zr_2O_{12}$ interface stability and kinetics as a function of temperature and current density, *J. Power Sources*, 2016, **302**, 135–139.
17 H. Muramatsu, A. Hayashi, T. Ohtomo, S. Hama and M. Tatsumisago, Structural change of $Li_2S–P_2S_5$ sulfide solid electrolytes in the atmosphere, *Solid State Ionics*, 2011, **182**(1), 116–119.
18 C. Dietrich, R. Koerver, M. W. Gaultois, G. Kieslich, G. Cibin, J. Janek and W. G. Zeier, Spectroscopic characterization of lithium thiophosphates by XPS and XAS – a model to help monitor interfacial reactions in all-solid-state batteries, *Phys. Chem. Chem. Phys.*, 2018, **20**(30), 20088–20095.
19 Y.-C. Lu, E. J. Crumlin, G. M. Veith, J. R. Harding, E. Mutoro, L. Baggetto, N. J. Dudney, Z. Liu and Y. Shao-Horn, *In Situ* Ambient Pressure X-ray Photoelectron Spectroscopy Studies of Lithium–Oxygen Redox Reactions, *Sci. Rep.*, 2012, **2**(1), 715.

20 A. Schwöbel, R. Hausbrand and W. Jaegermann, Interface reactions between LiPON and lithium studied by *in situ* X-ray photoemission, *Solid State Ionics*, 2015, **273**, 51–54.

21 H.-J. Deiseroth, S.-T. Kong, H. Eckert, J. Vannahme, C. Reiner, T. Zaiß and M. Schlosser, Li_6PS_5X: A Class of Crystalline Li-Rich Solids With an Unusually High Li^+ Mobility, *Angew. Chem., Int. Ed.*, 2008, **47**(4), 755–758.

22 M. Chen and S. Adams, High performance all-solid-state lithium/sulfur batteries using lithium argyrodite electrolyte, *J. Solid State Electrochem.*, 2015, **19**(3), 697–702.

23 S. Yubuchi, S. Teragawa, K. Aso, K. Tadanaga, A. Hayashi and M. Tatsumisago, Preparation of high lithium-ion conducting Li_6PS_5Cl solid electrolyte from ethanol solution for all-solid-state lithium batteries, *J. Power Sources*, 2015, **293**, 941–945.

24 T. Cheng, B. V. Merinov, S. Morozov and W. A. Goddard, Quantum Mechanics Reactive Dynamics Study of Solid Li-Electrode/Li_6PS_5Cl–Electrolyte Interface, *ACS Energy Lett.*, 2017, **2**(6), 1454–1459.

25 C. Yu, L. van Eijck, S. Ganapathy and M. Wagemaker, Synthesis, structure and electrochemical performance of the argyrodite Li_6PS_5Cl solid electrolyte for Li-ion solid state batteries, *Electrochim. Acta*, 2016, **215**, 93–99.

26 M. Winter, The Solid Electrolyte Interphase – The Most Important and the Least Understood Solid Electrolyte in Rechargeable Li Batteries, *Z. Phys. Chem.*, 2009, **223**(10–11), 1395–1406.

27 R. Inada, S. Yasuda, H. Hosokawa, M. Saito, T. Tojo and Y. Sakurai, Formation and Stability of Interface between Garnet-Type Ta-doped $Li_7La_3Zr_2O_{12}$ Solid Electrolyte and Lithium Metal Electrode, *Batteries*, 2018, **4**(2), 26.

28 C. Wang, Y. Gong, B. Liu, K. Fu, Y. Yao, E. Hitz, Y. Li, J. Dai, S. Xu, W. Luo, E. D. Wachsman and L. Hu, Conformal, Nanoscale ZnO Surface Modification of Garnet-Based Solid-State Electrolyte for Lithium Metal Anodes, *Nano Lett.*, 2017, **17**(1), 565–571.

29 *Surface Analysis by Auger and X-Ray Photoelectron Spectroscopy*, D. Briggs and J. T Grant, IM Publications, Chichester, 2003.

30 T. Waldmann, A. Iturrondobeitia, M. Kasper, N. Ghanbari, F. Aguesse, E. Bekaert, L. Daniel, S. Genies, I. J. Gordon, M. W. Löble, E. De Vito and M. Wohlfahrt-Mehrens, Review—Post-Mortem Analysis of Aged Lithium-Ion Batteries: Disassembly Methodology and Physico-Chemical Analysis Techniques, *J. Electrochem. Soc.*, 2016, **163**(10), A2149–A2164.

31 S. Wang, X. Xu, X. Zhang, C. Xin, B. Xu, L. Li, Y.-H. Lin, Y. Shen, B. Li and C.-W. Nan, High-performance Li_6PS_5Cl-based all-solid-state lithium-ion batteries, *J. Mater. Chem. A*, 2019, **7**(31), 18612–18618.

32 S.-K. Otto, Y. Moryson, T. Krauskopf, K. Peppler, J. Sann, J. Janek and A. Henss, In-Depth Characterization of Lithium-Metal Surfaces with XPS and ToF-SIMS: Toward Better Understanding of the Passivation Layer, *Chem. Mater.*, 2021, **33**(3), 859–867.

33 Y. Zhu, J. G. Connell, S. Tepavcevic, P. Zapol, R. Garcia-Mendez, N. J. Taylor, J. Sakamoto, B. J. Ingram, L. A. Curtiss, J. W. Freeland, D. D. Fong and N. M. Markovic, Dopant-Dependent Stability of Garnet Solid Electrolyte Interfaces with Lithium Metal, *Adv. Energy Mater.*, 2019, **9**(12), 1803440.

34 J. G. Connell, T. Fuchs, H. Hartmann, T. Krauskopf, Y. Zhu, J. Sann, R. Garcia-Mendez, J. Sakamoto, S. Tepavcevic and J. Janek, Kinetic *versus*

Thermodynamic Stability of LLZO in Contact with Lithium Metal, *Chem. Mater.*, 2020, **32**(23), 10207–10215.

35 S. Wenzel, T. Leichtweiss, D. Kruger, J. Sann and J. Janek, Interphase formation on lithium solid electrolytes-an *in situ* approach to study interfacial reactions by photoelectron spectroscopy, *Solid State Ionics*, 2015, **278**, 98–105.

36 X. Shen, H. Wang and T. Yu, How do the electron beam writing and metal deposition affect the properties of graphene during device fabrication?, *Nanoscale*, 2013, **5**(8), 3352–3358.

37 J. C. Meyer, C. O. Girit, M. F. Crommie and A. Zettl, Hydrocarbon lithography on graphene membranes, *Appl. Phys. Lett.*, 2008, **92**(12), 123110.

38 R. E. Honig, Vapor Pressure Data for the More Common Elements, *RCA Rev.*, 1957, **18**, 195–204.

39 A. Cabrero-Vilatela, R. S. Weatherup, P. Braeuninger-Weimer, S. Caneva and S. Hofmann, Towards a general growth model for graphene CVD on transition metal catalysts, *Nanoscale*, 2016, **8**(4), 2149–2158.

40 R. S. Weatherup, 2D Material Membranes for *Operando* Atmospheric Pressure Photoelectron Spectroscopy, *Top. Catal.*, 2018, **61**(20), 2085–2102.

41 X. Li, Y. Zhu, W. Cai, M. Borysiak, B. Han, D. Chen, R. D. Piner, L. Colombo and R. S. Ruoff, Transfer of Large-Area Graphene Films for High-Performance Transparent Conductive Electrodes, *Nano Lett.*, 2009, **9**(12), 4359–4363.

42 A. Jablonski and C. J. Powell, *NIST Standard Reference Database 71: NIST Electron Inelastic-Mean-Free-Path Database Version 1.2*, 2021.

43 S. Tanuma, C. J. Powell and D. R. Penn, Calculations of electron inelastic mean free paths. IX. Data for 41 elemental solids over the 50 eV to 30 keV range, *Surf. Interface Anal.*, 2011, **43**(3), 689–713.

44 N. Fairley, *CasaXPS, Version 2.3.17*, 2021.

45 WaveMetrics, Igor Pro, Version 6.37, 2021.

46 C. F. Mallinson, J. E. Castle and J. F. Watts, Analysis of the Li KLL Auger Transition on Freshly Exposed Lithium and Lithium Surface Oxide by AES, *Surf. Sci. Spectra*, 2013, **20**(1), 113–127.

47 X. Wu, C. Villevieille, P. Novák and M. El Kazzi, Insights into the chemical and electronic interface evolution of $Li_4Ti_5O_{12}$ cycled in Li_2S–P_2S_5 enabled by *operando* X-ray photoelectron spectroscopy, *J. Mater. Chem. A*, 2020, **8**(10), 5138–5146.

48 S. Suga, A. Sekiyama, H. Fujiwara, Y. Nakatsu, T. Miyamachi, S. Imada, P. Baltzer, S. Niitaka, H. Takagi, K. Yoshimura, M. Yabashi, K. Tamasaku, A. Higashiya and T. Ishikawa, Do all nuclei recoil on photoemission in compounds?, *New J. Phys.*, 2009, **11**(7), 073025.

49 S. Suga and A. Sekiyama, High energy photoelectron spectroscopy of correlated electron systems and recoil effects in photoelectron emission, *Eur. Phys. J. Spec. Top.*, 2009, **169**, 227–235.

50 M. W. Chase Jr, NIST-JANAF Thermochemical Table, Fourth Edition, *J. Phys. Chem. Ref. Data, Monogr.*, 1998, **9**, 1–1951.

51 M. M. Markowitz, Alkali metal–water reactions, *J. Chem. Educ.*, 1963, **40**(12), 633.

52 N. V. Sidgwick, *The Chemical Elements and Their Compounds*, Oxford University Press, London, 1950, vol. I.

53 M.-C. Bay, M. Wang, R. Grissa, M. V. F. Heinz, J. Sakamoto and C. Battaglia, Sodium Plating from Na-β″-Alumina Ceramics at Room Temperature, Paving the Way for Fast-Charging All-Solid-State Batteries, *Adv. Energy Mater.*, 2020, **10**(3), 1902899.

54 R. Rajagopalan, Y. Tang, X. Ji, C. Jia and H. Wang, Advancements and Challenges in Potassium Ion Batteries: A Comprehensive Review, *Adv. Funct. Mater.*, 2020, **30**(12), 1909486.

55 E. Franke, H. Neumann, M. Zeuner, W. Frank and F. Bigl, Particle energy and angle distributions in ion beam sputtering, *Surf. Coat. Technol.*, 1997, **97**(1–3), 90–96.

56 K. K. Kammara, R. Kumar and F. S. Donbosco, Reconsideration of metal surface sputtering due to bombardment of high-energy argon ion particles: a molecular dynamics study, *Computational Particle Mechanics*, 2016, **3**(1), 3–13.

57 T. Asano, N. Uetake and K. Suzuki, Mean Atomic Velocities of Uranium, Titanium and Copper during Electron Beam Evaporation, *J. Nucl. Sci. Technol.*, 1992, **29**(12), 1194–1200.

58 K. Reichelt and X. Jiang, The preparation of thin films by physical vapour deposition methods, *Thin Solid Films*, 1990, **191**(1), 91–126.

59 M. Koseoglou, E. Tsioumas, D. Ferentinou, N. Jabbour, D. Papagiannis and C. Mademlis, Lithium plating detection using dynamic electrochemical impedance spectroscopy in lithium-ion batteries, *J. Power Sources*, 2021, **512**, 230508.

60 F. Katzer and M. A. Danzer, Analysis and detection of lithium deposition after fast charging of lithium-ion batteries by investigating the impedance relaxation, *J. Power Sources*, 2021, **503**, 230009.

61 P. Shafiei Sabet, G. Stahl and D. U. Sauer, Non-invasive investigation of predominant processes in the impedance spectra of high energy lithium-ion batteries with nickel–cobalt–aluminum cathodes, *J. Power Sources*, 2020, **472**, 228189.

62 L. G. Bulusheva, M. A. Kanygin, V. E. Arkhipov, K. M. Popov, Y. V. Fedoseeva, D. A. Smirnov and A. V. Okotrub, *In Situ* X-ray Photoelectron Spectroscopy Study of Lithium Interaction with Graphene and Nitrogen-Doped Graphene Films Produced by Chemical Vapor Deposition, *J. Phys. Chem. C*, 2017, **121**(9), 5108–5114.

63 K. Murakami, T. Kadowaki and J.-i. Fujita, Damage and strain in single-layer graphene induced by very-low-energy electron-beam irradiation, *Appl. Phys. Lett.*, 2013, **102**(4), 043111.

64 P. Ahlberg, F. O. L. Johansson, Z.-B. Zhang, U. Jansson, S.-L. Zhang, A. Lindblad and T. Nyberg, Defect formation in graphene during low-energy ion bombardment, *APL Mater.*, 2016, **4**(4), 046104.

65 J.-H. Chen, W. G. Cullen, C. Jang, M. S. Fuhrer and E. D. Williams, Defect Scattering in Graphene, *Phys. Rev. Lett.*, 2009, **102**(23), 236805.

66 R. F. Egerton, P. Li and M. Malac, Radiation damage in the TEM and SEM, *Micron*, 2004, **35**(6), 399–409.

67 D. Teweldebrhan and A. A. Balandin, Modification of graphene properties due to electron-beam irradiation, *Appl. Phys. Lett.*, 2009, **94**(1), 013101.

68 R. Havelund, M. P. Seah, A. G. Shard and I. S. Gilmore, Electron Flood Gun Damage Effects in 3D Secondary Ion Mass Spectrometry Imaging of Organics, *J. Am. Soc. Mass Spectrom.*, 2014, **25**(9), 1565–1571.
69 I. S. Gilmore and M. P. Seah, Electron flood gun damage in the analysis of polymers and organics in time-of-flight SIMS, *Appl. Surf. Sci.*, 2002, **187**(1–2), 89–100.
70 H. W. Conru and P. C. Laberge, Oil contamination with the SEM operated in the spot scanning mode, *J. Phys. E: Sci. Instrum.*, 1975, **8**(2), 136–138.
71 D. R. Baer, K. Artyushkova, H. Cohen, C. D. Easton, M. Engelhard, T. R. Gengenbach, G. Greczynski, P. Mack, D. J. Morgan and A. Roberts, XPS guide: charge neutralization and binding energy referencing for insulating samples, *J. Vac. Sci. Technol., A*, 2020, **38**(3), 031204.
72 A. C. Ferrari and J. Robertson, Interpretation of Raman spectra of disordered and amorphous carbon, *Phys. Rev. B*, 2000, **61**(20), 14095–14107.

Faraday Discussions

ROYAL SOCIETY OF CHEMISTRY

PAPER

New directions in the analysis of buried interfaces for device technology by hard X-ray photoemission

O. Renault, *[a] P.-M. Deleuze, [a] J. Courtin,[a] T. R. Bure, [a] N. Gauthier, [a] E. Nolot,[a] C. Robert-Goumet, [b] N. Pauly, [c] E. Martinez [a] and K. Artyushkova [d]

Received 29th November 2021, Accepted 18th January 2022

DOI: 10.1039/d1fd00110h

Photoelectron spectroscopy is a characterization technique which plays a key role in device technology, a field requiring, very often, a reliable and reproducible analysis of buried, critical interfaces. The recent advent of laboratory hard X-ray spectrometers opens new perspectives toward routine studies of technologically-relevant samples for the qualification of processes and materials. In this review, the status of hard X-ray photoelectron spectroscopy (HAXPES) implemented with chromium Kα excitation (5.414 keV) and applied to technological research in nanoelectronics is presented. After an account of the role of synchrotron HAXPES and the specific effects to care about at the practical level, different aspects are developed, first for illustrating the benefits of the technique through specific application cases in the field of resistive memories and power transistors. Then, we provide a status update on quantification in HAXPES, both from core-level intensities and inelastic background analysis. Finally, we present preliminary results in a novel analytical field, *operando* HAXPES, where a prototypical device is operated *in situ* during the laboratory HAXPES experiment, opening up the possibility of unravelling the mechanisms occurring at buried interfaces and governing device operation.

Introduction

The first metal–oxide–semiconductor field effect transistor (MOSFET) fabricated in an integrated fashion on a silicon substrate appeared at the end of the 1950s, following different contributions, and is today governing the operation of silicon chips. The complexity of the latter has steadily increased since then, due to

[a]*Univ. Grenoble-Alpes, CEA, Leti, 38000 Grenoble, France. E-mail: olivier.renault@cea.fr*

[b]*Université Clermont Auvergne, Clermont Auvergne INP, CNRS, Institut Pascal, F-63000 Clermont-Ferrand, France*

[c]*Université Libre de Bruxelles, Service de Métrologie Nucléaire (CP 165/84), 50 av. F. D. Roosevelt, B-1050 Brussels, Belgium*

[d]*Physical Electronics Inc., 18725 Lake Drive E, Chanhassen, MN 55317, USA*

a continuous down-scaling of the size of MOSFET devices (from a 20 μm channel length down to 5 nm and even less today). This has been known since the mid 1960s as the empirical Moore's law,[1] yielding the fabrication of chips with improved performances (higher operation frequency, lower power consumption) at a constant cost, with the transistor amount doubling every two years. The framework of this review is the applied surface science context within the particular field of device technology, which has seen a tremendous evolution over the past 25 years because of down-scaling and the necessary introduction of new materials and novel device architectures. In this field, characterization in general and surface/interface analysis in particular has always played and still plays a crucial role, both in the efficiency of the developments made and the success of the achievements performed. The latter basically rely on the processing of a series of well-controlled interfaces governing the electrical characteristics and operation of the device, and being more or less critical for its final properties and operation. This fact is often overlooked and was first realized by Herbert Krömer,[2] stating nearly twenty years after in his Nobel Lecture:[3] "The interface is the device". Krömer pioneered heterojunction devices, in which, indeed, the quality of epitaxially-grown junctions is a critical factor for obtaining devices with good performances.

Before arriving at the topic of this article, we would like to take some time to recall the crucial role of XPS in the early days of device technology, and from a short historical perspective: the example of the silicon dioxide/silicon interface of MOSFET devices is highlighted in order to illustrate the important role of photoelectron spectroscopy in the understanding and optimization of this critical interface. Amongst the key contributors to modern silicon electronics, M. J. Atalla played a particular role because he solved the critical problem of silicon surface states which were preventing current flowing along the MOSFET channel: he demonstrated that a controlled surface passivation process of silicon would decrease the surface states by a factor of 10–100, the carefully grown silicon oxide reducing and stabilizing many of the interface trap states.[4] Then, shortly after the advent of XPS, it was realized that the surface sensitivity and energy resolution of laboratory photoelectron spectroscopy enabled the detection and quantitative measurement of the Si sub-oxides at the SiO_2/Si interface involved in the formation of interface states.[5,6] In parallel, thanks to soft X-ray photoemission with synchrotron radiation excitation,[7,8] the complex structure of this interface was unravelled. Photoelectron spectroscopy, performed at various levels of sophistication, contributed therefore significantly to the progress of the research in micro- and nanoelectronics by setting a proper ground for further understanding and optimization of the SiO_2/Si interface. About 15 years ago, XPS became a metrology technique thanks to its accuracy and reproducibility in the determination of nm-thick overlayers on Si.[9] Device fabrication and integration proceeds with sequential stacking of patterned layers, over various and numerous (several hundreds) processing steps, such as deposition, lithography, etching, cleaning,… in order to produce on a silicon wafer the final assembly of interconnected nano-devices, along with the much larger test areas for characterization and metrology purposes. Keeping in mind that down-scaling results in reduced film thicknesses for the active materials in the order of the photoelectron mean free path, one understands that the samples to be analyzed in a user facility attached to a research and technology organization (RTO) maybe highly heterogeneous in the three dimensions. The interfaces of interest for characterization or metrology are most of the time deeply buried underneath metal electrodes or oxide layers and/or localized in small areas, which has been an increasing challenge for XPS performed at RTOs. Also, the diversification in logic device architectures and types is involving a large variety of materials and is driving the need for innovative ways of analyzing, by either improving existing techniques, or implementing novel ones, or even by combining and correlating analytical methods.[10–12] The recent advent of HAXPES and its applications in device technology is following this trend.

HAXPES for device technology

The early demonstrations of photoemission with hard X-rays were from Siegbahn[13] with laboratory sources (using first Mo Kα_1 and Mo Kα_2 radiations at 17.48 and 17.37 keV, respectively, to expel Cu 1s photoelectrons, then Cu Kα and Cu Kβ radiation to expel Mg 1s photoelectrons) and from Lindau *et al.*,[14] who with 8 keV photons delivered the Stanford Synchrotron Radiation Project. Then, HAXPES first expanded at synchrotron radiation facilities, and this trend is still increasing; according to a recent review providing extended historical details,[15] there are to date 24 HAXPES beamlines operational worldwide, mostly operating up to 12 keV photon energies, with some of them delivering up to 35 keV photons, mainly for gas phase studies.[16] HAXPES with synchrotron radiation has two key benefits in applied surface science: first, the high energy resolution, typically better than 250 meV, allowing accurate chemical-state fingerprinting, and second, the high brilliance of the photon beam compensating for the dramatic drop in the photoionization cross-section. For these reasons, it is expected that synchrotron HAXPES will continue to play an important role in many fields of materials science and engineering. For some time, and also quite early on, synchrotron HAXPES was used at Leti for studying multilayer systems for technologically-relevant device applications,[17–19] mainly for the characterization of the critical gate oxide interfaces in advanced planar CMOS architectures. Although synchrotron HAXPES does provide meaningful chemical state analysis at buried interfaces, the accessibility of such facilities remains poorly compatible with the time scales in force at RTOs. Also, HAXPES endstations remain poorly characterized in general towards accurate elemental quantification, thereby preventing reliable concentration figures being provided to the user. This is mainly because synchrotron sources are often designed to generate nearly-perfect plane-polarized X-rays, which has an impact on the angular electron distribution collected by the analyzer.[20,21]

Some years ago, HAXPES instruments operated with laboratory hard X-ray sources became available, allowing flexible XPS and HAXPES analysis on the same sample. These instruments are nowadays produced by different manufacturers with a broad range of photon energies enabled by different X-ray technologies: conventional X-ray sources from solid-state anode for silver Lα (2984 eV photon energy)[22] and chromium Kα (5415 eV)-based excitations,[23] and liquid-jet technology for gallium Kα excitation (9250 eV).[24] The X-ray bandwidths after monochromatization are 1.3, 0.6 and 0.45 eV, respectively, for Ag, Cr and Ga emissions, meaning that overall energy resolutions well below 1 eV are attainable in HAXPES with Cr and Ga-based technologies. To our knowledge, there will be about 60 laboratory HAXPES spectrometers installed and operated worldwide in 2022; this figure indicates how the technique is rapidly renewing the field of surface and interface analysis by photoemission.

The results presented here were obtained using a laboratory-based hard X-ray photoelectron spectrometer enabling the collection of XPS and HAXPES spectra from the same area, excited either with monochromated Al Kα or Cr Kα radiation. The instrument (Quantes dual X-ray microprobe, Physical Electronics, Inc.) was implemented recently in different fields.[11,25,26] It provides Al Kα X-ray spots ranging from 7.5 μm to 200 μm, while the available Cr Kα X-ray spot sizes are from 7.5 to 100 μm, the spatial coincidence of both spots being within 7 μm. The angle between the Cr source and the analyzer is 46.05°, while for the Al source this angle is 45°. The best overall energy resolutions available from this instrument are 0.55 eV and 0.77 eV with Al Kα and Cr Kα excitation, respectively. In HAXPES, the high excitation energy results in a more or less dramatic decrease of the photoionization cross-section depending on the considered electron transition, the intensity drop being however larger at higher photon energies.[15,27–29] For this reason, HAXPES is best implemented with high brilliance synchrotron sources. Regarding device technology in particular and interface analysis in general, there are several advantages of performing lab-scale HAXPES with a Cr Kα source. Firstly, the photon energy is not too high, thereby limiting the intensity drop due to the decreased cross-section and avoiding the need for high brilliance X-ray sources. The second advantage is that for some key elements, the sensitivity can be conserved with respect to XPS thanks to the analysis of high energy transitions; Fig. 1 illustrates this and provides the evolution at increased excitation energy of the photoionization cross-section for the main transitions of key elements in device technology. The cross-section values are taken from tabulated data from Trzhashovskaya *et al.*,[28] obtained from the same atomic model than

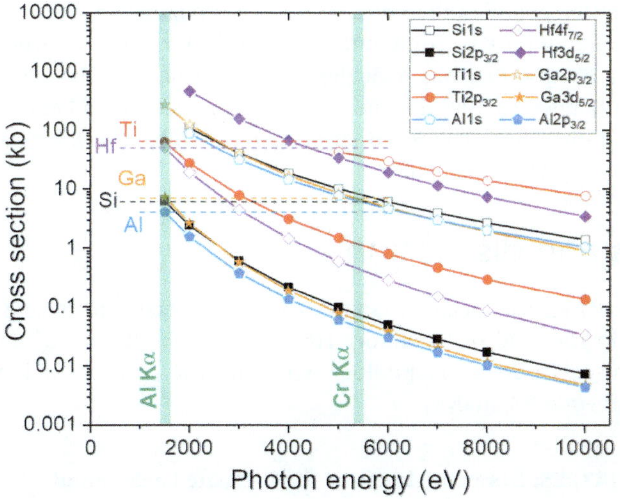

Fig. 1 Photoionization cross-section of the main photoelectric XPS and HAXPES lines for some of the key elements of semiconductor compounds, from 1 to 10 keV, taken from the tables of ref. 28.

Scofield used for the softer X-ray excitation case (Al Kα and Mg Kα),[29] but taking into account, in the frozen orbital approximation, the effect of the hole produced following the photoionization of the particular atomic sub-shell (both models being in agreement at low excitation energies). It is seen, for example, that the Si 1s (Ga 2p) cross-section upon Cr Kα photon excitation is nearly equal to the Si 2p (Ga 3d) cross-section upon Al Kα excitation. A similar conclusion can be drawn for low-energy, Al Kα-excited transitions of Hf, Al, and Ti compared to higher energy transitions of the same elements excited with Cr Kα photons. Therefore, as far as the main elements in force in device technology are concerned, HAXPES with Cr Kα enables the maintenance of the sensitivity of conventional XPS, without necessitating an increased brilliance of the photon source which may be harmful to some samples, *e.g.* oxide materials.[30,31]

A third advantage is related to the HAXPES-specific core-level recoil shift. This recoil effect arises from the momentum transfer between the hard X-ray photon and the high-energy expelled photoelectron. The recoil energy is proportional to the square of the momentum transfer and inversely proportional to the mass of the emitter atom. It is therefore larger for light elements and increases with photon energy. Kayanuma has reported recoil shifts of around 0.3 eV at the same excitation energy for C 1s in graphite.[32] Recoil shifts of 0.16 eV were reported for the Si 2p core level in HAXPES of V_3Si compounds excited with 8 keV photons,[33] which are significant. Core level energy shifts in semiconductors have many origins, including the depth-dependent effect occurring at interfaces where a potential drop exists because of band bending at the semiconductor interface or vacancy distribution in oxides; therefore, it is essential to determine whether it is valid or not to neglect this shift specific to HAXPES. The recoil energy also has a distribution which results in a broadening of the HAXPES core-levels. A short list of recoil shifts and broadening provided by Kövér[34] indicates that for elements heavier than Si excited by 5 keV photons, the recoil shift in Cr Kα HAXPES is <0.1 eV, while the recoil energy broadening is <0.2 eV. For lighter elements of technological importance, such as Li and C, as well as for heavier elements at higher excitation energies, these figures are not negligible any more. Therefore, as far as elements of pure or compound semiconductors are concerned, the recoil effect in Cr Kα HAXPES has a negligible impact on the interpretation of the complex, entangled core-level shifts as frequently encountered at semiconductor surfaces and interfaces. The recoil effect happens to be of greater importance in the field of gas phase HAXPES.[35]

Results and discussion

In the following, we focus on some selected aspects illustrating interesting directions in the field of laboratory HAXPES for the analysis of materials and interfaces used in device technology, stressing the concern of improving the reliability of HAXPES analysis.

Lab-scale HAXPES: from core-levels to the inelastic background

Interplay between XPS and HAXPES core-level analysis in capped V_2O_3 films. Vanadium oxides have attracted much attention for catalysis[36] and microelectronic applications.[37] Among them, vanadium sesquioxide, V_2O_3, or more particularly Cr doped V_2O_3, is of great interest in the field of resistive memories (ReRAM) because of its Mott transition. The fabrication of such novel devices with optimal performances requires the analysis of a buried electrode/V_2O_3 interface in a reliable way. Here, we present the first study of such an interface combining XPS and HAXPES. Vanadium can have a large number of valence states: V^{5+}, V^{4+}, V^{3+} and V^{2+}, leading to stoichiometric V_2O_5, VO_2, V_2O_3 and VO, respectively. Consequently, it is important to be able to characterize the nature of the phase of these oxides. To this aim, XPS is a suitable technique as it can provide chemical information because the binding energy of the V 2p core-level increases with the oxidation state of vanadium. Vanadium oxides have been extensively studied by XPS over the past decades.[38–42] Nevertheless, the large disparity in the reported binding energy values of the V 2p level prevents its use directly to determine the vanadium oxide phase. Additionally, Zimmerman *et al.*[39] showed that the V 2p core-levels in V_2O_5, VO_2 and V_2O_3 exhibit satellite peaks which overlap with the O 1s line. Because of this interference, it is not reliable to calculate the intensity ratio I_O/I_V to determine the oxidation state. Hence, Mendialdua *et al.*[38] have shown that the most reliable method to determine the oxidation state of vanadium is to measure the energy difference between the O 1s and V 2p lines while using the binding energy of O 1s = 530.0 eV as the energy reference. Therefore, this method has been employed to determine the nature of the vanadium oxide phase in this study.

In the application case presented here, the sample consists of 110 nm of V_2O_3 (with 4% of V substituted by Cr) capped with a 10 nm-thick layer of Si_3N_4 (SiN). The aim is to check if the vanadium oxide has indeed been deposited in the form of V_2O_3 as expected. In XPS, with an Al Kα excitation, the inelastic mean free path of V 2p electrons in SiN is calculated to be 2.4 nm with the TPP-2M method.[43] Hence, the maximum probing depth is only around 7 nm, which means that it is mandatory to partially sputter the capping layer before recording an meaningfull V 2p signal. However, Ar^+ sputtering can potentially damage the underlying vanadium oxide and therefore lead to spurious results.[41,42] Thus, it would be valuable to achieve the non-destructive characterization of the buried layer. To this end, HAXPES can provide a deeper probing depth, allowing information from the buried layer to be obtained without the need to sputter the top layer. Indeed, when using Cr Kα radiation, the inelastic mean free path of V 2p photoelectrons in SiN increases to 8.6 nm, which gives a probing depth of around 25 nm.

The sample was analyzed by both XPS and HAXPES. Before the XPS measurements, the sample was sputtered by monoatomic Ar^+ ions (1 keV) for 1.5 minutes, corresponding to the abrasion necessary to start to observe a V 2p signal. For the HAXPES experiment, no sputtering was performed prior to the measurements. The spectra are presented in Fig. 2. The first thing to note is regarding the O 1s peak shape, consisting of two components, one at high binding energy, related to O bonded to oxidized SiN, and another at a lower binding energy, corresponding to O in the vanadium oxide lattice. It is worth pointing out that the relative intensity of VO_x compared to that of oxidized SiN is higher in HAXPES, meaning that despite the absence of sputtering, HAXPES is able to probe deeper than XPS into the vanadium oxide layer. The second observation concerns the energy separation between O 1s and V 2p, which is equal to 15 eV in both XPS and HAXPES experiments. The O 1s–V 2p energy spacing is

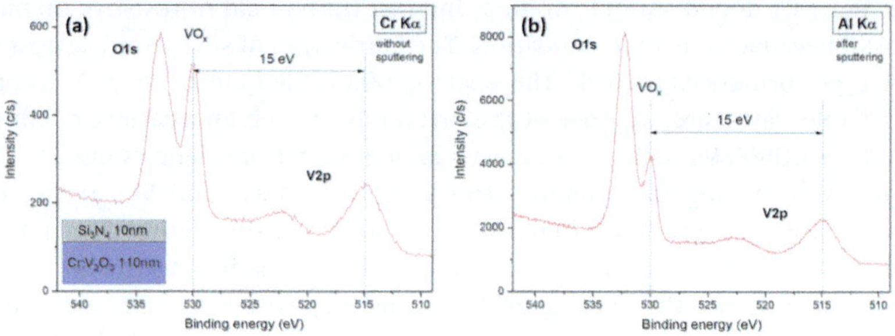

Fig. 2 (a) HAXPES and (b) XPS O 1s and V 2p core level spectra recorded on a V_2O_3 sample capped with a 10 nm-thick SiN overlayer. The overall energy resolution is 1.17 eV.

a reliable metric to determine the vanadium chemical state, and it has been demonstrated[42] that this value is in the range of the characteristic O 1s–V 2p energy separation values for the V_2O_3 phase, as expected here. Therefore, thanks to HAXPES, it is unambiguously shown that the sputtering before XPS did not modify the structure of the vanadium oxide layer thanks to the protective role played by the capping. However, HAXPES allows the measurement to be made and this conclusion to be drawn without any prior surface preparation or potential subsequent surface modifications. Additionally, with HAXPES, about 15 nm of the V_2O_3 film are probed, while only the very first nanometers are accessible with XPS. This gives the possibility to investigate the in-depth homogeneity of the film.

HAXPES of GaN-based devices and its benefits. Gallium nitride is an excellent candidate for transistors working at high frequencies and voltages (650 V) designed for high power electronics, due to a large bandgap (3.4 eV), high thermal stability and high electron mobility (>1500 $cm^2\ V^{-1}\ s^{-1}$). This mobility is enhanced by the use of AlGaN/GaN heterostructures, yielding the formation of a two-dimensional electron gas (2DEG), thereby enabling the fabrication of "high electron mobility transistors" (HEMTs). These transistors are so-called "normally-on", because the 2DEG is depleted only when a negative threshold voltage (V_{th}) is applied to the structure. For safety and reliability reasons, "normally-off' HEMTs, with a positive V_{th}, are desired for power electronics. They can be obtained by removing the 2DEG located underneath the gate with several technical solutions. One of them consists of etching the AlGaN layer under the gate to create a metal–oxide–semiconductor (MOS) stack, giving rise to the so-called MOSc-HEMTs (see Fig. 11). These transistors are "normally-off", with charge carriers moving along the channel only when a positive V_{th} is applied to the gate.

HAXPES is of particular interest for optimizing GaN HEMTs for the power electronic market because the layer thickness of the gate oxide necessarily increases well beyond the sub-5 nm values encountered in logic transistors. To get normally-off HEMTs, we have chosen to develop a technology based on the full recess approach, *e.g.* based on the full removal of the AlGaN layer below the gate. The dielectric and metal gate are then directly deposited on the GaN surface obtained after etching.[44] A schematic of the final device is presented in Fig. 3.

The final electrical properties of the transistors strongly depend on the quality of the gate oxide/GaN interface. This interface is currently being optimized by improving the technological processes, such as GaN etching, GaN surface

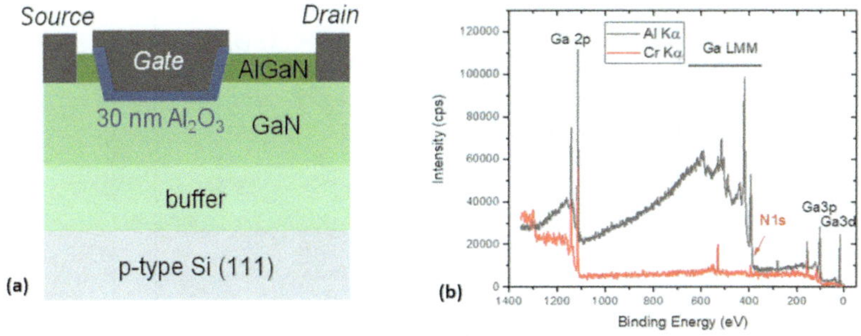

Fig. 3 (a) Schematic description of the normally-off GaN MOSc-HEMT architecture developed at CEA-Leti. (b) Survey spectra measured at 45° for GaN with Al Kα and Cr Kα sources.

passivation and alumina deposition. The development of these processes needs a precise chemical characterization of the Al_2O_3/GaN interface to investigate their impact on the GaN surface oxidation or contamination as well as the GaN stoichiometry. XPS has often be used to investigate the buried Al_2O_3/GaN interface but with Al_2O_3 thicknesses lower than 5 nm due to the extreme surface sensitivity of the technique.[44] Here, we present the first example of HAXPES as a promising technique to investigate a more realistic Al_2O_3/GaN interface with a thick gate oxide, without the need for extensive sputtering,[45] or to retrieve meaningful information regarding band bending.[46] Another interest of HAXPES for GaN-based structures, illustrated in Fig. 3(b) for Cr Kα excitation, is to get rid of the critical overlap between the N 1s core level and the Ga LMM Auger lines when using a standard Al Kα source. By using a Cr Kα source, and provided that reliable RSF are available (see the next section), a precise quantification of the nitrogen content is possible based on the N 1s peak intensity free from any other signal. The Al_2O_3 thickness should be ideally as close as possible to that used in the MOSc-HEMTs devices (30 nm). In order to see which thickness could be compatible with the use of a Cr Kα source (5.415 eV), we have run HAXPES measurements with increasing Al_2O_3 thicknesses from 5 to 30 nm.

Fig. 4 shows the Ga 2p signal recorded at a photoelectron take-off angle of 45° through increasing Al_2O_3 thicknesses. The peak intensity decreases when increasing the alumina thickness, with nearly no signal detected at 30 nm. This result is consistent with a sampling depth of 17 nm at 45° for Ga 2p photoelectrons across Al_2O_3, corresponding to an inelastic mean free path of 8 nm.[47] These spectra were measured with charge compensation and binding energy correction was done with respect to the Al 1s peak at 1562 eV.[48] Note that to investigate the GaN composition, the Ga 3d core level typically measured in XPS is here replaced by the Ga 2p core level because of a cross-section higher by one order of magnitude at 5.4 keV (Fig. 1). More investigations have been conducted to extract chemical information from HAXPES measurements, in particular from the decomposition of the Ga 2p peak, which provides a good example of how useful reference spectra can be for increasing the reliability of the analysis. A detailed analysis of the Ga 2p core level can bring some information about the GaN oxidation at the Al_2O_3/GaN interface. We first performed HAXPES measurements

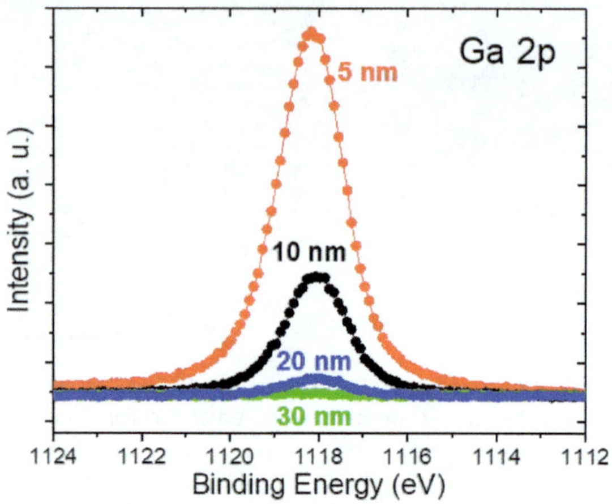

Fig. 4 Comparison of the Cr Kα HAXPES Ga $2p_{3/2}$ peak intensity for different Al_2O_3 thicknesses.

on reference samples, such as GaN and Ga_2O_3, to identify the binding energy of both the Ga–N (1117.6 eV) and Ga–O (1118.3 eV) chemical environments. The energy shift between these two chemical states was measured to be ~0.7 eV, as shown in Fig. 5(a). Due to the low intensity of the C 1s peak in HAXPES, the binding energy was referenced with respect to the N 1s peak for GaN (397.3 eV) and the O 1s peak for Ga_2O_3 (531.2 eV), considering the peak positions recorded with the Al Kα source.

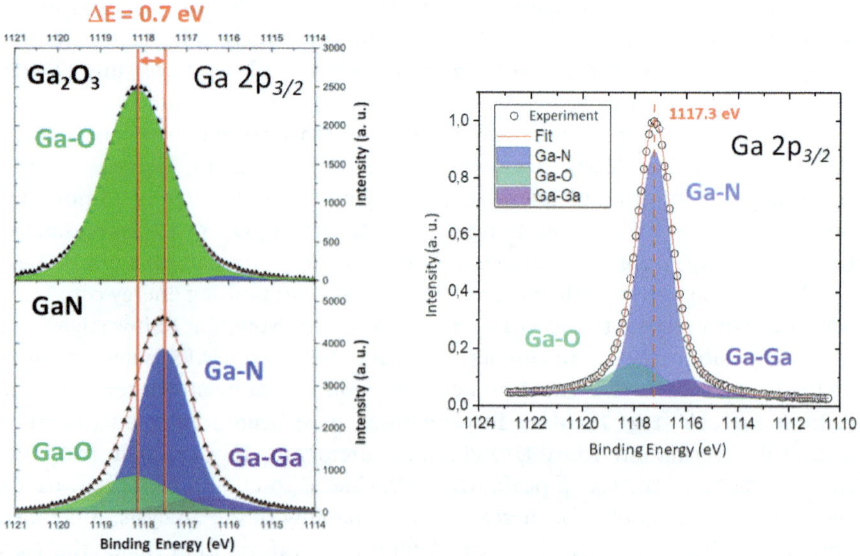

Fig. 5 Ga 2p HAXPES level spectra measured at a 45° take-off angle for the Ga_2O_3 and GaN bulk reference samples (left) and for the 15 nm-thick Al_2O_3/GaN sample (right).

For measurements on technologically relevant samples, the fitting of the Ga 2p core level is performed by considering as the external input the relative core level shift measured on the bulk references, yielding three Voigt functions assigned to the Ga–O, Ga–N and Ga–Ga chemical states; the energy shift between the Ga–O and Ga–N components is fixed at 0.7 eV, as deduced from the reference samples. An example of the Ga 2p decomposition is given in Fig. 5(b). On other types of GaN surfaces, the component of the higher energy is attributed to N-GaO states;[49] this is not expected here on epitaxially-grown GaN films, and examining the N 1s does not show any oxidized nitrogen contribution. Following the procedure using the bulk reference samples, it appears that a third Ga–Ga component for the Ga 3d core level is needed for the fit, similarly to previous works.[50] This methodology will be used in the next future to investigate the GaN oxidation at the Al_2O_3/GaN interface upon different cleaning, etching and deposition processes used for transistor fabrication.

Quantification in HAXPES

Quantification from core-level intensities. Device technology requires accurate determinations of film thicknesses, elemental concentrations, and distributions. Because of its key role in the qualification of processes and materials, there is a great demand for XPS and HAXPES regarding accurate elemental quantification. However, one needs to bear in mind that behind the formalism of quantitative analysis in photoelectron spectroscopy, there are important assumptions: (i) unpolarized X-rays are used for excitation, (ii) the sample is amorphous, and (iii) the elemental distribution within the probed volume is homogeneous. Quantitative XPS is most often performed by measurement of relative peak intensities. For this purpose, it is necessary to have relative sensitivity factors (RSF) to normalize the integrated peak intensities from the contribution of various physical factors such as differential photoionization cross-section, inelastic mean-free path as well as instrumental factors such as X-ray flux, analyzer acceptance angle and transmission function, detector efficiency. These RSF can be determined empirically from reference materials[51] or theoretically derived from first-principle calculations,[52,53] but for repeatable and reproducible experiments, it is important to select an appropriate set of RSFs in agreement with the acquisition conditions and the instrumental calibration.[20] However, the extensive work conducted over the years to develop reliable RSF databases has mostly considered the case of XPS with Al Kα X-ray excitation.[54] Higher energy X-rays give rise to different RSFs, mainly because of the significant decrease of photoionization cross-sections, which has the largest influence on the signal intensity. Some contributions have been made to determine the sensitivity factors for lab-based HAXPES instruments using Ag Lα photons[55,56] but RSFs for Cr Kα photons have not been reported yet. Therefore, in an effort towards the development of a reliable RSF database for Cr Kα HAXPES, we have started an extensive work, in collaboration with Physical Electronics, to empirically determine these sensitivity factors. The purpose of this section is to provide a status update for this work and account for the open issues coming out. The first step consisted in the determination of the pure-element sensitivity factors (PERSFs) S_i^{Ep}, which can be obtained by measuring the peak intensity of the selected element I_{ref} normalized by the peak intensity of a key material I_{key}, sputter clean Ag foil in our case, so that:

$$S_i^{Ep} = \frac{I_{ref}}{I_{key}}$$

When considering PERSF, matrix effects are not considered, namely differences in atomic densities in compound-specific inelastic mean free paths, along with elastic scattering effects. Atomic density was shown to be the largest contribution to the matrix correction,[57] though, given the large difference in kinetic energy with HAXPES, the effects of the matrix on the inelastic mean free path should be considered.

Elastic scattering remains less pronounced in HAXPES than in XPS,[58] being negligible above a photon energy of 3 keV,[59] and in any case, existing equations for correcting elastic scattering do not take into account the matrix and are valid for transition energies below kinetic energies of 2300 eV. Table 1 shows a comparison between PERSFs measured for a series of pure elements and theoretical RSFs. The careful, and fairly long, measurement protocol ensures in particular that core-level intensities are stable from the beginning to the end of the process, by checking the Ag 2p and Ag 3d intensities. In the calculation of theoretical RSFs, both dipolar and non-dipolar parameters are considered, as required in HAXPES; the dipolar approximation, which yields a simplification of the formalism for describing the transition probability in photoemission, neglects the spatial variations of the vector potential of the photon field over the characteristic length of the electron states, which is no longer the case at a shorter excitation wave-length. From a practical point of view, accounting for non-dipole effects enables us to better describe the photoelectron angular distribution, which is directly related to the collection of the photoelectron intensity in a particular experimental geometry. Non-dipolar contributions to the angular distribution become significant for photon energies higher than a few keV.[21,34] Regarding inelastic mean free paths, these are calculated according to the TPP-2M formula for the average matrix sample, while the elastic–electron corrections were calculated by a method based on a procedure[60] from the International Standard Organization. The differential photoionization cross-section was calculated according to Verner *et al.* who consider, unlike Scofield and Trzhashovskaya's data, a non-relativistic treatment of the electron states.[61] From Table 1, for most of the elements measured, an agreement within less than 15–20% is found, while for others, such as Ti and Cr, there is a huge discrepancy. Regarding Ti, this may be explained by the difficulty in maintaining a clean Ti surface during the measurements, while for Cr, the deviation looks to be transition-dependent with no obvious reason. Hence, the second part of this work aims to confirm the validity of the empirically derived

Table 1 Deviations between theoretical RSFs and PERSFs in the Cr Kα HAXPES of some pure elements

Photoelectric line	Al 1s	Al 2p	Si 1s	Si 2p	Ti 1s	Ti 2p	S 2s	S 2p	Cr 2s	Cr 2p	Co 2p
PERSF − RSF% deviation	12.8	2.4	0.9	16.0	−77.0	−79.0	−6.50	17.0	−67.0	23.0	−18.0

PERSF when applied to different compounds containing these elements and to correct them with matrix effects if necessary. This vast effort is still ongoing, but we were able to show that the empirical PERSFs give accurate quantification in aluminum compounds, such as Al_2O_3 (ref. 62) and AlN,[63] or silicon compounds, like SiC[64] and Si_3N_4.[65]

In Table 2, we provide a complementary account of the best agreements found for some compounds measured, showing the elemental concentrations as determined from the quantified HAXPES spectra. The quantification is performed either from the PERSFs or from the theoretical RSFs, and is shown against the expected or measured composition values determined by an independent method. It is seen that the agreement between both elemental compositions is dependent on the set of RSFs used, and also that the set of RSF to be used is compound-dependent. For TiS_2, the best agreement comes from theoretical RSFs, in line with the large deviation observed with PERSFs. This points to the fact that great care is needed at this stage of the work when considering composition determinations in Cr Kα HAXPES. However, these data show how useful and required the consideration of PERSF is for accurate composition measurements. It is, however, necessary, in the near future, to come up with a complete set of updated theoretical RSFs including recent cross-section data calculated for HAXPES;[27,66] indeed, it has been shown previously that the considered atomic model may, for specific transitions, yield large discrepancies.[67] It is also clear how important the constitution of an extended database of PERSFs accounting for matrix effects is.

Inelastic background analysis. As mentioned above, there are several important assumptions behind the model enabling the derivation of reliable relative elemental concentrations from core-level intensities in photoelectron spectroscopy, the most important being the homogeneity of the probed volume. As pointed out many years ago,[68] this is frequently not totally true in practical samples, which yield errors in the quantification determined from peak intensities or in the determination of layer thicknesses. As the volume probed in HAXPES is increased compared to that of XPS, there is therefore a larger probability of performing incorrect quantification. The quantitative analysis of the shape of the photoelectron loss signal, so-called inelastic background analysis,[69] has proven to be of great help for overcoming this difficulty and increasing the accuracy of the quantification of samples of practical interest. It can be implemented through different levels of sophistication, from a simple visual inspection of survey spectra

Table 2 Comparison between the HAXPES-derived best compositions of some compounds (in bold: use of PERSFs, otherwise use of theoretical RSFs) and their true composition

Compound	CoCr alloy	TiS_2	$MoSe_2$
HAXPES composition	**86% Co (2p)**	33% Ti (2p) 67% S (2p)	**35% Mo (2p)**
	17% Cr (2s)	31% Ti (1s) 69% S (1s)	**65% Se (2s)**
True composition (technique used)	80% Co	33% Ti	33% Mo
	20% Cr	67% S	67% Se
	(RBS)	(Raman[a])	(Bulk reference)

[a] Characteristic vibrational modes of crystalline, stoichiometric TiS_2 are detected.

to more complex analyses of the background shape.[70] Quite early on, it was realized how useful background analysis may be in HAXPES for probing much deeper into the bulk; indeed, the characteristic loss signal originates from typical depths up to eight times the inelastic mean free path, and therefore enables the probing of deeply buried interfaces. A first demonstration of this fact was done with the analysis of a sub-nm thick La layer buried underneath a 50 nm overlayer.[71] Then, several refinements of the technique were investigated (still using synchrotron-based HAXPES with photon energies in the 6–15 keV range), on the one hand toward increasing the accuracy of the depth distribution determination,[72,73] and on the other hand toward enabling the analysis of complex cases encountered in material science,[72,74] in both cases dealing with interfaces buried up to 70 nm below the surface. It was shown that, even without the use of a reference sample, accuracies of around 10–15% in the determination of the depth distribution of buried layers were possible. This figure was shown to decrease down to 5% thanks to an analysis performed with a reference. Since then, background analysis in HAXPES has started to be applied in different fields.[75]

Here, we introduce the inelastic background analysis of Cr Kα HAXPES spectra, toward the goal of its more widespread use for metrology purposes, and following some recent refinements performed in the field of routine analysis.[76] The samples are bilayer stacks of technologically-relevant oxides grown on Si by atomic layer deposition: HfO_2 and Al_2O_3, with the 2.5 nm-thick HfO_2 layer being buried underneath 10, 15 or 20 nm Al_2O_3, as determined from X-ray fluorescence. The Al 1s (Hf 3d) inelastic mean free path is 7.4 nm (7.2 nm) in alumina. In Fig. 6, we present superimposed survey spectra of the three samples recorded at a take-off angle of 45°. One observes close Al and Si XPS transitions in the high kinetic energy region of the spectra, while the intense Al 1s transition from the surface alumina layer predominates in the HAXPES region. In this latter region, the Al 1s, Hf 3d and Si 1s core levels are of interest, because, unlike the XPS core levels, they

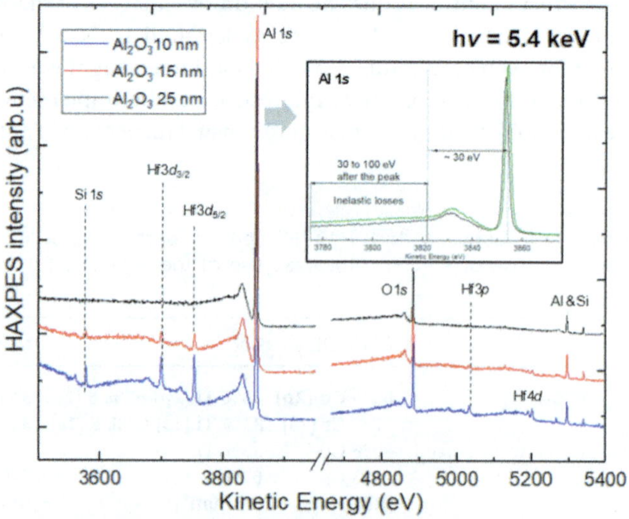

Fig. 6 HAXPES survey spectra of HfO_2 and Al_2O_3 bilayer structures on Si.

remain well separated from each other, which is suitable for background analysis. A visual inspection of the Al 1s background in the kinetic energy range, indicated in the inset, (after the first plasmon loss) is informative about its location at the surface of the stack, because the background steadily decreases at lower energy. The decreasing rate of the background signal (taken with respect to the signal at 3900 eV before the no-loss peak) is, qualitatively at this stage, indicative of the relative thicknesses of the alumina surface layer, with a smaller thickness for the blue spectrum and a greater thickness for the black spectrum.

Conversely, for the Si 1s background, a monotonic increase is observed away from the main peak, which is indicative of a deeply buried Si layer in the probed HAXPES volume. For the Hf 3d, we consider a smaller energy range, as presented in Fig. 7, where the spectra are now compared to the one of a thick, single layer of HfO_2. We can clearly observe that for the bilayer spectra, the inelastic background is nearly flat from 1740 eV and above in the case of the thinner overlayer, while it is steadily increasing over an energy range which is directly related to the thickness of the alumina overlayer. In contrast, the Hf 3d background for the single, thick HfO_2 layer ($z = 3 \times$ IMFP) decreases from 1740 eV. For the thicker overlayer, the Hf 3d main peak is no longer detected, however its constantly increasing background signal is an indirect fingerprint of a deeply buried Hf-containing layer within the HAXPES probed volume. As for previous works, we may quantitatively analyze the background signals of the Si 1s, Al 1s, Hf 3d and O 1s core levels in order to retrieve the elemental distributions.

Such an analysis is shown in Fig. 8 for the four elements of interest in one of the bilayer sample. As input parameters, we considered the effective inelastic mean free paths[73,76,77] and universal inelastic scattering cross-section. which were found to best model the background signal. We observe a good agreement between the overall depth distribution of O (Fig. 8(a)) and that of Si (Fig. 8(c)). The

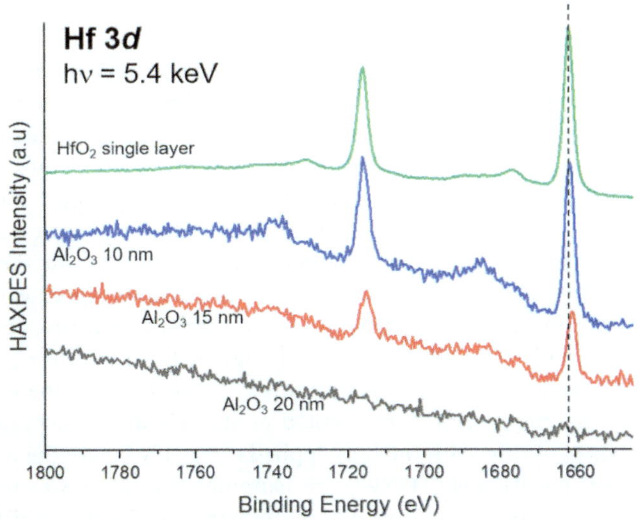

Fig. 7 Hf 3d core level and inelastic background for the bilayer structure and a thick, pure HfO_2 layer.

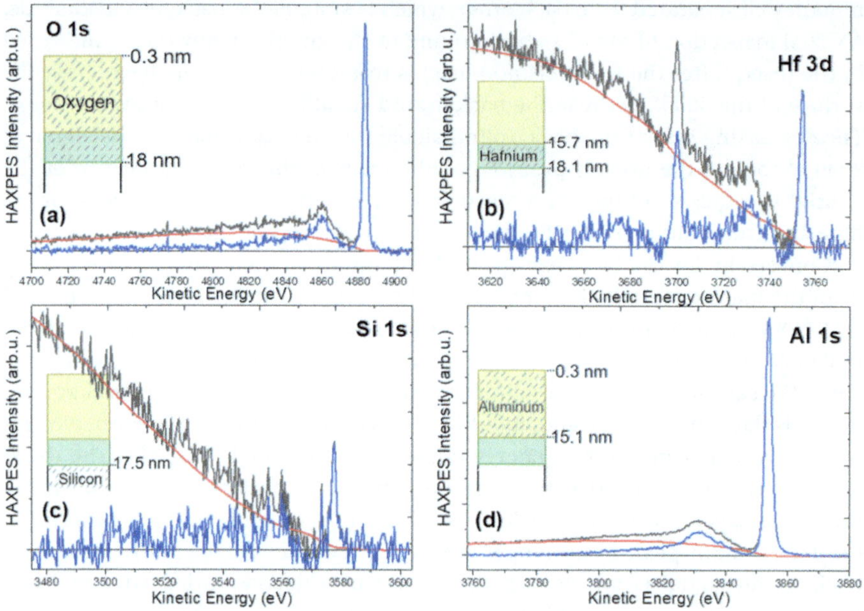

Fig. 8 Background analysis of the Al_2O_3 15 nm/HfO_2 2.5 nm bilayer sample. Black: experimental spectrum; red: calculated background; blue: subtracted spectrum.

thickness of the HfO_2 buried layer (Fig. 8(b)), as well as that of the alumina surface layer (Fig. 8(d)), are consistent with the nominal values determined by X-ray fluorescence within less than 10%. The analysis demonstrates the ability of the background analysis of Cr Kα HAXPES spectra to determine accurate layer thicknesses, paving the way to routine measurements.

Operando lab-scale HAXPES

XPS and HAXPES analyze materials and interfaces in the static mode, *i.e.* without any consideration regarding the mechanisms prevailing during the operation of the corresponding device fabricated from these structures. This means that one misses crucial information about intermediate states (*e.g.*, at a non-zero applied field) at interfaces, which may help in unravelling the physical mechanisms governing device operation. For example, in ferroelectric oxide-based high-density capacitors, it is desirable to determine to which extent the interface between the electrode and the oxide film, formed upon specific deposition conditions, contributes to degrading the overall polarization properties of the structure.[78] *Operando* HAXPES is a relatively new field in which a prototypical device is analyzed while an external stimulus is applied during the experiment, however until recently it was implemented exclusively with synchrotron radiation[78–82] and most of time with the bias applied externally before the experiment. Here, we report on *operando* HAXPES measurements with Cr Kα excitation on a prototypical memory device (100 × 100 μm active area) fabricated from a Sr-doped lanthanum manganite oxide as the active material, showing reversible resistance changes upon biasing, also know as resistive switching (RS). There are a large number of oxides presenting RS behavior, such as $SmNiO_3$,[83] $SrTiO_3$ [84] and $La_{1-x}Sr_xMnO_3$.[85,86] The RS phenomena have different origins[87] (interfacial or filamentary-like) and it can be difficult to determine the exact mechanism associated with RS. *Operando* HAXPES turns out to be a particularly efficient method to study RS, since it makes it possible to investigate the chemical modifications under bias and that it can be used as a local probe of the potential distribution at interfaces.[81]

We investigated, using lab-scale *operando* HAXPES, the interface between a Ti electrode and a p-type semiconducting, 20 nm-thick $La_{0.8}Sr_{0.2}MnO_3$ (001) film. The $La_{0.8}Sr_{0.2}MnO_3$ (001) film was grown on an $LaAlO_3$ (001) substrate by pulsed-injection metal organic chemical vapor deposition. The thickness of $La_{0.8}Sr_{0.2}MnO_3$ (001), determined from by X-ray reflectometry, was equal to 21 nm. The electrode was a 10 nm-thick (as determined by a quartz microbalance) Ti layer deposited under UHV by using an e-gun evaporator. In order to avoid the oxidation of titanium, we deposited a 2 nm-thick Pt film on the latter. A thin Pt overlayer was deposited for the measurements (*I–V* curves and *operando* HAXPES). Photoemission measurements were carried out using monochromatic Cr Kα sources at room temperature. The photoelectron kinetic energy was measured by employing a hemispherical analyzer. Our set-up allowed us to apply a voltage during photoemission measurements. Fig. 9(a) shows the *I–V* measurements at room temperature. This curve shows a hysteretic behavior during the sweep, in agreement with previous studies on LSMO/metal structures.[86] From the *I–V* curve, it is seen that the application of positive (negative) voltage leads to a RS effect from a low resistance state (high resistance state) to high resistance state (low resistance state). Thus, it appears that RS is reversible and reproducible in our experiment. Fig. 9(b) shows the temporal evolution of the system in the low resistance state (LRS) and the high resistance state (HRS). This figure shows that the two resistance states are stable with time. Fig. 10(a) shows a schematic representation of the system studied by *operando* HAXPES. We applied a voltage

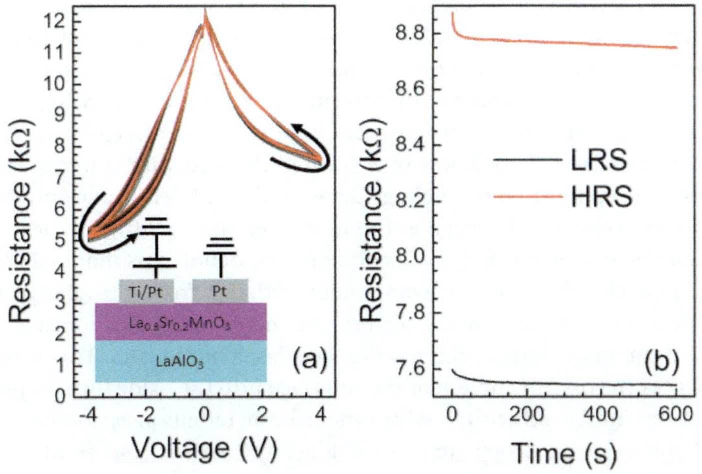

Fig. 9 Current–voltage characteristics (50 sweeps) acquired at room temperature for the Ti/LSMO interface (a) and resistance at −1.5 V as a function of time for the LRS and HRS (b).

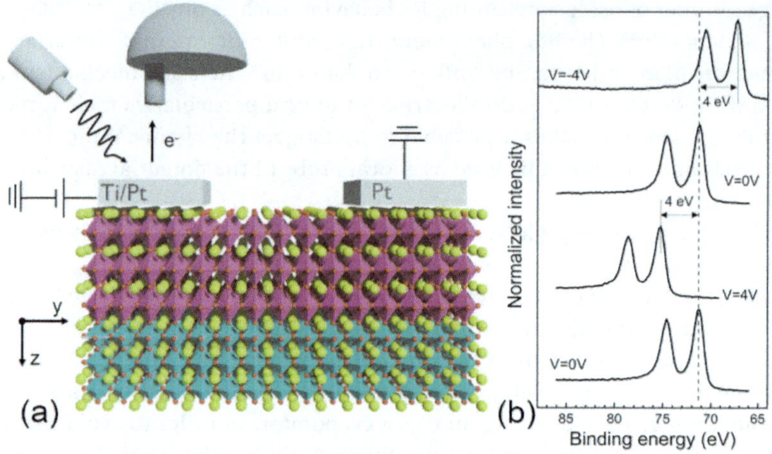

Fig. 10 Schematic representation of the study of the device by HAXPES (a) and Pt 4f core level spectra for several values of applied voltage (b). The spectra were collected at room temperature, normal emission and at a photon energy 1486.6 eV (Al Kα). Vertical black lines give the binding energy of Pt $4f_{7/2}$ component.

between the Ti/Pt and the Pt electrode *in situ* during the photoemission measurements. Fig. 10(b) shows the effect of applied voltage on the Pt 4f core level spectra (Ti capping layer). We find that the binding energy of the Pt 4f core level follows the applied voltage, which validates the experimental protocol.

Fig. 11(a and b) show the Ti 2p and La $3d_{5/2}$ core-level spectra of for different values of applied voltage between the Ti/Pt and Pt electrode at normal emission and excited with Cr Kα radiation. First, we focus on the measurements performed at zero bias after switching either into the HRS or the LRS. At zero bias, there is no contribution of the potential profile at the interface to the overall broadening of the spectra, therefore the core level lineshape is only indicative of chemical changes. The Ti 2p core level envelope was decomposed into three components (Ti^0, TiO and TiO_2). The Ti^0 component was fitted by an asymmetric Lorentzian line shape according to its metallic behavior, whereas the TiO and TiO_2 components were fitted by Gaussian–Lorentzian line shapes. We found that the titanium electrode was partially oxidized, in agreement with its expected high chemical affinity with oxygen. The important point here is that the magnitude of the TiO component increases (by a factor of 2.5) when the structure is in the HRS. This shows that the application of a voltage between the Ti/Pt electrode and the LSMO results in an exchange of oxygen at the interface. The La $3d_{5/2}$ core level spectra have been decomposed using three components, denoted as main, cf_B and cf_{AB}. The cf_B and cf_{AB} are satellite components arising from the charge transfer between the valence band of the ligand and the $4f^0$ orbital of ionized La.[88] We found that the magnitude of the satellite decreases in the HRS. This is related to a decrease of the oxygen content at the interface into the oxide film, as previously observed on similar structures with synchrotron radiation *operando* HAXPES[81] and XPS micro-spectroscopy after *ex situ* biasing.[89] This is consistent with the Ti oxidation at the interface observed in a more direct way from the Ti 2p spectra. Moreover, as shown in Fig. 11(b), an energy shift ($\sim$0.3 eV) between the HRS and

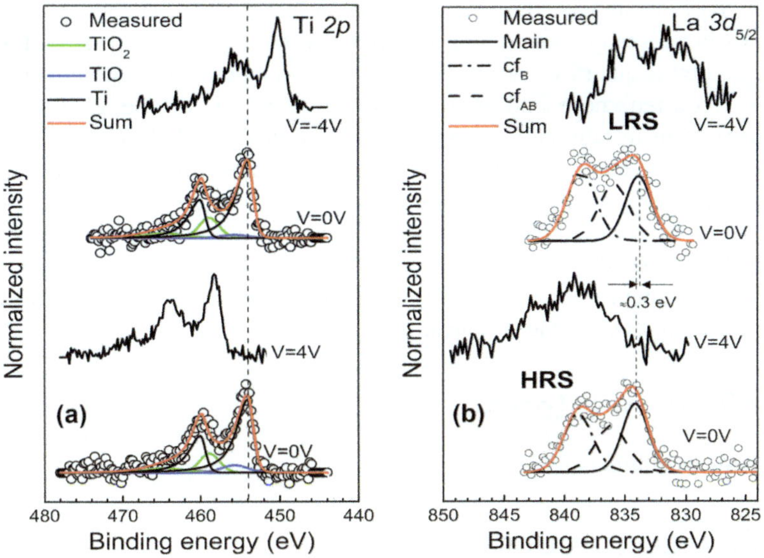

Fig. 11 Ti 2p (a) and La $3d_{5/2}$ (b) core level spectra for different applied voltage values. The Cr Kα HAXPES spectra were collected at room temperature, normal emission. The vertical dashed line in the La $3d_{5/2}$ core level spectra gives the energy of the main component. Shirley background subtraction was applied to all spectra.

the LRS is found. This latter is attributed to a displacement of the Fermi level toward the conduction band, consistent with a decrease of the hole density. In fact, oxygen vacancies act as electron donors. Overall, the photoemission measurements show unambiguously that the RS effect is due to an oxygen exchange between Ti and LSMO. In order to understand more precisely the operating modes of the device, we studied the HAXPES spectra obtained under bias. We found that the Ti 2p and La $3d_{5/2}$ core levels strictly follow the applied voltage. Therefore, the resistance of the TiO_x electrode is much lower than that of LSMO. The resistance of LSMO is the dominant resistance that limits the current flow. Moreover, we found that upon biasing, the La $3d_{5/2}$ spectra showed asymmetrical broadening. This is due to a potential distribution associated to a voltage drop in this layer, which needs to be accounted for by proper modelling (work in progress).

Conclusions

This review provided the current status of hard X-ray photoelectron spectroscopy (HAXPES) implemented with chromium Kα excitation (5.414 keV) as applied to technological research in device technology. It was shown from the different application cases in resistive oxide memories and GaN-based power transistors how HAXPES enables increasing the reliability of photoemission in this field with the analysis of samples closer to real device cases. The quantification of HAXPES spectra from core level intensities remains to be improved, accounting for updated cross-sections and matrix effects, based on encouraging first results in experimental sensitivity factors. However, it must be kept in mind that with greater probing depths, the probability of probing heterogeneous volumes in HAXPES increases in general and in applied surface science in general; for this reason, suitable quantification techniques, such as inelastic background analysis, will be required more and more. Finally, the novel technique of *operando* photoemission in laboratory HAXPES was introduced and shown to be a local probe of specific changes occurring at buried interfaces during device operation. Another HAXPES analytical method not mentioned here but of interest for spatially-resolved measurements is photoemission microscopy, the development of which in HAXPES is ongoing.[77,90] It is believed that laboratory HAXPES will continue to bring important benefits to the field of device technology and more generally in applied surface science.

Conflicts of interest

There are no conflicts to declare.

Author contributions

OR: conceptualization, validation, original draft – writing, review & editing, visualization ; P. M. D., E. M., T. R. B., K. A.: methodology, formal analysis, writing (results section); J. C.: conceptualization, methodology, formal analysis, investigation, writing (original draft, results section: *operando* measurements); C. R. G., N. P., E. N. : original draft – review and editing.

Acknowledgements

This work, carried out on the Platform For Nanocharacterisation (PFNC), was supported by the "Recherches Technologiques de Base" program of the French National Research Agency (ANR). We thank Carlos Moncasi (LMGP, CNRS and Grenoble Alpes Univ.) for providing the thin films used for the fabrication of the prototypical ReRAM devices studied by *operando* HAXPES. The authors are also grateful to L. Vauche for providing the GaN-based samples, with special thanks to L. De Brucker and P. Fernandes Paes Pinto Rocha for their help regarding data analysis of the alumina/GaN samples.

References

1 G. E. Moore, *Technical Digest 1975: International Electron Devices Meeting*, 1975, vol. 11.
2 H. Kröemer, *Surf. Sci.*, 1983, **132**(1–3), 543.
3 H. Kröemer, *ChemPhysChem*, 2001, **2**(8–9), 490.
4 B. Lojek, *History of Semiconductor Engineering*, Springer Science & Business Media, 2007, vol. 321.
5 J. M. Hill, D. G. Royce, C. S. Fadley, L. F. Wagner and F. J. Grunthaner, *Chem. Phys. Lett.*, 1976, **44**(2), 225.
6 P. J. Grunthaner, M. H. Hecht, F. J. Grunthaner and N. M. Johnson, *J. Appl. Phys.*, 1987, **61**(2), 629.
7 G. Hollinger and F. J. Himpsel, *Appl. Phys. Lett.*, 1984, **44**(1), 93.

8 F. J. Himpsel, F. R. McFeely, A. Taleb-Ibrahimi, J. A. Yarmoff and G. Hollinger, *Phys. Rev. B: Condens. Matter Mater. Phys.*, 1988, **38**(9), 6084.
9 C. R. Brundle, G. Conti and P. Mack, *J. Electron Spectrosc. Relat. Phenom.*, 2010, **178–179**, 433.
10 P. S. Johnson, I. Boukahil, F. J. Himpsel, K. L. Kearns, J. H. Kang, J.-C. Lin, A. Leugers, G. Meyers, S. Mukhopadhyay, D. H. K. Jackson and T. F. Kuech, *J. Phys. Chem. C*, 2016, **120**(3), 1366.
11 P. A. W. van der Heide, V. Spampinato, A. Franquet, C. Zborowski, T. Conard, J. Ludwig, K. Paredis, W. Vandervorst, A. Pirkl and E. Niehuis, *Surf. Interface Anal.*, 2020, **52**(12), 786.
12 D. P. B. Meunier, R. Rodriguez-Lamas, M. Boudard, O. Chaix-Pluchery, E. Martinez, N. Chevalier, C. Jimenez, M. Burriel and O. Renault, *ACS Appl. Electron. Mater.*, 2019, **1**(5), 675.
13 K. Siegbahn, *Nucl. Instrum. Methods Phys. Res., Sect. A*, 2005, **547**(1), 1.
14 I. Lindau, P. Pianetta, S. Doniach and W. E. Spicer, *Nature*, 1974, **250**(5463), 214.
15 K. Curran, N. K. Fernando, P. Bhatt, O. Fredrik, L. Johansson, A. Lindblad, H. Rensmo, L. Zendejas Medina, R. Lindblad, S. Siol, L. P. H. Jeurgens, C. Cancellieri, K. Rossnagel, K. Medjanik, G. Schönhense, M. Simon, A. X. Gray, S. Nemšák, P. Lömker, C. Schlueter and A. Regoutz, *J. Phys.: Condens. Matter*, 2021, **33**(23), 233001.
16 M. Oura, T. Gejo, K. Nagaya, Y. Kohmura, K. Tamasaku, L. Journel, M. N. Piancastelli and M. Simon, *New J. Phys.*, 2019, **21**(4), 043015.
17 J. Rubio-Zuazo, E. Martinez, P. Batude, L. Clavelier, F. Soria, A. Chabli and G. R. Castro, *AIP Conf. Proc.*, 2007, **931**(1), 329.
18 E. Martinez, O. Renault, L. Clavelier, C. Le Royer, J.-M. Hartmann, V. Loup, S. Minoret, V. Cosnier, Y. Campidelli, J. Zegenhagen and T.-L. Lee, *J. Vac. Sci. Technol., B: Microelectron. Nanometer Struct.–Process., Meas., Phenom.*, 2007, **25**(1), 86.
19 E. M. R. Boujamaa, F. Pierre, O. Renault, B. Detlefs, J. Zegenhagen, S. Baudot, M. Gros-Jean, F. Bertin and C. Dubourdieu, *Appl. Surf. Sci.*, 2015, **335**, 71.
20 A. G. Shard, *J. Vac. Sci. Technol., A*, 2020, **38**(4), 041201.
21 M. Novák, N. Pauly and A. Dubus, *J. Electron Spectrosc. Relat. Phenom.*, 2012, **185**(1–2), 4.
22 K. Yates and R. H. West, *Surf. Interface Anal.*, 1983, **5**(4), 133.
23 K. Kobayashi, M. Kobata and H. Iwai, *J. Electron Spectrosc. Relat. Phenom.*, 2013, **190**, 210.
24 R. Anna, M. Mascheck, T. Wiell, S. K. Eriksson, C. Liljenberg, K. Tetzner, A. Benjamin, D. Williamson, D. O. Scanlon and P. Paul, *Rev. Sci. Instrum.*, 2018, **89**(7), 073105.
25 S. Siol, J. Mann, J. Newman, T. Miyayama, K. Watanabe, P. Schmutz, C. Cancellieri and L. P. H. Jeurgens, *Surf. Interface Anal.*, 2020, **52**(12), 802.
26 O. Renault, E. Martinez, C. Zborowski, J. Mann, R. Inoue, J. Newman and K. Watanabe, *Surf. Interface Anal.*, 2018, **50**(11), 1158.
27 M. B. Trzhaskovskaya and V. G. Yarzhemsky, *At. Data Nucl. Data Tables*, 2019, **129–130**, 101280.
28 M. B. Trzhaskovskaya, V. I. Nefedov and V. G. Yarzhemsky, *At. Data Nucl. Data Tables*, 2001, **77**(1), 97.
29 J. H. Scofield, *J. Electron Spectrosc. Relat. Phenom.*, 1976, **8**(2), 129.

30 T. H. Fleisch and G. J. Mains, *J. Chem. Phys.*, 1982, **76**(2), 780.
31 A. B. Ahmad Kayani, S. Kuriakose, M. Monshipouri, F. Abdul Khalid, S. Walia, S. Sriram and M. Bhaskaran, *Small*, 2021, **17**(32), 2100621.
32 Y. Kayanuma, Hard X-ray Photoelectron Spectroscopy (HAXPES), in *Springer Series in Surface Science*, ed, J. C. Woicik, 2015, vol. 59.
33 S. Suga, S. Itoda, A. Sekiyama, H. Fujiwara, S. Komori, S. Imada, M. Yabashi, K. Tamasaku, A. Higashiya, T. Ishikawa, M. Shang and T. Fujikawa, *Phys. Rev. B: Condens. Matter Mater. Phys.*, 2012, **86**(3), 035146.
34 L. Kövér, *J. Electron Spectrosc. Relat. Phenom.*, 2010, **178–179**, 241.
35 E. Kukk, D. Céolin, O. Travnikova, R. Püttner, M. N. Piancastelli, R. Guillemin, L. Journel, T. Marchenko, I. Ismail, J. Martins, J. P. Rueff and M. Simon, *New J. Phys.*, 2021, **23**(6), 063077.
36 I. E. Wachs and B. M. Weckhuysen, *Appl. Catal., A*, 1997, **157**(1–2), 67; J. Haber, *Catal. Today*, 2009, **142**(3–4), 100.
37 R. B. Darling and S. Iwanaga, *Sadhana*, 2009, **34**(4), 531.
38 J. Mendialdua, R. Casanova and Y. Barbaux, *J. Electron Spectrosc. Relat. Phenom.*, 1995, **71**(3), 249.
39 R. Zimmermann, R. Claessen, F. Reinert, P. Steiner and S. Hüfner, *J. Phys.: Condens. Matter*, 1998, **10**(25), 5697.
40 M. Demeter, M. Neumann and W. Reichelt, *Surf. Sci.*, 2000, **454–456**, 41; N. Kamakura, M. Taguchi, K. Yamamoto, K. Horiba, A. Chainani, Y. Takata, E. Ikenaga, H. Namatame, M. Taniguchi, M. Awaji, A. Takeuchi, K. Tamasaku, Y. Nishino, D. Miwa, T. Ishikawa, Y. Ueda, K. Kobayashi and S. Shin, *J. Electron Spectrosc. Relat. Phenom.*, 2005, **144–147**, 841; G. Beamson, N. Moslemzadeh, P. Weightman and J. F. Watts, *J. Electron Spectrosc. Relat. Phenom.*, 2008, **162**(1), 19.
41 G. Silversmit, D. Depla, H. Poelman, G. B. Marin and R. De Gryse, *J. Electron Spectrosc. Relat. Phenom.*, 2004, **135**(2–3), 167.
42 E. Hryha, E. Rutqvist and L. Nyborg, *Surf. Interface Anal.*, 2012, **44**(8), 1022.
43 S. Tanuma, C. J. Powell and D. R. Penn, *Surf. Interface Anal.*, 1988, **11**(11), 577.
44 L. Vauche, C. Antoine, E. Martinez, M.-C. Roure, C. Le Royer, S. Bécu, R. Gwoziecki and M. Plissonnier, *ACS Appl. Electron. Mater.*, 2021, **3**(3), 1170.
45 W. Lisowski, E. Grzanka, J. W. Sobczak, M. Krawczyk, A. Jablonski, R. Czernecki, M. Leszczyński and T. Suski, *J. Alloys Compd.*, 2014, **597**, 181.
46 S. Ueda, *Appl. Phys. Express*, 2018, **11**(10), 105701.
47 S. Tanuma, C. J. Powell and D. R. Penn, *Surf. Interface Anal.*, 2011, **43**(3), 689.
48 G. Kobayashi, Y. Irii, F. Matsumoto, A. Ito, Y. Ohsawa, S. Yamamoto, Y. Cui, J.-Y. Son and Y. Sato, *J. Power Sources*, 2016, **303**, 250.
49 H. Mehdi, G. Monier, P. E. Hoggan, L. Bideux, C. Robert-Goumet and V. G. Dubrovskii, *Appl. Surf. Sci.*, 2018, **427**, 662.
50 H. Kim, D. H. Kim and B. J. Choi, *Appl. Phys. A: Mater. Sci. Process.*, 2017, **123**(12), 800; M. Mishra, T. C. Shibin Krishna, N. Aggarwal, M. Kaur, S. Singh and G. Gupta, *Phys. Chem. Chem. Phys.*, 2015, **17**(23), 15201.
51 C. D. Wagner, L. E. Davis, M. V. Zeller, J. A. Taylor, R. H. Raymond and L. H. Gale, *Surf. Interface Anal.*, 1981, **3**(5), 211.
52 M. P. Seah, *Surf. Interface Anal.*, 1980, **2**(6), 222.
53 M. P. Seah, I. S. Gilmore and S. J. Spencer, *Surf. Interface Anal.*, 2001, **31**(8), 778.
54 C. Richard Brundle and B. V. Crist, *J. Vac. Sci. Technol., A*, 2020, **38**(4), 041001.

55 M. J. Edgell, R. W. Paynter and J. E. Castle, *J. Electron Spectrosc. Relat. Phenom.*, 1985, **37**(2), 241.
56 A. G. Shard, J. D. P. Counsell, D. J. H. Cant, E. F. Smith, P. Navabpour, X. Zhang and C. J. Blomfield, *Surf. Interface Anal.*, 2019, **51**(7), 763.
57 S. Tanuma and T. Kimura, *J. Surf. Anal.*, 2003, **10**, 163.
58 C. S. Fadley, Hard X-ray Photoelectron Spectroscopy (HAXPES), in *Springer Series in Surface Science*, ed. J. C. Woicik, 2016, vol. 59, p. 1.
59 C. Weiland, A. K. Rumaiz, P. Pianetta and J. C. Woicik, *J. Vac. Sci. Technol., A*, 2016, **34**(3), 030801.
60 ISO, *ISO 18118:2015, Surface chemical analysis – Auger electron spectroscopy and X-ray photoelectron spectroscopy – Guide to the use of experimentally determined relative sensitivity factors for the quantitative analysis of homogeneous materials*, 2015.
61 D. A. Verner, G. J. Ferland, K. T. Korista and D. G. Yakovlev, *J. Surf. Anal.*, 1996, **465**, 487.
62 P.-M. Deleuze, E. Martinez, K. Artyushkova and O. Renault, *Surface Science Spectra*, 2022, **29**, 014003.
63 P.-M. Deleuze, E. Martinez, K. Artyushkova and O. Renault, *Surface Science Spectra*, 2022, **29**, 014004.
64 P.-M. Deleuze, E. Martinez, K. Artyushkova and O. Renault, *Surface Science Spectra*, 2022, **29**, 014006.
65 P.-M. Deleuze, E. Martinez, K. Artyushkova and O. Renault, *Surface Science Spectra*, 2022, **29**, 014007.
66 V. G. Yarzhemsky and M. B. Trzhaskovskaya, *At. Data Nucl. Data Tables*, 2021, **139**, 101387.
67 M. B. Trzhaskovskaya and V. K. Nikulin, *Phys. Rev. B: Condens. Matter Mater. Phys.*, 2007, **75**(17), 177104.
68 S. Tougaard, *J. Vac. Sci. Technol., A*, 1996, **14**(3), 1415.
69 S. Tougaard, *J. Electron Spectrosc. Relat. Phenom.*, 2010, **178–179**, 128.
70 S. Tougaard, *J. Vac. Sci. Technol., A*, 2021, **39**(1), 011201.
71 P. Risterucci, O. Renault, E. Martinez, B. Detlefs, V. Delaye, J. Zegenhagen, C. Gaumer, G. Grenet and S. Tougaard, *Appl. Phys. Lett.*, 2014, **104**(5), 051608.
72 P. Risterucci, O. Renault, C. Zborowski, D. Bertrand, A. Torres, J. P. Rueff, D. Ceolin, G. Grenet and S. Tougaard, *Appl. Surf. Sci.*, 2017, **402**, 78.
73 C. Zborowski, O. Renault, A. Torres, Y. Yamashita, G. Grenet and S. Tougaard, *Appl. Surf. Sci.*, 2018, **432**, 60.
74 C. Zborowski, O. Renault, A. Torres, C. Guedj, Y. Yamashita, S. Ueda, G. Grenet and S. Tougaard, *J. Appl. Phys.*, 2018, **124**(8), 085115.
75 B. F. Spencer, S. Maniyarasu, B. P. Reed, D. J. H. Cant, R. Ahumada-Lazo, A. G. Thomas, C. A. Muryn, M. Maschek, S. K. Eriksson, T. Wiell, T. L. Lee, S. Tougaard, A. G. Shard and W. R. Flavell, *Appl. Surf. Sci.*, 2021, **541**, 148635.
76 C. Zborowski and S. Tougaard, *Surf. Interface Anal.*, 2019, **51**(8), 857.
77 O. Renault, C. Zborowski, P. Risterucci, C. Wiemann, G. Grenet, C. M. Schneider and S. Tougaard, *Appl. Phys. Lett.*, 2016, **109**(1), 011602.
78 I. Gueye, G. Le Rhun, O. Renault, D. Cooper, D. Ceolin, J.-P. Rueff and N. Barrett, *Appl. Phys. Lett.*, 2017, **111**(3), 032906.
79 Y. S. Lin, F. Zeng, S. G. Tang, H. Y. Liu, C. Chen, S. Gao, Y. G. Wang and F. Pan, *J. Appl. Phys.*, 2013, **113**(6), 064510.

80 M. Sowinska, T. Bertaud, D. Walczyk, S. Thiess, P. Calka, A. Lambert, C. Walczyk and T. Schroeder, *J. Appl. Phys.*, 2014, **115**(20), 204509.
81 B. Meunier, E. Martinez, R. Rodriguez-Lamas, D. Pla, M. Burriel, M. Boudard, C. Jiménez, J.-P. Rueff and O. Renault, *J. Appl. Phys.*, 2019, **126**(22), 225302.
82 B. Meunier, E. Martinez, R. Rodriguez-Lamas, D. Pla, M. Burriel, C. Jimenez, Y. Yamashita and O. Renault, *ACS Appl. Electron. Mater.*, 2021, **3**(12), 5555–5562.
83 J. Shi, S. D. Ha, Y. Zhou, S. Frank and S. Ramanathan, *Nat. Commun.*, 2013, **4**(1), 2676.
84 K. Szot, W. Speier, G. Bihlmayer and R. Waser, *Nat. Mater.*, 2006, **5**(4), 312.
85 C. Moreno, C. Munuera, S. Valencia, F. Kronast, X. Obradors and C. Ocal, *Nano Lett.*, 2010, **10**(10), 3828; J. C. Gonzalez-Rosillo, R. Ortega-Hernandez, J. Jareño-Cerulla, E. Miranda, J. Suñe, X. Granados, X. Obradors, A. Palau and T. Puig, *J. Electroceram.*, 2017, **39**(1–4), 185.
86 R. Ortega-Hernandez, M. Coll, J. Gonzalez-Rosillo, A. Palau, X. Obradors, E. Miranda, T. Puig and J. Suñe, *Microelectron. Eng.*, 2015, **147**, 37.
87 J. S. Lee, S. Lee and T. W. Noh, *Appl. Phys. Rev.*, 2015, **2**(3), 031303; S. Bagdzevicius, K. Maas, M. Boudard and M. Burriel, *J. Electroceram.*, 2017, **39**(1–4), 157.
88 M. F. Sunding, K. Hadidi, S. Diplas, O. M. Løvvik, T. E. Norby and A. E. Gunnæs, *J. Electron Spectrosc. Relat. Phenom.*, 2011, **184**(7), 399.
89 B. Meunier, D. Pla, R. Rodriguez-Lamas, M. Boudard, O. Chaix-Pluchery, E. Martinez, N. Chevalier, C. Jiménez, M. Burriel and O. Renault, *ACS Appl. Electron. Mater.*, 2019, **1**(5), 675.
90 S. Tougaard, *J. Surf. Anal.*, 2017, **24**(2), 107.

Faraday Discussions

PAPER

Characterization of buried interfaces using Ga Kα hard X-ray photoelectron spectroscopy (HAXPES)†

B. F. Spencer, *[a] S. A. Church,[b] P. Thompson,[c] D. J. H. Cant,[d] S. Maniyarasu,[b] A. Theodosiou,[e] A. N. Jones,[e] M. J. Kappers,[f] D. J. Binks, [b] R. A. Oliver,[f] J. Higgins,[g] A. G. Thomas,[a] T. Thomson, [c] A. G. Shard [d] and W. R. Flavell *[b]

Received 28th January 2022, Accepted 28th February 2022
DOI: 10.1039/d2fd00021k

The extension of X-ray photoelectron spectroscopy (XPS) to measure layers and interfaces below the uppermost surface requires higher X-ray energies and electron energy analysers capable of measuring higher electron kinetic energies. This has been enabled at synchrotron radiation facilities and by using lab-based instruments which are now available with sufficient sensitivity for measurements to be performed on reasonable timescales. Here, we detail measurements on buried interfaces using a Ga Kα (9.25 keV) metal jet X-ray source and an EW4000 energy analyser (ScientaOmicron GmbH) in the Henry Royce Institute at the University of Manchester. Development of the technique has required the calculation of relative sensitivity factors (RSFs) to enable quantification analogous to Al Kα XPS, and here we provide further substantiation of the Ga Kα RSF library. Examples of buried interfaces include layers of memory and energy materials below top electrode layers, semiconductor heterostructures, ions implanted in graphite, oxide layers at metallic surfaces, and core–shell nanoparticles. The use of an angle-resolved mode enables depth profiling from the surface into the bulk, and is complemented with surface-sensitive XPS. Inelastic background modelling allows the

[a]*Henry Royce Institute, Photon Science Institute, Department of Materials, School of Natural Sciences, The University of Manchester, Manchester, M13 9PL, UK. E-mail: ben.spencer@manchester.ac.uk*

[b]*Henry Royce Institute, Photon Science Institute, Department of Physics and Astronomy, School of Natural Sciences, The University of Manchester, Manchester, M13 9PL, UK. E-mail: wendy.flavell@manchester.ac.uk*

[c]*Department of Computer Science, School of Engineering, The University of Manchester, Manchester, M13 9PL, UK*

[d]*Surface Technologies, Chemical and Biological Sciences Department, National Physical Laboratory, Hampton Road, Teddington, TW11 0LW, UK*

[e]*The Nuclear Graphite Research Group, The University of Manchester, Oxford Road, Manchester, M13 9PL, UK*

[f]*Department of Materials Science & Metallurgy, University of Cambridge, 27 Charles Babbage Road, Cambridge, CB3 0FS, UK*

[g]*Rolls Royce PLC, London, UK*

† Electronic supplementary information (ESI) available. See DOI: 10.1039/d2fd00021k

extraction of information about buried layers at depths up to 20 times the photoelectron inelastic mean free path.

1. Introduction

Hard X-ray photoelectron spectroscopy (HAXPES) extends the sampling depth of X-ray photoelectron spectroscopy (XPS) below the topmost surface for the detection of bulk-like materials and buried interfaces by increasing photoelectron kinetic energies and, thus, escape depths.[1] The recent development of laboratory-based HAXPES instruments is expected to lead to an accelerated uptake of this technique by a broad range of researchers; currently, commercial systems using monochromated Ag Lα (2.98 keV), Cr Kα (5.41 keV), and Ga Kα (9.25 keV) are available, with X-ray energies higher than those of traditional lab-based XPS systems, which typically use monochromated Al Kα X-ray sources (1.49 keV).[2] Here, we demonstrate measurements made using a Ga Kα HAXPES instrument, with the highest photon energy currently available from a commercial lab-based photoelectron spectrometer, manufactured by ScientaOmicron GmbH[3] in the Henry Royce Institute at the University of Manchester. These measurements include the detection of layers buried below 200 nm of organic material using inelastic background modelling,[4] where an angle-resolved mode of the EW4000 analyser was also used to obtain depth profiles from the maximum sampling depth towards the surface. Relative sensitivity factors (RSFs) were developed, including the important recent calculations of non-dipole contributions by Trzhaskovskaya and Yarzhemsky,[5,6] and these were verified using standard materials.[4] More recently, the instrument was used to depth profile through mixed-cation perovskite materials, where these RSFs were again utilised and, notably, HAXPES provided a greater sensitivity to the dilute elements Cs and Rb than can be achieved using XPS.[7] We have also combined HAXPES with other destructive depth-profiling techniques, where it was found that HAXPES is able to sample below the damage layer created by plasma etching in glow discharge optical emission spectroscopy (GDOES), a technique which has traditionally been challenging to calibrate.[8] This demonstrates that HAXPES is sensitive to the material below the altered layer on an ion-etched surface, and therefore offers measurement of the unmodified chemical composition prior to ion-induced damage and preferential sputtering. As all forms of etching (using monoatomic and more recent cluster sources) potentially alter material systems,[9] an important benefit of HAXPES analysis is to provide a minimally-destructive measurement of the material below the topmost surface.

Work to date has demonstrated that, by using primary photoelectron peaks in the elastic limit (where photoelectrons leave the sample for detection without loss of energy), Ga Kα HAXPES can increase the sampling depth, defined as three times the effective electron attenuation length, from a maximum of $\sim$10 nm (measuring C 1s in graphite using Al Kα X-rays, where the inelastic mean free path of electrons is calculated using the well-established TPP-2M formula) to $\sim$51 nm, approximating the effective attenuation length as the inelastic mean free path of electrons.[10] Use of the inelastic background has been demonstrated to further increase this depth-profiling capability, in enabling information to be extracted from a greater *inelastic* sampling depth (even when primary photoelectron peaks are no longer detected) of up to twenty times the effective electron attenuation length (potentially corresponding to more than hundreds of nm).[4]

In this paper, we describe the measurement of buried interfaces using core-level photoelectron peaks, angle-resolved measurements, and inelastic background analysis, through five case studies: $In_xGa_{1-x}N$/GaN heterostructures, probing Mn_xAl_{1-x} below a top capping layer, measuring below the surface oxide passivation layer in zirconium, Cs^+ ions implanted in graphite, and measuring through thicker shelled polymer core/shell nanoparticles. These case studies together demonstrate the potential for characterizing buried interfaces using laboratory-based HAXPES.

2. Experimental

The ScientaOmicron HAXPES spectrometer has been detailed previously, and the instrument at the Royce Institute is shown in Fig. 1.[4,11] In brief, the instrument is comprised of a Ga metal jet X-ray source (Excillum),[12] a bespoke monochromator, and an EW4000 electron energy analyser capable of measuring photoelectrons with kinetic energies up to 12 keV. The liquid metal jet provides a constantly refreshed anode material, enabling the use of a 70 kV electron gun at 250 W. The monochromator removes additional emission lines to isolate Ga Kα_1 X-rays (9.2517 keV photon energy),[13] as well as focussing the X-ray beam to *ca.* 50 × 50 μm at the sample position. This yields a relatively high X-ray flux in the laboratory environment, *ca.* three orders of magnitude greater than traditional lab sources for XPS, such as monochromated Al Kα sources (1.486 keV photon energy).[11] This,

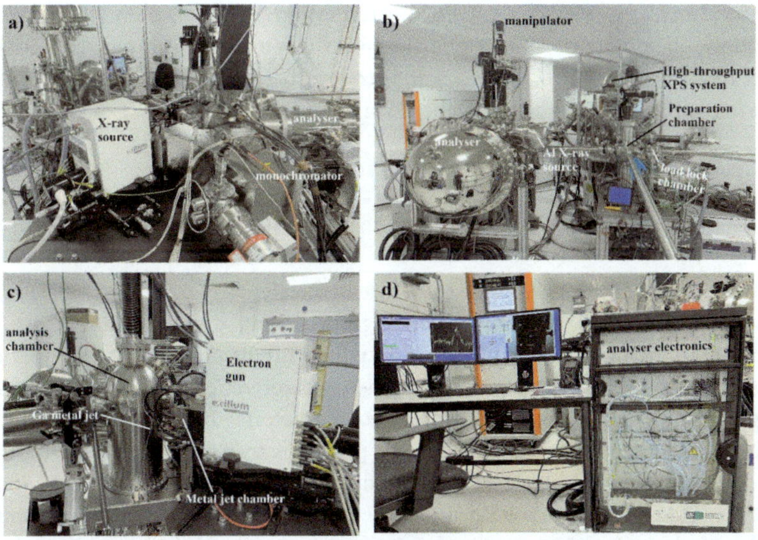

Fig. 1 Photographs of the HAXPES system (ScientaOmicron GmbH) at the Henry Royce Institute, (a) the X-ray source and monochromator, (b) the analyser, preparation chamber and separate XPS system (linked *via* ultra-high vacuum transfer lines), (c) the metal jet capillary and X-ray generation chamber, and (d) the analyser and instrument electronics racks.

crucially, enables HAXPES measurements to be undertaken on reasonable time-scales, analogous to spectrometers for XPS, overcoming sensitivity issues. The relative sensitivity to some photoelectron peaks (particularly for light elements) is heavily influenced by the photoionization cross section,[4] which in some cases drops by up to three orders of magnitude as the photon energy is increased from 1.5 to 9 keV.[6]

The metal jet, monochromator, and analysis chamber are isolated vacuum chambers; the base pressure in the analysis chamber is 1×10^{-10} mbar. Insulated material may be measured using an electron flood source (PREVAC) to replenish electrons at the surface, and differential charging may be avoided by floating the sample (detaching the sample manipulator from the ground). An Al Kα X-ray source is also mounted on the analysis chamber for the comparison of HAXPES with traditional surface-sensitive XPS. The HAXPES instrument is connected *via* ultra-high-vacuum transfer lines to a high-throughput ESCA2SR spectrometer with mon-oatomic Ar^+ ion etching (10 mA, 2–5 kV, FOCUS GmbH) and a cluster-ion-etching source (GCIB-10S, Ionopktica), as well as an ultra-violet source for UPS measurements (typically generating He II photons at 40.81 eV energy),[14] and a preparation chamber with a variety of ports available for additional instrumentation.

The EW4000 analyser may be operated in transmission mode or using angle-resolved modes, AR45 (with a total angular range of 45°), AR56 (total angular range of 56°), or AR60 (total angular range of 60°). The analyser entrance slit has a large acceptable angle of 60°, and the angular modes are used to transmit photoelectrons emitted at each angle directly to the 2D detector after they pass through the entrance slit, whereby the *y*-axis on the charge-coupled device detector image becomes the photoemission angle (which may then be converted to sampling depth).[4] This enables angle-resolved information to be captured in one shot as the sample is held in one position. The sampling depth, d_S, is defined as three times the electron effective attenuation length, which in the case of negligible elastic scattering is equivalent to three times the electron inelastic mean free path (IMFP, λ_i), and is reduced with increasing the photoemission angle, ϑ, with respect to the surface normal, according to[4]

$$d_S = 3\lambda_i \cos\vartheta, \tag{1}$$

where a reduction in sampling depth by 50% is achieved at a photoemission angle of 60°.

The angle-resolved mode may therefore be particularly useful for samples with lateral microstructure (such as patterning), or if homogeneity is a concern, as it avoids small changes in sample position which might occur as the sample tilt is adjusted. A sputter-cleaned polycrystalline gold reference sample was measured using the angle-resolved mode in order to characterise both the relative detector response and the natural decline in the photoelectron intensity with increasing photoemission angle (extracted *via* peak fitting the Au $3d_{5/2}$ peak at each angle in the 2D angle-resolved data with a simple Gaussian), so that normalized angular profiles of photoelectron intensity (again extracted *via* peak fitting a photoelectron peak) may be generated by dividing the measured angle-resolved photoelectron intensity by the Au $3d_{5/2}$ reference function. Angle-resolved measurements, however, may be limited by the roughness of the sample, as well as the amount of elastic scattering.[4]

Energy resolution down to approximately 0.5 eV is achievable, as measured using the Fermi edge on a gold reference sample,[3] and the resolution and count rate may be adjusted with a combination of entrance slit width adjustment and the choice of analyser pass energy.[3] For the data presented here, survey spectra were measured using a 1.5 mm slit width and 500 eV pass energy. High resolution spectra were acquired with a 1.5 mm slit width and 100 eV pass energy, a 0.8 mm slit with 200 eV pass energy, or a 0.3 mm slit width and 500 eV pass energy (which all achieve a similar resolution with differing count rates). In transmission mode, the complete binding energy (BE) range is measurable (that is, up to kinetic energies ~9.25 keV) at all pass energies, however, for the angle-resolved modes, the upper kinetic energy range is limited by the pass energy; for the complete BE range to be measured, 500 eV pass energy is required.

3. Results and discussion

3.1. InGaN/GaN heterostructures

$In_xGa_{1-x}N$/GaN quantum well (QW) heterostructures are important material systems for light emitting diodes.[15] Their optical properties are influenced by competing radiative and non-radiative recombination mechanisms; these are affected by strong internal electric fields in these structures (up to MV cm^{-1}).[16] The optical properties of these heterostructures may be altered by engineering the material structure, such as variations in the In content of the $In_xGa_{1-x}N$ QWs, the width of the QWs, the width of the GaN capping layer, and the inclusion of a doped underlayer.[17] Indeed, research into this material system continues because the internal quantum efficiency drops significantly with increasing emission wavelength (the 'green gap'),[17] a process which has been attributed to intrinsic field effects, alloy fluctuations and nonradiative defects.[18,19]

The internal electric fields lead to shifts in the electronic bands (band bending) at the interfaces, because the Fermi level is pinned at the surface, and therefore a measurement of band bending in a depth-resolving fashion would provide useful insights into the electronic structure within the stack, which is a crucial factor in determining device performance. XPS has been used to measure changes in the band bending at the surface with laser photoexcitation, by measuring changes in the binding energy of elemental core levels after carrier creation with time-resolution and chemical specificity.[20–23] Gaining depth resolution in such measurements for depletion region widths in excess of typical XPS sampling depths requires an extension of the sampling depth enabled by HAXPES, as demonstrated on GaN by Ueda using 5.95 keV X-rays at the BL15XU end station at SPring-8.[24] Similarly, Narita *et al.* measured changes in the band bending at the GaN surface after etching-induced damage, using 7.94 keV X-rays at the BL46XU end station.[25] In both studies, the photoemission angle, ϑ, was varied to reduce the sampling depth by cos ϑ, according to eqn (1).

Here, we demonstrate HAXPES measurements using the 9.25 keV laboratory source, on a set of $In_xGa_{1-x}N$/GaN heterostructures (a QW typically 3 nm thick, with varying In content, on top of 2.5 μm GaN, with a 7.5 nm GaN capping layer) on (0001) sapphire substrates, prepared by metal–organic chemical vapor deposition and characterised using high-resolution transmission electron microscopy and electron energy loss spectroscopy.[26,27] Fig. 2 shows normal emission HAXPES (Fig. 2(a)) and Al Kα XPS survey spectra (Fig. 2(b)) for a typical QW structure, as

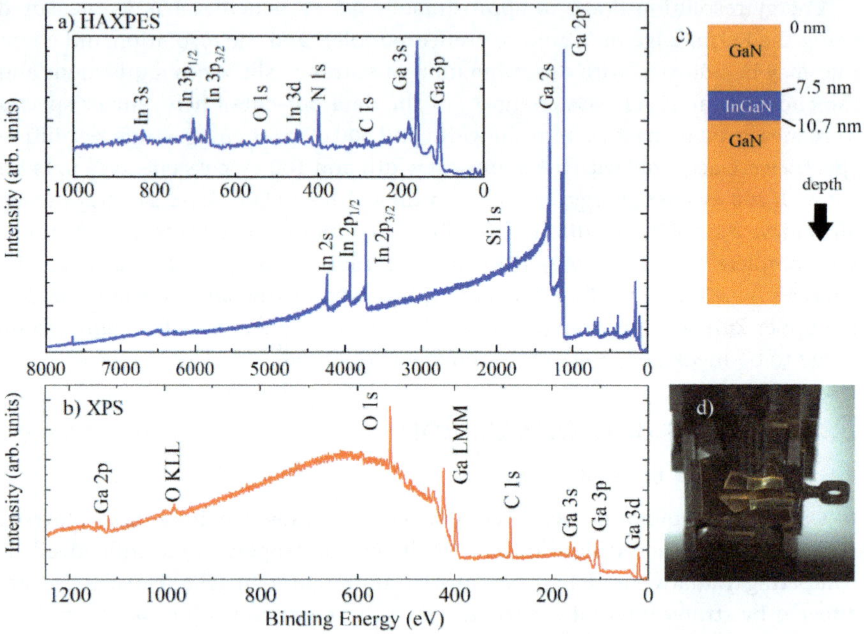

Fig. 2 (a) HAXPES (with the low BE region shown in the inset) and (b) Al Kα XPS survey spectra for an InGaN/GaN heterostructure ($In_{0.19}Ga_{0.81}N$, 3.2 nm thick), as shown in the schematic in (c). (d) Photograph of samples on sample plate on the manipulator.

illustrated in Fig. 2(c), containing 19% In ($In_{0.19}Ga_{0.81}N$) with a 3.2 nm thick QW. XPS measurements were taken at normal emission, and HAXPES using grazing incidence with a photoemission angle of 4° in transmission mode (Fig. 2(d)). For Al Kα XPS, there is an unfortunate overlap of the N 1s and Ga LMM signals, which is removed at different photon energies and has also recently been demonstrated using a Cr Kα laboratory HAXPES system.[28] No peaks associated with In are observed using XPS, which is dominated by surface contamination containing carbon and oxygen, whereas In 3s, p, d are observed with HAXPES, along with the deeper core levels In 2s and 2p.

HAXPES was performed on a set of QWs with nominal $In_xGa_{(1-x)}N$ compositions of $x = 0.05$, 0.15, 0.19, and 0.25, with QW widths of 2.8 nm, 2.9 nm, 3.2 nm and 3.3 nm, respectively, all with 7.5 nm GaN capping layers. Because the samples were prepared on insulating sapphire substrates, the electron flood source charge neutraliser was required to remove charging under the X-rays. Energy referencing of the HAXPES spectra was performed using the Ga $2p_{3/2}$ peak at 1117.8 eV, expected for GaN, and measured with XPS after energy referencing using C 1s at 284.8 eV. This gave BE positions for In $3d_{5/2}$ at ~444 eV and N 1s at ~397 eV, also as expected.[29] The use of C 1s for charge referencing in HAXPES (when it is measurable) is not recommended due to non-negligible recoil effects for light atoms at high photoelectron kinetic energies.[30]

Calculations of atomic concentrations were performed using established relative sensitivity factors (RSFs) for the Ga Kα instrument,[4] which have now been refined and expanded by Cant *et al.* into an important new tool which allows

Table 1 Atomic concentrations (equivalent homogeneous concentrations) of Ga, N, and In, measured using the In 2p (left hand side) and In 3d (right hand side) core levels for In. Using In 3d, the stoichiometry of $In_xGa_{(1-x)}N$ was calculated using a layer calculation (final column)

	Measured using In 2p			Measured using In 3d			
Nominal x for $In_xGa_{(1-x)}N$	Ga at%, ±0.3 at%	N at%, ±1 at%	In at%, ±0.1 at%	Ga at%, ±0.3 at%	N at%, ±1 at%	In at%, ±0.1 at%	Calculated x for $In_xGa_{(1-x)}N$
0.05	50.8	49	0.3	50.7	49	0.5	0.08 ± 0.02
0.15	55.2	44	0.5	54.9	44	1.1	0.15 ± 0.03
0.19	55.3	44	0.7	55.0	43	1.3	0.17 ± 0.03
0.25	53.2	46	0.9	52.5	45	2.0	0.26 ± 0.04

researchers to calculate RSFs for all atomic core levels measured with X-ray photon energies in the range 1.5–10 keV and in any instrument geometry.[31] Calculations were carried out using both the In $2p_{3/2}$ core level and the In 3d core level, as shown in Table 1, using high resolution core level spectra obtained at grazing incidence in transmission mode. Errors were estimated using relative changes in at% calculated using a range of acceptable background fits (as judged by eye), as well as the associated RSF.

As expected, the measured In content, calculated as the equivalent homogeneous concentration, increases with known In concentration within the QW and buried layer, and in all cases is much lower than the nominal value because of attenuation through the top GaN capping layer. The kinetic energy of In 3d photoelectrons is more than 3.3 keV greater than that of In 2p photoelectrons, and thus is representative of a greater sampling depth. Calculated IMFPs for GaN are approx. 7 nm and 11 nm for In $2p_{3/2}$ and In $3d_{5/2}$ respectively,[10] meaning that sampling depths are >20 and >30 nm, respectively (eqn (1)). Table 1 shows that the In content calculated using In 3d is approximately double that calculated using In 2p. Analysis using the straight line approximation (detailed in the ESI†),[32–34] the known structure (Fig. 2(c)) and the measured In 3d and Ga 2p intensities generates QW compositions in agreement with the nominal concentrations, as shown in Table 1. The In content measured from In 2p is lower than that measured from In 3d, primarily because of differential attenuation through the contamination overlayer, which is evident in the XPS spectrum. Surface contamination was therefore a concern, and so, prior to angle-resolved measurements, the samples were sputter etched in a vacuum using monoatomic Ar^+ ions (2 kV, 10 mA emission, with a large beam size of ~1 cm for 5 minutes). XPS showed reduction of the C and O content by >90% without Ar implantation (evidenced by a lack of signal from Ar 2p); HAXPES then showed no signal from O or C. It is known that HAXPES is not sensitive to an ion-induced damage layer at the surface.[8]

Angle-resolved measurements were performed on two of the heterostructures with In stoichiometries of $x = 0.15$ and $x = 0.19$ (QW widths of 2.9 and 3.2 nm respectively). Measurements were carried out using the AR56 mode of the EW4000 analyser, where the usable angular range (to neglect any slit edge effects) is approximately 40°. The sample was tilted so that the angle between the analyser entrance plane and the surface normal was 40°, meaning that photoemission angles in the range 20–60° were captured in a single shot.[4] Minimal depth-resolved information is gained in the range 0–20°; grazing incidence measurements measured at a photoemission angle of 4° provide an additional measurement at the maximum sampling depth.

Fig. 3 shows 2D angle-resolved measurements (photoelectron intensity at each BE plotted against photoemission angle) of Ga 2p, N 1s, In $2p_{3/2}$ for a $In_{0.19}Ga_{0.81}N$ QW. The Au $3d_{5/2}$ image shows the drop in photoelectron counts with a higher photoemission angle, as expected. Higher photoemission angles provide a smaller sampling depth (at 60°, a reduction of the maximum sampling depth by a factor of 0.5). Even though these raw data are influenced by the gradual information depth change, there is clearly a greater intensity of In at a lower photoemission angle (*i.e.* deeper in the sample) because the In atoms are buried. There is additional structure to the Ga 2p and N 1s intensity variation with angle that is not apparent in the Au signal. This is likely to be due to elastic scattering effects, such as diffraction, as expected for these single crystal materials.[35,36] The angular mode of the analyser has allowed these effects to become evident, whereas measurements in the transmission mode of the EW4000 analyser do not.

The 2D data may be converted to a set of spectra at different angles by summing ranges of angles, as previously demonstrated.[4] Here, each spectrum at every angle (>260 angles across the useful 40° range) was peak fitted with a single Gaussian, after a simple linear background subtraction across a ~10 eV binding energy range across the photoelectron peak (judged as satisfactory given the signal-to-noise level of the individual spectra) to extract the BE and peak intensity for the N 1s, Ga $2p_{3/2}$, In $2p_{3/2}$ and Au $3d_{5/2}$ photoelectron peaks, and plots of intensity *vs.* angle and BE *vs.* angle were compiled. The N, Ga, and In intensity plots were then normalized using the response from the gold reference to remove the natural loss of photoelectron signal with angle, as well as the analyser/detector response.

Fig. 4 shows the BE position, which is defined as the mean of the peak Gaussian fit, of Ga $2p_{3/2}$, N 1s and In $2p_{3/2}$ for these two heterostructures as a function of the photoemission angle, where a greater angle indicates a lower

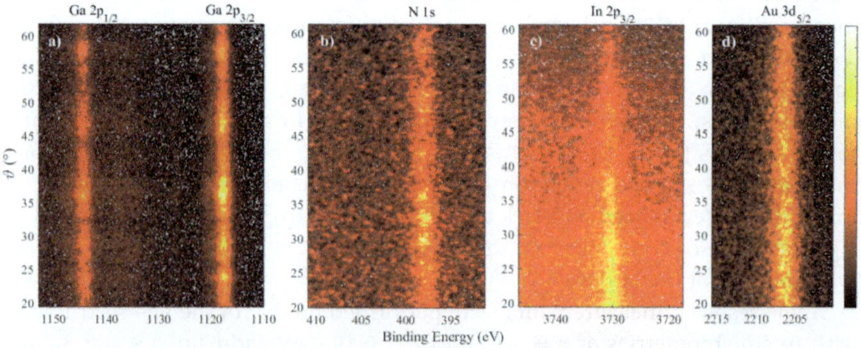

Fig. 3 2D angle-resolved data for an InGaN/GaN heterostructure (In fraction of 19%) and a gold reference sample, measured using the AR56 detection mode on the EW4000 analyser. (a) Ga 2p, (b) N 1s, (c) In $2p_{3/2}$, and (d) Au $3d_{5/2}$ from the reference sample. Increasing photoemission angle indicates increasing surface sensitivity. The intensity colour bar (arb. units) is shown on the right-hand side.

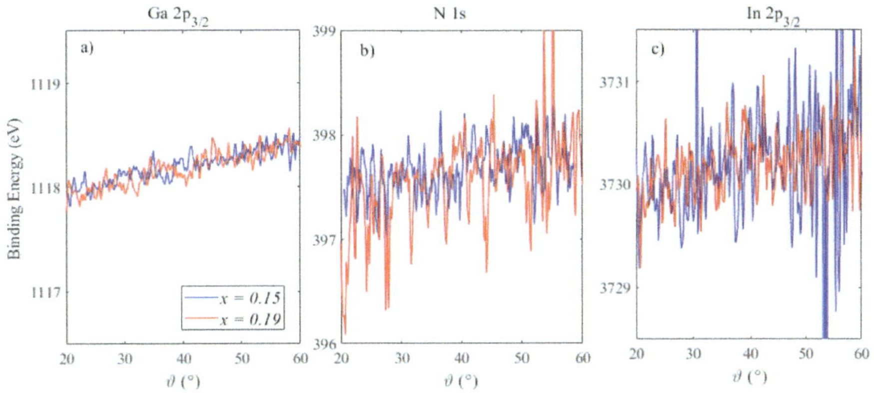

Fig. 4 Angle resolved HAXPES measurements of the binding energy peak positions of (a) Ga $2p_{3/2}$, (b) N 1s and (c) In $2p_{3/2}$ in two InGaN/GaN heterostructures containing 15% In (blue lines) and 19% In (red lines). Increasing the photoemission angle provides greater surface sensitivity. The binding energy scale ranges have been fixed for comparison.

sampling depth. The Ga $2p_{3/2}$ peak shifts to a higher BE by ~0.7 eV in both samples, indicative of the surface band bending in the n-type GaN capping layer.[24] The magnitude of this BE shift is relatively large over the sampling depths probed here, but this is in agreement with previous work and reflects the relatively high electric fields present in these structures.[24,25] The N 1s BE position shows a slightly reduced BE shift, however, the relatively low cross section of N 1s compared to Ga 2p means that the signal-to-noise level, even after extracting the parameters *via* peak fitting, is poor. The slopes of these BE variations were extracted, and found to be similar for Ga $2p_{3/2}$ (12 and 13 meV $^{\circ -1}$ for $x = 0.15$ and 0.19, respectively) and N 1s (11 and 18 meV $^{\circ -1}$ for $x = 0.15$ and 0.19, respectively), with a much larger associated error for N 1s. The In $2p_{3/2}$ peak may also show some shift in BE position, as much as Ga and N (~11 meV $^{\circ -1}$). However the signal-to-noise level deteriorates with increasing photoemission angle, as less In is detected from the buried layer as the sampling depth is reduced, meaning it is difficult to draw conclusions from photoemission angles above ~40°. The full-width at half maximum (FWHM) of the fitted peaks was also found to increase with photoemission angle; for Ga $2p_{3/2}$ the increase was from ~1.3 eV at 20° to ~1.5 eV at 60°, and the other peaks increased by a similar amount (15%). We have therefore assessed band bending through the GaN capping layer and QW and, although the equilibrium band bending will be larger than the measured BE shifts (as the depletion layer is not fully probed), the measurements demonstrate that band bending within a structure may be measured using angle-resolved HAXPES. Further work in this area will include using laser photoexcitation, delivered into the ultra-high-vacuum chamber *via* calcium fluoride windows for high ultra-violet transmission, to create carriers in the QW as changes to the band bending are monitored.

The photoelectron peak intensities (the area below the Gaussian peak fit to the spectra), normalized using the gold reference sample response, for Ga $2p_{3/2}$, N 1s and In $2p_{3/2}$ are shown in Fig. 5. Undulations are particularly clear for Ga $2p_{3/2}$, as observed in the 2D plot in Fig. 3, and these are not removed after normalising

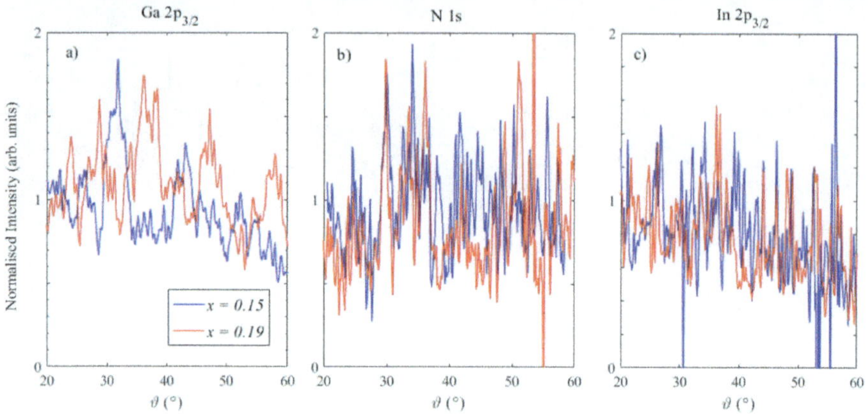

Fig. 5 Angle-resolved HAXPES photoelectron peak intensities (normalized using a gold reference sample) for (a) Ga $2p_{3/2}$, (b) N 1s and (c) In $2p_{3/2}$ in two InGaN/GaN heterostructures containing 15% In (blue lines) and 19% In (red lines). A higher photoemission angle indicates a lower sampling depth.

using the gold reference response, indicating that they arise from elastic scattering effects, including photoelectron diffraction.[37–39] The structure of the normalised Ga $2p_{3/2}$ intensities is different in the two samples, which suggests that the azimuthal angles with respect to the crystal planes were different (as expected, since the samples were arbitrarily mounted on the sample plate). Further development may enable the extraction of structural information: for instance, a manipulator with azimuthal rotation could enable diffractograms to be generated which can then be modelled using multiple scattering cluster (MSC) simulations.[37–39] Indeed, it is well known that using higher photoelectron kinetic energies (>500 eV) simplifies the analysis of diffractograms due to the so-called "forward focussing" effect, in which emission is enhanced along densely packed atomic planes and rows of atoms corresponding to low-index crystal directions.[38,40]

Disregarding these scattering effects, Ga and N core levels do not show any notable trend in intensity with photoemission angle. The In $2p_{3/2}$ intensity plots, in contrast, show a reduction at higher photoemission angles, as expected for a buried layer. These angle-dependent intensity plots may also be used to extract information about the depth distribution of the elements through modelling in software packages such as QUASES-ARXPS.[41–43] Analysis of the In angle-resolved data again indicated that some additional contamination was present at the surface, and also that the GaN capping layer was not fully covering the QW structure below, which is expected due to the potential removal of some of the capping layer with ion etching, as well as the presence of V-shaped pits on the surface.

3.2. Characterisation of thin films with capping layers: MnAl for spintronics

Extending the development of spintronics to new devices will require discovery and innovation in magnetic thin film materials.[44] The requirements are exacting because the materials must possess high magnetisation and Curie temperatures, very high magnetic anisotropy and, for applications involving magnetisation dynamics, a very low damping parameter. Creating materials with these desirable properties requires a deep understanding of physical properties, including composition, crystal and grain structure, and film thickness.[44] These then inform the understanding of magnetic, electrical and dynamic properties. Of the potential candidate materials for future spintronic technologies, binary alloys with an $L1_0$ structure show considerable promise. The $L1_0$ structure is a distorted fcc structure, so-called face centred tetragonal (fct), where alternating layers of the two constituent atoms form the desired structure, leading to a large magneto-crystalline anisotropy along the tetragonal axis. As alternating layers of atoms are a prerequisite, it follows that $L1_0$ binary alloys can only form for equiatomic compositions. A number of equiatomic binary alloys are known to form the $L1_0$ structure (FePt, FePd, NiFe, MnAl, CoPt),[45] of which the most well-known is FePt.[46]

One $L1_0$ binary alloy with potential for spintronic applications is MnAl. Neither Mn or Al show ferromagnetic ordering, however, when combined under the correct thermodynamic conditions (deposition and annealing temperatures) and an appropriate substrate template, the alloy provides magnetic properties similar to those of FePt.[47,48] Such properties are normally only found in materials containing 4f elements such as Pt. In the case of MnAl, these desirable magnetic properties are only obtained over a small (~10%) atomic composition window due to a complex phase diagram, hence a detailed understanding of composition is essential in developing these thin film materials.[44] In recent years, there have been relatively few reports on the fabrication of thin film MnAl due to the difficulties in creating the appropriate thermodynamic conditions and ensuring the correct stoichiometry. Here, we show the capability of HAXPES to provide key compositional information in films that are only 50 nm thick. The thin films were fabricated by magnetron sputtering, co-deposited from elemental Mn and Al targets, to understand phase formation of a ~50 nm MnAl film. To prevent contamination and oxidation, a 5 nm capping layer of Ta was deposited on top of the film. The film and capping layer thicknesses are routinely measured using X-ray reflection measurements, and standard protocols have been developed to enable consistent depositions.

Although the capping layer is very thin, it is thick enough to prevent characterisation with traditional Al Kα XPS, where the sampling depth (defined as three times the inelastic mean free path of electrons) for the Mn 2p core level is *ca.* 4.5 nm in Ta,[10] meaning a 5 nm capping layer is sufficient to prevent sampling of the MnAl material below. The MnAl films are themselves thin at ~50 nm, which in turn limits characterisation with other techniques, such as inductively-coupled plasma (ICP) etching[49] and bulk-sensitive X-ray fluorescence.[50] HAXPES with Ga Kα X-rays increases the XPS sampling depth to 27.9 nm for Ta using Mn 2p,[10] and the equivalent sampling depth through MnAl is estimated to be similar (33 nm),[10] meaning that the MnAl layer is easily measured. The sampling depth does not exceed the total film thickness, hence the calculated relative sensitivity factors for each core level,[4,31] assuming a 'thick' sample (thickness > sampling depth), may be used to accurately measure the atomic concentrations of Mn and Al in the thin film. For this material system, HAXPES is therefore an ideal probe of the atomic makeup of the thin films beneath a capping layer.

A set of MnAl thin films was fabricated, as detailed above, where the power applied to each target was varied, as detailed in Table 2. This set of spectra for one

Table 2 Sample details listing the Mn and Al target powers used for fabrication, the expected Mn atomic composition calculated from the ratio of the target powers, and the measured Mn composition, both using the Mn 1s and Al 1s core levels, and the Mn 2p and Al 2s core levels for comparison

Sample	Mn power (W)	Al power (W)	Expected Mn composition (at%)	Mn at%, Mn 1s, Al 1s, ±0.2 at%	Mn at%, Mn 2p, Al 2s, ±0.6 at%
A	40	40	50.00	50.1	51.5
B	40	55	42.11	27.0	38.8
C	40	80	33.33	33.4	27.8
D	30	30	50.00	53.4	49.8[a]
E	30	45	40.00	40.3	46.8[a]
F	30	60	33.33	31.5	25.2[a]
G	20	25	44.44	47.8	37.3[a]
H	20	35	36.36	25.9	24.8[a]
I	20	45	30.77	34.7	29.0

[a] For samples D–H, Mn 2p and Al 2s were not measured with high resolution; the intensity was instead extracted from the survey spectra, and so the error associated with these compositions is significantly greater (estimated to be ~1.0–2.0 at%)

sample required an accumulation time of *ca.* 2.5 hours, demonstrating that the flux of the Ga Kα X-ray source is sufficient to enable measurements with similar throughput times to those of traditional Al Kα laboratory instruments. The survey spectrum in Fig. 6(a) shows that, while the intensity of Ta photoelectron peaks is still relatively high, the Mn and Al peaks are easily measurable. These conductive samples did not require the use of a charge neutraliser and, hence, energy calibration was performed using a gold reference sample. The Mn $2p_{3/2}$ peak is at BE 638.5 eV, as expected for the metallic phase of Mn,[51] as well as MnAl alloys,[52] and the spectra indicate little (if any) oxidation, as expected when the film has been capped with Ta. The Mn 1s peak is at 6537.5 eV BE. The Al 2p peak is obscured by Ta 5s, hence the Al 2s peak was measured, and was found to be at 117.0 eV BE, also as expected for metal phase Al,[53] with the Al 1s peak position at 1558.8 eV, again as expected for metal phase Al, in agreement with measurements by Castle *et al.* using Si Kα X-rays.[54]

Now that HAXPES technology has developed, particularly in the laboratory environment where a variety of X-ray sources are available,[2] an update of reference data will increase confidence in the chemical states identified using deeper core levels. The Ta $3d_{5/2}$ photoelectron peak is present at 1731.8 eV BE, and is in agreement with the literature.[55] Here, however, an additional peak is present at an approx. 4.4 eV higher BE and assigned to the oxide, accounting for *ca.* 16% of the signal. As Ta readily oxidises, this is expected given that the Ta capping layer was exposed to the atmosphere. This chemical shift is similar to that observed for Ta $4f_{7/2}$ in the metallic phase and in Ta_2O_5 using XPS.[56,57] While BE positions for deeper core levels than those used in XPS (BE > 1500 eV) are less widely available, it is anticipated that chemical shifts may generally be consistent with those from traditional XPS measurements. Additional peaks across the range 3500–4200 eV BE in the survey spectrum are assigned to Mn Auger peaks, but again, a definitive assignment from the literature is lacking as they occur at relatively high kinetic

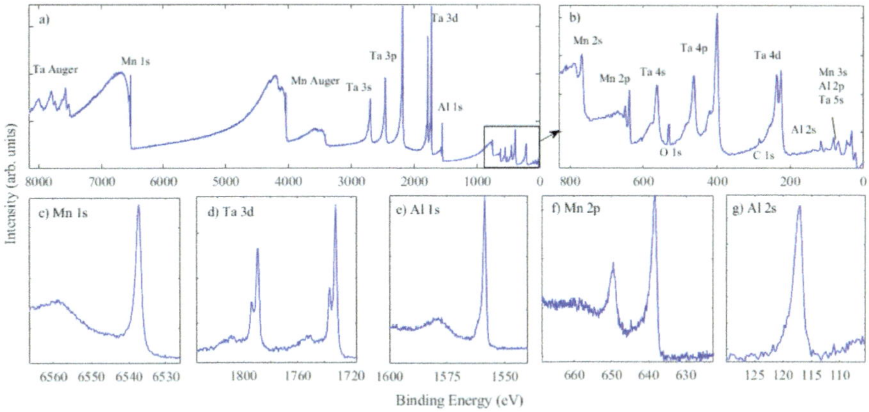

Fig. 6 (a) HAXPES survey spectrum, with the inset of (a) shown in (b) showing the low binding energy region, and high resolution spectra of (c) Mn 1s, (d) Ta 3d, (e) Al 1s, (f) Mn 2p and (g) Al 2s from a co-deposited MnAl thin film capped with a *ca.* 5 nm Ta surface layer. Measurements were performed with grazing incidence in transmission mode.

energies (>5 keV). Measurement of these Auger peak positions, again with further reference data, may also enable chemical state information to be extracted using HAXPES, using the Auger parameter, as for XPS,[58,59] and as recently demonstrated using HAXPES.[60]

Table 2 shows that there is generally good agreement between the measured composition and the expected composition (based on the ratio of the Mn and Al target powers, anticipating a linear relationship between the composition and target). Errors were estimated using the calculated relative changes in at%, as previously detailed in section 3.1 of the Results.

Two notable discrepancies are samples B and H, which also show relatively high inelastic backgrounds (*i.e.*, lower intensity primary photoelectron peaks), particularly for Mn 1s, as shown in Fig. 7. The survey spectra for these two samples show a higher amount of Ta than those of the other samples, meaning the capping layer was thicker in these samples. The Mn 1s photoelectrons have a lower kinetic energy than Al 1s (and are thus associated with a lower sampling depth); it may be in these cases the use of Mn 1s is not appropriate, and indeed for sample B, the measured composition using Mn 1s and Al 1s core levels is 27% Mn, much lower than the expected 42%. As can be seen from Table 2, use of the Mn 2p and Al 2s signals (at much higher and similar kinetic energies) gives a composition of 39% Mn, which is not significantly different from the expected value. Sample H, however, shows a similar Mn composition measured with both sets of core levels (~25%), lower than the expected value of 36%. These data were extracted using the survey spectrum, with a much greater associated error. Overall, for the majority of samples with thinner capping layers, there is good agreement between the Mn compositions extracted using the two sets of core levels at higher and lower binding energies, which adds to the confidence in the calculated RSFs.

These results demonstrate the capability to measure material composition and chemical states below top electrode layers in electrical devices. They also demonstrate that capping vacuum-deposited materials with a metal layer at the

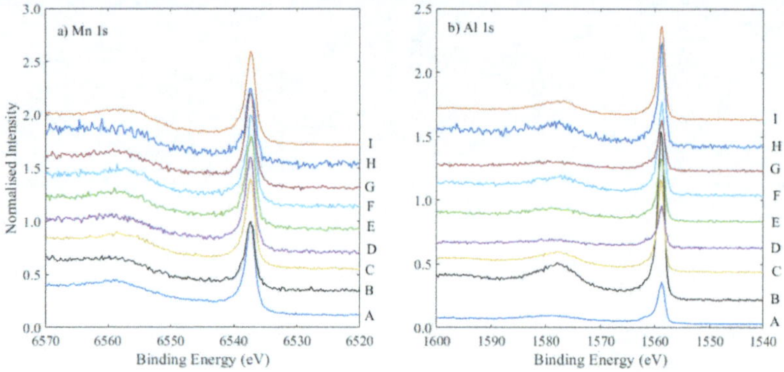

Fig. 7 (a) Mn 1s, and (b) Al 1s spectra, normalised to the Mn 1s peak intensity and offset on the y-axis, for a set of samples A–I, as labelled, where the powers of the Mn and Al targets were varied, as listed in Table 2.

surface can prevent atmospheric exposure without prohibiting the analysis of the material of interest *via* photoelectron spectroscopy with hard X-rays.

3.3. The metal oxide interface in zirconium

The surface of zirconium, like many metals, exhibits an oxide passivation layer. Zirconium dioxide itself has been extensively studied for applications in fuel cells, gas sensors, as a coating, and in metal-oxide semiconductor devices.[61] Metallic Zr can readily form a surface oxide layer even at low temperatures and in relatively low oxygen environments,[62] therefore the characterisation of the electronic properties of zirconium and its oxides is important for a number of sectors. Here, we examine a piece of untreated Zr metal, left exposed to the atmosphere for a prolonged period, measured using surface sensitive XPS and HAXPES.

Fig. 8 shows the survey spectra of untreated Zr metal measured using HAXPES and XPS, with the inset showing the Zr LMM Auger features at ~1850 eV kinetic energy and an enhanced view of the region below 1200 eV BE. The relative changes in the photoionisation cross sections are clear here, shown by the relative intensity of Zr 3d compared with the Zr 3p and 3s intensities in the XPS and HAXPES spectra: the relative intensity of the Zr 3d photoelectron line is diminished in the HAXPES spectrum due to a relative decrease in the photoionization cross section. The HAXPES survey spectrum shows much greater sensitivity to deeper core levels; the intensities of Zr 2s and 2p are an order of magnitude greater than those of 3s, 3p, and 3d.

Fig. 9 shows the high resolution spectra of Zr 3d, Zr 2p, O 1s and the valence band regions of the untreated Zr sample. As expected, the surface sensitive XPS spectrum of Zr 3d shows one spin–orbit split chemical species, with Zr $3d_{5/2}$ at 183.5 eV BE associated with ZrO_2.[62] HAXPES, however, shows two species; a second doublet is present at *ca.* 4.4 eV lower binding energy (Zr $3d_{5/2}$ at 179.1 eV), associated with metallic Zr.[62] Similarly, the Zr $2p_{3/2}$ peak is comprised of two chemical species (the doublet pair is not shown as the spin–orbit splitting of Zr 2p is relatively large). The BEs of the metallic and oxide species are 2221.9 and 2226.2 eV, respectively, meaning a chemical shift of 4.3 eV, similar to the Zr 3d

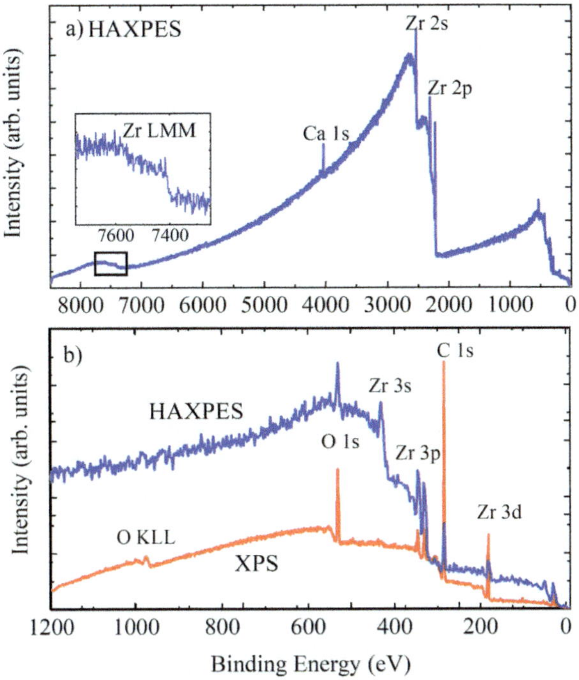

Fig. 8 (a) HAXPES survey spectrum of an untreated piece of Zr metal (inset shows the Zr LMM Auger region), and (b) overlay of HAXPES (blue line) and XPS survey (red line) spectra up to 1200 eV binding energy.

chemical shift. References for deeper core levels such as Zr $2p_{3/2}$ are very few, but the metallic BE position measured here is the same within error as that measured by Huschka *et al.* using Ag Lα and Ti Kα X-rays (not monochromated) on a Kratos

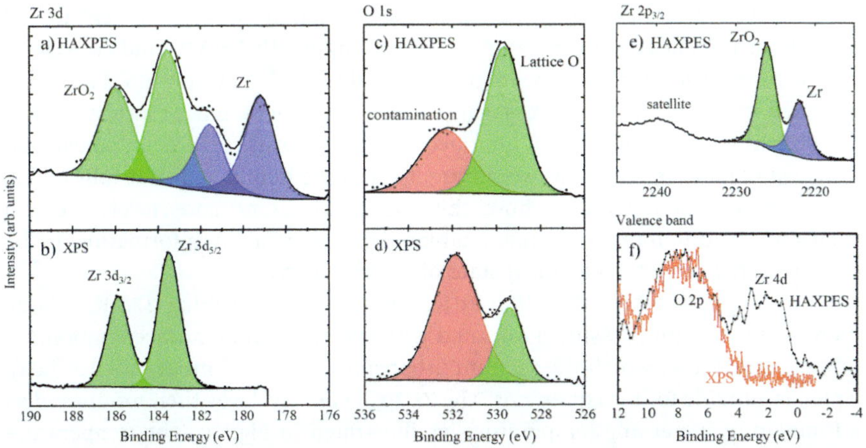

Fig. 9 Zr 3d spectra of an untreated piece of Zr metal, measured using (a) HAXPES (9.25 keV) and (b) XPS (1.486 keV), O 1s spectra measured with (c) HAXPES and (d) XPS, (e) Zr $2p_{3/2}$ spectra measured using HAXPES, and (f) valence band spectra measured using HAXPES (black points and line) and XPS (red points and line).

XSAM 800 spectrometer from Zr foil, cleaned *in situ* by scratching (2222.2 ± 0.4 eV).[55] The O 1s spectra show a greater proportion of lattice oxygen (measured at a lower binding energy, here ~529 eV) when measured with HAXPES, because there is lower sensitivity to oxygen-containing contamination (such as hydroxides and oxidised carbon species) at the surface. Finally, the valence band spectra show that HAXPES, sampling the metallic Zr below the oxide passivation layer, measures the Zr 4d core level at a low binding energy.[61,63]

The measured ratio of metallic to oxidised zirconium changes depending on whether Zr 3d or 2p photoelectrons are used, due to a change in the sampling depth of the photoelectrons, as the kinetic energy difference is *ca.* 2 keV. According to the TPP-2M formula,[64] and assuming that the majority of the photoelectrons are travelling through ZrO_2, the sampling depths for Zr 3d and 2p are 32 and 26 nm, respectively (calculated using eqn (1)). In comparison, XPS using Zr 3d has a calculated sampling depth of 7 nm. A simple calculation may be made to obtain the oxide layer thickness, using the ratio of metallic : oxide Zr 3d or $2p_{3/2}$ signals, an assumed stoichiometry of the oxide layer (Zr : O = 1 : 2), and assuming both a uniform and abrupt overlayer and identical shake-up intensity for all core-level peaks. The Strohmeier equation gives the relation for the oxide layer thickness, *d*:[65]

$$d = \lambda_{\mathrm{ox}} \cos\vartheta \ln\left[\frac{N_{\mathrm{m}}}{N_{\mathrm{ox}}}\frac{\lambda_{\mathrm{m}}}{\lambda_{\mathrm{ox}}}\frac{I_{\mathrm{ox}}}{I_{\mathrm{m}}} + 1\right], \tag{2}$$

where λ_{ox} and λ_{m} are the inelastic mean free paths (IMFP) of photoelectrons through the oxide and metal, respectively, and $N_{\mathrm{m}}/N_{\mathrm{ox}}$ is the ratio of the number density of metal atoms in the metal and oxide layers. ϑ is the photoelectron angle with respect to the surface normal (0° for normal emission), and $I_{\mathrm{ox}}/I_{\mathrm{m}}$ is the ratio of oxide to metal intensities from the photoelectron spectra. The ratio of the metal and oxide intensities from both 3d and $2p_{3/2}$ core levels may be compared (using the appropriate IMFPs in each case) to check for consistency. The oxide layer thicknesses calculated are 15.5 nm (using Zr 2p) and 13.8 nm (using Zr 3d). These results are consistent within the expected absolute error of these calculations of 10–20%, mainly associated with error in the calculated IMFPs.[65] The calculated oxide layer thickness is greater than the sampling depth for XPS and consistent with the observation of only oxide in XPS. In addition, an abrupt interface (as is assumed in eqn (2)) is not necessarily expected, *i.e.*, there may be a thin layer of sub-stoichiometric oxide at the interface, or uneven oxide thickness, so calculations using measurements at two different sampling depths may not fully agree. This example demonstrates how the use of different core levels, several thousand eV different in energy, enables depth-dependent information to be obtained, including the chemical state of the elements.

The sample was heated *in situ*; at temperatures approaching 373 K, a large increase in vacuum pressure associated with the removal of surface contamination occurred, after which the vacuum dropped to <2 × 10^{-8} mbar (2 × 10^{-6} Pa), which is suitable for measurement. The Zr $2p_{3/2}$ core level was then monitored as a function of increasing temperature, as illustrated in Fig. 10. The temperature was monitored using a K-type thermocouple attached to the base of the manipulator, and for temperatures over 493 K, a further calibration between the measured and sample temperature was required as provided by the manufacturer of the instrument. There is little change in the relative intensity of oxide and

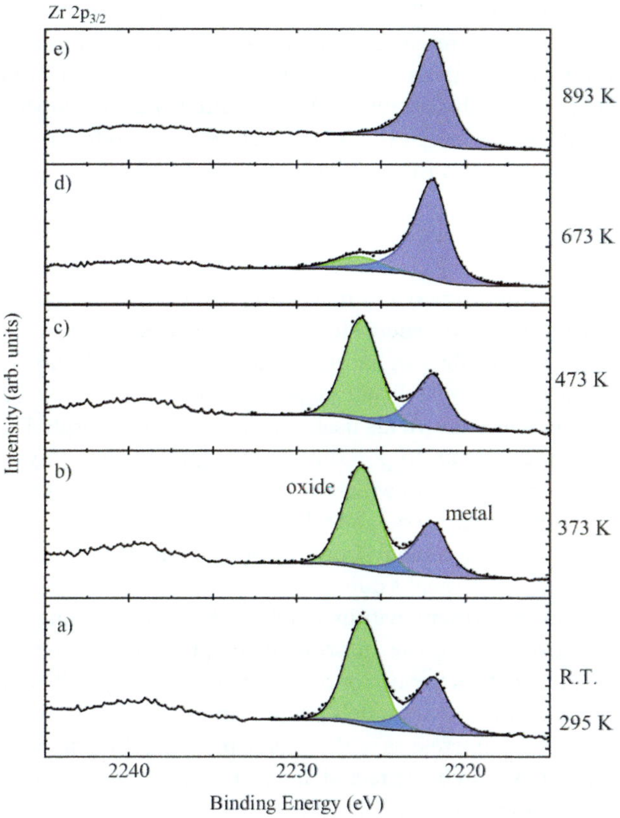

Fig. 10 Zr $2p_{3/2}$ spectra of a piece of untreated Zr metal which was heated from room temperature (a) to 893 K (e).

metal features up to temperatures of 473 K, but at 673 K the oxide is largely removed, and completely removed by 893 K. This confirms that the high BE species is associated with the oxide passivation layer. The complete removal of the oxide overlayer also allowed the extraction of the asymmetric metallic peak shape required for peak fitting and quantification of the data from the oxidised films.

3.4. Ion implantation in graphite

Implantation of charged species into the matrix of a material is an area of interest in many research fields.[66–69] Here, we examine highly oriented pyrolytic graphite (HOPG) implanted with Cs^+ ions. Caesium is a by-product of uranium fission and is known to be potentially mobile under certain conditions,[70,71] and HOPG is commonly used as a nuclear graphite simulant.[72] Therefore, sensitive methods to measure caesium distributions are useful. A sample of the simulant HOPG was implanted with 50 keV Cs^+ ions and then exposed to the atmosphere, as described elsewhere.[73] The ion beam was 50 μm in diameter and was rastered across the surface of the HOPG. SRIM calculations (stopping and range of ions in matter)[74] suggested that the ions should implant as a Gaussian distribution centred below the surface of the HOPG, at a depth of *ca.* 20 nm, and, consistent with this, initial measurements with Al Kα XPS detected a small amount of Cs.[73] However, subsequent XPS imaging revealed that the rastering had not yielded a uniform distribution of Cs across the sample, possibly implying lateral variations in the ion implantation depth distribution.[73] Insight into variations in the ion implantation depth distribution across the HOPG surface was therefore required and was enabled using HAXPES and inelastic background modelling using the Cs 2s and 2p core levels.

Inelastic background modelling, pioneered by Sven Tougaard,[75] and implemented in the QUASES-Tougaard software package,[76] makes use of the energy loss structure to lower kinetic energy from a primary photoemission peak. Inspection of the inelastic background (where electrons are scattered with associated energy loss) can be used to provide a signature of the morphology of that element below the surface: indeed, a simple visual inspection of the core levels and the background in a survey spectrum yields useful insight into the morphology.[77] A recent publication also provides an important practical guide to carrying out this analysis.[78] The ultimate depth from which information may be extracted depends on the measurable kinetic energy range of this energy loss structure, *i.e.*, the width of the range to high binding energy from a primary photoelectron emission peak that is unperturbed by any peaks due to other elements.[79] For XPS, where multiple core levels are often present within a relatively small binding energy range, together with Auger peaks, the information depth can be somewhat limited. However, HAXPES enables the use of deeper core levels, typically separated from other photoemission peaks by much greater energy intervals.[4] We have shown that, even after the complete loss of the elastic photoelectron signal, the depth of a buried element may still be extracted at up to 20 times the electron attenuation length, corresponding to hundreds of nm.[4]

This inelastic background modelling is well established and has been refined to take into account elastic scattering effects, layered structures with greatly differing cross sections and electron attenuation lengths.[80,81] Accuracy and depth resolution is generally improved using reference samples.[4,82] Background analysis also has the potential to correct for the effect of photoelectron attenuation through a gaseous atmosphere in near ambient pressure (NAP) XPS measurements.[83] Here, we simply demonstrate how dramatically the inelastic background is influenced by the morphology of the sample, using a sample of HOPG that has been implanted with Cs^+ ions.

Measurements were performed at grazing incidence in transmission mode, across the Cs 2s, 2p region. A range of several hundred eV to the high kinetic energy of the Cs $2p_{3/2}$ peak was used to carefully remove the preceding background, and the spectrum was measured over several hundred eV to the lower kinetic energy of the Cs 2s and 2p peaks. The inelastic mean free path of electrons was calculated to be 8.9 nm using the TPP-2M formula,[10] using the QUASES-TPP2M calculator tool, assuming that the photoelectrons are mainly travelling through graphite. Analysis was performed using the QUASES-Tougaard-Analyze software, which is applicable when the elastic photoelectron peaks are observable.[78] Several elemental distribution models are available in the software, including an exponential profile into or out of the bulk material, however, here it was found that a simple buried layer model was sufficient to adequately model the background shapes, and this provided insight into the caesium distributions following implantation.

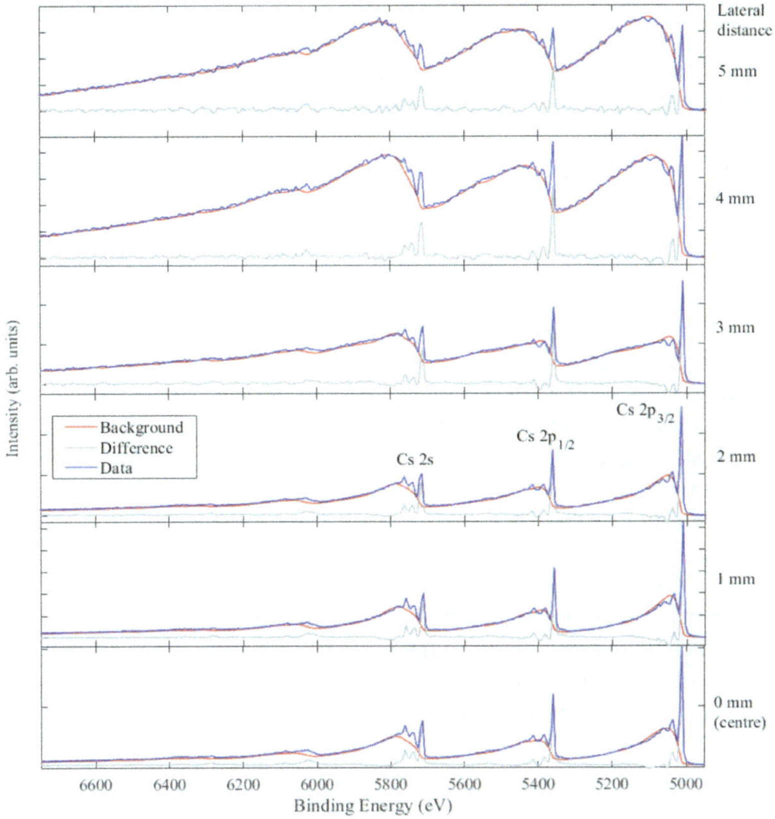

Fig. 11 HAXPES inelastic background modelling of Cs^+ ions implanted in HOPG for distances up to 5 mm away from the centre of the sample, as labelled. Data are shown by blue lines, the background model is shown by red lines, and the difference is shown by green lines. The backgrounds were modelled as buried layers (thicknesses ranging from 10–20 nm) at a starting depth of 9 nm (0 mm from the sample centre), 15 nm (1 mm), 17 nm (2 mm), 20 nm (3 mm), 22 nm (4 mm) and 30 nm (5 mm).

Fig. 11 shows the Cs 2s, 2p spectra taken at six positions across the HOPG sample, from the centre of the 10 mm × 10 mm sample to 5 mm away from the centre laterally in one direction, which was very close to the edge of the HOPG sample. The Cs $2p_{3/2}$ core level peak is at 5011.0 eV BE, and the Cs $3d_{5/2}$ core level signal was measured at 724.8 eV BE. The relative intensities of the primary peaks (from elastic processes) and secondary background features (from inelastic scattering) depend on the depth distribution and sampling depth associated with the peak. From a visual inspection of the inelastic backgrounds, it is clear that the depth distribution of Cs atoms embedded in the graphite changes with lateral distance away from the centre of the sample. Only at 4 mm and 5 mm away from the centre do the Cs peaks show large rising inelastic backgrounds to low KE from each photoelectron peak that exceed the primary photoelectron peak intensity, indicating that the Cs is heavily buried. The process of ion implantation is expected to damage the graphitic carbon matrix, particularly when high concentrations of ions are implanted; that damage may lead to the loss of graphite at the surface during implantation and may also allow for an enhancement in the mobility of the Cs ions after implantation. The amount of Cs measured (using the Cs $2p_{3/2}$, C 1s, and O 1s photoelectron peaks) varied significantly from *ca.* 11 at% (equivalent homogeneous concentration) at the centre, down to *ca.* 1 at% 5 mm away from the centre.

The changes in the spectra close to the sample centre suggest some removal of surface material due to implantation. The background models used to fit the data were all generated with a buried layer with a thickness of 10–20 nm. In order to model the possible effects of implantation, the starting depth of this buried layer was varied and optimized in order to adequately fit the data. It was found that from 0 to 5 mm from the sample centre, the best fits were obtained for buried layer starting depths of 9 nm, 15 nm, 17 nm, 20 nm, 22 nm, and 30 nm, respectively, consistent with SRIM calculations and a varying degree of lost surface materials, except for the measurement at the sample edge, where the calculated starting depth exceeded the SRIM model expectation. Alternatively, some ion migration could be occurring through the damaged carbon matrix at room temperature, or there is a combination of these effects; further investigation is planned.

Inelastic background analysis therefore provides another important tool for the characterisation of buried interfaces; overall, this analysis, where applicable, is more efficient than analogous angle-resolved measurements, as detailed in section 3.1 of the Results. However, for high accuracy, great care must be taken when determining the input parameters used in the analysis.[78]

3.5. Detection of nanoparticle cores under thick shells

Nanoparticles commonly have an internal structure which typically consists of layers, or shells, around a functional core. The purpose of the shell material may be protective, functional, or accidental. Examples of protection include the passivation of luminescent quantum dots for optoelectronic and bioimaging or a layer forming a diffusion barrier in drug-eluting particles where the active ingredient is centrally located. Functional coatings are usually designed to interact with the particle environment to aid particle dispersion, prevent adsorption of molecules (for example polyethylene glycol coatings to reduce protein attachment), or to provide targeted adhesion. The measurement of the average coating thickness and information on the elemental depth distributions in a population of particles can be performed using XPS, however the data analysis must consider the geometry of the particles. For large particles with a uniform coating, the signal intensity from the underlying material is approximately 2/3 that of a flat surface with the same coating thickness.[84] This implies that the information depth for particles should be estimated as twice the effective attenuation length (EAL) of electrons, rather than three times, as is typical for flat surfaces. For nanoparticles, where the dimension of the particle is similar to the EAL, there is a further drop in intensity from the core material, particularly for thick shells. With some knowledge of particle sizes and the internal distribution of materials in the particles, XPS data can be analysed to provide shell thicknesses and compositions.[85–88]

Due to the constraints described above on the depth information from XPS, HAXPES may be a suitable technique for the analysis of particles with small cores and thick shells. A series of reference nanoparticles with a PTFE core and PMMA shell have been analysed by a wide range of methods, including XPS.[89,90] For these particles, which possess a 47 nm diameter core and nominal shell thicknesses of 4.5 nm, 12 nm, 23.5 nm and 35.5 nm, Al Kα XPS was only able to reliably detect the F 1s peak from the PTFE core material for the three smallest particles. In the case of the nominal 12 and 23.5 nm shell thicknesses, this detection was possible because of the non-central location of the cores within these particles.[89] For the largest particle type (with a shell thickness of 35.5 nm), the core could not be detected and sputter depth profiling using argon cluster ions was necessary to obtain detail on the internal distribution of the materials.[91]

Fig. 12(a) and (b) demonstrate that the core material may be detected for particles with nominal shell thicknesses of 12 nm, 23.5 nm and 35.5 nm using Ga Kα HAXPES. For the thickest shell, the F 1s signal is close to the detection limit; equivalent homogeneous concentrations of the F at% were calculated to be 24.4,

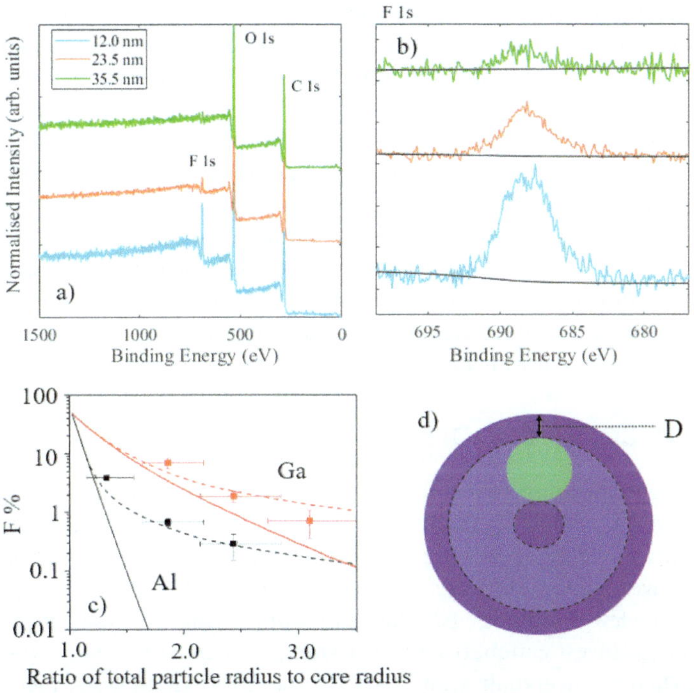

Fig. 12 (a) Survey spectra and (b) high resolution F 1s spectra for 12.0 nm, 23.5 nm and 35.5 nm nominal shell thickness PTFE-PMMA core–shell particles (cyan, orange and green lines, respectively). (c) Measured F 1s at% and modelled data. Solid lines indicate the expected fluorine at% for a centralised core, and dashed lines indicate the expected fluorine at% for a core displaced 3 nm from the particle surface, and points indicate measured values. Black lines and points are data corresponding to Al Kα X-rays, whereas red lines and points correspond to Ga Kα X-rays, as indicated. (c) Schematic diagram showing the morphology used for modelling; intensities are averaged for a core located anywhere within the shaded area, where *D* is the core displacement from the surface.

7.2 and 2.2 at%, respectively. In this case, it must be considered that the PTFE core is buried, on average, under more than 30 nm of PMMA and the volume fraction of PTFE in the core–shell particles is less than 5%, therefore this is a significant achievement for a light element with a low photoionisation cross-section. The graph in Fig. 12(c) illustrates the measured fluorine atomic fractions from XPS and HAXPES for these samples. Solid lines represent the expected intensities with a centrally located core and the dashed lines are the expected orientation-averaged intensities for a non-central core located 3 nm beneath the particle surface (as illustrated in the schematic in Fig. 12(d), where the displacement below the surface is labelled "D"). This 'off-centre' model is based upon transmission electron microscopy analysis,[89] and approximately describes the XPS and HAXPES data. Divergences indicate that for smaller particles, the cores are slightly closer to the surface than 3 nm, but for the largest particle type the cores are somewhat deeper.

4. Conclusions

These results showcase measurements on a wide range of material systems containing buried interfaces, with characterisation of the buried layers enabled by hard X-ray photoelectron spectroscopy, extending the traditional XPS sampling depth below the topmost surface. The HAXPES-Lab system from ScientaOmicron GmbH uses an EW4000 analyser with a wide acceptance angle, which also enables angle-resolved measurements, capturing >40° angular information in one shot. The capabilities of the instrument may be usefully combined with advanced analysis techniques, including the modelling of the angle-resolved data, and use of the inelastic background. Taken in combination, these approaches extend the depth-profiling capabilities of HAXPES smoothly from the XPS regime to hundreds of nm below the surface. It is therefore anticipated that research utilising HAXPES will continue to accelerate, especially now that high-throughput measurements are enabled in the laboratory environment, and these methods will be applicable to a wide range of advanced materials research.

Author contributions

BFS: conceptualization, data curation, formal analysis, funding acquisition, investigation, methodology, software, validation, visualization, writing – original draft, writing – review & editing. SAC: conceptualization, data curation, formal analysis, investigation, software, validation, visualization, writing – original draft, writing – review & editing. PT: data curation, formal analysis, investigation, methodology, investigation, resources, writing – original draft, writing – review & editing. DJHC: conceptualization, data curation, formal analysis, methodology, resources, software, validation, visualization, writing – original draft, writing – review & editing. SM: investigation, methodology, validation, writing – review & editing. AT: conceptualization, investigation, methodology, resources, writing – original draft, writing – review & editing. ANJ: conceptualization, resources, supervision, writing – review & editing. MJK: resources. DJB: conceptualization, resources, supervision, validation, writing – review & editing. RAO: conceptualization, resources, supervision, validation, writing – review & editing. JH: conceptualization, data curation, methodology, resources, validation, writing –

review & editing. AGT: conceptualization, supervision, writing – review & editing. TT: conceptualization, resources, validation, supervision, writing – original draft, writing – review & editing. AGS: conceptualization, data curation, formal analysis, funding acquisition, methodology, software, supervision, visualization, writing – original draft, writing – review & editing. WRF: conceptualization, funding acquisition, methodology, supervision, validation, writing – review & editing.

Conflicts of interest

There are no conflicts to declare.

Acknowledgements

This work was supported by the Henry Royce Institute for Advanced Materials, funded through the EPSRC grants EP/R00661X/1, EP/P025021/1 and EP/P025498/1, and the National Measurement System of the UK Department of Business, Energy and Industrial Strategy (BEIS).

Notes and references

1 J. Woicik, *Hard X-ray Photoelectron Spectroscopy (HAXPES)*, Springer International Publishing, Cham, 2016.

2 C. Kalha, N. K. Fernando, P. Bhatt, F. O. L. Johansson, A. Lindblad, H. Rensmo, L. Z. Medina, R. Lindblad, S. Siol, L. P. H. Jeurgens, C. Cancellieri, K. Rossnagel, K. Medjanik, G. Schonhense, M. Simon, A. X. Gray, S. Nemsak, P. Lomker, C. Schlueter and A. Regoutz, *J. Phys.: Condens. Matter*, 2021, **33**, 233001.

3 A. Regoutz, M. Mascheck, T. Wiell, S. K. Eriksson, C. Liljenberg, K. Tetzner, B. A. D. Williamson, D. O. Scanlon and P. Palmgren, *Rev. Sci. Instrum.*, 2018, **89**, 073105.

4 B. F. Spencer, S. Maniyarasu, B. P. Reed, D. J. H. Cant, R. Ahumada-Lazo, A. G. Thomas, C. A. Muryn, M. Maschek, S. K. Eriksson, T. Wiell, T. L. Lee, S. Tougaard, A. G. Shard and W. R. Flavell, *Appl. Surf. Sci.*, 2021, **541**, 148635.

5 M. B. Trzhaskovskaya and V. G. Yarzhemsky, *At. Data Nucl. Data Tables*, 2019, **129–130**, 101280.

6 M. B. Trzhaskovskaya and V. G. Yarzhemsky, *At. Data Nucl. Data Tables*, 2018, **119**, 99–174.

7 S. Maniyarasu, J. C. Ke, B. F. Spencer, A. S. Walton, A. G. Thomas and W. R. Flavell, *ACS Appl. Mater. Interfaces*, 2021, **13**, 43573–43586.

8 M. Bouttemy, S. Béchu, B. F. Spencer, P. Dally, P. Chapon and A. Etcheberry, *Coatings*, 2021, **11**, 702.

9 A. Theodosiou, B. F. Spencer, J. Counsell and A. N. Jones, *Appl. Surf. Sci.*, 2020, **508**, 144764.

10 H. Shinotsuka, S. Tanuma, C. J. Powell and D. R. Penn, *Surf. Interface Anal.*, 2015, **47**, 871–888.

11 A. Regoutz, M. Mascheck, T. Wiell, S. K. Eriksson, C. Liljenberg, K. Tetzner, B. A. D. Williamson, D. O. Scanlon and P. Palmgren, *Rev. Sci. Instrum.*, 2018, **89**, 73105.

12 D. H. Larsson, P. A. C. Takman, U. Lundström, A. Burvall and H. M. Hertz, *Rev. Sci. Instrum.*, 2011, **82**, 123701.

13 A. Thompson, D. Attwood, E. Gullikson, M. Howells, K.-J. Kim, J. Kirz, J. Kortright, I. Lindau, Y. Liu, P. Pianetta, A. Robinson, J. Scofield, J. Underwood, G. Williams and H. Winick, *X-ray Data Booklet*, Lawrence Berkeley National Laboratory, Berkeley, CA 94720, United States, 2009.

14 Q. Lian, M. Z. Mokhtar, D. Lu, M. Zhu, J. Jacobs, A. B. Foster, A. G. Thomas, B. F. Spencer, S. Wu, C. Liu, N. W. Hodson, B. Smith, A. Alkaltham, O. M. Alkhudhari, T. Watson and B. R. Saunders, *ACS Appl. Mater. Interfaces*, 2020, **12**, 18578–18589.

15 P. Dawson, S. Schulz, R. A. Oliver, M. J. Kappers and C. J. Humphreys, *J. Appl. Phys.*, 2016, **119**, 181505.

16 V. Fiorentini, F. Bernardini, F. Della Sala, A. Di Carlo and P. Lugli, *Phys. Rev. B: Condens. Matter Mater. Phys.*, 1999, **60**, 8849–8858.

17 S. A. Church, G. M. Christian, R. M. Barrett, S. Hammersley, M. J. Kappers, M. Frentrup, R. A. Oliver and D. J. Binks, *J. Phys. D: Appl. Phys.*, 2021, **54**, 475104.

18 M. Auf Der Maur, A. Pecchia, G. Penazzi, W. Rodrigues and A. Di Carlo, *Phys. Rev. Lett.*, 2016, **116**, 027401.

19 G. Kusch, E. J. Comish, K. Loeto, S. Hammersley, M. J. Kappers, P. Dawson, R. A. Oliver and F. C. P. Massabuau, *Nanoscale*, 2022, **14**, 402–408, DOI: 10.1039/d1nr06088k.

20 B. F. Spencer, M. J. Cliffe, D. M. Graham, S. J. O. Hardman, E. A. Seddon, K. L. Syres, A. G. Thomas, F. Sirotti, M. G. Silly, J. Akhtar, P. O'Brien, S. M. Fairclough, J. M. Smith, S. Chattopadhyay and W. R. Flavell, *Faraday Discuss.*, 2014, **171**, 275–298.

21 B. F. Spencer, M. J. Cliffe, D. M. Graham, S. J. O. Hardman, E. A. Seddon, K. L. Syres, A. G. Thomas, F. Sirotti, M. G. Silly, J. Akhtar, P. O'Brien, S. M. Fairclough, J. M. Smith, S. Chattopadhyay and W. R. Flavell, *Surf. Sci.*, 2015, **641**, 320–325.

22 B. F. Spencer, D. M. Graham, S. J. O. Hardman, E. A. Seddon, M. J. Cliffe, K. L. Syres, A. G. Thomas, S. K. Stubbs, F. Sirotti, M. G. Silly, P. F. Kirkham, A. R. Kumarasinghe, G. J. Hirst, A. J. Moss, S. F. Hill, D. A. Shaw, S. Chattopadhyay and W. R. Flavell, *Phys. Rev. B: Condens. Matter Mater. Phys.*, 2013, **88**, 195301.

23 B. F. Spencer, M. A. Leontiadou, P. C. J. Clark, A. I. Williamson, M. G. Silly, F. Sirotti, S. M. Fairclough, S. C. E. Tsang, D. C. J. Neo, H. E. Assender, A. A. R. Watt and W. R. Flavell, *Appl. Phys. Lett.*, 2016, **108**, 091603.

24 S. Ueda, *Appl. Phys. Express*, 2018, **11**, 105701.

25 T. Narita, D. Kikuta, N. Takahashi, K. Kataoka, Y. Kimoto, T. Uesugi, T. Kachi and M. Sugimoto, *Phys. Status Solidi A*, 2011, **208**, 1541–1544.

26 G. M. Christian, S. Schulz, M. J. Kappers, C. J. Humphreys, R. A. Oliver and P. Dawson, *Phys. Rev. B*, 2018, **98**, 155301.

27 D. M. Graham, A. Soltani-Vala, P. Dawson, M. J. Godfrey, T. M. Smeeton, J. S. Barnard, M. J. Kappers, C. J. Humphreys and E. J. Thrush, *J. Appl. Phys.*, 2005, **97**, 103508.

28 A. Vanleenhove, C. Zborowski, I. Vaesen, I. Hoflijk and T. Conard, *Surf. Sci. Spectra*, 2021, **28**, 014006.

29 DOI: 10.1088/0031-8949/22.

30 Y. Kayanuma, in *Hard X-ray Photoelectron Spectroscopy (HAXPES)*, ed. J. Woicik, Springer International Publishing, 2016, pp. 175–195, DOI: 10.1007/978-3-319-24043-5_8.
31 D. J. H. Cant, B. F. Spencer, W. R. Flavell and A. G. Shard, *Surf. Interface Anal.*, 2022, **54**, 442–454.
32 P. L. J. Gunter, O. L. J. Gijzeman and J. W. Niemantsverdriet, *Appl. Surf. Sci.*, 1997, **115**, 342–346.
33 L. P. Kazansky, I. A. Selyaninov and Y. I. Kuznetsov, *Appl. Surf. Sci.*, 2012, **259**, 385–392.
34 M. P. Seah, J. H. Qiu, P. J. Cumpson and J. E. Castle, *Surf. Interface Anal.*, 1994, **21**, 336–341.
35 P. J. Cumpson, *J. Electron Spectrosc. Relat. Phenom.*, 1995, **73**(1), 25–52.
36 V. I. Nefedov, *J. Electron Spectrosc. Relat. Phenom.*, 1999, **100**, 1–15.
37 A. Chassé and T. Chassé, *J. Phys. Soc. Jpn.*, 2018, **87**, 061006.
38 L. Despont, D. Naumović, F. Clerc, C. Koitzsch, M. G. Garnier, F. J. Garcia De Abajo, M. A. Van Hove and P. Aebi, *Surf. Sci.*, 2006, **600**, 380–385.
39 F. J. García De Abajo, M. A. Van Hove and C. S. Fadley, *Phys. Rev. B: Condens. Matter Mater. Phys.*, 2001, **63**, 075404.
40 H. A. Aebischer, T. Greber, J. Osterwalder, A. P. Kaduwela, D. J. Friedman, G. S. Herman and C. S. Fadley, *Surf. Sci.*, 1990, **239**, 261–264.
41 T. S. Lassen, S. Tougaard and A. Jablonski, *Surf. Sci.*, 2001, **481**, 150–162.
42 S. Tougaard, *Surf. Interface Anal.*, 1998, **26**, 249–269.
43 S. Tougaard and A. Jablonski, *Surf. Interface Anal.*, 1997, **25**, 404–408.
44 J. Cui, M. Kramer, L. Zhou, F. Liu, A. Gabay, G. Hadjipanayis, B. Balasubramanian and D. Sellmyer, *Acta Mater.*, 2018, **158**, 118–137.
45 A. Edström, J. Chico, A. Jakobsson, A. Bergman and J. Rusz, *Phys. Rev. B: Condens. Matter Mater. Phys.*, 2014, **90**, 014402.
46 B. S. D. C. S. Varaprasad, M. Chen, Y. K. Takahashi and K. Hono, *IEEE Trans. Magn.*, 2013, **49**, 718–722.
47 X. Zhang, L. L. Tao, J. Zhang, S. H. Liang, L. Jiang and X. F. Han, *Appl. Phys. Lett.*, 2017, **110**, 252403.
48 S. Zhao, T. Hozumi, P. Leclair, G. J. Mankey and T. Suzuki, *IEEE Trans. Mag.*, 2015, **51**, 2101604.
49 Z. Luo, S. Shao and T. Wu, *Microelectron. Eng.*, 2021, **242–243**, 111530.
50 R. Sitko, *Spectrochim. Acta, Part B*, 2009, **64**, 1161–1172.
51 C. J. Jenks, S. L. Chang, J. W. Anderegg, P. A. Thiel and D. W. Lynch, *Phys. Rev. B: Condens. Matter Mater. Phys.*, 1996, **54**, 6301–6306.
52 C. Navío, M. Villanueva, E. Céspedes, F. Mompeán, M. García-Hernández, J. Camarero and A. Bollero, *APL Mater.*, 2018, **6**, 101109.
53 P. M. Th, M. Van Attekum and J. M. Trooster, *Phys. Rev. B: Condens. Matter Mater. Phys.*, 1978, **18**, 3872–3883.
54 J. E. Castle, L. B. Hazell and R. D. Whitehead, *J. Electron Spectrosc. Relat. Phenom.*, 1976, **9**, 247–250.
55 W. Huschka, D. Ross, M. Maier and E. Umbach, *J. Electron Spectrosc. Relat. Phenom.*, 1988, **46**, 273–276.
56 E. Atanassova, T. Dimitrova and J. Koprinarova, *Appl. Surf. Sci.*, 1995, **84**, 193–202.
57 P. Prieto, L. Galán and J. M. Sanz, *Appl. Surf. Sci.*, 1993, **70–71**, 186–190.
58 J. E. Castle, *Surf. Interface Anal.*, 2022, **54**(4), 455–464.

59 C. D. Wagner and A. Joshi, *J. Electron Spectrosc. Relat. Phenom.*, 1988, **47**, 283–313.
60 S. Siol, J. Mann, J. Newman, T. Miyayama, K. Watanabe, P. Schmutz, C. Cancellieri and L. P. H. Jeurgens, *Surf. Interface Anal.*, 2020, **52**, 802–810.
61 Z. Azdad, L. Marot, L. Moser, R. Steiner and E. Meyer, *Sci. Rep.*, 2018, **8**, 16251.
62 I. Bespalov, M. Datler, S. Buhr, W. Drachsel, G. Rupprechter and Y. Suchorski, *Ultramicroscopy*, 2015, **159**, 147–151.
63 M. Magnuson, S. Schmidt, L. Hultman and H. Högberg, *Phys. Rev. B*, 2017, **96**, 195103.
64 H. Shinotsuka, S. Tanuma, C. J. Powell and D. R. Penn, *Surf. Interface Anal.*, 2015, **47**, 1132.
65 B. R. Strohmeier, *Surf. Interface Anal.*, 1990, **15**, 51–56.
66 H. Ryssel and I. Ruge, *Ion Implantation*, John Wiley & Sons Ltd., Chichester, UK, 1986, ISBN 10 047110311X, ISBN 13 9780471103110.
67 D. N. Jamieson, C. Yang, T. Hopf, S. M. Hearne, C. I. Pakes, S. Prawer, M. Mitic, E. Gauja, S. E. Andresen, F. E. Hudson, A. S. Dzurak and R. G. Clark, *Appl. Phys. Lett.*, 2005, **86**, 202101.
68 E. A. Scott, K. Hattar, J. L. Braun, C. M. Rost, J. T. Gaskins, T. Bai, Y. Wang, C. Ganski, M. Goorsky and P. E. Hopkins, *Carbon*, 2020, **157**, 97–105.
69 A. Theodosiou, A. F. Carley and S. H. Taylor, *J. Nucl. Mater.*, 2010, **403**, 108–112.
70 R. B. I. Evans, W. Davis, Jr. and A. L. Sutton, Jr., *Cesium Diffusion in Graphite*, Office of Scientific and Technical Information (OSTI), 1980.
71 B. Torstenfelt, *Radiochim. Acta*, 1986, **39**, 97–104.
72 K. Wen, J. Marrow and B. Marsden, *J. Nucl. Mater.*, 2008, **381**, 199–203.
73 A. Theodosiou, B. F. Spencer, J. Counsell, P. Ouzilleau, Z. He and A. N. Jones, 2022, submitted.
74 J. F. Ziegler, M. D. Ziegler and J. P. Biersack, *Nucl. Instrum. Methods Phys. Res., Sect. B*, 2010, **268**, 1818–1823.
75 S. Tougaard, *Surf. Sci.*, 1985, **162**, 875–885.
76 S. Tougaard, *Software Packages to Characterize Surface Nano-Structures by Analysis of Electron Spectra*, http://www.quases.com.
77 S. Tougaard, *Surf. Interface Anal.*, 2018, **50**, 657–666.
78 S. Tougaard, *J. Vac. Sci. Technol., A*, 2021, **39**, 011201.
79 S. Tougaard, *J. Electron Spectrosc. Relat. Phenom.*, 2010, **178–179**, 128–153.
80 P. Risterucci, O. Renault, C. Zborowski, D. Bertrand, A. Torres, J. P. Rueff, D. Ceolin, G. Grenet and S. Tougaard, *Appl. Surf. Sci.*, 2017, **402**, 78–85.
81 C. Zborowski and S. Tougaard, *Surf. Interface Anal.*, 2019, **51**, 857–873.
82 C. Zborowski, O. Renault, A. Torres, Y. Yamashita, G. Grenet and S. Tougaard, *Appl. Surf. Sci.*, 2018, **432**, 60–70.
83 S. Tougaard and M. Greiner, *Appl. Surf. Sci.*, 2020, **530**, 147243.
84 A. G. Shard, J. Wang and S. J. Spencer, *Surf. Interface Anal.*, 2009, **41**, 541–548.
85 D. J. H. Cant, Y.-C. Wang, D. G. Castner and A. G. Shard, *Surf. Interface Anal.*, 2016, **48**, 274–282.
86 H. Kalbe, S. Rades and W. E. S. Unger, *J. Electron Spectrosc. Relat. Phenom.*, 2016, **212**, 34–43.
87 C. J. Powell, W. S. M. Werner, H. Kalbe, A. G. Shard and D. G. Castner, *J. Phys. Chem. C*, 2018, **122**, 4073–4082.
88 A. G. Shard, *J. Phys. Chem. C*, 2012, **116**, 16806–16813.

89 D. J. H. Cant, C. Minelli, K. Sparnacci, A. Müller, H. Kalbe, M. Stöger-Pollach, W. E. S. Unger, W. S. M. Werner and A. G. Shard, *J. Phys. Chem. C*, 2020, **124**, 11200–11211.

90 A. Müller, K. Sparnacci, W. E. S. Unger and S. Tougaard, *Surf. Interface Anal.*, 2020, **52**, 770–777.

91 Y. Pei, D. J. H. Cant, R. Havelund, M. Stewart, K. Mingard, M. P. Seah, C. Minelli and A. G. Shard, *J. Phys. Chem. C*, 2020, **124**, 23752–23763.

Faraday Discussions

ROYAL SOCIETY OF CHEMISTRY

DISCUSSIONS

Buried interfaces: general discussion

Damien Aureau, Thorsten Bartels-Rausch, Mark A. Buckingham, Thierry Conard, Martina Dell'Angela, Wendy Flavell, Joshua S. Gibson, Georg Held, Mark Isaacs, Claus F. P. Kastorp, Alenka Krizan, Robert Lindsay, Patrick Lömker, David Morgan, Ayomide Osundare, Robert Palgrave, Yiwen Pei, Olivier Renault, Adam Roberts, Günther Rupprechter, Alexander Shard, Francine Solal, Ben Spencer and Robert Weatherup

DOI: 10.1039/d2fd90026b

Adam Roberts opened the discussion: We refer to hard and soft X-ray photoelectron spectroscopy (HAXPES and SOXPES, respectively). There does not appear to be a consensus within the photoemission community as to a definition of the photon energies associated with these acronyms. Is it possible to define the energies considered to be hard X-rays in the context of X-ray photoemission spectroscopy?

Joshua S. Gibson replied: I agree that this is an ambiguous term. People have also defined "tender" X-rays, which have an energy between soft and hard X-rays. Researchers have stated that silver-based lab sources are hard, compared to the Al source, so could "hard" be defined relative to the Al Kα source?

Alexander Shard remarked: Regarding the definition of HAXPES, this is currently under consideration in ISO TC201 and the literature definition from ref. 1 is currently used, where an X-ray energy of 2140 eV is the minimum HAXPES energy.

1 C. Weiland, A. K. Rumaiz, P. Pianetta and J. C. Woicik, *J. Vac. Sci. Technol., A*, 2016, **34**, 030801.

Georg Held said: In synchrotron radiation sources, hard X-ray beamlines are usually those which use crystal monochromators (typically Si(111) crystals), whereas soft X-ray beamlines use grating monochromators. The lowest energy that can be reached with Si(111) is around 2000 eV; typical plane grating monochromators can reach photon energies up to around 2500 eV. So there is some overlap between the two energy ranges.

Olivier Renault replied: Thanks for the comment and the sound reference. The value of 2 keV is about what HAXPES analysts have in mind, I guess.

Wendy Flavell commented: In the context of the value of HAXPES in being able to access deeper core levels, I can give a specific example. We have been studying alkali-metal doped perovskite light harvesters containing a small amount (a few %) of Rb and Cs dopants. The role of both (particularly Rb) in stabilising the structures has been debated. For Rb, measurement of the 2p core level at 1805 eV binding energy using a Ga Kα source yielded a sensitivity about 18 times greater than measuring the Rb 3d core level with XPS. The sensitivity to Cs, using the 2p core level and the Ga Kα source, was about 4.5 times greater than measuring using the 3d core level and an Al Kα source. We were able to depth profile the Rb distribution more accurately than had been possible before, and to build up a model for the role of Rb in the materials.[1]

While we are trying to define our territory, could I ask: what do we mean by '*operando*'?

1 S. Maniyarasu, J. C.-R. Ke, B. F. Spencer, A. S. Walton, A. G. Thomas and W. R. Flavell, *ACS Appl. Mater. Interfaces*, 2021, **13**, 43573–43586.

Joshua S. Gibson replied: *Operando* means "in operation", so in our situation, when charging and discharging a battery over a 1 to 20 hour period, *operando* can be considered to be observing changes with the evolution of time corresponding to the amount of charge passed. There is then potential for discussion around the available time resolution of XPS; the fastest you can collect core levels with reasonable signal is 2 minutes, then a full set of core level spectra, say 5 core levels, can take over 10 minutes. If you are probing rapidly evolving processes, I suppose that one could argue that 10 minute temporal resolution is not really *operando* spectroscopy. A solution to this could be using an approach similar to the multiple ion detection mode in mass spectrometry, where you sit at one, or cycle between a few, fixed binding energies and hope that your binding energies don't shift. I think that with the development of brighter sources, there will be coupled techniques that allow for the better mitigation of beam damage. This discussion of temporal measurement applies to the cycling of a battery, while in general, as mentioned, *operando* relates to being able to probe the device on a timescale that is suitable for its use.

Olivier Renault replied: It is a good and key question, and it's true that a precise definition of what *operando* means is highly desirable. First of all, looking at the literature simply by checking the occurrence of "*operando*" and "*in situ*" in conjunction with "photoelectron" in databases, it can be seen that the field of *operando* studies is much more recent (from 2002, and steadily growing) than the one related to *in situ* studies, which started in the late 1960s. However, it is becoming quite clear that *operando* studies are very specific and should not be confused with *in situ* works.

Literally, *operando* is the Latin gerund meaning "in operation" or "being operated" and, therefore, the wording "*in operando*" very often employed is incorrect and only "*operando*" should be used. Concerning how one defines *operando* experiments, and as seen from different papers and from a more personal perspective (also shared in my group and amongst colleagues), there is a clear relationship with the operation of a device in which the property of the material being studied is playing a key role, as opposed to an "*in situ*" experiment, where one is "only" looking how a specific property of the material itself varies with an external stimulus applied on this material while the measurement is performed (temperature, pressure, humidity, *etc.*). Consequently, both kind of experiment have the same important dynamic character and carry a far greater reach regarding applications but the degree of sophistication and how the experiment is representative of realistic conditions differ very much. I would therefore classify *operando* photoelectron spectroscopy experiments only in the following cases:

(1) The material analyzed is integrated in a device which is operated during the experiment conceptualized in such a way that the spectra may be trustfully and unambiguously correlated to the operation conditions. The device may be an electrochemical cell,[1] a capacitor,[2] a memory device,[3] an organic LED,[4] a transistor device,[5] a solar cell, a sensor, *etc.* This means, *e.g.* for electronic or electrochemical devices, that the electrical/electrochemical characteristic needs, if possible, to be recorded inside the spectrometer and that this characteristic needs to be stable during the whole experiment. At the very least, it is mandatory to record the electrical characteristic *ex situ* on the same device analyzed during the experiment after checking its stability with time. Within this field, there is plenty of room, especially for solar cell devices being studied under realistic illumination conditions.

(2) Experimental conditions that are indicative of real, realistic conditions for the material being analyzed. This is the case, for instance, in the field of catalysis or of gas sensing, where a particular gas at a particular pressure interacts with the surface of the material. I would propose that such experimental conditions would correspond to "*operando*" and not "*in situ*", for the sake of clarity.

With the proposed definition, it looks like time-resolved studies without the application of an external, specific stimulus, are going into a particular category, because only fundamental processes are being investigated.

I hope this view, which should of course be further debated, discussed and agreed upon, will find some degree of consensus within our community. The reliability and trustfulness of *operando* experiments are key and are becoming necessary in the perspective of a more widespread use of such experiments, which happen to be absolutely essential for accelerating the optimization of materials in order to help us facing the big challenges of energy transition and climate change.

1 A. Benayad, J. E. Morales-Ugarte, C. C. Santini and R. Bouchet, *J. Phys. Chem. A*, 2021, **125**, 1069–1081.
2 I. Gueye, G. Le Rhun, O. Renault, D. Cooper, D. Ceolin, J.-P. Rueff and N. Barrett, *Appl. Phys. Lett.*, 2017, **111**, 032906.
3 B. Meunier, E. Martinez, R. Rodriguez-Larnas, D. Pla, M. Burriel, C. Jimenez, Y. Yamashita and O. Renault, *ACS Appl. Electron. Mater.*, 2021, **3**, 5555–5562.
4 J. Ikeuchi, H. Hamamtsu, T. Miyamoto, S. Tanaka, Y. Yamashita, H. Yoshikawa and S. Ueda, *J. Appl. Phys.*, 2015, **118**, 085308.
5 P. Kumar, C. Leroux, F. Domengie, E. Martinez, V. Loup, D. Guiheux, Y. Morand, J.-M. Pedini, C. Tabone, F Gaillard and G. Ghibaudo, *2018 IEEE International Electron Devices Meeting (IEDM)*, 2018, 17.6.1–17.6.4, DOI: 10.1109/IEDM.2018.8614554.

Günther Rupprechter commented: The core of the *operando* approach is to use a second independent method for simultaneous measurements of a process. Measurements with only one method, even if under relevant conditions, do not qualify as *operando* (see *e.g.* ref. 1 and references therein).

1 G. Rupprechter, *Small*, 2021, **17**, 2004289.

Francine Solal said: The definition of *operando* experiments is when you look at materials when they are working/functional.

Olivier Renault replied: I totally agree. *Operando* experiments look at materials in their real operating conditions using the functional property they've been designed for.

Günther Rupprechter comments: That is not correct. For operando, one must make sure that the materials work properly (*e.g.* for catalysts by applying GC or MS product analysis). Just setting working conditions is not sufficient.

David Morgan opened the discussion of the paper by Joshua S. Gibson: It is mentioned that the electron charge compensation system can cause the migration of lithium to the surface of the material. Can you therefore be confident that, if the charge compensation system is used during analysis, the Li spectra you observe is a true representation of the material under analysis, or do you look for changes in the Li spectra by recording this elemental region first and again at the end to assess any changes, as one may do in a degradation study, for example?

Joshua S. Gibson replied: With these studies, we explicitly did not use the charge compensation system during analysis for a number of reasons, one being, as you say, the migration of lithium through the solid electrolyte, and the second being the observed surface damage. With the electron irradiation experiment, this was being monitored continually during irradiation, giving us confidence that what we were observing was being induced by the electron neutraliser drawing through the Li ions. In the absence of the electron neutraliser, we did not see any Li metal plating on the surface, suggesting that it was not an X-ray induced effect. The main disadvantage of using the electron neutraliser to control the lithium plating is the difficulty in controlling the neutraliser current in response to the surface potential. *Operando* approaches, using a potentiostat to control the potential, will help to overcome this.

Damien Aureau commented: You present a nice following of graphene flakes upon different treatments (Fig. 3 in the paper, https://doi.org/10.1039/d1fd00118c) by Raman spectroscopy. I was wondering if you also collected photoemission spectra (C 1s or O 1s, for instance) for the different stages in order to study chemical modifications of the flakes?

Joshua S. Gibson responded: We didn't see any particularly obvious differences in our surfaces, although for the samples that we put into the Raman, we needed to ensure that there was only a very thin Li layer to prevent signal attenuation. Over longer sputter deposition timescales on the Li_6PS_5Cl solid electrolyte, we saw damage occurring through the loss of carbon species, which we think is more likely to be physical damage, rather than chemical modification. Future work into studying the modification of the graphene flakes with increased deposition times could be of interest to look into chemical modification, although you might find, especially in the instance of sputtering, that you simply remove the graphene from your surface.

Mark A. Buckingham enquired: Lithium electrodes have a few decomposition pathways, such as dendrite formation. Do you think your electrodes would be more durable than normally synthesised electrodes? Have you looked at any long-term studies on these electrodes?

Joshua S. Gibson replied: It depends on your definition of long term, with battery cycles going on for up to and over 1000 cycles. This is something that we have not been able to do, partly due to instrumental limitations with high demand for the equipment, meaning that we cannot perform *in situ* cycling and currently have no storage spaces for cycling. These longer term cycling studies are definitely of interest to observe the longer term degradation processes, and they are something we are aiming to implement in future, but right now are not something that we can undertake.

Mark A. Buckingham said: Thank you for your excellent work, it was very interesting. I am curious to know if you have had a chance to test your *in situ* synthesised electrodes compared to typically synthesised electrodes?

Joshua S. Gibson responded: This is not something we have looked into at present, however it is something that will be of value in understanding the differences in surface structure and benefits of *in situ* electrode preparation compared to *ex situ* approaches.

Robert Palgrave queried: There is a lot of battery work going on looking at ions other than lithium, for example, sodium, potassium and magnesium. Would the issues highlighted here be relevant to those other studies using other ions?

Joshua S. Gibson responded: We expect that these findings are incredibly relevant to all alkali and alkali-earth metals, due to the high levels of reactivity of all of these metals. Like lithium, the other alkali metals have quite large inelastic mean free paths for photoelectrons, so studies into sodium and potassium ion battery interfaces are on the horizon, in collaboration with some other groups.

Francine Solal commented: Battery researchers tend to press the electrode on top of the electrolyte. Do you plan to study this real-world interface, and how do you expect they could differ from the one you studied?

Joshua S. Gibson replied: Pressing the electrode allows application of stack pressure, a parameter that people often vary in order to achieve and maintain good contact between the electrode and electrolyte, which in turn affects the ability to cycle at higher current densities without creating voids at the lithium metal–solid electrolyte interface. Through our evaporation method, we inherently get good, intimate contact between the electrode and electrolyte to begin with. With our thin film setup, we are limited in the maximum achievable current density, through void formation at high current densities in the absence of stack pressure. It would be difficult to apply a stack pressure whilst still being able to look through the electrode. The only option that springs to mind in being able to apply stack pressure in this way at present is the use of thin film membranes, however we would need to be careful not to burst these fragile membranes through the application of stack pressure.

Mark Isaacs commented: Was the choice to analyse these materials by Raman spectroscopy after air exposure a decision of purely experimental design concern? Do you think that the materials could have been affected by this air exposure?

Joshua S. Gibson answered: Unfortunately, our Raman spectrometer is in air, with no method of inertly transferring the Li coated graphene to the Raman. We did perform some other studies where we evaporated thicker lithium layers onto graphene, then washed them off in ethanol (which contains trace amounts of water to react the lithium away) and saw that the graphene remained undamaged, suggesting that it is chemically stable in contact with lithium.

Alexander Shard commented: Did you calculate the kinetic energy of sputtered lithium and backscattered argon using SRIM or another software to help understand the sputter deposition induced damage?

Joshua S. Gibson answered: We did not use any modelling for the energy profile of the sputtered material, but instead looked at the literature, finding that although the most likely energy is about 5 to 10 eV for a sputtered particle, there is a high energy tail up to 50 eV. When we consider these energies, it is maybe not surprising that sputter induced damage is observed, however this appears to have been overlooked in previous work for lithium deposition on solid electrolytes. This highlights the importance and need for the less energetic evaporation when studying interfacial species, which we advocate for in our paper (https://doi.org/10.1039/d1fd00118c).

Robert Weatherup opened the discussion of the paper by Ben Spencer: All of the papers mention recoil effects, what is the best approach to charge referencing when dealing with light elements?

Ben Spencer responded: For conductive samples, a gold reference sample electrically connected to the surface may be used for charge referencing, however, if a charge neutraliser is used for insulating samples, then a new value for the BE of the light element photoelectron peaks may be used, which takes into account the recoil effect. This is very straightforward to calculate using the conservation of momentum of a single atom, but it has been suggested that this may vary depending on the lattice and so may not be universally applicable.[1] Charge referencing is therefore still unresolved in HAXPES analysis and we, as a community, need to address this and develop standard procedures.

1 Y. Takata, Y. Kayanuma, M. Yabashi, K. Tamasuka, Y. Nishino, D. Miwa, Y. Harada, K. Horiba, S. Shin, S. Tanaka, E. Ikenaga, K. Kobayashi, Y. Senba, H. Ohashi and T. Ishikawa, *Phys. Rev. B*, 2007, **75**, 233404.

Robert Weatherup asked: For systems where heavy elements aren't available, could using carbon or other light elements with positions corrected based on calculated recoil shifts be appropriate?

Ben Spencer replied: In principle, this could be applied using a simple conservation of momentum for a single atom, but the influence of the lattice is not necessarily negligible; further experimental and theoretical work is required to validate whether this approach can be used as a standard procedure.

Joshua S. Gibson responded: Recoil effects were observed and discussed within our work here (https://doi.org/10.1039/d1fd00118c), especially for lithium, being such a light element. At a photon energy of 6.6 keV, the recoil shift on lithium was found to be approximately 0.5 eV. In this work, we perform our charge calibration to the Cl $2p_{3/2}$ at 198.5 eV, which is much heavier and only exhibits a negligible (<0.1 eV) recoil shift.

Olivier Renault answered: It is a good point, and I guess this may be a reasonable option because of the atomic character of the recoil shift a given excitation energy. Concerning energy-referencing using adventitious carbon, there is the intrinsic difficulty of the weak surface sensitivity of HAXPES which complicates the procedure. However, it also worth keeping in mind the numerous pitfalls of the conventional approach of binding energy referencing using the C 1s peak, as recently reviewed extensively by Greczynski *et al.* (see ref. 1).

1 Greczynski, *et al.*, *Prog. Mater. Sci.*, 2020, **107**, 100591.

Georg Held said: The recoil effect in the emission of a high-energy photoelectron is, in principle, very similar to the emission of a high-energy photon (typically > 10 keV) from a radioactive nucleus. In the latter case, if the emitter is incorporated in a solid, the Mössbauer effect is observed, *i.e.* a certain fraction of the emitted photons are emitted without recoil because the crystal lattice, rather than the nucleus, provides the counter-momentum to the photons. A similar mechanism seems to be ignored in the discussion of recoil in high-energy photoelectron emission. Why is this justified? (For energies of 10 keV, the momentum of an electron is about 10 times that of a photon.)

Joshua S. Gibson answered: In our study presented here (https://doi.org/10.1039/d1fd00118c), we were able to see photoelectron peak shifts which do correspond quite well to those predicted from the recoil shift, based on the momentum transfer equation, for both lithium and also in carbon. In this case, it appears that the emitter is the atom, rather than the crystal/grain of the cold pressed powder. The recoil shifts we observed differed for a particular atom at a particular photon energy; for example, the Li recoil shifts we observed were about 0.5 eV when using a 6.6 keV photon, and 0.15 eV when using a 2.2 keV photon. These observed shifts in the Li 1s photon peak equate to the values calculated using a momentum transfer equation, with the atom as the emitter.

Olivier Renault responded: The concern of the paper regarding the recoil shift was to investigate whether or not it would be neglected at a photon energy of 5.4

keV exciting elements heavier than silicon, and rather from an analytical perspective. This is indeed the case, but is no longer valid for lighter elements and at photon energies larger than 8 keV. Kayanuma, in the 2016 HAXPES book edited by Woicik (Chapter 8),[1] provided a detailed treatment of the recoil shift, indeed mentioning a similarity with the Mössbauer effect in solids. Quoting Kayanuma p. 181: "In the Mössbauer effect of solids, a part of the recoil momentum is accepted by the rigid crystal lattice, and the spectrum of the emitted x-ray now consists of a sharp recoilless line and its phonon sidebands. In the case of recoil effect in the core-level XPS, the zero-phonon line is indiscernible in many cases because of the lifetime broadening of the core-excited states, unlike the Mössbauer effect."

1 Y. Kayanuma, in *Hard X-ray Photoelectron Spectroscopy (HAXPES)*, ed. J. Woicik, Springer Series in Surface Sciences, Springer, 2016, Chapter 8, pp. 175–195.

Martina Dell'Angela asked: Do the recoil effects influence the photoelectron diffraction patterns?

Ben Spencer replied: Recoil effects are only expected to be non-negligible (with KE losses > 0.1–0.2 eV) for small atoms, so we do not expect any influence in the spectra from Ga, however there could be some influence in the spectra for N; this requires further investigation.

Robert Palgrave enquired: Do the currently used models of recoil in HAXPES take into account the lattice structure of the material, or is this not relevant?

Ben Spencer responded: Recoil in HAXPES may be treated simply as a conservation of momentum of a single atom, where at 8 keV kinetic energy, the C 1s recoil is calculated as being 0.36 eV. The recoil energy, however, will be dissipated in the lattice *via* the excitation of phonons in solids (see the book chapter by Y. Kayanuma[1] and Takata *et al.*[2]). It is suggested that dissipation in the lattice will also influence the line shape in photoemission, meaning the recoil effect, measured indirectly, could be used as a spectroscopic tool. The theoretical treatment of these effects is not arbitrary, however, so it is still unclear how this information can be extracted practically, especially for complex systems. And further work is required to measure variations in recoil energies caused by the lattice structure to assess how great that influence might be and thus how relevant it is.

1 Y. Kayanuma, in *Hard X-ray Photoelectron Spectroscopy (HAXPES)*, ed. J. Woicik, Springer, Cham, 2016, Chapter 8, pp. 175–195.
2 Y. Takata, Y. Kayanuma, M. Yabashi, K. Tamasuka, Y. Nishino, D. Miwa, Y. Harada, K. Horiba, S. Shin, S. Tanaka, E. Ikenaga, K. Kobayashi, Y. Senba, H. Ohashi and T. Ishikawa, *Phys. Rev. B*, 2007, **75**, 233404.

Claus F. P. Kastorp asked: Inelastic background analysis is no doubt a strong tool for determining the structure of buried species, but how well do you have to know the sample chemically speaking in order to do a useful analysis?

Ben Spencer replied: Indeed, the more that is known about the sample, the more accurate the inelastic background modelling performed can be, and combined measurements of reference materials (*e.g.*, a pure sample of a buried material) increase that accuracy, as we have demonstrated.[1] However, this does not mean that semi-unknown or completely unknown samples cannot utilise this modelling. If a photoelectron peak is present in the survey spectrum, and a sufficient energy loss function is accessible (a region of a higher BE/lower KE after the peak) without being interrupted by other peaks, an analysis to assess the morphology of that element within the material matrix can be performed. However, deeply buried layers, where the photoelectron peak is lost, will have a much more subtle influence on the survey spectrum and so may not be detected; here, knowledge of that material is required and measurement of a pure reference sample can enable modelling of that background (again, as demonstrated in ref. 1).

1 B. F. Spencer, S. Maniyarasu, B. P. Reed, D. J. H. Cant, R. Ahumada-Lazo, A. G. Thomas, C. A. Muryn, M. Maschek, S. K. Eriksoon, T. Weill, T.-L. Lee, S. Tougaard, A. G. Shard and W. R. Flavell, *Appl. Surf. Sci.*, 2021, **541**, 148635.

Olivier Renault commented: The chemical composition of your sample will be essentially provided by the survey spectrum, which contains already significant qualitative information, besides the nature of the elements present in the probed volume, namely, a rough depth distribution of the elements. Tougaard recently provided useful guidelines for looking quickly into this aspect by a visual inspection of the shape of the background peak of XPS spectra.[1] These can be directly extended to the case of HAXPES.

1 S. Tougaard, *Surf. Interface Anal.*, 2018, **50**, 657–666.

Robert Weatherup enquired: Is there anything we can learn from inelastic background fitting that we can't learn from X-ray reflectometry?

Ben Spencer answered: X-ray reflectometry is an excellent technique for measuring thin film thicknesses, and is an important technique to verify inelastic background modelling results. The main difference is that the inelastic background is measured alongside the photoelectron spectra in a vacuum, yielding the chemical information.

Alexander Shard commented: Regarding the comparison of HAXPES with X-ray reflectometry (XRR) for buried layer measurements, there are a number of advantages. XRR requires the sample to be flat and uniform over a large area and have layers with differences in electron density. Inelastic background analysis of HAXPES can be applied to samples with topography, layers with interfacial roughness, and layers that are distinguished by their elemental composition but which have similar electron densities. HAXPES can therefore be used to measure materials that XRR cannot.

Robert Lindsay continued the discussion of the paper by Ben Spencer: For the data in Fig. 5 of the paper (https://doi.org/10.1039/d1fd00118c), you mention that they are normalised using the gold reference sample response. Could you describe this process? Also, was the azimuthal orientation of the sample varied to examine the impact of this parameter on the diffracted signal.

Ben Spencer replied: By the gold reference response, we mean the dependence on the photoelectron signal with angle. Photoelectron counts naturally drop off at a higher emission angle, and the response of the analyser may additionally add artefacts to the angular response. Using a thick, very flat, vacuum-cleaned, homogeneous gold reference sample allows the response function to be measured, removing any influence of the sample morphology. We then divide the angle-resolved data from the samples by the angular response to normalise the angular data.

The azimuthal orientation was not varied, and the samples were randomly mounted on the sample plates, hence the response for two of the samples was different. Varying the azimuthal angle would allow a diffractogram to be generated and this is planned in future work.

Thierry Conard remarked: In your angle resolved data (shown in Fig. 3 of the paper, https://doi.org/10.1039/d1fd00118c), did you use the same normalisation for all peaks? Did you use any peak energy dependence for the angle signature?

Ben Spencer replied: Yes we used the same normalisation (using a clean, thick, homogeneous gold reference sample) for all the photoelectron peaks here. We found no noticeable difference using Au $3d_{5/2}$ (2206 eV BE) and Au $4f_{7/2}$ (84 eV BE), but Au 3d was used due to a much greater signal-to-noise ratio. As such, no peak energy dependence was used.

Adam Roberts communicated: With respect to the interpretation of the data presented in Fig. 3 of the paper (https://doi.org/10.1039/d1fd00118c), the paper explains that the 2D data is converted to a set of spectra (>260 over the useful 40° angular range). If I assume a step (pixel) step size of 0.2 eV, there would be 200 pixels in the energy direction, such that the Ga 2p image has 52 000 pixels. I would like to know the acquisition time of the set of images, allowing an estimate of the dwell time per pixel. It would be instructive to see a representative spectrum for each core-level with the curve fit, as this data is used to generate Fig. 4 and 5 of the paper. This would then lead to a discussion of the ability to use multivariate analysis techniques to partition the data from noise in the images and then using only the data abstract factors to reconstruct the spectral information. This technique has been applied to spectra-from-images data acquired using a spherical mirror analyser.[1]

1 J. Walton and N. Fairley, *Surf. Interface Anal.*, 2008, **40**, 478–481.

Ben Spencer communicated in reply: The total acquisition time for obtaining the set of angle-resolved data (yielding the 2D images) was approximately 18 hours. For Ga 2p, this gives an estimated dwell time of approximately 0.5 s per pixel (the BE step was 0.25 eV). Example curve fits are presented in Fig. 1 for Ga 2p, N 1s and In 2p. Here, we have performed a simple curve fitting (as we were mainly interested in the peak position) at every angle in order to demonstrate the capability, although more commonly one would sum sets of angular spectra together to provide an improved signal-to-noise ratio for a more careful peak fitting (as demonstrated in ref. 1) with a shortened acquisition time. I agree that

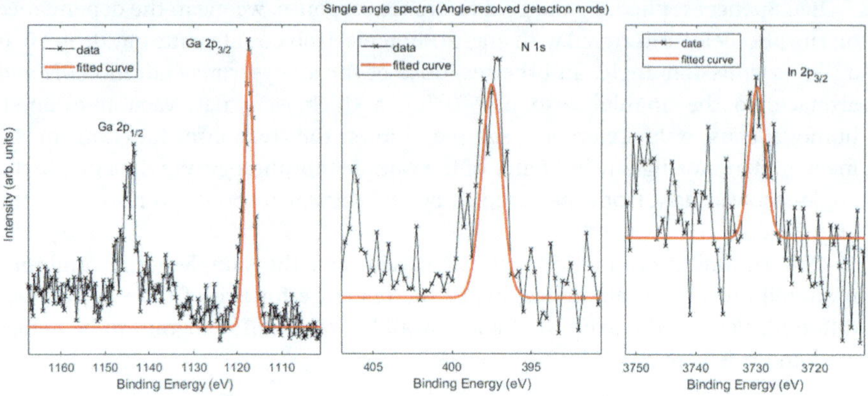

Fig. 1 Example curve fits for Ga 2p, N 1s and In 2p.

multivariate analysis techniques would be very useful to refine the analysis of the 2D angle-resolved data, especially when different chemical states are expected.

1 B. F. Spencer, S. Maniyarasu, B. P. Reed, D. J. H. Cant, R. Ahumada-Lazo, A. G. Thomas, C. A. Muryn, M. Maschek, S. K. Eriksson, T. Wiell, T.-L. Lee, S. Tougaard, A. G. Shard and W. R. Flavell, *Appl. Surf. Sci.*, 2021, **541**, 148635.

Francine Solal said: In Fig. 4 of the paper (https://doi.org/10.1039/D1FD00118C), the angular dependency of the binding energy with respect to the emission angle is attributed to the dependency of band bending on the escape depth. As this is an integration on the complete escape depth for a given angle, a rigid shift is unexpected. In addition, the increase of the binding energy with angle is counter intuitive considering that the semiconductor is n-type.

Ben Spencer responded: The rigid shift observed was attributed to the fact that these "depth profiles" are cumulated depth profiles, you are always measuring from the surface inwards and what we are changing is how far we are extracting the information from. As such, the depth dependence of band bending was expected to be distorted, however not necessarily into straight, or rigid, lines. The GaN is expected to be n-type, and so yes, the shift direction is also unexpected here. The samples were grown on sapphire substrates, and so a charge neutraliser (an electron flood source in the vacuum chamber) was required. Further experimental work is thus required to verify the materials and whether the charge neutralisation system is influencing the material and results.

Yiwen Pei remarked: HAXPES is a useful tool to reveal the chemistry and structure of PTFE–PMMA core–shell nanoparticles with difference shell thickness. Do you think it would be beneficial to combine HAXPES with a sputtering process to provide additional information on these materials?

Ben Spencer answered: Absolutely, the combination of sputtering processes with HAXPES analysis will assist in benchmarking both the sputtering process (*e.g.* etching rates) and the non-destructive HAXPES measurement of buried layers (for example, for materials with well established etch rates).

Damien Aureau opened a general discussion of the paper by Olivier Renault: In Fig. 11.b of the paper (https://doi.org/10.1039/d1fd00110h), the spectra shown at +4 V and −4 V seem to be too noisy to obtain reliable information. They would deserve to be counted longer in a slightly extended range to confirm the tendencies. Otherwise, references 62 to 65 are exactly identical.

Olivier Renault answered: The statistics of the spectra (especially the La) reflect the low cross-section of low binding energy core-levels excited with a hard X-ray source. We also would remind you that in such an *operando* experiment, the same core-level needs to be recorded several times at different applied voltages, so one needs to optimize the overall duration of the experiment. It is unfortunately not possible to find an intense, high-binding energy core-level available with appropriate acquisition conditions, and HAXPES is, much more than XPS, always a trade-off between probing depth and intrinsic signal intensity. As stated in the paper (https://doi.org/10.1039/d1fd00110h), we worked on spectra acquired at 0 V in order to suppress the peak asymmetry arising from the heterogeneous potential distribution, but of course the issue of the counting statistics is the same.

Alenka Krizan asked: Have you tried to fit Ti 2p under bias (Fig. 11a of the paper, https://doi.org/10.1039/d1fd00110h)? Could you extract the ratio between the resistance of the Ti layer and LSMO layer by investigating the peak broadening in XPS?

Olivier Renault replied: As stated at the end of the section dedicated to the *operando* measurements, we have measured both Ti and La core-levels under bias, but we have not shown a proper peak fitting because this requires the peak asymmetry arising from the potential drop to be quantitatively removed, which was not available at the time of writing. A first-order analysis tells us, however, that the La 3d and Ti 2p shifts strictly follow the applied bias, meaning that the LSMO resistance is limiting the current flow. The Ti 2p spectra under bias do not show the peak asymmetry, thereby confirming this conclusion.

Francine Solal asked: Are you expecting to be able to relate the asymmetry of the La peak to the precise voltage drop in LSMO and oxygen location?

Olivier Renault answered: Yes. This is an ongoing work which is expected to be published this year. Several approaches have been published already, including the pioneering one from Lim *et al.* in the Chambers group,[1] where built-in electric fields across the epitaxial $SrNbTiO_3$/Si(0001) interface are quantitatively derived from the simulation of HAXPES core-level asymmetry.

1 Z. H. Lim, N. F. Quackenbush, A. N. Penn, M. Chrysler, M. Bowden, Z. Zhu, J. M. Ablett, T.-J. Lee, J. M. LeBeau, J. C. Woicik, P. V. Sushko, S. A. Chambers and J. H. Ngal, *Phys. Rev. Lett.*, 2019, **123**, 026805.

Mark Isaacs addressed Ben Spencer and Olivier Renault: Regarding the quantification of overlayers/buried interfaces using the inelastic background scattering methodology – does the resolution of this technique scale with photon energy, *i.e.* would you obtain more sensitive information from the top 10 or so nanometers of a layered structure if you were to use Al Kα X-rays for the close-surface region, with HAXPES being used for deeper investigation – or is HAXPES equally effective if investigating deep interfaces *vs.* near-surface structures?

Ben Spencer answered: We estimate that the effective maximum sampling depth (the depth from which information may be extracted using inelastic background modelling) is around 20 times the inelastic mean free path of electrons,[1] so approximately 6–7 times greater than the elastic sampling depth. So with Al Kα X-rays, the inelastic sampling depth is of course less than that using HAXPES, but in principle should allow information over several tens of nm to be extracted.

If the sampling depth is much greater than the morphological feature (*e.g.* for a 10 nm top layer, when your inelastic sampling depth is > 200 nm), the inelastic background features will be more difficult to observe in the survey spectrum. A strong background feature after a photoelectron peak indicates that it is buried, but it will appear "less buried" if your sampling depth extends much deeper into the sample. So in this way, yes, using Al Kα X-rays may make the background signatures of structures over 10–20 nm more obvious. The main problem with applying the technique to Al Kα XPS is that the accuracy is dependent on the range over which the energy loss function (to a higher BE after a photoelectron peak) can be tracked, and up to 1000 eV, there are usually many overlapping peaks, and for Al Kα XPS, many different Auger peaks, which limits that energy loss function range. And importantly, in principle, the background analysis is applicable to any morphological structure sampled (and, indeed, comparing HAXPES with Al Kα XPS can add confidence to your findings).

1 B. F. Spencer, S. Maniyarasu, B. P. Reed, D. J. H. Cant, R. Ahumada-Lazo, A. G. Thomas, C. A. Muryn, M. Maschek, S. K. Eriksson, T. Wiell, T.-L. Lee, S. Tougaard, A. G. Shard and W. R. Flavell, *Appl. Surf. Sci.*, 2021, **541**, 148635.

Olivier Renault responded: This is totally true. The inelastically scattered photoelectrons originate from between 3 times and up to 8 times the mean free path, so we have qualitatively the same dependence on the excitation energy regarding the probing depth than for core-level electrons.

Ayomide Osundare addressed Olivier Renault: In comparison to using a synchrotron, what are the possible challenges one could face while using lab-based HAXPES for resistive switches?

Olivier Renault responded: The main challenges are related to the analytical performances of lab-scale HAXPES with Cr Kα excitation compared to synchrotron: a limited energy resolution (see figures provided in the text) and poorer counting statistics, meaning significantly increased acquisition times. This is the counterpart of the much enhanced accessibility (increasing analytical throughput) and instrumental reliability for quantification.

Thorsten Bartels-Rausch addressed Olivier Renault and Ben Spencer: I'm impressed by the background analysis and modelling. Have you applied this to – or do you see the possibility to do so in future – near-ambient pressure experiments, where gas-phase scattering adds further complexity?

Ben Spencer responded: The application of inelastic background modelling to near-ambient pressure experiments has been demonstrated by Tougaard and Greiner,[1] where spectra are corrected for the distortion caused by the gas; the sample becomes a buried layer, and an effective "overlayer" thickness is related to the gas pressure. Indeed, I expect further application of this analysis to NAP-XPS experiments both in the laboratory and at synchrotron facilities in the future, and it is something we are exploring.

1 S. Tougaard and M. Greiner, *Appl. Surf. Sci.*, 2020, **530**, 147243.

Olivier Renault responded: The effect of electron scattering in the gas phase and the associated specific distortions of XPS spectra recorded at high pressure has been recently addressed, see for instance ref. 1 and 2. A method has been proposed to correct the spectra from this effect, following the inelastic background subtraction performed in conventional XPS or in HAXPES, as shown in our paper, and enabling easier interpretation of NAP-XPS spectra, which is applicable to all gases. In any case, the conclusions derived from the careful Ti 2p and La 3d peak fitting show that an internal consistency is reached and a physical agreement is found; therefore, we do not think that the counting statistics here are a big issue.

1 S. Tougaard and M. Greiner, *Appl. Surf. Sci.*, 2020, **530**, 147543.
2 L. Pielsticker, R. Nicholls, S. Beeg, C. Hartwig, G. Klihm, R. Schögl, S. Tougaard and M. Greiner, *Surf. Interface Anal.*, 2021, **53**, 605–617.

Patrick Lömker said: Using inelastic background modelling is a useful technique for films, as they have a (mostly) well defined morphology. The potential of using gas phase inelastic scattering to model the thickness and pressure of an overlayer seems appealing but I am wary of the unknowns therein, *i.e.* non-uniform distribution of the pressure close to the aperture into the (AP-)XPS spectrometer. These must be calculated. See *e.g.* ref. 1.

1 P. Amann, D. Degerman, M.-T. Lee, J. D. Alexander, M. Shipilin, H.-Y. Wang, F. Cavalca, M. Weston, J. Gladh, M. Blom, M. Björkhage, P. Löfgren, C. Schlueter. P. Loemker, K. Ederer, W. Drube, H. Noei, J. Zehetner, H. Wentzel, J. Åhlund and A. Nilsson, *Rev. Sci. Instrum.*, 2019, **90**, 103102.

Conflicts of interest

There are no conflicts to declare.

Faraday Discussions

ROYAL SOCIETY OF CHEMISTRY

PAPER

Photoelectron spectra of water and simple aqueous solutions at extreme conditions

Zifan Ye,†[a] Cunzhi Zhang†[a] and Giulia Galli*[abc]

Received 10th January 2022, Accepted 27th January 2022
DOI: 10.1039/d2fd00003b

Determining the electronic structure of aqueous solutions at extreme conditions is an important step towards understanding chemical bonding and reactions in water under pressure (*P*) and at high temperature (*T*). We present calculations of the photoelectron spectra of water and a simple solution of NaCl under pressure at conditions relevant to the Earth's interior (11 GPa and 1000 K). We combine first-principles and deep-potential molecular dynamics with electronic structure calculations with dielectric-dependent hybrid functionals. These functionals are defined with a fraction of exact exchange determined from the dielectric constant of the liquid computed in extreme conditions. We find a broadening of the spectra relative to ambient conditions, particularly prominent in the merging of the two main peaks below the onset of the spectra. Furthermore we find an overall red shift at high pressure and temperature, which is however not constant over the whole energy range and varies between 1.1 and 2.4 eV. Our results also show that the anion energy levels are closer to the valence band maximum of the liquid than at ambient conditions, indicating that as *P* and *T* are increased, the defect levels of Cl^- and OH^- in water may eventually lie below the valence band maximum of water. Finally, we characterize the ionization potential of hydrated species deriving from rapid water dissociation, *e.g.* hydrated hydroxide and hydronium, and we elucidate the electronic states associated with proton transfer events at high pressure. Our results represent a first, important step in predicting the electronic properties of solutions in super-critical conditions.

I. Introduction

Water-rich fluids are important constituents of the Earth's crust and mantle[1–3] and they play a central role in many fundamental processes, including metasomatism,[4] transport of oxidized carbon[5,6] and ultimately in the evolution of the continental crust.[3] The presence of so called deep-earth aqueous fluids has been

[a]*Pritzker School of Molecular Engineering, The University of Chicago, Chicago, IL 60637, USA. E-mail: gagalli@uchicago.edu*
[b]*Department of Chemistry, The University of Chicago, Chicago, IL 60637, USA*
[c]*Materials Science Division, Argonne National Laboratory, Lemont, IL 60439, USA*
† These two authors contributed equally, listed in alphabetical order.

reported in aqueous pockets in diamond,[7] metamorphic minerals[8,9] and stable hydrous silicate minerals under pressure (P),[10] and inferred from seismic data[11] and conductivity measurements.[12,13]

While most of the current knowledge of deep-earth aqueous fluids comes from geophysical models[14–23] and from vibrational spectroscopic measurements, progress has been made in recent years in understanding the properties of these fluids at the microscopic level, using first principles simulations. The latter have been used to investigate fundamental properties at high P and T (HPT) including the dielectric constant of water,[24,25] ionic speciation,[5,6] ion solvation[26,27] and ionic conductivity of simple solutions.[28–31] However, relatively little is known about the electronic structure[32] of aqueous fluids under extreme conditions, which is important to understand, for instance, electron and proton transfer processes in redox reactions occurring[33,34] in water and solutions in contact with rocks in the Earth's interior.

Many electronic structure probes used at ambient conditions are difficult to employ at high temperature and pressure. For example, photoelectron (PE) spectroscopy has been widely used to probe the electronic properties of hydrogen bonded liquids, providing information on their occupied energy levels,[35–37] including the ionization potentials (IP) of solvated ions,[38–41] *e.g.* hydrated OH^-, H_3O^+, Cl^-, Na^+, thus elucidating their structural and chemical environment. However, obtaining PE spectra for HPT water and solutions is a challenging task not only because of the reactivity of water at high P and T, but also for the difficulty in preparing super-critical liquids in contact with vacuum and detecting PE signals with high time-resolution in the presence of rapid thermal expansion.[42–44]

First principles theoretical methods based on quantum mechanical calculations have been applied to predict the PE spectra of several aqueous solutions at ambient P and T (APT),[40,41,45–47] providing results in excellent agreement with experiments. In particular, recent investigations have shown that a combination of first-principles molecular dynamics (FPMD) and density functional theory calculations with dielectric-dependent hybrid (DDH) functionals[48] is a robust, predictive approach to study photoelectron spectra of aqueous fluids.[40,41] It is hence interesting to explore how to predict PE under extreme conditions and gain insight into the electronic properties of aqueous fluids at high P and T.

Here, we generalize the approach recently used to compute PE spectra at ambient conditions to the study of aqueous fluids under pressure, at high temperature. We carry out first principles simulations with semilocal and hybrid functionals, and we consider two specific systems at conditions (11 GPa and 1000 K) relevant to the Earth upper mantle, where pressure can reach ~13 GPa and temperature ~1700 K;[49,50] in particular we investigate pure water and a 0.68 M NaCl solution, whose structural and bonding properties have been previously studied.[27,31] We report PE spectra on an absolute scale, by using a technique to refer energy levels to vacuum which is appropriate for hot compressed fluids; we discuss results obtained with dielectric-dependent hybrid functionals, and with a fraction of exact exchange determined for the specific conditions studied here.[32] We also analyze the contribution of dissociated species to the PE spectra, in particular hydroxide (OH^-), hydronium (H_3O^+) ions and H_4O_2, which play an important role in proton transfer processes and acid–base chemistry,[51–53] and contribute to the increased ionic conductivity[31] of water under extreme conditions.

The rest of the paper is organized as follows: the methods used here are described in the next section, followed by a discussion of our results in Section 3. Our conclusions are presented in Section 4.

II. Methods

A. First-principles simulations and electronic structure calculations

We consider the FPMD trajectories generated at 11 GPa and 1000 K by Rozsa *et al.* and Zhang *et al.*[27,31] to model pure water and a 0.68 M NaCl solution, respectively. These trajectories were obtained with the Qbox[54] code, employing the Perdew–Burke–Ernzerhof (PBE)[55] functional, with 64 water molecules to model pure water and 126 water molecules and 1 NaCl formula to model the solution. The length of the trajectory is ~240 (300) ps for pure water (NaCl solution). We note that in the case of the solution, the Na–Cl distance was constrained to be <6 Å in ref. 27 to sample ion-pair configurations. From the FPMD trajectories we extracted 500 (285) snapshots by every ~0.5 (~1) ps for pure water (NaCl solution), which we then used to compute the electronic density of state (EDOS), as well as the projected EDOS, with the Quantum ESPRESSO[56,57] code, using both the PBE semi-local and dielectric dependent hybrid (DDH) functionals.[48] The DDH functional was previously shown to provide an excellent description of PE spectra at ambient conditions, in good agreement with experimental measurements.[40,41] In order to model HPT conditions, we used a DDH functional with a fraction of exact exchange (42%) equal to the inverse of the macroscopic dielectric constant of water at 11 GPa and 1000 K, which was determined to be 2.37 in a recent study.[32] In our electronic structure calculations, we used norm-conserving pseudopotentials[58] to describe the interactions between core and valence electrons, a plane-wave basis set with a kinetic energy cutoff of 60 Ry and only the Γ-point to sample the Brillouin zone.

To predict PE spectra comparable to experiments, the calculation of absolute energies referenced to vacuum is required, and the strategy chosen to do so in our work is described in detail in the next section.

B. Evaluation of absolute orbital energies

While eigenvalue and total energy differences are well defined, absolute energies obtained with the plane-wave pseudopotential method for periodic systems are not.[40,45,59] In order to obtain absolute energies $\tilde{\varepsilon}_i$, *i.e.* to refer the electronic energies to vacuum, we evaluate:

$$\tilde{\varepsilon}_i = \varepsilon_i - \Delta V - V_{bulk} \tag{1}$$

where ε_i denotes Kohn–Sham eigenvalues computed for either bulk water or a NaCl solution in periodic plane-wave calculations. The term V_{bulk} is the average electrostatic potential for bulk water samples, which can be easily computed following the procedure outlined in ref. 60. Consistent with the results reported in ref. 40, we found that the value of V_{bulk}, 0.81 eV, is insensitive to the choice of the energy functional. The term ΔV is the difference between the electrostatic potential in the bulk region of a water sample interfaced with vacuum, and the vacuum region. In previous calculations of PE spectra at ambient conditions, the term ΔV was obtained by carrying out MD simulations of water in contact with vacuum using empirical potentials and then carrying out electronic structure calculations on the trajectories generated for the water/vacuum slab. However, this scheme is not applicable at HPT, where hot, compressed water in contact with a vacuum region would immediately expand. Hence, here we adopt a different approach to compute ΔV that we call *Slice* and that we compare in Fig. 1 with other approaches adopted in the literature.

Fig. 1 illustrates three possible strategies to obtain ΔV, which are summarized below.

• *Vacuum*: a water sample at a chosen, initial density is in contact with vacuum within a periodically repeated slab that is equilibrated using MD at a given temperature. This scheme was used for calculations of PE spectra at APT and is applicable if the density of water is such that immediate evaporation is not observed, over the time scale of the simulations. Hence it is not applicable to simulate water at high P and T, where the sample would tend to expand into vacuum over short time scales.

• *Wall*: similar to the previous approach, a water sample is in contact with vacuum within a periodically repeated slab that is equilibrated using MD at given temperature. However in this case a repulsive potential is added to the Hamiltonian so as to maintain the water sample at a density corresponding to the desired pressure. Such a repulsive potential prevents water from evaporating into the vacuum portion of the slab. This method can in principle be used at any P–T conditions, however the results may be affected by the choice of the specific form of the repulsive potential and the approach requires the use of relatively large systems to obtain a converged electrostatic potential.

• *Slice*: a relatively large bulk water model is equilibrated in a periodically repeated supercell at desired P and T conditions. A slice of the sample is then extracted after equilibration and placed in contact with vacuum, with no further MD simulations of the whole slab. To avoid dangling bonds at the interface, when taking the slice, we retain the H–O covalent bonds. This scheme is applicable to any P–T conditions, provided a slice of sufficient thickness is chosen.

To reduce the computational cost in preparing the samples required to obtain ΔV, we carried out simulations with empirical potentials first, for benchmarking purposes, and then we used the deep-MD potential[61,62] implemented in the

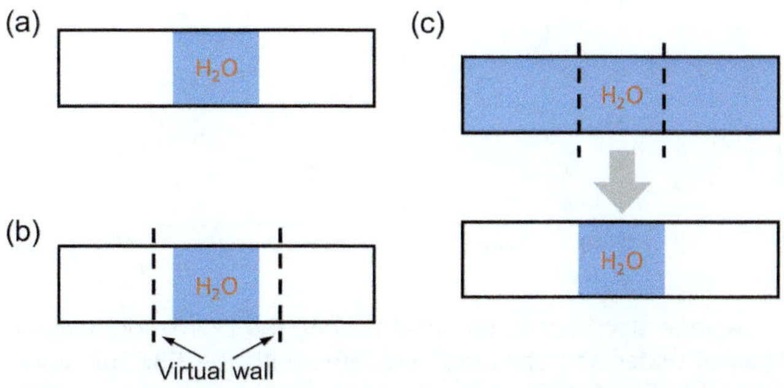

Fig. 1 Strategies to generate water samples in contact with vacuum: *Vacuum* (a), *Wall* (b) and *Slice* (c) (see text). The three rectangles schematically represent a slab used and periodically repeated in our pseudopotential-plane-wave calculations, and the white and blue regions denote vacuum and a water sample, respectively.

DeePMD-kit package.[63] The deep-MD potential (DP)[64] is trained on first principle data obtained with the SCAN functional[65] at HPT and allows for the proper description of dissociated species at extreme conditions, unlike empirical potentials. In addition, using the DP potential, simulations with cells larger than those affordable with FPMD can be carried out and longer time scales can be sampled with first-principle accuracy, as described below.

In order to assess the robustness of our methodology, we first carried out simulations with the TIP3P[66] potential and we compared results obtained with the *Vacuum* and *Slice* procedures at ambient conditions; we also compared our results with those previously reported in the literature. We then further compared the results obtained with the *Wall* and *Slice* methods at HPT. MD simulations were carried out in the *NVT* ensemble with the LAMMPS package.[67,68] A timestep of 2 (1) fs was used for TIP3P simulations at APT (11 GPa & 1000 K). When using the *Vacuum* method, our samples consisted of 108 water molecules interfaced with vacuum in a cell of dimensions 12.77 × 12.77 × 80 Å. When using the *Slice* method, we equilibrated water samples with 435 and 683 molecules at APT and 11 GPa & 1000 K, respectively, corresponding to the densities of 1.0 and 1.57 g cm^{-3} in a cell of dimensions 12.77 × 12.77 × 80 Å. The thickness of a slice was chosen to be 20 (18) Å at APT (11 GPa & 1000 K). When adopting the *Wall* method, we considered 300 water molecules at 11 GPa & 1000 K, and the *wall/lj*126 scheme implemented in the LAMMPS code. The distance between two virtual walls simulating a repulsive potential in the Hamiltonian was set to be 39 Å. The repulsive part of Lennard-Jones interactions between the wall and water was described by using carbon–oxygen Lennard-Jones parameters[69] to mimic the interaction with water oxygen atoms only. For all the procedures adopted here, the MD simulations with the TIP3P potential were performed for 4 ns following a 2 ns equilibration, with 200 equally-spaced snapshots extracted every 20 ps.

The results for the average potential obtained with the PBE functional on the configurations extracted from our MD simulations are presented in Fig. 2. Our

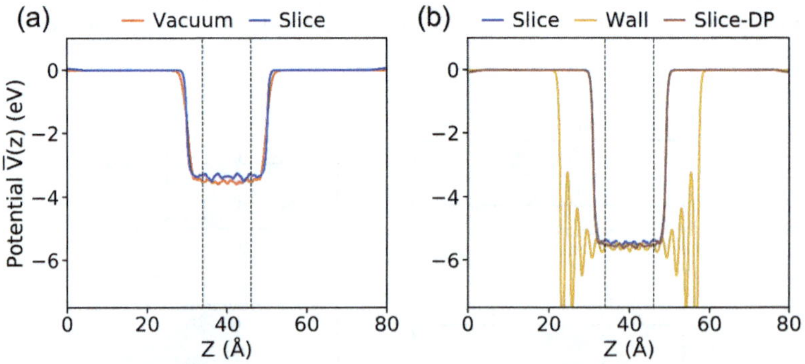

Fig. 2 Plane-average electrostatic potential $\bar{V}$ along the z direction perpendicular to the interface between water and vacuum, calculated with the PBE functional at ambient conditions (a) and 11 GPa & 1000 K (b). The vacuum level is set to zero. The equilibrium configurations were generated using molecular dynamics simulations and the TIP3P potential except for the case denoted as Slice-DP, for which we used the deep-MD potential.[61] See Fig. 1 for the definition of different procedures. The dashed lines define the region used for averaging the electrostatic potential.

results with the *Vacuum* method are nearly identical with those reported using the same scheme in the literature: 3.53 eV *versus* 3.54 eV.[40] When using the *Slice* method we obtain a value of 3.36 eV and we consider the difference relative to the result with the *Vacuum* method (0.17 eV) to be negligible, given that the position of the valence band maximum (VBM) of water is at approximately 10 eV (ref. 39–41 and 46) relative to vacuum. The difference of 0.17 eV may be ascribed to the specific surface of the chosen slice (the statistical error in our evaluation of ΔV is smaller than 0.1 eV).

At 11 GPa and 1000 K, as reported in Fig. 2(b), ΔV is computed to be 5.43 (*Slice*) and 5.58 (*Wall*) eV. Again, we consider the difference of 0.15 eV to be small for the purpose of our study.

Next we compared the results obtained with the *Slice* method and the TIP3P potential with those obtained with the DP potential. When using the DP potential, we performed 210 independent MD simulations (with a time step of 0.25 fs) each of 13 ps long, at the end of which we extracted configurations for our electronic structure calculations. We obtained $\Delta V = 5.53$ eV, in good agreement with the value computed with TIP3P, indicating that the value of ΔV depends rather weakly on subtle structural differences in the liquid samples. These comparisons between different approaches gave us confidence that the *Slice* method is a reliable technique to use both at APT and HPT.

Therefore, in the following we compute ΔV with the *Slice* method and the DP potential to obtain the spectra at HPT and to discuss the physical properties of water and the NaCl solution. We note that when using the DDH functional, we could only carry out calculations of ΔV for a small number of configurations (5), much smaller than with the PBE functional. Hence we estimated the difference between ΔV computed with PBE and DDH for those limited number of configurations, and we applied the same difference between the DDH and PBE result (5.31 *vs.* 5.53 eV) to all eigenvalues obtained at the DDH level of theory for the bulk samples.

III. Results and discussion

A. Photoelectron spectra of water and NaCl solution at high pressure and temperature

The PE spectra are evaluated as the EDOS of single-particle eigenvalues referred to vacuum, as an approximation to the electronic binding energies (BE). In Fig. 3, we compare low and high pressure spectra of the NaCl solution computed at the PBE (left hand side) and DDH (right hand side) level of theory. The positions in energy of peaks' maxima are summarized in Table 1.

The four major peaks (labeled, in order of increasing BE as: $2a_1$, $1b_2$, $3a_1$, and $1b_1$) are notably broadened at HPT, with the $3a_1$ and $1b_1$ features strongly overlapping with each other, consistent with larger structural fluctuations in the HPT samples. The DDH results show that the relative spacing of the water $2a_1$ band with respect to the $1b_1$ peak is decreased by ~1.1 eV at high *P*, while that of $1b_2$ and $3a_1$ bands are less affected at extreme conditions. Interestingly, calculations with the DDH functional predict a reduction of the BE by roughly 1.1–2.4 eV at high pressure, depending on the character of electronic states; however calculation with the PBE or standard hybrid functionals such as PBE0 (ref. 70 and 71) yield negligible differences between the band positions in the PE spectra at high

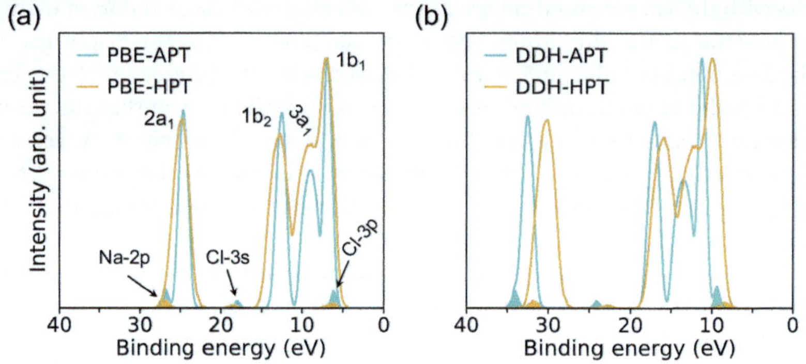

Fig. 3 Photoelectron spectra of a 1 M NaCl solution at ambient conditions (APT)[40] and a 0.68 M NaCl solution at high pressure conditions (HPT) (11 GPa & 1000 K) obtained with the PBE (a) and DDH (b) functionals. The intensities are rescaled to those of the water $1b_1$ peak; the shaded areas show the distribution of ionization potentials of solvated Cl^- and Na^+ ions.

Table 1 Electron binding energy (eV) of NaCl solution at ambient conditions (APT) and high pressure and temperature (HPT) (11 GPa & 1000 K) obtained with the PBE and DDH functionals. See Fig. 3 for the labeling of the electronic states

Method	2p (Na^+)	$2a_1$	3s (Cl^-)	$1b_2$	$3a_1$	$1b_1$	3p (Cl^-)
PBE-APT[a]	26.90	24.71	18.06	12.63	9.03	6.99	6.20
DDH-APT[a]	34.04	32.57	24.15	17.14	13.56	11.34	9.46
PBE-HPT[b]	27.02	24.81	18.62	12.84	8.98	7.12	6.28
DDH-HPT[b]	31.92	30.18	22.93	15.98	12.01	10.02	8.41

[a] Results at APT of a 1 M NaCl solution from ref. 40. [b] Results at 11 GPa & 1000 K of a 0.68 M NaCl solution.

and low pressure. This finding is important and highlights the need to correctly take into account screening effects at HPT and correspondingly adjust the amount of exact exchange used to define hybrid functionals (~60% in DDH at APT;[40,41] ~42% in DDH at the HPT conditions of this study, corresponding to the dielectric constants $\varepsilon_\infty = 1.78$ and 2.37, respectively).

The ionization potentials (IPs) of hydrated Cl^- and Na^+ at APT and HPT are indicated in Fig. 3 as shaded areas. Similar to the BE of water electronic states, the IPs of ions are underestimated by PBE, relative to DDH, and DDH results show a red shift with increasing P and T which is not reproduced by the PBE functional. In addition, when using the DDH functional, the relative positions of IPs referenced to water is slightly different at low and high P. In particular, the separation between the 3p-level of Cl^- and the $1b_1$ peak of water is decreased by 0.27 eV, suggesting a modified electronic structure near the band edges of water at high P, which in turn may affect the chemical reactivity of solutions under extreme conditions. Our results hint at a possible tendency of anion levels to move below the valence band of water as a function of increasing pressure, at high T. We note that the BE of water and IP of ions are not sensitive to the Cl^-–Na^+ distance, hence the results reported here should be representative of weakly concentrated solutions.

B. Water dissociation

Fig. 4 compares the spectra obtained at HPT for water and the solution. We find that the effects of ions on the electronic properties of water are minor at the concentrations considered here (0.68–1 M);[41] for example, the centers of the $1b_1$ and $1b_2$ peaks obtained in HPT water and in the solutions with the DDH functional differ by only 0.03 and 0.13 eV, respectively.

In our FPMD simulations, we directly observe frequent dissociation events at HPT, indicating that water itself becomes an electrolyte. The short-lived dissociated products[27,31] are characterized using a cutoff distance for O–H bonds of 1.25 Å. A detailed analysis reveals that about 2% of water molecules are dissociated, into species including solvated HO^-, H_3O^+ and H_4O_2. These species play a critical role in proton transfer events and acid–base reactions in water.

As shown in Fig. 5(a), the signal coming from ionic species resulting from water dissociation spans the entire spectrum range. The signal is weak and expected to become more prominent at higher pressure and temperature, where the concentration of dissociated species will increase. The contributions to the overall PE signal of hydrated HO^- and H_3O^+ are presented in Fig. 5(b) and (c), and it has been identified by carrying out projected EDOS calculations. Similar to ambient conditions, the solvated HO^- level is just above the VBM of water, and it originates from two degenerate lone-pair orbitals ($1e_1$) localized on the oxygen atom. The solvated H_3O^+ levels lie instead close to the high-energy edge of the $1b_2$ feature of bulk water and originates from the ionization from two degenerate covalent orbitals with mainly p-character (1e). However the IPs of hydrated HO^- and H_3O^+ at HPT, peaking at 8.9 and 17.75 eV, are red-shifted relative to ambient conditions, where they are centered at ~9.2 and ~20 eV (~9.99 and ~19.01 eV) in the experiments of ref. 39 (calculated by many-body perturbation approach[46]). In addition, the separation between the $1e_1$ level of

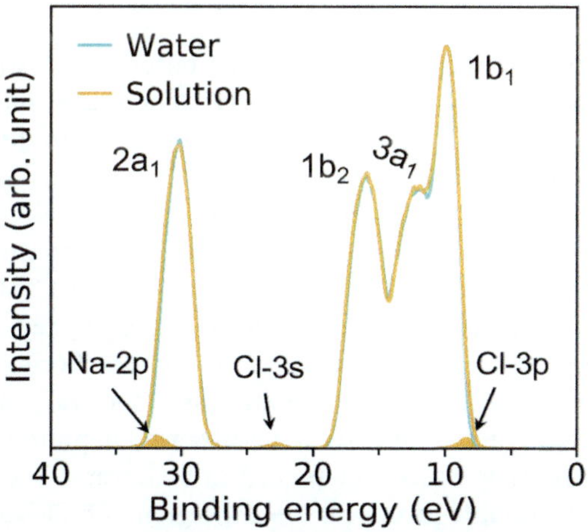

Fig. 4 Photoelectron spectra of water and a 0.68 M NaCl solution obtained with the DDH functional at 11 GPa & 1000 K. The intensities of the spectra are rescaled to the water $1b_1$ peak; the shaded areas indicate the ionization potential of solvated Cl^- and Na^+ ions.

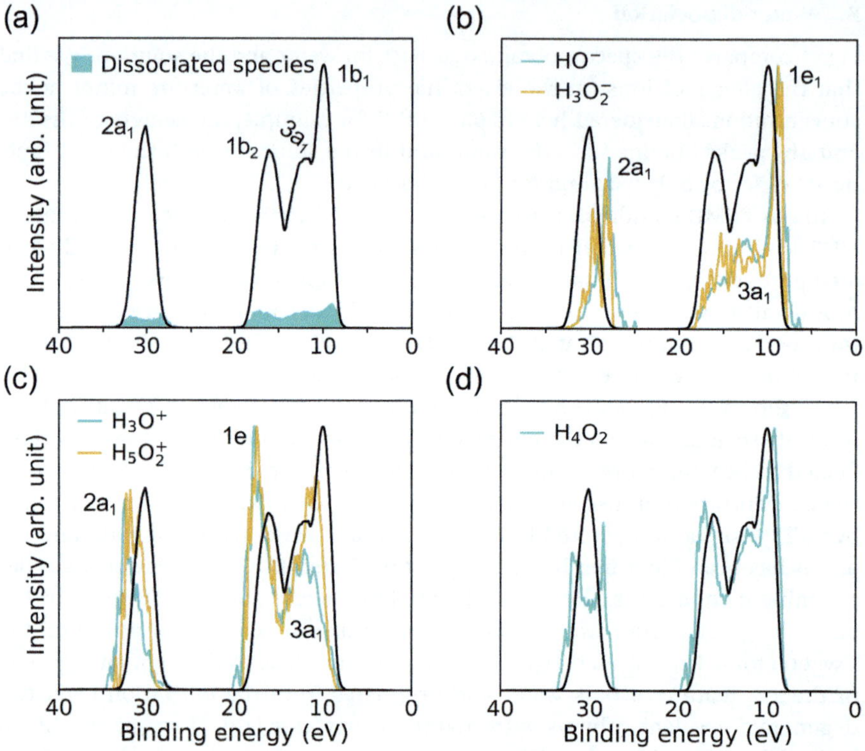

Fig. 5 Signals of dissociated water species contributing to the photoelectron spectra (PE) of water under pressure, at high temperature, calculated with the DDH functional at 11 GPa & 1000 K. The full spectrum of bulkwater is represented by the black line. (a) The shaded areas (cyan) represent the total PE spectrum from all dissociated species (magnified five times for clarity). The intensities in (b) HO^- and $H_3O_2^-$, (c) H_3O^+ and $H_5O_2^+$, (d) H_4O_2 are magnified between ~100 (*e.g.* HO^- and H_4O_2) and ~1000 (*e.g.* $H_3O_2^-$) times.

HO^- and $1b_1$ peak of water is ~1.91 eV at APT based on experiment,[39] while we find that at HPT it is 1.12 eV, when using DDH. Hence our simulations show that similar to the case of the Cl^-, the level of HO^- gets closer to the VBM of water as pressure is increased.

To understand the electronic states involved in proton transfer events at high pressure, we plot the IP distributions for three hydrated intermediate complexes $H_3O_2^-$, $H_5O_2^+$ and H_4O_2 in Fig. 5(b–d). With respect to HO^-, the IP of $H_3O_2^-$ is blueshifted with a slightly enhanced signal near the $1b_2$ region of bulk water, similar to what is found in proton transfer events at APT.[46] Compared to H_3O^+, we observe a red shift in the IP of $H_5O_2^+$ with increased spectral weights near the $1b_1$ region of bulk water, again similar to what is observed in proton transfer events at APT.[46] For solvated H_4O_2 cluster referenced to bulk water, the major change at high pressure is the double-peak near the $2a_1$ region of bulk water, giving rise to the $2a_1$ peaks with lower or higher IP in the resulting species HO^- and H_3O^+. The results reported here for ionization potentials of dissociated species in water may help, at least qualitatively, to understand proton transfer reactions at high pressure.

IV. Conclusions

In summary, we proposed a new computational scheme to evaluate photoelectron spectra of water and simple aqueous solutions at high P and T (HPT). By combining deep-MD and first-principles simulations, and electronic structure calculations with hybrid functionals, we computed the absolute orbital energies of water and a 0.68 M NaCl solution at 11 GPa and 1000 K, which are conditions relevant to the earth mantle. We found a broadening of the spectra relative to ambient conditions, particularly prominent for the merging of the $3a_1$ and $1b_1$ peaks of water. In addition we found an overall red shift of the HPT spectra relative to ambient conditions, which is however not constant over the whole energy range and varies between 1.1 and 2.4 eV. Our results also show that the anion energy levels are closer to the VBM of water at high pressure and temperature and we speculate that as P and T are increased, the "defect levels" of Cl^- and OH^- in water may eventually lie below the valence band maximum. We also characterized the IP of hydrated species deriving from rapid water dissociation, *e.g.* hydrated hydroxide and hydronium, and elucidated the electronic states associated with proton transfer events at high pressure. Our results represent a first step in predicting the electronic properties of solutions in super-critical conditions. Finally we note that the prediction of hybrid and PBE functionals for the electronic properties of the APT and HPT fluids are markedly different; in addition PBE predicts no notable changes between high and low pressure spectra, at variance with dielectric dependent hybrid functionals, when used with a screening fraction computed for the specific HPT conditions investigated here.

Conflicts of interest

There are no conflicts to declare.

Acknowledgements

This work was completed with resources provided by the University of Chicago's Research Computing Center. Z. Y. acknowledges support from the Advanced Materials for Energy-Water Systems Center, an Energy Frontier Research Center funded by the US Department of Energy, Office of Science, Basic Energy Sciences. C. Z. acknowledges support from Midwest Integrated Center for Computational Materials as part of the Computational Materials Sciences Program funded by the US Department of Energy, Office of Science, Basic Energy Sciences. We thank Marcello Puligheddu for useful discussions and Linfeng Zhang and Roberto Car for providing the deep-MD potential.

References

1 K. Bucher and I. Stober, *Geofluids*, 2010, **10**, 241.
2 A. Liebscher, *Geofluids*, 2010, **10**, 3.
3 C. E. Manning, *Annu. Rev. Earth Planet. Sci.*, 2018, **46**, 67.
4 A. Putnis and H. Austrheim, *Geofluids*, 2010, **10**, 254.
5 D. Pan and G. Galli, *Sci. Adv.*, 2016, **2**, e1601278.
6 D. Pan and G. Galli, *Nat. Commun.*, 2020, **11**(1), 421.

7 O. Tschauner, S. Huang, E. Greenberg, V. Prakapenka, C. Ma, G. Rossman, A. Shen, D. Zhang, M. Newville, A. Lanzirotti, *et al.*, *Science*, 2018, **359**, 1136.
8 J. Touret, *Lithos*, 2001, **55**, 1.
9 B. W. Yardley and R. J. Bodnar, *Geochem. Perspect.*, 2014, **3**, 1.
10 M. Nishi, T. Irifune, J. Tsuchiya, Y. Tange, Y. Nishihara, K. Fujino and Y. Higo, *Nat. Geosci.*, 2014, **7**, 224.
11 G. C. Beroza and S. Ide, *Annu. Rev. Earth Planet. Sci.*, 2011, **39**, 271.
12 T. J. Shankland and M. E. Ander, *J. Geophys. Res. Solid Earth*, 1983, **88**, 9475.
13 R. Hyndman and P. Shearer, *Geophys. J. Int.*, 1989, **98**, 343.
14 H. C. Helgeson and D. H. Kirkham, *Am. J. Sci.*, 1976, 276.
15 J. C. Tanger and H. C. Helgeson, *Am. J. Sci.*, 1988, **288**, 19.
16 D. A. Sverjensky, B. Harrison and D. Azzolini, *Geochim. Cosmochim. Acta*, 2014, **129**, 125.
17 D. Dolejš and C. Manning, *Geofluids*, 2010, **10**, 20.
18 C. E. Manning, *Rev. Mineral. Geochem.*, 2013, **76**, 135.
19 J. W. Johnson, E. H. Oelkers and H. C. Helgeson, *Comput. Geosci.*, 1992, **18**, 899.
20 D. Fernandez, A. Goodwin, E. W. Lemmon, J. Levelt Sengers and R. Williams, *J. Phys. Chem. Ref. Data*, 1997, **26**, 1125.
21 J.-M. Huizenga, *Lithos*, 2001, **55**, 101.
22 D. Dolejš, *Rev. Mineral. Geochem.*, 2013, **76**, 35.
23 D. I. Foustoukos, *Chem. Geol.*, 2016, **447**, 183.
24 R. Hou, Y. Quan and D. Pan, *J. Chem. Phys.*, 2020, **153**, 101103.
25 D. Pan, L. Spanu, B. Harrison, D. A. Sverjensky and G. Galli, *Proc. Natl. Acad. Sci.*, 2013, **110**, 6646.
26 V. Rozsa and G. Galli, *J. Chem. Phys.*, 2021, **154**, 144501.
27 C. Zhang, F. Giberti, E. Sevgen, J. J. de Pablo, F. Gygi and G. Galli, *Nat. Commun.*, 2020, **11**(1), 3037.
28 A. F. Goncharov, N. Goldman, L. E. Fried, J. C. Crowhurst, I.-F. W. Kuo, C. J. Mundy and J. M. Zaug, *Phys. Rev. Lett.*, 2005, **94**, 125508.
29 M. French, T. R. Mattsson and R. Redmer, *Phys. Rev. B: Condens. Matter Mater. Phys.*, 2010, **82**, 174108.
30 M. French, S. Hamel and R. Redmer, *Phys. Rev. Lett.*, 2011, **107**, 185901.
31 V. Rozsa, D. Pan, F. Giberti and G. Galli, *Proc. Natl. Acad. Sci. U. S. A.*, 2018, **115**, 6952.
32 D. Pan, Q. Wan and G. Galli, *Nat. Commun.*, 2014, **5**(1), 3919.
33 R. A. Marcus, *J. Chem. Phys.*, 1956, **24**, 966.
34 R. A. Marcus and N. Sutin, *Biochim. Biophys. Acta Rev. Bioenerg.*, 1985, **811**, 265.
35 B. Winter and M. Faubel, *Chem. Rev.*, 2006, **106**, 1176.
36 B. Winter, *Nucl. Instrum. Methods Phys. Res. A*, 2009, **601**, 139.
37 R. Seidel, S. Thurmer and B. Winter, *J. Phys. Chem. Lett.*, 2011, **2**, 633.
38 B. Winter, R. Weber, I. V. Hertel, M. Faubel, P. Jungwirth, E. C. Brown and S. E. Bradforth, *J. Am. Chem. Soc.*, 2005, **127**, 7203.
39 B. Winter, M. Faubel, I. V. Hertel, C. Pettenkofer, S. E. Bradforth, B. Jagoda-Cwiklik, L. Cwiklik and P. Jungwirth, *J. Am. Chem. Soc.*, 2006, **128**, 3864.
40 A. P. Gaiduk, M. Govoni, R. Seidel, J. H. Skone, B. Winter and G. Galli, *J. Am. Chem. Soc.*, 2016, **138**, 6912.
41 T. A. Pham, M. Govoni, R. Seidel, S. E. Bradforth, E. Schwegler and G. Galli, *Sci. Adv.*, 2017, **3**, e1603210.

42 O. Link, E. Vöhringer-Martinez, E. Lugovoj, Y. Liu, K. Siefermann, M. Faubel, H. Grubmüller, R. B. Gerber, Y. Miller and B. Abel, *Faraday Discuss.*, 2009, **141**, 67.
43 E. Vöhringer-Martinez, O. Link, E. Lugovoy, K. Siefermann, F. Wiederschein, H. Grubmüller and B. Abel, *Phys. Chem. Chem. Phys.*, 2014, **16**, 19365.
44 T. Gladytz, B. Abel and K. R. Siefermann, *Phys. Chem. Chem. Phys.*, 2015, **17**, 4926.
45 T. A. Pham, C. Zhang, E. Schwegler and G. Galli, *Phys. Rev. B: Condens. Matter Mater. Phys.*, 2014, **89**, 060202.
46 C. W. Swartz and X. Wu, *Phys. Rev. Lett.*, 2013, **111**, 087801.
47 D. Opalka, T. A. Pham, M. Sprik and G. Galli, *J. Chem. Phys.*, 2014, **141**, 034501.
48 J. H. Skone, M. Govoni and G. Galli, *Phys. Rev. B: Condens. Matter Mater. Phys.*, 2014, **89**, 195112.
49 D. J. Frost, *Elements*, 2008, **4**, 171.
50 A. B. Thompson, *Nature*, 1992, **358**, 295.
51 M. Chen, L. Zheng, B. Santra, H.-Y. Ko, R. A. DiStasio Jr, M. L. Klein, R. Car and X. Wu, *Nat. Chem.*, 2018, **10**, 413.
52 M. Tuckerman, K. Laasonen, M. Sprik and M. Parrinello, *J. Chem. Phys.*, 1995, **103**, 150.
53 D. Marx, A. Chandra and M. E. Tuckerman, *Chem. Rev.*, 2010, **110**, 2174.
54 F. Gygi, *IBM J. Res. Dev.*, 2008, **52**, 137.
55 J. P. Perdew, K. Burke and M. Ernzerhof, *Phys. Rev. Lett.*, 1996, **77**, 3865.
56 P. Giannozzi, S. Baroni, N. Bonini, M. Calandra, R. Car, C. Cavazzoni, D. Ceresoli, G. L. Chiarotti, M. Cococcioni, I. Dabo, *et al.*, *J. Phys.: Condens. Matter*, 2009, **21**, 395502.
57 P. Giannozzi, O. Andreussi, T. Brumme, O. Bunau, M. B. Nardelli, M. Calandra, R. Car, C. Cavazzoni, D. Ceresoli, M. Cococcioni, *et al.*, *J. Phys.: Condens. Matter*, 2017, **29**, 465901.
58 D. Hamann, M. Schlüter and C. Chiang, *Phys. Rev. Lett.*, 1979, **43**, 1494.
59 R. Sundararaman and Y. Ping, *J. Chem. Phys.*, 2017, **146**, 104109.
60 C. G. Van de Walle and R. M. Martin, *Phys. Rev. B: Condens. Matter Mater. Phys.*, 1986, **34**, 5621.
61 L. Zhang, J. Han, H. Wang, R. Car and E. Weinan, *Phys. Rev. Lett.*, 2018, **120**, 143001.
62 L. Zhang, J. Han, H. Wang, W. A. Saidi, R. Car and E. Weinan, *Proceedings of the 32nd International Conference on Neural Information Processing Systems, NIPS18*, Curran Associates Inc., Red Hook, NY, USA, 2018, pp. 4441–4451.
63 H. Wang, L. Zhang, J. Han and E. Weinan, *Comput. Phys. Commun.*, 2018, **228**, 178.
64 L. Zhang, H. Wang, R. Car and E. Weinan, *Phys. Rev. Lett.*, 2021, **126**, 236001.
65 J. Sun, A. Ruzsinszky and J. P. Perdew, *Phys. Rev. Lett.*, 2015, **115**, 036402.
66 D. J. Price and C. L. Brooks III, *J. Chem. Phys.*, 2004, **121**, 10096.
67 A. P. Thompson, H. M. Aktulga, R. Berger, D. S. Bolintineanu, W. M. Brown, P. S. Crozier, P. J. in't Veld, A. Kohlmeyer, S. G. Moore, T. D. Nguyen, *et al.*, *Comput. Phys. Commun.*, 2021, 108171.
68 S. Plimpton, *J. Comput. Phys.*, 1995, **117**, 1.
69 G. Hummer, J. C. Rasaiah and J. P. Noworyta, *Nature*, 2001, **414**, 188.
70 J. P. Perdew, M. Ernzerhof and K. Burke, *J. Chem. Phys.*, 1996, **105**, 9982.
71 C. Adamo and V. Barone, *J. Chem. Phys.*, 1999, **110**, 6158.

Faraday Discussions

ROYAL SOCIETY OF CHEMISTRY

PAPER

Predicting core electron binding energies in elements of the first transition series using the Δ-self-consistent-field method†

J. Matthias Kahk *[a] and Johannes Lischner[b]

Received 29th November 2021, Accepted 8th March 2022

DOI: 10.1039/d1fd00103e

The Δ-Self-Consistent-Field (ΔSCF) method has been established as an accurate and computationally efficient approach for calculating absolute core electron binding energies for light elements up to chlorine, but relatively little is known about the performance of this method for heavier elements. In this work, we present ΔSCF calculations of transition metal (TM) 2p core electron binding energies for a series of 60 molecular compounds containing the first row transition metals Ti, V, Cr, Mn, Fe and Co. We find that the calculated TM $2p_{3/2}$ binding energies are less accurate than the results for the lighter elements with a mean absolute error (MAE) of 0.73 eV compared to experimental gas phase photoelectron spectroscopy results. However, our results suggest that the error depends mostly on the element and is rather insensitive to the chemical environment. By applying an element-specific correction to the binding energies the MAE is reduced to 0.20 eV, similar to the accuracy obtained for the lighter elements.

1 Introduction

Theoretical calculations of core electron binding energies are important for aiding the interpretation of experimental core level X-ray photoelectron spectra. In recent years, it has been shown that the all-electron Δ-Self-Consistent-Field (ΔSCF) method[1] can yield accurate absolute core electron binding energies in both free molecules and periodic solids, provided that computational details such as the choice of the exchange–correlation functional, the treatment of relativistic effects, basis set convergence, and size convergence in periodic calculations are

[a]*Institute of Physics, University of Tartu, W. Ostwaldi 1, 50411 Tartu, Estonia. E-mail: juhan.matthias.kahk@ut.ee*

[b]*Department of Materials, and the Thomas Young Centre for Theory and Simulation of Materials, Imperial College London, London SW7 2AZ, UK*

† Electronic supplementary information (ESI) available: Full numerical results, core-enhanced numerical basis sets used for the atoms with a core hole, an evaluation of the basis sets used in this work, relative to uncontracted Gaussian basis sets, sample control.in files, and the relaxed structures of all molecules. See DOI: 10.1039/d1fd00103e

carefully considered.[2–5] It has also been shown that eigenvalue self-consistent full-frequency GW calculations can yield accurate absolute core electron binding energies in free molecules.[6]

However, thus far, almost all theoretical studies of core electron binding energies have focused on a narrow subset of elements in the periodic table. In particular, the prediction of 1s core electron binding energies in the elements carbon to fluorine has attracted the most attention.[2–24] In addition, a few studies have considered the 2p core levels of elements of the third row of the periodic table.[2,3,7–11]

Much less work has been carried out for systems containing transition metal atoms. A few studies applied the ΔSCF approach to atomic systems: all subshells of neutral atoms with Z between 2 and 106 were considered in ref. 25, and 2s and 2p binding energies in isolated transition metal ions were computed in ref. 26. Moreover, methods that go beyond the conventional ΔSCF approach, where the wavefunction of the core hole state is expressed as a combination of multiple Slater determinants, have recently been used to study Ti 2p photoemission from $SrTiO_3$, and Fe 2p photoemission from FeO and Fe_2O_3 by Bagus *et al.*[27–29] These studies are primarily concerned with explaining the satellite structures and peak broadenings observed in experimental measurements. Finally, we note that ΔSCF calculations based on Hartree–Fock theory and complete active space SCF (CASSCF) calculations of Zn 2s and 2p core electron binding energies in the Zn atom, small molecular compounds of Zn, and small clusters of Zn metal and ZnO were reported by Rößler *et al.* in ref. 30 and 31.

In this work, we examine the performance of the all-electron ΔSCF method based on density functional theory (DFT) for predicting $2p_{3/2}$ binding energies in molecules containing first row transition metals. In particular, we carry out calculations for 60 different molecules containing Ti, V, Cr, Mn, Fe and Co using the formalism introduced in ref. 2 and 3 and compare the calculated core electron binding energies to experimental values from gas phase photoelectron spectroscopy. We find that the discrepancy between calculated and measured binding energies is larger than for the lighter elements. Importantly, however, our results suggest that the error depends predominantly on the element and is rather insensitive to the atom's chemical environment. By correcting for this systematic error in the calculated binding energies, highly accurate absolute binding energies can be obtained.

2 Methods

All core electron binding energies reported in this work have been calculated using the ΔSCF method, in which the binding energy is obtained as the total energy difference between the N-electron ground state and the $N-1$ electron state with a core hole. The total energy of the $N-1$ electron state is calculated by converging the self-consistent field in the presence of an explicit core hole, which is achieved by imposing a non-Aufbau-principle occupation of the Kohn–Sham eigenstates at each SCF iteration. The Maximum Overlap Method (MOM) is used to keep track of the core hole, in case the energy ordering of the core orbitals changes between successive SCF iterations.[32,33] The calculations have been performed using the formalism that was originally introduced in ref. 2. In particular, we use Density Functional Theory (DFT) with the Strongly Constrained, Appropriately Normed

(SCAN) exchange–correlation functional[34] and the spin-unrestricted open shell Kohn–Sham method. In order to examine the effect of the exchange–correlation functional on the accuracy of the calculated binding energies, additional calculations are performed using the PBE and PBE0 functionals.[35,36] Scalar relativistic effects are accounted for *via* the scaled Zeroth Order Regular Approximation (scaled ZORA).[37–39] To enable direct comparison with experimental measurements for the $2p_{3/2}$ core electron binding energies, we subtract 1/3 of the experimental spin–orbit splitting from the calculated scalar-relativistic binding energy. We have used spin–orbit splittings of 6.4 eV, 7.7 eV, 9.7 eV, 11.2 eV, 13.1 eV, and 15.1 eV for the elements Ti, V, Cr, Mn, Fe, and Co respectively.[40]

DFT calculations were carried out using the all-electron electronic structure program FHI-aims, in which the Kohn–Sham eigenstates are represented in terms of atom-centered basis functions defined on a numerical grid.[41–43] For the atoms with a core hole, the default "Tier" basis sets of FHI-aims were augmented with additional functions in order to facilitate the relaxation of all remaining core and valence electrons in the presence of a core hole. These core-augmented basis sets are provided in the ESI† which also contains further details on basis set convergence. For the remaining atoms, the default basis sets of FHI-aims at the "tight" level were employed.[41] All ΔSCF calculations were performed at the relaxed molecular geometries. For geometry relaxation, the SCAN functional, the "atomic ZORA" treatment of relativistic effects, and the default "tight" basis sets of FHI-aims were used.[41] The integration grids were tightened as follows: the radial multiplier was set to 8, and the outer angular grid was increased by one step.[41,44] Geometries were relaxed using the Broyden–Fletcher–Goldfarb–Shanno (BFGS) algorithm until the forces on all atoms were less than 5×10^{-3} eV Å^{-1}.[43,45] Sample input files are provided in the ESI.†

3 Results and discussion

A comparison of the calculated and experimental core electron binding energies for a dataset of 60 molecular transition metal compounds containing Ti, V, Cr, Mn, Fe, and Co is presented in Fig. 1 and 2 and Table 1. Here, we only consider systems with a closed-shell ground state with most systems being organometallic compounds that follow the 18-electron rule. A full list of all molecules as well as all numerical results are provided in the ESI.† All experimental values have been taken from the compilation of gas phase X-ray Photoelectron Spectroscopy (XPS) data by Jolly and co-workers.[46]

Fig. 1 shows the results for all molecules in one graph. It can be seen that the scalar relativistic all-electron ΔSCF method with the spin–orbit corrections correctly captures the trend in $2p_{3/2}$ binding energy with increasing atomic number across the series.

A closer look at the results for each individual element, see Fig. 2, indicates that the absolute values of the calculated core electron binding energies are always smaller than the measured value (with only one exception). The mean signed errors of the calculated core electron binding energies are −1.17 eV for Ti $2p_{3/2}$, −1.05 eV for V $2p_{3/2}$, −0.94 eV for Cr $2p_{3/2}$, −0.65 eV for Mn $2p_{3/2}$, −0.48 eV for Fe $2p_{3/2}$, and −0.45 eV for Co $2p_{3/2}$. These errors are significantly greater than the mean signed errors found for the 1s levels in the elements boron to fluorine,

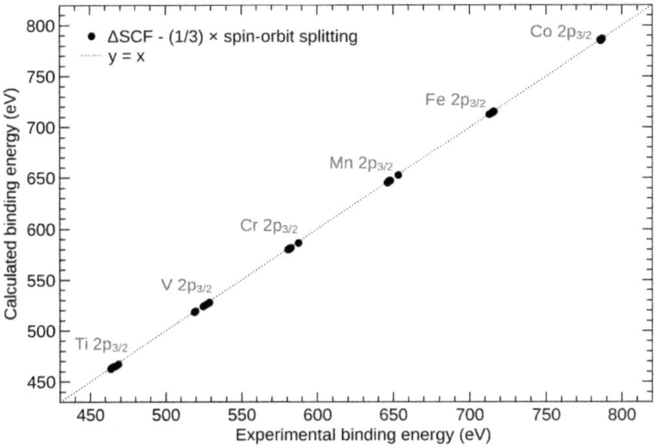

Fig. 1 A comparison of the calculated core electron binding energies from scalar relativistic ΔSCF calculations with spin–orbit correction to experimental values from gas phase photoelectron spectroscopy. The data for 60 molecular compounds containing 3d transition metals Ti, V, Cr, Mn, Fe and Co are shown together in one graph.

and the $2p_{3/2}$ levels in the elements silicon to chlorine obtained in our previous study,[2] which were all smaller than 0.35 eV, and typically below 0.15 eV.

However, Fig. 2 also suggests that the errors of the absolute $2p_{3/2}$ binding energies depend mostly on the element, but not significantly on the chemical

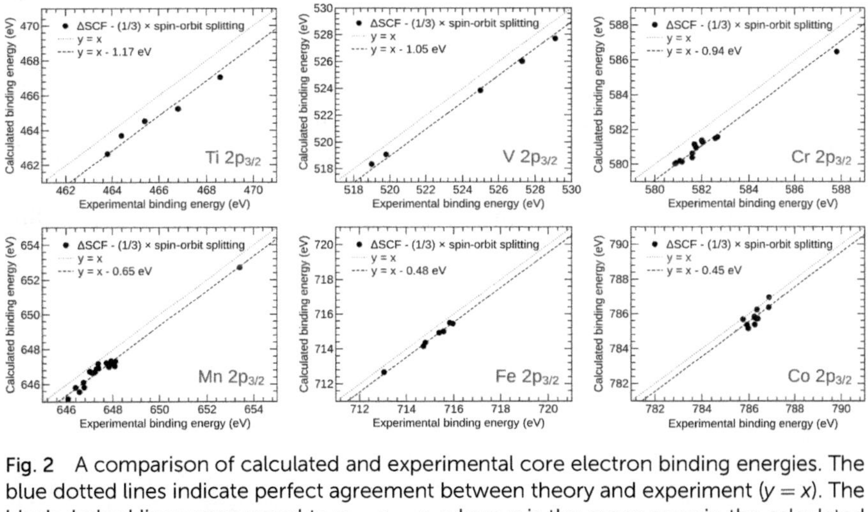

Fig. 2 A comparison of calculated and experimental core electron binding energies. The blue dotted lines indicate perfect agreement between theory and experiment ($y = x$). The black dashed lines correspond to $y = x - a$, where a is the mean error in the calculated core electron binding energies for that element. The R^2 values for the black dashed lines are 0.94 for Ti $2p_{3/2}$, 0.99 for V $2p_{3/2}$, 0.98 for Cr $2p_{3/2}$, 0.98 for Mn $2p_{3/2}$, 0.99 for Fe $2p_{3/2}$, and 0.64 for Co $2p_{3/2}$. The weaker correlation between the calculated and the experimental binding energies for the Co $2p_{3/2}$ core level stems principally from the fact that the entire spread of the experimental values is just over 1 eV, which is of a comparable magnitude to the accuracy of the calculated relative binding energies (0.20 eV for the entire dataset). The mean absolute error in the calculated Co $2p_{3/2}$ binding energies after the application of the element-dependent shift is still relatively small: 0.24 eV.

Table 1 The mean errors and mean absolute errors of the calculated $2p_{3/2}$ binding energies with and without the element-specific correction. The mean errors of the corrected binding energies are not shown because they are zero by construction. All values are given in electronvolts

	Ti $2p_{3/2}$	V $2p_{3/2}$	Cr $2p_{3/2}$	Mn $2p_{3/2}$	Fe $2p_{3/2}$	Co $2p_{3/2}$	Entire dataset
Number of molecules	5	5	13	19	7	11	60
Mean error (uncorrected)	−1.17	−1.05	−0.92	−0.65	−0.48	−0.45	−0.73
Mean absolute error (uncorrected)	1.17	1.05	0.92	0.65	0.48	0.47	0.73
Mean absolute error (corrected)	0.32	0.28	0.19	0.20	0.07	0.24	0.20

environment of the transition metal atom. In particular, it can be seen that the results are well described by a shifted line $y = x - a$ (black dashed lines in Fig. 2), where y and x denote the calculated and measured binding energies, respectively, and a is the magnitude of the systematic error in the calculated core electron binding energies. Of course, the value of a is equal to the mean signed error for each element (given in the previous paragraph). In other words, accurate absolute binding energies can be obtained once an element-specific correction is added to the calculated ΔSCF binding energies. It should be noted that some of the corrections used in this work (*e.g.* for Ti and V) are currently derived from a relatively small number of datapoints. This is due to the limited availability of gas phase reference data for these elements. Future studies where a greater variety of compounds are studied, including solids and surface species, should be able to improve on this aspect.

The accuracy of the calculated binding energies with and without the element-specific systematic corrections is analyzed in Table 1 and Fig. 3. Before the application of the corrections, the mean error for the whole dataset is −0.73 eV and the mean absolute error is 0.73 eV. In contrast, the mean absolute error in the binding energies after the application of the element-specific correction is just 0.20 eV. The distribution of errors of the corrected binding energies is shown in Fig. 3. In almost all cases, errors smaller than 0.5 eV are found. One trend that can be noted is that even after the application of the corrections, core electron binding

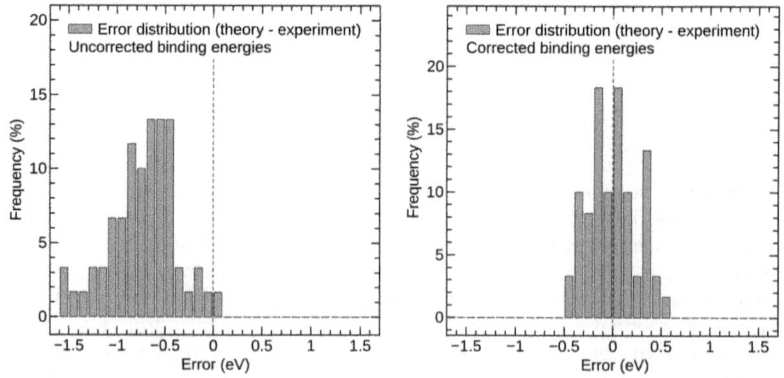

Fig. 3 Histograms showing the distribution of errors in the calculated binding energies with and without the element-specific corrections.

energies of metal atoms in very high oxidation states, bonded to several fluorine or oxygen atoms, are typically underestimated by a few tenths of an eV. The errors for the Ti $2p_{3/2}$ level in TiF_4, the V $2p_{3/2}$ level in VF_5, the V $2p_{3/2}$ level in VOF_3, the Cr $2p_{3/2}$ level in CrO_2Cl_2, and the Mn $2p_{3/2}$ level in MnO_3Cl are −0.39 eV, −0.35 eV, −0.24 eV, −0.43 eV, and −0.05 eV respectively. A similar tendency was observed in ref. 2 for nonmetals in high oxidation states.

In previous studies we have found that the same computational formalism, based on the SCAN functional, yields accurate absolute core electron binding energies in lighter elements, without the need for any empirical correction factors. For example, for a reference set of 103 1s and 2p binding energies of the elements B–F and Si–Cl the calculated binding energies have a mean absolute error of just 0.16 eV compared to results of gas phase photoemission measurements.[2] In addition, for a set of 15 core electron binding energies of the elements Li–Si in solids, the method yields a mean absolute error of 0.24 eV.[3] In light of the excellent agreement of the calculated absolute core electron binding energies for lighter elements, one naturally wonders why significant element-specific corrections are needed to obtain accurate absolute binding energies for the $2p_{3/2}$ core levels of first row transition metals. We have carefully considered the dependence of the calculated binding energies on the basis set used in the calculation, and found that basis set incompleteness cannot account for the observed systematic errors (see ESI† for details). Therefore, we believe that the two most important sources of error are the treatment of relativistic effects and the choice of the exchange–correlation functional.

Our treatment of relativistic effects using scalar-relativistic DFT (followed by subtracting one third of the experimentally measured atomic spin–orbit splitting to obtain $2p_{3/2}$ binding energies) is well justified for the 1s levels in elements of the second period of the periodic table and also for the 2p levels of the elements silicon to chlorine even though the magnitude of the spin–orbit splittings are already significant in these systems (ranging from 0.4 eV for Si to 1.6 eV for Cl). In contrast, for the 2p levels of the elements titanium to cobalt, the spin–orbit splittings are much greater, ranging from 6.4 eV for Ti to 15.1 eV for Co and therefore the applied spin–orbit correction shifts the binding energies by several eV. It is possible that our treatment of spin–orbit coupling as a post-SCF correction is no longer reliable for these elements and future studies should examine whether fully relativistic ΔSCF calculations, in which the energy of the final state with a $2p_{3/2}$ core hole can be calculated directly, yield more accurate absolute core electron binding energies.

As an alternative to a fully relativistic ΔSCF calculation, one could attempt to correct the 2p core electron binding energy from a scalar relativistic ΔSCF calculation using the spin–orbit splitting from a fully relativistic ground state calculation of the same compound. Such a scheme could capture the dependence of the spin–orbit splitting on the chemical environment of the atom and therefore lead to more accurate relative core electron binding energies. However, it is unlikely that such a scheme could correct for the systematic errors in the $2p_{3/2}$ binding energies observed in this work (in which atomic experimental values for the spin–orbit splittings are used). In order to correct the systematic errors, the difference between the calculated and experimental spin–orbit splittings would have to be three times larger than the systematic error in the $2p_{3/2}$ binding energies, *i.e.* about 3 eV if the systematic error is about 1 eV. However, our

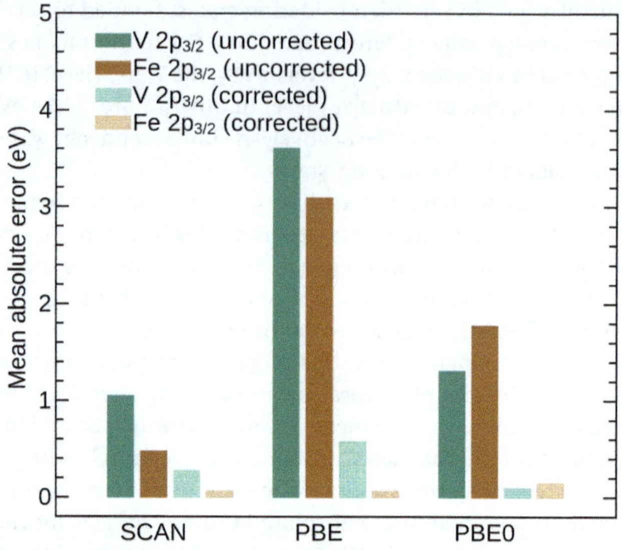

Fig. 4 A comparison of the mean absolute errors in calculated V $2p_{3/2}$ and Fe $2p_{3/2}$ core electron binding energies from calculations using three different exchange–correlation functionals: SCAN, PBE, and PBE0. For each functional, the raw theoretical results exhibit a relatively large systematic error – once this systematic error is accounted for, good agreement with experiment is observed.

preliminary tests indicate that for the compounds considered in this work, calculated spin–orbit splittings typically lie within a few tenths of an eV of the atomic experimental values. Moreover, if the calculated spin–orbit splitting were very different from the experimental value, it would not be possible for the calculated positions of the $2p_{3/2}$ and $2p_{1/2}$ peaks to be accurate at the same time, as their separation is of course equal to the splitting parameter.

In order to examine the influence of the exchange–correlation functional on the accuracy of the calculated binding energies, we have repeated the calculations for two of the elements (V and Fe) using the PBE and PBE0 functionals. The results of these calculations are summarized in Fig. 4 and ESI Fig. 1.† We find that the binding energies obtained using PBE and PBE0 exhibit even larger systematic errors than those obtained using SCAN. In particular, the systematic errors in the calculated V $2p_{3/2}$ and Fe $2p_{3/2}$ core electron binding energies are 3.61 eV and 3.1 eV for PBE, and 1.31 eV and 1.78 eV for PBE0. As before, the agreement between theory and experiment can be dramatically improved by eliminating these systematic errors *via* element- and functional-specific corrections. The mean absolute errors in the corrected V $2p_{3/2}$ and Fe $2p_{3/2}$ binding energies are 0.58 eV and 0.07 eV for PBE, and 0.10 eV and 0.15 eV for PBE0.

4 Conclusions

Previous studies have established the ΔSCF method as a highly accurate and computationally affordable method for calculating core-electron binding energies. However, thus far, the vast majority of theoretical work has focused on elements of the 2nd and 3rd periods of the periodic table, whereas experimentally, core level spectra of all stable elements, except for H and He, are studied. In this work, we have evaluated the suitability of the ΔSCF method for predicting $2p_{3/2}$ core electron binding energies in compounds of the first row transition metals Ti, V, Cr, Mn, Fe, and Co.

We have found that a formalism based on scalar-relativistic DFT calculations using the SCAN functional systematically underestimates absolute TM $2p_{3/2}$ binding energies in these compounds, and that the errors in the calculated absolute core electron binding energies are significantly larger than the errors found for lighter elements using the same method.[2] However, we have also found, that for the dataset of 60 molecules considered in this work, the calculated relative binding energies are still highly accurate, with a mean absolute error of 0.20 eV compared to experiment. Moreover, we have shown that with the use of element-specific corrections, the prediction of accurate absolute TM $2p_{3/2}$ core electron binding energies in closed shell compounds of first row transition metals is also possible.

Together with the recent work on extending the ΔSCF method to periodic solids,[3] these results lay the foundations for the use of ΔSCF calculations for predicting core electron binding energies in a wide range of experimentally relevant systems. Further theoretical work is required to examine the possibility of performing constrained fully relativistic DFT calculations with the non-Aufbau principle occupation of eigenstates that would allow the total energies of core hole states, where the core hole resides in a spin–orbit split p-, d-, or f-orbital, to be calculated directly.

Conflicts of interest

There are no conflicts to declare.

Acknowledgements

This project has received funding from the European Union's Horizon 2020 research and innovation programme under grant agreement No 892943. JMK acknowledges support from the Estonian Centre of Excellence in Research project "Advanced materials and high-technology devices for sustainable energetics, sensorics and nanoelectronics" TK141 (2014-2020.4.01.15-0011). This work used the ARCHER2 UK National Supercomputing Service *via* J. L.'s membership of the HEC Materials Chemistry Consortium of UK, which is funded by EPSRC (EP/L000202).

Notes and references

1 P. S. Bagus, *Phys. Rev.*, 1965, **139**, A619–A634.
2 J. M. Kahk and J. Lischner, *Phys. Rev. Mater.*, 2019, **3**, 100801.
3 J. M. Kahk, G. S. Michelitsch, R. J. Maurer, K. Reuter and J. Lischner, *J. Phys. Chem. Lett.*, 2021, **12**, 9353–9359.
4 N. Pueyo Bellafont, G. Álvarez Saiz, F. Viñes and F. Illas, *Theor. Chem. Acc.*, 2016, **135**, 35.
5 D. Hait and M. Head-Gordon, *J. Phys. Chem. Lett.*, 2020, **11**, 775–786.
6 D. Golze, L. Keller and P. Rinke, *J. Phys. Chem. Lett.*, 2020, **11**, 1840–1847.

7 T. Ozaki and C.-C. Lee, *Phys. Rev. Lett.*, 2017, **118**, 026401.
8 Y. Takahata and A. D. S. Marques, *J. Electron Spectrosc. Relat. Phenom.*, 2010, **178–179**, 80–87.
9 T. Aoki and K. Ohno, *J. Phys.: Condens. Matter*, 2018, **30**, 21LT01.
10 T. Zhu and G. K.-L. Chan, *J. Chem. Theory Comput.*, 2021, **17**, 727–741.
11 M. Walter, M. Moseler and L. Pastewka, *Phys. Rev. B*, 2016, **94**, 041112.
12 J. M. Kahk and J. Lischner, *Phys. Chem. Chem. Phys.*, 2018, **20**, 30403–30411.
13 A. Regoutz, M. S. Wolinska, N. K. Fernando and L. E. Ratcliff, *Electron. Struct.*, 2020, **2**, 044005.
14 D. Golze, J. Wilhelm, M. J. van Setten and P. Rinke, *J. Chem. Theory Comput.*, 2018, **14**, 4856–4869.
15 G. Cavigliasso and D. P. Chong, *J. Chem. Phys.*, 1999, **111**, 9485–9492.
16 D. P. Chong, *Reviews of Modern Quantum Chemistry*, WORLD SCIENTIFIC, 2002, pp. 1106–1139.
17 J. Liu, D. Matthews, S. Coriani and L. Cheng, *J. Chem. Theory Comput.*, 2019, **15**, 1642–1651.
18 M. Lembinen, E. Nõmmiste, H. Ers, B. Docampo-Álvarez, J. Kruusma, E. Lust and V. B. Ivaništšev, *Int. J. Quantum Chem.*, 2020, **120**, e26247.
19 I. Tolbatov and D. M. Chipman, *Theor. Chem. Acc.*, 2014, **133**, 1560.
20 C.-H. Hu and D. P. Chong, *Chem. Phys. Lett.*, 1996, **262**, 733–736.
21 X. Zheng and L. Cheng, *J. Chem. Theory Comput.*, 2019, **15**, 4945–4955.
22 N. Pueyo Bellafont, F. Viñes and F. Illas, *J. Chem. Theory Comput.*, 2016, **12**, 324–331.
23 N. A. Besley, A. T. B. Gilbert and P. M. W. Gill, *J. Chem. Phys.*, 2009, **130**, 124308.
24 B. P. Klein, S. J. Hall and R. J. Maurer, *J. Phys.: Condens. Matter*, 2021, **33**, 154005.
25 K.-N. Huang, M. Aoyagi, M. H. Chen, B. Crasemann and H. Mark, *At. Data Nucl. Data Tables*, 1976, **18**, 243–291.
26 D. Clark, B. Cromarty and A. Sgamelotti, *J. Electron Spectrosc. Relat. Phenom.*, 1980, **19**, 303–309.
27 P. S. Bagus, C. J. Nelin, C. R. Brundle and S. A. Chambers, *J. Phys. Chem. C*, 2019, **123**, 7705–7716.
28 P. S. Bagus, C. J. Nelin, C. R. Brundle, B. Vincent Crist, N. Lahiri and K. M. Rosso, *J. Chem. Phys.*, 2020, **153**, 194702.
29 P. S. Bagus, C. J. Nelin, C. R. Brundle, B. V. Crist, N. Lahiri and K. M. Rosso, *J. Chem. Phys.*, 2021, **154**, 094709.
30 N. Rößler and V. Staemmler, *Phys. Chem. Chem. Phys.*, 2003, **5**, 3580–3586.
31 N. Rössler, K. Kotsis and V. Staemmler, *Phys. Chem. Chem. Phys.*, 2006, **8**, 697–706.
32 A. T. B. Gilbert, N. A. Besley and P. M. W. Gill, *J. Phys. Chem. A*, 2008, **112**, 13164–13171.
33 G. S. Michelitsch and K. Reuter, *J. Chem. Phys.*, 2019, **150**, 074104.
34 J. Sun, A. Ruzsinszky and J. Perdew, *Phys. Rev. Lett.*, 2015, **115**, 036402.
35 J. P. Perdew, K. Burke and M. Ernzerhof, *Phys. Rev. Lett.*, 1996, **77**, 3865–3868.
36 C. Adamo and V. Barone, *J. Chem. Phys.*, 1999, **110**, 6158–6170.
37 C. Chang, M. Pelissier and P. Durand, *Phys. Scr.*, 1986, **34**, 394–404.
38 E. van Lenthe, E. J. Baerends and J. G. Snijders, *J. Chem. Phys.*, 1994, **101**, 9783–9792.

39 S. Faas, J. Snijders, J. van Lenthe, E. van Lenthe and E. Baerends, *Chem. Phys. Lett.*, 1995, **246**, 632–640.
40 A. Thompson, D. Attwood, J. Kirz, M. Gullikson, M. Howells, J. Kortright, Y. Liu, A. Robinson, J. Underwood, K.-J. Kim, I. Lindau, P. Pianetta, H. Winick, G. Williams and J. Scofield, *X-Ray Data Booklet*, Lawrence Berkeley National Laboratory, Berkeley, California, 2009, p. 94720.
41 V. Blum, R. Gehrke, F. Hanke, P. Havu, V. Havu, X. Ren, K. Reuter and M. Scheffler, *Comput. Phys. Commun.*, 2009, **180**, 2175–2196.
42 V. Havu, V. Blum, P. Havu and M. Scheffler, *J. Comput. Phys.*, 2009, **228**, 8367–8379.
43 F. Knuth, C. Carbogno, V. Atalla, V. Blum and M. Scheffler, *Comput. Phys. Commun.*, 2015, **190**, 33–50.
44 B. Delley, *J. Comput. Chem.*, 1996, **17**, 1152–1155.
45 J. D. Head and M. C. Zerner, *Chem. Phys. Lett.*, 1985, **122**, 264–270.
46 W. Jolly, K. Bomben and C. Eyermann, *At. Data Nucl. Data Tables*, 1984, **31**, 433–493.

Faraday Discussions

PAPER

Corrosion inhibition in acidic environments: key interfacial insights with photoelectron spectroscopy

Kiran Kousar,[a] Michael Dowhyj,[ab] Monika S. Walczak,[a] Thomas Ljungdahl,[c] Alexander Wetzel,[c] Hans Oskarsson,[c] Alex S. Walton,[bd] Paolo Restuccia,[e] Nicholas M. Harrison[e] and Robert Lindsay*[ab]

Received 29th November 2021, Accepted 6th January 2022
DOI: 10.1039/d1fd00106j

In many engineering scenarios, surface-active organic species are added to acidic solutions to inhibit the corrosion of metallic components. Given suitable selection, such corrosion inhibitors are highly effective, preventing significant degradation even in highly aggressive environments. Nevertheless, there are still considerable gaps in fundamental knowledge of corrosion inhibitor functionality, severely restricting rational development. Here, we demonstrate the capability of X-ray photoelectron spectroscopy (XPS), supported by *ab initio* modelling, for revealing key details of inhibited substrates. Attention is focussed on the corrosion inhibition of carbon steel through the addition of an exemplar imidazoline-based corrosion inhibitor (OMID) to aqueous solutions of both HCl and H_2SO_4. Most notably, it is demonstrated that interfacial chemistry varies with the identity of the acid. High resolution Fe 2p, O 1s, N 1s, and Cl 2p XPS spectra, acquired from well-inhibited carbon steel in 1 M HCl, show that there are two different singly protonated OMID species bound directly to the metallic carbon steel substrate. In sharp contrast, in 0.01 M H_2SO_4, OMID adsorbs onto an ultra-thin surface film, composed primarily of a ferric sulfate ($Fe_2(SO_4)_3$)-like phase. Such insight is essential to efforts to develop a mechanistic description of corrosion inhibitor functionality, as well as knowledge-based identification of next generation corrosion inhibitors.

Introduction

Electrochemical corrosion of metallic engineering components continues to be a *bête-noire* of our modern world, resulting in significant costs, both economic

[a]Corrosion and Protection Centre, Department of Materials, The University of Manchester, Sackville Street, Manchester, M13 9PL, UK. E-mail: robert.lindsay@manchester.ac.uk
[b]Photon Science Institute, The University of Manchester, Oxford Road, Manchester, M13 9PL, UK
[c]Nouryon, Hamnvägen 2, SE-444 85 Stenungsund, Sweden
[d]Department of Chemistry, The University of Manchester, Oxford Road, Manchester, M13 9PY, UK
[e]Department of Chemistry and Institute for Molecular Science and Engineering, Imperial College London, White City Campus, 80 Wood Lane, London W12 0BZ, UK

and beyond.[1] Given that this phenomenon is almost always thermodynamically inevitable, slowing material degradation down through controlling reaction kinetics is the best that can be achieved. To pursue this goal, corrosion engineers typically adopt one or more of three basic approaches, *i.e.*, material selection, coating application, or *operando* chemical treatment. In the latter category, one option is the addition of corrosion inhibitor species, that reduce corrosion by inducing a change in the surface termination of the metallic substrate.[2] This approach is well established, and highly effective in many real-world applications. There are, however, significant gaps in the fundamental understanding of corrosion inhibitor functionality, severely restricting rational development; selection of corrosion inhibitors at present is largely a trial-and-error exercise. In this contribution, it will be demonstrated that X-ray photoelectron spectroscopy (XPS), supported by *ab initio* modelling, can be exploited to gain fresh insight into inhibited interfaces in acidic environments, offering the possibility of more knowledge-based development of next generation corrosion inhibitors.

In aqueous acidic solution, organic surface-active species are invariably employed as corrosion inhibitors.[3,4] As illustrated in Fig. 1(a), it is typically assumed that these species function through binding to the metallic substrate, forming either a monolayer or bilayer.[4] It should be emphasised, however, that this cartoon is rather speculative, with relatively little direct experimental evidence acquired to date, *e.g.*, the nature of the surface of the metallic substrate is generally unknown. XPS is one technique that can be applied to begin to remedy this situation, as it can deliver chemically sensitive information about such interfaces. Here, XPS is employed to characterise inhibited interfaces formed on carbon steel in acidic solutions in the presence of an exemplar

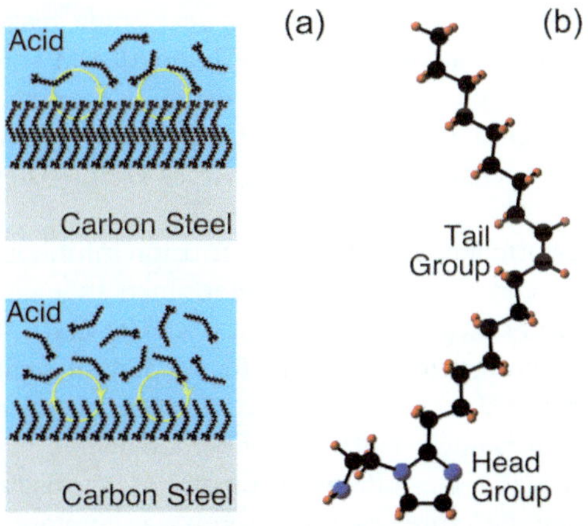

Fig. 1 (a) Cartoons of organic corrosion inhibitor species forming a monolayer (bottom) or bilayer (top) on a metallic substrate (carbon steel) in acidic solution. The yellow circles with arrows indicate that the adsorbed layers are dynamic, *i.e.*, they dissipate in the absence of corrosion inhibitor in the bulk solution. (b) Ball and stick model of OMID, where the black, blue, and red spheres represent carbon, nitrogen, and hydrogen atoms, respectively. The head group, which is expected to be involved in substrate bonding, comprises a heterocyclic ring ($C_3N_2H_4$) and a pendant side group ($C_2H_4NH_2$). The tail group is an unbranched aliphatic chain ($C_{17}H_{33}$).

Table 1 Acidic solutions employed in this study

Acid solutions
1 M HCl
0.01 M HCl
1 M H_2SO_4
0.01 M H_2SO_4

imidazoline-based surfactant, which will henceforth be referred to as OMID; a ball and stick model of OMID is depicted in Fig. 1(b).

Previous studies have demonstrated that OMID is a very efficient corrosion inhibitor for carbon steel in acidic solution.[5–8] Moreover, we have already applied XPS to characterise the interaction of OMID with carbon steel in 1 M HCl, concluding that the well-inhibited interface comprises OMID bound to film-free substrate, *i.e.*, there is no surface oxide/salt phase.[9] Furthering this work, in the following sections we present a study exploring the impact of varying the composition of the acidic solution (see Table 1 for a list of the solutions) on both the corrosion inhibition performance of OMID at the critical micelle concentration (CMC),† and the nature of the interface derived from XPS data; we focus on measurements at the CMC, as it is expected that OMID should perform optimally at this concentration.[4] Density functional theory (DFT) calculations have been undertaken to aid data analysis/interpretation.

Experimental

A single batch of carbon steel samples (10 mm × 10 mm × 1.5 mm), purchased from RCSL Corrosion Monitoring Ltd, were used as substrates. For preparation of the acidic solutions listed in Table 1, analytical grade reagents from Fischer Scientific were employed, namely 10.2 M HCl and 18.4 M H_2SO_4. OMID (97.7% purity) was synthesised in our laboratories; see ref. 8 for details of the synthetic route. The CMC of OMID in each acidic solution was determined through pendant drop surface tensiometry, using a Krüss GmbH Drop Shape Analyzer, DSA100.

To assess the performance of OMID as a corrosion inhibitor for carbon steel in each acidic solution, corrosion rates were determined after 4 h of immersion at $T \sim 298$ K, in both the absence and presence of OMID (at CMC); 4 h was adopted as the standard immersion period, as it was found that all samples had achieved a steady state by this point. A combination of electrochemical techniques, *i.e.*, linear polarisation resistance (LPR) and potentiodynamic polarisation (PDP),[10] were used to provide a quantitative estimate of corrosion rate. For these measurements, a standard three-electrode setup was employed in a 1 L glass cell containing the solution of interest; LPR (PDP) data were collected with a computer-controlled potentiostat (Ivium Compactstat), using a scan rate of 10 mV min^{-1} (120 mV min^{-1}) over the range −10 mV to +10 mV (−250 mV to +250 mV) relative to open circuit potential (OCP). Before insertion of a carbon steel

† The critical micelle concentration (CMC) is the concentration of surfactant above which micelles (aggregates of surfactant molecules) start to form.

sample (working electrode) into the glass cell, it was mounted in epoxy resin, ground with a series of SiC papers (240 grit, 400 grit, 600 grit, and 800 grit), washed with ethanol and de-ionized water, and dried.

XPS spectra were recorded using a Kratos Axis Ultra spectrometer, equipped with an Al Kα X-ray source ($h\nu = 1486.6$ eV), a 165 mm hemispherical energy analyser, and an electron flood gun; samples were exposed to low energy electrons (≤ 3 eV) during data acquisition to minimize charge accumulation. Spectra were recorded at normal emission ($\theta_E = 0°$), using pass energies of 80 eV and 20 eV for overview spectra and higher resolution single core level spectra, respectively. The metallic Fe $2p_{3/2}$ peak at a binding energy (BE) of 706.7 eV was used for calibration (± 0.1 eV).[11]

Carbon steel samples for XPS analysis were ground with a series of SiC papers (400 grit, 600 grit, 800 grit, 1200 grit, 2400 grit, 4000 grit), and then polished with diamond paste (1 μm) to a mirror finish. Subsequently, each sample was rinsed with acetone and de-H_2O, and then dried. To avoid air exposure, a nitrogen filled glove box directly attached to the XPS instrument, as shown in Fig. 2(a), was used for sample immersion. Following solution exposure for 4 h, the sample was retrieved, blown dry with nitrogen to minimise physical surface deposition of non-volatile solution constituents, and inserted directly into the XPS load-lock chamber. Implementation of this sample handling procedure, which is reported in detail in ref. 13, is essential for mitigating post-immersion substrate oxidation, which can lead to misinterpretation of inhibited interface chemistry. Fig. 2(b) and (c) demonstrate the advantage of this approach through comparison of Fe 2p and O 1s core level XPS spectra from highly inhibited carbon steel samples formed in 1 M HCl in the presence of the corrosion inhibitor 2-

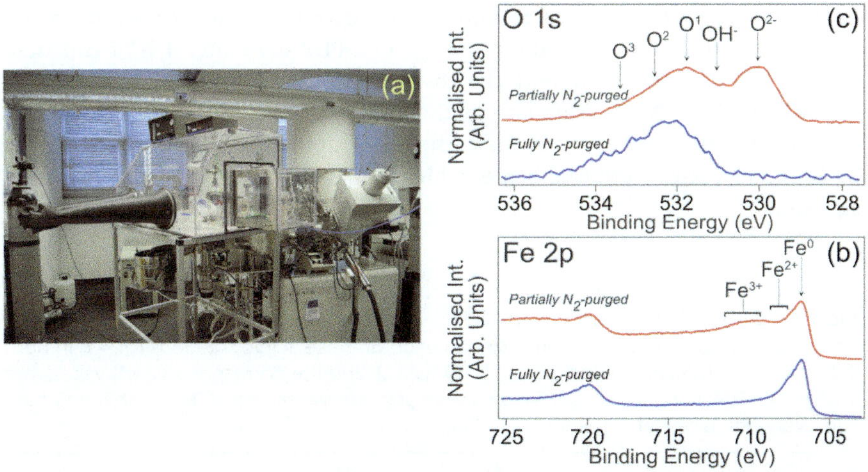

Fig. 2 (a) Photograph of N_2-filled glove box attached to load lock of XPS spectrometer to allow transfer of samples without post-immersion air exposure. This setup was employed for all XPS data presented in this study. (b) Fe 2p XPS spectra ($h\nu = 1486.6$ eV, $\theta_E = 0°$) acquired from carbon steel samples that had been immersed in 1 M HCl inhibited with MBI.[12,13] The blue (red) spectrum is from a sample transferred through a fully (partially) N_2-purged glove box.[12,13] (c) O 1s XPS spectra acquired under analogous conditions; O^1, O^2 and O^3 arise from adsorbed OH (O^1) and an adventitious carbon layer (O^1–O^3).

mercaptobenzimidazole (MBI).[12,13] Blue (red) spectra were recorded from samples that had been removed from solution into a fully (partially) N_2-purged glove box. The partially N_2-purged data display features consistent with interface oxidation, *i.e.*, signals assigned to O^{2-}/OH^- and $Fe^{2+/3+}$, which are absent from the fully N_2-purged data, indicating that post-immersion oxidation can have a significant bearing on XPS core level profiles.

Concerning *ab initio* DFT modelling, calculations have been undertaken on both isolated OMID head group (HG-OMID) molecules and a 5-layer slab of (110) oriented Fe decorated with chlorine atoms, using the CRYSTAL 17 program[14,15] that expands molecular and crystal orbitals as a linear combination of atom centered Gaussian functions. For molecules, electronic exchange and correlation were approximated using the hybrid HSE06 functional, which describes the strong on-site electronic interactions of electrons localised in molecular orbitals well. Quadruple valence basis sets with polarisation function were used for the C, N, and H atoms, *i.e.*, 6-31111++G** for C and N, and 31111++G** for H. Following energy minimisation through geometry optimisation, projected electronic density of states (PDOS) and proton affinities (PAs) were computed. The 2D periodic slab calculations of the surface were performed as previously outlined in ref. 9, with electron exchange and correlation approximated with the PBE exchange–correlation functional that describes the delocalised states of the metal adequately. All electron basis sets with polarisation were used for the Fe and Cl atoms, *i.e.*, 8-64111d41G quadruple valence for Fe and 86-311G triple valence with the addition of a diffuse d-symmetry polarisation function for Cl. Again, following geometry optimisation, the PDOS was computed.

Results and discussion

Table 2 shows the CMC value determined for OMID in each acidic solution, along with the corrosion rate (CR) derived from LPR/PDP data after 4 h of substrate (carbon steel) immersion in both the absence ([OMID] = 0 mM) and presence of OMID ([OMID] = CMC); a standard methodology was employed to determine CRs in mm per year.[10] Furthermore, the corrosion inhibitor efficiency (η%) displayed by OMID in each solution is listed, which was determined from the following equation:

Table 2 CMC for OMID in each of the acidic solutions considered in this study. Corrosion rate of carbon steel following immersion in each of these acidic solutions for 4 h in both the absence (0 mM) and presence of OMID (CMC), and the corresponding η% values, are also listed. The numbers in bold indicate values consistent with OMID exhibiting high corrosion inhibition performance

Solution	OMID CMC (mM)	CR, 0 mM [OMID] (mm per year)	CR, CMC [OMID] (mm per year)	η% (%)
1 M HCl	0.18 ± 0.03	3.75 ± 0.07	**0.11 ± 0.01**	**97.1 ± 0.3**
0.01 M HCl	5.90 ± 1.50	1.96 ± 0.12	0.28 ± 0.03	85.7 ± 1.8
1 M H_2SO_4	0.22 ± 0.05	52.71 ± 0.80	1.09 ± 0.07	**97.9 ± 0.1**
0.01 M H_2SO_4	0.70 ± 0.12	4.43 ± 0.70	**0.20 ± 0.01**	**95.5 ± 0.8**

$$\eta\% = \frac{U - I}{U} \times 100,$$

where U and I are the uninhibited and inhibited corrosion rates, respectively.[2] This figure of merit for corrosion inhibitor performance is commonly used for selection purposes in engineering applications, with species displaying $\eta\% \geq 95\%$ typically being identified as high performing.[16]

Surveying the $\eta\%$ values in Table 2, it is evident that OMID is a high performing corrosion inhibitor (*i.e.*, $\eta\% \geq 95\%$) for carbon steel in 3 out of the 4 acidic solutions considered. Moreover, the absolute inhibited corrosion rate is ≤0.2 mm per year for two of these solutions, which we specify as an additional condition for a high performing corrosion inhibitor in this study.

Focusing on the two well-inhibited solutions, namely 1 M HCl with OMID at CMC and 0.01 M H_2SO_4 with OMID at CMC, which will henceforth be referred to as 1MHCl/OMID and 0.01MH_2SO_4/OMID, respectively, Fig. 3 shows overview XPS spectra acquired from carbon steel samples subject to 4 h of immersion in each of these solutions; a spectrum acquired from a polished carbon steel substrate is also displayed. Concerning the latter spectrum, features originating from photoemission out of Fe 2p, O 1s, and C 1s core levels are clearly visible. The Fe signal originates from the carbon steel substrate, O from both a surface oxidic film and adsorbed species, and C is a result of adventitious surface contamination.[9,12]

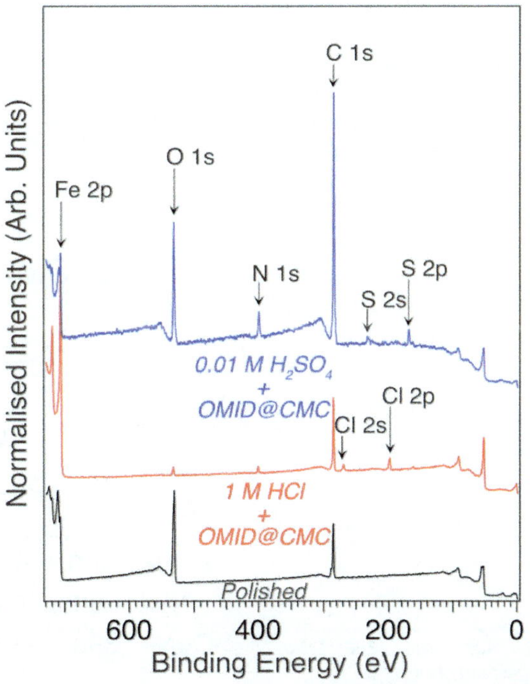

Fig. 3 Overview XPS spectra ($h\nu = 1486.6$ eV, $\theta_E = 0°$) acquired from carbon steel subsequent to immersion for 4 h in either 1MHCl/OMID (red line) or 0.01MH_2SO_4/OMID (blue line). The bottom-most spectrum (black line) was acquired from a polished carbon steel substrate.

Immersion in either acidic solution induces modification of the overview spectral profile. One common change is the appearance of a N 1s feature, which is consistent with the adsorption of OMID, as it contains three nitrogen atoms (see Fig. 1). Fig. 4(a) shows corresponding higher resolution N 1s core level XPS spectra; the data shown in red were acquired from carbon steel following immersion in 1MHCl/OMID, and those shown in blue following immersion in 0.01MH_2SO_4/OMID. Evidently, OMID adsorption is not identical in the two cases, as the spectra differ significantly.

To aid interpretation of such N 1s XPS data, we presented in ref. 8 a N 1s core level spectrum acquired from a multilayer of the molecule depicted in Fig. 4(b), which is the OMID head group (HG-OMID). It was demonstrated that the core level signal was composed of three components (BEs = 398.9 eV, 399.6 eV, and 400.4 eV) of equal intensity, consistent with the three chemically distinct N atoms in OMID, *i.e.*, imine (>C=N–), primary amine (–NH_2), and tertiary amine (–N<). In accord with earlier work,[17–19] the following BE assignments were proposed: >C=N– to BE = 398.9 eV, –NH_2 to BE = 399.6 eV, and –N< to BE = 400.4 eV.

In order to corroborate these suggested N 1s XPS BE attributions, an *ab initio* DFT calculation has been undertaken of isolated HG-OMID. Following geometric optimisation, the binding energy of each N 1s core level was extracted from computed PDOS. The values obtained are compared to the experimental XPS BEs in Fig. 4(b). As expected, given that the calculation is performed for the ground state of the molecule and single particle eigenvalues used to approximate the excitation energy, the absolute values differ. There is excellent agreement, however, in terms of relative energies, which are expressed in terms of binding energy shifts (ΔBE) relative to –N< (ΔBE = 0 eV). This result suggests that the N 1s XPS assignments are reasonable.

Using the ΔBE data summarised in Fig. 4(b) as a starting point, Fig. 5 (lower spectrum) shows the best fit to the N 1s core level spectrum acquired from

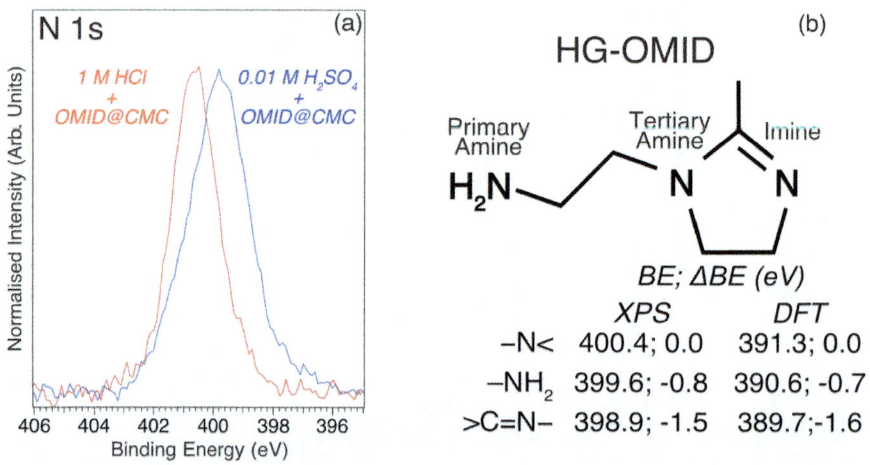

Fig. 4 (a) N 1s core level XPS spectra ($h\nu = 1486.6$ eV, $\theta_E = 0°$) acquired from carbon steel following immersion for 4 h in either 1MHCl/OMID (red line) or 0.01MH_2SO_4/OMID (blue line). (b) Skeletal model of OMID head group (HG-OMID), and BEs, plus ΔBEs relative to –N<, for each nitrogen atom, derived from either XPS or DFT calculations.

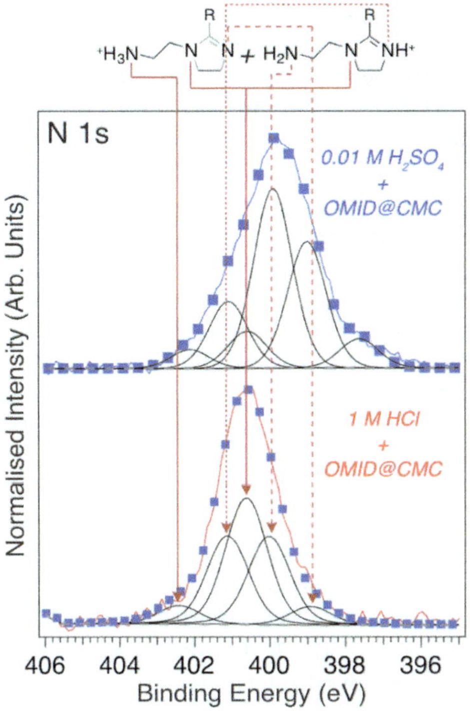

Fig. 5 N 1s core level XPS spectra ($h\nu = 1486.6$ eV, $\theta_E = 0°$) acquired from carbon steel following immersion for 4 h in either 1MHCl/OMID (lower spectrum) or 0.01MH_2SO_4/OMID (upper spectrum). Blue markers show best fits to the experimental spectra (red/blue lines), using GL(30) (black lines) and Shirley-type (broken grey lines) functions. Peak assignments, consistent with OMID being adsorbed on carbon steel in 1MHCl/OMID in the two $OMIDH^+$ forms shown at the top of this figure, are also indicated. R = $C_{17}H_{33}$.

a carbon steel sample post immersion in 1MHCl/OMID; a combination of Gaussian–Lorentzian (GL(*p*))‡ line shape functions and a Shirley-type background function were employed to model the photoelectron peaks and secondary electron background, respectively. 5 GL(30) functions, each having a full width at half maximum (FWHM) of ~1.2 eV, were required to achieve a satisfactory fit to the data. As discussed previously[8,9] and illustrated in Fig. 5, this fit can be reconciled with the adsorption of OMID in two different singly protonated ($OMIDH^+$) forms, where either the primary amine ($-NH_2 \rightarrow -NH_3^+$) or imine ($>C{=}N- \rightarrow >C{=}NH^+-$) is protonated. We note that OMID is expected to be doubly protonated, $-NH_3^+$ plus $>C{=}NH^+-$, in bulk solution in both 1 M HCl and 0.01 M H_2SO_4;[8] the tertiary amine group ($-N<$) remains unprotonated. Hence, adsorption results in a reduction in the degree of protonation, *i.e.*, $OMID2H^+ \rightarrow OMIDH^+$.

From the relative intensity of the $-NH_2$ and $>C{=}NH^+-$ components in Fig. 5 (lower spectrum), it can be concluded that the majority of OMID adsorbed on carbon steel in 1MHCl/OMID possesses a protonated imine group ($>C{=}NH^+-$).

‡ *p*, which lies in the range $0 \leq p \leq 100$, describes the degree of Gaussian–Lorentzian mixing, where GL(100) (GL(0)) is a pure Lorentzian (Gaussian) line shape.

Interestingly, this result agrees with the PA deduced from DFT total energy differences for protonation of each N in HG-OMID (HG-OMID → HG-OMIDH$^+$). For optimised HG-OMID geometries it was found that (>C=N-) displayed the greatest PA, *i.e.*, PA(>C=N) ~ 713 kJ mol^{-1} > PA(–NH$_2$) ~ 613 kJ mol^{-1} > PA(–N<) ~ 582 kJ mol^{-1}, and so >C=NH$^+$– is expected to be the dominant protonated group for OMIDH$^+$. We point out that the *ab initio* modelling is performed on isolated species and so takes no account of either solvent effects or OMID–substrate/OMID–solution interactions, but is nevertheless consistent with interpretation of the XPS data.

Turning to the N 1s XPS spectrum recorded from carbon steel subsequent to immersion in 0.01MH$_2$SO$_4$/OMID, we have attempted to reproduce the experimental data with the same 5 GL(30) functions used for 1MHCl/OMID. To achieve the fit shown in Fig. 5 (upper spectrum), the BE of each peak was allowed to shift by up to ~0.3 eV relative to those determined for 1MHCl/OMID, and an additional GL(30) component was required (BE = 397.6 eV). Assuming these peaks correspond to the assignments above, the most intense component is a result of photoemission from the –NH$_2$ group, which is not consistent with adsorption of the two OMIDH$^+$ species illustrated in Fig. 5. If only these two species were adsorbed, then the peak attributed to –N< (BE = 400.7 eV) would always be the most intense, as it is common to both OMIDH$^+$ species.

Further effort to interpret the N 1s spectrum for 0.01MH$_2$SO$_4$/OMID has involved both consideration of other adsorbed species, *e.g.*, OMID2H$^+$ and OMID, as well as reassignment of components to other N atoms. To date, no physically reasonable solution has been found, indicating that the adsorption of OMID onto carbon steel in 0.01MH$_2$SO$_4$/OMID is significantly more complex than in 1MHCl/OMID.

The information derived from the N 1s spectra in Fig. 5 is summarised in Fig. 6. As illustrated, these data do not reveal anything about the nature of the substrate terminations. To begin to gain insight into this aspect of the interfaces, we return to the overview spectra in Fig. 3. Comparing the spectrum of the

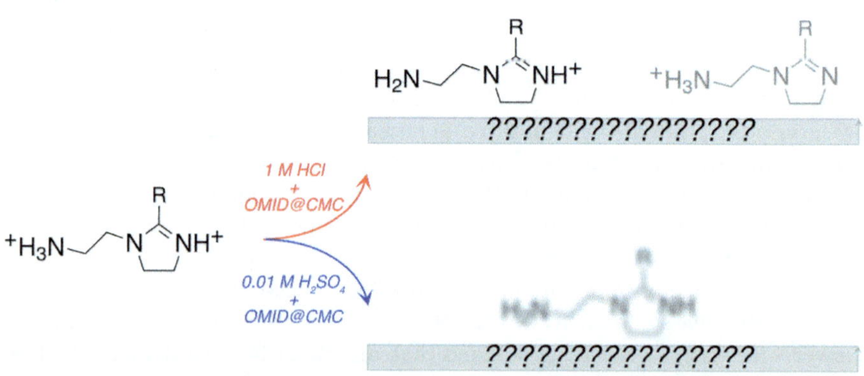

Fig. 6 Cartoon summarising information obtained from N 1s XPS spectra acquired from carbon steel following immersion for 4 h in either 1MHCl/OMID or 0.01MH$_2$SO$_4$/OMID. For 1MHCl/OMID, two OMIDH$^+$ species are concluded to be adsorbed, with the dominant one containing >C=NH$^+$–. For 0.01MH$_2$SO$_4$/OMID, the exact nature of the adsorbed OMID remains ill-defined. R = C$_{17}$H$_{33}$.

polished carbon steel to that of the sample immersed in 1MHCl/OMID, there is a clear reduction in O 1s signal, suggesting depletion of the surface oxidic layer. Moreover, features assigned to Cl XPS core levels (2s and 2p) appear, which almost certainly arise from the interaction of the substrate with Cl^-(aq) anions. Concerning the sample exposed to $0.01MH_2SO_4$/OMID, an intense O 1s feature is apparent, as well as S 2s and S 2p peaks, which suggest the presence of substrate bound SO_4^{2-} anions.

To facilitate more detailed understanding of the substrate terminations, higher resolution Fe 2p, O 1s, Cl 2p and S 2p core level XPS spectra, and corresponding best fits, acquired from carbon steel following immersion in either 1MHCl/OMID or $0.01MH_2SO_4$/OMID are shown in Fig. 7; a Cl 2p (S 2p) spectrum is not shown for $0.01MH_2SO_4$/OMID (1MHCl/OMID) as there was no signal above background. For the two common core levels (Fe 2p and O 1s), there are clear differences between the two acids, suggesting that not only does the adsorbed state(s) of OMID vary with solution chemistry, but so too does the substrate termination.

Concentrating on the 1MHCl/OMID data in Fig. 7, the Fe 2p spectrum is consistent with an oxide/salt-free metallic termination, as only features associated with Fe^0 are visible; a single Gaussian-convoluted Lorentzian based asymmetric (LF) function[20] has been used to fit the Fe $2p_{3/2}$ feature. 3 GL(30) components, labelled O^1, O^2, and O^3, have been employed to fit the O 1s signal,

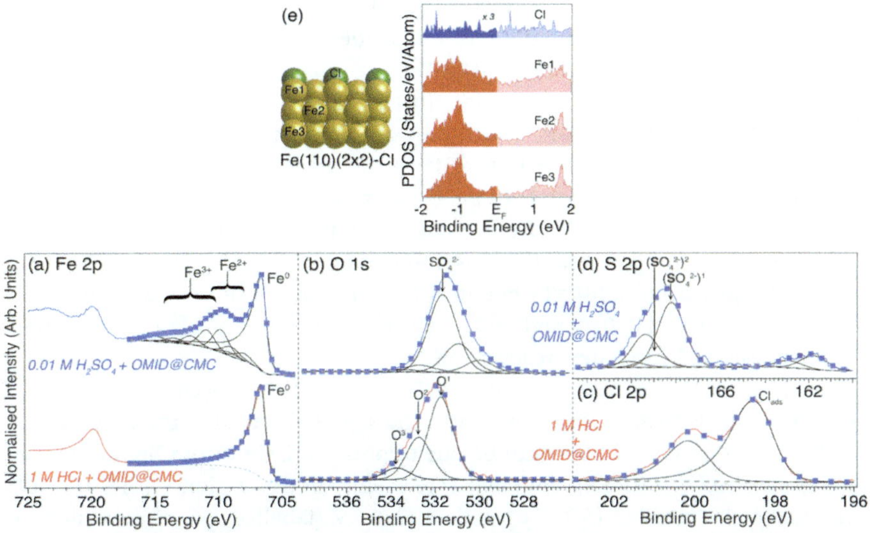

Fig. 7 (a) Fe 2p, (b) O 1s, (c) Cl 2p and (d) S 2p core level XPS spectra acquired from carbon steel following immersion for 4 h in either 1MHCl/OMID (lower row) or $0.01MH_2SO_4$/OMID (upper row). Spectra have been normalised such that minima are 0, and maxima are 1. Blue markers show best fits to the experimental spectra (solid red/blue lines), using GL (black lines), LF (black lines), and Shirley-type (broken grey lines) functions. Peak assignments are also indicated. For the Cl 2p and S 2p spin–orbit split doublets ($2p_{1/2}$ and $2p_{3/2}$), the BE differences and intensity ratios are consistent with expected values.[11] (e) On the left, side view of the structure of ordered (2 × 2) overlayer of Cl atoms on Fe(110) that was optimised in DFT calculations to produce PDOS for Cl, Fe1, Fe2, and Fe3. The corresponding PDOS are plotted on the right; both Fe and Cl display finite PDOS around E_F.

which we have previously ascribed to adsorbed OH (O^1) and adventitious RC_xO_y (O^1, O^2, and O^3) species.[9] The Cl 2p spectrum exhibits a single spin–orbit doublet, with the Cl $2p_{3/2}$ component located at BE ~ 198.6 eV, which has been interpreted as resulting from surface adsorbed Cl atoms (Cl_{ads}).[9,12]

Reflecting on our fitting of the Cl 2p spectrum, a notable detail is that two asymmetric LF line shape functions have been used, rather than GL functions. Such line shapes are more typically associated with the XPS core levels of metallic materials, *e.g.*, Fe 2p of carbon steel, as per above. In metals, the origin of this asymmetry depends upon there being a finite density of states at the Fermi level (E_F). Upon photoexcitation of a core level, this electronic structure gives rise to a weighted continuum of final states through electron–hole pair generation, and so a spectral profile with a tail to higher BE.[21] As argued in ref. 12, analogous line shapes are expected for Cl_{ads}, as there is appreciable hybridisation between substrate Fe^0 and Cl_{ads} resulting in a finite PDOS located on Cl_{ads} in the vicinity of the E_F. The existence of these PDOS is demonstrated in Fig. 7(e), which shows the results of DFT calculations performed for a (2 × 2) ordered overlayer of atomic chlorine on Fe(110).

Moving to the Fe 2p spectrum (Fig. 7(a)) of carbon steel following immersion in $0.01MH_2SO_4$/OMID, the profile is substantially different to that of the sample exposed to 1MHCl/OMID. As annotated, there are features that can be assigned to the presence of both Fe^{2+} and Fe^{3+}, with the latter being the major cationic species; the presented $Fe2p_{3/2}$ fit comprises a LF line shape function for Fe^0, along with a series of GL functions to describe the $Fe^{2+/3+}$ multiplet states and satellites.[22,23] Evidently, in sharp contrast to the inhibited interface formed in 1MHCl/OMID, the substrate termination is not oxide/salt-free metallic carbon steel following exposure to $0.01MH_2SO_4$/OMID.

Comparing the O 1s spectra (Fig. 7(b)) of carbon steel immersed in either 1MHCl/OMID or $0.01MH_2SO_4$/OMID, they are far from identical. For the latter, constraining each GL(30) function to have the same FWHM (*i.e.* ~1.5 eV), six components were required to achieve a satisfactory fit to the experimental data. Unique assignment of each of these peaks has proven to be unfeasible at this juncture, but we propose that the most intense signal, labelled SO_4^{2-} (BE ~ 531.7 eV), originates from substrate bound SO_4^{2-} anions.[24,25] In addition, the peaks located at BEs of ~530.0 eV and ~531.0 eV are likely to be associated with surface iron oxides and hydroxides, respectively.[9,12]

As regards the existence of surface SO_4^{2-}, the S 2p spectrum in Fig. 7(d) supports this deduction. More explicitly, the signal observed in the range 166 eV $\lesssim$ BE $\lesssim$ 171 eV is well known to be due to photoemission from highly oxidised sulfur species.[24,25] Two S 2p doublets were used to fit this region, with S $2p_{3/2}$ BEs of ~168.2 eV (labelled $(SO_4^{2-})^1$) and ~169.0 eV (labelled $(SO_4^{2-})^2$), which are concluded to be both due to surface bound SO_4^{2-}.[24,25] The reason for two SO_4^{2-} related doublets is uncertain but may be due to significant heterogeneity at the carbon steel surface. It should be noted that reduction of S to a lower formal oxidation state is a less likely explanation, as the BE difference would almost certainly be larger.[25] There is also a lower intensity doublet in Fig. 7(d) located at much lower BE (S $2p_{3/2}$ BE ~ 161.8 eV), which is attributed to surface sulfides, potentially arising from solution contaminants or possibly a SO_4^{2-} reduction reaction.

Merging our analysis of the Fe 2p, O 1s, and S 2p core level XPS spectra acquired from carbon steel following immersion in $0.01MH_2SO_4$/OMID, we posit the substrate termination depicted in Fig. 8. OMID is adsorbed onto an ultra-thin oxide/salt film, composed primarily of a ferric sulfate ($Fe_2(SO_4)_3$)-like phase, which is formed atop the carbon steel. Also shown in Fig. 8 is the inhibited interface concluded to form on carbon steel in 1MHCl/OMID. In this case, $OMIDH^+$ adsorbs onto an oxide/salt-free carbon steel surface decorated with chlorine atoms (Cl_{ads}).

Regarding the above results, to our knowledge, these are the first data to explicitly demonstrate that the identity of the aqueous acid can determine the chemistry of a well-inhibited interface. Hence, speculation about the origin of this variation is of value. With regards to the surface termination being a ferric sulfate like layer in $0.01MH_2SO_4$/OMID, one potentially salient point is that carbon steel is protected from corrosion by a layer of ferrous sulfate in highly concentrated H_2SO_4 ($\gtrsim$ 9 M);[26] no analogous protective iron chloride film forms on carbon steel in concentrated HCl. On this basis, in $0.01MH_2SO_4$/OMID, perhaps sorption of OMID limits dissolution of the ferric sulfate through formation of a protective hybrid surface film, thus inhibiting corrosion of the carbon steel.

Concerning 1MHCl/OMID, one topic of interest is the role of the adsorbed chlorine atoms, *i.e.*, 'Is Cl_{ads} essential to corrosion inhibition or simply a spectator adsorbate?'. Referring to Table 2, the $\eta\%$ of OMID at CMC in 0.01 M HCl is ~86%, which is a decrease of about 11% compared to 1 M HCl. This trend suggests that Cl_{ads} does play a role in the adsorption/performance of OMID as a corrosion inhibitor in HCl. To test this hypothesis, we have determined $\eta\%$ for carbon steel in a solution consisting of OMID at CMC, 0.01 M HCl and 1 M NaCl, so that the total Cl^- concentration is 1.01 M, which is similar to that in 1 M HCl. It was found that $\eta\% = 94.6 \pm 1.8\%$, suggesting that Cl_{ads} is required to ensure that OMID

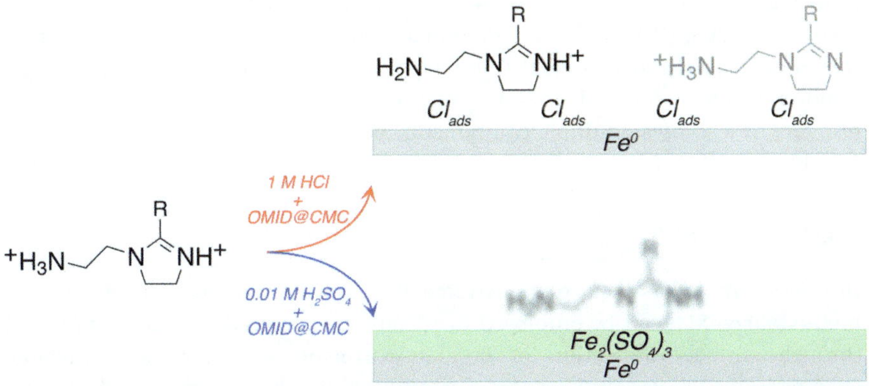

Fig. 8 Cartoon summarising interpretation of N 1s, Fe 2p, O 1s, Cl 2p, and S 2p XPS spectra acquired from carbon steel following immersion for 4 h in either 1MHCl/OMID or $0.01MH_2SO_4$/OMID. For 1MHCl/OMID, two $OMIDH^+$ species are concluded to be adsorbed onto an oxide/salt-free surface, decorated with chlorine atoms. For $0.01MH_2SO_4$/OMID, the exact nature of the adsorbed OMID remains ill-defined, but it is adsorbed onto an ultra-thin oxide/salt film, composed primarily of a ferric sulfate ($Fe_2(SO_4)_3$)-like phase. $R = C_{17}H_{33}$.

exhibits excellent corrosion inhibition in hydrochloric acid. We note that contrary to this variation in OMID performance with HCl concentration, Table 2 indicates that there is little loss in $\eta\%$ as the concentration of H_2SO_4 is reduced from 1 M to 0.01 M. This distinction is probably a consequence of a rather different inhibition mechanism, as discussed above.

Finally, we would like to make some comments on the approach adopted for both acquisition and analysis/interpretation of the presented XPS data. For the former, the importance of using the N_2-filled glove box for sample transfer into the XPS facility, to avoid post-immersion oxidation, cannot be over emphasised. Similar measures are apparently not taken in many other related studies, including some conducted quite recently,[27–29] which almost certainly leads to flawed interface characterisation. It should be borne in mind, however, that even using the glove box, the spectra have been acquired *ex situ*, and interface modification remains a concern, *e.g.*, adventitious carbon, which is almost inevitable, may modify corrosion inhibitor adsorption geometry, if not the chemistry. Furthermore, the process of removing a sample from solution may strip off weakly bound corrosion inhibitors, *e.g.*, those forming the outer part of an adsorbed bilayer (Fig. 1(a)).

As to data analysis/interpretation, our goal is to achieve a physically/chemically reasonable description of the inhibited interfaces through fitting of the spectral profiles with model line shape functions. We attempt to assign each peak to a specific state, although this can prove intractable on occasion, *e.g.*, the N 1s peak acquired in 0.01MH_2SO_4/OMID (Fig. 5). Given the complexity of the systems studied, in our opinion, this lack of clarity, does not invalidate the effort, but indicates that other techniques are likely required to complete characterisation, *e.g.*, X-ray absorption spectroscopy. In addition, atomic scale modelling may also facilitate more complete interface elucidation. Here, we have already demonstrated that *ab initio* DFT calculations can be used to support the assignment of core level peaks for HG-OMID, as well as predict the most favourable protonated states (HG-OMID → HG-OMIDH$^+$). Furthermore, DFT calculations have allowed us to justify the use of an asymmetric line shape function for fitting Cl 2p features assigned to Cl_{ads} (Fig. 7(c)). As highlighted in another recent study,[30] we suggest that proper selection of model line shape functions will become increasingly important moving forward, as system complexity increases, *e.g.*, in near ambient pressure XPS studies, multiple components are commonly required to fit an XPS core level spectrum.

Conclusions

In this study, it has been demonstrated that, through application of rigorous methodology, XPS can be employed to effectively characterise interfaces formed through surface interaction of organic corrosion inhibitors with metallic substrates in acidic solutions. Most notably, it has been determined that for corrosion inhibition of carbon steel by OMID, the nature of the well-inhibited interface is dependent on the acid. In 1 M HCl, it is concluded that OMID, assisted by the presence of Cl_{ads}, adsorbs as two different singly protonated species onto an oxide/salt-free carbon steel surface. In sharp contrast, in 0.01 M H_2SO_4, OMID adsorbs onto an ultra-thin surface film, composed primarily of a ferric sulfate ($Fe_2(SO_4)_3$)-like phase.

Conflicts of interest

There are no conflicts of interest to declare.

Acknowledgements

Funding and technical support from AkzoNobel and Nouryon through a collaboration with The University of Manchester is acknowledged. KK and MD are appreciative of funding from the EPSRC (EP/L01680X/1) through the Materials for Demanding Environments (M4DE) Centre for Doctoral Training (CDT). Furthermore, we are grateful for additional funding and technical support from bp through the bp International Centre for Advanced Materials (bp-ICAM) and EPSRC through the prosperity partnership (EP/G036850/1).

References

1 G. Koch, J. Varney, N. Thompson, O. Moghissi, M. Gould and J. Payer, *NACE 2016 Impact Study, International Measures of Prevention Application, and Economics of Corrosion Technologies Study*, 2016.
2 R. Lindsay and S. B. Lyon, in *Shreir's Corrosion, vol. 4, Management and Control of Corrosion*, ed. R. A. Cottis, M. J. Graham, R. Lindsay, S. B. Lyon, J. A. Richardson, J. D. Scantlebury and F. H. Stott, Elsevier, Amsterdam, 2010, pp. 2891–2899.
3 M. Finšgar and J. Jackson, *Corros. Sci.*, 2014, **86**, 17–41.
4 Y. Zhu, M. L. Free, R. Woollam and W. Durnie, *Prog. Mater. Sci.*, 2017, **90**, 159–223.
5 O. Olivares-Xometl, N. V. Likhanova, M. A. Domínguez-Aguilar, J. M. Hallen, L. S. Zamudio and E. Arce, *Appl. Surf. Sci.*, 2006, **252**, 2139–2152.
6 O. Olivares-Xometl, N. V. Likhanova, R. Martínez-Palou and M. A. Domínguez-Aguilar, *Mater. Corros.*, 2009, **60**, 14–21.
7 M. Yadav, D. Behera and U. Sharma, *Arabian J. Chem.*, 2016, **9**, S1487–S1495.
8 K. Kousar, T. Ljungdahl, A. Wetzel, M. Dowhyj, H. Oskarsson, A. S. Walton, M. S. Walczak and R. Lindsay, *J. Surfactants Deterg.*, 2020, **23**, 225–234.
9 K. Kousar, M. S. Walczak, T. Ljungdahl, A. Wetzel, H. Oskarsson, P. Restuccia, E. A. Ahmad, N. M. Harrison and R. Lindsay, *Corros. Sci.*, 2021, **180**, 109195.
10 R. A. Cottis, in *Shreir's Corrosion, vol. 2, Corrosion in Liquids, Corrosion Evaluation*, ed. R. A. Cottis, M. J. Graham, R. Lindsay, S. B. Lyon, J. A. Richardson, J. D. Scantlebury and F. H. Stott, Elsevier, Amsterdam, 2010, pp. 1341–1373.
11 ThermoFisher Scientific, https://www.thermofisher.com/uk/en/home/materials-science/learning-center/periodic-table.html, accessed November 20, 2021.
12 P. Morales-Gil, M. S. Walczak, R. A. Cottis, J. M. Romero and R. Lindsay, *Corros. Sci.*, 2014, **85**, 109–114.
13 M. S. Walczak, P. Morales-Gil, T. Belashehr, K. Kousar, P. Arellanes Lozada and R. Lindsay, *J. Visualized Exp.*, 2017, 55163.
14 R. Dovesi, V. Saunders, C. Roetti, R. Orlando, C. Zicovich-Wilson, F. Pascale, B. Civalleri, K. Doll, N. Harrison, I. Bush, P. D'Arco, M. Llunell, M. Causà,

Y. Noël, L. Maschio, A. Erba, M. Rerat and S. Casassa, *CRYSTAL17 User's Manual*, University of Torino, 2017.

15 R. Dovesi, A. Erba, R. Orlando, C. M. Zicovich-Wilson, B. Civalleri, L. Maschio, M. Rérat, S. Casassa, J. Baima, S. Salustro and B. Kirtman, *Wiley Interdiscip. Rev.: Comput. Mol. Sci.*, 2018, **8**, e1360.

16 M. S. Walczak, P. Morales-Gil and R. Lindsay, *Corros. Sci.*, 2019, **155**, 182–185.

17 J. S. Stevens, S. J. Byard and S. L. M. Schroeder, *Cryst. Growth Des.*, 2010, **10**, 1435–1442.

18 J. S. Stevens, A. C. de Luca, M. Pelendritis, G. Terenghi, S. Downes and S. L. M. Schroeder, *Surf. Interface Anal.*, 2013, **45**, 1238–1246.

19 J. S. Stevens, S. J. Byard, C. C. Seaton, G. Sadiq, R. J. Davey and S. L. M. Schroeder, *Phys. Chem. Chem. Phys.*, 2014, **16**, 1150–1160.

20 N. Fairley, *CasaXPS Manual 2.3.15 Spectroscopy 1.3*, Casa Software Ltd, 2009.

21 G. Wertheim and P. Citrin, in *Topics in Applied Physics, Photoemission in Solids I, General Principles*, ed. M. Cardona and L. Ley, Springer-Verlag, Berlin, Heidelberg, New York, 1978, pp. 197–236.

22 P. Gupta and S. K. Sen, *Phys. Rev. B: Solid State*, 1975, **12**, 15–19.

23 A. P. Grosvenor, B. A. Kobe, M. C. Biesinger and N. S. McIntyre, *Surf. Interface Anal.*, 2004, **36**, 1564–1574.

24 M. Descostes, F. Mercier, N. Thromat, C. Beaucaire and M. Gautier-Soyer, *Appl. Surf. Sci.*, 2000, **165**, 288–302.

25 M. Wahlqvist and A. Shchukarev, *J. Electron Spectrosc. Relat. Phenom.*, 2007, **156–158**, 310–314.

26 Z. Panossian, N. L. de Almeida, R. M. F. de Sousa, G. de S. Pimenta and L. B. S. Marques, *Corros. Sci.*, 2012, **58**, 1–11.

27 J. Haque, V. Srivastava, M. A. Quraishi, D. Singh Chauhan, H. Lgaz and I.-M. Chung, *Corros. Sci.*, 2020, **172**, 108665.

28 A. Espinoza-Vázquez, F. J. Rodríguez-Gómez, G. E. Negrón-Silva, R. González-Olvera, D. Ángeles-Beltrán, M. Palomar-Pardavé, A. Miralrio and M. Castro, *Corros. Sci.*, 2020, **174**, 108853.

29 L. Zhu, J. Fan, H. Huang, L. Guo, M. Zhu, X. Zheng and I. B. Obot, *J. Mol. Struct.*, 2021, **1243**, 130852.

30 C. R. O'Connor, M. A. van Spronsen, M. Karatok, J. Boscoboinik, C. M. Friend and M. M. Montemore, *J. Phys. Chem. C*, 2021, **125**, 10685–10692.

Faraday Discussions

ROYAL SOCIETY OF CHEMISTRY

PAPER

Resonant X-ray photoelectron spectroscopy: identification of atomic contributions to valence states†

J. M. Seymour,[a] E. Gousseva,[a] R. A. Bennett,[a] A. I. Large,[b] G. Held,[b] D. Hein,[c] G. Wartner,[c] W. Quevedo,[c] R. Seidel,[c] C. Kolbeck,[d] C. J. Clarke,[e] R. M. Fogarty,[f] R. A. Bourne,[g] R. G. Palgrave,[h] P. A. Hunt[i] and K. R. J. Lovelock*[a]

Received 7th December 2021, Accepted 16th February 2022

DOI: 10.1039/d1fd00117e

Valence electronic structure is crucial for understanding and predicting reactivity. Valence non-resonant X-ray photoelectron spectroscopy (NRXPS) provides a direct method for probing the overall valence electronic structure. However, it is often difficult to separate the varying contributions to NRXPS; for example, contributions of solutes in solvents or functional groups in complex molecules. In this work we show that valence resonant X-ray photoelectron spectroscopy (RXPS) is a vital tool for obtaining atomic contributions to valence states. We combine RXPS with NRXPS and density functional theory calculations to demonstrate the validity of using RXPS to identify atomic contributions for a range of solutes (both neutral and ionic) and solvents (both molecular solvents and ionic liquids). Furthermore, the one-electron picture of RXPS holds for all of the closed shell molecules/ions studied, although the situation for an open-shell metal complex is more complicated. The factors needed to obtain a strong RXPS signal are investigated in order to predict the types of systems RXPS will work best for; a balance of element electronegativity and bonding type is found to be important. Additionally, the dependence of RXPS spectra on both varying solvation environment and varying local-covalent bonding is probed. We find that RXPS is a promising fingerprint method for identifying species in solution, due to the spectral shape having a strong dependence on local-covalency but a weak dependence on the solvation environment.

[a]*Department of Chemistry, University of Reading, Reading, UK. E-mail: k.r.j.lovelock@reading.ac.uk*

[b]*Diamond Light Source, Didcot, Oxfordshire, UK*

[c]*Helmholtz-Zentrum Berlin für Materialien und Energie (HZB), Berlin, Germany*

[d]*Fritz-Haber-Institut der Max-Planck-Gesellschaft, Berlin, Germany*

[e]*School of Chemistry, University of Nottingham, Nottingham, UK*

[f]*Department of Materials, Imperial College London, London, UK*

[g]*Institute of Process Research and Development, Schools of Chemistry and Chemical and Process Engineering, University of Leeds, Leeds, UK*

[h]*Department of Chemistry, University College London, London, UK*

[i]*School of Chemical and Physical Sciences, Victoria University of Wellington, Wellington, New Zealand*

† Electronic supplementary information (ESI) available. See DOI: 10.1039/d1fd00117e

1. Introduction

Theories linking electronic structure to reactivity are well-established.[1–5] For many applications which revolve around electron donation in the liquid phase, *e.g.* electrochemical energy storage and catalysis,[6–10] knowledge of the occupied electronic states is vital for understanding and predicting reactivity. Experimental measures of the occupied valence states (OVSs)[11] and calculations of the occupied molecular orbitals (OMOs)[1–5,12] for gas phase molecules and for solids/adsorbates on solids are widely available. The highest OVS (HOVS) from experiments and highest OMO (HOMO) from calculations are often found to be important for explaining reactivity, but deeper OVSs/OMOs at larger binding energy (E_B) have also been found to be critical.[13–15] Therefore, knowledge of both E_B and the atomic contributions to the OVSs/OMOs is crucial to understanding and predicting reactivity for the liquid phase. Ideally, this question would be answered using experimental data, but measuring every solute/solvent combination is not possible; calculations offer a potential for screening solute/solvent combinations. E_B(HOMO) calculated on gas phase species has been suggested as a suitable descriptor for choosing solutes (*e.g.* redox mediators) and solvents for battery electrolytes,[6,7,16,17] although gas phase calculations do not include solvation effects.[18] Calculations on models of the liquid phase[8,19] or where solvent effects are accounted for,[20] have been published, but capturing solvation with calculations is very challenging, both in terms of the methods and the system size. Therefore, benchmark experimental data is required for validation of these calculations.

Information about the electronic structure can be obtained from valence non-resonant X-ray photoelectron spectroscopy (NRXPS) (Fig. 1b). Currently there is little liquid phase experimental data for OVSs compared to the amount of experimental data available for the gas phase and solids, especially for solvents other than water, as solutions pose significant challenges. Given the volatile nature of most solutions, liquid jet sample set-ups are usually required,[21–24] although UHV-compatible ionic liquids (ILs, liquids containing only ions and no neutral molecules) have also been studied.[25] A further challenge is when contributions from the solute and solvent overlap; contributions from the solvent dominate, making OVSs from the solute particularly hard to identify. There are a small number of solutions where the solute HOVS was clearly the sample HOVS, meaning that NRXPS could be used to measure E_B for the solute HOVS, *e.g.* NaI in water.[26] For certain solutes, the solute HOVS has been found to overlap with the solvent HOVS; therefore, the solute E_B(HOVS) can be measured with very careful valence NRXPS measurements and subtraction of the solvent valence NRXPS contribution. However, E_B for most other solute OVSs cannot be readily measured, *e.g.* imidazole in water.[27] Little valence NRXPS exists for solutes with large HOVS E_B due to the significant experimental challenges.

A comprehensive description of liquid-phase electronic structure can potentially be obtained from valence resonant XPS (RXPS, Fig. 1c and d), which is element-specific and orbital-specific, and at times site-specific.[28–78] In particular, contributions from overlapping OVSs can be identified using valence RXPS, which is very important for complex systems such as solutions. RXPS has been used for many years to study readily vaporised small molecules, *e.g.* water, N_2.[28–32] In the

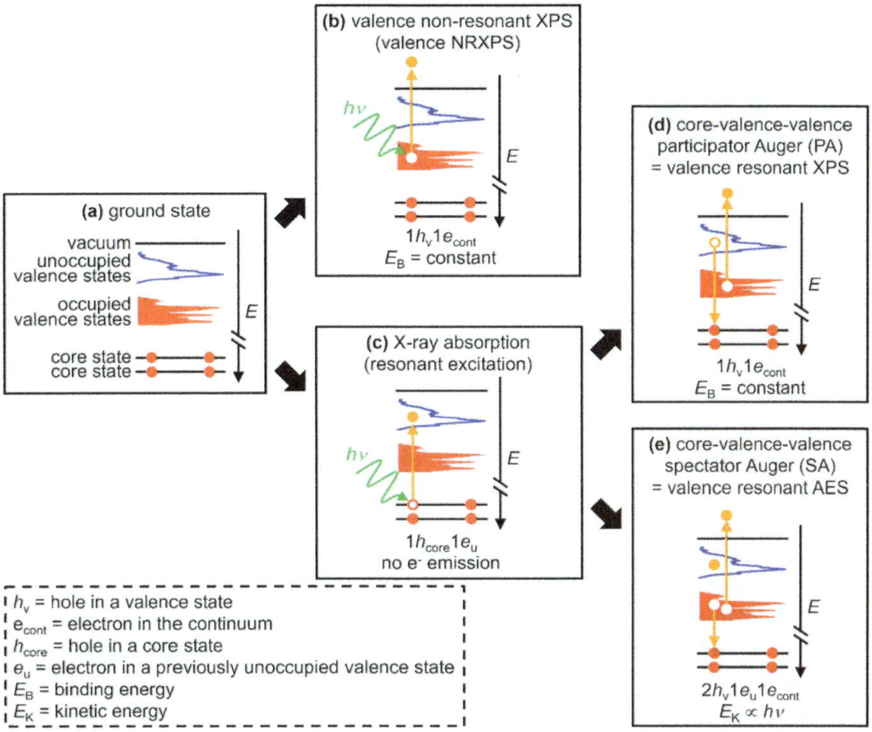

Fig. 1 Sketches explaining the different electron dynamics events featured in this article. (a) Ground state. (b) Valence non-resonant XPS (NRXPS). (c) X-ray absorption (resonant excitation). (d) Core–valence–valence participator Auger (PA) transition; X-ray absorption followed by a valence PA transition = valence resonant XPS (RXPS). (e) Core–valence–valence spectator Auger (SA) transition; X-ray absorption followed by a valence SA transition = valence resonant Auger electron spectroscopy (RAES). The text at the bottom of each sketch gives the final state of each process and the approximate electron kinetic energy (E_K) or binding energy (E_B). $1e_{cont}$ = electron in the continuum. $1e_u$ = electron in a previously unoccupied valence state. Element- and orbital-specificity are achieved through studying different element core state X-ray absorptions followed by RXPS for the same samples; site-specificity can be achieved for certain samples by studying two or more different X-ray absorptions followed by RXPS for the same element of the same sample.

condensed phase, RXPS has been used to study many solids: metal salts/complexes,[33–35] inorganic salts,[36–45] small organic molecules,[46–52] protein,[53] polymers,[54–59] organic perovskites.[60,61] Furthermore, RXPS has been used at the solid–adsorbate interface.[62–66] More recently, RXPS has been used to study liquids, both for the solvent (water,[67] ILs[68,69]) and more commonly for solutes in water (H_2O_2,[67] metal complexes[70–75] such as Fe^{3+} in aqueous solution[70] and Ti^{3+} in aqueous solution,[71] and nanoparticles[76,77]).

Valence RXPS is best pictured as a two-step process. (i) X-ray absorption (XA) by a core state to give a $1h_{core}1e_u$ state ($1h_{core}$ represents the core-hole and $1e_u$ represents an electron in a previously unoccupied valence state, Fig. 1c), where the absorption cross-section is large for dipole allowed transitions (*e.g.* N 1s → N 2p π* or Co $2p_{3/2}$ → Co 3d). (ii) A core–valence–valence participator Auger (PA) transition as the de-excitation step; the excited valence electron refills the core-hole (*i.e.* participating in the process through direct recombination) and an electron from an OVS is ejected (Fig. 1d). The final state for valence PA emission is one hole in the valence, $1h_v$ (Fig. 1d), 1+ relative to the ground state (Fig. 1a), the same final state as for valence NRXPS (Fig. 1b). Hence, XA followed by valence PA emission is known as valence RXPS and can be interpreted in terms of a one-electron picture, *i.e.* the frozen orbital approximation.

The selection rules are different for valence NRXPS and valence RXPS, with the selection rules well-known for valence NRXPS (explained by the Gelius method[79,80]) but unclear for valence RXPS.[31,52,55] One of the requirements for a PA transition is that the OVS (in the ground state) involved in the PA process has good overlap with the core-hole. Therefore, features due to RXPS appear as resonant enhancements in the intensity of valence NRXPS contributions and normally appear at constant binding energy, E_B; the same E_B as for valence NRXPS from the same OVS.[34,56] Using a combination of results from RXPS (*i.e.* PA transitions), valence NRXPS and ground state calculations to obtain the partial density-of-states (pDoS), it has been shown that RXPS can be used to identify atomic contributions to OVSs for a number of relatively simple systems, *e.g.* pyridine,[46] pyrazine,[46] s-triazine,[46] butadiene,[51] furan,[49] pyrrole,[49] benzene,[48] and also the polymer poly(4-hydroxystyrene) (PHS),[55] *i.e.* the one-electron picture held. Leading on from these findings, RXPS has been used to identify the HOVS as originating from: (a) DNA bases and not in the backbone of DNA duplexes,[57] and (b) the cation (and not the anion) for nine ILs (out of 60 ILs studied).[68,69] However, most RXPS studies have been limited to organic molecules containing π-bonding.

To probe the OVSs, intense RXPS transitions are ideal. Therefore, it is important to determine what structures give intense RXPS transitions. At $E_B >$ ~20 eV (depending on the sample and element edge studied) transitions are detected, *e.g.* spectator Auger (SA) transitions, where the final state is $2h_v1e_u$ with one-electron ejected to the continuum (Fig. 1e). For these SA transitions the one-electron picture does not hold.[49] SA transitions compete with PA transitions to fill the core-hole; large PA : SA intensity ratios are best for probing OVSs. In general, valence SA features are more intense than valence PA features.[31,36,43–45,49,50,52,54,55] For X-ray absorption for core state → π* *versus* X-ray absorption for core state → σ*, organic molecules with π* showed much larger PA : SA intensity ratios than organic molecules with no π* and only σ* (which generally showed no PA intensity),[50,54] *e.g.* ethylene *versus* cyclohexane.[50] For different elements, much larger PA : SA intensity ratios were obtained for the B 1s → π* absorption for solid BN and solid B_2O_3 than for the N 1s → π* absorption for solid BN.[43–45] For solid pyrrole the PA : SA intensity ratios were 1 : 10 for the C 1s → π* absorption and 1 : 50 for the N 1s → π* absorption.[49] Furthermore, the PA : SA intensity ratios were 1 : 10 for the C 1s → π* absorption for solid furan, polystyrene and poly(4-hydroxystyrene).[49,54,55] For the gas phase of four simple O-containing organic molecules (formic acid, acetaldehyde, acetic acid and methyl formate), the PA : SA intensity ratio was small even for O 1s → π* absorption.[31] Lastly, no PA intensity was observed at the F 1s absorption edge for solid CaF_2 and fluorinated nanotubes.[36,78]

Studies have investigated whether valence RXPS spectra gave fingerprint contributions that represent functional groups within a sample. For solid $M[NO_3]$ (where $M = Li^+, Na^+$) the $[NO_3]^-$ anion was essentially quasi-isolated, as the RXPS spectra were broadly independent of the counterion identity.[41,42] Conversely, for four gas phase small C=O containing organic molecules (formic acid, acetaldehyde, acetic acid and methyl formate), the RXPS spectra did not match; RXPS was deemed too sensitive to the chemical bonding to enable identification of specific functional groups in these molecules.[31]

In this article, we investigate the liquid phase electronic structure of both solutes and solvents using valence RXPS and DFT calculations. The key molecules and ions studied are given in Fig. 2. A full list of samples studied using valence RXPS is given in ESI Table S1.† Four samples were molecular liquids (water, ethanol = EtOH, acetonitrile = MeCN and propylene carbonate = PC), 11 samples were solutes in molecular liquids (water and MeCN), and eight samples were neat ILs. Coordinates for the molecules/ions for which DFT calculations were performed are given in ESI Section 2;† to capture the liquid phase, a polarisable continuum model was used (see Section 2.6 for more details). Separation of solute and solvent contributions to the valence electronic structure was achieved using valence RXPS. The factors that underpin valence RXPS intensity were probed. It is challenging to determine the factors for many of the solutes studied here due to the dominance of solvent contributions; hence, we mainly focused in this part on neat solvents, *i.e.* molecular liquids and ILs. By combining knowledge from: (i) E_B

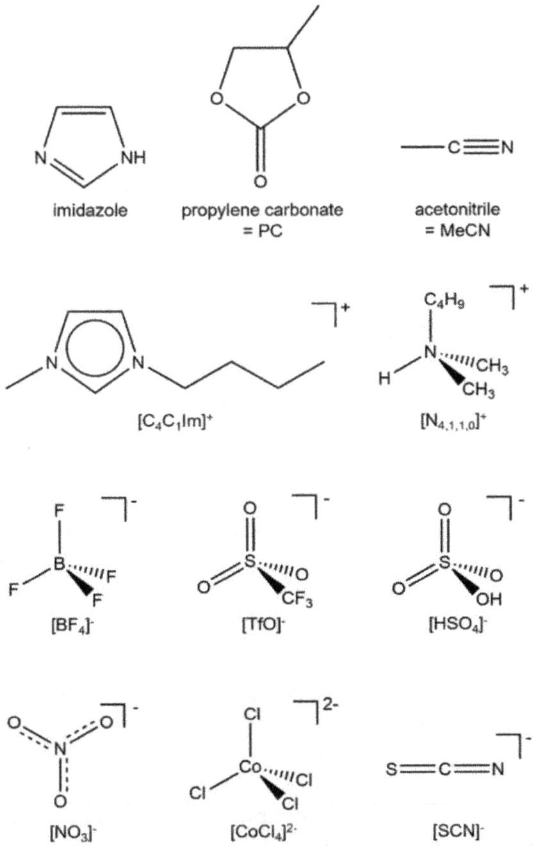

Fig. 2 The key molecules, cations and anions studied.

matches from valence RXPS and NRXPS and (ii) comparisons of valence RXPS traces to pDOS from calculations, we demonstrate that the one-electron picture for valence XPS holds for closed shell molecules/ions in the liquid phase. The cyano-based anions are ideal for testing the one-electron picture, as these anions gave sharp and well-defined features in valence NRXPS. Lastly, insights into bonding were gained by making comparisons between experimental RXPS traces. The use of valence RXPS traces as fingerprints for both intramolecular and intermolecular bonding/interactions is investigated; this knowledge is important to understand interactions of ions/molecules with both other solutes and surfaces.

2. Experimental

2.1 Sample synthesis

Details of sample synthesis/preparation are given in ESI Section 1.†

2.2 Lab XPS apparatus

Laboratory-based XPS for the IL $[C_4C_1Im][BF_4]$ was carried out using a Thermo K-alpha spectrometer utilising Al Kα radiation ($h\nu = 1486.6$ eV), as detailed in ref. 68. No impurities were observed after Ar^+ ion sputtering (ESI Fig. S21†).

2.3 Synchrotron XPS apparatus

The measurement for 15 solutions was performed on the U49/2-PGM 1 beamline with SOL^3PES end-station[81] at BESSY II (Germany). Solutes were weighed and mixed with a corresponding mass of solvent to achieve the desired concentration. Liquid-jet apparatus was used. XP and RXP spectra were acquired using a Scienta Omicron R4000 HIPP-2 hemispherical electron analyser.

The measurements for two ILs ($[C_8C_1Im]_2[CoCl_4]$ and $[C_8C_1Im]_2[Co(NTf_2)_4]$) were performed on the B07-C (VerSoX) beamline at Diamond Light Source (UK).[82] A thin film (less than 0.1 mL, essentially so that no drop could be observed) of the IL sample was placed on a tantalum sample holder. The "T-pot" analyser chamber was used, which allowed for rastering of the sample continually, perpendicular to the analyser entrance nozzle during X-ray irradiation. Rastering allowed for the maximum X-ray flux of B07, as beam damage did not occur during the raster experiments (unlike during static experiments, see ref. 69). XP and RXP spectra were acquired using a PHOIBOS 150 NAP hemispherical electron analyser supplied by SPECS. No sputtering was applied to the samples prior to measurement (no contaminants were observed by NRXPS).

The measurements for five ILs ($[C_8C_1Im][C(CN)_3]$, $[C_4C_1Im][SCN]$, $[P_{6,6,6,14}][NO_3]$, $[N_{2,1,1,0}][TfO]$ and $[N_{4,1,1,0}][HSO_4]$) were performed on the I311 beamline at MAX-lab (Sweden). The apparatus used at MAX-lab is explained in ref. 68.

The angle between the polarisation vector of the incoming X-rays and the analyser was different for the experimental set-ups used here (see ESI Table S1† for the angle each sample was run at). However, this angle is unimportant for valence RXPS, and will only affect the relative intensity of peaks in valence NRXPS.

2.4 Producing RXPS traces

Valence RXPS traces were produced using the subtraction procedure outlined in ref. 68. Briefly, traces of valence RXPS plus valence NRXPS had the valence NRXPS contributions subtracted. In some cases, this subtraction method gave negative features, due to either complications due to 2nd order light (*e.g.* Fig. 8c) or RXPS contributions being much lower intensity than the NRXPS contributions (*e.g.* Fig. 6d and 7e). The latter scenario causes a problem when a small percentage increase in the NRXPS contribution at the absorption edge (*i.e.* the $h\nu$ at which the RXPS contribution occurs) relative to the $h\nu$ at which only NRXPS contribution occurs that is used for subtraction, leading to negative peaks in the subtracted RXPS trace.

2.5 Charge referencing methods for XP spectra

The valence NRXPS features were at constant E_B with respect to $h\nu$ for all experiments presented here, see Fig. 3a and ESI Fig. S2–S20.† Therefore, any changes in E_B for resonant features relative to non-resonant features were not due to sample charging.

For the 11 aqueous solutions, all RXPS spectra were charge referenced to $E_B(1b_1) = 11.33$ eV by fitting the valence NRXPS [83] measured at $h\nu$ just below the

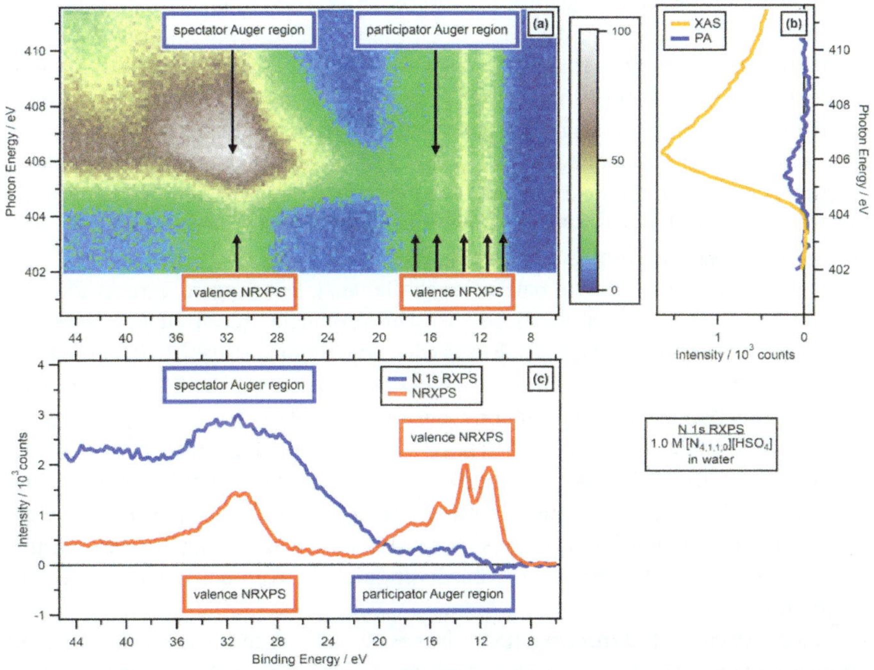

Fig. 3 N 1s valence RXPS for 1.0 M $[N_{4,1,1,0}][HSO_4]$ in water. (a) Heat map of photon energy against binding energy for the N 1s valence RXPS. (b) Partial electron yield N 1s NEXAFS spectra (spectator Auger region plus participator Auger region = XAS, and participator Auger region only). (c) Valence NRXPS below the N 1s absorption edge ($h\nu = 403.0$ eV) and N 1s valence RXPS ($h\nu = 405.6$ eV). The valence RXPS trace was produced using the procedure outlined in ref. 68. All electron spectra were charge referenced using procedures outlined in Section 2.5.

relevant absorption edge. No charge referencing was applied to RXPS for neat MeCN, neat EtOH and neat PC, given no literature E_B values could be found for valence NRXPS. For 0.5 M $[C_4C_1Im][OAc]$ in MeCN, the O 1s RXPS was charge referenced to the O 1s RXPS spectrum for 1.3 M $[C_4C_1Im][OAc]$ in water to allow for visual comparisons. All IL spectra were effectively charge referenced to $E_B(C_{alkyl}\ 1s) = 289.58$ eV.[84] Further small E_B shifts were then applied to allow visual comparisons to the RXPS measured in molecular solvents.

2.6 Calculation methods

DFT calculations were carried out at the B3LYP-D3(BJ)/6-311+G(d,p) level; further details can be found in ESI Section 2.† Structures were optimised as individual, isolated molecules/ions. A generalised solvation model has been employed (SMD), and the parameters employed are reported in ESI Table S2.† Coordinates for the molecules/ions studied are given in ESI Section 10.†

3. Results

3.1 Separating solute and solvent contributions to OVSs

3.1.1 Identifying solute contributions to OVSs. For the 15 ions/molecules studied here, our approach of identifying solute contributions using valence RXPS worked very well, despite the presence of very intense NRXPS solvent/counterion contributions, *e.g.* Fig. 4, 6, 7 and 8. For some solutes, the RXPS contributions were more intense than the solvent NRXPS contributions, *e.g.* C 1s RXPS for 0.5 M $[C_4C_1Im]I$ in water (ESI Fig. S14†). For other solutes, it was more challenging to distinguish solute contributions from solvent contributions. For example, for N 1s RXPS of 1.0 M $[N_{4,1,1,0}][HSO_4]$ in water a weak but detectable PA feature is visible on the heat map at 404 eV < $h\nu$ < 407 eV (Fig. 3b) and 19 eV > E_B > 10 eV (Fig. 3a and c); the NRXPS contributions from water were larger than N 1s RXPS contributions from $[N_{4,1,1,0}]^+$ (Fig. 3a and c).

O 1s RXPS for solutes is a particular challenge, given oxygen features in many common solvents, especially those with the properties required for successful liquid jet XPS experiments, *e.g.* solvents with large relative permittivity. O 1s RXPS for $[OAc]^-$ in water worked very well, the O 1s → π* X-ray absorption energy for $[OAc]^-$ ($h\nu$ ~ 533 eV) was significantly lower than the O 1s → σ* X-ray absorption energies for water ($h\nu$ > 534 eV). Therefore, the PA contributions for $[OAc]^-$ could be separated from the PA contributions for water. RXPS maps and traces for a number of key solvents (water, ethanol, propylene carbonate, acetonitrile) are given in ESI Section 4.† This data is a useful resource, especially for determining whether solute contributions can be distinguished from solvent contributions using RXPS.

3.1.2 Factors determining RXPS intensity. In the competition between SA transitions and PA transitions to fill the core-hole, SA features were more intense than valence PA features (Fig. 4), matching the literature.[31,36,43–45,49,50,52,54,55] For example, for N 1s RXPS of 1.0 M $[N_{4,1,1,0}][HSO_4]$ in water, very strong features due to SA transitions were observed at 45 eV > E_B > 20 eV relative to the features due to PA transitions (Fig. 3a and c).

For the anions $[BF_4]^-$ and $[TfO]^-$ the F 1s RAES/RXPS spectra gave very low intensity features due to PA transitions at 24 eV > E_B > 0 eV (Fig. 4a), giving very

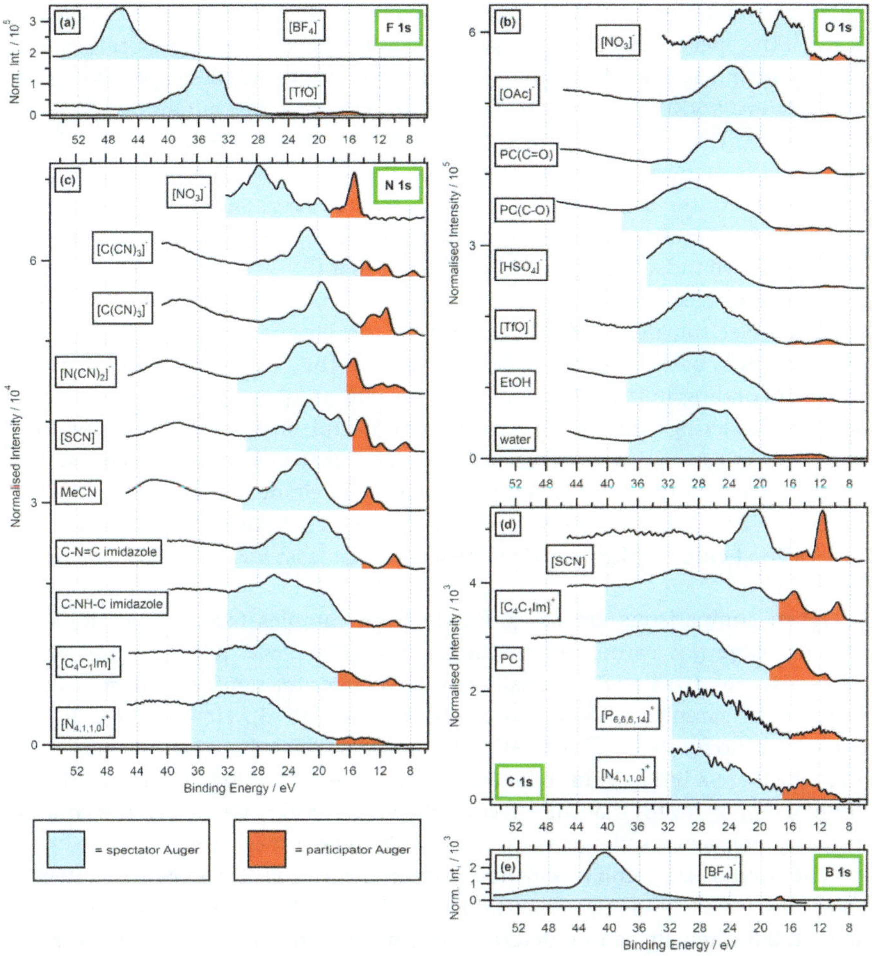

Fig. 4 Valence RXPS at different absorption edges. Shaded regions are shown for estimated spectator Auger (SA, in light blue) and participator Auger (PA, in red) transition contributions; backgrounds have not been included, so the SA region in particular will give an overestimate of the true SA area. (a) F 1s valence RXPS for 0.5 M [C_4C_1Im][TfO] in water ($h\nu = 690.6$ eV) and 0.5 M [C_4C_1Im][BF_4] in water ($h\nu = 696.4$ eV). (b) O 1s valence RXPS for [$P_{6,6,6,14}$][NO_3] ($h\nu = 531.9$ eV), 1.3 M [C_4C_1Im][OAc] in water ($h\nu = 532.7$ eV), propylene carbonate (PC, $h\nu = 533.3$ eV for C=O absorption and $h\nu = 536.2$ eV for C–O absorption), [$N_{4,1,1,0}$][HSO_4] ($h\nu = 536.8$ eV), [$N_{2,2,1,0}$][TfO] ($h\nu = 535.3$ eV), ethanol (EtOH, ($h\nu = 535.4$ eV)) and water ($h\nu = 535.0$ eV). (c) N 1s valence RXPS for [$P_{6,6,6,14}$][NO_3] ($h\nu = 405.2$ eV), [C_8C_1Im][$C(CN)_3$] ($h\nu = 399.0$ eV and $h\nu = 400.0$ eV), 0.5 M [C_4C_1Im][$N(CN)_2$] in water ($h\nu = 400.1$ eV), 0.5 M [C_4C_1Im][SCN] in water ($h\nu = 399.8$ eV), acetonitrile (MeCN, ($h\nu = 400.1$ eV)), 1.8 M imidazole in water ($h\nu = 400.2$ eV for C–N=C absorption and $h\nu = 401.9$ eV for C–NH–C absorption), 0.5 M [C_4C_1Im]I in water ($h\nu = 402.1$ eV) and 3.3 M [$N_{4,1,1,0}$][HSO_4] in water ($h\nu = 405.6$ eV). (d) C 1s valence RXPS for 0.5 M K[SCN] in water ($h\nu = 287.5$ eV), 0.5 M [C_4C_1Im]I in water ($h\nu = 287.0$ eV), propylene carbonate (PC, $h\nu = 290.5$ eV), [$P_{6,6,6,14}$][NO_3] ($h\nu = 287.6$ eV) and [$N_{4,1,1,0}$][HSO_4] ($h\nu = 288.5$ eV). (e) B 1s valence RXPS for 0.5 M [C_4C_1Im][BF_4] in water ($h\nu = 201.4$ eV). The valence NRXPS contributions were subtracted using the procedure outlined in ref. 68. All valence XP spectra were charge referenced using procedures outlined in Section 2.5. The upper E_B limit for the participator Auger region was chosen based on a combination of the relative intensity compared with other features from participator/spectator transitions and variation in peak E_B with respect

small PA : SA intensity ratios. For the anion $[SCN]^-$, three cations and PC, the C 1s RAES/RXPS spectra gave much larger PA : SA intensity ratios than observed for F 1s (Fig. 4d and a). The PA : SA intensity ratios for O 1s RAES/RXPS spectra (Fig. 4b) and N 1s RAES/RXPS spectra (Fig. 4c) were, on average across all samples, between the two extremes of F 1s RAES/RXPS and C 1s RAES/RXPS. The correlation between the element electronegativity and PA : SA intensity ratio holds for B 1s RAES/RXPS for the solids BN and B_2O_3, where large PA : SA intensity ratios were observed.[43–45] However, for B 1s RAES/RXPS for $[BF_4]^-$ the PA : SA intensity ratio was very small indeed (Fig. 4e). It is expected that $[BF_4]^-$ is a special case, as the difference in electronegativity is so large between B and F.[85] Boron had little localised electron density/OVSs, leading to the observed weak PA signal; this observation is backed up by the boron pDoS for $[BF_4]^-$ (Fig. 7e), demonstrating that the bonding in the individual ion/molecule needs to be taken into account when considering the PA : SA intensity ratio and electronegativity. For most samples, the PA : SA intensity ratios follow the order from largest to smallest: B 1s > C 1s > N 1s > O 1s > F 1s. This trend matches the element electronegativity,[85] as noted for four gas phase small organic molecules.[31] Overall, electronegativity gives a good guide to the PA : SA intensity ratio, at least for the period 2 elements from B to F.

When comparisons are made for different samples focusing on the same element edge (*i.e.* carbon *versus* carbon, nitrogen *versus* nitrogen and oxygen *versus* oxygen), the PA : SA intensity ratio was larger for samples with π^* valence states than samples with σ^* valence states (Fig. 4b–d). However, the PA : SA intensity ratio for $[N_{4,1,1,0}]^+$ N 1s → σ^* RAES/RXPS (Fig. 4c) was significantly larger than the PA : SA intensity ratio for PC(C=O) O 1s → π^* RAES/RXPS (Fig. 4b). This finding demonstrates that the effect on the PA : SA intensity ratio of the bonding type was not nearly so large as the effect of the electronegativity.

The overall RXPS enhancement is captured with the RXPS : NRXPS intensity ratio for any E_B region (Fig. 5, 6b and ESI Table S4†). For the data in Fig. 5 and 6b, the NRXPS intensity can be determined from the off-resonant signal below the absorption edge; the RXPS intensity can be determined from the on-resonant signal. By far the largest RXPS : NRXPS intensity ratios were observed for N 1s RXPS for MeCN (Fig. 5a and ESI Fig. S6†), *e.g.* for the HOVS of MeCN the RXPS : NRXPS intensity ratio was ~100 (ESI Table S4†). C_{ring} 1s RXPS for $[C_4C_1Im]^+$ (*i.e.* the three carbon atoms in the imidazolium ring) and N 1s RXPS for $[C(CN)_3]^-$ both gave RXPS : NRXPS intensity ratios of ~20 (Fig. 5a, b and ESI Table S4†). All other RXPS : NRXPS intensity ratios were below 10 (ESI Table S4†). In particular, the RXPS : NRXPS intensity ratios were in the range of 1 to 2 for functional groups with little or no π^* bonding and only σ^* bonding, *e.g.* O 1s RXPS for water gave a RXPS : NRXPS intensity ratio of ~2 (Fig. 5c and ESI Table S4†).

Overall, the RXPS : NRXPS intensity ratios did not match to the PA : SA intensity ratios. For example, for MeCN N 1s RXPS the RXPS : NRXPS intensity ratio was the largest value by far, but for MeCN N 1s RXPS the PA : SA intensity

to $h\nu$. The upper E_B limit for the spectator Auger region was chosen based on the Auger dip (where measured, otherwise the upper limit of the measurement); the lower E_B limit for the spectator Auger region was the same as the upper limit for the participator Auger region.

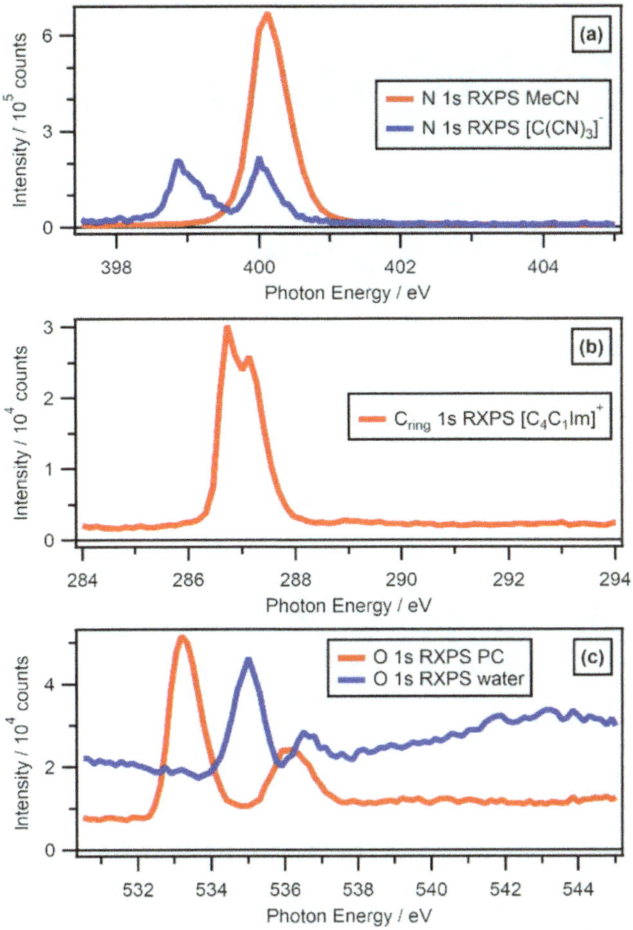

Fig. 5 Partial electron yield NEXAFS spectra. (a) N 1s edge for acetonitrile (MeCN, 12.6 eV > E_B > 11.2 eV) and 0.5 M $[C_4C_1Im][SCN]$ in water (9.4 eV > E_B > 7.4 eV). (b) C 1s edge for 0.5 M $[C_4C_1Im]I$ in water (10.0 eV > E_B > 9.2 eV). (c) O 1s edge for propylene carbonate (PC, 11.4 eV > E_B > 10.0 eV) and water (12.4 eV > E_B > 10.8 eV).

ratio was less than a number of other samples, *e.g.* $[SCN]^-$ C 1s RXPS. Therefore, it appears that the RXPS : NRXPS intensity ratio was determined by a combination of the NEXAFS absorption cross-section and the competition between PA and SA transitions. However, it is difficult to draw wide-ranging conclusions given the dataset of RXPS : NRXPS intensity ratios available; especially problematic is the sparse data focused on functional groups with only σ* bonding (and only those at the F 1s or O 1s edges, where the PA : SA intensity ratios are very small).

3.2 Demonstrating the one-electron picture holds for valence RXPS

3.2.1 Valence NRXPS *versus* valence RXPS: E_B comparisons. Comparing valence NRXPS and N 1s RXPS for $[SCN]^-$, there were subtle differences in E_B for the main features (Fig. 6a and c). The peak maxima for the features at $E_B \sim 7.5$ eV and $E_B \sim 13$ eV were shifted by $E_B \sim +0.2$ eV for N 1s RXPS compared to NRXPS;

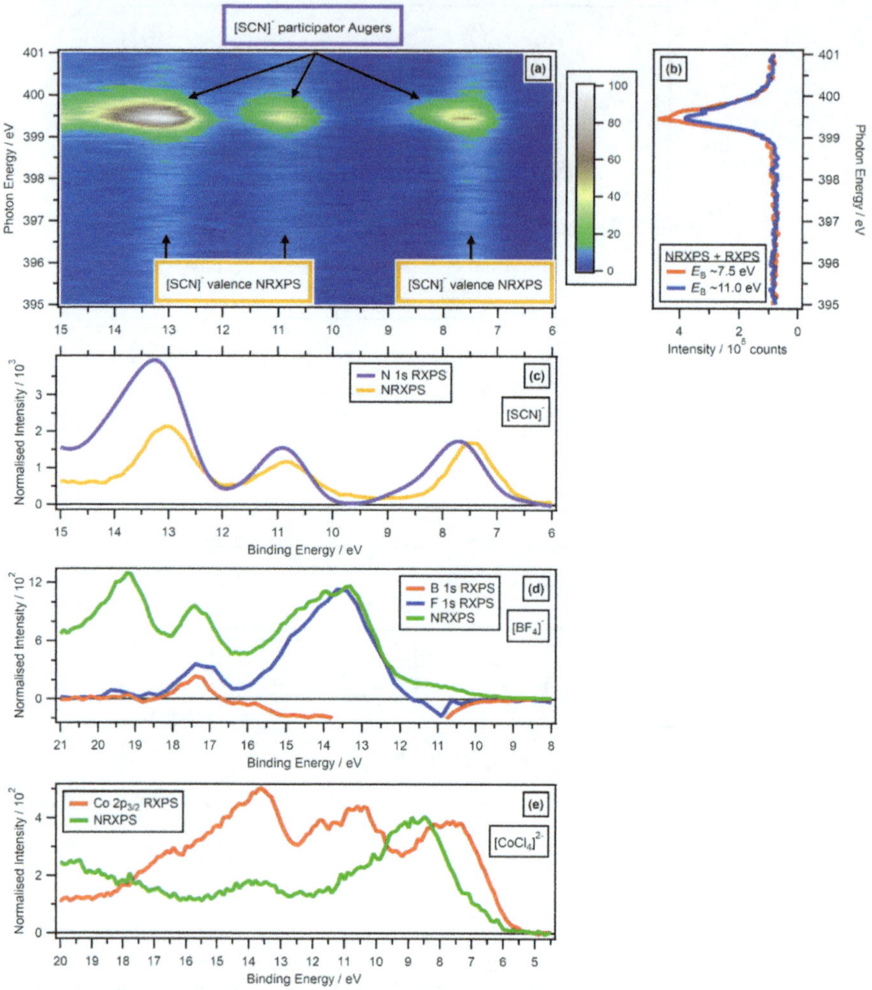

Fig. 6 Comparing the valence RXPS and valence NRXPS for the anion N 1s → π* absorption for $[C_4C_1Im][SCN]$: (a) N 1s valence RXPS heat map for $[C_4C_1Im][SCN]$; (b) partial electron yield N 1s NEXAFS spectra for two regions ($E_B \sim 7.5$ eV and $E_B \sim 11.0$ eV). (c) N 1s valence RXPS for $[C_4C_1Im][SCN]$ ($h\nu = 399.5$ eV), compared to the valence NRXPS for $[C_4C_1Im][SCN]$ ($h\nu = 1486.6$ eV). The $[C_4C_1Im]^+$ contribution to the valence NRXPS was negligible at $h\nu = 1486.6$ eV.[68,69] (d) B 1s valence RXPS ($h\nu = 201.4$ eV) and F 1s valence RXPS ($h\nu = 696.4$ eV) for 0.5 M $[C_4C_1Im][BF_4]$ in water, compared to the valence NRXPS for $[C_4C_1Im][BF_4]$ ($h\nu = 1486.6$ eV). (e) Co $2p_{3/2}$ valence RXPS for $[C_8C_1Im]_2[CoCl_4]$ ($h\nu = 774.9$ eV), compared to the valence NRXPS for $[C_8C_1Im]_2[CoCl_4]$ ($h\nu = 773.0$ eV, intensity multiplied by 4). The valence NRXPS contributions to valence RXPS were subtracted using the procedure outlined in ref. 68. All electron spectra were charge referenced using procedures outlined in Section 2.5.

however, the peak maxima for the features at $E_B \sim 11$ eV matched well for RXPS and NRXPS (Fig. 6c). Small E_B deviations between NRXPS and C 1s RXPS of about 0.1 eV to 0.2 eV have previously been observed for thin films of pure organic model molecules;[34] the explanation was occupation of vibrational sub-levels in the final state of C 1s RXPS.[56] We presume these vibrations gave similar E_B shifts for dilute solutes in solution.

For $[C(CN)_3]^-$, two distinct N 1s(C≡N) → π* absorptions were observed at $h\nu = 399.0$ eV and $h\nu = 400.0$ eV, both of which originate from the same core state, as the three nitrogen atoms in $[C(CN)_3]^-$ are equivalent (see ref. 86 for more details). The two N 1s RXPS traces for $[C(CN)_3]^-$ were very similar (Fig. 4c); furthermore, the SA features were broadly similar in terms of feature shape (Fig. 4c). Unsurprisingly, the OVSs that contributed to the two N 1s RXPS traces were essentially the same. For the two N 1s RXPS contributions at different $h\nu$ for $[C(CN)_3]^-$, the feature at lowest E_B from the HOVS was different by $E_B \sim 0.15$ eV, most likely due to subtle differences in final state vibrational contributions for the two PA transitions. Furthermore, for $[C(CN)_3]^-$ mismatches were observed between the NRXPS and N 1s RXPS features for the HOVS; $E_B \sim 0.2$ eV for $h\nu =$ 399.0 eV and $E_B \sim 0.35$ eV for $h\nu = 400.0$ eV (ESI Fig. S8e†). Again, these differences must be due to final state vibrational contributions in the PA transitions not present in the valence NRXPS.

For 0.5 M $[C_4C_1Im][BF_4]$ in water, signals from valence NRXPS for the solute were dwarfed by signal from the solvent (ESI Fig. S11†). Therefore, comparisons are made between the valence NRXPS for neat IL $[C_4C_1Im][BF_4]$, which were dominated by anion contributions (in this case, F 2p contributions) due to photoionisation cross-section effects.[69,87] These comparisons show a very good visual match at 18 eV > E_B > 11 eV between the valence NRXPS for neat IL $[C_4C_1Im][BF_4]$ and both B 1s RXPS and F 1s RXPS for $[BF_4]^-$ (Fig. 6d).

For the IL $[C_8C_1Im]_2[CoCl_4]$, valence NRXPS and Co $2p_{3/2}$ RXPS showed significant differences in both E_B for the main features and the spectral shapes (Fig. 6e). However, there were far more features observed in Co $2p_{3/2}$ RXPS than in valence NRXPS, particularly at 18 eV > E_B > 12 eV. The kinetic energy E_K was ~764 eV for the Co $L_3M_{4,5}L_{4,5}$ Auger transition from NRXPS,[69] which gives an apparent $E_B \sim 11$ eV at $h\nu = 775$ eV, meaning that features due to SA transitions would be expected in the same E_B region that the resonant features were detected for $[CoCl_4]^{2-}$. However, all of the resonant features remained at approximately constant E_B when $h\nu$ was increased, strongly suggesting that PA transitions dominated over SA transitions. It is possible that for the RXPS features at 12 eV > E_B > 6 eV the one-electron picture held, but determining that is beyond the scope of this article. However, for the features at 18 eV > E_B > 12 eV the one-electron picture certainly did not hold.

These comparisons for closed shell ions all demonstrate that the one-electron picture broadly held for valence RXPS. The lack of matching in terms of E_B for the NRXPS and RXPS features, strongly indicates that the one-electron picture failed for open shell complexes such as $[CoCl_4]^{2-}$.

3.2.2 Valence RXPS traces *versus* pDoS calculations. For almost all samples studied here, and all samples that contained π bonding, comparisons of RXPS traces and pDoS from calculations showed excellent matches (Fig. 7 and ESI Section 9†), *e.g.* for seven different inorganic anions that contain π-bonding, five different organic solutes that contain π-bonding, and two inorganic anions that contain no π-bonding and only σ-bonding in the relevant functional groups, $[BF_4]^-$ (Fig. 7e) and $[TfO]^-$ (ESI Fig. S24b†).

The most impressive matches of RXPS traces and pDoS was for samples for which more than one RXPS trace was recorded, *i.e.* PC, $[C_4C_1Im]^+$, imidazole,

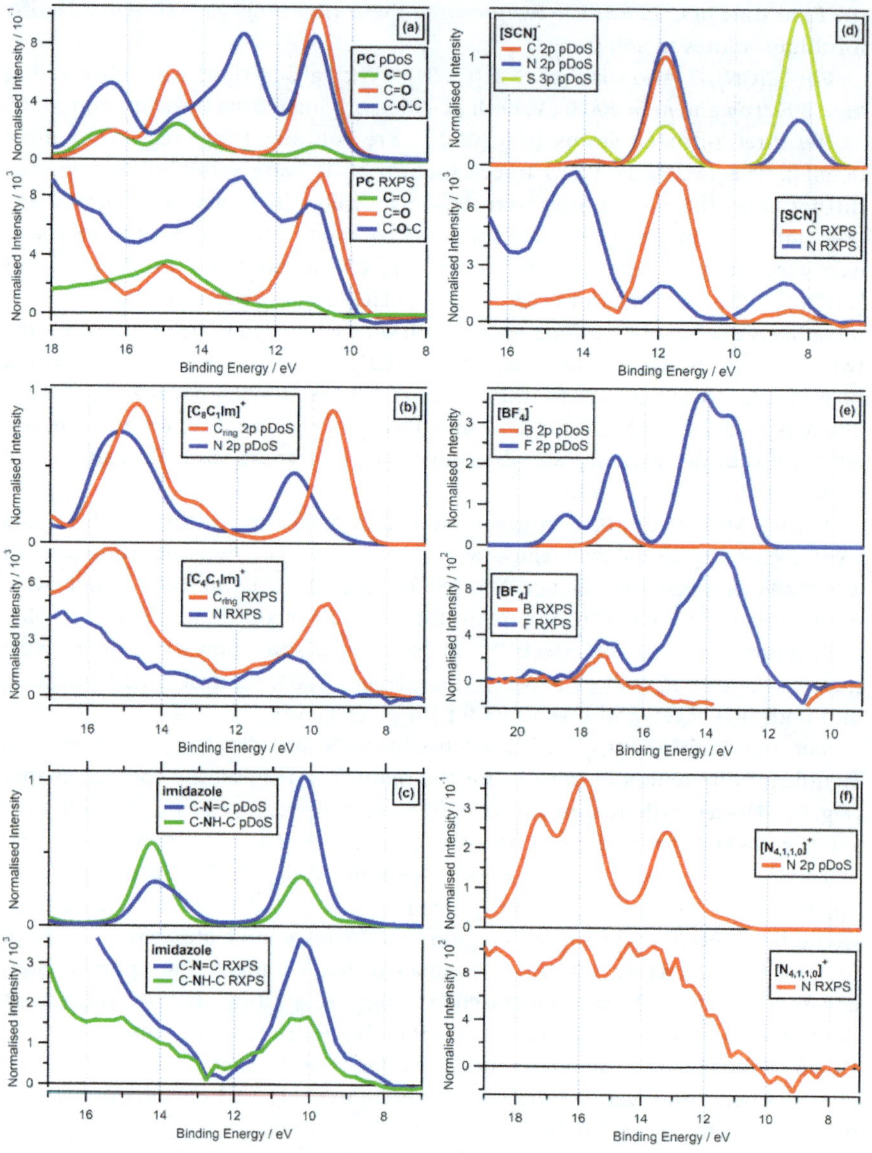

Fig. 7 Valence RXPS at different absorption edges compared to molecule/lone ion SMD pDoS calculations. (a) O 1s valence RXPS traces for propylene carbonate (PC) at both the C=O and C–O–C absorption edges ($h\nu = 533.3$ eV and $h\nu = 536.2$ eV respectively), and C 1s valence RXPS trace for propylene carbonate (PC) at the C=O absorption edge ($h\nu = 290.5$ eV), compared to C=O O pDoS, C–O–C O pDoS and C=O C pDoS calculated for PC in an SMD. (b) N 1s valence RXPS ($h\nu = 402.1$ eV) and C 1s valence RXPS ($h\nu = 287.0$ eV) traces for 0.5 M $[C_4C_1Im]I$ in water, compared to C_{ring} 2p (*i.e.* the three carbon atoms in the imidazolium ring) and N 2p pDoS calculated for lone $[C_8C_1Im]^+$ in an SMD. (c) N 1s valence RXPS traces for 1.8 M imidazole in water at both the C–N=C and C–NH–C absorption edges ($h\nu = 400.2$ eV and $h\nu = 401.9$ eV respectively), compared to C–N=C and C–NH–C N pDoS calculated for imidazole in an SMD. (d) N 1s valence RXPS trace for 0.5 M $[C_4C_1Im][SCN]$ in water ($h\nu = 399.8$ eV) and C 1s valence RXPS trace for 0.5 M K[SCN] in water ($h\nu = 287.5$ eV), compared to C 2p, N 2p and S 3p pDoS calculated for lone $[SCN]^-$ in an SMD. (e) B 1s valence RXPS ($h\nu = 201.4$ eV) and F 1s valence RXPS ($h\nu = 696.4$ eV) traces

$[SCN]^-$, $[BF_4]^-$ and $[NO_3]^-$ (Fig. 7 and ESI Section 9†). The visual match for PC was remarkable, as a comparison we showed that the pDoS matched the very small (but reliable, given they are measured in the same O 1s experiment with no charging effects observed) shift in E_B for the lowest E_B features for C=O and C–O–C (Fig. 7a). Furthermore, the match of the E_B peak separation for the features at lowest E_B for carbon and nitrogen for $[C_4C_1Im]^+$ was also very noteworthy; from both experimental RXPS and pDoS the E_B peak separation was ~1.0 eV (Fig. 7b), further emphasising the excellent match of the cheap, simple calculations and experimental RXPS. Lastly, for $[BF_4]^-$ F 1s RXPS and F 2p pDoS gave excellent visual matches for the two main features at $E_B \sim 17$ eV and $E_B \sim 13$ eV (Fig. 7e) and $[BF_4]^-$ B 1s RXPS and B 2p pDoS gave an excellent visual match at $E_B \sim 17$ eV (Fig. 7e).

For three different organic cations ($[N_{4,1,1,0}]^+$, $[N_{2,2,1,0}]^+$ and $[P_{6,6,6,14}]^+$) and both water and ethanol, which all contain only σ-bonding, there were reasonable matches of valence RXPS and pDoS calculations (see ESI Section 9†), but far less satisfactory than the other systems explained previously here. For all five of these ions/molecules, the RXPS traces showed a broad feature with no clear peaks, unlike all of the ion/molecules with π bonding (and $[BF_4]^-$ and $[TfO]^-$ too) which showed clear, readily identifiable peaks due to PA transitions. Furthermore, for all five of these ions/molecules the relevant pDoS showed clear, readily identifiable peaks; hence, the matches were reasonable but clearly poorer than other systems studied here. For example, N 1s RXPS for $[N_{4,1,1,0}]^+$ gave a broader feature than the N 2p pDoS (Fig. 7f), agreeing with observations for a perovskite with an organic, protic ammonium cation.[60] The small RXPS : NRXPS intensity ratios are probably not the explanation, otherwise all ions/molecules that contain only σ-bonding would show the same reasonable matches. All five of the ions/molecules that gave reasonable matches all had hydrogen atoms covalently bonded to the atoms studied using RXPS, *e.g.* one N–H bond for $[N_{4,1,1,0}]^+$. Therefore, final state vibrational contributions were likely a significant contribution to the broad experimental RXPS traces. Furthermore, it is possible that a range of ground state liquid phase environments not captured by the single calculated conformer (*i.e.* conformational flexibility and/or hydrogen bonding) also contributed for some ions/molecules. To untangle this problem further, RXPS calculations including a core-hole may be required, which are very challenging for condensed systems, particularly for solutes; these types of calculations have only been achieved for a small number of systems, with a focus on gas phase molecules with few atoms (*e.g.* formic acid,[31] HNCO,[88] CH_4,[89] and pyrimidine[89]).

$[SCN]^-$ N 1s RXPS and N 2p pDoS gave excellent visual matches for the three main features at lowest E_B (Fig. 7d). Furthermore, $[SCN]^-$ C 1s RXPS and C 2p pDoS also gave excellent visual matches for the three main features at lowest E_B (Fig. 7d). One small difference is the feature at $E_B \sim 8$ eV, which has an experimental C 1s RXPS contribution but not matching C 2p pDoS contribution, also found by other calculations.[90,91] This mismatch was most likely due to subtle

for 0.5 M $[C_4C_1Im][BF_4]$ in water, compared to B 2p and F 2p pDoS calculated for lone $[BF_4]^-$ in an SMD. (f) N 1s valence RXPS trace for 3.3 M $[N_{4,1,1,0}][HSO_4]$ in water ($h\nu = 405.6$ eV), compared to N 2p pDoS calculated for lone $[N_{4,1,1,0}]^+$ in an SMD. The valence NRXPS contributions were subtracted using the procedure outlined in ref. 68. All electron spectra were charge referenced using procedures outlined in Section 2.5.

differences of experiments and calculations, *i.e.* the HOMO and HOMO−1 for $[SCN]^-$ have no C 2p contribution but significant N 2p and S 3p contributions, and it is these N 2p and S 3p contributions that most likely gave rise to the feature in the experimental C 1s RXPS data for $[SCN]^-$. Essentially, how the density is assigned in pDoS calculations and the selection rules for RXPS do not quite match, which is not a surprising observation; furthermore, given the lack of success establishing selection rules for valence RXPS[31,52,55] and the difficulty performing RXPS calculations[31,88,89] solving this mismatch will not be straightforward.

Overall, these excellent matches of experimental valence RXPS and pDoS demonstrate that the one-electron picture of valence RXPS holds for these closed shell molecules/ions. Moreover, these excellent matches support the use of RXPS to identify atomic contributions to OVSs.

3.3 Insights into bonding

3.3.1 Element-, orbital- and site-specific comparisons of valence RXPS: insights into bonding. For four of the ions/molecules studied here ($[BF_4]^-$, $[SCN]^-$, $[NO_3]^-$, PC), RXPS traces recorded for more than one element gave excellent E_B matches of features/peaks (Fig. 7 and ESI Fig. S24d†). For $[BF_4]^-$ B 1s RXPS and F 1s RXPS both gave a feature at $E_B \sim 17$ eV (Fig. 7e), an excellent visual match; these features must have originated from the same OVSs (or degenerate states). The N 1s and C 1s RXPS traces for $[SCN]^-$ matched very well in terms of the E_B of the three main features (Fig. 7d). The N 1s and O 1s RXPS traces for $[NO_3]^-$ matched very well in terms of the E_B of the three main features (ESI Fig. S24d†). The PC C 1s RXPS trace at the C 1s → π* absorption at $h\nu = 290.5$ eV (most likely due to C=O) matched well in terms of E_B to the O 1s NRXPS traces for C=O and C–O–C; all three RXPS traces gave features at $E_B \sim 11$ eV and 15 eV (Fig. 7a). All these excellent matches strongly demonstrate that the one-electron picture held for these systems.

For both $[C_4C_1Im]^+$ and imidazole (which both have rings in their structure), the same element gave two core state → π* NEXAFS absorptions, C 1s for $[C_4C_1Im]^+$ and N 1s for imidazole (ESI Fig. S14b and S8b†). For both $[C_4C_1Im]^+$ and imidazole, the two absorptions were caused by two different core states,[92–95] and involved absorption to the same π* previously unoccupied valence state (see ESI Section 8† for more details). For both $[C_4C_1Im]^+$ and imidazole, the RXPS traces recorded for the two core state → π* NEXAFS absorptions gave excellent matches of features/peaks and were essentially identical (ESI Fig. S14d† and 7c respectively), demonstrating that the same OVSs contributed to both RXPS traces. It can be concluded that the same OVSs had similarly good overlap with different core-holes in the ring structures, pointing towards delocalisation of the valence states over the ring structures.

3.3.2 RXPS spectra as fingerprints: effect of the surroundings. A visual comparison of the peak E_B separation and relative peak intensity for ions in different solvation environments shows a good match for all six different ions (Fig. 8). For example, $[OAc]^-$ in water and $[OAc]^-$ in MeCN shows the same complex shape (Fig. 8c). Moreover, for $[C_nC_1Im][A]$ where the anion $[A]^-$ was varied the N 1s RXPS gave almost identical peak E_B separation and relative peak intensity, *e.g.* for $[C_nC_1Im]^+$ the E_B separation between the two main features was

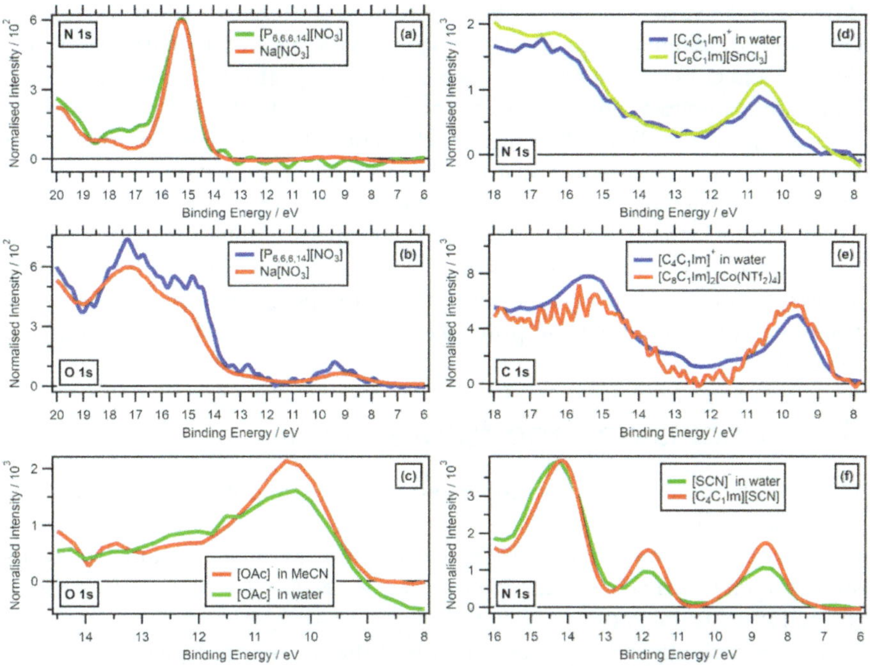

Fig. 8 Valence RXPS at different absorption edges. (a) N 1s valence RXPS traces for solid $Na[NO_3]$ ($h\nu = 405.0$ eV, data from ref. 41) and $[P_{6,6,6,14}][NO_3]$ ($h\nu = 405.0$ eV). (b) O 1s valence RXPS traces for solid $Na[NO_3]$ ($h\nu = 532.0$ eV, data from ref. 41) and $[P_{6,6,6,14}][NO_3]$ ($h\nu = 405.2$ eV). (c) O 1s valence RXPS traces for 1.3 M $[C_4C_1Im][OAc]$ in water ($h\nu = 532.7$ eV) and 0.5 M $[C_4C_1Im][OAc]$ in MeCN ($h\nu = 532.7$ eV). (d) N 1s valence RXPS traces for 0.5 M $[C_4C_1Im]I$ in water ($h\nu = 402.1$ eV) and $[C_8C_1Im][SnCl_3]$ ($h\nu = 401.8$ eV, data taken from ref. 69). (e) C 1s valence RXPS traces for 0.5 M $[C_4C_1Im]I$ in water ($h\nu = 287.0$ eV) and $[C_8C_1Im]_2[Co(NTf_2)_4]$ ($h\nu = 285.5$ eV, data taken from ref. 69). (f) N 1s valence RXPS traces for 0.5 M $[C_4C_1Im][SCN]$ in water ($h\nu = 399.8$ eV) and $[C_4C_1Im][SCN]$ ($h\nu = 399.5$ eV). The valence NRXPS contributions were subtracted using the procedure outlined in ref. 68. All electron spectra were charge referenced using procedures outlined in Section 2.5.

~6 eV for both $[C_4C_1Im]^+$ in water and $[C_8C_1Im][SnCl_3]$ (Fig. 8d).[68,69] Furthermore, the match for $[C_nC_1Im]^+$ for C_{ring} 1s RXPS (*i.e.* the three carbon atoms in the imidazolium ring) in two different solvation environments was also good (Fig. 8e). Therefore, the solvation environment of these ions in solution (and in certain solids, *e.g.* $Na[NO_3]$ in ref. 41) did not strongly affect the E_B separation of the OVSs, *i.e.* either the covalent bonding of the ions in question was not strongly affected by the solvation environment, or the covalent bonding of the ions was all equally affected. Furthermore, these observations demonstrate that ions will give fingerprint RXPS spectra, independent of the solvation environment.

While varying solvation environments appear to have little effect on RXPS, small changes in the local covalent environment have been found to lead to significant differences in RXPS for a small number of organic molecules in the gas phase.[31] The four ions/molecules studied here that contain at least one cyano functional group ($[SCN]^-$, $[N(CN)_2]^-$, $[C(CN)_3]^-$ and MeCN), the N 1s RXPS spectra were very different (Fig. 4c), matching the literature findings. Other examples where the same/similar functional group in two different samples gave very different RXPS spectra include: N 1s RXPS for $[C_nC_1Im]^+$ *versus* imidazole (Fig. 4c), O 1s RXPS for ethanol *versus* water (Fig. 4b), O 1s RXPS for PC(C=O) *versus* $[OAc]^-$ (Fig. 4b) and O 1s RXPS for $[HSO_4]^-$ *versus* $[TfO]^-$ (Fig. 4b). These findings match literature examples of small gas phase organic molecules.[31]

Broadly, it has been shown that changing the covalent, intermolecular bonding of an ion/molecule had a significant impact on the RXPS, but changing the intramolecular solvation environment had far less of an impact on the RXPS. In ref. 31, the extreme sensitivity of RXPS to changes in intermolecular covalent bonding was seen as a downside, as the technique could not be used for structural analysis of functional groups. However, the RXPS spectra published here are the beginnings of a dataset of fingerprint RXPS spectra of commonly studied ions and molecules.

4. Conclusions and future work

RXPS can be used to identify atomic contributions to OVSs. This approach is particularly valuable for complicated systems, *e.g.* solutes at low concentration. Closed shell ions/molecules with π bonding are particularly suited to RXPS studies, both in terms of giving high intensity RXPS features, but also giving well-defined RXPS features too.

Very importantly, all evidence shows that the one-electron picture holds for valence RXPS for closed-shell ions/molecules. Valence RXPS data matches well in terms of E_B to valence NRXPS, with the small differences that sometimes appear due to final state vibrational contributions. This finding demonstrates that the one-electron picture of RXPS holds for closed shell ions/molecules. However, for $[CoCl_4]^{2-}$ the valence RXPS and valence NRXPS were very different, demonstrating that the one-electron picture of RXPS does not hold for open shell ions/molecules for all PA transitions, although the one-electron picture may hold for certain PA transitions. RXPS matched very well with pDoS for a wide range of samples and element edges. Therefore, RXPS data provides an excellent benchmark for validation of ground state calculations of OVSs. This finding reinforces that the one-electron picture of RXPS holds for closed shell ions/molecules. Furthermore, insight can be gained into bonding by comparisons of RXPS and pDoS.

Our experimental set-ups are well-suited to measuring small signals, *e.g.* PA transitions for fluorine, due to the constantly refreshed sample environment, by liquid jet or rastering respectively. Significantly, a measurable PA signal was observed for every molecule/ion studied here. For example, for F-containing anions, a F 1s PA signal was observed, whereas no F 1s PA signal was observed for solid CaF_2.[36,38] As far as we are aware, this represents the first measurement of an F 1s PA transition. Furthermore, the F 1s PA signal for both $[BF_4]^-$ and $[TfO]^-$ matched very well to calculated pDoS data.

Two guidelines for ions/molecules to give large RXPS signal are: (i) the element studied having low electronegativity, (ii) π bonding involving the element studied. For example, F 1s valence RXPS for $[TfO]^-$ gave a small RXPS signal, even though the SO_3 functional group contains π bonding, because the CF_3 functional group contains only σ bonding. Therefore, O 1s valence RXPS of water as a solute will be challenging given the small and broad RXPS signal; conversely, water is a good solvent for studying O 1s valence RXPS of solutes dissolved in water. Overall, to obtain high quality data for RXPS for solutes with the same element as the solvent, the ideal is solutes that give large RXPS signal and solvents that give small RXPS.

An area of RXPS studies that has potential is studying functional groups of ions/molecules that contain only σ bonding. RXPS measurements of *e.g.* alkylamines and alkanes to determine their PA : SA intensity ratios would be instructive, to determine the relative importance of low electronegativity elements *versus* the bonding type. Furthermore RXPS for functional groups which only have σ bonding, *e.g.* tetraalkylammonium, alkanes, ethers, to determine the relative importance of covalent bonds to hydrogen for RXPS.

RXPS traces were used to demonstrate the strong impact of covalent bonding on RXPS, but the weak impact of non-covalent intermolecular interactions on RXPS. The dataset of fingerprint RXPS spectra of commonly studied ions and molecules presented here will prove very useful for understanding interactions of these ions and molecules with surfaces. Many simple molecules have been studied as adsorbates on surfaces for many years,[96] along with larger adsorbates,[62] but most of the complex ions presented here in particular are yet to be studied as adsorbates using RXPS.

Conflicts of interest

There are no conflicts to declare.

Acknowledgements

KRJL acknowledges support from a Royal Society University Research Fellowship (URF\R\150353). JMS acknowledges support from a Royal Society University Research Fellowship Enhancement Award (RGF\EA\180089). EG acknowledges support from a Royal Society Research Grant for Research Fellows (RGF\R1\180053). AL acknowledges part funding for a studentship from Johnson Matthey plc. DH, GW, WQ, and RS thank the Deutsche Forschungsgemeinschaft (German Research Foundation, DFG) for support through an Emmy-Noether grant (SE 2253/3-1). MAX-lab are thanked for awarding beamtime on I311. Measurements were carried out at the SOL^{3}PES instrument (beamline U49-2_PGM-1) at the BESSY II electron storage ring operated by the Helmholtz-Zentrum Berlin für Materialien und Energie. We acknowledge Diamond Light Source for time on Beamline B07 (VERSOX) under Proposals SI24304 and SI25929. The research leading to this result has been supported by the project CALIPSOplus under the Grant Agreement 730872 from the EU Framework Programme for Research and Innovation HORIZON 2020. Bernd Winter is acknowledged for inspirational support and advice. Matthijs Van Spronsen is acknowledged for beamtime support on Proposal SI25929. The authors are grateful to Jacek Osiecki and Karsten Handrup (MAX-lab, Sweden) for assistance with experiments.

References

1 I. Fleming, *Molecular Orbitals and Organic Chemical Reactions*, Wiley, Chichester, 2010.

2 K. Fukui, *Angew. Chem., Int. Ed. Engl.*, 1982, **21**, 801–809.

3 R. Hoffmann, *Angew. Chem., Int. Ed. Engl.*, 1982, **21**, 711–724.

4 R. G. Pearson, *Inorg. Chem.*, 1988, **27**, 734–740.

5 R. G. Pearson, *J. Org. Chem.*, 1989, **54**, 1423–1430.

6 K. Yoshida, M. Nakamura, Y. Kazue, N. Tachikawa, S. Tsuzuki, S. Seki, K. Dokko and M. Watanabe, *J. Am. Chem. Soc.*, 2011, **133**, 13121–13129.

7 H. D. Lim, B. Lee, Y. Zheng, J. Hong, J. Kim, H. Gwon, Y. Ko, M. Lee, K. Cho and K. Kang, *Nat. Energy*, 2016, **1**, 16066.

8 X. D. Ren, P. Y. Gao, L. F. Zou, S. H. Jiao, X. Cao, X. H. Zhang, H. Jia, M. H. Engelhard, B. E. Matthews, H. P. Wu, H. Lee, C. J. Niu, C. M. Wang, B. W. Arey, J. Xiao, J. Liu, J. G. Zhang and W. Xu, *Proc. Natl. Acad. Sci. U. S. A.*, 2020, **117**, 28603–28613.

9 S. E. Denmark and G. L. Beutner, *Angew. Chem., Int. Ed.*, 2008, **47**, 1560–1638.

10 E. Vedejs and S. E. Denmark, *Lewis Base Catalysis in Organic Synthesis*, Wiley-VCH, Weinheim, 2016.

11 P. J. Linstrom and W. G. Mallard, *NIST Chemistry WebBook*, NIST Standard Reference Database Number 69, DOI: 10.18434/T4D303, accessed 6 December 2021.

12 F. A. Bulat, J. S. Murray and P. Politzer, *Comput. Theor. Chem.*, 2021, **1199**, 113192.

13 W. Liu, Y. D. Jiang, K. H. Dostert, C. P. O'Brien, W. Riedel, A. Savara, S. Schauermann and A. Tkatchenko, *Sci. Adv.*, 2017, **3**, e1700939.

14 J. T. Li, S. Yang, J. C. Ren, G. R. Su, S. Li, C. J. Butch, Z. G. Ding and W. Liu, *J. Phys. Chem. Lett.*, 2019, **10**, 6755–6761.

15 T. D. Spivey and A. Holewinski, *J. Am. Chem. Soc.*, 2021, **143**, 11897–11902.

16 X. D. Ren, L. F. Zou, S. H. Jiao, D. H. Mei, M. H. Engelhard, Q. Y. Li, H. Y. Lee, C. J. Niu, B. D. Adams, C. M. Wang, J. Liu, J. G. Zhang and W. Xu, *ACS Energy Lett.*, 2019, **4**, 896–902.

17 R. A. Miranda-Quintana and J. Smiatek, *Electrochim. Acta*, 2021, **384**, 138418.

18 P. Peljo and H. H. Girault, *Energy Environ. Sci.*, 2018, **11**, 2306–2309.

19 E. R. Fadel, F. Faglioni, G. Samsonidze, N. Molinari, B. V. Merinov, W. A. Goddard, J. C. Grossman, J. P. Mailoa and B. Kozinsky, *Nat. Commun.*, 2019, **10**, 3360.

20 V. Pande and V. Viswanathan, *J. Phys. Chem. Lett.*, 2019, **10**, 7031–7036.

21 B. Winter and M. Faubel, *Chem. Rev.*, 2006, **106**, 1176–1211.

22 B. Winter, *Nucl. Instrum. Methods Phys. Res., Sect. A*, 2009, **601**, 139–150.

23 T. Fransson, Y. Harada, N. Kosugi, N. A. Besley, B. Winter, J. J. Rehr, L. G. M. Pettersson and A. Nilsson, *Chem. Rev.*, 2016, **116**, 7551–7569.

24 R. Seidel, B. Winter and S. E. Bradforth, *Annu. Rev. Phys. Chem.*, 2016, **67**, 283–305.

25 K. R. J. Lovelock, I. J. Villar-Garcia, F. Maier, H. P. Steinrück and P. Licence, *Chem. Rev.*, 2010, **110**, 5158–5190.

26 M. N. Pohl, E. Muchová, R. Seidel, H. Ali, S. Sršeň, I. Wilkinson, B. Winter and P. Slavíček, *Chem. Sci.*, 2019, **10**, 848–865.

27 B. Jagoda-Cwiklik, P. Slavíček, D. Nolting, B. Winter and P. Jungwirth, *J. Phys. Chem. B*, 2008, **112**, 7355–7358.

28 M. N. Piancastelli, R. F. Fink, R. Feifel, M. Bassler, S. L. Sorensen, C. Miron, H. Wang, I. Hjelte, O. Björneholm, A. Ausmees, S. Svensson, P. Salek, F. K. Gel'mukhanov and H. Agren, *J. Phys. B: At., Mol. Opt. Phys.*, 2000, **33**, 1819–1826.

29 I. Hjelte, M. N. Piancastelli, R. F. Fink, O. Björneholm, M. Bassler, R. Feifel, A. Giertz, H. Wang, K. Wiesner, A. Ausmees, C. Miron, S. L. Sorensen and S. Svensson, *Chem. Phys. Lett.*, 2001, **334**, 151–158.

30 E. E. Rennie, U. Hergenhahn, O. Kugeler, A. Rüdel, S. Marburger and A. M. Bradshaw, *J. Chem. Phys.*, 2002, **117**, 6524–6532.

31 U. Hergenhahn, A. Rüdel, K. Maier, A. M. Bradshaw, R. F. Fink and A. T. Wen, *Chem. Phys.*, 2003, **289**, 57–67.

32 P. Bolognesi, P. O'Keeffe, Y. Ovcharenko, L. Avaldi and V. Carravetta, *J. Chem. Phys.*, 2012, **136**, 154308.

33 A. Kikas, R. Ruus, A. Saar, E. Nommiste, T. Kaambre and S. Sundin, *J. Electron Spectrosc. Relat. Phenom.*, 1999, **101**, 745–749.

34 H. Peisert, I. Biswas, L. Zhang, B. E. Schuster, M. B. Casu, A. Haug, D. Batchelor, M. Knupfer and T. Chasse, *J. Chem. Phys.*, 2009, **130**, 194705.

35 R. H. Temperton, N. W. Rosemann, M. Y. Guo, N. Johansson, L. A. Fredin, O. Prakash, K. Warnmark, K. Handrup, J. Uhlig, J. Schnadt and P. Persson, *J. Phys. Chem. A*, 2020, **124**, 1603–1609.

36 T. Tiedje, K. M. Colbow, D. Rogers and W. Eberhardt, *Phys. Rev. Lett.*, 1990, **65**, 1243–1246.

37 E. Kukk, S. Aksela, H. Aksela, E. Nommiste, A. Kikas, A. Ausmees and M. Elango, *Phys. Rev. B: Condens. Matter Mater. Phys.*, 1994, **50**, 9079–9085.

38 H. Aksela, E. Kukk, S. Aksela, A. Kikas, E. Nommiste, A. Ausmees and M. Elango, *Phys. Rev. B: Condens. Matter Mater. Phys.*, 1994, **49**, 3116–3123.

39 M. Elango, A. Kikas, A. Maiste, R. Ruus and A. Saar, *Phys. B*, 1995, **208**, 47–48.

40 A. S. Vinogradov, A. B. Preobrajenski, S. L. Molodtsov, S. A. Krasnikov, R. Szargan, A. Knop-Gericke and M. Havecker, *Chem. Phys.*, 1999, **249**, 249–258.

41 A. B. Preobrajenski, A. S. Vinogradov, S. L. Molodtsov, S. K. Krasnikov, T. Chasse, R. Szargan and C. Laubschat, *Phys. Rev. B: Condens. Matter Mater. Phys.*, 2002, **65**, 205116.

42 A. B. Preobrajenski, A. S. Vinogradov, S. L. Molodtsov, S. A. Krasnikov, R. Szargan and C. Laubschat, *Chem. Phys. Lett.*, 2003, **368**, 125–131.

43 A. B. Preobrajenski, A. S. Vinogradov, E. Kleimenov, A. Knop-Gericke, S. A. Krasnikov, R. Szargan and N. Mårtensson, *Phys. Scr.*, 2005, **T115**, 1071–1073.

44 A. B. Preobrajenski, A. S. Vinogradov and N. Mårtensson, *J. Electron Spectrosc. Relat. Phenom.*, 2005, **148**, 59–64.

45 H. Shimada, N. Matsubayashi, M. Imamura, T. Sato, Y. Kakudate, S. Fujiwara and A. Nishijima, *J. Electron Spectrosc. Relat. Phenom.*, 1996, **79**, 211–214.

46 R. Dudde, M. L. M. Rocco, E. E. Koch, S. Bernstorff and W. Eberhardt, *J. Chem. Phys.*, 1989, **91**, 20–28.

47 D. Menzel, G. Rocker, H. P. Steinrück, D. Coulman, P. A. Heimann, W. Huber, P. Zebisch and D. R. Lloyd, *J. Chem. Phys.*, 1992, **96**, 1724–1734.

48 T. Porwol, G. Domotor, H. J. Freund, R. Dudde, C. M. Liegener and W. Vonniessen, *Phys. Scr.*, 1992, **T41**, 197–207.

49 M. Mauerer, P. Zebisch, M. Weinelt and H. P. Steinrück, *J. Chem. Phys.*, 1993, **99**, 3343–3352.

50 W. Wurth and D. Menzel, *J. Electron Spectrosc. Relat. Phenom.*, 1993, **62**, 23–31.

51 S. L. Sorensen, S. J. Osborne, A. Ausmees, A. Kikas, N. Correia, S. Svensson, A. N. deBrito, P. Persson and S. Lunell, *J. Chem. Phys.*, 1996, **105**, 10719–10724.

52 J. J. Gallet, F. Bournel, S. Kubsky, G. Dufour, F. Rochet and F. Sirotti, *J. Electron Spectrosc. Relat. Phenom.*, 2002, **122**, 285–295.
53 D. V. Vyalikh, V. V. Maslyuk, A. Bluher, A. Kade, K. Kummer, Y. S. Dedkov, T. Bredow, I. Mertig, M. Mertig and S. L. Molodtsov, *Phys. Rev. Lett.*, 2009, **102**, 098101.
54 J. Kikuma and B. P. Tonner, *J. Electron Spectrosc. Relat. Phenom.*, 1996, **82**, 41–52.
55 J. J. Gallet, F. Jolly, F. Rochet, F. Bournel, G. Dufour, P. A. Avila, F. Sirotti and P. Torelli, *J. Electron Spectrosc. Relat. Phenom.*, 2002, **122**, 11–25.
56 R. Friedlein, S. L. Sorensen, A. Baev, F. Gel'mukhanov, J. Birgerson, A. Crispin, M. P. de Jong, W. Osikowicz, C. Murphy, H. Agren and W. R. Salaneck, *Phys. Rev. B: Condens. Matter Mater. Phys.*, 2004, **69**, 125204.
57 H. S. Kato, M. Furukawa, M. Kawai, M. Taniguchi, T. Kawai, T. Hatsui and N. Kosugi, *Phys. Rev. Lett.*, 2004, **93**, 086403.
58 L. Wang, K. Q. Zhang, X. Y. Gao, X. Y. Liu and A. T. S. Wee, *Appl. Phys. Lett.*, 2006, **89**, 013902.
59 M. Gliboff, D. Sulas, D. Nordlund, D. W. deQuilettes, P. D. Nguyen, G. T. Seidler, X. S. Li and D. S. Ginger, *J. Phys. Chem. C*, 2014, **118**, 5570–5578.
60 G. J. Man, C. M. Sterling, C. Kamal, K. A. Simonov, S. Svanstrom, J. Acharya, F. O. L. Johansson, E. Giangrisostomi, R. Ovsyannikov, T. Huthwelker, S. M. Butorin, P. K. Nayak, M. Odelius and H. Rensmo, *Phys. Rev. B*, 2021, **104**, L041302.
61 M. Kot, K. Wojciechowski, H. Snaith and D. Schmeisser, *Chem.–Eur. J.*, 2018, **24**, 3539–3544.
62 J. Schnadt, P. A. Brühwiler, L. Patthey, J. N. O'Shea, S. Södergren, M. Odelius, R. Ahuja, O. Karis, M. Bässler, P. Persson, H. Siegbahn, S. Lunell and N. Mårtensson, *Nature*, 2002, **418**, 620–623.
63 J. J. Gallet, F. Bournel, S. Carniato, G. Dufour, F. Rochet, S. Rangan and F. Sirotti, *Surf. Sci.*, 2007, **601**, 552–561.
64 F. Petraki, H. Peisert, I. Biswas, U. Aygul, F. Latteyer, A. Vollmer and T. Chasse, *J. Phys. Chem. Lett.*, 2010, **1**, 3380–3384.
65 C. Sauer, M. Wiessner, A. Scholl and F. Reinert, *New J. Phys.*, 2015, **17**, 043016.
66 C. Sauer, M. Wiessner, A. Scholl and F. Reinert, *Phys. Rev. B*, 2016, **93**, 045101.
67 S. Thürmer, I. Unger, P. Slavíček and B. Winter, *J. Phys. Chem. C*, 2013, **117**, 22268–22275.
68 R. M. Fogarty, R. G. Palgrave, R. A. Bourne, K. Handrup, I. J. Villar-Garcia, D. J. Payne, P. A. Hunt and K. R. J. Lovelock, *Phys. Chem. Chem. Phys.*, 2019, **21**, 18893–18910.
69 J. M. Seymour, E. Gousseva, A. I. Large, C. J. Clarke, P. Licence, R. M. Fogarty, D. A. Duncan, P. Ferrer, F. Venturini, R. A. Bennett, R. G. Palgrave and K. R. J. Lovelock, *Phys. Chem. Chem. Phys.*, 2021, **23**, 20957–20973.
70 S. Thürmer, R. Seidel, W. Eberhardt, S. E. Bradforth and B. Winter, *J. Am. Chem. Soc.*, 2011, **133**, 12528–12535.
71 R. Seidel, K. Atak, S. Thürmer, E. F. Aziz and B. Winter, *J. Phys. Chem. B*, 2015, **119**, 10607–10615.
72 S. S. N. Lalithambika, K. Atak, R. Seidel, A. Neubauer, T. Brandenburg, J. Xiao, B. Winter and E. F. Aziz, *Sci. Rep.*, 2017, **7**, 40811.
73 S. S. N. Lalithambika, R. Golnak, B. Winter and K. Atak, *Inorg. Chem.*, 2019, **58**, 4731–4740.

74 K. D. Mudryk, R. Seidel, B. Winter and I. Wilkinson, *Phys. Chem. Chem. Phys.*, 2020, **22**, 20311–20330.
75 R. H. Temperton, W. Quevedo, R. Seidel, J. Uhlig, J. Schnadt and P. Persson, *Phys. Rev. Res.*, 2021, **3**, 033030.
76 H. Ali, R. Seidel, M. N. Pohl and B. Winter, *Chem. Sci.*, 2018, **9**, 4511–4523.
77 H. Ali, R. Seidel, A. Bergmann and B. Winter, *J. Mater. Chem. A*, 2019, **7**, 6665–6675.
78 A. Y. Klyushin, M. M. Brzhezinskaya, A. V. Generalov, R. Puttner and A. S. Vinogradov, *Fullerenes, Nanotubes, Carbon Nanostruct.*, 2010, **18**, 600–604.
79 U. Gelius and K. Siegbahn, *Faraday Discuss.*, 1972, **54**, 257–268.
80 U. Gelius, *J. Electron Spectrosc. Relat. Phenom.*, 1974, **5**, 985–1057.
81 R. Seidel, M. N. Pohl, H. Ali, B. Winter and E. F. Aziz, *Rev. Sci. Instrum.*, 2017, **88**, 073107.
82 G. Held, F. Venturini, D. C. Grinter, P. Ferrer, R. Arrigo, L. Deacon, W. Q. Garzon, K. Roy, A. Large, C. Stephens, A. Watts, P. Larkin, M. Hand, H. C. Wang, L. Pratt, J. J. Mudd, T. Richardson, S. Patel, M. Hillman and S. Scott, *J. Synchrotron Radiat.*, 2020, **27**, 1153–1166.
83 S. Thürmer, S. Malerz, F. Trinter, U. Hergenhahn, C. Lee, D. M. Neumark, G. Meijer, B. Winter and I. Wilkinson, *Chem. Sci.*, 2021, **12**, 10558–10582.
84 G. Greczynski and L. Hultman, *Prog. Mater. Sci.*, 2020, **107**, 100591.
85 L. C. Allen, *J. Am. Chem. Soc.*, 1989, **111**, 9003–9014.
86 R. M. Fogarty, R. P. Matthews, M. T. Clough, C. R. Ashworth, A. Brandt-Talbot, P. J. Corbett, R. G. Palgrave, R. A. Bourne, T. W. Chamberlain, T. Vander Hoogerstraete, P. B. J. Thompson, P. A. Hunt, N. A. Besley and K. R. J. Lovelock, *Phys. Chem. Chem. Phys.*, 2017, **19**, 31156–31167.
87 J. J. Yeh and I. Lindau, *At. Data Nucl. Data Tables*, 1985, **32**, 1–155.
88 F. Holzmeier, T. J. A. Wolf, C. Gienger, I. Wagner, J. Bozek, S. Nandi, C. Nicolas, I. Fischer, M. Guhr and R. F. Fink, *J. Chem. Phys.*, 2018, **149**, 034308.
89 G. Grell and S. I. Bokarev, *J. Chem. Phys.*, 2020, **152**, 074108.
90 S. E. Bradforth, E. H. Kim, D. W. Arnold and D. M. Neumark, *J. Chem. Phys.*, 1993, **98**, 800–810.
91 S. Rangan, J. Viereck and R. A. Bartynski, *J. Phys. Chem. B*, 2020, **124**, 7909–7917.
92 C. Ehlert, M. Holzweber, A. Lippitz, W. E. S. Unger and P. Saalfrank, *Phys. Chem. Chem. Phys.*, 2016, **18**, 8654–8661.
93 I. J. Villar-Garcia, E. F. Smith, A. W. Taylor, F. L. Qiu, K. R. J. Lovelock, R. G. Jones and P. Licence, *Phys. Chem. Chem. Phys.*, 2011, **13**, 2797–2808.
94 M. J. Thomason, C. R. Seabourne, B. M. Sattelle, G. A. Hembury, J. S. Stevens, A. J. Scott, E. F. Aziz and S. L. M. Schroeder, *Faraday Discuss.*, 2015, **179**, 269–289.
95 D. Nolting, N. Ottosson, M. Faubel, I. V. Hertel and B. Winter, *J. Am. Chem. Soc.*, 2008, **130**, 8150–8151.
96 O. Bjorneholm, A. Sandell, A. Nilsson, N. Martensson and J. N. Andersen, *Phys. Scr.*, 1992, **T41**, 217–225.

Faraday Discussions

ROYAL SOCIETY OF CHEMISTRY

DISCUSSIONS

Future directions: general discussion

Rosa Arrigo, Loren Ban, Thorsten Bartels-Rausch, Philip R. Davies, Samuel Eyley, Wendy Flavell, Giulia Galli, Shaoliang Guan, Samuel Hall, Georg Held, Juhan Kahk, Roxy Lee, Robert Lindsay, Johannes Lischner, Kevin Lovelock, Anders Nilsson, David Payne, Olivier Renault, Alexander Shard and Sefik Suzer

DOI: 10.1039/d2fd90027k

Anders Nilsson opened a discussion of the paper by Giulia Galli: When we are comparing the electronic structure of water at ambient conditions with high pressure conditions there can be other changes than only the formation of different species. The basic structure can change quite a bit so that high density becomes more dominating. Can this aspect be part of understanding the electronic structure change?

Giulia Galli responded: To really reply to this question we should have simulations (which we do not have yet) as a function of pressure and really have trends as a function of pressure. Hopefully we will have those in the next few months but they are not yet available to us.

Anders Nilsson said: There is a shift in the $3a_1$ towards the $1b_1$ orbital peak when pressure is applied. Could this be the result of forming a disordered band-like structure in a condensed phase and could there be an increased overlap between orbitals of neighboring molecules with increasing pressure?

Giulia Galli replied: Yes indeed, I think this is what we are observing in our simulations. We are now examining other conditions and trying to figure out how this overlap depends on pressure in more detail.

Loren Ban commented: Thank you for this nice presentation and the great work, I believe that calculated binding energy spectra of liquids are an important ingredient for further progress in understanding condensed phase electronic structure. These calculations will definitely be very valuable to experimentalists looking at similar systems.

You show in Fig. 3 of the paper (https://doi.org/10.1039/d2fd00003b) that the binding energy spectrum of water at HPT is notably different from that at APT – both in terms of peak widths and peak positions. For the latter, you find an overall redshift of the HPT spectra. You also show that ionization energies of

Na/Cl ions are redshifted by a similar amount. All of these are very different ionization processes in terms of the charge in the initial and final states (for water neutral → cation, for chloride anion → neutral, for sodium cation → doubly charged cation). Therefore, it seems interesting that the redshift occurs with a similar magnitude for all these species. Can you elaborate on the origin of this redshift and the influence of the dielectric constant? Can it naively be related to specific initial or final state effects?

And finally, regarding the difference between PBE and DDH functionals – could you also calculate adiabatic ionization energies from your approach? By calculating these, I expect that the shifts in ionization energies would be very different for the above mentioned species, and you could possibly compare them to a simple thermodynamical model (Born equation, solvation free energies, see *e.g.* ref. 1).

1 B. Winter, R. Weber, I. V. Hertel, M. Faubel, P. Jungwirth, E. C. Brown and S. E. Bradforth, *J. Am. Chem. Soc.*, 2005, **127**, 7203–7214.

Giulia Galli answered: I need to clarify which dielectric constant one is talking about.

The dielectric constant ε_0 (low frequency dielectric constant) decreases at high pressure and high *T*. We have computed it in previous work and at the conditions studied here it becomes on the order of 35–40 (to be compared with 78 at ambient conditions). Hence the solvation properties of salts in water are affected in terms of the relative stability of contact ion pairs and separated pairs (also studied in previous work). So this is one effect of the low frequency dielectric constant.

Then there is the effect of the high frequency dielectric constant that increases a bit under pressure (that would be ε_∞) and that enters as input in the definition of the hybrid functional DDH. It is because of the change in dielectric constant (ε_∞) that we observe a red shift.

But overall the picture is complex because solvation properties do change (due to ε_0) but also electronic properties and ε_∞ change. And we are seeing the effect of these combined complex changes.

Johannes Lischner remarked: As you describe, your calculations contain two steps: first, you carried out *ab initio* calculations to identify a structural model and then you carry out excited s-state simulations to obtain spectra. I am wondering how sensitive the spectra are to the results of the first step, *i.e.* the structural model, and how sensitive the structural model is to the employed exchange–correlation functional.

Giulia Galli replied: Unfortunately we only have partial answers to these important questions.

The sensitivity of the spectra to the functionals, given a structural model, is important and, for example, semi-local functionals would not be able to yield the same differences between high and low pressure that a hybrid functional yields. This point is discussed a bit in our paper (https://doi.org/10.1039/d2fd00003b).

There is a much weaker dependence of the spectra on the structural model "within reason". What I mean is, for structural models that at least qualitatively give you similar solvation properties, then the spectra turn out to be very similar. But unfortunately these differences are not so trivial to quantify.

Loren Ban commented: In Section III.B of your paper (https://doi.org/10.1039/d2fd00003b) you present a nice analysis of the short-lived products arising from water dissociation events. The calculated photoelectron spectra (Fig. 5 of the paper) of OH^- and H_3O^+, as well as their hydrated analogues, will in my opinion become very useful for identifying their contributions in future photoelectron spectroscopy experiments.

In addition to water dissociation induced here by HPT conditions, photodissociation of water also leads to interesting short-lived products. One of the important photodissociation channels includes the formation of H_3O^+, an OH radical and a hydrated electron. I was wondering if you had considered looking at electron solvation in your calculations. Would you expect that OH radicals or hydrated electrons could appear at HPT conditions?

Giulia Galli responded: Great question! We do want to look at electron solvation but have not yet done so. It is an interesting and important question. I very much expect that solvated electrons would appear and I also expect the solvation properties to be rather different from those at ambient conditions and this is something we hope to be doing in the near future.

Loren Ban asked: In a follow-up to the previous question, can your calculations give you an insight into the spatial extent of specific orbitals? Do you observe that orbital extent increases in dissociation events or are the orbitals localized to a certain molecule?

A similar concept was reported in Boero *et al.*[1] looking at unoccupied orbitals that were assigned to the solvated electron (so called wet electron).

1 M. Boero, K. Terakura, T. Ikeshoji, C. C. Liew and M. Parrinello, *J. Chem. Phys.*, 2001, **115**, 2219.

Giulia Galli answered: Yes our simulations can indeed give insight into localization properties of orbitals (or spatial extent) and we do see delocalization (which I believe is what you are hinting at) in the presence of dissociative events. We can also characterize dissociative events by looking at the centers of maximally localized Wannier functions.

Loren Ban commented: For the solvated H_4O_2 cluster, you observe a double-peak feature near the $2a_1$ region of bulk water. Is this a feature specific to HPT water or would it be present in ambient water as well? Do you know what the origin of this splitting is – is it a solvation effect or would it also be observed in gas phase water dimers? The gas phase water dimer has been studied by photoelectron spectroscopy (*e.g.* ref. 1–3), but unfortunately there seems to be no detailed information on the inner valence $2a_1$ orbital.

1 P. A. Pieniazek, J. VandeVondele, P. Jungwirth, A. I. Krylov and S. E. Bradforth, *J. Phys. Chem. A*, 2008, **112**, 6159–6170.
2 S. Barth, M. Ončák, V. Ulrich, M. Mucke, T. Lischke, P. Slavíček and U. Hergenhahn, *J. Phys. Chem. A*, 2009, **113**, 13519–13527.

3 S. Hartweg, B. L. Yoder, G. A. Garcia, L. Nahon and R. Signorell, *Phys. Rev. Lett.*, 2017, **118**, 103402.

Giulia Galli responded: Unfortunately we could not do the same analysis for ambient water due to the time scale of our simulations. What I mean is the following: the reason why we can identify these species (which are very short lived) at high pressure and high temperature is because they form and dissolve over the short time scales of our simulations (orders of ps). At ambient conditions one would require much longer simulations to see autoionization and we cannot perform such long simulations. However we could carry out a sort of *gedanken* experiment and put those species in an ambient condition sample and ask the question about which signal they would give rise to in photoelectron spectra. That would be an interesting computational experiment to do.

I suspect that would be a solvation effect but I cannot be 100% sure.

Anders Nilsson opened a discussion of the paper by Juhan Matthias Kahk and Johannes Lischner: It is important when computing binding energy shift in terms of difference in total energy between ground and core hole states to know whether these are adiabatic or vertical energies. Please comment on your calculations in this regard.

Juhan Kahk replied: In this *Faraday Discussions* article (https://doi.org/10.1039/d1fd00103e) and in previous work, I have calculated vertical ionization energies, *i.e.* the energies of both the ground state and the core hole state are evaluated at the relaxed geometry of the ground state.

Anders Nilsson said: Experimentally we see vertical binding energy shifts and often we assume that the geometry in the final state corresponds to a frozen case. However, sometimes the vibrational envelope in the spectrum can be complex and the peak becomes highly asymmetrical. Do you think that in these special cases it would be good to also compute the whole line shape of the XPS peaks?

Juhan Kahk answered: I certainly agree that in many cases it is important to go beyond the prediction of vertical ionization potentials, and to compute the whole lineshape of XPS peaks taking into account both vibrational and electronic effects that contribute to peak asymmetry.

In the ΔSCF method, the calculation of the core hole state has approximately the same cost as the calculation of the ground electronic state, and the forces on atoms as well as the second derivatives of the energy can in principle be computed analytically. Therefore, the potential energy surfaces can be mapped efficiently in both the initial and the final electronic states. However, the prediction of vibrational peak broadening or vibrational satellites is still challenging.

For small molecules, *e.g.* H_2O, a fully quantum mechanical treatment of molecular vibrations has been used to very accurately model the vibrational fine structure of the O 1s photoemission peak in ref. 1. However, it would be challenging to apply the same approach to larger systems with many more vibrational degrees of freedom.

The prediction of vibrational peak broadening in XPS of ionic crystals has been demonstrated in ref. 2, based on an approach where vertical ionization potentials are calculated at the classical turning points of a specific vibrational mode along the ground state potential energy surface. However, it remains to be seen whether this approach that largely ignores the quantum mechanical nature of vibrations would be able to correctly reproduce the asymmetric peak shapes that are seen in some materials.

In short, I believe that the prediction of vibrational effects on lineshapes in core level XPS is presently still an open question.

1 R. Sankari, M. Ehara, H. Nakatsuji, Y. Senba, K. Hosokawa, H. Yoshida, A. De Fanis, Y. Tamenori, S. Aksela and K. Ueda, *Chem. Phys. Lett.*, 2003, **380**, 647–653.
2 C. J. Nelin, P. S. Bagus, M. A. Brown, M. Sterrer and H.-J. Freund, *Angew. Chem., Int. Ed.*, 2011, **50**, 10174–10177.

Sefik Suzer remarked: You have introduced the spin–orbit-coupling in an *ad hoc* manner, which is understandable, considering the complexity of using full relativistic effects. However, due to the relativistic effects, there are additional differences imparted to the $P_{1/2}$ and $P_{3/2}$ spinors, such as their radial distribution and/or bonding properties. For example it is known that $P_{1/2}$ spinors are more compacted and less bonding when compared to their $P_{3/2}$ counterparts. As a result, I am asking, if you were to make a set of similar calculations for the very same systems, but for the experimental $2P_{1/2}$ peaks, whether or not your calculations would be reflecting the very same trends? It would be valuable for the XPS community if such a discussion was included in your final manuscript.

Juhan Kahk responded: Indeed, the radial wavefunctions of the $2p_{1/2}$ and $2p_{3/2}$ orbitals will be different. The neglect of this in the present work is likely one of the reasons why the calculated absolute $2p_{3/2}$ core electron binding energies do not agree as well with experiment, as they do for lighter elements where the magnitude of spin–orbit coupling is much smaller, or indeed for 1s levels.

That said, in first row transition metals, both the $2p_{1/2}$ and $2p_{3/2}$ levels can be safely considered core-like, and in compounds they largely retain their atomic like character. The 2p electrons in 3d transition metals do not participate in the formation of chemical bonds to any significant extent.

Much less experimental data is available for $2p_{1/2}$ binding energies than for $2p_{3/2}$ binding energies in the literature – it is for this reason that the $2p_{3/2}$ binding energies were used for comparisons between theory and experiment.

Samuel Hall commented: The mean error for the selection of transition metals decreases as you go across the period and I want to know if there is any reason explaining this.

Juhan Kahk answered: There are in principle three likely sources of error in the absolute binding energies: the exchange–correlation functional, the numerical implementation of the methods including the basis sets, and the treatment of relativistic effects. Finally, of course, one can in general challenge the validity of modelling the core hole state using occupancy constrained DFT, although recent progress in orbital optimized excited state mean field methods would suggest that there is at least strong empirical evidence of the robustness of this approach.

Careful tests have been conducted to examine the numerical accuracy of the calculations (see the ESI of the paper, https://doi.org/10.1039/d1fd00103e), and we have concluded that this, including the basis sets, cannot be responsible for the systematic errors observed.

The remaining two factors are the exchange–correlation functional, and the approximate treatment of spin and spin–orbit coupling. Unfortunately, it is very difficult to disentangle these effects, and to therefore rationalize why the systematic errors are larger for some elements than for others. In the future, fully relativistic ΔSCF calculations of core electron binding energies should shed more light on this issue.

Georg Held addressed Juhan Kahk and Johannes Lischner: All calculations presented in the paper (https://doi.org/10.1039/d1fd00103e) use molecules. Can the same accuracy also be achieved for solids?

Johannes Lischner responded: We have recently carried out a set of benchmark calculations on solids using periodic Δ-SCF calculations[1] and we have found a similar accuracy to that for molecular compounds.

1 J. M. Kahk, G. S. Michelitsch, R. J. Maurer, K. Reuter and J. Lischner, *J. Phys. Chem. Lett.*, 2021, **12**(38), 9353–9359.

Juhan Kahk answered: Our recent proof-of-principle study of applying the ΔSCF method to periodic solids[1] yielded promising results: MAE = 0.24 eV for 15 core levels. However, more work is still required to better understand the strengths and limitations of the ΔSCF method when applied to solids and surface species.

1 J. M. Kahk, G. S. Michelitsch, R. J. Maurer, K. Reuter and J. Lischner, *J. Phys. Chem. Lett.*, 2021, **12**(38), 9353–9359.

Georg Held addressed Johannes Lischner and Juhan Kahk: Can final state effects, such as spin–orbit splitting and satellites, be included in the formalism used for the calculations presented in the paper (https://doi.org/10.1039/d1fd00103e)?

Johannes Lischner responded: These are more difficult to capture and would certainly require extension of the current modelling framework. Inclusion of spin–orbit splitting is possible through higher level treatments of relativistic effects. Vibrational or electronic satellites could in principle be modelled as well by combining the Δ-SCF method with methods to describe electron–phonon and electron–plasmon interactions.

Juhan Kahk replied: At present, the prediction of ground state spin–orbit splittings is relatively straightforward, but the computational optimization of final states, where a core hole is created in a spin–orbit-split level, *e.g.* a $2p_{1/2}$ or $2p_{3/2}$ level, is to the best of my knowledge presently only possible for atomic systems and small molecules in the electronic structure code Dirac. However, it is likely that in the next few years other implementations of fully relativistic ΔSCF calculations will also become available.

Philip R. Davies asked: Were there any patterns in the effects of the ligands on your calculations?

Juhan Kahk answered: A general trend in ΔSCF calculations of core electron binding energies using the SCAN functional seems to be that core electron binding energies are slightly underestimated for atoms in very high oxidation states, *i.e.* those bonded to multiple F or O atoms.

In our previous study, where core electron binding energies in the elements B–F and Si–Cl were considered, the six highest errors in the calculated binding energies were −1.04 eV for B 1s in BF_3, −0.84 eV for S $2p_{3/2}$ in SF_6, −0.75 eV for Si 2p in SiF_4, −0.68 eV for P $2p_{3/2}$ in $P(CF_3)_3$, −0.63 eV for P $2p_{3/2}$ in PF_3, and −0.61 eV for P $2p_{3/2}$ in POF_3.[1]

In this work, after the application of the element specific corrections, the TM $2p_{3/2}$ core electron binding energies in TiF_4, VF_5, VOF_3, CrO_2Cl_2, and MnO_3Cl are still underestimated by −0.39 eV, −0.35 eV, −0.24 eV, −0.43 eV, and −0.05 eV respectively.

1 J. M. Kahk and J. Lischner, *Phys. Rev. Mater.*, 2019, **3**, 100801(R).

Thorsten Bartels-Rausch opened a discussion of the paper by Robert Lindsay: This is a question on the speciation of the adsorbed chlorine, which is described as atomic (page 12, line 8). My first question is then, where did the electron end up?

Secondly, by de-convoluting the Cl 2p, as shown in Fig. 8 of the paper (https://doi.org/10.1039/d1fd00106j), using 2 features for the protonated and unprotonated chlorine we have recently suggested that at the air–ice interface HCl adsorbs molecularly to the ice surface and dissociates only after entering the interfacial layer where the chloride can form sufficient numbers of hydrogen bonds to stabilise the ion and so favour dissociation energetically. The spectroscopic data thus revealed a Janus-type behaviour of the acid: the physisorbed form at the surface and dissociation only taking place deeper within the air–ice interface.[1] The binding energy shift of unprotonated and protonated chloride on ice goes back to the work by P. Parent, J. Lasne, G. Marcotte and C. Laffon[2] where Fig. 3 impressively shows the two features in the Cl 2p spectra on ice. This brings me to my question: would there be any indication of HCl and/or Cl^- being present at the Fe-surface in the presence of the OMID?

1 X. Kong, A. Waldner, F. Orlando, L. Artiglia, T. Huthwelker, M. Ammann and T. Bartels-Rausch, *J. Phys. Chem. Lett.*, 2017, **8**, 4757–4762.
2 P. Parent, J. Lasne, G. Marcotte and C. Laffon, *Phys. Chem. Chem. Phys.*, 2011, **13**, 7142–7148.

Robert Lindsay replied: Concerning our description of the adsorbed chlorine atoms as atomic, we are using the term to distinguish between isolated surface adsorbates and chlorine as a constituent of a surface phase (*e.g.*, iron chloride salt layer), rather than the explicit charge state of the adsorbed chlorine. In terms of the fate of the electron on the chlorine anion, it will most likely be shared with the substrate during surface bond formation.

As regards the presence of molecular HCl at the surface, we suggest that this species is very unlikely to be present. In contrast to the study of Kong *et al.*, where

HCl is dosed onto the surface from the gas phase, we are adsorbing species from aqueous HCl, in which the HCl (a strong acid) will already be essentially fully dissociated, *i.e.*, there is no obvious source of molecular HCl. Furthermore, our measurements are conducted at room temperature, and so any molecular HCl would desorb into the XPS vacuum chamber. We note that in the absence of OMID, we have concluded that the surface is terminated by a thin layer of iron chloride.[1]

1 K. Kousar, M. S. Walczak, T. Ljungdahl, A. Wetzel, H. Oskarsson, P. Restuccia, E. A. Ahmad, N. M. Harrison and R. Lindsay, *Corros. Sci.*, 2021, **180**, 109195.

Samuel Eyley commented: In both nitrogen 1s spectra in Fig. 5 of the paper (https://doi.org/10.1039/d1fd00106j) the same binding energy is assigned to the tertiary amine nitrogen of the imidazoline moiety irrespective of the protonation state of the imidazoline. One might expect charge delocalization from the protonated imine functionality across the two nitrogens of the imidazoline resulting in similar binding energies for the two nitrogens. If this is the case, what would the impact be on the ratio of protonated species on the surface and did you consider using DFT to determine the binding energies for the protonated species?

Robert Lindsay responded: Our fitting/interpretation of the N 1s spectrum is based on work presented in ref. 1. In that paper, they fit N 1s XPS spectra acquired from both a theophylline/maleic acid co-crystal and a theophylline/5-sulfosalicylic acid salt (see Fig. 2 in ref. 1). The theophylline exhibits an imidazole ring, which becomes protonated in the salt, *i.e.*, $>$C=N– → $>$C=NH$^+$–. They fit their N 1s XPS data such that this protonation leads to a BE shift of +2.3 eV for this nitrogen atom, with the BE of the other N atom in the ring ($>$N–H) being unaffected, *i.e.*, it is concluded to remain at a constant BE. Assuming that the above does not pertain to our system, but rather, as you suggest, that the two nitrogen atoms in the imidazoline ring should exhibit the same BE following protonation, we have explored the impact of this possibility on the fitting of the N 1s data. To date, we have not been able to come up with a physically reasonable solution. As regards DFT of the protonated species, effort is concentrating on modelling the interaction of these species with the substrate. These calculations are ongoing, but may deliver further insight into the interface, as substrate bonding may produce N 1s BE shifts that will provide an alternative interpretation of the N 1s profile.

1 J. S. Stevens, S. J. Byard, C. C. Seaton, G. Sadiq, R. J. Davey and S. L. M. Schroeder, *Phys. Chem. Chem. Phys.*, 2014, **16**, 1150–1160.

Wendy Flavell remarked: In the context of the S 2p spectra in Fig. 7 of the paper (https://doi.org/10.1039/d1fd00106j), you have identified 2 sulphate species. Sulphur oxidises very easily through many oxidation states from S(0) to sulphate in wet conditions, so I wondered if you have some sulphite (SO_3^{2-}) at the surface? This is seen at the surface of PbS quantum dots as they degrade in wet environments, and transforms to sulphate over time (*e.g.* ref. 1).

1 D. J. H. Cant, K. L. Syres, P. J. B. Lunt, H. Radtke, J. Treacy, P. J. Thomas, E. A. Lewis, S. J. Haigh, P. O'Brien, K. Schulte, F. Bondino, E. Magnano and W. R. Flavell, *Langmuir*, 2015, **31**, 1445–1453.

Robert Lindsay replied: In the current study (https://doi.org/10.1039/d1fd00106j), the BE difference between the two S $2p_{3/2}$ peaks is ~0.8 eV (BEs are ~169 eV and ~168.2 eV). We assign both of these features to SO_4^{2-}, as we would expect SO_3^{2-} to have a lower S $2p_{3/2}$ BE. In the paper by Cant *et al.*, the BE difference of the S $2p_{3/2}$ peaks for SO_4^{2-} and SO_3^{2-} is ~1.7 eV (from Table 1 of ref. 1), which is essentially double what is found in our study.

1 D. J. H. Cant, K. L. Syres, P. J. B. Lunt, H. Radtke, J. Treacy, P. J. Thomas, E. A. Lewis, S. J. Haigh, P. O'Brien, K. Schulte, F. Bondino, E. Magnano and W. R. Flavell, *Langmuir*, 2015, **31**, 1445–1453.

Georg Held commented: The C 1s signal in Fig. 3 of the paper (https://doi.org/10.1039/d1fd00106j) is all assigned to adventitious carbon. Given that the sample is carbon steel and that the inhibitor is an organic molecule with a high carbon content, can the carbon signal from these two sources be completely excluded?

Robert Lindsay answered: To clarify the statement in the paper (https://doi.org/10.1039/d1fd00106j), the C 1s signal is all assigned to adventitious carbon only for the polished carbon steel sample (bottommost spectrum in Fig. 3 of the paper). In the presence of OMID at CMC, there will also be a contribution from the carbon atoms in the adsorbed molecules. As regards a contribution to the C 1s spectrum from carbon in the carbon steel substrate, this carbon is expected to primarily exist as cementite (Fe_3C), which should produce a peak at BE ~283.5 eV. We do not observe such a feature in any of the data presented in this paper, although we have found a very low intensity peak in this region for poorly inhibited samples. This peak can be attributed to carbon in Fe_3C, which becomes more evident on the surface following loss of the Fe matrix due to corrosion; the Fe_3C phase does not corrode.

Rosa Arrigo remarked: In the N 1s spectrum in Fig. 5 of the paper (https://doi.org/10.1039/d1fd00106j), there is a component at around 398 eV, do you have any idea about the nature of this species? The BE of C=N bonds is commonly found at 398.5 eV similar to your findings. However below this binding energy only Fe–N bonds in nitride were reported but generally at a much lower BE.

Robert Lindsay responded: Currently, we are uncertain as to the origin of the N 1s component at BE = 397.6 eV. We do agree that it is located at a lower BE than expected for nitrogen in common organic compounds. Further efforts are underway to allow identification of this feature.

Rosa Arrigo said: The peak at 402 eV in the N 1s spectrum in Fig. 5 of the paper (https://doi.org/10.1039/d1fd00106j), assigned to quaternary nitrogen, looks a bit high in binding energy, would you exclude oxygen bound to the nitrogen atom?

What is the rationale/strategy used in this study for peak assignment?

Robert Lindsay answered: Reference N 1s XPS data from a multilayer of the OMID head group (see ref. 1) was used as a starting point for peak assignments. The multilayer N 1s spectrum was fitted with 3 peaks, which were assigned to the three chemically different nitrogen atoms in the molecule, *i.e.*, imine, tertiary amine, and primary amine. In addition, data presented in ref. 2 were used to corroborate the BE shifts of these peaks induced by protonation. The BE shift associated with protonation of $-NH_2$ to $-NH_3^+$ (+2.4 eV) in our work is consistent with that indicated by Stevens *et al.* The absolute BE for the $-NH_3^+$ group in our work (402.4 eV) is somewhat higher than that reported by Stevens *et al.* (401.9 eV), but this difference may result from the different environments (*i.e.*, surface adsorbate and bulk organic salts/crystals, respectively). We note that we cannot entirely exclude that the peak at BE = 402.4 eV is due to N bonded to O. However, we suggest that it is unlikely, as the only O 1s signal present is almost certainly due to adventitious deposits, and so oxygen is not an intrinsic component of the interface.

1 K. Kousar, T. Ljungdahl, A. Wetzel, M. Dowhyj, H. Oskarsson, A. S. Walton, M. S. Walczak and R. Lindsay, *J. Surfactants Deterg.*, 2020, **23**, 225–234.
2 J. S. Stevens, S. J. Byard, C. C. Seaton, G. Sadiq, R. J. Davey and S. L. M. Schroeder, *Phys. Chem. Chem. Phys.*, 2014, **16**, 1150–1160.

Samuel Eyley remarked: The nitrogen 1s spectrum of the 1 M HCl + OMID sample in Fig. 5 of the paper (https://doi.org/10.1039/d1fd00106j) contains a large number of fitted component peaks, while the overall peak profile is relatively featureless. Other than the peak positions from Fig. 4(b), were any constraints on parameters such as peak area used to stabilize the fitting of this peak and if so, what was the basis for these constraints?

Robert Lindsay replied: Our general philosophy in terms of fitting XPS core levels is to fit with the fewest number of components, each having the same fwhm (unless there is a good reason not to do so), that enables a physically reasonable interpretation of the interface chemistry. This approach was employed for the N 1s spectrum (Fig. 5 of the paper, https://doi.org/10.1039/d1fd00106j) acquired for 1 M HCl/OMID, using N 1s XPS data from a multilayer of the OMID head group (see ref. 1) as a starting point. The multilayer N 1s spectrum was fitted with 3 peaks, which were assigned to the three chemically different nitrogen atoms in the molecule, *i.e.*, imine, tertiary amine, and primary amine. Attempts to fit the spectrum in Fig. 5 with only these 3 components, whilst maintaining physically reasonable intensity ratios, failed to produce an acceptable fit. Given that it was also established (ref. 1) that OMID is doubly protonated in 1 M HCl, we considered different combinations of protonated species, both singly and doubly protonated during fitting; we employed data presented in ref. 2 to aid peak assignments. Using this approach, and ensuring that the component intensities were physically reasonable, we produced the fit displayed in Fig. 5.

1 K. Kousar, T. Ljungdahl, A. Wetzel, M. Dowhyj, H. Oskarsson, A. S. Walton, M. S. Walczak and R. Lindsay, *J. Surfactants Deterg.*, 2020, **23**, 225–234.
2 J. S. Stevens, S. J. Byard, C. C. Seaton, G. Sadiq, R. J. Davey and S. L. M. Schroeder, *Phys. Chem. Chem. Phys.*, 2014, **16**, 1150–1160.

Sefik Suzer commented: In your paper (https://doi.org/10.1039/d1fd00106j), you have written that you had used HCl and H_2SO_4, both in 1 M and 0.01 M concentrations, but in the figures, you mostly compare 1 M HCl and 0.01 M H_2SO_4. Is there a reason for not comparing them at equal molarities?

What are the typical conditions (*e.g.* pH) in a corrosive environment?

Robert Lindsay answered: It was decided to compare data from samples immersed in 1 M HCl and 0.01 M H_2SO_4, as we wanted to compare well-inhibited systems. Here, we define a well-inhibited system as having an inhibition efficiency $\geq$ 95% and an absolute inhibited corrosion rate of $\leq$ 0.2 mm y^{-1}. At lower inhibition efficiencies/higher absolute corrosion rates, it is more likely that the XPS data may not be representative of a well-inhibited substrate, complicating interpretation.

In engineering scenarios, acid concentration can vary widely. For example, for acidizing applications in oilfield operations, 8 M HCl can be used.

Wendy Flavell said: You comment in your conclusions on the significant difference in chemistry between the two systems; with sulphuric acid you see an oxidised film at the surface, which is not present in the case of hydrochloric acid. Could this be because sulphuric acid is an oxidising acid, and hydrochloric acid isn't?

Robert Lindsay replied: Firstly, we note that both HCl and H_2SO_4 are oxidising the substrate, *i.e.*, acidic protons formed through dissociation in solution act as an oxidising agent for Fe^0 to Fe^{2+}. As an "oxidising acid", it could be that the $SO_4{}^{2-}$ anion in H_2SO_4 is also acting as an oxidising agent. However, given the concentrations of sulfuric acid employed in this study (https://doi.org/10.1039/d1fd00106j), this reaction will almost certainly not be significant; much more concentrated H_2SO_4 is required. Hence, the differences observed between HCl and H_2SO_4 cannot, in our opinion, be simply attributed to H_2SO_4 being an "oxidising acid".

Kevin Lovelock asked: For the N 1s XPS shown in Fig. 5 of the paper (https://doi.org/10.1039/d1fd00106j), when the imidazoline is protonated on the ring, have you checked to see if the two ring nitrogen atoms are equivalent in XPS? Your XPS fitting model shows them as relatively different. In the literature, for the closely related cationic heterocycle imidazolium with one alkyl and one proton substituent, the N 1s XPS gave one symmetric peak with a relatively narrow width,[1] suggesting that the two nitrogen atoms were equivalent. Moreover, imidazolinium cations have also been depicted in the literature with delocalisation over the N–C–N bonds, suggesting that the two nitrogen atoms are equivalent.[2]

1 R. M. Fogarty, R. G. Palgrave, R. A. Bourne, K. Handrup, I. J. Villar-Garcia, D. J. Payne, P. A. Hunt and K. R. J. Lovelock, Electron spectroscopy of ionic liquids: experimental identification of atomic orbital contributions to valence electronic structure, *Phys. Chem. Chem. Phys.*, 2019, **21**(35), 18893–18910.
2 A. L. Chong, M. Forsyth and D. R. MacFarlane, Novel imidazolinium ionic liquids and organic salts, *Electrochim. Acta*, 2015, **159**, 219–226.

Robert Lindsay responded: As discussed in reply to Samuel Eyley, our fitting/interpretation of the N 1s data is based on ref. 1, where they conclude that the two ring N atoms remain inequivalent following protonation of the imine (>C=

N– → >C=NH$^+$–); we have not succeeded in obtaining a physically reasonable fit, as yet, assuming that the two ring atoms are equivalent.

Examining the comprehensive study of Fogarty *et al.*, the pertinent N 1s spectra, as you write, are entirely consistent with the two ring N atoms being equivalent following protonation. This result apparently contradicts Stevens *et al.* and so needs to be investigated further. Potentially, it may be system related, *e.g.*, bulk crystalline solid *versus* bulk ionic liquid, or the chemistry of the surface region probed by XPS may not be representative of the bulk for either one or the other. In terms of the adsorbed OMID, it may also be that substrate interactions should be considered to interpret the N 1s profile. Currently, we have not explored this possibility, as supporting evidence (*e.g.*, DFT calculations) would be required to move beyond mere speculation.

1 J. S. Stevens, S. J. Byard, C. C. Seaton, G. Sadiq, R. J. Davey and S. L. M. Schroeder, *Phys. Chem. Chem. Phys.*, 2014, **16**, 1150–1160.

Kevin Lovelock commented: For the S 2p XPS in Fig. 7d of the paper (https://doi.org/10.1039/d1fd00106j), did you consider the formation of the anion $[HSO_4]^-$? XPS of both $[HSO_4]^-$ and $[SO_4]^{2-}$ have been published for H_2SO_4 dissolved in water.[1] $[HSO_4]^-$ anions have also been studied in ionic liquids using XPS.[2]

1 A. M. Margarella, K. A. Perrine, T. Lewis, M. Faubel, B. Winter and J. C. Hemminger, Dissociation of sulfuric acid in aqueous solution: determination of the photoelectron spectral fingerprints of H_2SO_4, HSO_4^-, and SO_4^{2-} in water, *J. Phys. Chem. C*, 2013, **117**(16), 8131–8137.
2 R. M. Fogarty, R. Rowe, R. P. Matthews, M. T. Clough, C. R. Ashworth, A. Brandt, P. J. Corbett, R. G. Palgrave, E. F. Smith, R. A. Bourne, T. W. Chamberlain, P. B. J. Thompson, P. A. Hunt and K. R. J. Lovelock, Atomic charges of sulfur in ionic liquids: experiments and calculations, *Faraday Discuss.*, 2018, **206**, 183–201.

Robert Lindsay answered: You make an excellent suggestion. Examining the paper by Margarella *et al.*, the S 2p BE difference for SO_4^{2-} and HSO_4^- is ~0.7 eV. This BE difference is consistent with that we report (*i.e.*, ~0.8 eV) for the two S 2p features that we associate with two different states of SO_4^{2-}, due to significant surface heterogeneity. On this basis, it is plausible that these two states are a result of surface SO_4^{2-} and HSO_4^-, *i.e.*, the salt layer consists of both ferric sulfate and ferric hydrogensulfate. Further work is required to confirm this suggestion.

Shaoliang Guan said: I'm very fascinated by the naturally adsorbed Cl atoms on Fe that show the asymmetric characteristic. We've tried Cl adsorption on Cu single crystals with HCl in various ways, including gas doping and electrochemical methods, but never observed such a phenomenon. Is it due to the different metal system and why is that? Also, if a potential is applied on your HCl/Fe system, what would you expect to see regarding the changes in the Cl adsorbed layer? Thanks.

Robert Lindsay responded: As you suggest, I believe that the difference in Cl 2p line shape is a result of a change in substrate, *i.e.*, Cu *versus* Fe. This conclusion is based on a study of the O 1s XPS line shape for ordered overlayers of atomic oxygen on Ni(100) and Cu(100) by Tillborg *et al.*[1] They demonstrate that the asymmetry of the core level peak can vary significantly with substrate, as a result of changes in the location of the adsorbate/substrate hybrid (2p–3d) states relative to the Fermi level. Concerning the impact of varying the potential on interface termination, including changes in Cl adsorbed layer, I am uncertain as to how it would vary in detail. However, I would expect there to be substantial variation as you polarise the sample anodically/cathodically relative to the open circuit potential.

1 H. Tillborg, A. Nilsson, T. Wiell, N. Wassdahl, N. Mårtensson and J. Nordgren, *Phys. Rev. B: Condens. Matter Mater. Phys.*, 1993, **47**, 16464.

David Payne remarked: The need to have excellent HAXPES reference spectra is of great importance, but it is often difficult to persuade review panels to give beam time for this work. How can this be overcome?

Robert Lindsay replied: I agree that it is difficult to get synchrotron beam time for such work, as I have already had experience of this issue. One approach may be to organise the community to prepare a long-term proposal, which would deal systematically with a wide range of samples. Alternatively, lab-based sources may also be employed for such data acquisition.

Olivier Renault communicated: What is the energy resolution of the XPS analysis? Values of the analyzer pass energy are stated in the Experimental section of the paper (https://doi.org/10.1039/d1fd00106j), but do not tell us anything about the effective contribution of the electron spectrometer to the overall energy resolution (photon band width included). This is important to know regarding the fitting presented in Fig. 5 of the paper, in which some components look fairly close. The question is therefore: how does the smallest energy separation of the N 1s chemical state components compare with the effective energy resolution of the experiment?

Robert Lindsay communicated in reply: The instrumental energy resolution is regularly assessed through acquisition of Ag 3d spectra. Under conditions analogous to those employed for collection of higher-resolution data in the paper (*i.e.*, monochromatic Al Kα source and an analyser pass energy of 20 eV), the fwhm of the Ag $3d_{5/2}$ peak was found to be $\sim$0.7 eV. In comparison, the smallest BE separation of the N 1s components in Fig. 5 of the paper (https://doi.org/10.1039/d1fd00106j) is $\sim$0.4 eV, with the next smallest being 0.7 eV; it should be noted that a fwhm of $\sim$1.2 eV was employed for each of the N 1s components during fitting.

Roxy Lee opened a discussion of the paper by Kevin Lovelock: Do you get element specific information from the resonant spectra, and if you summed the resonant spectra for all elements does it at all resemble the non-resonant spectra?

Kevin Lovelock answered: Yes, resonant XPS is element specific. Measuring resonant XPS for all contributions of an ion or molecule is very challenging, *e.g.* alkyl chains give relatively weak C 1s RXPS signals.[1] The transition probabilities for resonant XPS and non-resonant XPS are different. Therefore, for samples where one manages to record resonant XPS for all elements (*e.g.* the $[SCN]^-$ anion in our article, https://doi.org/10.1039/d1fd00117e), the resonant XPS and non-resonant XPS spectra are very unlikely to match. Moreover, the transition probabilities for non-resonant XPS are pretty well understood, *i.e.* the Gelius method. In contrast, for resonant XPS the transition probabilities are still not well understood; there are relatively few papers including resonant XPS calculations, especially for relatively large molecules (see ref. 2 and 3 below as excellent examples).

1 R. M. Fogarty, R. G. Palgrave, R. A. Bourne, K. Handrup, I. J. Villar-Garcia, D. J. Payne, P. A. Hunt and K. R. J. Lovelock, *Phys. Chem. Chem. Phys.*, 2019, **21**, 18893–18910.
2 F. Holzmeier, T. J. A. Wolf, C. Gienger, I. Wagner, J. Bozek, S. Nandi, C. Nicolas, I. Fischer, M. Guhr and R. F. Fink, *J. Chem. Phys.*, 2018, **149**, 034308.
3 G. Grell and S. I. Bokarev, *J. Chem. Phys.*, 2020, **152**, 074108.

Wendy Flavell remarked: In your conclusions, you comment on the guidelines for strong resonant XPS. One of these is low electronegativity. What do you think is the physical picture behind this?

Kevin Lovelock responded: This explanation came about from a discussion with Prof. Reinhold Fink (Tübingen). For a bond A–B, where we assume that atom A has lower electronegativity than atom B, the bonding orbital will be mainly located on atom B, and consequently the antibonding orbital will be mainly located on atom A. For resonant XPS, an electron will be promoted from a core level into the antibonding orbital located primarily on atom A, meaning that atom A will be more likely to have resonant XPS transitions occurring (*i.e.* participator Auger transitions) than atom B.

Johannes Lischner commented: Your paper (https://doi.org/10.1039/d1fd00117e) concludes that the one-electron picture is appropriate for the interpretation of your experiments. What would be the signals of a breakdown of this picture and has such a breakdown been observed for any of the systems studied in your experiments?

Kevin Lovelock replied: In our article (https://doi.org/10.1039/d1fd00117e) we briefly comment on resonant XPS for the open-shell metal complex anion $[CoCl_4]^{2-}$, and how the one-electron picture holds at lower binding energies (12 eV > E_B > 6 eV) but fails at higher binding energies (18 eV > E_B > 12 eV), judged by a visual match (for the one-electron picture holding) and mismatch (for the one-electron picture failing, *i.e.* spectator Auger transitions occurring) to the non-resonant XP spectrum. We have further resonant XPS data for similar open-shell halometallate anions, but we decided to focus on samples where the one-electron picture held in this article. The challenges of resonant XPS for open-shell samples are discussed in more detail in ref. 1.

1 R. H. Temperton, M. Y. Guo, G. D'Acunto, N. Johansson, N. W. Rosemann, O. Prakash, K. Wärnmark, J. Schnadt, J. Uhlig and P. Persson, *Sci. Rep.*, 2021, **11**, 22144.

Juhan Kahk asked: In your studies of resonant photoemission from liquids, did you ever see evidence of interference between direct photoemission and participator Auger decay in the form of the Fano lineshape in a plot of intensity *vs.* incident photon energy?

Kevin Lovelock responded: We have not seen any evidence of a Fano lineshape. It should be noted that the experiments we perform are relatively tricky, both in terms of recording the data for liquid samples and for subtracting the non-resonant XPS contributions, so any small features may be too small to be observed.

Wendy Flavell said: In your presentation you commented that the reactivity of molecules is explained by their electron distribution. How do you see the atomic character of the density of states influencing reactivity?

Kevin Lovelock answered: There are a number of theories linking density of states and reactivity, and the most well-known is probably by Fukui.[1] All revolve around knowledge of the occupied density of states at low binding energy, *i.e.* what we call the highest occupied valence state from our experiments. Determining the partial density of states for relatively large molecules and ions is particularly important for understanding reactivity, given that the highest occupied valence state could be localised on different parts of the molecule. For example, in our article we show for the anion $[TfO]^-$ ($[CF_3SO_3]^-$), the three fluorine atoms do not contribute significantly to the occupied valence states near the highest occupied valence state (see Fig. S19 in the ESI of the paper, https://doi.org/10.1039/d1fd00117e). Therefore, the fluorine atoms in $[TfO]^-$ are very unlikely to be good electron donors and be the primary drivers of the reactivity for the $[TfO]^-$ anion – this result will not surprise an organic chemist, but it is a good demonstration of the power of resonant XPS. The nitrogen partial density of states for imidazolium is a similar case, where the ring carbon atoms dominate the highest occupied valence state for imidazolium (Fig. 7b of the paper).

1 K. Fukui, *Angew. Chem., Int. Ed. Engl.*, 1982, **21**, 801–809.

Alexander Shard opened a general discussion: Can the panel comment on progress in the calculation of shake-up positions and intensities?

Juhan Kahk responded: For calculating the positions and intensities of shake-up satellites (and other satellites, *e.g.* those caused by plasmon excitation), genuine excited state calculations need to be performed, unlike in the ΔSCF method, in which the energy of the lowest energy core hole state is obtained from an effective ground state calculation. Therefore, the computational cost of calculations of photoemission satellites is considerably higher than the cost of predicting binding energies under the assumption that the fully screened final state is reached.

Nevertheless, considerable progress has been made in recent years. For the latest on the development of theoretical methods, I would refer to the work of J. J. Rehr, and J. J. Kas, *e.g.* ref. 1–4 and others.

Myself, I have recently used the GW + cumulant approach that is already quite well established for predicting satellite structures in core and valence level photoelectron spectra of the layered oxide $PdCoO_2$ and W metal.[5,6]

1 F. D. Vila, J. J. Rehr, J. J. Kas, K. Kowalski and B. Peng, *J. Chem. Theory Comput.*, 2020, **16**, 6983–6992.
2 J. J. Rehr and J. J. Kas, *J. Vac. Sci. Technol. A*, 2021, **39**, 060401.
3 F. D. Vila, J. J. Kas, J. J. Rehr, K. Kowalski and B. Peng, *Front. Chem.*, 2021, **9**, 734945.
4 J. J. Kas, J. J. Rehr and T. P. Devereaux, *arXiv*, 2022, preprint, arXiv:2107.10409v2, https://arxiv.org/abs/2107.10409v2.
5 Q. Lu, H. Martins, J. M. Kahk, G. Rimal, S. Oh, I. Vishik, M. Brahlek, W. C. Chueh, J. Lischner and S. Nemsak, *Comm. Phys.*, 2021, **4**, 143.
6 C. Kalha, L. E. Ratcliff, J. J. Gutiérrez Moreno, S. Mohr, M. Mantsinen, N. K. Fernando, P. K. Thakur, T.-L. Lee, H.-H. Tseng, T. S. Nunney, J. M. Kahk, J. Lischner and A. Regoutz, *Phys. Rev. B*, 2022, **105**, 045129.

Johannes Lischner commented: The conference organizers have placed talks containing theoretical work into the "Future Directions" session indicating that a closer coupling of theory and experiments is an important step for the interpretation of future photoelectron spectroscopy experiments. In your opinion, what needs to be done to accelerate this process and ensure closer collaborations between theorists and experimentalists?

Robert Lindsay replied: In my opinion, fruitful collaboration between experimentalists and theorists is founded on early and close interaction, *i.e.*, getting the theory group on board at the start of a project, where a clear target is to use modelling to aid interpretation of XPS data. Approaching a modeller at a later stage of the research can result in fragmented interactions, and likely poorer outcomes.

Giulia Galli responded: First I think the value of validating the theory and methods we use with controlled experiments needs to be recognized. No theory is exact, no first principles method is exact. The level of theory for a given problem and property should be validated with controlled experiments, *i.e.* on very well defined samples, before proceeding to predictions. If the value of validation is recognized as important, then experimentalists would want to work with and for theorists (and not only ask theorists for 'service') because they would recognize that a good validated theory would then be very important to interpret their experiments and then to have predictions of new things that then experimentalists could try.

In short: if both parts recognize the value of the other part to solve a problem (instead of just producing a number to be written in a paper, to publish the paper faster), then that is a good start. More easily said than done, of course.

Juhan Kahk answered: Firstly, I very much share the sentiment expressed in the question. A lot of time and money is spent on experimental photoelectron spectrometry – this is particularly true for state-of-the-art applications such as endstations at synchrotrons and ambient pressure spectrometers. For these instruments to be able to deliver on their promise to yield revolutionary insights into processes such as heterogeneous catalysis or environmental degradation, it is critical that we are able to extract as much useful information from the recorded spectra as possible. Theoretical modelling can certainly help in this regard.

Several things could be done to accelerate the uptake of theoretical methods for the analysis of experimental spectra. For example:

It is important to understand to what extent the results obtained using existing theoretical methods can be trusted. Benchmarking studies of well characterized reference systems are highly required to answer this question.

The ability of existing first-principles methods to deal with transition metal compounds with an open d-shell is still somewhat limited. Further theoretical developments are required.

In general, modelling the surface chemistry in a complex system, and coming up with reasonable candidate structures for photoemission calculations is a time-consuming process. If theoretical simulations are to be used to aid the interpretation of experimental spectra, it will have to be someone's job to perform these simulations. In the ideal scenario, in future funding applications for next generation photoemission instruments, the need for theoretical modelling will be foreseen at the stage of applying for funding, and dedicated resources for hiring computational researchers will be requested as part of the project.

Georg Held said: Using synchrotrons we have been able to measure photoemission and Auger spectra as a function of photon energy for some time. There is a lot to be learned from such experiments, *e.g.* about charge transfer, chemical identity, defects, *etc.*, but these are often difficult to interpret accurately. This is a challenge which can only be met in close collaboration between experimentalists and theoreticians.

Conflicts of interest

There are no conflicts to declare.

Faraday Discussions

PAPER

Photo-induced lattice distortion in 2H-$MoTe_2$ probed by time-resolved core level photoemission

R. Costantini,[a] F. Cilento,[b] F. Salvador,[a] A. Morgante,[ac] G. Giorgi,[de] M. Palummo[f] and M. Dell'Angela[*a]

Received 29th November 2021, Accepted 2nd March 2022
DOI: 10.1039/d1fd00105a

The technological interest in $MoTe_2$ as a phase engineered material is related to the possibility of triggering the 2H–1T′ phase transition by optical excitation, potentially allowing for an accurate patterning of metallic areas into a semiconducting canvas *via* laser irradiation. In this paper, we investigate the photo-induced modifications of a bulk 2H-$MoTe_2$ crystal by means of time-resolved X-ray photoemission spectroscopy. We observe that in the microsecond timescale, the core levels shift to higher kinetic energies due to surface photovoltage fields, while in the sub-nanosecond range, the photoemission peaks shift in the opposite direction. With the support of DFT calculations, we ascribe the latter effect to the deformation of the lattice in the out-of-plane direction, which is along the pathway for the 2H–1T′ phase transition. Our data indicate an intermediate lattice excitation state with a measured lifetime in the order of 600 ps, in which the displacement of Mo and Te atoms causes the Te 4d electrons to shift towards higher binding energies.

Introduction

Transition metal dichalcogenides (TMDs) have been widely explored for the development of innovative nanoscale devices. TMDs are compounds with layered structures, with each layer consisting of a three atom unit (MX_2) with a transition metal atom (M) sandwiched between two chalcogen layers (X = S, Se, Te). The multiple combinations of transition metal and chalcogen atoms in TMDs result in a large range of electronic properties, which also depend on the crystal phase

[a]*CNR-IOM Laboratorio TASC, SS 14 Km 163.5, I-34149 Trieste, Italy. E-mail: dellangela@iom.cnr.it*
[b]*Elettra Sincrotrone Trieste S.C.p.A., Strada Statale 14 Km 163.5, I-34149 Trieste, Italy*
[c]*Department of Physics, University of Trieste, Via A. Valerio 2, I-34127 Trieste, Italy*
[d]*Department of Civil and Environmental Engineering (DICA), University of Perugia, Via G. Duranti 93, I-06125 Perugia, Italy*
[e]*CNR-SCITEC, I-06123 Perugia, Italy*
[f]*Dipartimento di Fisica, INFN, Università di Roma "Tor Vergata", Via della Ricerca Scientifica 1, 00133 Roma, Italy*

and on the sample thickness.[1] The indirect gap of bulk TMDs becomes direct when thinned down to monolayers, and the material character can change from semiconducting to (semi-)metallic if the trigonal prismatic (1H/2H) phase is distorted into an octahedral (1T/1T′) structure.

Molybdenum ditelluride ($MoTe_2$) is a prototypical TMD. It presents two crystal structures at room temperature: the semiconducting hexagonal phase (2H or α-phase) and the semimetallic monoclinic phase (1T′ or β-phase). In its 2H phase, $MoTe_2$ has attracted significant interest from a technological point of view due to its bandgap in the near infrared region (0.9 eV in the bulk and 1.1 eV in the monolayer), which makes it a potential alternative to silicon for optoelectronic applications. Very recently, carrier multiplication has been demonstrated in 2H-$MoTe_2$, making it a promising material for the development of light harvesting devices.[2] The 1T′ phase of $MoTe_2$ is metastable, since a barrier of around 0.7 eV (ref. 3 and 4) impedes the reconversion of the system to the 2H phase, making this material a promising template for relevant phase engineering applications.[5,6] Laser induced phase patterning has in fact been successfully applied to $MoTe_2$ to fabricate an ohmic heterophase homojunction.[7] The use of optical excitations allows for a precise transformation of the structural properties of the substrate with high area and phase selectivity, which is not achievable with traditional thermal and strain treatments.

The microscopic nature of the 2H–1T′ photo-induced phase transition in $MoTe_2$ is still debated. Several studies have recently tackled the photoexcited carrier dynamics in 2H-$MoTe_2$ by means of optical, terahertz, electron and X-ray spectroscopies.[8–11] It has been shown that photoexcitation is followed by thermalization of the lattice within 2 ps and the subsequent cooling process of the phonons occurs in the 100 ps timescale.[10] MeV ultrafast electron diffraction experiments have shown that by tuning the excitation wavelength, it is possible to control the evolution of the phonon populations by coupling to zone-center or zone-edge vibrational modes and, therefore, to control the structural dynamics.[12] More recently, a theoretical work has demonstrated that the photo-induced transition from 2H to 1T′ in $MoTe_2$ monolayers is purely triggered by electronic excitation, which causes the softening of the A''_2 phononic mode, provided that the pump energy is above 1.96 eV.[13]

In this work, we study the structural dynamics of an optically excited 2H-$MoTe_2$ single crystal by means of time-resolved X-ray photoemission spectroscopy (tr-XPS). We use laser pulses with a photon energy of 2.4 eV as a pump and synchrotron X-ray pulses as a probe. We show that the optical excitation results in the accumulation of electrons in the top layer, generating a long-lived surface photovoltage (SPV), which causes the shift of the photoelectron spectra to lower binding energies, as compared to those of the unpumped system. Conversely, on the sub-nanosecond timescale, we observe a transient shift of the Te 4d core level to positive binding energies, which cannot be ascribed to the SPV. We used DFT calculations to simulate the deformation of the lattice along the A''_2 mode of the monolayer, corresponding to an out-of-plane displacement of Mo and Te atoms moving in opposite directions. The theoretical model shows that an atomic displacement of the order of 0.01 Å suffices to cause a ∼10 meV positive binding energy shift, in agreement with the experimental data. The fingerprints of this transient distortion are spectroscopically inequivalent to those of the irreversible 2H-to-1T′ transition. We therefore suggest that this core level shift is due to a different and reversible structural distortion of the irradiated sample, which recovers in the 600 ps timescale.

Experimental details

Materials and methods

The 2H-$MoTe_2$ crystal (>99.995% purity) was purchased from HQ Graphene (Groningen, the Netherlands) and cleaved in an ultra-high vacuum (base pressure in the experimental chamber <3×10^{-10} mbar). All measurements were carried out at the ANCHOR-SUNDYN endstation of the ALOISA beamline at the Elettra synchrotron in Trieste, Italy.[14] 2.4 eV (515 nm) pulses from an Yb-doped yttrium aluminum garnet (Yb:YAG) fiber laser (Tangerine HP, Amplitude Systèmes, Pessac, France) were used as a pump to excite the samples. The laser was operated at 385.67 kHz and was used to pump one out of every three hybrid X-ray pulses of the storage ring (1.157 MHz), as detailed in ref. 15. The X-ray and optical beams were impinging on the sample in a quasi-collinear geometry, and the spatial overlap was preliminarily determined on a YAG crystal mounted on the sample holder. XPS spectra in the pump on/pump off series were acquired with a photon energy of 400 eV and a pass energy of 20 eV. As such scans were acquired using the multibunch radiation of Elettra, the pump-induced effects were averaged on a 2.6 μs delay range, *i.e.*, the separation between consecutive laser pulses. By acquiring the scans in pump off/pump on sequences, we ensured the reversibility of the effect and excluded radiation damage. Time-resolved spectra were acquired with the pump laser synchronized to 150 eV synchrotron probe pulses. The temporal coincidence of the pump and probe pulses was calibrated by measuring the surface photovoltage shift on a silicon crystal. Considering the significant difference in pulse duration for the pump (~300 fs) and the probe (~100 ps) pulses, the latter dominates the cross-correlation, and therefore the temporal resolution in time-resolved XPS measurements.

Theoretical methodology

Geometric structures were optimized and single-point calculations of the $MoTe_2$ monolayers (see Fig. 1) were performed by means of density functional theory (DFT), as implemented in the VASP code,[16–19] within the generalized gradient approximation (GGA) of Perdew–Burke–Ernzerhof (PBE).[20] The ion–electron interactions were described by the projector augmented wave (PAW) method.[21] A plane-wave basis set energy cutoff of 600 eV was employed. Also, the zero damping DFT-D3 (Becke–Johnson) method of Grimme was employed to include the long-range interactions.[22] As an initial model, we employed the three-atom unit cell (1 Mo, 2 Te) of the ML, whose Brillouin zone (BZ) was sampled with a Γ-centered k-point mesh of $24 \times 24 \times 1$. A sufficient amount of vacuum was added along the non-periodic direction to avoid any possible spurious interaction among replicas. Any possible residual dipole along the same non-periodic direction was properly corrected. Optimizations were carried out until residual forces were below 0.005 eV $Å^{-1}$. The optimized lattice parameters and angles of the ML are $a = b =$ 3.4996 Å and $\alpha = \beta = 90°$ and $\gamma = 120°$.

We then calculated the changes in the core level energies according to the final state effect scheme, which consisted of isolating one atom and creating a core

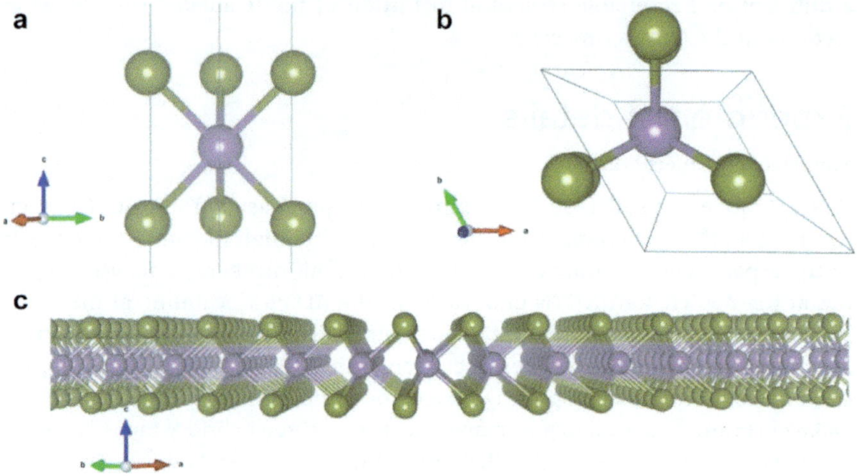

Fig. 1 Lateral (a) and top (b) view of the optimized structure of the ML of $MoTe_2$. (c) Lateral view of the supercell of the same ML of $MoTe_2$. Purple: Mo; gold: Te.

hole and injecting the electron back into the empty states;[23] in our calculations, we focused on Te 4d electrons. Convergence of the calculations was checked up to 8×8 supercells (128 Te & 64 Mo atoms).

Results and discussion

We measured the effects of the optical excitation on the 2H-$MoTe_2$ crystal core levels, from the sub-nanosecond to the microsecond timescale. We first acquired photoemission spectra in the Mo 3d and Te 4d ranges by optically exciting the sample with 2.4 eV photons and detecting the photoelectrons generated by the whole multibunch radiation of the Elettra synchrotron. The pump-induced effects in this case are averaged over the time between consecutive laser pulses, as described in the Methods section. If compared to the actual time-resolved XPS scans, which are presented later in the text, this acquisition method ensures a higher count rate and can be used to quickly assess the presence of long-lived effects and the entity of the possible radiation damage caused by the laser (not detected in this experiment). Fig. 2 shows the Mo 3d and Te 4d core levels as a function of the applied laser fluence. The spectra were acquired by alternating pump-off and pump-on scans while increasing the laser power to examine whether any irreversible changes had been introduced. Below 0.4 mJ cm^{-2} all pump-induced effects are completely reversible and consist of: (i) a rigid shift of the Mo 3d and Te 4d lines to lower (higher) binding (kinetic) energies, (ii) the appearance of a weak component ~0.6 eV below the Mo $3d_{5/2}$ peak, as indicated by the black arrow and (iii) the reduction of the Te 4d peak intensity. By switching the laser off, the spectrum returns to its original shape, thus, we only show the first of the unpumped spectra for clarity. Above 0.4 mJ cm^{-2}, the appearance of the novel component (ii) becomes persistent even after switching off the optical excitation. The energy position of (ii) is compatible with Mo $3d_{5/2}$ photoelectrons from Mo atoms in the 1T′ phase.[7,24] Concurrently, we would expect its Mo $3d_{3/2}$

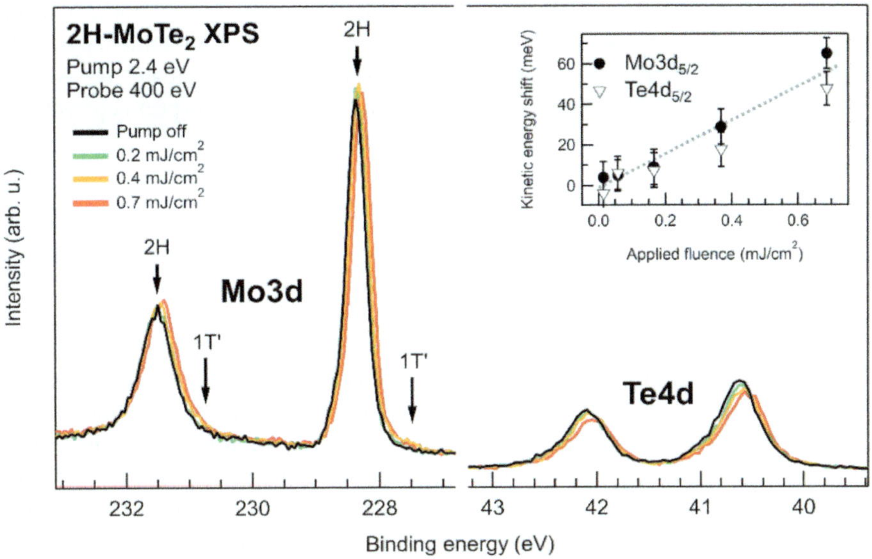

Fig. 2 Mo 3d and Te 4d photoemission lines as a function of the applied laser fluence. The pump photon energy is 2.4 eV and the probe photon energy is 400 eV. The spectra shift consistently to lower binding energies with increasing fluence due to the surface photovoltage caused by the long-lived charge carriers. The components of the Mo 3d core level assigned to the 2H and 1T′ phases are also labelled. The inset shows the energy shift of Mo $3d_{5/2}$ and Te $4d_{5/2}$ as a function of the applied fluence with respect to the laser off spectra.

counterpart and a Te 4d doublet with a ~0.6 eV lower binding energy with respect to the original Te 4d lines; due to the larger intrinsic width of the Mo $3d_{3/2}$ and Te 4d peaks and the weak intensity of the 1T′ component (~2.5% with respect to the pristine 2H-$MoTe_2$, as derived from the fits), such contributions do not emerge as clearly as for Mo $3d_{5/2}$ in the current dataset. The slight broadening of the Te 4d lines, however, is consistent with the appearance of a chemically inequivalent component near the 2H related peak. Moreover, the decreasing intensity of Te 4d (iii) as a function of the increasing pump fluence is a further proof of a partial 2H-to-1T′ transition, since the 1T′ phase is also characterized by Te vacancies.[7,24] Indeed, the diffusion and ordering of Te vacancies is one of the proposed mechanisms for the 2H-to-1T′ phase transition.[11,25]

The last optically induced effect to be discussed is the spectral shift to higher kinetic energies (i). Due to the semiconducting properties of 2H-$MoTe_2$, we ascribe this shift to the surface photovoltage arising after the separation of the photogenerated excitons. The direction of the shift indicates that the material is a p-type semiconductor; even though the material is intrinsically n-type, it has been shown that exposure to air and the presence of adsorbates converts $MoTe_2$ into a p-type semiconductor,[26] in agreement with our observations. It was also shown that in order to restore the original n-type conductance of the material, an annealing of the crystal at 350 K in a vacuum is necessary.[26] This fact led us to conclude that the pump fluences used in our experiments, which were all conducted at room temperature, did not cause a significant temperature increase in the sample.

The shift amplitude as a function of the applied fluence is shown in the inset of Fig. 2. The trend appears to be linear, despite the expected dependency of SPV induced shifts on the laser power being logarithmic.[27] To evaluate the excited carrier density in $MoTe_2$, we should consider the saturable absorption correction, which reduces the absorption coefficient according to:

$$\alpha = \frac{\alpha_0}{1 + \frac{I}{I_S}}$$

where α and α_0 are the absorption coefficients with and without saturable absorption effects, respectively, I is the incident peak intensity and I_S is the saturation peak intensity. The latter quantity has been reported for 2H-$MoTe_2$ and a 300 fs pulse at 515 nm,[28] and amounts to ~60 GW cm^{-2}. In our experiments, the peak power at 0.7 mJ cm^{-2} fluence is ~2 GW cm^{-2}, which gives the lowest α/α_0 ratio of 0.97; we can therefore neglect saturable absorption effects in the evaluation of the excited carrier density. From the optical properties of $MoTe_2$ (ref. 29), we find a reflectivity $R = 0.44$ at normal incidence and a penetration depth of $\sim 2 \times 10^{-4}$ cm, which gives an upper limit for the excited carrier density of $\sim 5 \times 10^{18}$ cm^{-3} with an applied fluence of 0.7 mJ cm^{-2}, assuming one excited electron per absorbed photon. For photon energies larger than twice the band gap (~1 eV for bulk $MoTe_2$ (ref. 30)) quantum yields of 1.5–2 have been observed,[31] raising the upper limit of the excited carrier density to $\sim 1 \times 10^{19}$ cm^{-3} in our measurements. Considering that 2H-$MoTe_2$ has 18 valence electrons per unit cell (*i.e.*, in a volume of $\sim 1.7 \times 10^{22}$ cm^3), we get a maximum excitation density[32] of 0.01%. Moreover, if we assume a uniform absorption in the interaction volume and a monolayer thickness of 0.7 nm, we obtain an average of $\sim 7 \times 10^{11}$ cm^{-2} carriers per layer, which is lower than the Mott threshold of $\sim 1 \times 10^{13}$ cm^{-2} reported for other TMDs.[33,34] On one hand, these low numbers are in agreement with the observed SPV linear trend; on the other hand, this estimate disagrees with the requirement of a threshold concentration of carriers for the 2H-to-1T′ transition to occur,[13] since the fingerprints of both the Te vacancies and the 1T′ component in the Mo 3d region become noticeable at 0.4 mJ cm^{-2}. A plausible explanation for this unexpected behaviour is that higher concentrations of carriers are accumulated at the surface of the material after the separation of the excitons that are generated in the interaction volume, therefore locally triggering the phase transition. This mechanism has been recently observed in 2H-MoS_2, in which the transition to the 1T metallic phase is reversible.[35]

The measurements presented in Fig. 2 characterize the photoexcited system in the microsecond range, therefore they define the SPV corrected position of the core levels acquired in the sub-nanosecond XPS measurement. The concept is summarized in Fig. 3. The static XPS spectrum acquired without optical excitation (a) serves as a reference for the absolute binding energy positions of the core levels. The measurement has then been repeated while illuminating the sample with increasing optical fluence, but without synchronizing the pump and probe sources (b). In these measurements, the pump-induced effects are averaged over 2.6 μs, which is the time separation between two consecutive laser pulses; as we have shown in Fig. 2, the dominant effect is the SPV shift to lower binding energies. Finally, we performed the time-resolved scans by synchronizing the laser to the hybrid pulses of the Elettra storage ring[14] and progressively varying the delay between the two (c). We observed

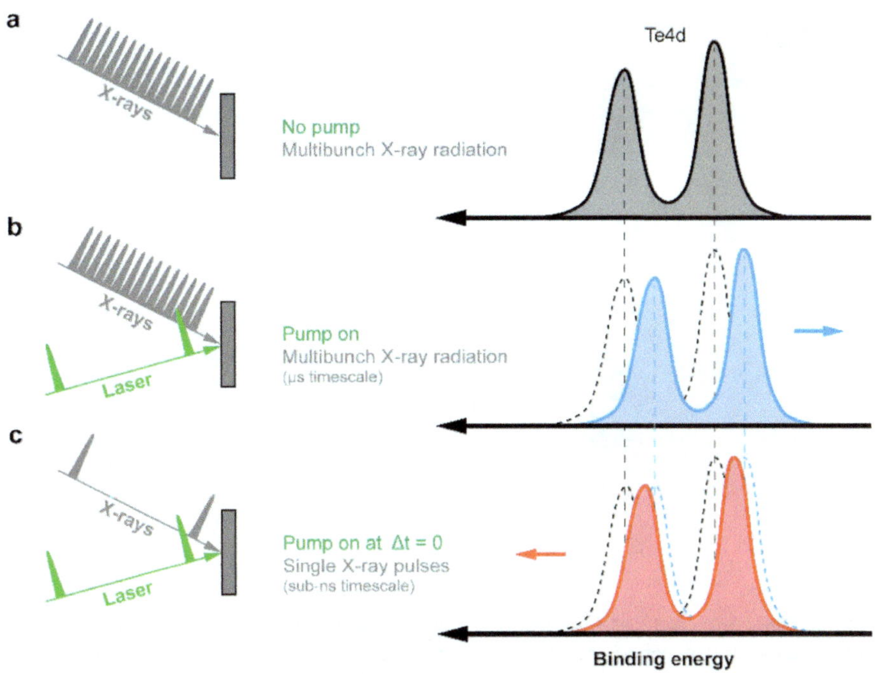

Fig. 3 Scheme of the acquisition modes. (a) The reference spectrum of the Te 4d multiplet is acquired without optical excitation (black) to define the reference binding energy of the peaks. (b) A series of spectra as a function of the applied laser fluence (blue) is acquired using the multibunch radiation as a probe, in order to detect the average optically induced modification in the system in the microsecond range. The peaks shift to lower binding energies due to the surface photovoltage, as indicated by the blue arrow. (c) The time-resolved XPS experiment is performed using the single X-ray pulses as a probe, unveiling the photo-induced dynamics in the sub-nanosecond range (red). The red arrow indicates that the spectral shift in this case is towards positive binding energies, suggesting that the origin of the effect is different from the photovoltage effect that is present at longer delay times.

that when the pump and probe pulses are temporally overlapped, the core levels shift to higher binding energies which, interestingly, is the opposite direction of the SPV effect occurring at longer delay times.

These results are shown in Fig. 4 for the Te $4d_{5/2}$ core level. In particular, the system is excited with 2.4 eV laser pulses at a 0.7 mJ cm^{-2} fluence and probed with 150 eV synchrotron pulses; the laser repetition rate being 1/3 of the synchrotron revolution frequency allows for the parallel acquisition of the photoelectron signal generated by the pumped synchrotron pulse, and a reference signal generated by the subsequent X-ray pulse, *i.e.*, ~0.9 μs after the optical pump, which is used for normalization (when the timescales of the observed effect are short enough). In Fig. 4a, we refer to the former as the "pumped" signal (red trace) and to the latter as the "unpumped" signal (blue trace). In this case, the unpumped trace is equivalent to a "laser on" spectrum at the same pump fluence, therefore it shifted to lower binding energies according to the SPV effect, while the pumped trace displays the dynamics occurring on top of the SPV on shorter timescales. In the bottom panel of Fig. 4a, we also present the false color map of

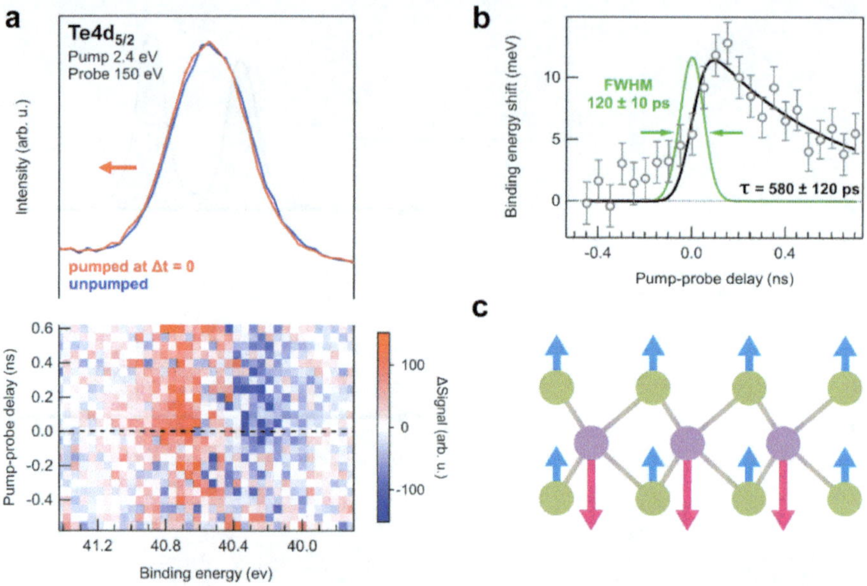

Fig. 4 Time-resolved XPS measurements on the Te $4d_{5/2}$ core level. (a) The unpumped (blue trace) and pumped (red trace) spectra are shown, revealing a rigid shift of the photoemission peak to higher binding energies. A false color map of the difference between the pumped and unpumped signal as a function of the pump–probe delay is also shown, evidencing the shift of the center of mass of the peak after the photoexcitation. (b) Te 4d binding energy shift (circles) obtained by fitting the experimental data for each delay point. The data is fitted by the convolution of a Gaussian peak and an exponential decay (black trace). The Gaussian (green trace) FWHM is held to 120 ps to account for the cross correlation between the pump and probe pulses. The exponential decay yields a time constant of 580 ± 120 ps. (c) Sketch of the A''_2-like displacements applied in the $MoTe_2$ layer to simulate the chemical shift observed experimentally. Note that the Mo displacement is twice that of Te. Purple: Mo; gold: Te.

the difference between the pumped and unpumped scans for each delay point, from which the shift of the center of mass of the Te $4d_{5/2}$ peak to higher binding energies clearly emerges. In Fig. 4b we show the binding energy shift (circles) as a function of the pump–probe delay that has been obtained by fitting the Te 4d multiplet with a Doniach–Sunjic pair. In particular, the magnitude of the shift is the difference between the binding energies of the pumped and unpumped data at each delay point, meaning that the shift is corrected for any eventual photon energy drift of the synchrotron beam that may occur over the scan acquisition time. The same procedure has been used to derive the broadening and the intensity change of the Te $4d_{5/2}$ peak, revealing no significant variations; the photo-induced effects in the sub-nanosecond range therefore consist of a rigid shift of the core levels to higher binding energies. To derive the lifetime of this effect, we fitted the data points of Fig. 4b using the convolution of and exponential decay with a Gaussian function, to account for the instrumental response function. In particular, the Gaussian FWHM of the fit is held to 120 ps since this value should correspond to the cross-correlation between the pump and probe pulses, which has been independently determined by measuring the SPV dynamics on a Si crystal under identical experimental conditions. The recovery dynamics of the binding energy shift derived from the fits are therefore well described by a single exponential decay with $\tau = 580 \pm 120$ ps.

As already mentioned, the most interesting aspect of this sub-nanosecond effect is the direction of the shift, which is opposite to the one of the SPV induced shift. In fact, if we suppose that the sub-nanosecond shift is also due to the SPV, we would expect it to be in the same direction of the shift measured in the microsecond range. Even assuming a SPV that has not completely relaxed in the time elapsed between consecutive pump pulses, we should measure a positive (in the case of a p-doped semiconductor) binding energy shift also around time zero.[36–38] We also exclude vacuum space-charge effects[39,40] because the number of secondary electrons photoemitted by the pump pulse is negligible, and because the typical fingerprint of such an effect is a positive kinetic energy shift. The nature of the shift observed in Fig. 4 must therefore be different. Considering the timescales of hundreds of picoseconds, a plausible hypothesis is that lattice related effects are involved. Ultrafast electron diffraction measurements have recently revealed how hot carriers can couple to high-energy phonons, leading to lattice displacements;[12] long-lived lattice heat effects have also been detected in transient X-ray absorption measurements,[8,9] and lattice cooling time constants in the order of 200 ps have been measured by transient optical absorption.[10]

We performed DFT calculations to examine the possibility that the sub-nanosecond shift in our measurement is indeed due to the modification of the chemical environment of the Te 4d electrons that occurs as a consequence of lattice distortion. As an example, the thermally induced lattice expansion in a Ta(100) crystal was experimentally observed to cause a ~30 meV increase in the Ta 4f binding energy.[41] In our case, we model the lattice distortion along the A''_2 phononic mode of the monolayer, displacing Mo and Te atoms in opposite directions along the *c* axis (see Fig. 4c). This is the mode dominating the distortion for photoexcitation with pump photon energies near 2.4 eV,[13] although we acknowledge that in the real system, other phononic modes may also contribute to the lattice distortion. We also underline that in our study, the DFT calculations have been performed on monolayer 2H-$MoTe_2$ and not on a bulk sample, as in the experiment. Despite the different electronic and vibrational properties of monolayer and bulk $MoTe_2$, both undergo the same 2H–1T′ structural transition and we assume that the direction of the core level binding energy shifts of the distorted structure is maintained. The results of our simulations are reported in Table 1. In detail, we observe that the calculated binding energy difference (ΔBE) of the Te 4d

Table 1 Mo and Te displacements along the *c* axis and relative binding energy variations (ΔBE) compared to the unperturbed initially optimized case

Displacement along the *c* axis (Å)		
Mo	Te	ΔBE (eV)
−0.025	+0.0125	0.03
−0.05	+0.025	0.06
−0.10	+0.05	0.11
−0.15	+0.075	0.14

electrons increases linearly with the atomic displacement, compared to the unperturbed initially optimized lattice positions. It is worth mentioning that, since our simulations are statically mimicking the A''_2 phononic mode, we consider relevant displacements up to 0.15 Å for Mo (0.075 Å for Te); above this threshold, we observe a sensitive stretch of the Mo–Te bond, which is a fingerprint of an incipient bond breaking (thus, far stronger than a phononic vibrational mode), which is not relevant for our analysis.

Our model is in agreement with the experimental findings and shows that a slight structural distortion can indeed cause a positive binding energy shift of the Te 4d level for tens of meVs; a linear fit of the data reported in Table 1 gives that a ~0.01 Å shift for Mo atoms is required to reproduce the ~10 meV shift observed in Fig. 4. In particular, this can be due to a combination of the initial state (density of states) and final state (core–hole screening) contributions, but a more thorough analysis is required to quantify the weight of both effects on the total shift observed. Interestingly, the direction of the transient core level shift reported above is in the opposite direction with respect to the position of the Te 4d line in 1T′-$MoTe_2$, as the core levels in the latter phase are found at a ~0.6 eV lower binding energy with respect to the 2H phase.[7,42] Recent calculations by Peng *et al.*[13] demonstrate that the excited carrier density threshold for the photo-induced phase transition to occur in $MoTe_2$ is in the order of 10^{14} cm^{-2}, which is more than two orders of magnitude higher than the excitation density in our experiment; moreover, the authors show that, since the 2H-to-1T′ transition is a result of phonon softening, the transition should occur within the timescales of the lattice vibration periods (<1 ps), which is too fast to be probed with our current setup. We therefore suggest that the transient modification of the photoelectron spectra reported in this paper is rather due to a lattice distortion with a ~600 ps time constant, supporting the theoretical hypothesis of the existence of intermediate phases along the 2H–1T′ transition.[4,43]

Conclusions

We have investigated the photo-induced dynamics of a 2H-$MoTe_2$ crystal by means of time-resolved X-ray photoemission spectroscopy. We have shown that the optical excitation generates a surface photovoltage shift of the photoemission spectra to lower binding energies, which persists for microseconds. When observing the sub-nanosecond dynamics of the system, we observed a transient shift of the Te 4d core level to higher binding energies instead, which we assigned to structural deformations in the out-of-plane direction; our DFT calculations show that an atomic displacement in the order of 0.01 Å is enough to cause a positive binding energy shift of tens of meVs, in agreement with the experiment. These results show that high-resolution time-resolved photoemission, combined with theoretical simulations, can provide valuable information not only on electronic and chemical modifications of photoexcited systems, but also on lattice distortions and phase transitions.

Author contributions

Conceptualization: MD; data curation: RC and MD; formal analysis: GG and MP; investigation: RC and MD; resources: FC and FS; visualization: RC; writing –

original draft: RC and MD; writing – review & editing: RC, FC, AM, GG, MP and MD.

Conflicts of interest

There are no conflicts to declare.

Acknowledgements

MD and RC acknowledge support from the SIR grant SUNDYN [Nr RBSI14G7TL, CUP B82I15000910001] of the Italian Ministry of Education University and Research MIUR. The work has been supported by EUROFEL MIUR Progetti Internazionali and PRIN 2017-FERMAT (Nr 2010KFY7XF). MP and GG acknowledge the availability of high-performance computing resources at Cineca under Iscra-B and Iscra-C initiatives. GG thanks the Dipartimento di Ingegneria Civile e Ambientale of the University of Perugia for allocated computing time within the project "Dipartimenti di Eccellenza 2018–2022". MP acknowledges support from the Physics Department Research Project TESLA 2021.

Notes and references

1 Q. H. Wang, K. Kalantar-Zadeh, A. Kis, J. N. Coleman and M. S. Strano, *Nat. Nanotechnol.*, 2012, **7**, 699–712.
2 W. Zheng, M. Bonn and H. I. Wang, *Nano Lett.*, 2020, **20**, 5807–5813.
3 A. Krishnamoorthy, L. Bassman, R. K. Kalia, A. Nakano, F. Shimojo and P. Vashishta, *Nanoscale*, 2018, **10**, 2742–2747.
4 A. V. Kolobov, P. Fons and J. Tominaga, *Phys. Rev. B*, 2016, **94**, 094114.
5 K.-A. N. Duerloo, Y. Li and E. J. Reed, *Nat. Commun.*, 2014, **5**, 4214.
6 D. Voiry, A. Mohite and M. Chhowalla, *Chem. Soc. Rev.*, 2015, **44**, 2702–2712.
7 S. Cho, S. Kim, J. H. Kim, J. Zhao, J. Seok, D. H. Keum, J. Baik, D. Choe, K. J. Chang, K. Suenaga, S. W. Kim, Y. H. Lee and H. Yang, *Science*, 2015, **349**, 625–628.
8 A. Britz, A. R. Attar, X. Zhang, H.-T. Chang, C. Nyby, A. Krishnamoorthy, S. H. Park, S. Kwon, M. Kim, D. Nordlund, S. Sainio, T. F. Heinz, S. R. Leone, A. M. Lindenberg, A. Nakano, P. Ajayan, P. Vashishta, D. Fritz, M.-F. Lin and U. Bergmann, *Struct. Dyn.*, 2021, **8**, 014501.
9 A. R. Attar, H. T. Chang, A. Britz, X. Zhang, M. F. Lin, A. Krishnamoorthy, T. Linker, D. Fritz, D. M. Neumark, R. K. Kalia, A. Nakano, P. Ajayan, P. Vashishta, U. Bergmann and S. R. Leone, *ACS Nano*, 2020, **14**, 15829–15840.
10 Z. Chi, H. Chen, Q. Zhao and Y.-X. Weng, *J. Chem. Phys.*, 2019, **151**, 114704.
11 Z. Wang, X. Li, G. Zhang, Y. Luo and J. Jiang, *ACS Appl. Mater. Interfaces*, 2017, **9**, 23309–23313.
12 A. Krishnamoorthy, M. F. Lin, X. Zhang, C. Weninger, R. Ma, A. Britz, C. S. Tiwary, V. Kochat, A. Apte, J. Yang, S. Park, R. Li, X. Shen, X. Wang, R. Kalia, A. Nakano, F. Shimojo, D. Fritz, U. Bergmann, P. Ajayan and P. Vashishta, *Nano Lett.*, 2019, **19**, 4981–4989.
13 B. Peng, H. Zhang, W. Chen, B. Hou, Z.-J. Qiu, H. Shao, H. Zhu, B. Monserrat, D. Fu, H. Weng and C. M. Soukoulis, *npj 2D Mater. Appl.*, 2020, **4**, 14.

14 R. Costantini, M. Stredansky, D. Cvetko, G. Kladnik, A. Verdini, P. Sigalotti, F. Cilento, F. Salvador, A. De Luisa, D. Benedetti, L. Floreano, A. Morgante, A. Cossaro and M. DellAngela, *J. Electron Spectrosc. Relat. Phenom.*, 2018, **229**, 7–12.

15 R. Costantini, R. Faber, A. Cossaro, L. Floreano, A. Verdini, C. Hättig, A. Morgante, S. Coriani and M. DellAngela, *Commun. Phys.*, 2019, **2**, 56.

16 G. Kresse and J. Hafner, *Phys. Rev. B: Condens. Matter Mater. Phys.*, 1993, **48**, 13115–13118.

17 G. Kresse and J. Furthmüller, *Phys. Rev. B: Condens. Matter Mater. Phys.*, 1996, **54**, 11169–11186.

18 G. Kresse and J. Hafner, *Phys. Rev. B: Condens. Matter Mater. Phys.*, 1994, **49**, 14251–14269.

19 G. Kresse and J. Furthmüller, *Comput. Mater. Sci.*, 1996, **6**, 15–50.

20 J. P. Perdew, K. Burke and M. Ernzerhof, *Phys. Rev. Lett.*, 1996, **77**, 3865–3868.

21 P. E. Blöchl, *Phys. Rev. B: Condens. Matter Mater. Phys.*, 1994, **50**, 17953–17979.

22 S. Grimme, J. Antony, S. Ehrlich and H. Krieg, *J. Chem. Phys.*, 2010, **132**, 154104.

23 L. Köhler and G. Kresse, *Phys. Rev. B: Condens. Matter Mater. Phys.*, 2004, **70**, 165405.

24 Y. Tan, F. Luo, M. Zhu, X. Xu, Y. Ye, B. Li, G. Wang, W. Luo, X. Zheng, N. Wu, Y. Yu, S. Qin and X.-A. Zhang, *Nanoscale*, 2018, **10**, 19964–19971.

25 C. Si, D. Choe, W. Xie, H. Wang, Z. Sun, J. Bang and S. Zhang, *Nano Lett.*, 2019, **19**, 3612–3617.

26 J. Liu, Y. Wang, X. Xiao, K. Zhang, N. Guo, Y. Jia, S. Zhou, Y. Wu, Q. Li and L. Xiao, *Nanoscale Res. Lett.*, 2018, **13**, 291.

27 S. Tanaka, S. D. More, J. Murakami, M. Itoh, Y. Fujii and M. Kamada, *Phys. Rev. B: Condens. Matter Mater. Phys.*, 2001, **64**, 155308.

28 K. Wang, Y. Feng, C. Chang, J. Zhan, C. Wang, Q. Zhao, J. N. Coleman, L. Zhang, W. J. Blau and J. Wang, *Nanoscale*, 2014, **6**, 10530–10535.

29 A. R. Beal and H. P. Hughes, *J. Phys. C: Solid State Phys.*, 1979, **12**, 881–890.

30 A. J. Grant, T. M. Griffiths, G. D. Pitt and A. D. Yoffe, *J. Phys. C: Solid State Phys.*, 1975, **8**, L17–L23.

31 J.-H. Kim, M. R. Bergren, J. C. Park, S. Adhikari, M. Lorke, T. Frauenheim, D.-H. Choe, B. Kim, H. Choi, T. Gregorkiewicz and Y. H. Lee, *Nat. Commun.*, 2019, **10**, 5488.

32 R. Costantini, A. Morgante and M. DellAngela, *J. Electron Spectrosc. Relat. Phenom.*, 2022, **254**, 147141.

33 B. Radisavljevic and A. Kis, *Nat. Mater.*, 2013, **12**, 815–820.

34 A. Chernikov, C. Ruppert, H. M. Hill, A. F. Rigosi and T. F. Heinz, *Nat. Photonics*, 2015, **9**, 466–470.

35 N. L. A. N. Sorgenfrei, E. Giangrisostomi, R. M. Jay, D. Kühn, S. Neppl, R. Ovsyannikov, H. Sezen, S. Svensson and A. Föhlisch, *Adv. Mater.*, 2021, **33**, 2006957.

36 S. Tokudomi, J. Azuma, K. Takahashi and M. Kamada, *J. Phys. Soc. Jpn.*, 2008, **77**, 014711.

37 D. Bröcker, T. Gießel and W. Widdra, *Chem. Phys.*, 2004, **299**, 247–251.

38 A. Shavorskiy, S. Neppl, D. S. Slaughter, J. P. Cryan, K. R. Siefermann, F. Weise, M.-F. Lin, C. Bacellar, M. P. Ziemkiewicz, I. Zegkinoglou, M. W. Fraund, C. Khurmi, M. P. Hertlein, T. W. Wright, N. Huse, R. W. Schoenlein,

T. Tyliszczak, G. Coslovich, J. Robinson, R. A. Kaindl, B. S. Rude, A. Ölsner, S. Mähl, H. Bluhm and O. Gessner, *Rev. Sci. Instrum.*, 2014, **85**, 093102.

39 S. Hellmann, K. Rossnagel, M. Marczynski-Bühlow and L. Kipp, *Phys. Rev. B: Condens. Matter Mater. Phys.*, 2009, **79**, 035402.

40 M. DellAngela, T. Anniyev, M. Beye, R. Coffee, A. Föhlisch, J. Gladh, S. Kaya, T. Katayama, O. Krupin, A. Nilsson, D. Nordlund, W. F. Schlotter, J. A. Sellberg, F. Sorgenfrei, J. J. Turner, H. Öström, H. Ogasawara, M. Wolf and W. Wurth, *Struct. Dyn.*, 2015, **2**, 025101.

41 D. Riffe, W. Hale, B. Kim and J. Erskine, *Phys. Rev. B: Condens. Matter Mater. Phys.*, 1996, **54**, 17118–17122.

42 J. R. Shallenberger, R. Katz and Z. Mao, *Surf. Sci. Spectra*, 2021, **28**, 024001.

43 H. H. Huang, X. Fan, D. J. Singh, H. Chen, Q. Jiang and W. T. Zheng, *Phys. Chem. Chem. Phys.*, 2016, **18**, 4086–4094.

Faraday Discussions

ROYAL SOCIETY OF CHEMISTRY

PAPER

Surface photovoltage dynamics at passivated silicon surfaces: influence of substrate doping and surface termination†

Debora Pierucci, ‡[ab] Mathieu G. Silly, *[a] Heloise Tissot, §[ab] Philippe Hollander,[a] Fausto Sirotti ¶[a] and Francois Rochet [b]

Received 29th November 2021, Accepted 25th February 2022

DOI: 10.1039/d1fd00107h

We have monitored the temporal evolution of the band bending at controlled silicon surfaces after a fs laser pump excitation. Time-resolved surface photo-voltage (SPV) experiments were performed using time resolved photoemission spectroscopy with time resolution of about 30 ns. To disentangle the influence of doping and surface termination on SPV dynamics, we compare the results obtained on two surface terminations: the water saturated (H,OH)–Si(001) surface and the thermally oxidized Si(001) one. The SPV dynamics were explored as a function of laser fluence and as a function of time for the two surface terminations at given doping levels. The return to equilibrium involves a characteristic time in the 0.1 μs to 10 μs range, depending on the surface termination and bulk doping. Exploring several laser fluences, different SPV regimes were found for the two surface terminations at given doping levels. For low laser fluence the SPV dynamic follows the commonly accepted thermionic model. At higher fluence, the SPV signal reaches a saturation value, and if the fluence is further increased, the decay time of the SPV increases and can no longer be explained by a thermionic model alone.

I. Introduction

Since the 40s and the discovery of the silicon based p–n junction, junction transistor and integrated circuit, the silicon semiconductor has a dominant

[a]Synchrotron SOLEIL, L'Orme des Merisiers, Saint-Aubin, BP 48, F-91192 Gif-sur-Yvette, France. E-mail: Mathieu.Silly@Synchrotron-Soleil.fr

[b]Sorbonne Universités, UPMC Univ. Paris 06, and CNRS UMR 7614, Laboratoire de Chimie Physique Matière et Rayonnement (LCPMR), F-75005, Paris, France

† Electronic supplementary information (ESI) available. See DOI: 10.1039/d1fd00107h

‡ Current address: Université Paris-Saclay, CNRS, Centre de Nanosciences et de Nanotechnologies, 91120, Palaiseau, Paris, France.

§ Current address: Aix-Marseille Univ, CNRS, CINaM, Marseille, France.

¶ Current address: Physique de la Matière Condensée, CNRS and Ecole Polytechnique, IP Paris, F-91128 Palaiseau, France.

position in microelectronics.[1] Silicon has established itself as one of the major semiconducting materials thanks to its high structural quality with suitable electronic properties such as band gap (1.12 eV), mobility, capability to be n or p doped and passivated by forming an insulating SiO_2 layer. Silicon also holds high interest in photonics[2] as a photodetector and light emitter, photocatalysis,[3] and solar cell applications.[4] Silicon based solar cells with an efficiency up to 26.7% represented about 95% of the worldwide production in 2020.[5] Minority charge carrier lifetime is a key parameter for device efficiency.[6] Surface photovoltage is a powerful technique as a non-destructive and contact-less technique to determine the minority charge carrier lifetime.[7] Transient surface photovoltage measurement based on time resolved photoemission spectroscopy (TR-PES) has been successfully used to characterize various materials and to combine the determination of the chemistry, band alignment and charge carrier lifetime. TR-PES has provided evidence of different excitation and relaxation processes depending on the nature of the substrate,[8] but also the influence of surface termination on the deexcitation pathway.[9] Even if minority charge carrier lifetime in semiconductors plays an important role, the parameters controlling it remain mainly unexplored. Band alignment and charge transfer at the surface and interface involve different parameters intimately imbricated.

At semiconductor surfaces and interfaces, a spatially extended space-charge layer (SCL) may be present. This behavior distinguishes semiconductors from metals, and it is due to the low densities of free carriers in non-degenerately doped semiconductors. This results in large screening lengths, and in the existence of electronic surface and interface states within the band gap that may become charged.

A semiconductor surface that possesses electronic surface states usually represents a perturbation in the local charge balance. Depending on the type of surface states (donor or acceptor) and on the position of the Fermi level at the surface, the surface states may carry charges. The position of the Fermi level at the surface is determined by the condition of charge neutrality where the charge at the surface Q_{ss} (a charge per unit area) is compensated by an opposite charge inside the semiconductor.

Electronic states within the band gap determine the position of the Fermi level at semiconductor surfaces/interfaces. The charge transfer to the surface leads to a band bending (qV_{BB}) within a space SCL, whose thickness ranges typically from nm to hundreds of nm, thickness depending on the doping level. The band bending direction depends on the bulk doping and is downward for a p-type sample and upward for an n-type (see Fig. 1).

The band bending, in absolute value, can be diminished or even nullified after adsorption of photons with an energy greater than E_g. In the case of silicon for a light of $\lambda = 800$ nm ($h\nu = 1.55$ eV) the absorption depth is around 10 μm,[11,12] electron–hole pairs are created not only near the surface, but also into the bulk of the sample. In most of the bulk there is no electric field, so that the concentration of electrons is the same as the concentration of holes. In the SCL, however, the electric field separates the electron–hole pairs. On p-type samples (with downward band banding), the electric field drives the electrons (minority carrier) toward the surface and the holes toward the bulk. Just the opposite occurs for an n-type sample. As a result, there will be a non-equilibrium accumulation of electrons at the surface of a p-type sample, and a depletion of electrons at the

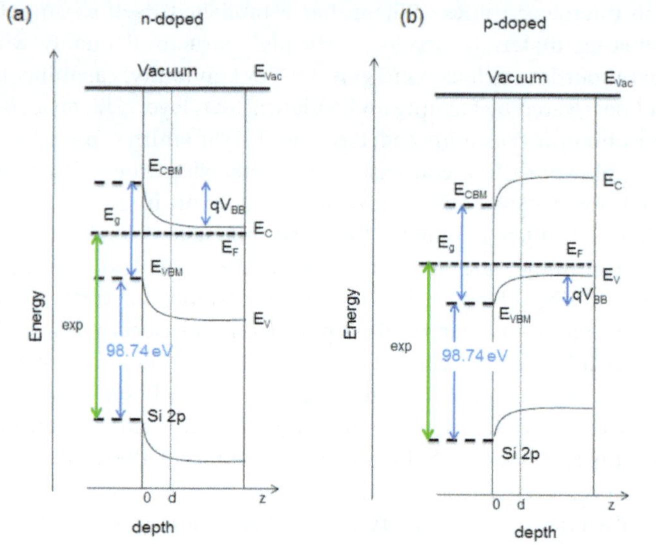

Fig. 1 Schematic view of band bending in (a) n-doped and (b) p-doped silicon as a function of the depth z. Through XPS we can measure the Si 2p binding energy (BE). The distance between the Si $2p_{3/2}$ and the valence band maximum (VBM) is fixed[10] and equal to 98.74 eV. Knowing the silicon band gap (1.12 eV) we can calculate the band bending qV_{BB} and the surface charge density.

surface of an n-type sample. The accumulation of carriers of one charge and the depletion of those of the other charge in the near surface region generates a voltage, referred to as surface photo-voltage (SPV), lowering the band bending at the surface. Equilibrium is restored by transport of majority carrier from the bulk to the surface *via* thermionic emission over the surface potential barrier.[13–15]

The SPV effect on semiconductor surface has been attracting considerable interest in surface science, for a review see Kronik and Shapira.[7]

Besides electrical methods, photoemission spectroscopy is also one of the most powerful tools to study the SPV effect, especially synchrotron radiation core-level photoemission spectroscopy (measured in surface sensitive conditions) combined with pulsed-laser illumination that creates photo-excited carriers. The time profile of the SPV effect can be directly obtained following the core level energy shift under laser excitation. After the pioneering work of Long *et al.*,[16] several time-resolved core level photoemission studies combining synchrotron radiation with pulsed-laser,[15,17,18] including free-electron laser,[19,20] have been performed to study the SPV effect on clean and oxidized Si(100)-2 × 1 and Si(111)-7 × 7 surfaces. From these studies, it was deduced that the SPV dynamics are influenced by various parameters, depending on bulk and surface properties of Si.

The most important bulk parameter is the Fermi level position at the surface/interface, which determines the band bending at the surface. From the existing literature, it was still unclear how the time scale of the SPV phenomenon is related to the different bulk (*e.g.* doping level) and surface parameter (*e.g.* surface termination). To disentangle the influence of surface termination and bulk doping on SPV dynamics, a systematic study for four different doping levels [from highly doped ($N_{D/A} \sim 10^{19}$ cm^{-3}) to low doped ($N_{D/A} \sim 10^{14}$ cm^{-3}) n (phosphorus) and p (boron) types], and two surface terminations [$Si(001)/SiO_2$ interface and water-terminated Si surface ((H,OH)–Si(001)-2 × 1)] was performed. These two surface terminations exhibit similar defects responsible for the positioning of the Fermi level at the surface: a trivalent silicon atom in the $^{\bullet}Si{\equiv}Si_3$ configuration. This "dangling bond" defect is called P_b in the case of $Si(001)/SiO_2$ (ref. 21 and 22) and isolated dangling bonds (IDBs) in the case of the water-reacted surface.[23,24] These defects are responsible for the presence of gap states at the $Si(001)/SiO_2$ and (H,OH)–Si(001)-2 × 1 surfaces and as a consequence of a residual surface band bending $|qV_{BB}|$. Before the SPV dynamics study these two surfaces were well characterized (see Section III).

The SPV process can be schematized in three steps (Fig. 2). The first step is the generation of carriers by photo-excitation (Fig. 2(a)). In the case of Si we have an energy gap of 1.12 eV. If we excite the system with a photon energy of 1.55 eV (λ = 800 nm), the main process is the "band-to-band adsorption". Valence electrons are excited into the conduction band (super-band gap excitation). The penetration depth in silicon for this wavelength is about 10 µm.[11,12] This ensures a uniform excitation in the SCL (for high doping the SCL is ~5 nm and is ~700 nm for low doping, see Table 1).

The second step is the separation of the photo-excited carriers in the subsurface electric field. In the case of n-type Si (Fig. 2) the electric field drives the holes toward the surface and the electron in the bulk (Fig. 2(a)). In the steady-state, an excess of positive charge builds up at the surface (Fig. 2(b)), producing a potential difference (the SPV = $|qV_{BB} - qV|$ where qV_{BB} is the band bending in the dark and qV is the new band bending under illumination at a time t) between the surface and the bulk, which reduces the band bending.

The third step is the restoration of the equilibrium by transport of electron from the bulk to the surface through a thermionic emission over the surface potential barrier (like the case of Schottky barrier in metal–semiconductor contacts[13]).

The time scales of the first and the second steps are in the ps–ns. The third step is characterized by a much longer time scale ranging from ns to µs. We can easily understand that the SPV decay is not a simple exponential. The potential barrier at the surface is itself a function of time. The decay rate is highest just after the

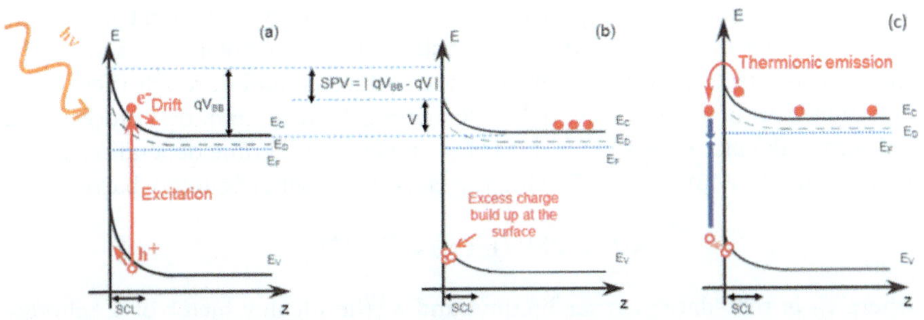

Fig. 2 Evolution of the surface photo-voltage effect from excitation (a) to excess charge build up (b) and the following relaxation through (c) the thermionic effect. E_C and E_V are the energy of the conduction and valence band edge in the bulk, respectively. E_D donor level, E_F Fermi level. Figure adapted from ref. 25.

laser pulse, when the bands have been flattened: bulk majority carriers can be easily emitted thermionically to the surface and recombine with the excess surface charge. At later times when the band bending has been restored the barrier is higher and the characteristic decay time becomes larger, ultimately reaching the dark value of τ_D, which is the lifetime of charge carriers for SPV = 0. It is said that the SPV decelerates continuously.

As a consequence, following Long *et al.*,[16] Hamers *et al.*[14,26,27] and Bröcker *et al.*[15] the amount of charge carriers at the surface is connected to SPV by the expression (valid for $kT \ll \mathrm{SPV} \ll |qV_{BB}|$):

$$\frac{\mathrm{SPV}}{kT}\mathrm{e}^{\frac{\mathrm{SPV}}{k_B T}} = \frac{n_p}{n_0}\mathrm{e}^{|qV_{BB}|/k_B T} \tag{1}$$

where n_0 is the doping carrier concentration, n_p is the photoexcited carrier concentration $\left(\frac{n_p}{n_0} = \Delta_p\right)$, qV_{BB} is the equilibrium band bending, k_B is the Boltzmann constant and T the temperature ($k_BT = 25$ meV for $T = 300$ K).

Eqn (1) can also be read as a formula connecting laser intensity and SPV if the laser intensity is assumed to be proportional to the excited carrier density. In particular in the limit of $kT \ll \mathrm{SPV} \ll |qV_{BB}|$) the laser intensity dependence of SPV can be approximated by:

$$\mathrm{SPV} = A\ln\left(1 + \frac{n_p}{n_0}\right) \tag{2}$$

Since n_p is proportional to laser intensity,[15,26,27] eqn (2) can be rewritten as:

$$\mathrm{SPV}(I) = A\ln(1 + \gamma I) \tag{3}$$

where $A = \alpha k_B T$. α is a proportionality factor and has been interpreted to be equivalent to the ideality factor in a Schottky diode. Following this model α is 1 for a pure Schottky barrier. Depending on the interface properties α can take values[7] between 0.5 and 2, but larger values are found in the literature: $\alpha = 3.5 \pm 1.5$ (Ogawa *et al.*[25]) and $\alpha = 9.7$ (Pietzsch *et al.*[28]). γ is another material parameter. Eqn (3) is valid only if n_p is small compared to the total carrier density n_0 (low laser intensity). At much higher intensities a saturation of the SPV signal is observed[15] (even though the flat band condition is not reached) and SPV becomes nearly independent of intensity.

However, the excited charged carriers will recombine. As shown in Fig. 2(c), the electrons have to approach the holes and overcome a potential barrier. The transport of electrons from the bulk to the surface is done through thermionic emission over the surface potential barrier where the recombination process with holes can take place. This process leads to an electron lifetime depending on the surface potential (SPV)[14,15] (following the Hecht's model of Schottky barrier[13]):

$$\tau(t) = \tau(\mathrm{SPV}) = \tau_D\,\mathrm{e}^{-\mathrm{SPV}/\alpha' k_B T} \tag{4}$$

where τ_D is the "dark" carrier lifetime and α' the ideality factor of a Schottky diode. The SPV dependent decay time leads to a decay equation for the surface charge given by:[15]

$$\frac{d\Delta_p}{dt} = -\frac{\Delta_p}{\tau(\mathrm{SPV})} \tag{5}$$

Combining eqn (1) and (5) and assuming $\alpha' = \alpha$ the differential equation is solved by:[15]

$$\mathrm{SPV}(t) = -\alpha kT \ln(1 - (1 - e^{-\mathrm{SPV}_0/\alpha k_B T})e^{-t/\tau_D}) \tag{6}$$

where SPV_0 is the maximum value of SPV. Using eqn (6) to fit the experimental data, the relevant time scale τ_D was determined for the different Si doping and for the two surface terminations.

II. Experimental

The doping and surface termination dependence of the surface photo-voltage (SPV) dynamics was studied using two different doping species, boron and phosphorus:

- n doped with an N_D (phosphorus) concentration of 8.2×10^{14} cm^{-3} ($5.5\ \Omega \times$ cm) and 2.2×10^{19} cm^{-3} ($0.003\ \Omega \times$ cm).
- and p doped with an N_A (boron) concentration of 7.5×10^{14} cm^{-3} ($18\ \Omega \times$ cm) and 4×10^{19} cm^{-3} ($0.003\ \Omega \times$ cm).

And two different Si surface terminations, (H,OH)–Si(001)-2 $\times$ 1 surface and thermally oxidized Si(001). The steady states of the samples are fully characterized in dark conditions using high resolution photoemission spectroscopy. From the measurement of the binding energy of the Si 2p core level we can measure the Fermi level position ($E_F - E_{VBM}$) and the band bending qV_{BB} (upward in the case of phosphorus-doped sample and downward in the case of boron-doped samples) at the surface (and from the band bending the width d of the depletion region as $= \sqrt{\frac{2\varepsilon\varepsilon_0|V_{BB}|}{qN_D}}$, where ε is relative permittivity (for Si $\varepsilon \sim 12$) and ε_0 the vacuum permittivity) as given in eqn (7) and (8) respectively:

$$(E_F - E_{VBM}) = \mathrm{BE}_{\mathrm{Si2p}_{3/2}} - 98.74\ \mathrm{eV} \tag{7}$$

where 98.74 is energy distance between the valence band maximum (VBM) and the Si $2p_{3/2}$.[29]

$$qV_{BB} = (E_F - E_V) - (E_F - E_{VBM}) \tag{8}$$

where $(E_F - E_V)$ is calculated as $E_F - E_V = kT\ln\left(\frac{N_V}{N_A}\right)$ (where the effective density of states in the valence band is $N_V = 1.04 \times 10^{19}$ cm^{-3} and N_A is the density of acceptor). We recall that E_F is defined only at thermodynamic equilibrium.

It is well known that the synchrotron radiation can induce SPV on Si sample and this phenomenon increases as the temperature decreases.[30–32] For this reason, we performed test experiments by varying the X-ray flux (for a fixed photon energy we detuned the undulator gap). For all the samples, with an X-ray spot of 300 μm $\times$ 200 μm and a photon energy of 150 eV, it has been checked that no SPV

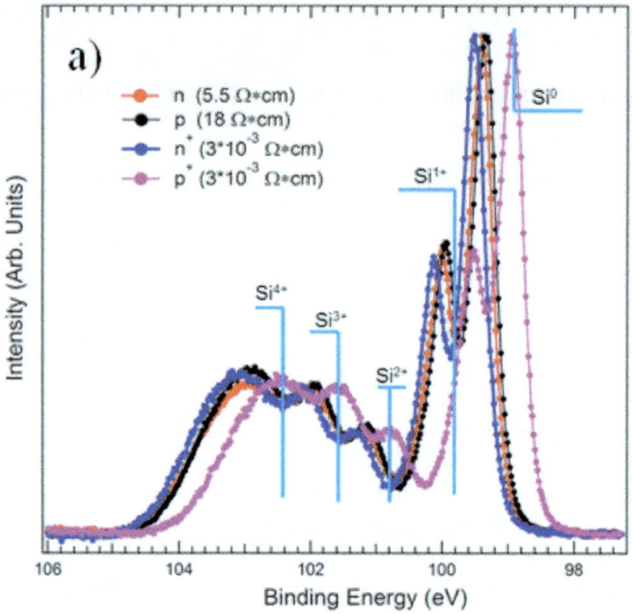

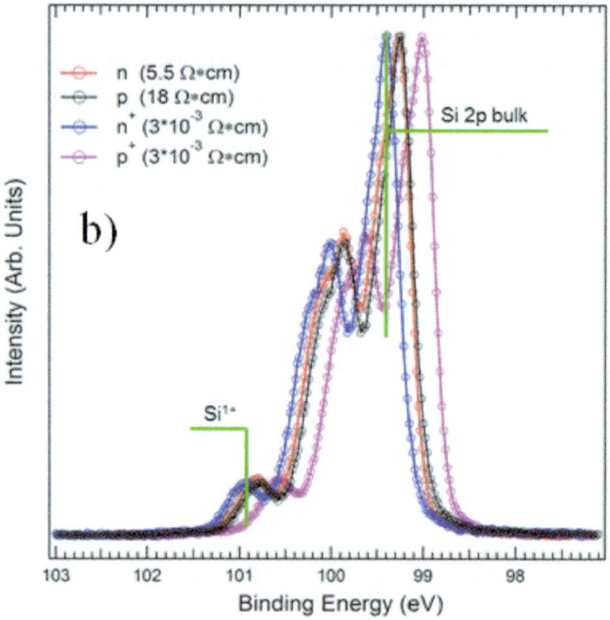

Fig. 3 (a) XPS Si 2p spectrum ($h\nu = 150$ eV) of the Si(001)/SiO_2 interface for different silicon doping: pink curve p^+ sample ($N_A = 4 \times 10^{19}$ cm^{-3} and $\rho = 3 \times 10^{-3}$ $\Omega \times$ cm), black curve p-sample ($N_A = 7.5 \times 10^{14}$ cm^{-3} and $\rho = 18$ $\Omega \times$ cm), blue curve n^+ sample ($N_D = 2.25 \times 10^{19}$ cm^{-3} and $\rho = 3 \times 10^{-3}$ $\Omega \times$ cm), red curve n sample ($N_D = 8.20 \times 10^{14}$ cm^{-3} and $\rho = 5.5$ $\Omega \times$ cm). At Si(001)/SiO_2 interface the oxidation is incomplete, instead of having only Si^{4+} as in SiO_2, at the interface Si is also found in its Si^{1+}, Si^{2+} and Si^{3+} oxidation states depending on the O atoms surrounding the Si atom. Equivalent degrees of oxidation are present for all the samples. (b) XPS Si 2p spectrum ($h\nu = 150$ eV) of the water-saturated Si(001)-2 × 1 surface for different silicon doping: pink curve p^+ sample ($N_A = 4 \times 10^{19}$ cm^{-3} and $\rho = 3 \times 10^{-3}$ $\Omega \times$ cm), black curve p-sample ($N_A = 7.5 \times 10^{14}$ cm^{-3} and $\rho = 18$ Ω effect was induced by X-rays. These experimental conditions were kept for all SPV experiments.

Photoelectron spectroscopy experiments were performed at the TEMPO beamline[33] at SOLEIL synchrotron radiation source. The time-resolved data were obtained using stroboscopic pump (laser) and probe (synchrotron radiation (SR)) method. The laser pulse duration was about 60 fs. The pumping laser was set to provide a photon energy ($h\nu$) of 1.51 eV with a repetition rate varying between 14 and 94 kHz. Details of pump probe experiments are given in ref. 34. The data acquisition system is described in detail in Bergeard *et al.*[35]

All the important experimental parameters characterizing the samples are collected in Table 1 for the two surface terminations Si(001)/SiO_2 and (H,OH)–Si(001)-2 $\times$ 1.

III. Results

In the case of thermal oxide Si(001)/SiO_2 (Fig. 3(a)), we see the Si bulk contribution, three suboxides Si^{1+}, Si^{2+}, Si^{3+} and the SiO_2 component Si^{4+}. In the notation Si^{n+}, n denotes the number of oxygen ligands around the silicon atom. Note the additivity of the chemical shifts (about 0.9 eV per ligand).[36,37] For water saturated (H,OH)–Si(001) (Fig. 3(b)) only the first oxidation state is present (Si^{1+}). The Si^{1+} component presents a surface core-level shift (SCLS) of $\sim$ +0.9 eV from the bulk line (Si 2p bulk).[24] We note the absence of the higher oxidation states. The doping changes the density and sign of the charge present at the interface/surface. The Fermi level position and the band bending are given in Table 1.

The first part of the study is focused on the third step of the SPV dynamics (Fig. 2(c)): the restoration of the equilibrium by transport of charges (electron or hole) from the bulk to the surface through a thermionic emission over the surface potential barrier. As shown in previous studies by Bröcker *et al.*[15] this process covers a time scale of tens of microseconds. For each sample, the measuring frequency is chosen in order to recover the equilibrium condition before the arrival of the next laser excitation. The sampling time is chosen in relation to the measured time interval and is set between 10 and 30 ns. To study the SPV dynamics, the evolution of the position in energy of the Si 2p core level spectra is collected under laser illumination as a function of time with the electron detection in fixed mode: one photoemission spectrum is obtained in one single shot from the detector image. The measurement is done at photon energy of 150 eV in order to be highly sensitive to the surface (analyzer pass energy PE = 50 eV), for a laser energy of 1.55 eV (λ = 800 nm, super-band gap excitation).

In Fig. 4, the Si 2p photoemission intensity measured as a function of the time between two laser pulses is presented in a color scale. When the laser pulse hits the surface, a sudden variation in Si 2p kinetic energy happens due to the SPV (pink curve in Fig. 4). The system relaxes back to its steady state (purple curve in Fig. 4).

$\times$ cm), blue curve n^+ sample ($N_D = 2.25 \times 10^{19}$ cm^{-3} and $\rho = 3 \times 10^{-3}$ $\Omega \times$ cm), red curve n sample ($N_D = 8.20 \times 10^{14}$ cm^{-3} and $\rho = 5.5$ $\Omega \times$ cm). Only the first degree of oxidation of Si is present corresponding to Si–OH species.

Table 1 Summary of the properties as a function of the doping. (H,OH)–Si(001)-2 × 1 and Si(001)/SiO_2 characteristic parameters (columns) obtained for the four substrates' characteristics (lines). The substrates are defined by the resistivity (Ω × cm), doping (cm^{-3}) calculated from the resistivity,[11] ($E_F - E_V$) calculated in the Maxwell–Boltzmann approximation equation. For each surface termination Si $2p_{3/2surf}$ is measured with photoemission, ($E_F - E_{VBM}$) calculated using eqn (7), qV_{BB} (eV) (upward in the case of phosphorus-doped sample and downward in the case of boron-doped sample) calculated from eqn (8), depletion region (nm) calculated from equation $d = \sqrt{\frac{2\varepsilon\varepsilon_0|V_{BB}|}{qN_D}}$

Substrate			Surface →	Si(001)/SiO_2				H,OH–Si(001)-2 × 1			
Type	ρ (Ω × cm)	Doping (cm^{-3})	Calc. $E_F - E_{VB}$ (eV)	Si $2p_{3/2}$ exp (eV)	Exp. $E_F - E_{VB}$ (eV)	qV_{BB} (eV)	Depl. region (nm)	Si $2p_{3/2}$ exp (eV)	Exp. $E_F - E_{VB}$ (eV)	qV_{BB} (eV)	Depl. region (nm)
n^+	0.003	2.25×10^{19}	1.12	99.53	0.79	+0.33	4.4	99.41	0.67	+0.45	5.1
n	5.5	8.20×10^{14}	0.85	99.36	0.62	+0.23	607	99.27	0.53	+0.32	716
p	18	7.5×10^{14}	0.25	99.34	0.60	−0.35	783	99.26	0.52	−0.27	688
p^+	0.003	4×10^{19}	0	99.0	0.26	−0.26	3	99.02	0.28	−0.28	3

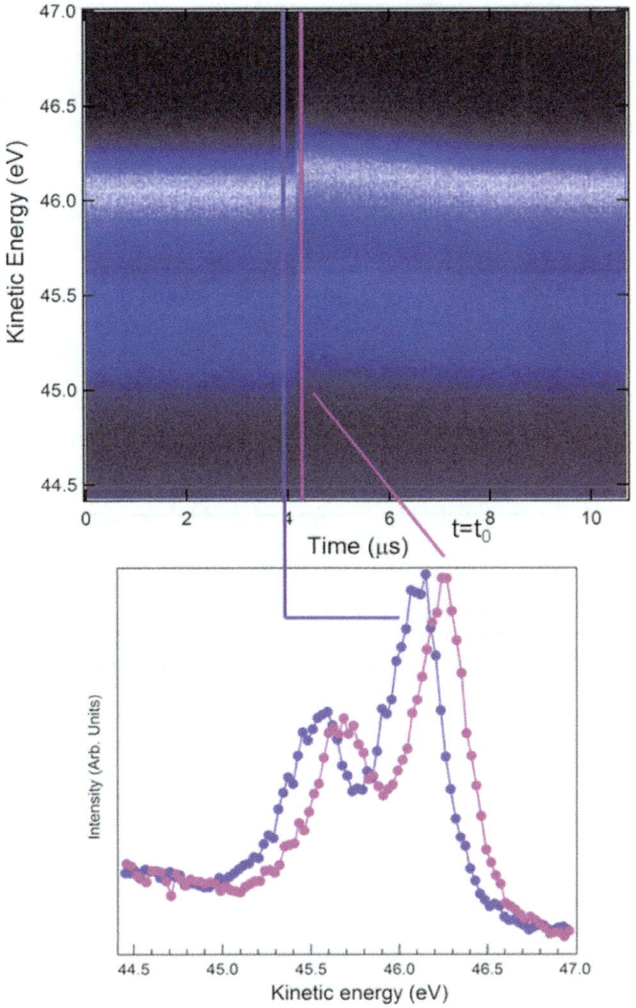

Fig. 4 Temporal evolution of Si 2p core level (Si/SiO_2 p-sample $N_A = 7.5 \times 10^{14}$ cm^{-3} (18 $\Omega \times$ cm)) as a function of time (10 μs time window). Photon energy of 150 eV and laser energy of 1.55 eV with a repetition rate of $f \sim 94$ kHz. Pink curve: energy shift of the Si 2p spectrum due to SPV effect, purple curve: Si 2p spectrum in unperturbed state. Acquisition time for each spectrum 10 ns.

In the example shown in Fig. 4, the temporal evolution of the Si 2p core level is measured for 10 μs ($f \sim 94$ kHz) with an acquisition time of 10 ns per spectrum. The reversibility of the dynamic process ensures the possibility to repeat the acquisition until sufficient statistics for the experiment were obtained. The time resolution is limited by the time of flight spread of the electrons having the same kinetic energy but following different trajectories in the analyzer. This time spread depends on the kinetic energy of the photoelectron and on the pass energy of the analyzer. During the experiment, a PE = 50 eV is used with a kinetic energy window centered at around 50 eV. The FWHM of the spread of the electrons in these conditions is 30 ns.

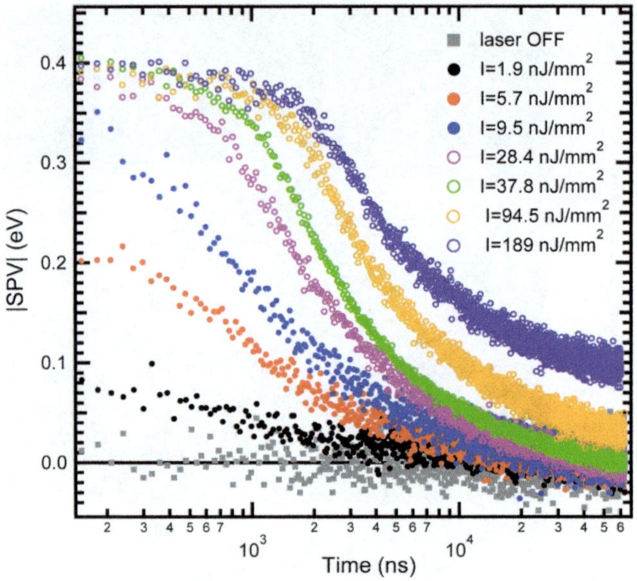

Fig. 5 n-doped (H,OH)–Si(001)-2 × 1 surface, absolute value of the SPV shift as a function of time for seven different laser fluences. For each curve, we extract the maximum of the SPV shift with respect to the value of the binding energy of Si 2p in the dark (laser OFF value set as zero of the energy scale, black line). Laser OFF baseline is determined from measurement without laser (solid square). Experimental decays are plotted in solid circles for laser fluence below the saturation of the SPV signal and in open circles above the saturation.

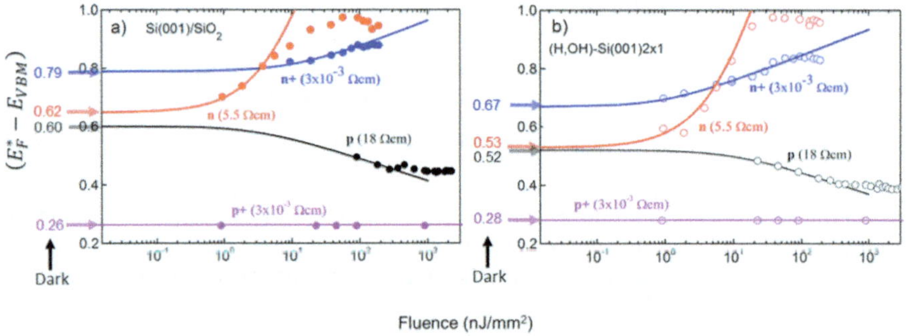

Fig. 6 SPV shift as a function of laser fluence I (nJ mm^{-2}) (a) Si(001)/SiO$_2$ sample (b) (H,OH)–Si(001)-2 × 1 for different silicon doping: pink circles p$^+$ sample ($N_A = 4 \times 10^{19}$ cm^{-3} and $\rho = 3 \times 10^{-3}\ \Omega \times$ cm), black circles p-sample ($N_A = 7.5 \times 10^{14}$ cm^{-3} and $\rho = 18\ \Omega \times$ cm), blue circles n$^+$ sample ($N_D = 2.25 \times 10^{19}$ cm^{-3} and $\rho = 3 \times 10^{-3}\ \Omega \times$ cm), red circles n sample ($N_D = 8.20 \times 10^{14}$ cm^{-3} and $\rho = 5.5\ \Omega \times$ cm). The solid-colored lines are the results of the fitting procedure using eqn (10).

An example of temporal evolution of the Si 2p binding energy for the n-doped (H,OH)–Si(001)-2 × 1 surface is shown in Fig. 5 for different laser fluences. The SPV shift is referenced with respect to the binding energy of the Si 2p in the dark.

Similar measurements are performed for the different surface termination and bulk doping (not shown here).

The maxima of the SPV signals obtained immediately after the laser excitation for all the samples are summarized in Fig. 6 as a function of laser fluence (I). In particular, the data points in Fig. 6 ((a) Si(001)/SiO_2 samples and (b) (H,OH)–Si(001)-2 × 1 samples) are the binding energy shift translated, to show them as the variation of the steady-state Fermi level E_F^* with respect to the VBM ($E_F^* - E_{VBM}$) (following eqn (7)).

The fluence scale is logarithmic; the threshold for observing a modification in the band bending is strongly dependent on the doping. The lower fluence to induce photo-voltage is obtained for the n-doped sample (1 nJ mm^{-2}). 10 nJ mm^{-2} are needed to observe a corresponding signal for the n$^+$ doping. The direction of the SPV is reversed for the p doped sample (as expected, due the opposite sign of the band bending with respect to the n-doped sample) and the binding energy shift is observed for higher fluences than those needed for the n doped samples.

The data were fitted following Long *et al.*,[16] Hamers *et al.*[14,26,27] and Bröcker *et al.*[15] in the framework of the thermionic model, with eqn (9):

$$\text{SPV}(I) = \alpha k_B T \ln(1 + \gamma I) \tag{9}$$

In the present case, $(E_F^* - E_{VBM}) = \pm\text{SPV} + c$ where c is a constant (the SPV is multiplied by +1 for n samples and −1 for p ones). Then:

$$(E_F^* - E_{VBM})(I) = \pm\alpha k_B T \ln(1 + \gamma I) + c \tag{10}$$

Constant c is the value of the Fermi level in the gap in the dark ($E_F - E_{VBM}$) (Table 1). The values of c are reported on the left axis of Fig. 6(a) and (b). The results of the fitting are presented in Fig. 6 as continuous lines having the same colours of the corresponding data points. All the measurements were taken at room temperature ($k_B T = 25$ meV).

The found α and γ values for the different surface terminations and substrate doping are collected in Table 2.

From the comparison in Fig. 6, the thermionic model well represents the variation of the Fermi level at the surface ($E_F^* - E_{VBM}$) over a wide intensity range, but for high fluence (between 10–450 nJ mm^{-2} depending on the doping of the sample) a saturation of the SPV signal is observed[15] (even if the theoretical value of flat band condition is not reached) and SPV becomes

Table 2 Values of α and γ obtained by fitting the experimental data measured at low fluence, corresponding to the thermionic relaxation process, with eqn (10), for the Si(001)/SiO_2 and (H,OH)–Si(001)-2 × 1 surfaces for three different doped substrates

	Si(001)/SiO_2		(H,OH)–Si(001)-2 × 1	
Doping type	α	γ (mm^2 nJ^{-1})	α	γ (mm^2 nJ^{-1})
n$^+$	1.5	0.09	1.5	0.97
n	13.7	0.15	13.7	0.15
p	1.27	0.30	1.27	0.10

nearly independent of intensity. For the n doped sample a reduction of the SPV signal is observed for higher laser fluence which is related to a strong modification of the time dependent SPV signal (see Fig. 8). The α parameter does not depend on the surface termination but only on the doping of the sample. α should be 1 for an ideal Schottky barrier, depending on the material properties α could take values[7] between 0.5–2. In our case only for the n^+ and p samples we found values of α of 1.5 and 1.27. In the case of n doped samples an α value of 13.7 is necessary to reproduce the observed behavior. High values are also present in the literature.[25,28] The γ (a proportional factor) value seems not to have a particular dependence on doping nor on surface termination (Bröcker *et al.*[15] found a value of 2 $mm^2\ nJ^{-1}$ and Ogawa *et al.*[25] $10 \pm 10\ cm^2\ \mu J^{-1}$). For the p^+-doped sample no SPV was observed over the full range of explored fluence. This result can come from the small dimension of the depletion region (3 nm). When the depletion region is very small the thermionic emission is not the only process responsible for recombination of carriers but tunneling recombination can also be present. A previous study on Schottky barriers[38] showed that at low dopant concentration, only recombinations due to thermionic emission play a role in the recombination. At high dopant concentration, tunneling takes over as a dominant recombination mechanism leading to a strong decrease in the SPV effect. Hecht [31] showed, for a metal–GaAs (100) interface, that for a doping level above 10^{17} cm^{-3} a complete neutralization of the SPV by the tunneling is presented. Also Mulhollan *et al.*[39] showed, in the case of p-doped GaAs, that the SPV effect decreases with the increase in doping, and goes to zero at a doping level of 5×10^{19} cm^{-3}. Ogawa *et al.*[25] have shown that the characteristic time of the tunneling is very small in the range between 0.1 ns to a maximum value of 20 ns, below the resolution time of 30 ns in the stroboscopic mode used in this study. Short decay time can explain the absence of SPV signal for the p^+ sample. On the other hand, a SPV signal is measured for the highly doped n^+ sample (with a reduced amplitude with respect to the n sample). The width of the space charge layer calculated from the band bending and the dopant concentration is 10 to 20 Å thicker for the p^+ than for n^+ (Table 1). This may be sufficient to reduce considerably the tunnel component.

It is clear from the results shown in Fig. 6 that for laser fluence lower than 450 nJ mm^{-2} the description of the SPV dynamics is in excellent agreement with the thermionic model. In Table 3, the maximum laser fluences allowed in order to describe the SPV dynamics with the thermionic model as a function of doping are summarized. From Fig. 6, it appears that the maximum fluence value is only related to the doping of the sample and not to the surface termination. For the

Table 3 Maximum laser fluence allowed in order to describe the SPV dynamics with the thermionic model

Doping label	Resistivity	Doping (cm^{-3})	Max fluence (nJ mm^{-2})
n^+	$3 \times 10^{-3}\ \Omega$ cm	$N_D = 2.25 \times 10^{19}$	94.6 nJ mm^{-2}
n	5.5 Ω cm	$N_D = 8.20 \times 10^{14}$	9.5 nJ mm^{-2}
p	18 Ω cm	$N_A = 7.5 \times 10^{14}$	450 nJ mm^{-2}

phosphorus-doped sample we see the higher the doping level, the higher the maximum laser fluence needed to saturate the SPV.

When the laser fluence exceeds the values reported in Table 3 the SPV effect saturates. In the n doped sample at laser fluence above the saturation, a modification in shape of the curves and longer decay time are observed (Fig. 5). The curves cannot be approximated by the thermionic model (eqn (10)), the system leaves the domain of validity of the thermionic model. The thermionic deexcitation process is not efficient enough to relax the total amount of photogenerated charges. It has been shown by Long *et al.*[16] that, at high fluence, bulk recombination processes become dominant.

IV. SPV analysis as a function of time

The possibility to tune the laser fluence and adapt it to the sample characteristics is indispensable to describe the observed variations in the core level binding energy shifts. To explain the amplitude of SPV *versus* fluence and describe its time dependence the thermionic model is only valid at low fluence. Consequently, we applied this model to describe the time dependent curves, when the SPV shift is less than 70% of the maximum SPV at saturation. In this case we have used eqn (11) to fit the experimental results and determine for each sample the value of τ_D, the dark carrier lifetime.

$$\mathrm{SPV}(t) = -\alpha kT\ln(1-(1-\mathrm{e}^{-\mathrm{SPV}_0/\alpha k_BT})\mathrm{e}^{-t/\tau_D}) \tag{11}$$

This fitting procedure was applied to the full set of samples. In Fig. 7, we show the comparison of the SPV energy shift as a function of time and laser fluence obtained for n-doped Si ($Si(001)/SiO_2$) and p-doped Si ($Si(001)/SiO_2$) surfaces. The time dependence of the surface photo-voltage signal measured for three laser fluences are presented in the right panel. The continuous line is obtained with eqn (11) by selecting proper values of the α and τ_D parameters.

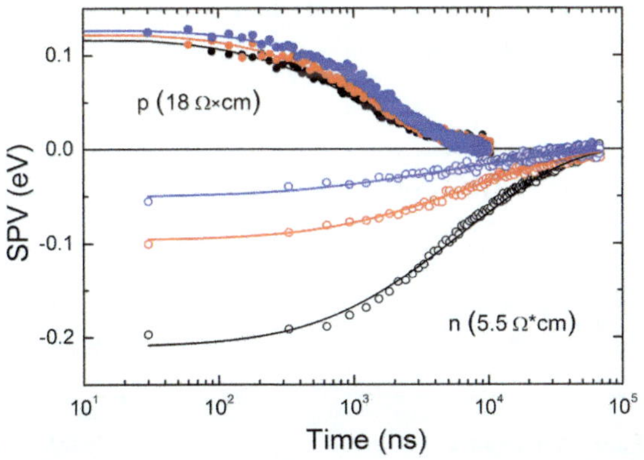

Fig. 7 SPV as a function of time (log scale) of n-doped $Si(001)/SiO_2$ ($N_D = 8.20 \times 10^{14}$ cm^{-3}, 5.5 Ω cm) and p-doped-$Si(001)/SiO_2$ ($N_A = 7.5 \times 10^{14}$ cm^{-3}, 18 Ω cm) samples; black, red and blue lines are obtained with the thermionic fit equation. Black circles $I = 0.9$ nJ mm^{-2}, red circles $I = 1.9$ nJ mm^{-2}, pink circles $I = 3.7$ nJ mm^{-2}. Empty black circles $I =$ 90 nJ mm^{-2}, empty red circles $I = 180$ nJ mm^{-2}, empty blue circles $I = 450$ nJ mm^{-2}.

All the values of α and τ_D for the different samples are reported in Fig. 8 (sample n doped $N_D = 8.20 \times 10^{14}$ cm^{-3} and $\rho = 5.5\ \Omega \times$ cm, sample n$^+$ doped $N_D = 2.25 \times 10^{19}$ cm^{-3} and $\rho = 3 \times 10^{-3}\ \Omega \times$ cm, sample p-doped $N_A = 4 \times 10^{19}$ cm^{-3} and $\rho = 18 \times 10^{-3}\ \Omega \times$ cm).

The trend for α and τ_D as a function of the laser fluence is presented in Fig. 8. When the laser fluence increases α increases and τ_D decreases. This agrees with the concept of the thermionic process: as the fluence increases the SPV increase corresponds to a reduction of the band bending at the surface. The potential barrier that the charges have to overcome in order to recombine is lowered and then the decay time becomes shorter. As a function of doping τ_D varies from tens of µs in the case of low phosphorus doped samples to hundreds of ns in the case of highly doped ones.

The reduction of τ_D with the increase of the doping is consistent with the established carrier lifetime increase with resistivity in Si.[40] It has to be noticed that in the case of the highly phosphorus doped sample the SPV is reduced with respect to the case of low doping. This is in agreement with the previous study in GaAs of Mulhollan *et al.*[39] and it is attributed to the reduction of the depletion region thickness. When the depletion region is thin the charges can tunnel through the barrier and increase the recombination efficiency. The limit is to reach the condition of no detectable SPV for high doping and very thin depletion region (as seen for the p$^+$ sample).

In agreement with Ogawa *et al.*,[25] τ_D does not show a strong dependence on surface termination for low boron doped samples (p doped). On the contrary, the effect of the surface termination is not negligible for the phosphorus doped samples, n doped ($N_D = 8.20 \times 10^{14}$ cm^{-3}) and n$^+$ doped ($N_D = 2.25 \times 10^{19}$ cm^{-3}) silicon surfaces. For the n$^+$ doped sample, the decay time observed for the thermally oxidized surface is three times faster than the decay time measured for the water terminated surface. In opposition, for the n doped sample, the decay time observed for the thermally oxidized surface is the longest observed in

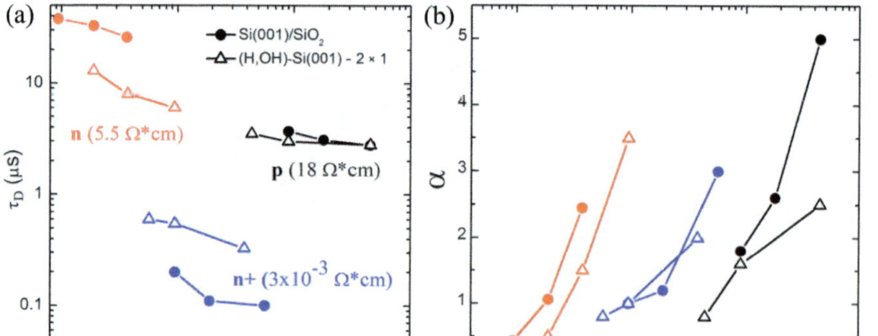

Fig. 8 Values of (a) τ_D and (b) α as a function of laser fluence (nJ mm^{-2}); colored circles Si(001)/SiO$_2$ samples, colored triangles (H,OH)–Si(001)-2 × 1 samples, for different silicon doping: black p-sample; blue n$^+$ sample; red n sample. This figure is done with small fluences. The variation of the dark SPV value is then small. The fitting curves are obtained with the thermionic model.

the experiments and is twice the decay time observed for the water terminated silicon for the same doping level. In the case of the n doped sample, this behavior can be attributed to the modification of the doping level in the vicinity of the surface because of the sample preparation procedure (the water saturated surface is obtained at RT, the silicon oxide is fabricated at a sample temperature of ~500 °C). This doping is also the one exhibiting the highest sensitivity to the incoming laser radiation: SPV is observed for laser fluence as low as 1 nJ mm^{-2}. The longest decay time is also observed for the water terminated surface. The decay time observed for phosphorus doped samples presents large discrepancies depending on the surface termination when the decay time measured for the boron doped samples seems independent of the surface termination.

V. Conclusion

The charge carriers dynamics at the Si(001)/SiO_2 and (H,OH)–Si(001)-2 × 1 surfaces have been studied as a function of Si doping by time-dependent core level spectroscopy in the ns to microsecond range. At low fluence we also give more detailed information on the SPV dynamics as a function of time and laser fluence than the previous studies.[15,17–20] The study as a function of laser fluence has allowed to define for each Si doping, the range of laser fluence in agreement with the thermionic model. The measured SPV value is well reproduced by the model and the fitting parameters can be applied to both surface terminations. The threshold fluence values are dependent on the Si doping level: the thermionic model is valid only up to 10 nJ mm^{-2} for the n doped sample (5.5 Ω × cm) and to 500 nJ mm^{-2} for the p doped one (18 Ω × cm). This thermionic fluence range is followed at higher fluence by a saturation region where the measured SPV energy shift remains constant. A third region at still higher photon fluence has been evidenced, where the increase of laser fluence induces modification in SPV relaxation time. The relaxation process leaves the regime of validity of the thermionic model.

The SPV energy shift increases with the laser fluence while the decay time τ_D decreases. Moreover, a reduction of the τ_D is observed while increasing the doping level. For example, in the case of phosphorus doped samples, from (n-sample) $N_D = 8.20 \times 10^{14}$ cm^{-3} to (n^+ sample) $N_D = 2.25 \times 10^{19}$ cm^{-3}, τ_D decreases from tens of microseconds to hundreds of ns.

In agreement with Ogawa *et al.*[25] no strong dependence of SPV dynamics is observed on surface termination for the highly doped phosphorus sample and low boron doped samples. The surface termination is important only for the n doped ($N_D = 8.20 \times 10^{14}$ cm^{-3}) silicon surfaces. The decay time τ_D observed for the thermally oxidized surface is the longest we have seen during our experiments and is the double of the decay time observed for the water terminated silicon with the same doping level.

With this study, a complete characterization of the time τ_D necessary to the system to recover its equilibrium situation has been performed. For each doping, the time interval during which an accumulation of excited holes or electrons is present at the surface has been determined. This is a relevant and usually unknown parameter in photochemistry. Indeed supra-band gap illumination is used to accelerate the grafting rate of π-bonded molecules on silicon surfaces, to create a stable monomolecular layer on top of them.[41–43] In many cases the reaction rate is extremely small in the dark. Holes are believed to trigger the reaction by rendering the surface silicon electron-poor, which in turn favors a nucleophilic attack by the π-bonded molecule. The lifetime of the excited holes at the surface is indeed a crucial parameter,[44] as the longer this time, the higher the reactivity of Si with this class of molecule.

Conflicts of interest

There are no conflicts to declare.

References

1 J. T. Clemens, *Bell Labs Tech. J.*, 1997, **2**, 76–102.

2 M. J. Deen and P. K. Basu, *Silicon Photonics: Fundamentals and Devices*, Wiley, 2012.

3 P. Lv, C. Xu and B. Peng, *ACS Omega*, 2020, **5**, 6358–6365.

4 C. Kamaraki, M. T. Klug, T. Green, L. Miranda Perez and C. Case, *Appl. Phys. Lett.*, 2021, **119**, 070501.

5 S. Philipps and W. Warmuth, *Photovoltaics Report 2022*, Fraunhofer ISE, PSE Projects GmbH, https://www.ise.fraunhofer.de/en/publications/studies/photovoltaics-report.html.

6 R. Jayakrishnan, S. Gandhi and P. Suratkar, *Mater. Sci. Semicond. Process.*, 2011, **14**, 223–228.

7 L. Kronik and Y. Shapira, *Surf. Sci. Rep.*, 1999, **37**, 1–206.

8 B. F. Spencer, D. M. Graham, S. J. O. Hardman, E. A. Seddon, M. J. Cliffe, K. L. Syres, A. G. Thomas, S. K. Stubbs, F. Sirotti, M. G. Silly, P. F. Kirkham, A. R. Kumarasinghe, G. J. Hirst, A. J. Moss, S. F. Hill, D. A. Shaw, S. Chattopadhyay and W. R. Flavell, *Phys. Rev. B: Condens. Matter Mater. Phys.*, 2013, **88**, 195301.

9 M. Ogawa, S. Yamamoto, K. Fujikawa, R. Hobara, R. Yukawa, S. Yamamoto, S. Kitagawa, D. Pierucci, M. G. Silly, C.-H. Lin, R.-Y. Liu, H. Daimon, F. Sirotti, S.-J. Tang and I. Matsuda, *Phys. Rev. B: Condens. Matter Mater. Phys.*, 2013, **88**, 165313.

10 F. J. Himpsel, G. Hollinger and R. A. Pollak, *Phys. Rev. B: Condens. Matter Mater. Phys.*, 1983, **28**, 7014–7018.

11 S. M. Sze, *Physics of Semiconductor Devices*, New York, 2nd edn, 1981.

12 A. J. Sabbah and D. M. Riffe, *J. Appl. Phys.*, 2000, **88**, 6954.

13 M. H. Hecht, *Phys. Rev. B: Condens. Matter Mater. Phys.*, 1991, **43**, 12102–12105.

14 R. J. Hamers and D. G. Cahill, *J. Vac. Sci. Technol., B: Microelectron. Nanometer Struct.–Process., Meas., Phenom.*, 1991, **9**, 514–518.

15 D. Bröcker, T. Gießel and W. Widdra, *Chem. Phys.*, 2004, **299**, 247–251.

16 J. P. Long, H. R. Sadeghi, J. C. Rife and M. N. Kabler, *Phys. Rev. Lett.*, 1990, **64**, 1150–1161.

17 W. Widdra, D. Bröcker, T. Gießel, I. Hertel, W. Krüger, A. Liero, F. Noack, V. Petrov, D. Pop, P. Schmidt, R. Weber, I. Will and B. Winter, *Surf. Sci.*, 2003, **543**, 87–94.

18 M. Ogawa, S. Yamamoto, Y. Kousa, F. Nakamura, R. Yukawa, A. Fukushima, A. Harasawa, H. Kondoh, Y. Tanaka, A. Kakizaki and I. Matsuda, *Rev. Sci. Instrum.*, 2012, **83**, 023109.

19 M. Marsi, M. E. Couprie, L. Nahon, D. Garzella, T. Hara, R. Bakker, M. Billardon, A. Delboulbé, G. Indlekofer and A. Taleb-Ibrahimi, *Appl. Phys. Lett.*, 1997, **70**, 895.
20 M. Marsi, R. Belkhou, C. Grupp, G. Panaccione, A. Taleb-Ibrahimi, L. Nahon, D. Garzella, D. Nutarelli, E. Renault, R. Roux, M. E. Couprie and M. Billardon, *Phys. Rev. B: Condens. Matter Mater. Phys.*, 2000, **61**, 5070–5073.
21 J. P. Campbell and P. M. Lenahan, *Appl. Phys. Lett.*, 2002, **80**, 1945–1947.
22 A. Stirling, A. Pasquarello, J. Charlier and R. Car, *Phys. Rev. Lett.*, 2000, **85**, 2773–2776.
23 D. Pierucci, J. J. Gallet, F. Bournel, F. Sirotti, M. G. Silly, H. Tissot, A. Naitabdi and F. Rochet, *J. Phys. Chem. C*, 2016, **120**, 21631–21641.
24 J.-J. Gallet, F. Bournel, F. Rochet, U. Köhler, S. Kubsky, M. G. Silly, F. Sirotti and D. Pierucci, *J. Phys. Chem. C*, 2011, **115**, 7686–7693.
25 M. Ogawa, S. Yamamoto, K. Fujikawa, R. Hobara, R. Yukawa, S. Yamamoto, S. Kitagawa, D. Pierucci, M. G. Silly, C.-H. Lin, R. Y. Liu, H. Daimon, F. Sirotti, S.-J. Tang and I. Matsuda, *Phys. Rev. B: Condens. Matter Mater. Phys.*, 2013, **88**, 165313.
26 R. J. Hamers and K. Markert, *J. Vac. Sci. Technol., A*, 1990, **8**, 3524–3530.
27 R. J. Hamers and K. Markert, *Phys. Rev. Lett.*, 1990, **64**, 1051–1054.
28 A. Pietzsch, A. Föhlisch, F. Hennies, S. Vijayalakshmi and W. Wurth, *Appl. Phys. A: Mater. Sci. Process.*, 2007, **88**, 587–592.
29 F. J. Himpsel, B. S. Meyerson, F. R. McFeely, J. F. Morar, A. Taleb-Ibrahimi and J. A. Yarmoff, in *Proceedings of Enrico Fermi School on "Photoemission and Adsorption Spectroscopy of Solids and Interfaces with Synchrotron Radiation"*, ed. M. Campana and R. Rosei, Amsterdam, 1992.
30 J. E. Demuth, W. J. Thompson, N. J. DiNardo and R. Imbihl, *Phys. Rev. Lett.*, 1986, **56**, 1408–1411.
31 M. H. Hecht, *Phys. Rev. B: Condens. Matter Mater. Phys.*, 1990, **41**, 7918–7922.
32 N. Barrett, L. F. Zagonel, O. Renault and A. Bailly, *J. Phys.: Condens. Matter*, 2009, **21**, 314015.
33 F. Polack, M. Silly, C. Chauvet, B. Lagarde, N. Bergeard, M. Izquierdo, O. Chubar, D. Krizmancic, M. Ribbens, J.-P. Duval, C. Basset, S. Kubsky, F. Sirotti, R. Garrett, I. Gentle, K. Nugent and S. Wilkins, *AIP Conf. Proc.*, 2010, **1234**, 185–188.
34 M. G. Silly, T. Ferté, M. A. Tordeux, D. Pierucci, N. Beaulieu, C. Chauvet, F. Pressacco, F. Sirotti, H. Popescu, V. Lopez-Flores, M. Tortarolo, M. Sacchi, N. Jaouen, P. Hollander, J. P. Ricaud, N. Bergeard, C. Boeglin, B. Tudu, R. Delaunay, J. Luning, G. Malinowski, M. Hehn, C. Baumier, F. Fortuna, D. Krizmancic, L. Stebel, R. Sergo and G. Cautero, *J. Synchrotron Radiat.*, 2017, **24**, 886–897.
35 N. Bergeard, M. G. Silly, D. Krizmancic, C. Chauvet, M. Guzzo, J. P. Ricaud, M. Izquierdo, L. Stebel, P. Pittana, R. Sergo, G. Cautero, G. Dufour, F. Rochet and F. Sirotti, *J. Synchrotron Radiat.*, 2011, **18**, 245–250.
36 F. Jolly, F. Rochet, G. Dufour, C. Grupp and A. Taleb-Ibrahimi, *J. Non-Cryst. Solids*, 2001, **280**, 150–155.
37 F. J. Himpsel, F. R. McFeely, A. Taleb-Ibrahimi, J. A. Yarmoff and G. Hollinger, *Phys. Rev. B: Condens. Matter Mater. Phys.*, 1988, **38**, 6084.
38 A. Bauer, M. Prietsch, S. Molodtsov, C. Laubschat and G. Kaindl, *Phys. Rev. B: Condens. Matter Mater. Phys.*, 1991, **44**, 4002–4005.

39 G. A. Mulhollan, A. V. Subashiev, J. E. Clendenin, E. L. Garwin and R. E. Kirby, *Phys. Lett. A*, 2001, **282**, 309–318.
40 R. J. Hamers and D. G. Cahill, *Appl. Phys. Lett.*, 1990, **57**, 2031.
41 J. M. Buriak, *Chem. Rev.*, 2002, **102**, 1271–1308.
42 L. A. Huck and J. M. Buriak, *Langmuir*, 2012, **28**, 16285–16293.
43 L. A. Huck and J. M. Buriak, *J. Am. Chem. Soc.*, 2012, **134**, 489–497.
44 X. Wang, R. E. Ruther, J. A. Streifer and R. J. Hamers, *J. Am. Chem. Soc.*, 2010, **132**, 4048–4049.

Faraday Discussions

ROYAL SOCIETY OF CHEMISTRY

PAPER

Time-dependent photoemission from droplets: influence of size and charge on the photophysics near the surface†

Loren Ban, * Hanchao Tang, Bruce L. Yoder and Ruth Signorell *

Received 29th November 2021, Accepted 10th January 2022

DOI: 10.1039/d1fd00108f

Photoemission from submicrometer droplets containing a mixture of dioctyl phthalate and dioctyl sebacate was investigated by femtosecond and nanosecond photoionization. Photoelectron spectra recorded after ionization with single 4.7 eV femtosecond or nanosecond laser pulses showed marked differences between the two cases. These differences were attributed to ionization of long-lived states which only occurred within the duration of the nanosecond pulse. The tentative assignment of the long-lived states to dioctyl phthalate triplet states is discussed. A nanosecond–femtosecond pump–probe scheme using 4.7 eV (pump) and 3.1 eV (probe) pulses was used to investigate the decay dynamics of these long-lived states. The dynamics showed an accelerated decay rate at higher dioctyl phthalate concentrations. Furthermore, the dependence of the decay dynamics on droplet size and charge was investigated. The decay of the long-lived states was found to be faster in smaller droplets as well as in neutral droplets compared with both positively and negatively charged droplets. Possible mechanisms to explain these observations and the dominance of contributions from the droplets surface are discussed.

Introduction

Photoelectron spectroscopy (PES) of droplets and particles has been established as a versatile spectroscopic method to investigate aerosol samples of nanometer and micrometer size.[1–6] Due to their large surface-to-volume ratio, studies of aerosol interfaces are of great importance to understand their physics and chemistry.[7,8] In the context of atmospheric aerosol studies, PES has been used to investigate heterogeneous reactivity at droplet surfaces.[9–11] This was done by using high-energy radiation (in the soft X-ray and near-edge X-ray regimes) which yields surface-sensitive information due to short experimental probing depths

Department of Chemistry and Applied Biosciences, ETH Zürich, Vladimir-Prelog Weg 2, CH-8093 Zürich, Switzerland. E-mail: lban@ethz.ch; rsignorell@ethz.ch

† Electronic supplementary information (ESI) available: Additional information on the experimental and theoretical methods. See DOI: 10.1039/d1fd00108f

limited to a few nanometers by inelastic electron scattering. As a result, it was shown that chemical gradients are formed at the droplet surface as a result of diffusion-limited mixing in the droplet, therefore altering the chemical kinetics compared to bulk systems.[9,12]

Droplet PES also offers insight into fundamental properties of the condensed phase such as electron transport scattering[13–18] or the electronic structure.[6,19–22] In a recent publication,[18] we illustrated how a charged droplet–vacuum interface influences such properties. For example, the study of low-energy electron escape from charged droplet–vacuum interfaces can provide detailed information on the shape of the interface potential. Quantum effects make low-energy electrons particularly sensitive to the height and width of the interface potential.[18] Furthermore, a charged interface leads to characteristic shifts in the measured photoelectron spectra which need to be taken into account for precise determination of electronic structure in droplet PES.[18]

In this work, we extend our previous study of the droplet–vacuum interface and examine its influence on the photophysics near the surface. As a model system, we chose bis(2-ethylhexyl) phthalate (DEHP) dissolved in bis(2-ethylhexyl) sebacate (DEHS) droplets, where both components are liquid and their low volatility offers good control over the interface properties. The study was motivated by the observation that resonant photoionization of DEHP droplets with 4.7 eV light leads to significantly different photoelectron spectra depending on whether nanosecond or femtosecond laser pulses were used. To get at the root of this phenomenon, we performed nanosecond–femtosecond pump–probe experiments and show that following excitation at 4.7 eV, DEHP survives in long-lived states. Measurements of the electron yield decay dynamics showed these long-lived states to decay on a microsecond timescale. The dynamics turned out to depend on the DEHP concentration as well as on the droplet's size (500 nm and 1200 nm in diameter) and charge state (neutral and charge states of up to about $\pm 20e$). These results suggest that surface effects could play a role in the decay of the long-lived states. Although still at an early stage, this work illustrates the prospects of droplet PES for time-resolved studies of photoinduced processes in submicrometer systems where surface effects can lead to enhanced reactivity.

Experimental and theoretical methods

The experimental setup used in this work is shown in Fig. 1. It consists of an air-side configuration used to prepare the aerosol sample, and a droplet photoelectron spectrometer. Here we briefly describe the key elements. Further details can be found in the ESI (Section 1).†

Aerosol sample preparation

The aerosol is prepared by atomization of liquid bis(2-ethylhexyl) phthalate (DEHP) and bis(2-ethylhexyl) sebacate (DEHS) mixtures, followed by size-selection in an Aerodynamic Aerosol Classifier (AAC) based on their aerodynamic diameter D_a.[23] A neutral aerosol sample is generated by removing charged droplets from the aerosol flow with an electrostatic precipitator (ESP). Positively and negatively charged aerosol samples are generated by replacing the ESP with a unipolar diffusion charger.[18] The aerosol flow is then directed to the photoelectron

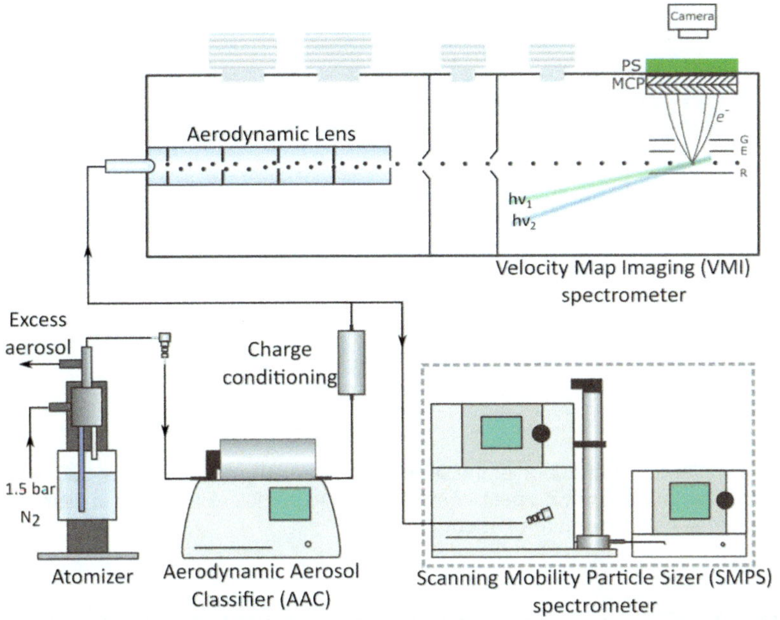

Fig. 1 Sketch of the experimental setup consisting of the aerosol generation part (atomizer, AAC, charge conditioning), photoelectron spectrometer (top) and in-parallel particle sizing by the SMPS. The droplets were transferred to vacuum *via* the aerodynamic inlet device and ionized by the laser pulses ($h\nu_{1,2}$) in between the repeller (R) and extractor (E) plates. Photoelectrons were detected in a velocity map imaging configuration (R, E and ground G plates) by the imaging detector (microchannel plates MCP, phosphor screen PS, camera).

spectrometer. A Scanning Mobility Particle Sizer (SMPS) monitors droplet size and charge in parallel to the photoionization experiment, as shown in Fig. 1.

Typical droplet size and charge distributions obtained from the SMPS are shown in Fig. 2 and 3. The size-selected ($D_a = 500$ nm) sample shows a single peak with a full-width at half-maximum of about 90 nm. The corresponding charge distributions for positively and negatively charged samples are shown in Fig. 3. The average droplet charge states of +15*e*, +20*e* and −17*e* are obtained, with a standard deviation of a few charge units.

Photoelectron spectrometer

The second part of the aerosol flow enters the photoelectron spectrometer through an aerodynamic lens (ADL), whose design is based on ref. 24. The ADL transmits a wide range of particle diameters in the submicrometer range and is similar to those used in our previous studies.[13,16,18,25] The droplet beam is intersected by femtosecond ($h\nu_1$) and nanosecond ($h\nu_2$) laser pulses in the photoelectron spectrometer. The femtosecond pulses have photon energies of either 3.1 eV or 4.7 eV and nominal durations of <70 fs, with peak intensities of ~1–35 GW cm^{-2}. The nanosecond pulses have a photon energy of 4.7 eV, a pulse duration of ~7 ns and peak intensities of ~100 kW cm^{-2}. For pump-probe experiments performed during this work, the instrument response function (IRF) is well approximated by the temporal profile of the nanosecond pulse.

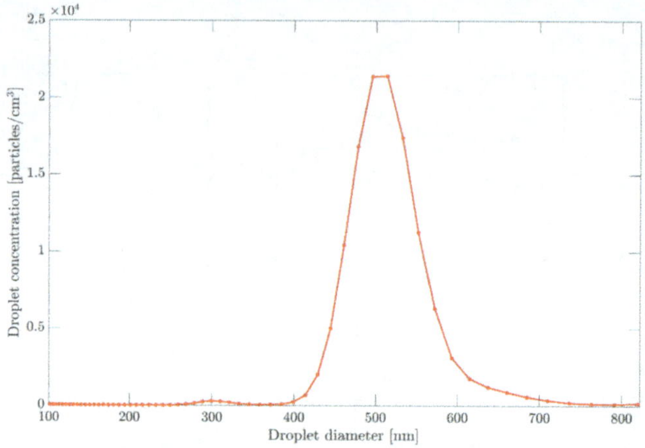

Fig. 2 Droplet size distribution of the DEHP/DEHS aerosol size-selected at 500 nm by the AAC. The distribution shows a peak corresponding to particles of 500 nm in diameter with a full width at half-maximum of about 90 nm.

The photoelectron spectrometer is operated in a Velocity Map Imaging (VMI) configuration, with two different detection modes employed in this work. In the first part of the paper, we use the camera to record angle-resolved photoelectron images. Photoelectron kinetic energy (eKE) spectra are retrieved by image reconstruction along the laser propagation axis[1] with MEVIR[26] and background-corrected by subtraction of an image collected in the absence of the droplet beam (with laser beams on). In the second part of the paper, we decoupled the electron signal directly from the imaging detector and recorded electron time-of-flight (eTOF) spectra to retrieve total photoelectron yields. Recording photoelectron yields requires the collection of fewer laser shots to achieve a reasonable

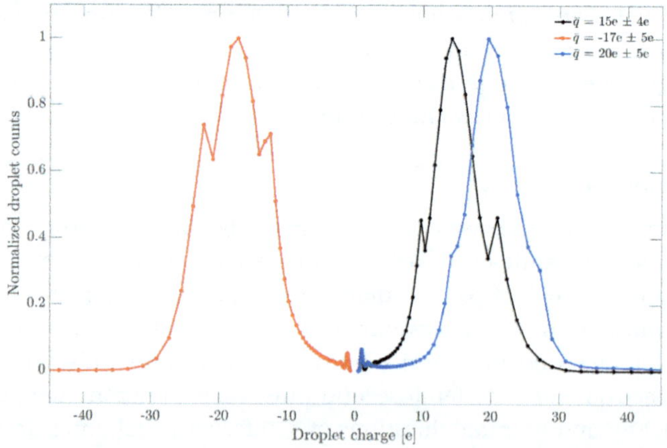

Fig. 3 Charge distributions of droplets with diameter of 500 nm after passing through the unipolar diffusion charger at positive and negative polarity. The average charge states (of elementary charge units) are shown in the legend. These charge states correspond to the measurements presented in the last section of the paper.

signal-to-noise ratio and was therefore used to perform pump–probe studies over a wide range of time delays (see below).

DFT calculations

To interpret the experimental data, we need information on the electronic energy levels relevant for our system. To the best of our knowledge, no detailed calculations exist in the literature for either liquid or gaseous DEHP. As a basis for our analysis, we performed density functional theory (DFT) calculations on a model system to obtain information on the electronic level structure that could be related to the experimental data. The chosen model system was a gas-phase dibutyl phthalate (DBP) molecule whose electronic structure should be similar to DEHP[27] (Fig. 4) with the advantage of providing a smaller and thus a more tractable system for quantum chemical calculations. The calculations were further simplified by constraining the conformational space that can be covered by the aliphatic chains. To estimate the vertical ionization energies (VIE) and adiabatic ionization energies (AIE) in the liquid droplet, we performed additional calculations using a polarizable continuum model. The calculations were compared with the limited information on phthalates (see above) and other similar aromatic systems, for which more detailed information is available in the literature (*i.e.* benzene and benzoic acid). Details of the calculations and the comparisons can be found in the ESI (Section 2).†

Results and discussion – neutral droplets

Single-pulse multiphoton ionization at 4.7 eV

In Fig. 5 we compare photoelectron images obtained by single-pulse photoionization of DEHP droplets with nanosecond (panel a) and femtosecond (panel b) pulses. Neutral droplets with a diameter of 500 nm and a DEHP concentration of 1.5 M were used. The signal in the photoelectron spectra originates from resonant multiphoton ionization (REMPI) of DEHP (see below). The solvent, DEHS, does not significantly contribute to the signal since it has no aromatic moiety and non-resonant multiphoton ionization is inefficient at the laser intensities used. This was confirmed by additional measurements of pure DEHS droplets at equal laser intensities.

The images are asymmetric along the laser propagation axis, with signal found on the image half away from the incident laser direction (single-headed arrow in Fig. 5). This is a characteristic feature of VMI PES of weakly-absorbing droplets that show so-called nanofocusing.[1,5,28] A clear difference between the ns and fs images can be seen, with the signal in the nanosecond case extending towards larger radii, *i.e.*, higher electron kinetic energies.

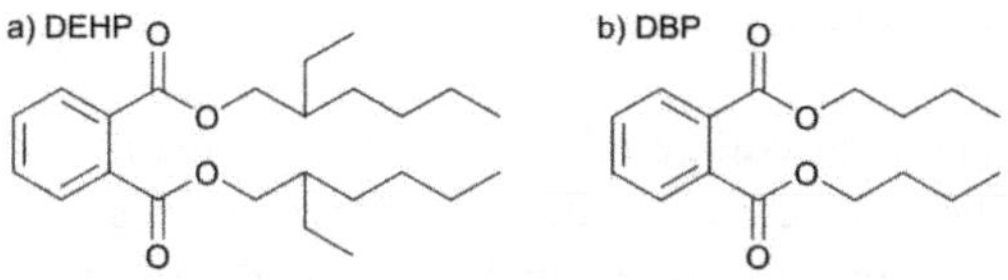

Fig. 4 Chemical structures of DEHP (a) and DBP (b).

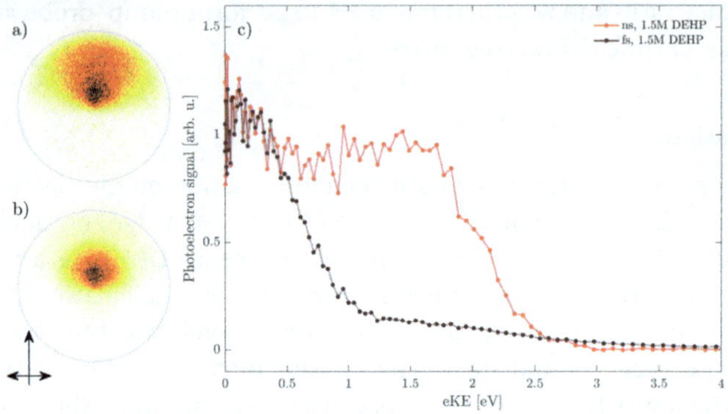

Fig. 5 Droplet photoelectron images generated by nanosecond (a) and femtosecond (b) ionization of neutral DEHP/DEHS droplets of 500 nm diameter and DEHP concentration of 1.5 M. The gray circles indicate an eKE of 2.4 eV. Light propagation (vertical) and polarization (horizontal) vectors are indicated in panel (b). Corresponding photoelectron spectra are shown in panel (c). The spectra are normalized to their respective signal at eKE ~ 0.2 eV.

For a more quantitative comparison, the corresponding photoelectron spectra are shown in Fig. 5c. The spectra are normalized to their respective signal at eKE ~ 0.2 eV. The nanosecond spectrum (red) shows two bands that extend up to electron kinetic energies (eKE) of ~3 eV. The spectrum is in good agreement with our previous study on pure DEHP droplets.[18] By contrast, the femtosecond spectrum (black) shows about one order of magnitude lower signal at eKE > 1 eV (relative to the signal at 0.2 eV) and the signal seems to extend to higher eKE (see difference at ~3 eV).

Gaussian component fitting of femto and nano spectra

To illustrate the differences more clearly, Fig. 6 shows the same spectra on a logarithmic *y*-axis. The femtosecond spectrum (dots in Fig. 6b) does indeed extend to significantly higher eKE than the nanosecond spectrum (dots in Fig. 6a). We analyzed the spectra in terms of a multi-component Gaussian fit to the experimental data. Rather than focusing on the peak maxima we compare high kinetic energy onsets, which we defined as $eKE^{\sigma} = eKE_0 + 3\sigma$, where eKE_0 is the Gaussian peak maximum and σ its standard deviation. Difficulties in using the peak maxima for the determination of vertical ionization energies (VIE) in condensed phase photoelectron spectroscopy have been previously assigned to fast excited-state dynamics[29,30] and inelastic electron scattering.[31,32] At the low eKE values measured in this work, inelastic electron scattering can additionally distort the peak shape away from a Gaussian, leading to an asymmetric tail toward lower eKEs.[14] Effects of electron scattering can in principle be accounted for by using a scattering model,[14] but scattering cross sections are not available for DEHP/DEHS mixtures.

Using eKE onsets (to be compared with the AIE) instead of peak maxima (to be compared with the VIE) circumvents some of these issues. The eKE onsets determined from the multi-component fits are qualitatively compared with the calculated ionization energies (IE, Table SI1).†

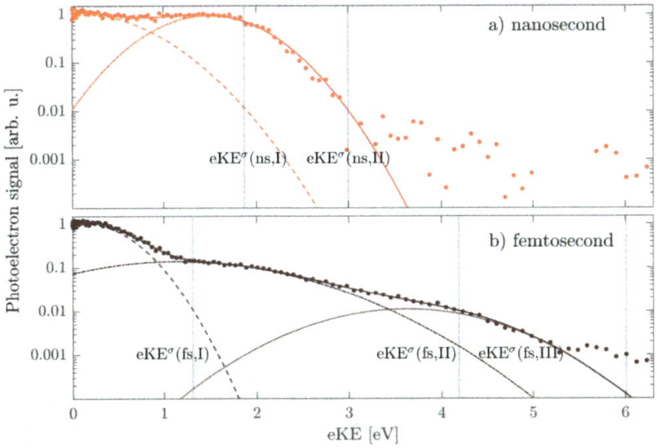

Fig. 6 Droplet photoelectron spectra (circles) from Fig. 5 shown on a logarithmic y-scale. The spectra are fitted by a multi-component Gaussian function (full-lines, components shown as dashed, dashed-dotted and dotted lines). Panel (a) shows nanosecond and panel (b) femtosecond ionization. Individual onset values are indicated by vertical gray dotted lines.

The resulting fits to the data are shown as full lines in Fig. 6, with dashed lines representing individual components. The nanosecond case (Fig. 6a) is well fitted by 2 component bands with onsets of $eKE^\sigma(ns,I) = 1.9$ eV and $eKE^\sigma(ns,II) = 3.0$ eV and similar peak amplitudes. In the femtosecond case (Fig. 6b), 3 components were necessary to fit the data with $eKE^\sigma(fs,I) = 1.3$ eV, $eKE^\sigma(fs,II) = 4.2$ eV and $eKE^\sigma(fs,III) = 6.0$ eV. The first component (fs,I) has about one and two orders of magnitude higher amplitude than components two (fs,II) and three (fs,III), respectively.

Band assignment and a simple energy diagram

We base our band assignment on the DFT calculations described in the ESI (Section 2).† Fig. 7 shows a simplified electronic energy diagram of a gas phase DBP molecule, which is assumed to be similar to a gas phase DEHP molecule. It includes vertical excitation energies of the singlet state manifold. The lowest excited state (S_1) lies at 3.2 eV above the ground state (S_0). Higher excited states (S_n) lie in the range of ~4.7 eV and enable resonant excitation at the photon energies used here. The gas phase VIE is calculated for a vertical transition to the lowest ionic state (D_0) and equals 9.0 eV. The condensed phase VIE is estimated by the transition to the ionic state D_0^{solv} and equals 8.1 eV. A condensed phase shift of 0.9 eV is reasonable considering the values determined for gas phase and condensed phase benzene.[33] We only account for the shift in the ionization energies, since energy levels of the neutral DEHP molecule will only be slightly shifted by the weakly polar ($\varepsilon \sim 4$)[34] solvent DEHS. The corresponding AIE equals 8.6 eV and 7.8 eV for the gas and condensed phase, respectively.

First, we consider the 1 + 1 REMPI process indicated by (1) in Fig. 7. Absorption of the first 4.7 eV photon leads to resonant excitation (blue arrow). Absorption of an additional photon is enough to reach the ionic state manifold (D_0,

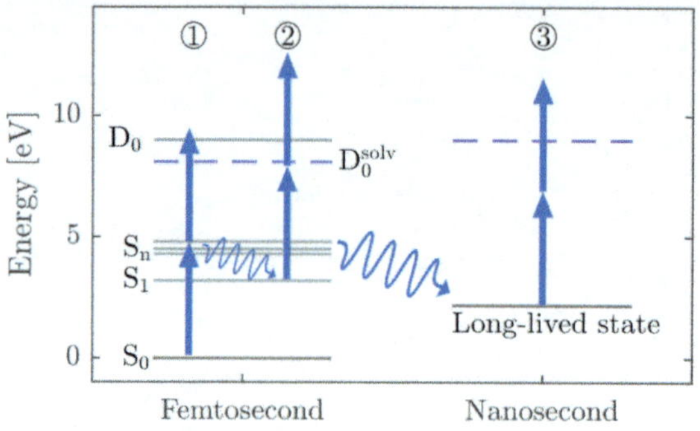

Fig. 7 A sketch of an energy level diagram of a DEHP molecule indicating 3 different processes that could explain the bands observed in the femtosecond (pathway 1 and 2) and nanosecond (pathway 1 and 3) spectra. The horizontal axis indicates the timescale at which these processes take place. Arrows indicate the 4.7 eV laser pulses. We suggest a 2-photon ionization process of the long-lived states on a nanosecond timescale (process 3).

D_0^{solv}) completing the 1 + 1 REMPI (1). The calculated condensed phase AIE (see ESI, Table SI1)† yields an eKE onset value of 1.6 eV in reasonable agreement with the experimentally determined eKE^{σ}(ns,I) of 1.9 eV and eKE^{σ}(fs,I) of 1.3 eV. Therefore, we assign these bands to a 1 + 1 REMPI from the ground state. The difference in the onsets of ~0.6 eV might be ascribed to the overlap of the (ns,I) band with the asymmetric low eKE tail of inelastically scattered electrons of the strong (ns,II) band. The fit accounts for this by increasing the width of the lower eKE Gaussian component, thereby shifting eKE^{σ}(ns,I) to correspondingly larger values. This shift is virtually absent in the fs spectrum because the (fs,II) band is much weaker than the (fs,I) band.

On the premise that ionization requires a minimum of 2 photons (process (1)), the difference between the ns and fs spectra could arise from the presence of higher-order (1 + n) REMPI. These processes are much more probable in the fs case due to the shorter pulse duration and, in turn, significantly higher peak intensities (GW cm^{-2} in the fs case, kW cm^{-2} in the ns case). Since bands originating from (1 + n) REMPI would be shifted to higher eKE by integer multiples of the photon energy, we can assign the eKE^{σ}(fs,III) band at 6.0 eV to 1 + 2 REMPI (eKE^{σ}(fs,III) − eKE^{σ}(fs,I) = 4.7 eV). Significantly weaker amplitude of the 1 + 2 REMPI band is reasonable since it includes an additional non-resonant step which has a significantly lower cross section.

However, bands eKE^{σ}(ns,II) at 3 eV and eKE^{σ}(fs,II) at 4.2 eV cannot be explained in this way. Their eKE is too high to be explained by the 1 + 1 REMPI and too low to be explained by the 1 + 2 REMPI. This indicates that the excited-state dynamics in the S_n manifold taking place within the pulse duration plays a role.[29,30] The effect of this dynamics will be very different in the ns case and the fs case. Only ultrafast (<70 fs) relaxation processes are expected to play a role in the fs case, while slower processes such as intersystem crossing can occur within the duration of the ns pulse. In both cases, all the states generated during these dynamics can absorb additional photons and be ionized. This picture is illustrated by the "ladder" and "ladder switching" models.[35] We propose that the bands eKE^{σ}(ns,II) and eKE^{σ}(fs,II) originate from distinct molecular states formed as a result of the dynamics initiated by single-photon excitation. These intermediate states are ionized by additional photons within the duration of the same ns or fs pulse.

The (fs,II) band can be assigned to 2-photon ionization of the S_1 state upon fast internal conversion of the S_n manifold, designated as process (2) in Fig. 7. The calculated values of the condensed phase AIE and the S_1 energy lead to an estimate of the eKE onset value of 4.8 eV, in reasonable agreement with the experimental onset $eKE^{\sigma}(fs,II) = 4.2$ eV. Part of the difference might be ascribed to geometry relaxation in the S_1 state not accounted for in our DFT calculations. Our assignment implies an internal conversion of the S_n manifold to S_1 state on a sub-100 fs timescale, which is not unlikely considering literature on similar systems.[36–38]

The long duration of the ns pulse (~7 ns), by contrast, also allows for slow relaxation processes to take place (*e.g.* intersystem crossing, photofragmentation and solvated electron formation). For the ns band we determined an onset $eKE^{\sigma}(ns,II) = 3.0$ eV which would lead to IE = 6.4 eV (2 photon ionization) or IE = 1.7 eV (1-photon ionization). The IE of 1.7 eV is rather low and could only be explained by weakly bound molecular states such as anions formed by electron attachment after 2-photon ionization by process (1). It might also be conceivable, that weakly-bound solvated electrons are formed by localization of a photoelectron generated in process (1) – analogous to the above-mentioned electron attachment. The low-polarity of the solvent DEHS ($\varepsilon \sim 4$),[34] however, makes these options appear unlikely. In addition, solvated electrons were not reported in previous studies on similar systems.[39] One might also think of anionic species formed along the fragmentation pathway of DEHP. However, DEHP fragments formed by UV dissociation which are reported in the literature include only neutral and cationic species which are unlikely to have such a low IE.[40–42]

Discarding major contributions from such 2 + 1 processes, *i.e.*, process (1) followed by single photon ionization, we are left with the possibility of a 1 + 2 process. We propose to assign the (ns,II) band to the 2-photon ionization of long-lived states formed after photoexcitation to the S_n manifold, as indicated by process (3) in Fig. 7. The strong contribution of the (ns,II) band to the spectrum indicates that the efficiency of this ionization channel is comparable to the direct (1 + 1) REMPI channel (process (1)). This implies that the long-lived states are formed with a high probability and that it has a comparable ionization cross section to the ground state.

Concentration dependence of single-pulse multiphoton ionization at 4.7 eV

Fig. 8 shows photoelectron images and the corresponding spectra for nanosecond ionization of pure DEHP (panel c and open circles) and of DEHP in different concentrations in DEHS (panel b and full circles for 1.5 M, panel a and crosses for 0.5 M). Images and spectra collected upon femtosecond ionization are shown in Fig. 9 for a similar range of DEHP concentrations (panel a and open circles for 1.5 M, panel b and full circles for 0.5 M, panel c and crosses for 0.1 M). All spectra are normalized to their signal at eKE ≈ 0.2 eV (see Fig. 5).

In the ns case, the intensity of the (ns,II) band relative to that of the (ns,I) band increases with decreasing concentration. By contrast, the fs spectra show similar relative band intensities for all concentrations. This illustrates that the (fs,II) and

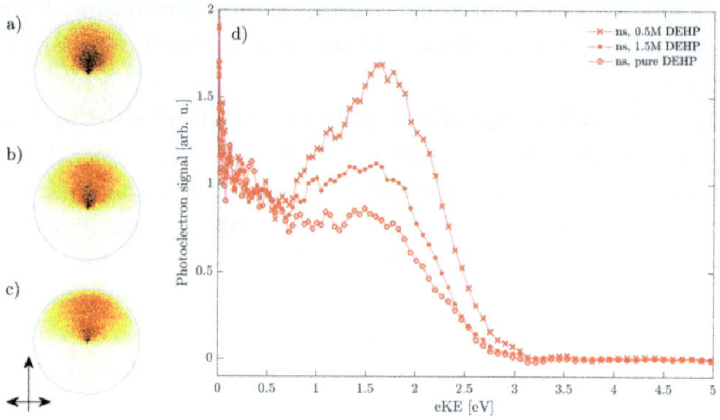

Fig. 8 Droplet photoelectron images ((a) 0.5 M DEHP, (b) 1.5 M DEHP, (c) pure DEHP) and spectra (d) from nanosecond ionization of neutral droplets with varying DEHP concentration. The single-headed arrow denotes the laser propagation direction and the double-headed arrow denotes the linear polarization of the electric field. The spectra are normalized to their respective signal at eKE ~ 0.2 eV and smoothed by a 3-point moving average.

(fs,III) bands are only minor spectral components whose relative contributions do not depend on the concentration of DEHP. This is consistent with the proposed origin of these bands in fast unimolecular relaxation (internal conversion) and higher order ionization.

The pronounced concentration dependence in the ns case (Fig. 8) suggests that the contribution of the (ns,II) band is sensitive to the surrounding DEHP molecules. The relative contribution of the (ns,II) band decreases at higher DEHP concentrations. This behavior suggests that the long-lived states are efficiently quenched at higher concentrations by additional DEHP molecules. While this simple mechanism qualitatively explains the data, we note that a complex interplay with additional mechanisms (*e.g.* bimolecular association to form a DEHP

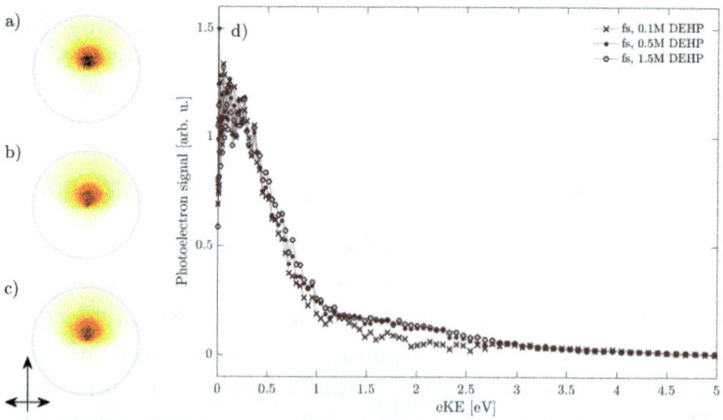

Fig. 9 Droplet photoelectron images ((a) 1.5 M DEHP, (b) 0.5 M DEHP, (c) 0.1 M DEHP) and spectra (d) from femtosecond ionization of neutral droplets with varying DEHP concentration. The single-headed arrow denotes the laser propagation direction and the double-headed arrow denotes the linear polarization of the electric field. The spectra are normalized to their respective signal at eKE ~ 0.2 eV.

dimer) could also be consistent with the experimentally observed concentration dependence (see below).

Nanosecond–femtosecond pump–probe VMI

To verify the assignment of the (ns,II) band to long-lived states formed upon photoexcitation, we turned to pump–probe studies. We used nanosecond 4.7 eV pulses to photo-excite (pump) the system followed by a femtosecond pulse, also at 4.7 eV, as a probe, with the pump–probe delay set to $\Delta t = 30$ ns. To determine whether the (ns,II) band also appears in the ns–fs pump–probe spectrum, the signal contributions from the single pulse excitations have to be removed from the VMI. This was achieved by subtracting an image obtained at $\Delta t = -20$ μs from the image at $\Delta t = 30$ ns. The image at long negative time delay serves as a background that contains only the respective single-pulse contributions (which do not depend on the time delay) but no pump–probe signal.

Fig. 10 compares the resulting pump–probe spectrum (black) with the single-pulse nanosecond case (red). The pump–probe spectrum (black) contains a clear contribution from a high-energy band that agrees well with the (ns,II) band of the single-pulse nanosecond experiment (red). This supports the assignment presented above, indicating that the long-lived states are generated during the ns pump pulse (7 ns) and persist for at least 30 ns before they are ionized by the fs-probe. The VIE and the eKE onset are in good agreement with the single-pulse nanosecond spectrum (red). While we cannot exclude that the long-lived states undergo further transformations after the ns-pump pulse, these transformations do not seem to affect their photoelectron spectrum significantly. Therefore, we conclude that the differences in photoionization by femtosecond and by nanosecond pulses (*i.e.* the (ns,II) band) are due to strong contributions from long-lived states formed during the nanosecond pulse.

A weak feature in the pump–probe spectrum at eKE $\sim$ 2.8 eV might be attributed to an artefact of the image reconstruction or the background

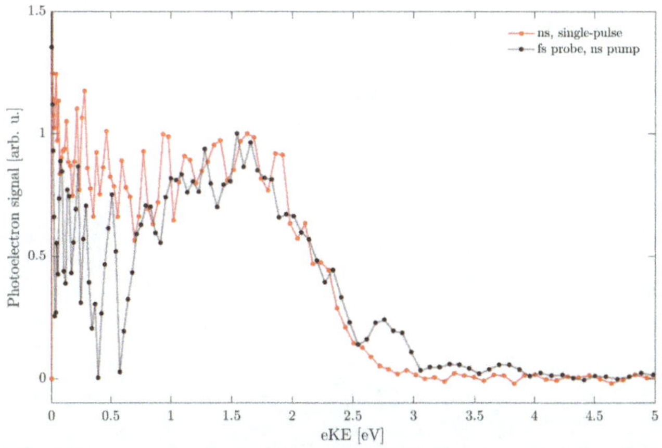

Fig. 10 Comparison of the nanosecond single-pulse photoelectron spectrum (red) with the pump–probe spectrum (black). Nanosecond (4.7 eV) pulses were used as a pump and femtosecond (4.7 eV) pulses as a probe, delayed by $\sim$30 ns. The spectra are normalized to their respective signal at eKE $\sim$ 1.75 eV.

subtraction. At low kinetic energies (eKE < 0.7 eV) the strong single-pulse femtosecond contribution (Fig. 5) in the background image causes significant scatter in the pump–probe signal after background subtraction so that meaningful analysis of this eKE range is currently impossible.

Total electron yield decay dynamics

To complement the photoelectron spectra, we performed pump–probe electron yield measurements to gain insight into the decay dynamics of the long-lived states. We used neutral droplets with a diameter of 500 nm composed of varying DEHP concentrations in DEHS. Nanosecond 4.7 eV pulses were used as a pump and time-delayed femtosecond 3.1 eV pulses as a probe. The photon energy of the fs-probe pulse was lowered so as to produce almost no single-pulse background. In this way, more reliable background subtraction and better signal-to-noise ratios were achieved, enabling measurements over a longer pump–probe delay range. We assumed that the same long-lived states are probed with both 4.7 eV and 3.1 eV pulses, which seems reasonable since 2-photon ionization is also feasible at lower photon energy (see Section 2 in the ESI).† Time-dependent total electron yields $Y(\Delta t)$ were extracted from the eTOF traces after correcting for the single-pulse background from the nanosecond and femtosecond pulses (see ESI, Section 3).†

Fig. 11 shows the decay dynamics of the electron yield $Y(\Delta t)$ for $t > 10$ ns (just outside of the instrument response function) as a function of the DEHP concentration. All data are normalized to the value $Y(\Delta t = 10 \text{ ns})$ of the measurement at the highest concentration (1.5 M). The experimental data (circles, shaded areas are the estimated signal uncertainties) show that the electron yield systematically increases with increasing DEHP concentration. As such this is no surprise as the number of absorbing DEHP molecules in the probe volume increases. However, this dependence is complicated by additional fast (Δt

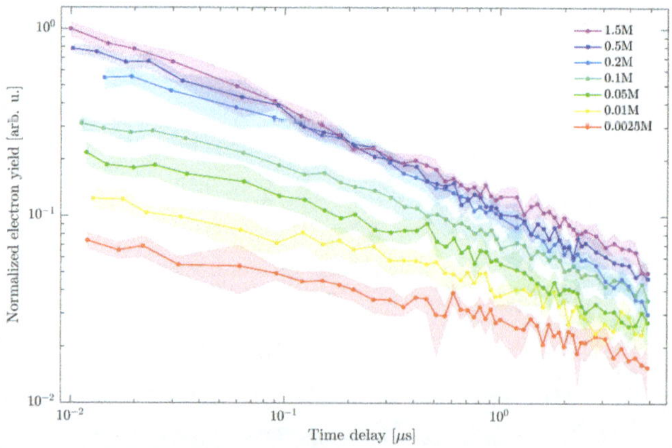

Fig. 11 Electron yield decay dynamics as a function of DEHP concentration extracted from the pump–probe eTOF traces. Shaded regions illustrate the estimated signal uncertainty and lines are connecting the measured data points to guide the eye. Data is normalized to the signal at 10 ns from the measurement with highest DEHP concentration (1.5 M).

< 10 ns) decay channels that become important at higher DEHP concentrations. The decay dynamics shows a concentration dependence, with faster signal decays at higher DEHP concentrations, evidenced in Fig. 11 by the decreasing spread of the yield traces on the logarithmic scale. Thus, the ratio of the yields at the latest and earliest delay times increases from $\frac{Y(5\ \mu s)}{Y(10\ ns)} = 0.06$ at a concentration of 1.5 M and to a value of 0.23 at a concentration of 0.0025 M. While the long-lived states invariably survive in significant amounts for at least 5 µs, their decay accelerates in more concentrated DEHP solutions.

Discussion of the long-lived states assignment

All the evidence so far speaks for the existence of long-lived states, but what are their nature? In the following we discuss the assignment of the long-lived states based on the presented experimental data, in particular the IE and decay dynamics. For the latter, we analyzed the data from Fig. 11 with two different phenomenological models (first order decay kinetics and diffusion-limited kinetics).

A commonly encountered long-lived state in the literature on similar aromatic systems is the triplet (T_1) state. For example, efficient and fast formation of the triplet T_1 state following excitation to the S_n manifold was reported for the benzoic acid monomer in the gas phase[43–45] and in acetonitrile.[38] In the latter, photoexcitation at 267 nm resulted in the formation of T_1 (~2.5 ps) with a large quantum yield (~0.65). No photofragments were observed. A T_1 lifetime of 9 µs was measured in N_2 saturated solutions and 0.1 µs in air saturated solutions. The photo-physics of DEHP was reported in ref. 46 following excitation with 20 ns, 266 nm pulses in dilute 1,1,2-trichlorotrifluoroethane solutions. The authors report that long-lived triplet states decay *via* a first-order decay process (~80 ns) in O_2 saturated solutions, while second-order triplet–triplet annihilation kinetics was reported in Ar-saturated solutions with half-lives of a few microseconds, leading to deactivation into the singlet manifold and to the formation of biradicals on a longer timescale (~20 µs). Similar results were found in smaller analogues of DEHP, such as dibutyl phthalate,[39] where triplet states were also suggested as primary photoproducts.

From the above, it seems reasonable to assume that a likely candidate for the long-lived states is the triplet (T_1) state. This assignment is supported by the calculated IE (5.4 eV for VIE and 4.6 eV for AIE, see ESI, Section 2)† which is in reasonable agreement with the experiment (Fig. 5 and 10). Contributions from other candidates, such as photo-fragments or solvated electrons do not appear plausible for the same reasons already given for the band assignment in the ns single pulse PES (see above).

Let us now consider the decay dynamics (Fig. 11) as a further test of the triplet state assignment. The photophysics of the triplet (T_1) state[47] includes various unimolecular processes (*e.g.* phosphorescence and intersystem crossing) and efficient quenching reactions by contaminants (most often O_2). The intrinsic T_1 lifetimes (in the absence of quenchers) are typically on the order of microseconds, so that the <5 µs decay (Fig. 11) must have another explanation. Provided quenchers are present in excess (or if the process is unimolecular), the quenching follows pseudo-first order kinetics, described by a single exponential decay with a rate constant equal to the sum of rate constants for all quenching reactions starting from the same initial state.

Therefore, in a first phenomenological approach we use an exponential decay model to analyze the data. A minimum of three independent exponential decay terms were required to obtain a reasonable fit to the data in Fig. 11

$$Y_{\text{fit}}(t) = a_1 \exp(-k_1 t) + a_2 \exp(-k_2 t) + a_3 \exp(-k_3 t)$$

where $(a_{1,2,3})$ are amplitudes that depend on initial concentrations. The fit performed on the data with different DEHP concentrations resulted in concentration-independent reaction rates $k_{1,2}$ while k_3 varied notably with DEHP concentration. The corresponding time constants are $k_1^{-1} = 30$ ns and $k_2^{-1} = 241$ ns. A range of $k_3^{-1} = 3.2$ μs to 6.3 μs was obtained for the concentration-dependent term. There is no systematic concentration dependence in the variations of k_3 and they still stay within the fit uncertainties. The observed systematic concentration dependence is therefore solely contained in the amplitudes $(a_{1,2,3})$, whose increase can be related to the increased number of absorbing DEHP molecules.

The need for three exponential decays suggests that the electron yield dynamics cannot be explained by triplet quenching processes only (*e.g.*, oxygen quenching). Either additional decay channels (*e.g.*, bimolecular processes) or additional states seem to contribute to the signal (*e.g.*, intermediate states or initial states other than the triplet). It is currently not possible to assign these three timescales to a physical reaction mechanism. Further experiments examining the effect of quenchers (*i.e.*, oxygen) on the decay kinetics are necessary since in our present experiment the oxygen content could not be determined or controlled. As a caveat, we note that a sufficient number of exponential decays can eventually describe any decay dynamics, even when physical mechanisms other than first-order reactions are at play.[48]

As an alternative picture let us consider a situation where the above-mentioned quenching reactions are absent and only bimolecular processes (self-quenching and triplet–triplet annihilation[46,47]) are observed on the time scale of the experiment. We found that the following ansatz leads to a reasonable fit to the data in Fig. 11

$$\frac{1}{Y(\Delta t)} = \frac{1}{Y(\Delta t = 0)} + 2a\Delta t + 4b\sqrt{\Delta t},$$

where $Y(\Delta t = 0)$, a and b are concentration-dependent fit parameters. This expression can be rationalized by a diffusion-limited triplet–triplet annihilation process[46,47] ($T_1 + T_1 \rightarrow$). Albeit producing a reasonable fit, the resulting fit parameters a and b are not fully consistent with the values expected for a diffusion-limited process. Further experiments are necessary to clarify whether this phenomenological approach can be physically justified.

Finally, we would like to note that our electron yield measurements cannot directly distinguish triplet state contributions from other intermediate states with a similar IE. In both approaches (exponential and diffusion-limited) the involvement of intermediate states would entail more complicated kinetic expressions to reproduce the data with a physically meaningful parameterization. For example, excimer (excited DEHP dimer) formation is one of the outcomes of the triplet–triplet annihilation process.[47] Since their IE is similar to the triplet state, the electron yield could well contain contributions from both species.

In conclusion, we cannot at this moment assign the electron yield decay dynamics to a precise physical mechanism. We analyzed the data considering two limiting cases, composite first order (exponential) kinetics and diffusion-limited kinetics. Neither model affords a conclusive analysis of the data in terms of physically interpretable parameters, possibly indicating that intermediate levels are involved in the observed "long-lived states" dynamics. Further kinetics experiments are necessary to obtain a clear picture, *e.g.*, by quantifying the influence of experimental parameters such as the oxygen concentration, the viscosity of the liquid or the temperature.

Results and discussion – size effects

In this section, we focus on the effect of droplet size on the electron yield decay dynamics. Fig. 12 shows results obtained from the ns-pump (4.7 eV) and fs-probe (3.1 eV) experiment on neutral droplets with a DEHP concentration of 0.2 M. The droplet diameter was varied from 500 nm to 1200 nm by adjusting the settings of the AAC. We also show additional measurements for even larger particles, for which we can only roughly estimate a size range. These droplets formed in the ADL at higher operating pressures, presumably in a coagulation process from a 500 nm droplet sample. Based on the observation of a strong light scattering signal, we qualitatively assign these particles to a size range of a few micrometers. The traces in Fig. 12 are normalized to the respective signal at ~10 ns in order to account for different number concentrations of droplets in the spectrometer as a consequence of different droplet generation conditions.

The electron yield shows a decay rate which decreases with increasing droplet diameter from 500 nm (black) to 1200 nm (red) to a few micrometers (green).

How can the observed size-dependence be explained? First, we need to consider the effects of inelastic electron scattering which determines the

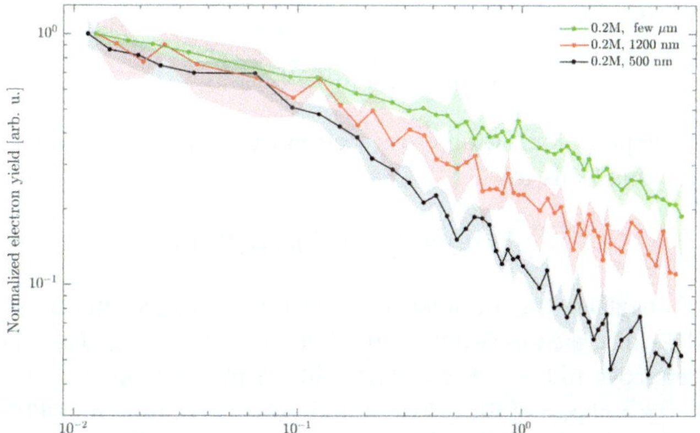

Fig. 12 Electron yield decay dynamics for different diameters of neutral droplets containing 0.2 M DEHP. Shaded regions illustrate the estimated signal uncertainty and lines are connecting the measured data points to guide the eye. Each individual trace is normalized to its signal at 10 ns.

experimental probing depth by limiting the distance from the surface from which photoelectrons can escape to vacuum and be detected. For water, the probing depth of <10 nm at an eKE of a few eV results from a value of ~3 nm for the inelastic mean-free path (IMFP).[14,31] Taking data on solid benzene[49,50] as a rough estimate for the IMFP in DEHP/DEHS mixtures we arrive at a range of ~1–10 nm. The probing depth is likely slightly higher than in water (a lower density of scatterers leads to a longer IMFP), but presumably does not exceed ~10 nm. Consequently, we can assume that our measurements (Fig. 12) are only sensitive to long-lived states that are present within a thin layer less than ~10 nm away from the surface. The range of diffusion in DEHP/DEHS droplets on a timescale of ~200 ns to ~2 μs lies between ~2 nm and ~6 nm ($x = \sqrt{2Dt}$, assuming $D = 10^{-2}$ $nm^2\ ns^{-1}$).[51] These distances are smaller than the probing depth and indicate that the long-lived states do not diffuse out of the layer defined by the probing depth within our pump–probe time-window.

In addition to the probing depth, the light intensity distribution inside the droplet must be taken into account. For droplets with diameters in the sub-micrometer range, interacting with visible or UV light, the internal light intensity distribution is not homogeneous,[1,5,13] leading to spatially dependent photoexcitation and photoionization probabilities. Therefore, it is important to examine whether the observed size-dependence could arise from the internal electric field magnitude or size-dependent spatial overlap of pump and probe pulses. The probability of forming long-lived states (we assume a triplet) in an infinitesimal volume dV is proportional to

$$[T_1(\mathrm{d}V)] \propto \Phi_T \sigma_S [S_0] I_{\mathrm{pump}}(\mathrm{d}V)$$

where $[T_1(\mathrm{d}V)]$ is the time-dependent concentration of the long-lived states. Φ_T is the yield for triplet formation from excited single states following single-photon excitation of the ground state (S_0) molecules with an absorption cross section σ_S by light with intensity $I_{\mathrm{pump}}(\mathrm{d}V)$. $I_{\mathrm{pump}}(\mathrm{d}V)$ is the space-dependent internal light intensity in the droplet volume element dV (see Fig. 13). Then, the electron signal $Y(\mathrm{d}V)$ generated by the probe laser (I_{probe}, assuming a 2-photon process) will be proportional to

$$Y(\mathrm{d}V) \propto \sigma_T [T_1(\mathrm{d}V)] I_{\mathrm{probe}}{}^2(\mathrm{d}V)$$

where σ_T is the photoionization cross-section for a 2-photon process. Neglecting all constants, this leads to

$$Y(\mathrm{d}V) \propto I_{\mathrm{pump}}(\mathrm{d}V) I_{\mathrm{probe}}{}^2(\mathrm{d}V)$$

This formula neglects the influence of inelastic electron scattering (see above).

We used the discrete-dipole approximation[52] to calculate $I_{\mathrm{pump}}(\mathrm{d}V)$ and $I_{\mathrm{probe}}(\mathrm{d}V)^2$ assuming refractive indices of $N = 1.52 + i0.001$ (4.7 eV) and $N = 1.49 + i0.0001$ (3.1 eV). For the real part of the refractive index we used values for DEHP[13] (DEHS is assumed to have the same value[53]). For the imaginary part, we scaled the DEHP values at the corresponding wavelength by 0.1 to account for a concentration of 0.2 M (DEHS does not significantly absorb at these wavelengths).

Fig. 13 shows the resulting internal light intensity distributions in the plane spanned by the laser propagation and polarization directions. The color scale

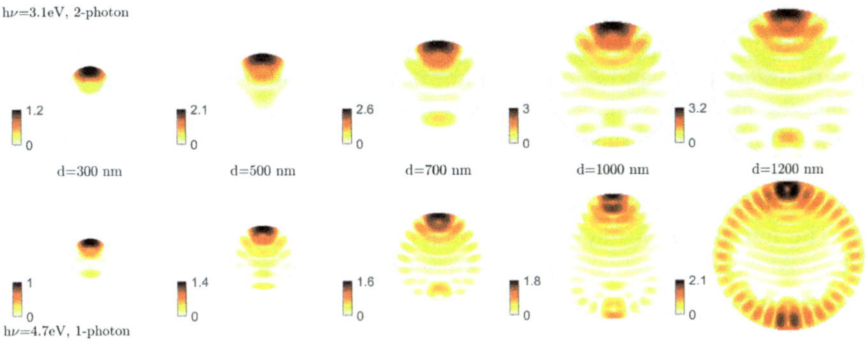

Fig. 13 ADDA simulations top row I_{probe}^2, bottom row I_{pump}. Light propagation direction is upwards and the polarization vector lies horizontally. The color axis is given as the internal intensity (I_{probe}^2, I_{pump}) relative to the incident intensity outside the droplet. The color scale is given on a logarithmic axis and signal below the input intensity (<1) is shown as white.

shows the logarithm of the internal light intensity relative to the incident light intensity. The effect of nano-focusing is clearly visible in the asymmetry of the distribution and the intensity enhancement which increases for larger particles (left to right). $I_{probe}(dV)^2$ (top row) shows a large light intensity enhancement with particle size, with maximum values increasing by almost 1 order of magnitude going from a particle diameter of 500 nm to 1000 nm (see color bars). By comparison, $I_{pump}(dV)$ increases by a factor of $\sim$2 in the same size range. The larger enhancement in the probe process ($I_{probe}(dV)^2$) reflects the dependence of the 2-photon ionization probability on the square of the intensity.

The most obvious effect of droplet size is the different magnitude of light intensity enhancement. Data in Fig. 12 are normalized to signal at $\sim$10 ns so that total relative signal at longer times should reflect the survival probability of the long-lived states since

$$\frac{Y(\Delta t)}{Y(\Delta t = 10\ \text{ns})} \approx \frac{[T_1(\Delta t)]}{[T_1]_0}$$

This assumption neglects the effects of electron scattering and additional transient states that could be ionized by the increased probe intensity inside the droplet. It is also assumed that the imaginary part of the refractive index remains small enough at the present DEHP concentration, so that its temporal variations can be neglected. Therefore, the only mechanism that could influence $[T_1(\Delta t)]$ is triplet–triplet annihilation due to its sensitivity to $[T_1]_0$. The increase in $[T_1]_0$ in larger droplets due to higher internal light intensity would lead to a faster signal decay, opposite of what we observed.

In addition, the spatial distribution of the light intensity needs to be considered. From Fig. 13, the spatial distribution of both pump and probe pulses seems qualitatively similar. To investigate the spatial distribution of the light intensity in more detail, we integrated the 3-dimensional internal light distributions over the whole solid angle (Ω)

$$I(r) = \iint I(r,\Omega) r^2 \mathrm{d}\Omega$$

where $I = I_{pump}$, $I = I_{probe}^2$ or $I = I_{pump}I_{probe}^2$, and r is the distance from the center of the droplet. Obtained integrals are shown in Fig. 14. Internal light intensity distributions I_{pump} and I_{probe}^2 were normalized to their corresponding maxima since the arbitrary scaling factor does not influence the result (see above). In larger droplets (d = 1000 nm and 1200 nm), oscillations of the internal light intensity distribution originate from Mie resonances. Data for these droplet sizes were therefore smoothed to average out the oscillations.

The internal light intensity distribution of pump pulses I_{pump} (Fig. 14a) is maximal close to the surface and extends further into the droplet interior for larger droplets. In the probe case, I_{probe}^2 (Fig. 14b), the distribution shows a pronounced maximum close to the surface and a weaker tail extending to the droplet interior for larger sizes. The product $I_{pump}I_{probe}^2$, which determines the spatial overlap of our pump–probe experiment is shown in Fig. 14c. It shows one dominant feature close to the droplet surface. For analyzed droplets diameters this feature is essentially size-independent up to ~30 nm away from the surface. This distance is larger than the probing depth (<10 nm) and we can therefore conclude that the size-dependent distribution of light inside the droplets plays no role on the observed decay dynamics.

From the above, it is clear that the observed increase in the electron survival probability with droplet size cannot be explained by diffusion out of the finite probing depth nor by variations in the internal light distribution. This leaves the influence of the droplet surface on the reactions taking place within a thin layer <10 nm from the surface as a possible cause for the observed size-dependence of the dynamics.

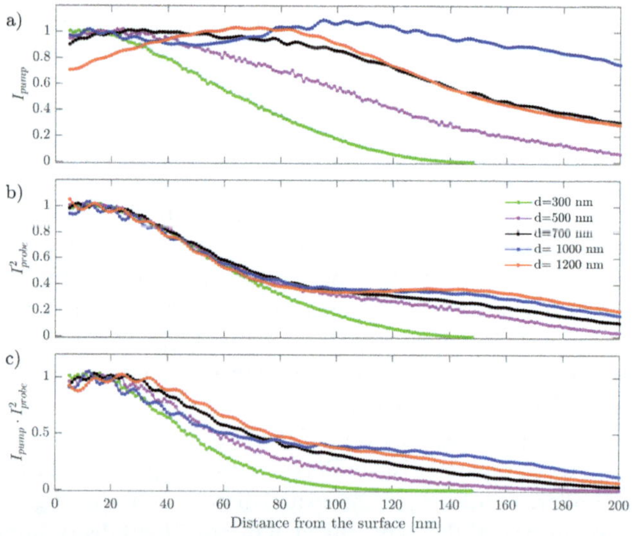

Fig. 14 Internal light intensity distribution integrated over the whole solid angle for droplets of different diameter (d). Data for d = 1000 nm and 1200 nm were smoothed to average out the oscillations coming from the Mie resonances. Panel (a) is a plot of the pump intensity (I_{pump}), panel (b) is a plot of the probe intensity squared (I_{probe}^2) and panel (c) is their product.

Accelerated reaction rates have been previously reported in micrometer sized droplets, but their physical origin is still being discussed.[54] A possible explanation can be proposed in the case of diffusion-limited reaction, where the reaction time is lowered by the decreased dimensionality at the droplet surface.[55] This constraint is relaxed as the droplet size increases and the reaction slows down, in qualitative agreement with our data. This hypothesis needs further experimental testing and must be interpreted with care. As highlighted in ref. 54, detailed and critical analysis of different effects is necessary before assigning the altered kinetics to a particular surface effect.

Results and discussion – charge effects

For the final section of this paper, we continue our discussion by presenting experiments on charged droplets. Surface charge has been proposed as a mechanism to explain accelerated reactivity observed in nanometer and micrometer sized droplets.[54,56]

In Fig. 15, we show the electron yield decay dynamics recorded on the droplets with a diameter of 500 nm with varying degrees of charging. The electron yield is normalized to its signal at ~10 ns. The charge states correspond to the measurements shown in Fig. 3. The droplet charge is assumed to be homogeneously distributed on the droplet surface. This assumption is supported by computational studies on electrospray ionization.[54,57,58] In this picture, the electric field magnitude E on the droplet surface of diameter d can be calculated from the droplet charge state ($q = ne$) by the simple model of a charged conductive sphere (see also ref. 54)

$$E = \frac{ne}{\pi \varepsilon_0 d^2}$$

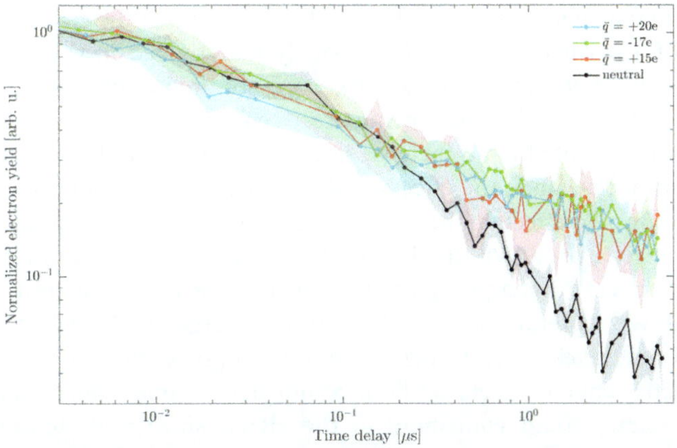

Fig. 15 Effect of charge on electron yield dynamics. Droplets of 500 nm in diameter with charge states of +15*e*, +20*e* and −17*e* (colored traces) show a significantly slower electron decay rate than neutral droplets of the same size (black). Shaded regions illustrate the estimated signal uncertainty and lines are connecting the measured data points to guide the eye. Each individual trace is normalized to its signal at 10 ns.

For the charge states in the present study this results in electric fields of about $3.5–4.6 \times 10^{-4}$ V nm^{-1} for a droplet with diameter of 500 nm carrying $n = 15–20$ charges.

The number density of the charged droplet sample was significantly lower than in the neutral case, which is reflected in the larger signal uncertainty. The temporal evolution of the electron yield of charged droplets clearly deviates from the neutral sample and leads to higher electron yield at times above ~200 ns. The electron yield is enhanced for both charge polarities and no clear dependence on the electric field magnitude was observed within the relatively large uncertainty.

First, it is important to discuss the nature of the droplet charging process since we are investigating excited state photophysics where quenching by impurities could have a strong effect. The droplets are charged by collisions with ions generated by the discharge in the gas flow. The species responsible for charging therefore depends on the polarity of the discharge, specific charge settings as well as the gas composition. In air, positive corona discharge primarily consists of small, protonated water clusters.[59] The situation is more complicated in the negative case,[60] where the generation of ozone leads to the presence of hydrated cluster anions of $OH^-$and NO_3^-. The number of absorbed ions directly determines the droplet charge, which at the charge states used in our experiment corresponds to a negligible concentration of these ions. The effects of ions on the photo-physics can therefore be safely neglected and cannot explain the difference from the neutral case. This is further confirmed by the fact that both polarities (different ions) show the same effect even though different ions would be formed.

But how then can the presence of charge influence measured photo-electron yields? Let us consider two general mechanisms by which an electric field can affect the electron yield. The first one applies to ionization processes that proceed *via* charge-pair states. In this case, the electric field can enhance the electron signal by facilitating charge separation and thus increasing the ionization probability. Examples of this mechanism are found in benzene excimers excited to charge-transfer states just below the ionization threshold.[33] Electric fields of about 10^{-3} V nm^{-1} produced a measurable increase in free charge generation.

The second mechanism works through the influence of the electric field on the diffusion of a particle. This has been widely investigated in the context of the recombination of charged particle pairs[61] (*e.g.* solvated electron) and of neutral-charged pairs.[62] For ion-pair geminate recombination,[61] electric fields of $\sim 10^{-2}$ V nm^{-1} resulted in an increase in the escape probability. However, in the case of neutral reactants, the effects of the electric field are not obvious.

The electric field effect on the electron yield could bring additional insight into the decay mechanism. Within the assumptions of our kinetic model, we can exclude the effect of charged particle diffusion since only neutral species are considered. We therefore propose that surface charge facilitates the charge-pair separation that takes place during the ionization process.

The electric charge effects could be a hint that excimer states, in addition to the triplet state, could contribute to the signal since their ionization often proceeds *via* a charge-separation mechanism.[33] This could be rationalized by a triplet–triplet annihilation mechanism which proceeds *via* an excimer intermediate ($T + T \leftrightarrows E \rightarrow$). Future studies should therefore extend the kinetic analysis to account for the possible effect of diffusion-limited monomer-excimer reactivity.[63,64] An intriguing observation is that our exponential kinetic model yields a time constant of ~240 ns, commensurate with the time delay after which the charged droplet electron yield departs from the neutral case in Fig. 15.

Conclusions

We performed a photoionization study of submicrometer droplets consisting of solutions of DEHP in DEHS and investigated the effect of pulse duration on the resonant photoionization process. When comparing single-pulse photoelectron spectra obtained upon ionization with 4.7 eV nanosecond and femtosecond pulses, we observed a pronounced difference. The presence of an intense higher kinetic energy band in the spectra of the nanosecond case was attributed to contributions from long-lived states populated within the duration of the nanosecond pulse. This assignment was confirmed by pump–probe photoelectron spectra, in which only the long-lived states contributions were observed. The relative contribution of the long-lived states decreased at higher DEHP concentrations, suggesting a more efficient quenching process in the presence of more DEHP molecules. The decay dynamics of the electron yield was measured by a pump–probe experiment using 4.7 eV nanosecond pump and 3.1 eV femtosecond probe pulses. The electron signal survived for at least 5 μs and showed an accelerated decay rate at higher DEHP concentrations, consistent with the single-pulse nanosecond spectra. We tentatively assign the long-lived states to a DEHP triplet, which is supported by the experimental photoelectron spectrum and DFT calculations. The decay dynamics were analyzed in terms of two limiting cases (1st order multi-exponential and diffusion-limited), but neither of these phenomenological approaches affords a conclusive interpretation of the fit parameters. The results might hint at the involvement of an additional intermediate state in the observed "long-lived states" dynamics.

We further investigated the influence of particle size and charge state on the decay dynamics of the electron yield. We observed a decrease of the decay rate with increasing droplet size. Because of the relatively low probing depth (<10 nm), the experiment preferentially probes the kinetics near the surface. Therefore, we tentatively attribute the decrease of the rate observed for larger droplets to an altered diffusion of the reactants near the surface. Further experiments are necessary before a detailed mechanism can be proposed. The decay rate of the electron yield also decreases in charged compared with neutral droplets. This effect does not depend on the charge polarity. We propose that the electric field on the surface of the droplets increases the ionization probability by facilitating charge-separation in the ionization process. This result hints that excimer states could be important intermediates in the long-lived states decay kinetics.

Although many questions remain open, the present study highlights the potential of droplet photoelectron imaging to gain insight into the photophysics and photochemistry in confined systems. The combination of pump–probe photoelectron spectroscopy with droplet beams of controlled size (submicrometer regime) and charge enables the study of kinetics at a surface with high time resolution.

Conflicts of interest

There are no conflicts to declare.

Acknowledgements

We would like to thank Dr Martin Fierz for his help and lending us devices for aerosol charging experiments. We thank David Stapfer, Dr Egor Chasovskikh and Dr Thomas E. Gartmann for technical contributions to the experimental setup, in particular our droplet photoelectron spectrometer. We also thank Daniel Zindel, Markus Steger and Dr Erich C. Meister for technical support regarding chemical and electronic aspects of the project. This project has received funding from the European Unions Horizon 2020 research and innovation program from the European Research Council under the Grant Agreement No. 786636, and the research was supported by the NCCR MUST, funded by the Swiss National Science Foundation (SNSF), through ETH-FAST, and through SNSF project No. 200020_200306.

Notes and references

1 L. Ban, B. L. Yoder and R. Signorell, *Annu. Rev. Phys. Chem.*, 2020, **71**, 315–334.

2 J. Shu, K. R. Wilson, M. Ahmed and S. R. Leone, *Rev. Sci. Instrum.*, 2006, **77**, 043106.

3 K. R. Wilson, H. Bluhm and M. Ahmed, in *Fundamentals and Applications in Aerosol Spectroscopy*, ed. R. Signorell and J. P. Reid, CRC Press, 2010, pp. 367–399.

4 E. Antonsson, H. Bresch, R. Lewinski, B. Wassermann, T. Leisner, C. Graf, B. Langer and E. Rühl, *Chem. Phys. Lett.*, 2013, **559**, 1–11.

5 K. R. Wilson, Z. Shengli, S. Jinian, E. Rühl, S. R. Leone, G. C. Schatz and M. Ahmed, *Nano Lett.*, 2007, **7**, 2014–2019.

6 C. C. Su, Y. Yu, P. C. Chang, Y. W. Chen, I. Y. Chen, Y. Y. Lee and C. C. Wang, *J. Phys. Chem. Lett.*, 2015, **6**, 817–823.

7 B. R. Bzdek and J. P. Reid, *J. Chem. Phys.*, 2017, **147**, 220901.

8 B. R. Bzdek, J. P. Reid and M. I. Cotterell, *Commun. Chem.*, 2020, **3**, 10–13.

9 M. I. Jacobs, B. Xu, O. Kostko, N. Heine, M. Ahmed and K. R. Wilson, *J. Phys. Chem. A*, 2016, **120**, 8645–8656.

10 E. Antonsson, M. Patanen, C. Nicolas, J. J. Neville, S. Benkoula, A. Goel and C. Miron, *Phys. Rev. X*, 2015, **5**, 1–7.

11 O. Sublemontier, C. Nicolas, D. Aureau, M. Patanen, H. Kintz, X. Liu, M. A. Gaveau, J. L. Le Garrec, E. Robert, F. A. Barreda, A. Etcheberry, C. Reynaud, J. B. Mitchell and C. Miron, *J. Phys. Chem. Lett.*, 2014, **5**, 3399–3403.

12 M. I. Jacobs, B. Xu, O. Kostko, A. A. Wiegel, F. A. Houle, M. Ahmed and K. R. Wilson, *J. Phys. Chem. A*, 2019, **123**, 6034–6044.

13 R. Signorell, M. Goldmann, B. L. Yoder, A. Bodi, E. Chasovskikh, L. Lang and D. Luckhaus, *Chem. Phys. Lett.*, 2016, **658**, 1–6.

14 R. Signorell, *Phys. Rev. Lett.*, 2020, **124**, 205501.

15 O. Kostko, M. I. Jacobs, B. Xu, K. R. Wilson and M. Ahmed, *J. Chem. Phys.*, 2019, **151**, 2–5.

16 S. Amanatidis, B. L. Yoder and R. Signorell, *J. Chem. Phys.*, 2017, **146**, 224204.

17 M. I. Jacobs, O. Kostko, M. Ahmed and K. R. Wilson, *Phys. Chem. Chem. Phys.*, 2017, **19**, 13372–13378.

18 L. Ban, T. E. Gartmann, B. L. Yoder and R. Signorell, *Phys. Rev. Lett.*, 2020, **124**, 13402.
19 K. R. Wilson, M. Jimenez-Cruz, C. Nicolas, L. Belau, S. R. Leone and M. Ahmed, *J. Phys. Chem. A*, 2006, **110**, 2106–2113.
20 K. R. Wilson, D. S. Peterka, M. Jimenez-Cruz, S. R. Leone and M. Ahmed, *Phys. Chem. Chem. Phys.*, 2006, **8**, 1884.
21 S. Tigrine, N. Carrasco, D. K. Bozanic, G. A. Garcia and L. Nahon, *Astrophys. J.*, 2018, **867**, 164.
22 S. Hartweg, G. A. Garcia, D. K. Božanić and L. Nahon, *J. Phys. Chem. Lett.*, 2021, **12**, 2385–2393.
23 F. Tavakoli and J. S. Olfert, *Aerosol Sci. Technol.*, 2013, **47**, 916–926.
24 X. Wang, F. E. Kruis and P. H. McMurry, *Aerosol Sci. Technol.*, 2005, **39**, 611–623.
25 M. Goldmann, J. Miguel-Sánchez, A. H. C. West, B. L. Yoder and R. Signorell, *J. Chem. Phys.*, 2015, **142**, 224304.
26 B. Dick, *Phys. Chem. Chem. Phys.*, 2014, **16**, 570–580.
27 J.-B. Du, Y.-L. Tang, Z.-W. Long, S.-H. Hu and T. Li, *Russ. J. Phys. Chem. A*, 2014, **88**, 819–822.
28 R. Signorell, M. Goldmann, B. L. Yoder, A. Bodi, E. Chasovskikh, L. Lang and D. Luckhaus, *Chem. Phys. Lett.*, 2016, **658**, 1–6.
29 A. Roy, R. Seidel, G. Kumar and S. E. Bradforth, *J. Phys. Chem. B*, 2018, **122**, 3723–3733.
30 F. Buchner, H. H. Ritze, J. Lahl and A. Lübcke, *Phys. Chem. Chem. Phys.*, 2013, **15**, 11402–11408.
31 D. Luckhaus, Y. Yamamoto, T. Suzuki and R. Signorell, *Sci. Adv.*, 2017, **3**, e1603224.
32 Y. I. Yamamoto, S. Karashima, S. Adachi and T. Suzuki, *J. Phys. Chem. A*, 2016, **120**, 1153–1159.
33 T. W. Scott, C. L. Braun and A. C. Albrecht, *J. Chem. Phys.*, 1982, **76**, 5195–5202.
34 C. Wohlfahrt, in *Static Dielectric Constants of Pure Liquids and Binary Liquid Mixtures*, ed. O. Madelung, Springer-Verlag, Berlin/Heidelberg, pp. 5–228.
35 R. Weinkauf, P. Aicher, G. Wesley, J. Grotemeyer and E. W. Schlag, *J. Phys. Chem.*, 1994, **98**, 8381–8391.
36 A. L. Thompson and T. J. Martínez, *Faraday Discuss.*, 2011, **150**, 293–311.
37 T. Suzuki, *Molecules*, 2014, **19**, 2410–2433.
38 C. Yang, H. Su, X. Sun and M. W. George, *J. Chem. Phys.*, 2012, **136**, 204507.
39 N. Liu, M.-H. Wu, G. Xu, Q. Yuan, L. Tang and J. Zhou, in *2009 International Conference on Energy and Environment Technology*, IEEE, 2009, vol. 3, pp. 455–458.
40 Y. A. Jeilani, B. H. Cardelino and V. M. Ibeanusi, *J. Am. Soc. Mass Spectrom.*, 2011, **22**, 1999–2010.
41 M. Lacko, P. Papp and Š. Matejčík, *J. Chem. Phys.*, 2018, **148**, 214305.
42 T. K. Lau, W. Chu and N. Graham, *Chemosphere*, 2005, **60**, 1045–1053.
43 J. Li, F. Zhang and W.-H. Fang, *J. Phys. Chem. A*, 2005, **109**, 7718–7724.
44 G. Meijer, M. S. De Vries, H. E. Hunziker and H. R. Wendt, *J. Phys.Chem.*, 1990, **94**, 4394–4396.
45 S. I. Kamei, H. Abe, N. Mikami and M. Ito, *J. Phys. Chem.*, 1985, **89**, 3636–3641.
46 G. Hizal, Q. Q. Zhu, C. H. Fischer, T. Majima and W. Schnabel, *J. Photochem. Photobiol., A*, 1992, **69**, 33–39.

47 J. B. Birks, *Photophysics of aromatic molecules*, Wiley-Interscience, London, New York, 1970.
48 D. J. S. Birch, A. D. Dutch, R. E. Imhof, B. Nadolski and I. Soutar, *J. Photochem.*, 1987, **38**, 239–254.
49 T. Goulet and J. P. Jay-Gerin, *Solid State Commun.*, 1985, **55**, 619–623.
50 T. Goulet and J.-P. Jay-Gerin, *Int. J. Radiat. Appl. Instrum., Part C*, 1986, **27**, 229–239.
51 K. R. Harris, B. Ganbold and W. S. Price, *J. Chem. Eng. Data*, 2015, **60**, 3506–3517.
52 M. A. Yurkin and A. G. Hoekstra, *J. Quant. Spectrosc. Radiat. Transfer*, 2011, **112**, 2234–2247.
53 L. De Lorenzi, M. Fermeglia and G. Torriano, *J. Chem. Eng. Data*, 1997, **42**, 919–923.
54 G. Rovelli, M. I. Jacobs, M. D. Willis, R. J. Rapf, A. M. Prophet and K. R. Wilson, *Chem. Sci.*, 2020, **11**, 13026–13043.
55 S. Mondal, S. Acharya, R. Biswas, B. Bagchi and R. N. Zare, *J. Chem. Phys.*, 2018, **148**, 244704.
56 S. Banerjee, E. Gnanamani, X. Yan and R. N. Zare, *Analyst*, 2017, **142**, 1399–1402.
57 L. Konermann, E. Ahadi, A. D. Rodriguez and S. Vahidi, *Anal. Chem.*, 2013, **85**, 2–9.
58 C. F. Chamberlayne and R. N. Zare, *J. Chem. Phys.*, 2020, **152**, 184702.
59 D. Y. H. Pui, S. Fruin and P. H. McMurry, *Aerosol Sci. Technol.*, 1988, **8**, 173–187.
60 K. Sekimoto and M. Takayama, *Int. J. Mass Spectrom.*, 2007, **261**, 38–44.
61 A. Mozumder, *J. Chem. Phys.*, 1974, **60**, 4300–4304.
62 S. Y. Reigh, K. J. Shin and M. Tachiya, *J. Chem. Phys.*, 2008, **129**, 1–12.
63 M. N. Berberan-Santos and J. M. G. Martinho, *J. Chem. Phys.*, 1991, **95**, 1817–1824.
64 K. Sienicki and M. A. Winnik, *J. Chem. Phys.*, 1987, **87**, 2766–2772.

Faraday Discussions

PAPER

Aligning time-resolved kinetics (TAP) and surface spectroscopy (AP-XPS) for a more comprehensive understanding of ALD-derived 2D and 3D model catalysts†

Evgeniy A. Redekop, *[a] Hilde Poelman, [b] Matthias Filez,‡[bc] Ranjith K. Ramachandran, [c] Jolien Dendooven, [c] Christophe Detavernier,[c] Guy B. Marin, [b] Unni Olsbye [a] and Vladimir V. Galvita [b]

Received 13th December 2021, Accepted 2nd February 2022

DOI: 10.1039/d1fd00120e

The spectro-kinetic characterization of complex catalytic materials, *i.e.* relating the observed reaction kinetics to spectroscopic descriptors of the catalyst state, presents a fundamental challenge with a potentially significant impact on various chemical technologies. We propose to reconcile the kinetic characteristics available from temporal analysis of products (TAP) pulse-response kinetic experiments with the spectroscopic data available from ambient pressure X-ray photoelectron spectroscopy (AP-XPS), using atomic layer deposition (ALD) to synthesize multicomponent model surfaces on 2D and 3D supports. The accumulated surface exposure to a key reactant (total number of collisions) is used as a common scale within which the results from the two techniques can be rigorously compared for microscopically-equivalent surfaces. This approach is illustrated by proof-of-principle TAP and AP-XPS experiments with $PtIn/MgO/SiO_2$ catalysts for alkane dehydrogenation at 800 K. Similarly to industrially-relevant Pt-based bimetallic catalysts on high-surface area supports, the initial period of coke accumulation on the surface resulted in gradually decreased conversion and increased selectivity towards propylene. We were able to monitor the process of coke deposition with both AP-XPS and TAP. The evolution of the C 1s photoelectron spectra is aligned on the common exposure scale with the evolution of the coke amounts deposited per Pt site during a multi-pulse TAP experiment. Moreover, TAP provided quantitative kinetic descriptors of propane consumption and product mean residence time within this common exposure scale. The challenges and opportunities presented by this novel tandem methodology are discussed in the context of catalysis research.

[a]*Centre for Materials Science and Nanotechnology (SMN), University of Oslo, Oslo, Norway. E-mail: evgeniyr@smn.uio.no*

[b]*Laboratory for Chemical Technology, Ghent University, Ghent, Belgium*

[c]*Conformal Coatings of Nanomaterials Group, Ghent University, Ghent, Belgium*

† Electronic supplementary information (ESI) available. See DOI: 10.1039/d1fd00120e

‡ Current affiliation.

Introduction

Heterogeneous catalysis proceeds through reactions that involve free molecules in the fluid phase, species adsorbed at the surface, and several (sub)surface layers of the catalytic material. Reaction kinetics and stability against deactivation are largely determined by the atomic and electronic configurations of the surfaces themselves and surface species that emerge *in situ* from the starting material under the influence of pretreatments and reaction environments (temperature, fluid phase composition, *etc.*). These interfacial gas/solid phenomena have been investigated by various methods for over a century, but the fundamental challenge of surface characterization in its reactive state persists to this day, especially with regard to linking the structure and performance of complex, multicomponent materials. Hypothetically, this challenge can be addressed by systematic, multiplexed spectro-kinetic studies that can quantitatively correlate the state of the surface, as captured by spectroscopy, with the intrinsic kinetic characteristics, as captured by the transformation rates of the fluid phase reactants. Critically, the model materials and characterization methods involved in such studies must be tailored to (i) capture the essential features of the reaction mechanism at the time scales of the catalytic cycle and potential side reactions and (ii) reveal how the structure–reactivity relationships evolve across wide ranges of the compositional, structural, and dynamic complexities. Since catalysis is an intrinsically kinetic phenomenon, time-resolved and transient methods in both gas-phase kinetics and surface spectroscopy are particularly attractive. Such time-resolved spectro-kinetic data could then be used to more readily discriminate hypothetical reaction mechanisms, benchmark theoretical models of surface reactions, and establish structure–performance relationships for the design of functional materials. Herein we explore the possibilities of combining transient temporal analysis of products (TAP) kinetic and *in situ* (near ambient pressure) X-ray photoelectron spectroscopy (AP-XPS) spectroscopic measurements, as viewed from two vantage points: the synthesis of equivalent materials amenable to the two techniques and aligning the range of gas exposures that these samples experience during typical TAP and AP-XPS experiments.

For model surfaces of single crystals under UHV conditions, the spectro-kinetic challenge has been addressed since the 1980s by combining molecular beam scattering (MBS) and temperature-programmed reaction spectroscopy (TPRS)[1] with various surface-sensitive spectroscopies,[2] whereby gaseous reactants are delivered by collimated molecular beams or transported by the free molecular flow and monitored by mass-spectrometry. Similar methods can also be applied to more complex planar model surfaces, such as thin films and nanoparticles deposited on planar substrates.[3]

For high-surface area and/or 3D supports, sub-second time-resolution and nuanced kinetic details can be achieved in temporal analysis of products (TAP) reactor systems,[4] which uniquely enable steady-state and non-steady-state experiments ranging from the well-defined low-pressure Knudsen flow regime to the ambient-pressure convective flow (and above). In the Knudsen flow regime, the nanomolar pulses used as gas phase perturbations allow for precise control over the gradual change of the surface composition under the influence of reactants, while simultaneously monitoring the intrinsic, *i.e.* solely determined by the chemical steps in the catalytic cycle, reaction kinetics.[5,6] Different pressure regimes can be compared for the same sample within the same device to derive deeper insights into the reaction-induced transformations.[7–9]

The gas pressure range sampled by the Knudsen flow TAP transients (0–100 Pa) is compatible with the soft and more gentle X-ray spectroscopies, including AP-XPS[10–12] and photoelectron-detected NEXAFS,[13] that can deliver highly surface-sensitive and chemically-specific *in situ* information about the catalyst. These techniques are also feasible at minute[14] to sub-second[15] time resolutions, providing fundamental mechanistic and kinetic insights into processes such as graphene growth on metal substrates[16,17] and many catalytically-relevant reactions,[18–20] although they have been applied primarily to model surfaces. It can be surmised that combined, or at least quantitatively reconciled, TAP and AP-XPS experiments with complex reactive surfaces are an intriguing prospect with considerable potential for materials discovery and reactivity understanding. There have been several reports employing TAP in combination with soft X-ray spectroscopic techniques[21–23] where spectroscopy has been used to characterize the limiting states of the catalyst surface. Quantitative time-resolved reconciliation has, to the best of our knowledge, not yet been fully demonstrated in the literature.

There are two primary challenges that must be resolved in order to achieve dynamic spectro-kinetic TAP/AP-XPS coupling:

(1) Microscopically-equivalent surfaces must be prepared on support materials that are compatible with the two techniques.

(2) Pretreatment protocols and gas-phase perturbations employed in specific TAP and AP-XPS instruments must be matched on a common scale to ensure that equivalent *in situ* states of the surface are accessible during both experiments.

The "materials gap" between TAP-compatible and planar, surface-science compatible samples is bridged only in certain fortuitous cases. For example, the kinetics of oxidative methanol coupling were qualitatively and quantitatively reconciled within the same microkinetic model on single crystal and nanoporous gold surfaces, largely due to the fact that both materials expose essentially equivalent surface terminations.[24] In general, special synthesis methods must be developed to re-create the same active sites on both 2D planar and 3D supports. Atomic beam deposition (ABD)[25] and magnetron sputtering[26] provide two examples of such synthetic versatility utilized for TAP measurements. However, both methods are mostly used for metal deposition, whereas more versatile synthesis platforms are needed to build complex, multicomponent surfaces, including oxide layers, for systematic catalytic studies of industrially-relevant catalysts.[27] Atomic layer deposition (ALD) can potentially offer a more flexible and precise method for tailored surface construction. ALD processes have been developed to prepare a wide variety of surfaces which mimic or even outperform industrially-relevant catalytic sites.[28–30] There have also been examples of successful ALD depositions of active ingredients into macroporous and even microporous supports.[31–34]

Before fully-coupled, simultaneous quantitative AP-XPS and gas-phase analysis within a well-defined transport regime becomes a reality, especially with sub-second time-resolution, a number of challenges can already be addressed with separate AP-XPS and TAP experiments. Given that materials present the same active sites and *in situ* pretreatment protocols are reproducible, the reaction

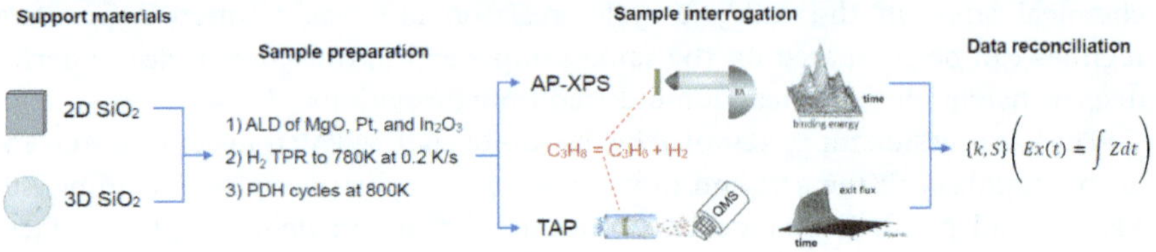

Fig. 1 Experimental and data reconciliation workflow for combined TAP/AP-XPS experiments with ALD-derived model catalysts.

kinetics observed in both experiments can be compared on a unified scale of kinetically-relevant quantities, such as the gas composition, the total number of gas–surface collisions, and the history of the surface evolution. Herein, we demonstrate the reconciliation of TAP and AP-XPS experiments on a relatively complex, yet tractable model material prepared on 2D (for AP-XPS) and 3D (for TAP) supports by the same sequence of ALD processes and *in situ* pretreatments. In particular, we monitor with AP-XPS the deposition of carbon on PtIn alloyed nanoparticles supported on MgO during propane dehydrogenation (PDH) and semi-quantitatively reconcile these data with changes in the intrinsic PDH kinetics monitored within the same exposure scale with TAP. Fig. 1 exemplifies this novel workflow.

PDH on supported Pt nano-catalysts doped with metals like Sn, In, and Ga constitutes a widely studied model system for probing structural and electronic effects in catalysis by alloys, including *in situ* surface restructuring.[35–40] Mounting evidence suggests that these dopant element alters the electronic structure of Pt surface sites *via* charge redistribution in such a way that the product alkene is less strongly bound, thus minimizing sequential reactions of deep dehydrogenation and eventual coking. The structural influence of the dopant is attributed to the preferential blocking of low-coordinated Pt sites and the decreased size of Pt ensembles, which are responsible for hydrogenolysis and coke formation. Various aspects of ALD-based models for Pt-based bimetallic catalysts were previously investigated in our research groups.[41–48] More recently, an ALD-derived PtIn catalyst was reported that achieved unprecedented catalytic performance by combining propane dehydrogenation with selective hydrogen combustion.[49] Although time-resolved XPS studies have been extensively reported for similar (single crystal) model materials,[16,17,50] our findings provide the first example in which AP-XPS and TAP data were evaluated on nanoscopically-equivalent, triple-component (Pt, In, and MgO) ALD model materials.

Experimental and numerical methods

Sample preparation and characterization

Model materials were prepared by applying sequential ALD deposition steps and *in situ* pretreatments on two different substrates: 2D – Si wafer for AP-XPS and 3D – polycrystalline quartz particles (SiO_2, $250 < d_p < 400$ μm) for TAP. In both cases, the outer surface is terminated by SiO_2, which can be partially hydroxylated to initiate ALD surface reactions but is likely to remain inert in high-temperature catalytic reactions. Furthermore, the SiO_2 layer plays a role in minimizing charging effects during XPS. First, a layer of MgO was deposited onto the SiO_2-terminated substrates using the $Mg(MeCp)_2$ and water ALD process. Next, Pt was deposited on top of the MgO layer using 50 ALD cycles of ($MeCpPtMe_3$) and ozone.[51] Finally, 40 ALD cycles of In_2O_3 were performed with [In(TMHD)(3)] and O_2 plasma.[52] For the 3D support, a bed of quartz particles was loaded into a molybdenum cup, placed on the sample heater, and subjected to the same ALD steps, but using quadrupled precursor and reactant pulse times to create an equivalent surface.[53] Upon reduction in a H_2/N_2 mixture at ambient pressure, while heating from RT to 780 K at 0.2 K s^{-1}, PtIn NPs were formed. These "as prepared" alloyed materials were further pre-treated *in situ* with multiple redox cycles before the main TAP and AP-XPS data collection. Post mortem XRD, EDX, and CO chemisorption were performed on the samples collected after the experiments to reveal the reaction-induced structural changes.

CO chemisorption at 323 K was performed on the used 3D-supported sample using an AutoChem II 2920 instrument. XRD measurements were performed using a Bruker D8 diffractometer with Cu Kα radiation. SEM images were taken on a Hitachi SU8230 field emission scanning electron microscope (FE-SEM). XRF measurements were performed using a Bruker Artax system with a Mo X-ray source (at an angle of 45° with the sample surface) and a silicon drift detector placed at an angle of 52° with the sample surface. The fluorescence signal was integrated over a period of 100 seconds.

AP-XPS experiments

In situ AP-XPS measurements were performed at the NAP-XPS experimental station attached to the TEMPO beamline[54] of the Soleil synchrotron radiation facility. The end-station comprised a load-lock for sample introduction, a catalysis chamber for ambient pressure pretreatments, a preparation chamber (not used here), and an analysis chamber containing a differentially-pumped SPECS Phoibos 150 NAP electron energy analyzer. The 2D Si-wafer sample was mounted on a standard sample holder, which allowed for resistive heating and temperature monitoring by a spot-welded thermocouple. Gases were introduced into the pretreatment and analysis chambers in the back-filled mode using SPECS gas-delivery system. Core photoelectrons from different elements were collected (see the ESI†). The main experiment was conducted with PDH at 800 K by introducing a reactant mixture into the analysis chamber (batch mode) as quickly as possible, while monitoring the accumulation of carbonaceous species on the surface of the catalyst in the C 1s spectral range. The nominal zero time of the experiment was recorded with the precision of several seconds based on the fast pressure measurement in the pre-lens region of the electron analyzer, which was calibrated with respect to the pressure in the chamber ahead of time. To maximize the surface sensitivity, transient experiments reported herein detected C 1s and Mg 2s photoelectrons with kinetic energies of 100 eV (20 eV pass energy, 0.05 eV step size, 0.05 s dwell time) and 295 eV (20 eV pass energy, 0.1 eV step size, 0.005 s dwell time), respectively, which were excited by the incidence photons of 385 eV. The energy scale was calibrated using the Mg 2s peak corresponding to the MgO substrate, as Mg 2p overlaps with Pt 5p. According to Schulze *et al.*,[55] Mg 2s is quite invariant to chemical shifts. Therefore, the energy spectrum was adjusted to match the binding energy (BE) of the Mg 2s photoelectron to a value of 89 eV.

Analysis of the collected spectra was performed by component de-convolution using casaXPS software.

TAP experiments

A TAP-3E reactor system at the University of Oslo (Mithra Technologies) was used to perform pulse-response experiments under the conditions of Knudsen flow. 280 mg of the 3D quartz-supported sample was loaded in the middle of a tubular, resistively-heated quartz microreactor (Ø 4 mm, 50 mm length) as a thick zone (20 mm) sandwiched between two zones of inert quartz (same $250 < d_p < 400$ μm size fraction). The temperature was measured in the middle of the catalyst bed by a K-type thermocouple. The sample was first pretreated in the flow of H_2 at 20 sccm, while ramping to 800 K at 5 K min^{-1}. After evacuation to the background pressure of 10^{-8} mbar at 800 K, a series of C_3H_8/Ne pulses (0.25/0.75 composition) was issued at a constant temperature of 800 K using high-speed pulsing valves (100 μs duration, $\sim 10^{-8}$ mol pulse size based on Ne calibration), while the following AMUs were monitored (one AMU per pulse) at the reactor exit by an Extrel quadrupole mass-spectrometer with 1 ms time resolution. The same number of PDH/oxidation/re-reduction cycles as in the AP-XPS experiments were performed before collecting the dataset discussed herein. The relative intensities of the overlapping fragments, as well as the absolute signal intensities, were calibrated by injecting C_3H_8/Ne, C_3H_6/Ne, CH_4/Ne, and H_2/Ne mixtures over an inert quartz bed. TAPFIT code from Ghent University was used to model the gas concentration transients at various positions of the packed bed.[56,57] Further analysis of experimental and simulated data was performed using custom python scripts and jupyter notebooks.

Results and discussion

Scales of exposure and criteria of equivalency for the catalyst states in TAP and AP-XPS experiments

The quantitative reconciliation of data from two dissimilar measurements, especially when different phases are monitored (*e.g.* the gas phase in TAP *vs.* the surface phase in AP-XPS), necessitates that data from both measurements are evaluated within a common standard scale of accessible surface states. So far as reaction kinetics are concerned, a surface state can be understood as a microstructural and electronic configuration of the material that results in particular observable kinetic properties when probed with a catalytic reaction. Surface states themselves emerge *in situ* and are dependent on the history of the surface's interactions with the gas phase. Ideally, both measurements that are to be reconciled should probe exactly the same kinetic surface state, which is challenging to achieve in two separate devices. In practice, equivalency between different surface states must be evaluated in terms of some phenomenological descriptor of the state, such as the surface coverage of a key reaction intermediate, the concentration of a poison (*e.g.* coke), or the average oxidation state of the surface. If both experiments begin from the same stable and reproducible initial condition, then the subsequent state of the surface can be controlled by reproducing well-defined treatments of gases and temperature. Therefore, in a broad range of cases, the integral (accumulated) exposure of a surface to gaseous reactants during an experiment is a suitable straightforward descriptor of its catalytic state.

In general, for multi-reactant, multi-pathway reactions involving reversible steps, a descriptor to scale the catalyst state $U(t)$ can be defined based on the overall mass balance of the experiment as an instantaneous difference between the uptake and release of certain moieties that are conserved in the reaction:

$$U(t) = \text{uptake}(t) - \text{release}(t) \tag{1}$$

This descriptor, termed surface storage, is generally distinct from the surface coverage of an individual surface intermediate and may reflect the transient accumulation of several substances on the surface. Surface storage has been used for the analysis of catalytic steps at the millisecond scale[58–60] and is the basis of the linear phenomenological rate-reactivity model (RRM).[61] However, in the absence of the transient mass balance, the uptake and release terms can be challenging to obtain. Herein, we consider a simplified situation in which certain assumptions about the reaction mechanism can be made:

(1) the reaction mechanism consists of the main catalytic cycle and a single side reaction,

(2) the side reaction is irreversible,

(3) the side reaction is relatively slow in comparison with the main catalytic cycle, at least on the time scale of the measurement.

This simplification is applicable to a wide range of industrially-relevant reactions, *e.g.* when a catalyst slowly deactivates through the parallel and/or sequential reactions of the reactant and product decomposition into coke. The surface storage of coke $U_C(t)$ in these cases (quantified in mol_C) is reduced to the residual uptake term in eqn (1), after a linear combination of the consumption and the release of all carbon-containing gaseous species has been taken into account. For an irreversible reaction, $U_C(t)$ can be expressed as a function of the integral gas exposure. Thus, a correlation can be established between the reaction kinetics k and a descriptor of the catalyst state U, which is monitored as a function of the integral, generally time-dependent exposure Ex:

$$k \sim f(U_C(\text{Ex}(t))) \tag{2}$$

The integral exposure of a surface to a reactant is given by:

$$\text{Ex}(t) = \int_0^t Z(t')\text{d}t', \tag{3}$$

where $Z(t)$ is the (generally) time-dependent impingement rate of gas molecules per unit surface area, $1/\text{m}^2\ \text{s}^{-1}$. $Z(t)$ can be expressed, assuming that the ideal gas law applies, through either pressure $P(t)$, Pa, or concentration $C(t)$, mol m^{-3}, using the Hertz–Knudsen equation:

$$Z(t) = \frac{P(t)}{\sqrt{2\pi k_B m T}} = \frac{C(t)N_A}{\sqrt{2\pi MRT}}, \tag{4}$$

where k_B is the Boltzmann constant, N_A is the Avogadro constant, T is temperature, m is the mass of a gas molecule, and M is the gas molecular mass. In the surface science literature, exposure is typically reported in Langmuirs, defined as $1L = 10^{-6}$ torr s (1.33 mbar s). For the experimental results herein, exposure is quantified in site-specific molar units, $mol_{collision}$ per site, provided that the surface density of the relevant sites, N_S, is quantified by independent measurements (*e.g.* chemisorption):

$$\mathrm{Ex}(t) = \frac{\int_0^t Z(t')\mathrm{d}t'}{N_A N_S}. \tag{5}$$

It is useful to evaluate and compare typical exposure histories experienced by a sample in TAP and AP-XPS experiments in order to improve the methodology for the quantitative reconciliation of the two techniques.

In a typical TAP experiment, approximately 1–10 nmol of a gas mixture is injected into the microreactor with each pulse, with the amount of catalyst sample in the microreactor and the gas composition varied according to the needs of the experiment. Usually, 5–100 mg of a high-surface area pelletized (250 $< d <$ 400 micron sieve fraction) material is packed into a 1–5 mm thin zone in the middle of the microreactor between two inert zones packed with quartz. The thin zone configuration with a very small length ratio, $l_{\mathrm{cat/linert}}$, maintains the spatial uniformity of the surface state within the sample,[62–64] but this requirement can practically be met only for reactions that are neither too fast nor too slow per amount of material. For intrinsically slower reactions or materials with a lower concentration of active sites per catalyst mass, TAP data with a reasonable signal-to-noise ratio can be obtained in a standard TAP experiment by increasing the thickness of the catalytic zone. Such data can be quantitatively interpreted through numerical modeling, albeit with additional uncertainty in the spatial sample uniformity. Ultimately, the optimal experimental conditions are determined by the interplay of the reaction kinetics, bed-scale transport, and detector sensitivity.

The maximal integral exposure of a catalytic sample to a gas during a TAP pulse-response experiment can be estimated by substituting the appropriately scaled concentration transient of an inert tracer into eqn (3), and this result is generally valid for reactions in which reactive gases irreversibly accumulate as surface species (*e.g.* coke). The diffusion-only concentration transients for this estimation can be generated for any set of operating conditions by numerically solving 1D reaction-diffusion equations of the standard TAP model.[59,65] Several parameters of this model and of the real TAP experiment can be varied in order to control the exposure: (i) the amount of reactant injected per pulse, which is proportional to the total pulse size and the reactant's partial pressure in the feed mixture, (ii) temperature, and (iii) the number of sequential pulses in the pulse train. Fig. 2a and b depict an example TAP transient in the units of concentration and pressure, respectively, in the middle of the catalytic zone during a typical single-pulse experiment (pulse size $N_p = 1$ nmol, 100% composition, $T = 800$ K). At its peak, the concentration reaches 3.8 mmol m^{-3}, which corresponds to the peak pressure of 25 Pa. The corresponding exposure history calculated using eqn (3) and space-averaged over the entire catalyst zone is shown in Fig. 2c. The final integral exposure of *ca.* 10^{-21} 1/m^2 corresponds well to that estimated by Wang *et al.* for similar TAP pulses in a shorter reactor.[66,67]

AP-XPS measurements are routinely performed in either back-filled chambers at static pressure or under a steady-state flow of reactants through a relatively small volume dedicated cell.[68–70] The discussion below is focused on the AP-XPS experiments with a static reactant pressure, although the impact of more

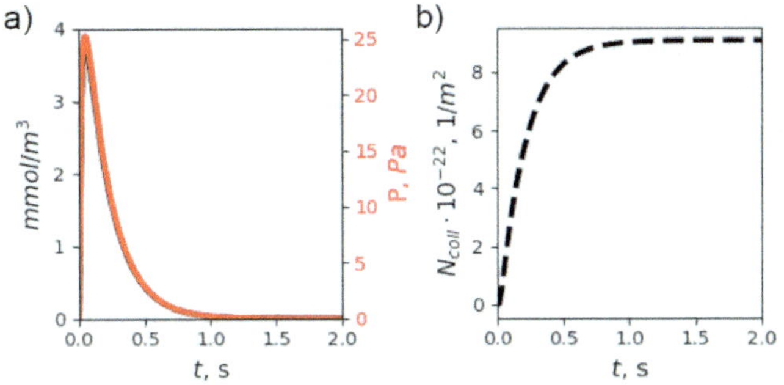

Fig. 2 Representative TAP pulse transients in the Knudsen flow regime (middle of the catalyst zone): (a) in the units of gas-phase concentration and pressure, and (b) in the units of exposure.

sophisticated control of the gas environment, including transients, is discussed in the outlook section of this paper. The accumulated exposure in an experiment with constant pressure is controlled by the partial pressure of the reactant and the time of exposure:

$$P(t) = p_i t. \tag{6}$$

In Fig. 3, the exposures (in total mol of collisions per m^2) for typical TAP and AP-XPS experiments are mapped over representative ranges of experimental conditions controlling the exposure, *i.e.* the pulse size, N_p, and the number of pulses for TAP and the partial pressure of the key reactant and the time of exposure for AP-XPS. The full set of model parameters used for the simulations can be found in the ESI.† Comparison of these maps shows that there is a significant overlap of accessible collision numbers between the two techniques, particularly in the limit of <0.1 mbar (<10 Pa) pressure in AP-XPS and relatively large TAP pulses (10^{-8} mol). Equipped with these estimations, we perform

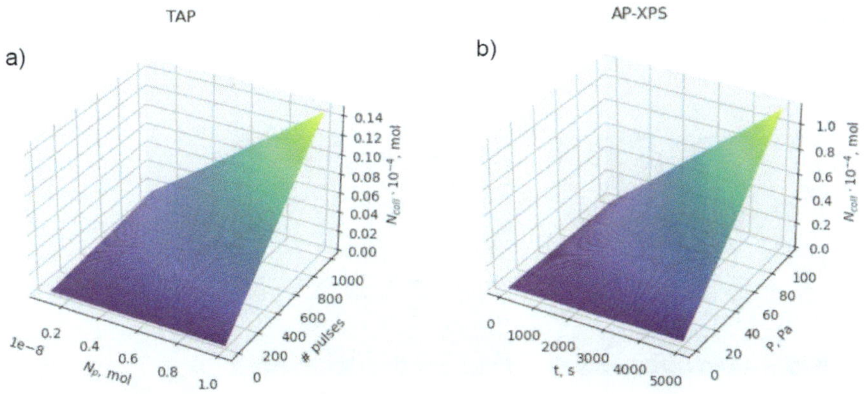

Fig. 3 Accessible ranges of exposure *vs.* the key experimental conditions: (a) TAP – exposure *vs.* the pulse intensity, N_p, and the total number of injected pulses, (b) AP-XPS – exposure *vs.* the partial pressure of the key reactant, *P*, and the time of exposure.

a proof-of-principle alignment of the exposure scale for a pair of ALD-derived model PtIn/MgO/SiO_2 samples.

Model materials

The main objective of the experiments described herein was to compare the kinetic information about the same model process, in this case, the initial stages of coke deposition on Pt-based catalysts during alkane dehydrogenation, as measured by the two complementary techniques of TAP and AP-XPS. Model bimetallic PtIn/MgO catalysts prepared *via* ALD contain relatively small amounts of In (Pt/(Pt + In) ratio of 0.86 based on XRF of a 2D sample) in comparison with the performance-optimized, high-surface area PtIn/Mg(Al,In)O_x catalysts.[40] XRF, SEM, XRD, and XPS (see the ESI†) reveal that the deposition of a MgO film resulted in a relatively smooth surface, while subsequent depositions of Pt and In_2O_3, followed by a temperature-programmed reduction to 780 K, initially resulted in the formation of clearly distinguishable *ca.* 10–20 nm metal nanoparticles. Previous reports on very similar catalysts derived from layered double hydroxides (LDH) and ALD-derived samples[36,71] suggest that after several cycles of reduction/dehydrogenation/oxidation/re-reduction treatments, Pt and In form compositionally and structurally stable alloyed NPs, the degree of sintering of which depends on various preparation factors, that display consistent kinetic behavior in propane dehydrogenation. In order to prepare such a stable starting surface, the as-prepared 2D and 3D samples were subjected to the standard sequence of (*ex situ* and *in situ*) cycling pretreatments before further characterization. According to the post mortem (after PDH reaction) characterization of the 2D sample (see the SEM and XRD in Fig. 4), the resulting surface is comprised of partially-sintered worm-like metal nanostructures with an average width of 20 nm and highly variable length in the order of hundreds of nm. The powder diffraction pattern of the freshly reduced (before) catalyst shows a broad feature at around 39°, which can be attributed to the (111) diffraction of the Pt_3In alloy (expected 2θ is 39.1°). However, a contribution of Pt (111) cannot be excluded (expected 2θ is 39.8°) and is even likely, based on the XRF quantification of Pt/(Pt + In) = 0.86 and previous characterization of similarly-constructed samples.[53] The broadness of the peak can also be explained by the morphology and size of the Pt_3In features, as observed in the SEM. The XRD pattern of the used catalysts shows a more intense

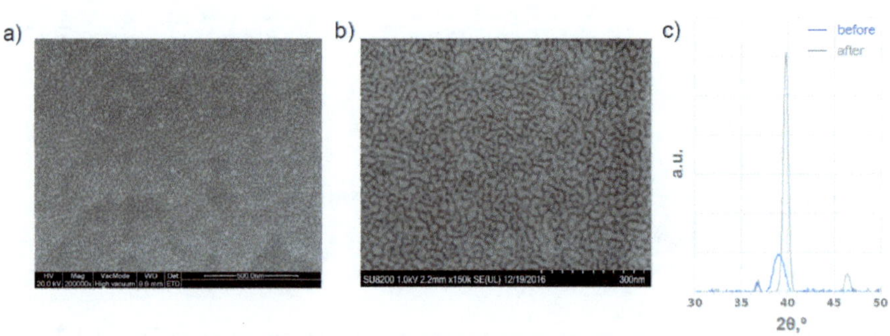

Fig. 4 SEM (a and b) and PXRD (c) characterization of the as-prepared and used ALD-derived 2D model PtIn/MgO/SiO_2 catalysts, respectively.

and narrow diffraction peak at around 39.8°, with an additional diffraction feature at around 46.5°. Both diffractions can be attributed to Pt, *i.e.* (111) and (200), respectively. Together with the XPS data discussed below, which clearly indicate the presence of PtIn alloy, the XRD suggest that the *in situ* surface state of the 2D sample at 800 K comprised sub-stoichiometric (In-poor) Pt_3In alloy, possibly with a more Pt-rich core. The total surface concentration of exposed Pt sites on the 3D sample was quantified by CO chemisorption (assuming a 1 : 1 adsorption stoichiometry) to be (0.2 $\mu mol\ g^{-1}$ for the 3D sample) 30 $\mu mol_{Pt}\ m^{-2}$, which is consistent with particles of several hundred nm. This estimate was used in the subsequent alignment of the exposure scale, assuming that the concentration of exposed Pt sites is comparable on both the 2D and 3D samples due to the high controllability of the ALD technique.

TAP experiments

As we have previously demonstrated with TAP experiments,[72] the accumulation of carbon deposits on a monometallic LDH-derived Pt/Mg(Al)O_x catalyst initially results in gradual catalyst deactivation, accompanied by increasing selectivity towards propene, with the subsequent release of methane and hydrogen. In comparison, analogous TAP pulse-response experiments with the model 3D ALD-derived PtIn/MgO/SiO_2 material revealed a generally similar, yet distinct, pattern of activity and selectivity. Initial experiments showed a much lower conversion of propane per catalyst loading, as expected for a non-porous material with a drastically lower concentration of Pt sites per available surface, both due to the low surface area of the quartz support and much larger size of the PtIn nanoparticles. To increase the conversion, the catalyst loading was extended to almost a half of the reactor volume, but was kept within the known limits of the central isothermal zone. The catalyst number, *i.e.* the ratio of injected gas molecules per pulse to the total number of surface sites, amounted to ~0.5 $mmol_{C_3H_8}\ mol_{Pt}^{-1}$ for the freshly reduced catalyst. Since the pulse size was maintained considerably lower than the amount of sites on the sample, the experiment was regarded as state-defining, in a sense that individual pulses did not drastically alter the state of the catalyst and the catalyst composition gradient along the reactor axis was modest.

A typical experiment started with a fully-reduced, carbon-free pretreated catalyst at 800 K. In Fig. 5a, the stacked sequences of millisecond-resolved responses of propane, propene, methane, and hydrogen during a typical pulse sequence are shown as functions of the pulse number. Based on quantitative MS calibration including overlapping fragmentation patterns, the overall mass balance for this experiment is reflected in Fig. 5b as a function of the injected propane amount. The conversion of propane gradually decreased from 0.7 to 0.3, and only a very small fraction was converted to propene (<0.1). Methane was detected only in trace amounts (not quantified further), and no other volatile products were observed in measurable quantities. It is reasonable to assume that the majority of missing carbon remained on the surface in the form of carbonaceous deposits (coke). In agreement with previous results for Pt-based catalysts,[72] the yield of propene increased towards the end of the experimental sequence, when a considerable amount of coke was already accumulated on the surface. The final coke content reached 35 mol_C/mol_{Pt}, in full agreement with the

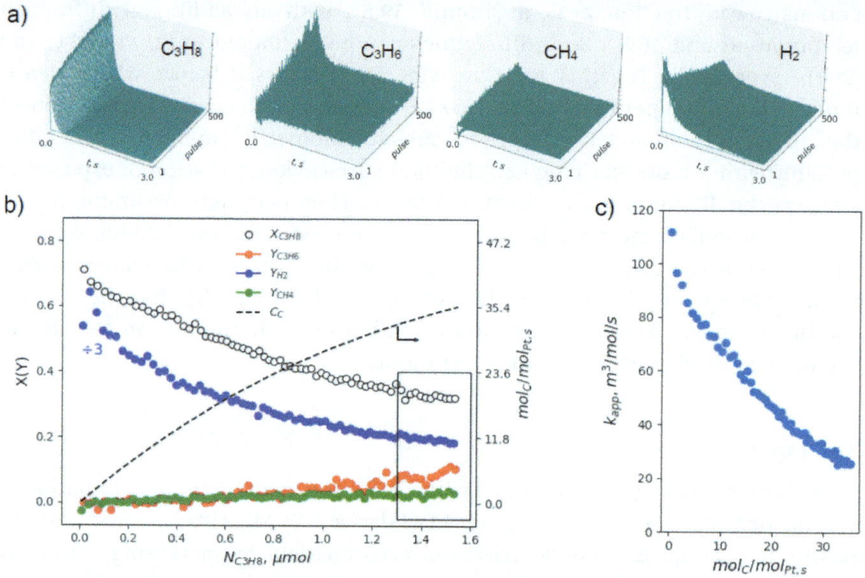

Fig. 5 TAP results for propane dehydrogenation at 800 K on the ALD-derived 3D PtIn/MgO/SiO_2 model catalyst: (a) transient responses of propane, propene, methane, and hydrogen as functions of time and pulse number within the sequence, (b) propane conversion and product yields *vs.* the amount of injected propane. The same panel also shows carbon accumulation on the surface in $mol_C/mol_{Pt,s}$ that was estimated from the mass-balance of the experiment. The rectangular area indicates the range of data used for the mean residence time analysis in Fig. 6. (c) the apparent rate coefficient of propane consumption *vs.* the amount of accumulated carbon.

numbers reported by Sun *et al.*[40] for PtIn/Mg(Al,In)O_x catalysts with low In content. Hydrogen production generally followed the same trend as propane conversion, but was quantified in much lower-than-stoichiometric amounts, assuming four mols of gaseous H_2 are released for every molecule of completely dehydrogenated propane. While a refinement of the hydrogen quantification was not attempted in this work, we note that a systematic calibration of TAP measurements, in principle, can provide an estimate of the residual H/C ratio in coke deposits based on the overall hydrogen balance of the experiment.

The height-normalized responses of neon, propane, propene, and hydrogen are compared in Fig. 6a to reveal clear differences in their temporal behavior. Their mean residence times inside the microreactor are shown in Fig. 6b, which were obtained from the height-normalized responses, F_{norm}, using first and second statistical moments:

$$\tau_r = \frac{\int tF_{\text{norm}}(t)\text{d}t}{\int F_{\text{norm}}(t)\text{d}t} \tag{7}$$

Here, the experimentally-observed values are compared to the expected diffusion-only residence times, *e.g.* inert τ_{Ne} corrected for the sqrt(T/M) dependency of the Knudsen diffusivity on the molecular mass, in order to delineate transport effects from kinetics. Only the last 85 pulses (rectangle in Fig. 6b) were

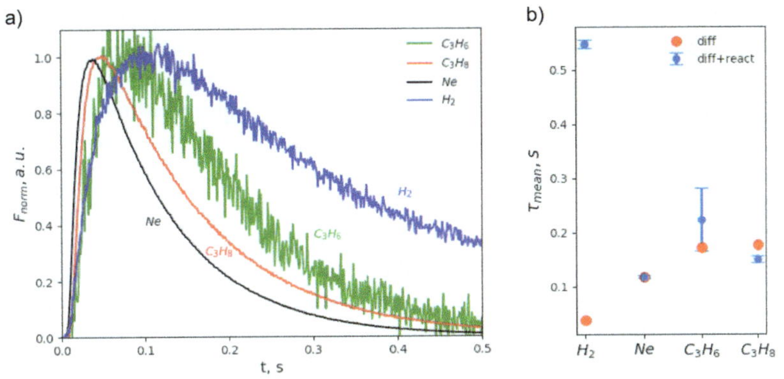

Fig. 6 Comparison of the mean reactor residence times of different gas molecules: (a) height-normalized transients and (b) moment-derived mean residence times.

considered to obtain these values, as propene generation was too low to be distinguished from noise earlier in the sequence. For propane, the mean residence time was on average lower than the diffusion-only value, as expected for an irreversibly-consumed gas. For propene, despite a rather large uncertainty, it can be concluded that the average mean residence time was slightly higher than the expected diffusion-only value. The scale and the opposite sign of the diffusion-normalized tau values with respect to the inert standard confirm that the calibrated signals at AMUs 29 and 42 indeed reflect the behavior of independent chemical species – propane and propene. The increased surface residence time of propene is consistent with its generation from propane by surface reactions, followed by its reversible re-adsorption. However, re-adsorbed propene on Pt catalysts is further transformed into coke,[73–75] which counter-currently lowers $\tau_{C_3H_8}$. Therefore, the magnitude of experimentally-observed $\tau_{C_3H_8}$ reflects the balance between its production, desorption, re-adsorption, and deep dehydrogenation rates. Finally, hydrogen exhibited a considerably longer mean residence time, which can be attributed to the combined effect of the primary hydrogen formation from propane dehydrogenation into propene, the parallel/sequential hydrogen formation from deep dehydrogenation of coke precursors, and the spillover of atomic hydrogen onto the MgO surface surrounding metal NPs. The latter process was detected by the increased mean residence time of hydrogen in TAP experiments on the LDH-derived Mg(Al,Ga)O surfaces.[76]

The activity of the catalyst towards consuming propane was quantified by estimating an apparent first order rate coefficient (k) from propane conversion within each pulse in the sequence. For this goal, the following reaction–diffusion equation was numerically solved for the gaseous propane concentration (C) within the packed bed:

$$\varepsilon\frac{\partial C}{\partial t} = D\frac{\partial^2 C}{\partial x^2} - \frac{m_{\mathrm{cat}}}{AL_{\mathrm{cat}}}kC_{\mathrm{Pt}}C, \tag{8}$$

where ε is the bed void fraction, D is the effective Knudsen diffusivity in $m^2\ s^{-1}$, m_{cat} is the catalyst loading in kg, A is the reactor cross-sectional area in m^2, x is the axial distance from the reactor inlet, L_{cat} is the length of the catalytic zone, and C_{Pt} is the nominal concentration of Pt sites in the sample in mol $kg_{\mathrm{cat}}{}^{-1}$. The dependency of the intrinsic rate coefficient on the accumulated amount of coke is plotted in Fig. 5c. The scale of this rate constant reflects the relatively poor activity of this specific model catalyst. After an initial sharp decline, the rate coefficient decreased nearly linearly with coke content, but the slight non-linearity suggests that in the limit of high coke coverage, the constant will retain a small, but finite value. Overall, TAP experiments on the 3D model sample provide a direct correlation between the amount of coke deposited on the catalyst surface and the kinetic parameters of propane dehydrogenation into propene and coke. Next, we recreate the process under comparable reaction conditions during AP-XPS experiments with an equivalent 2D sample in order to monitor the coke formation by surface spectroscopy.

AP-XPS experiments

The 2D sample supported on a Si wafer was pretreated *in situ* in a hydrogen atmosphere by ramping the temperature to 800 K at a rate of 5 K min^{-1}. Several dehydrogenation/regeneration cycles were performed at 800 K to stabilize the catalyst, thus mimicking the initial catalyst state in the TAP experiments. Deconvolution (fitting) of the core level Pt 4f and In 3d spectra for the prepared catalyst surface (see the ESI†) indicates that Pt is fully reduced and alloyed with In. Specifically, a doublet component of the Pt 4f spectra with the binding energies of 70.63 and 73.93 eV can be assigned to the Pt–In alloy, which has a lower binding energy compared to the doublet corresponding to pure metal Pt in the same sample (71.29/73.93 eV, respectively) as a result of electron donation from In. Indium is likely present in several oxidation states. However, a prominent narrow In $3d_{5/2}$ component at 443.24 eV and a smaller and broader component with a slightly higher BE at 443.8 eV can be confidently assigned to In^0 free metal and In^0 in the Pt–In alloy, respectively, in which In partially donates charge to Pt. The metallic and alloyed nature of In is also consistent with the $In_4M_{4,5}M_{4,5}$ Auger line with the Auger parameter of 853.31 eV. These observations are consistent with the common hypothesis of dopant-induced electronic effects being a major driver of enhanced catalytic properties of alloys in comparison with monometallic Pt. The C 1s signal is significant in most spectra of as-prepared and coked surfaces, but completely disappears upon high-temperature treatment in O_2, offering a suitable background for time resolved measurements.

Having verified the alloyed and reduced nature of the catalytic NPs, the freshly prepared carbon-free surface was rapidly subjected to a 0.1 mbar static pressure of either C_3H_8 or $C_3H_8 + H_2$ (1 : 1) mixture at 800 K, while C 1s and Mg 2s (reference) photoelectrons were collected at the kinetic energies of 100 and 295 eV, respectively, for maximum surface sensitivity. The rate of data acquisition was 70 s per each pair of sequentially recorded C 1s (BE 295–270 eV) and Mg 2s (BE 95–85 eV) spectral windows. The resulting transient C 1s signals energy-referenced to Mg 2s are depicted in Fig. 7. As expected, the C 1s signal located a few tenths of eV above 284 eV has gradually increased and shifted slightly to a higher BE due to the accumulation of propane dehydrogenation intermediates on the surface and the formation of carbon deposits. The final state of carbon was characterized in more detail by a high-resolution scan at the same temperature (see ESI†).

Our observations of a single, relatively narrow signal in the 284–285 eV BE range are consistent with many reports in the literature concerning sp^2 and sp^3

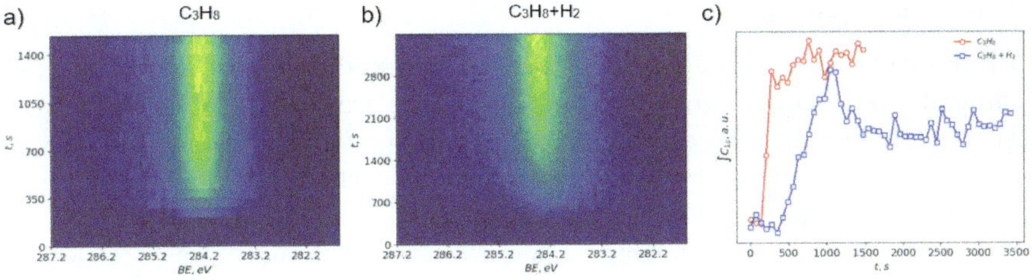

Fig. 7 *In situ* C 1s XPS spectra *vs.* time during the exposure of the ALD-derived 2D PtIn/MgO/SiO$_2$ model catalyst to 0.1 mbar of (a) pure C_3H_8 and (b) a 1 : 1 C_3H_8 : H_2 mixture. The integrated areas of the C 1s envelope (a.u.) are plotted *vs.* time in panel (c).

carbon deposition on model Pt surfaces[77–80] The quantitative interpretation of the C 1s spectra of carbonaceous deposits on complex materials can present considerable challenges.[81,82] However, based on the abundant literature referenced above, it is reasonable to hypothesize that the data in the figure reflect the kinetics of sp^2 and sp^3 carbon accumulation on the surface primarily due to the PDH reaction. The potential influence of the X-ray beam (damage) on the formation of the surface carbon from either the degradation of the underlying material or hydrocarbon reactions can be neglected for the results reported herein, at least at the relevant time scale of the experiment. This is supported by negligible C 1s signal observed for extended periods of time under vacuum and during the initial scans during/after the rapid propane introduction. In order to avoid over-interpretation of the C 1s signal in this proof-of-concept experiment, the integrated area of the entire C 1s envelope (linear background-corrected) was used to quantify the rate of carbon accumulation on the surface *vs.* time, including stable coke species, coke precursors, and minor contributions from all reaction intermediates.

Fig. 7c depicts the transient accumulation of carbon-containing species on the surface. For pure propane, the initial rapid growth of this envelope quickly approached a plateau, after which carbon coverage remained practically unchanged. In the presence of hydrogen, the initial rise in C 1s was delayed in time and occurred more slowly than in pure propane. After approximately 1000 s, the carbon content reached a well-defined maximum and then approached a steady-state plateau at the level slightly exceeding half of the plateau without hydrogen. The influence of hydrogen on the kinetics and stability of steady-state alkane dehydrogenation has been shown[35,83] to be beneficial for the overall process yield up to a 1.3 ratio, despite seemingly unfavorable thermodynamic factors. The difference in the kinetics of the initial carbon accumulation on our model catalysts with and without hydrogen is consistent with the removal of coke precursors by hydrogen, which has emerged as the leading explanation of the positive hydrogen effect in PDH.[74,75] This example illustrates that time-resolved AP-XPS in combination with quantitative kinetic analysis can offer a promising approach for discriminating mechanistic hypotheses in catalysis, which may have remained inconclusive, *e.g.* because of the limited absolute quantification potential of the C 1s envelope. Further progress in this direction requires the ability to monitor the evolution of surface species within the same range of exposure as in quantitative gas-phase kinetic measurements.

A common exposure scale

The amount of coke deposited on the model catalysts during TAP experiments (3D sample) can now qualitatively be compared to the accumulated C 1s signal from AP-XPS (2D sample) on a common scale of propane exposure per Pt site (Fig. 8). For the TAP dataset, the gas transport through the packed bed was simulated with realistic (experimental) dimensions using a fully-resolved three-zone diffusion model, from which the integral space-averaged exposure of the catalyst to propane (exposure per surface Pt atom) was estimated to be $\sim 2 \times 10^6$ $mol_{C_3H_8}/mol_{Pt,s}$. The C 1s signal was not calibrated due to the large experimental uncertainty of its intensity as a function of carbon content and is presented in the same figure on a separate *y*-scale in arbitrary units. However, even such semi-quantitative (*x*-axis only) data alignment between two dissimilar experiments demonstrates a surprising agreement in the rate of carbon accumulation when comparing TAP and AP-XPS transients for pure C_3H_8 with the AP-XPS transients for the C_3H_8 + H_2 mixture.

The concept of correlating AP-XPS data to the kinetics of carbon growth on the exposure scale has been successfully exploited in the context of graphene growth on model materials.[16,17,50] In comparison, our work introduces this concept to the field of the precise kinetic characterization of catalytic reactions. To fully appreciate the possibilities entailed by a fully quantitative data alignment, it should be emphasized that TAP offers a detailed view of reaction kinetics at several temporal scales – from sub-second differences in the mean residence times of species, through several seconds within each pulse, to the accumulated changes over hundreds of pulses – and all these characteristics could be related to the surface-sensitive spectral features probed by XPS for cases of technological and fundamental importance.

Our estimates provide only order of magnitude, first-order approximations to some of the conditions controlling the exposure experienced by the catalyst surface, some of which are poorly understood in general. For example, the actual partial pressure of reactants in front of the sample area probed by AP-XPS is

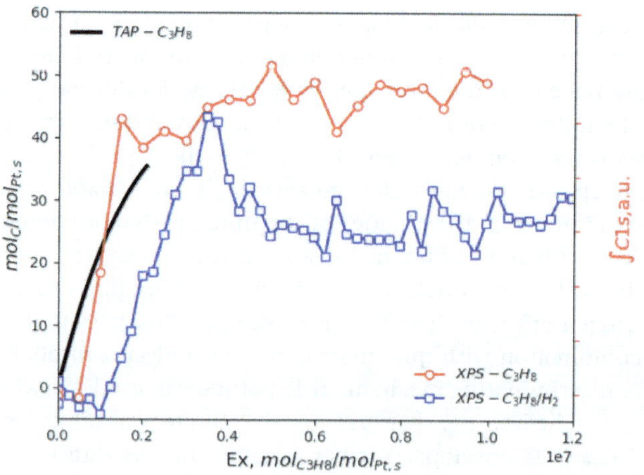

Fig. 8 TAP and AP-XPS datasets on carbon accumulation, reconciled on the common scale of gas exposure.

usually lower than the nominal pressure measured in the chamber due to the differential pumping through the analyzer.[84] Considerable progress has been made in improving the understanding and control of the gas transport in AP-XPS devices,[14,85–87] and further corrections can be devised for the exposure models in both devices. We envision that these advanced capabilities will enable more robust and flexible transient experiments to be performed in kinetic and spectroscopic devices with eventual data reconciliation in mind from the start of experimental planning and data acquisition.

The range of equivalent exposures achievable simultaneously in TAP and AP-XPS can be extended by considering longer and shorter time scales, which were not covered in this work. For example, progress in data acquisition and the wider accessibility of high-brilliance photon sources have drastically improved the signal to noise ratio accessible to time-resolved AP-XPS, increasing the potential time resolution of such measurements. Likewise, longer or drastically shorter[88] residence times can be achieved in TAP reactors with tailored diffusivity characteristics in order to implement experiments with much shorter or higher total exposure on a reasonable astronomical time scale. Undoubtedly, a more systematic use of quantitative measurements within the intermediate pressure range between Knudsen diffusion and atmospheric flow[89] will provide an even wider range of catalyst states accessible to both AP-XPS (up to 100 mbar) and TAP.

From the materials synthesis perspective, the main challenge is to achieve better control of ALD depositions on 2D and 3D surfaces, including macro- and meso-porous supports, thus expanding the range of model surface configurations that can be probed by the tandem TAP/AP-XPS experiments. Improved *in situ* stability of ALD-derived surfaces against sintering and composition changes is also highly desirable. Ultimately, fully combined and automated systems will be developed for the incremental synthesis, guided by transient kinetics, whereby the information is acquired from the gas- and surface-phase perspectives.

Conclusions

We have proposed the alignment of TAP and AP-XPS data on a common scale of the integral gas–surface collisions (exposure) in order to reconcile the kinetic characteristics available from TAP with spectroscopic signatures of surface species available from AP-XPS. The concept was demonstrated by aligning datasets for a series of multi-component, ALD-derived model catalysts consisting of PtIn bimetallic nanoparticles supported on a layer of MgO. These model surfaces were prepared by the same ALD steps on two supports: a planar SiO_2-terminated Si wafer as a 2D support for AP-XPS and non-porous, $250 < d_p < 400$ μm quartz (SiO_2) particles as a 3D support for TAP. Both surfaces were pretreated in the same way, and then the initial stages of the propane dehydrogenation reaction were monitored on these samples using AP-XPS and TAP, respectively.

TAP data indicate that carbon deposition on the surface of these catalysts resulted in gradual deactivation and an increase in propene selectivity. The AP-XPS experiment was repeated with an equimolar mixture of propane and hydrogen, which resulted in a delayed and slower rate of carbon growth. Importantly, the descriptors of carbon deposition from both measurements were aligned on a common scale of gas exposure per Pt site. More generally, our estimations of accessible exposure scales in TAP and AP-XPS predict that a vast variety of reaction kinetics can be probed by their tandem application. We hypothesize that in the future it will be possible to fully reconcile such time-dependent spectro-kinetic data within the mean-field microkinetic or other phenomenological models of transient reaction kinetics, such as the rate-reactivity model (RRM).[61] The current limitations of the tandem TAP/AP-XPS approach are discussed, and the most prominent challenges are summarized to provide motivation for future research and development.

Conflicts of interest

There are no conflicts to declare.

Acknowledgements

This work was funded by the Research Council of Norway (RCN) under Contract No. 272266 (TAPXPS project – a Nano2021 project). E. A. R. also acknowledges the Marie Curie International Incoming Fellowship granted by the European Commission (Grant Agreement No. 301703). E. A. R. and U. O. also acknowledge the Norwegian National Surface And Interface Characterisation Laboratory (NICE) for access to the TAP facility and Mr Christopher Affolter of Oslo, Norway for the SEM/EDX characterization. C. D. and V. V. G. acknowledge the Research Foundation – Flanders (FWO) for funding (G020911N, G032920N). The authors would like to express their sincere gratitude to the beamline scientists at the TEMPO beamline of the SOLEIL synchrotron (proposal 20141283), especially to Dr Jean-Jacques Gallet, and to Dr Esko Kokkonen of MAX-IV Laboratory for his advice on data reduction software.

References

1 M. P. D'Evelyn and R. J. Madix, Reactive Scattering from Solid Surfaces, *Surf. Sci. Rep.*, 1983, **3**(8), 413–495, DOI: 10.1016/0167-5729(83)90002-X.

2 W. Ho, Spectroscopy of Surface Kinetics and Reaction Mechanisms, *J. Phys. Chem.*, 1987, **91**(4), 766–779, DOI: 10.1021/j100288a006.

3 J. Libuda and H.-J. Freund, Molecular Beam Experiments on Model Catalysts, *Surf. Sci. Rep.*, 2005, **57**(7–8), 157–298, DOI: 10.1016/j.surfrep.2005.03.002.

4 J. T. Gleaves, G. Yablonsky, X. Zheng, R. Fushimi and P. L. Mills, Temporal Analysis of Products (TAP) – Recent Advances in Technology for Kinetic Analysis of Multi-Component Catalysts, *J. Mol. Catal. A: Chem.*, 2010, **315**(2), 108–134, DOI: 10.1016/j.molcata.2009.06.017.

5 S. O. Shekhtman, G. S. Yablonsky, J. T. Gleaves and R. Fushimi, "State Defining" Experiment in Chemical Kinetics – Primary Characterization of Catalyst Activity in a TAP Experiment, *Chem. Eng. Sci.*, 2003, **58**(21), 4843–4859, DOI: 10.1016/j.ces.2003.08.005.

6 S. O. Shekhtman, A. Goguet, R. Burch, C. Hardacre and N. Maguire, CO Multipulse TAP Studies of 2% Pt/CeO_2 Catalyst: Influence of Catalyst Pretreatment and Temperature on the Number of Active Sites Observed, *J. Catal.*, 2008, **253**(2), 303–311, DOI: 10.1016/j.jcat.2007.10.028.

7 Y. Schuurman and J. T. Gleaves, A Comparison of Steady-State and Unsteady-State Reaction Kinetics of n-Butane Oxidation over VPO Catalysts Using a TAP-

2 Reactor System, *Catal. Today*, 1997, **33**(1–3), 25–37, DOI: 10.1016/S0920-5861(96)00115-0.
8 R. Fushimi, S. O. Shekhtman, A. Gaffney, S. Han, G. S. Yablonsky and J. T. Gleaves, TAP Vacuum Pulse-Response and Normal-Pressure Studies of Propane Oxidation over MoVTeNb Oxide Catalysts, *Ind. Eng. Chem. Res.*, 2005, **44**(16), 6310–6319, DOI: 10.1021/ie049162k.
9 X. Zheng, J. T. Gleaves, G. S. Yablonsky, T. Brownscombe, A. Gaffney, M. Clark and S. Han, Needle in a Haystack Catalysis, *Appl. Catal., A*, 2008, **341**(1–2), 86–92, DOI: 10.1016/j.apcata.2008.02.040.
10 L. Nguyen, F. F. Tao, Y. Tang, J. Dou and X.-J. Bao, Understanding Catalyst Surfaces during Catalysis through Near Ambient Pressure X-Ray Photoelectron Spectroscopy, *Chem. Rev.*, 2019, **119**(12), 6822–6905, DOI: 10.1021/acs.chemrev.8b00114.
11 J. Schnadt, J. Knudsen and N. Johansson, Present and New Frontiers in Materials Research by Ambient Pressure X-Ray Photoelectron Spectroscopy, *J. Phys.: Condens. Matter*, 2020, **32**, 413003, DOI: 10.1088/1361-648X/ab9565.
12 A. F. Lee, V. Prabhakaran and K. Wilson, Surface X-Ray Studies of Catalytic Clean Technologies, *Chem. Commun.*, 2010, **46**(22), 3827–3842, DOI: 10.1039/c002853c.
13 J. Cao, A. Rinaldi, M. Plodinec, X. Huang, E. Willinger, A. Hammud, S. Hieke, S. Beeg, L. Gregoratti, C. Colbea, R. Schlögl, M. Antonietti, M. Greiner and M. Willinger, In Situ Observation of Oscillatory Redox Dynamics of Copper, *Nat. Commun.*, 2020, **11**(1), 3554, DOI: 10.1038/s41467-020-17346-7.
14 L. Artiglia, F. Orlando, K. Roy, R. Kopelent, O. Safonova, M. Nachtegaal, T. Huthwelker and J. A. van Bokhoven, Introducing Time Resolution to Detect Ce^{3+} Catalytically Active Sites at the Pt/CeO_2 Interface through Ambient Pressure X-Ray Photoelectron Spectroscopy, *J. Phys. Chem. Lett.*, 2017, **8**(1), 102–108, DOI: 10.1021/acs.jpclett.6b02314.
15 O. Höfert, C. Gleichweit, H.-P. Steinrück and C. Papp, Ultrafast X-Ray Photoelectron Spectroscopy in the Microsecond Time Domain, *Rev. Sci. Instrum.*, 2013, **84**(9), 093103, DOI: 10.1063/1.4821496.
16 F. Presel, H. Tetlow, L. Bignardi, P. Lacovig, C. A. Tache, S. Lizzit, L. Kantorovich and A. Baraldi, Graphene Growth by Molecular Beam Epitaxy: An Interplay between Desorption, Diffusion and Intercalation of Elemental C Species on Islands, *Nanoscale*, 2018, **10**(16), 7396–7406, DOI: 10.1039/c8nr00615f.
17 V. Johánek, V. Nehasil, T. Skála and N. Tsud, Carbon Chain Length Dependence of Graphene Formation via Thermal Decomposition of Alkenes on Pt(111), *J. Phys. Chem. C*, 2019, **123**(13), 7911–7921, DOI: 10.1021/acs.jpcc.8b07165.
18 C. J. Weststrate, D. Sharma, D. Garcia Rodriguez, M. A. Gleeson, H. O. A. Fredriksson and J. W. Niemantsverdriet, Mechanistic Insight into Carbon–Carbon Bond Formation on Cobalt under Simulated Fischer-Tropsch Synthesis Conditions, *Nat. Commun.*, 2020, **11**(1), 750, DOI: 10.1038/s41467-020-14613-5.
19 M. Wang, C. Zhou, N. Akter, W. T. Tysoe, J. A. Boscoboinik and D. Lu, Mechanism of the Accelerated Water Formation Reaction under Interfacial Confinement, *ACS Catal.*, 2020, **10**(11), 6119–6128, DOI: 10.1021/acscatal.9b05289.

20 R. Gao, A. Fernandez, T. Chakraborty, A. Luo, D. Pesquera, S. Das, G. Velarde, V. Thoréton, J. Kilner, T. Ishihara, S. Nemšák, E. J. Crumlin, E. Ertekin and L. W. Martin, Correlating Surface Crystal Orientation and Gas Kinetics in Perovskite Oxide Electrodes, *Adv. Mater.*, 2021, **33**(20), 2100977, DOI: 10.1002/adma.202100977.

21 K. Wiese, A. M. Abdel-Mageed, A. Klyushin and R. J. Behm, Dynamic Changes of Au/ZnO Catalysts during Methanol Synthesis: A Model Study by Temporal Analysis of Products (TAP) and Zn LIII near Edge X-Ray Absorption Spectroscopy, *Catal. Today*, 2019, **336**, 193–202, DOI: 10.1016/j.cattod.2018.11.074.

22 C. Breitkopf, S. Matysik and H. Papp, Selective Poisoning of Active Centers of Sulfated Zirconia Monitored by TAP, XPS, and DRIFTS, *Appl. Catal., A*, 2006, **301**(1), 1–8, DOI: 10.1016/j.apcata.2005.11.009.

23 J. Pasel, D. Schmitt, H. Hartmann, A. Besmehn, J. Dornseiffer, J. Werner, J. Mayer and R. Peters, Combined Near-Ambient Pressure Photoelectron Spectroscopy and Temporal Analysis of Products Study of CH4 Oxidation on Pd/γ-Al_2O_3 Catalysts, *Catal. Today*, 2021, **360**, 444–453, DOI: 10.1016/j.cattod.2019.12.026.

24 C. Reece, E. A. Redekop, S. Karakalos, C. M. Friend and R. J. Madix, Crossing the Great Divide between Single-Crystal Reactivity and Actual Catalyst Selectivity with Pressure Transients, *Nat. Catal.*, 2018, **1**(11), 852–859, DOI: 10.1038/s41929-018-0167-5.

25 R. Fushimi, J. T. Gleaves, G. Yablonsky, A. Gaffney, M. Clark and S. Han, Combining TAP-2 Experiments with Atomic Beam Deposition of Pd on Quartz Particles, *Catal. Today*, 2007, **121**(3–4), 170–186, DOI: 10.1016/j.cattod.2006.07.061.

26 X. Zheng, G. M. Veith, E. Redekop, C. S. Lo, G. S. Yablonsky and J. T. Gleaves, Oxygen and CO Adsorption on Au/SiO_2 Catalysts Prepared by Magnetron Sputtering: The Role of Oxygen Storage, *Ind. Eng. Chem. Res.*, 2010, **49**(21), 10428–10437, DOI: 10.1021/ie100547f.

27 R. Fushimi, X. Zheng, J. T. Gleaves, G. S. Yablonsky, A. Gaffney, M. Clark and S. Han, Techniques for Fabricating Nanoscale Catalytic Circuits, *Top. Catal.*, 2008, **49**(3–4), 167–177, DOI: 10.1007/s11244-008-9090-3.

28 K. Cao, J. Cai, X. Liu and R. Chen, Review Article: Catalysts Design and Synthesis via Selective Atomic Layer Deposition, *J. Vac. Sci. Technol., A*, 2018, **36**(1), 010801, DOI: 10.1116/1.5000587.

29 B. J. O'Neill, D. H. K. Jackson, J. Lee, C. Canlas, P. C. Stair, C. L. Marshall, J. W. Elam, T. F. Kuech, J. A. Dumesic and G. W. Huber, Catalyst Design with Atomic Layer Deposition, *ACS Catal.*, 2015, 1804–1825, DOI: 10.1021/cs501862h.

30 Z. Lu, O. Kizilkaya, A. J. Kropf, M. Piernavieja-Hermida, J. T. Miller, R. L. Kurtz, J. W. Elam and Y. Lei, Design and Synthesis of Model and Practical Palladium Catalysts Using Atomic Layer Deposition, *Catal. Sci. Technol.*, 2016, **6**(18), 6845–6852, DOI: 10.1039/c6cy00682e.

31 T. M. Onn, R. Küngas, P. Fornasiero, K. Huang and R. J. Gorte, Atomic Layer Deposition on Porous Materials: Problems with Conventional Approaches to Catalyst and Fuel Cell Electrode Preparation, *Inorganics*, 2018, **6**(1), 34, DOI: 10.3390/inorganics6010034.

32 D. Xu, J. Yin, Y. Gao, D. Zhu and S. Wang, Atomic-Scale Designing of Zeolite Based Catalysts by Atomic Layer Deposition, *ChemPhysChem*, 2021, **22**(13), 1287–1301, DOI: 10.1002/cphc.202100116.

33 S. P. Sree, J. Dendooven, T. I. Korányi, G. Vanbutsele, K. Houthoofd, D. Deduytsche, C. Detavernier and J. A. Martens, Aluminium Atomic Layer Deposition Applied to Mesoporous Zeolites for Acid Catalytic Activity Enhancement, *Catal. Sci. Technol.*, 2011, **1**(2), 218–221, DOI: 10.1039/c0cy00056f.

34 C. Detavernier, J. Dendooven, S. P. Sree, K. F. Ludwig and J. A. Martens, Tailoring Nanoporous Materials by Atomic Layer Deposition, *Chem. Soc. Rev.*, 2011, **40**(11), 5242–5253, DOI: 10.1039/c1cs15091j.

35 V. Galvita, G. Siddiqi, P. Sun and A. T. Bell, Ethane Dehydrogenation on Pt/Mg(Al)O and PtSn/Mg(Al)O Catalysts, *J. Catal.*, 2010, **271**(2), 209–219, DOI: 10.1016/j.jcat.2010.01.016.

36 E. A. Redekop, V. V. Galvita, H. Poelman, V. Bliznuk, C. Detavernier and G. B. Marin, Delivering a Modifying Element to Metal Nanoparticles via Support: Pt–Ga Alloying during the Reduction of Pt/Mg(Al,Ga)O_x Catalysts and Its Effects on Propane Dehydrogenation, *ACS Catal.*, 2014, **4**(6), 1812–1824, DOI: 10.1021/cs500415e.

37 Y. Nakaya, J. Hirayama, S. Yamazoe, K. Shimizu and S. Furukawa, Single-Atom Pt in Intermetallics as an Ultrastable and Selective Catalyst for PropaneDehydrogenation, *Nat. Commun.*, 2020, **11**(1), 2838, DOI: 10.1038/s41467-020-16693-9.

38 A. Virnovskaia, E. Rytter and U. Olsbye, Kinetic and Isotopic Study of Ethane Dehydrogenation over a Semicommercial Pt,Sn/Mg(Al)O Catalyst, *Ind. Eng. Chem. Res.*, 2008, **47**(19), 7167–7177, DOI: 10.1021/ie800361a.

39 A. Virnovskaia, S. Jørgensen, J. Hafizovic, Ø. Prytz, E. Kleimenov, M. Hävecker, H. Bluhm, A. Knop-Gericke, R. Schlögl and U. Olsbye, In Situ XPS Investigation of Pt(Sn)/Mg(Al)O Catalysts during Ethane Dehydrogenation Experiments, *Surf. Sci.*, 2007, **601**(1), 30–43, DOI: 10.1016/j.susc.2006.09.002.

40 P. Sun, G. Siddiqi, W. C. Vining, M. Chi and A. T. Bell, Novel Pt/Mg(In)(Al)O Catalysts for Ethane and Propane Dehydrogenation, *J. Catal.*, 2011, **282**(1), 165–174, DOI: 10.1016/j.jcat.2011.06.008.

41 J. Dendooven, R. K. Ramachandran, E. Solano, M. Kurttepeli, L. Geerts, G. Heremans, J. Rongé, M. M. Minjauw, T. Dobbelaere, K. Devloo-Casier, J. A. Martens, A. Vantomme, S. Bals, G. Portale, A. Coati and C. Detavernier, Independent Tuning of Size and Coverage of Supported Pt Nanoparticles Using Atomic Layer Deposition, *Nat. Commun.*, 2017, **8**(1), 1074, DOI: 10.1038/s41467-017-01140-z.

42 M. Filez, H. Poelman, R. K. Ramachandran, J. Dendooven, K. Devloo-Casier, E. Fonda, C. Detavernier and G. B. Marin, In Situ XAS and XRF Study of Nanoparticle Nucleation during O_3-Based Pt Deposition, *Catal. Today*, 2014, **229**, 2–13, DOI: 10.1016/j.cattod.2014.01.011.

43 R. K. Ramachandran, M. Filez, J. Dendooven, V. V. Galvita, H. Poelman, E. Solano, E. Fonda, G. B. Marin and C. Detavernier, Size- and Composition-Controlled Pt–Sn Bimetallic Nanoparticles Prepared by Atomic Layer Deposition, *RSC Adv.*, 2017, **7**(33), 20201–20205, DOI: 10.1039/c7ra01463e.

44 N. V. Srinath, A. Longo, H. Poelman, R. K. Ramachandran, J.-Y. Feng, J. Dendooven, M.-F. Reyniers and Vladimir. V. Galvita, In Situ XAS/SAXS

Study of Al_2O_3-Coated PtGa Catalysts for Propane Dehydrogenation, *ACS Catal.*, 2021, **11**(18), 11320–11335, DOI: 10.1021/acscatal.1c03146.

45 E. Solano, J. Dendooven, J.-Y. Feng, P. Brüner, M. M. Minjauw, R. K. Ramachandran, M. V. Daele, K. V. d. Kerckhove, T. Dobbelaere, A. Coati, D. Hermida-Merino and C. Detavernier, In Situ Study of the Thermal Stability of Supported Pt Nanoparticles and Their Stabilization via Atomic Layer Deposition Overcoating, *Nanoscale*, 2020, **12**(21), 11684–11693, DOI: 10.1039/d0nr02444a.

46 J. Dendooven, M. V. Daele, E. Solano, R. K. Ramachandran, M. M. Minjauw, A. Resta, A. Vlad, Y. Garreau, A. Coati, G. Portale and C. Detavernier, Surface Mobility and Impact of Precursor Dosing during Atomic Layer Deposition of Platinum: In Situ Monitoring of Nucleation and Island Growth, *Phys. Chem. Chem. Phys.*, 2020, **22**(43), 24917–24933, DOI: 10.1039/d0cp03563g.

47 R. K. Ramachandran, J. Dendooven and C. Detavernier, Controlled Synthesis of Fe–Pt Nanoalloys Using Atomic Layer Deposition, *Nanotechnology*, 2021, **32**(9), 095602, DOI: 10.1088/1361-6528/abc5f5.

48 R. K. Ramachandran, M. Filez, E. Solano, H. Poelman, M. M. Minjauw, M. Van Daele, J.-Y. Feng, A. La Porta, T. Altantzis, E. Fonda, A. Coati, Y. Garreau, S. Bals, G. B. Marin, C. Detavernier and J. Dendooven, Chemical and Structural Configuration of Pt-Doped Metal Oxide Thin Films Prepared by Atomic Layer Deposition, *Chem. Mater.*, 2019, **31**(23), 9673–9683, DOI: 10.1021/acs.chemmater.9b03066.

49 H. Yan, K. He, I. A. Samek, D. Jing, M. G. Nanda, P. C. Stair and J. M. Notestein, Tandem In2O3-Pt/Al2O3 Catalyst for Coupling of Propane Dehydrogenation to Selective H_2 Combustion, *Science*, 2021, **371**(6535), 1257–1260, DOI: 10.1126/science.abd4441.

50 A. Politano, L. Wang and G. Chiarello, Insight on Thermally Activated Hydrocarbon Dehydrogenation on the Pt_3Ni(111) Surface: From Adsorbed Hydrocarbons up to Graphene Formation, *J. Phys. Chem. C*, 2018, **122**(7), 3885–3892, DOI: 10.1021/acs.jpcc.7b11102.

51 J. Dendooven, R. K. Ramachandran, K. Devloo-Casier, G. Rampelberg, M. Filez, H. Poelman, G. B. Marin, E. Fonda and C. Detavernier, Low-Temperature Atomic Layer Deposition of Platinum Using (Methylcyclopentadienyl) Trimethylplatinum and Ozone, *J. Phys. Chem. C*, 2013, **117**(40), 20557–20561, DOI: 10.1021/jp403455a.

52 R. K. Ramachandran, J. Dendooven, H. Poelman and C. Detavernier, Low Temperature Atomic Layer Deposition of Crystalline In_2O_3 Films, *J. Phys. Chem. C*, 2015, **119**(21), 11786–11791, DOI: 10.1021/acs.jpcc.5b03255.

53 R. K. Ramachandran, J. Dendooven, M. Filez, V. V. Galvita, H. Poelman, E. Solano, M. M. Minjauw, K. Devloo-Casier, E. Fonda, D. Hermida-Merino, W. Bras, G. B. Marin and C. Detavernier, Atomic Layer Deposition Route To Tailor Nanoalloys of Noble and Non-Noble Metals, *ACS Nano*, 2016, **10**(9), 8770–8777, DOI: 10.1021/acsnano.6b04464.

54 F. Polack, M. Silly, C. Chauvet, B. Lagarde, N. Bergeard, M. Izquierdo, O. Chubar, D. Krizmancic, M. Ribbens, J. -P. Duval, C. Basset, S. Kubsky and F. Sirotti, TEMPO: A New Insertion Device Beamline at SOLEIL for Time Resolved Photoelectron Spectroscopy Experiments on Solids and Interfaces, *AIP Conf. Proc.*, 2010, **1234**(1), 185–188, DOI: 10.1063/1.3463169.

55 R. K. Schulze, M. A. Hill, R. D. Field, P. A. Papin, R. J. Hanrahan and D. D. Byler, Characterization of Carbonated Serpentine Using XPS and TEM, *Energy Convers. Manage.*, 2004, **45**(20), 3169–3179, DOI: 10.1016/j.enconman.2004.02.003.

56 R. Roelant, *Mathematical Determination of Reaction Networks from Transient Kinetic Experiments*, Universiteit Gent, 2010.

57 R. Roelant, *Model-Free Deduction of Connectivity Features of Reaction Networks from Thin-Zone-TAP-Data*, 2011.

58 E. A. Redekop, G. S. Yablonsky, D. Constales, P. A. Ramachandran, J. T. Gleaves and G. B. Marin, Elucidating Complex Catalytic Mechanisms Based on Transient Pulse-Response Kinetic Data, *Chem. Eng. Sci.*, 2014, **110**, 20–30, DOI: 10.1016/j.ces.2013.11.050.

59 E. A. Redekop, G. S. Yablonsky, D. Constales, P. A. Ramachandran, C. Pherigo and J. T. Gleaves, The Y-Procedure Methodology for the Interpretation of Transient Kinetic Data: Analysis of Irreversible Adsorption, *Chem. Eng. Sci.*, 2011, **66**(24), 6441–6452, DOI: 10.1016/j.ces.2011.08.055.

60 Y. Wang, M. R. Kunz, Z. Fang, G. Yablonsky and R. Fushimi, Accumulation Dynamics as a New Tool for Catalyst Discrimination: An Example from Ammonia Decomposition, *Ind. Eng. Chem. Res.*, 2019, **58**(24), 10238–10248, DOI: 10.1021/acs.iecr.9b01470.

61 G. S. Yablonsky, E. A. Redekop, D. Constales, J. T. Gleaves and G. B. Marin, Rate-Reactivity Model: A New Theoretical Basis for Systematic Kinetic Characterization of Heterogeneous Catalysts, *Int. J. Chem.Kinet.*, 2016, **48**(6), 304–317, DOI: 10.1002/kin.20988.

62 P. Phanawadee, S. O. Shekhtman, C. Jarungmanorom, G. S. Yablonsky and J. T. Gleaves, Uniformity in a Thin-Zone Multi-Pulse TAP Experiment: Numerical Analysis, *Chem. Eng. Sci.*, 2003, **58**(11), 2215–2227, DOI: 10.1016/s0009-2509(03)00080-0.

63 S. O. Shekhtman, G. S. Yablonsky, J. T. Gleaves and R. R. Fushimi, Thin-Zone TAP Reactor as a Basis of "State-by-State Transient Screening, *Chem. Eng. Sci.*, 2004, **59**(22–23), 5493–5500, DOI: 10.1016/j.ces.2004.09.015.

64 S. M. Pietrzyk, E. A. Redekop, G. S. Yablonsky and G. B. Marin, Precise Kinetic Measurements and Spatial Uniformity of Catalytic Beds, *Chem. Eng. Sci.*, 2015, **134**, 367–373, DOI: 10.1016/j.ces.2015.05.026.

65 G. S. Yablonsky, M. Olea and G. B. Marin, Temporal Analysis of Products: Basic Principles, Applications, and Theory, *J. Catal.*, 2003, **216**(1–2), 120–134, DOI: 10.1016/s0021-9517(02)00109-4.

66 L.-C. Wang, M. L. Personick, S. Karakalos, R. Fushimi, C. M. Friend and R. J. Madix, Active Sites for Methanol Partial Oxidation on Nanoporous Gold Catalysts, *J. Catal.*, 2016, **344**, 778–783, DOI: 10.1016/j.jcat.2016.08.012.

67 L.-C. Wang, C. M. Friend, R. Fushimi and R. J. Madix, Active Site Densities, Oxygen Activation and Adsorbed Reactive Oxygen in Alcohol Activation on NpAu Catalysts, *Faraday Discuss.*, 2016, **188**, 57–67, DOI: 10.1039/c5fd00161g.

68 J. Knudsen, J. N. Andersen and J. Schnadt, A Versatile Instrument for Ambient Pressure X-Ray Photoelectron Spectroscopy: The Lund Cell Approach, *Surf. Sci.*, 2016, **646**, 160–169, DOI: 10.1016/j.susc.2015.10.038.

69 G. Kerherve, A. Regoutz, D. Bentley, C. Hood, K. Feeley, S. Knight, A. Robson, C. Turner, N. Singh, J. Pontefract, J. Åhlund, J. M. Kahk, I. J. Villar-Garcia and D. J. Payne, Laboratory-Based High Pressure X-Ray Photoelectron

Spectroscopy: A Novel and Flexible Reaction Cell Approach, *Rev. Sci. Instrum.*, 2017, **88**(3), 033102, DOI: 10.1063/1.4975096.
70 F. Tao, Design of an In-House Ambient Pressure AP-XPS Using a Bench-Top X-Ray Source and the Surface Chemistry of Ceria under Reaction Conditions, *Chem. Commun.*, 2012, **48**(32), 3812–3814, DOI: 10.1039/c2cc17715c.
71 M. Filez, H. Poelman, E. A. Redekop, V. V. Galvita, K. Alexopoulos, M. Meledina, R. K. Ramachandran, J. Dendooven, C. Detavernier, G. Van Tendeloo, O. V. Safonova, M. Nachtegaal, B. M. Weckhuysen and G. B. Marin, Kinetics of Lifetime Changes in Bimetallic Nanocatalysts Revealed by Quick X-Ray Absorption Spectroscopy, *Angew. Chem., Int. Ed.*, 2018, **57**(38), 12430–12434, DOI: 10.1002/anie.201806447.
72 E. A. Redekop, S. Saerens, V. V. Galvita, I. P. González, M. Sabbe, V. Bliznuk, M.-F. Reyniers and G. B. Marin, Early Stages in the Formation and Burning of Graphene on a $Pt/Mg(Al)O_x$ Dehydrogenation Catalyst: A Temperature- and Time-Resolved Study, *J. Catal.*, 2016, **344**, 482–495, DOI: 10.1016/j.jcat.2016.10.023.
73 S. Saerens, M. K. Sabbe, V. V. Galvita, E. A. Redekop, M.-F. Reyniers and G. B. Marin, The Positive Role of Hydrogen on the Dehydrogenation of Propane on Pt(111), *ACS Catal.*, 2017, **7**(11), 7495–7508, DOI: 10.1021/acscatal.7b01584.
74 L. Xiao, Y.-L. Shan, Z.-J. Sui, D. Chen, X.-G. Zhou, W.-K. Yuan and Y.-A. Zhu, Beyond the Reverse Horiuti–Polanyi Mechanism in Propane Dehydrogenation over Pt Catalysts, *ACS Catal.*, 2020, **10**(24), 14887–14902, DOI: 10.1021/acscatal.0c03381.
75 Z. Lian, C. Si, F. Jan, S. Zhi and B. Li, Coke Deposition on Pt-Based Catalysts in Propane Direct Dehydrogenation: Kinetics, Suppression, and Elimination, *ACS Catal.*, 2021, **11**(15), 9279–9292, DOI: 10.1021/acscatal.1c00331.
76 M. Filez, E. A. Redekop, V. V. Galvita, H. Poelman, M. Meledina, S. Turner, G. V. Tendeloo, A. T. Bell and G. B. Marin, The Role of Hydrogen during Pt–Ga Nanocatalyst Formation, *Phys. Chem. Chem. Phys.*, 2016, **18**(4), 3234–3243, DOI: 10.1039/c5cp07344h.
77 O. Endo, F. Matsui, W.-J. Chun, M. Nakamura, K. Amemiya and H. Ozaki, Nanographene Growth from Benzene on Pt(111), *Surf. Sci.*, 2021, **711**, 121874, DOI: 10.1016/j.susc.2021.121874.
78 N. M. Rodriguez, P. E. Anderson, A. Wootsch, U. Wild, R. Schlögl and Z. Paál, XPS, EM, and Catalytic Studies of the Accumulation of Carbon on Pt Black, *J. Catal.*, 2001, **197**(2), 365–377, DOI: 10.1006/jcat.2000.3081.
79 Z. Paál, A. Wootsch, K. Matusek, U. Wild and R. Schlögl, Intentional Carbonization of Pt Black: A Model Spectroscopic and Catalytic Study, *Catal. Today*, 2001, **65**(1), 13–18, DOI: 10.1016/s0920-5861(00)00540-x.
80 A. Grüneis, K. Kummer and D. V. Vyalikh, Dynamics of Graphene Growth on a Metal Surface: A Time-Dependent Photoemission Study, *New J. Phys.*, 2009, **11**(7), 073050, DOI: 10.1088/1367-2630/11/7/073050.
81 R. Blume, D. Rosenthal, J.-P. Tessonnier, H. Li, A. Knop-Gericke and R. Schlögl, Characterizing Graphitic Carbon with X-Ray Photoelectron Spectroscopy: A Step-by-Step Approach, *ChemCatChem*, 2015, **7**(18), 2871–2881, DOI: 10.1002/cctc.201500344.
82 G. Greczynski and L. Hultman, C 1s Peak of Adventitious Carbon Aligns to the Vacuum Level: Dire Consequences for Material's Bonding Assignment by

Photoelectron Spectroscopy, *ChemPhysChem*, 2017, **18**(12), 1507–1512, DOI: 10.1002/cphc.201700126.

83 G. Siddiqi, P. Sun, V. Galvita and A. T. Bell, Catalyst Performance of Novel Pt/Mg(Ga)(Al)O Catalysts for Alkane Dehydrogenation, *J. Catal.*, 2010, **274**(2), 200–206, DOI: 10.1016/j.jcat.2010.06.016.

84 S. Blomberg, J. Zhou, J. Gustafson, J. Zetterberg and E. Lundgren, 2D and 3D Imaging of the Gas Phase Close to an Operating Model Catalyst by Planar Laser Induced Fluorescence, *J. Phys.: Condens. Matter*, 2016, **28**(45), 453002, DOI: 10.1088/0953-8984/28/45/453002.

85 A. Shavorskiy, J. Schnadt and J. Knudsen, Time Resolved Ambient Pressure X-Ray Photoelectron Spectroscopy, in *Ambient Pressure Spectroscopy in Complex Chemical Environments; ACS Symposium Series*, American Chemical Society, 2021, vol. 1396, pp. 219–248, DOI: 10.1021/bk-2021-1396.ch009.

86 J. Knudsen, T. Gallo, V. Boix, M. D. Strømsheim, G. D'Acunto, C. Goodwin, H. Wallander, S. Zhu, M. Soldemo, P. Lömker, F. Cavalca, M. Scardamaglia, D. Degerman, A. Nilsson, P. Amann, A. Shavorskiy and J. Schnadt, Stroboscopic Operando Spectroscopy of the Dynamics in Heterogeneous Catalysis by Event-Averaging, *Nat. Commun.*, 2021, **12**(1), 6117, DOI: 10.1038/s41467-021-26372-y.

87 E. A. Redekop, N. Johansson, E. Kokkonen, S. Urpelainen, F. Lopes da Silva, M. Kaipio, H.-E. Nieminen, F. Rehman, V. Miikkulainen, M. Ritala and U. Olsbye, Synchronizing Gas Injections and Time-Resolved Data Acquisition for Perturbation-Enhanced APXPS Experiments, *Rev. Sci. Instrum.*, 2021, **92**(4), 044101, DOI: 10.1063/5.0039957.

88 D. Wang and Z. Li, Time Resolution of the TAP Reactor, in *Studies in Surface Science and Catalysis*, ed. Froment G. F. and Waugh K. C., Elsevier, Reaction Kinetics and the Development and Operation of Catalytic Processes, 2001, vol. 133, pp. 553–557, DOI: 10.1016/s0167-2991(01)82011-5.

89 K. L. Gering, C. Baroi and R. R. Fushimi, Transport Modeling and Mapping of Pulsed Reactor Dynamics near and beyond the Onset of Viscid Flow, *Chem. Eng. Sci.*, 2018, **192**, 576–590, DOI: 10.1016/j.ces.2018.07.060.

Faraday Discussions

ROYAL SOCIETY OF CHEMISTRY

DISCUSSIONS

Time resolved surface analysis (kinetic and molecular time scales): general discussion

Rosa Arrigo, Damien Aureau, Loren Ban, Thorsten Bartels-Rausch, James J. C. Counter, Philip R. Davies, Martina Dell'Angela, Michael Dowhyj, Andrew Evans, Wendy Flavell, Georg Held, Claus F. P. Kastorp, Robert Lindsay, Anders Nilsson, Evgeniy Redekop, Olivier Renault, Günther Rupprechter, Mathieu G. Silly and Sefik Suzer

DOI: 10.1039/d2fd90029g

Wendy Flavell opened a general discussion of the paper by Martina Dell'Angela: You have very nicely separated out two separate timescales for the dynamics in this system. In the context of the SPV, clearly this scales with fluence, but (quite apart from the 'n/p' issue) naively I wondered if you expect to see an SPV shift at all in this system – it seems to be fairly two-dimensional to me – does it have a depletion layer at the surface that could give rise to an SPV shift?

Martina Dell'Angela answered: We thank you for the question. $MoTe_2$ is indeed a 2D layered material, but we are measuring a bulk crystal in the 2H phase, which is a semiconductor. Interlayer interactions are weak in this class of material, but the out-of-plane conductivity is not entirely suppressed (around 10^3 times lower than the in-plane conductivity), thus allowing for the formation of a depletion region and for the rise of photovoltage effects, as observed in other transition metal dichalchogenides, *e.g.*, MoS_2 (ref. 1 and 2) and WSe_2.[3]

1 Y. Liang, B.-H. Li, Z. Li, G. Zhang, J. Sun, C. Zhou, Y. Tao, Y. Ye, Z. Ren and X. Yang, *Mater. Today Phys.*, 2021, **21**, 100506.
2 N. L. A. N. Sorgenfrei, E. Giangrisostomi, R. M. Jay, D. Kühn, S. Neppl, R. Ovsyannikov, H. Sezen, S. Svensson and A. Föhlisch, *Adv. Mater.*, 2021, **33**, 2006957.
3 A. Jakubowicz, D. Mahalu, M. Wolf, A. Wold and R. Tenne, *Phys. Rev. B: Condens. Matter Mater. Phys.*, 1989, **40**, 2992–3000.

Wendy Flavell said: On a related point, are you able to control the doping so you could prepare n- and p-variants and see if the SPV shift switches direction?

Martina Dell'Angela answered: We thank for the question. We were not able to control the doping, however we have already planned experiments with a different glue to hold the samples that will allow heating and switching from p- to n-doping as in reference 1.

1 J. Liu, Y. Wang, X. Xiao, K. Zhang, N. Guo, Y. Jia, S. Zhou, Y. Wu, Q. Li and L. Xiao, *Nanoscale Res. Lett.*, 2018, **13**, 291.

Anders Nilsson asked: The shift you measure is very small. Is there any indication of the space-charge effects?

Martina Dell'Angela answered: The analysis of the vacuum space-charge effects on the measured photoemission line in pump–probe experiments is indeed very challenging. Here we exclude that the measured shifts are due to vacuum space-charge effects because the number of secondary electrons photoemitted by the pump pulse is very low (we measured drain currents in the picoampere range, while in space-charge conditions on metals the current is in the order of nanoampere). At the beginning of the experiments we measured the space-charge shift on a Au single crystal to calibrate the experimental conditions. As shown in the figure here, it presents a stronger positive (negative) kinetic (binding) energy shift as compared to the measured dynamics on $MoTe_2$, and does not show long-lived dynamics.

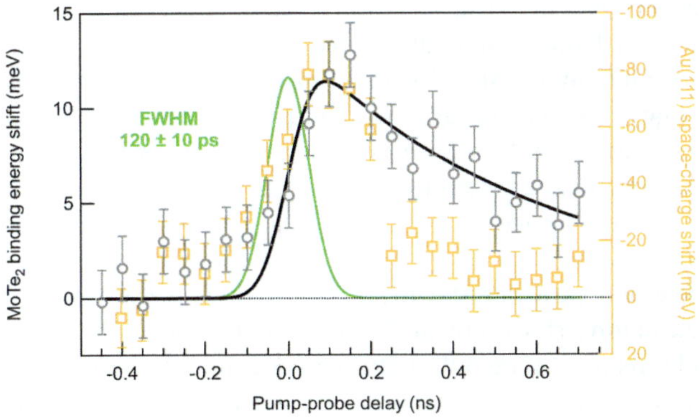

Anders Nilsson asked: Due you see any increase in the phonon temperature due to the HOMO–LUMO excitation? Can you apply something similar to the two-temperature model that is often used for metals?

Martina Dell'Angela replied: The simulation of the dynamics of the lattice temperature is a very interesting issue. As pointed out in the question we cannot apply the two-temperature model to a semiconductor. However, the temporal evolution of the lattice temperature could be estimated by solving the heat diffusion equation numerically with a source term that describes the Gaussian pump pulse. Such simulations are time demanding and we did not perform them. As described in ref. 1, we can obtain an estimate of the maximum temperature increase after each pulse and of the maximum temperature reached by the system after the irradiation at high repetition rates. Using a 2H-$MoTe_2$ thermal diffusivity of about 4 W mK^{-1} (average value in the literature) and considering the highest applied pump fluence, we expect that the temperature increase for each pulse in the hottest point is about 10 K and that it reaches 50 K after long irradiation of many pulses. In our data when comparing laser on-laser off spectra, we observe an additional broadening of the Mo 3d line of about 15 meV. Using the formula for the core level broadening in the high temperature approximation in ref. 2, such broadening corresponds to a temperature of 390 K, in good agreement with the expected 350 K. Unfortunately, we cannot resolve the temperature increase after each single pulse because it is beyond the experimental resolution: the time-resolved measurements are performed with high analyser pass energies to improve the count-rate.

1 Y. Xu, R. Wang, S. Ma, Lei Zhou, Y. R. Shen and C. Tian, *J. Appl. Phys*, 2018, **123**, 025301.
2 D. M. Riffe and G. K. Wertheim, *Phys. Rev. B: Condens. Matter Mater. Phys.*, 2000, **61**, 2302.

Claus F. P. Kastorp commented: On page 5 of your article (https://doi.org/10.1039/d1fd00105a), it is mentioned that the 1T′-phase is characterised by tellurium defects. Does that imply that the phase transition is induced by tellurium desorption? On that note, would an extended laser fluence be able to transition a larger part of the sample or merely destroy it?

Martina Dell'Angela replied: The details of the phase transition mechanism are still being debated. Although it is certain that the 1T′ phase is characterized by tellurium defects, it is not clear if the presence of vacancies alone may cause the transition. In ref. 1 the authors attempted to revert the 2H-to-1T′ transition by heating the sample in an excessive Te atmosphere with no successful outcome, suggesting that other factors are also contributing. In fact, calculations have shown that optical excitation both facilitates vacancy ordering and lowers the transition barrier. Ref. 2 instead demonstrates that the phase transition may occur even in the absence of Te vacancies and only requires changes in the electron densities.

Laser excitation provides the means for inducing the phase transition in a controllable manner: if the optical fluence is high enough, the beam can be used to irreversibly create a 1T′ metallic pattern inside the 2H semiconducting surface, paving the way for the design of novel nanoscale devices. As stated in the text, in our measurements we were careful to keep the fluence below such a threshold so as to avoid irreversible modification of the sample.

1 Y. Tan, Fang Luo, M. Zhu, X. Xu, Y. Ye, B. Li, G. Wang, W. Luo, X. Zheng, N. Wu, Y. Yu, Sh. Qin and X.-A. Zhang, *Nanoscale*, 2018, **10**, 19964–19971.
2 B. Peng, H. Zhang, W. Chen, B. Hou, Z.-J. Qiu, H. Shao, H. Zhu, B. Monserrat, D. Fu, H. Weng and C. M. Soukoulis, *npj 2D Mater. Appl.*, 2020, **4**, 14.

Claus F. P. Kastorp added: You mention that you have been unable to heat and clean the samples due to the glue that you have been using. Do you know if your samples are compatible with carbon glue, which is heat resistant?

Martina Dell'Angela answered: We thank you for the observation. We already planned further experiments with the samples glued using the UHV conductive expoxy from Acu-Glass that will allow annealing in UHV up to 150 °C to observe the change from p to n-doping.

Michael Dowhyj opened the discussion of the paper by Mathieu G. Silly: How do you verify and adjust overlap (both spatially and temporally) between pump pulse and X-ray pulse for these measurements?

Mathieu G. Silly replied: We first adjust the spatial overlap using an unencapsulated silicon photodiode (AXUVHS6A from Opto Diode Corporation) mounted on a transferable sample holder. The photocurrent of the diode is measured using a picoamperemeter (Keithley). The sensitive area of the diode is 0.025×0.025 mm^2, smaller than the size of the synchrotron and laser beams. This feature permits us to measure photocurrent maps by scanning the manipulator across the synchrotron and laser beam and measuring the photocurrent. This procedure allows us to determine the synchrotron beam position precisely and to perfectly overlap the laser beam on it. It is also possible to determine the spot size of the laser beam to determine the incident laser fluence.

When we want to use the temporal resolution of the synchrotron, the temporal overlap is then done using space charge effect on metals, we usually use a reconstructed gold surface, or surface photovoltage effect using a doped silicon sample. Photoelectrons coming from the isolated bunch in hybrid mode or bunches in height or single bunches operation mode, are collected for different delays between the isolated bunch of the synchrotron and the laser pulse. An energy shift of the core level arises when the synchrotron bunch and laser pulse are temporally overlapped which can be unambiguously determined. Space charge effect duration in metals is about 500 ps and the amplitude of the energy shift is about 500 meV.[1] The surface photovoltage rise time in silicon is about 150 ps and the amplitude of the SPV effect is about 100 meV.[2] Maximum space charge effect due to high energy shift is easier to determine but the effect is broader than the SPV effect which increases the uncertainty on the temporal reference T0.

1 *Ultrafast Magnetism I Proceedings of the International Conference UMC 2013 Strasbourg, France, October 28th–November 1st*, Springer Link, ed. J.-Y. Bigot, W. Hübner, T. Rasing and R. Chantrell, 2013, vol. 159, https://doi.org/10.1007/978-3-319-07743-7_97.

2 M. G. Silly, T. Ferté, M. A. Tordeux, D. Pierucci, N. Beaulieu, C. Chauvet, F. Pressacco, F. Sirotti, H. Popescu, V. Lopez-Flores, M. Tortarolo, M. Sacchi, N. Jaouen, P. Hollander, J. P. Ricaud, N. Bergeard, C. Boeglin, B. Tudu, R. Delaunay, J. Luning, G. Malinowski, M. Hehn, C. Baumier, F. Fortuna, D. Krizmancic, L. Stebel, R. Sergo and G. Cautero, *J. Synchrotron Radiat.*, 2017, **24**(4), 886–897.

Martina Dell'Angela added: The coarse temporal overlap between the laser and the X-rays at the sample is achieved by means of a reverse biased photodiode (Opto Diode, SXUV5) mounted into the experimental chamber. The precision of this measurement is however, limited by the response time of the photodiode, which is ~2 ns. The fine temporal overlap is obtained by measuring optical laser induced vacuum space charge effects in the photoemission of Au(111). Such effects are well established in photoemission measurements with pulsed sources like lasers, HHG and FELs. The Au 4f lines exhibit kinetic energy shifts that last for several hundred picoseconds. The duration of the transition from zero to the maximum kinetic energy shift is only limited by the X-ray pulse width and therefore determines time-zero within our experimental resolution. We described the temporal and spatial alignment procedures in ref. 1. Recently we upgraded the laser incoupling, and the laser and the X-ray beams enter the chamber almost collinear. The spatial overlap of the two beams is determined on the surface of a cerium doped-YAG ($Y_3Al_5O_{12}$) crystal mounted on the sample holder at the same height of the sample surface. The two samples can be reached by a vertical translation of the manipulator.

1 R. Costantini, M. Stredansky, D. Cvetko, G. Kladnik, A. Verdini, P. Sigalotti, F. Cilento, F. Salvador, A. De Luisa, D. Benedetti, L. Floreano, A. Morgante, A. Cossaro and M. Dell'Angela, *J. Electron Spectrosc. Relat. Phenom.*, 2018, **229**, 7–12.

Georg Held addressed Martina Dell'Angela and Mathieu G. Silly: Could you please discuss the factors that contribute to the time resolution of the analysers you use in your experiments?

Mathieu G. Silly answered: By measuring the photoelectrons coming from the isolated bunch in hybrid operation mode of the Synchrotron SOLEIL, which presents a FWHM of about 50 ps, we determined a temporal resolution of the analyzer for our experimental setup (pass energy 50 and central kinetic energy 50 eV) of about 30 ns. Even if the temporal resolution of the delay line detector is about 50 ps,[1] the temporal resolution of the analyzer is limited by the spread in energy of the electron in the analyzer hemisphere. Depending on the trajectory of the photoelectron in the hemisphere depending on the kinetic energy, the photoelectrons do not reach the detector at the same time. A complete study of the temporal resolution of our analyzer has been performed and has shown that the temporal resolution is comprised between 30 and 70 ns depending on the central kinetic energy and pass energy use.[2] To carry out time resolved experiments with higher temporal resolution, we have to use the temporal resolution of the synchrotron. The laser repetition rate and detector collection have to be synchronized to the radiofrequency of the machine. By applying a delay between the laser and the synchrotron it is possible to go down to 10 ps in low alpha operation mode.[3]

1 G. Cautero, R. Sergo, L. Stebel, P. Lacovig, P. Pittan, M. Predonzani and S. Carrato, *Nuclear Instruments and Methods in Physics Research Section A: Accelerators, Spectrometers, Detectors and Associated Equipment*, 2008, **595**(2), 447–459.
2 N. Bergeard, M. G. Silly, D. Krizmancic, C. Chauvet, M. Guzzo, J. P. Ricaud, M. Izquierdo, L. Stebel, P. Pittana, R. Sergo, G. Cautero, G. Dufour, F. Rochet and F. Sirotti, *J. Synchrotron Rad.*, 2011, **18**, 245–250.
3 M. G. Silly, T. Ferté, M. A. Tordeux, D. Pierucci, N. Beaulieu, C. Chauvet, F. Pressacco, F. Sirotti, H. Popescu, V. Lopez-Flores, M. Tortarolo, M. Sacchi, N. Jaouen, P. Hollander, J. P. Ricaud, N. Bergeard, C. Boeglin, B. Tudu, R. Delaunay, J. Luning, G. Malinowski, M. Hehn, C. Baumier, F. Fortuna, D. Krizmancic, L. Stebel, R. Sergo and G. Cautero, *J. Synchrotron Radiat.*, 2017, **24**(4), 886–897.

Martina Dell'Angela answered: In our experiments, due to the significant difference between laser and synchrotron pulse durations, the latter is the main contributor to the temporal resolution (*i.e.*, the pump–probe cross-correlation), which equals roughly 120 ps. The time resolution of the detection system is an important issue, but in the pulse discrimination scheme employed it only affects the measurements indirectly: the time-of-flight (TOF) spread of the electrons crossing the hemispheres determines the minimum separation between consecutive pulses that allows a clear discrimination of the electrons photoemitted by a specific probe pulse. The TOF spread is a function of both kinetic and pass energy settings, with higher energies yielding a narrower spread, but a worse energy resolution; for a 150 mm analyzer it can be brought down to around 1 ns with pass energies greater than 200 eV, up to several nanoseconds with lower pass energies (see ref. 1). When working with the hybrid pulse of the Elettra synchrotron, the separation between the single bunch, acquired for the time-resolved signal, and the nearest pulse of the multibunch train is in the order of 50–70 ns, which allows us to effectively use pass energies down to 20 eV and obtain energy resolutions better than 0.2 eV, depending on the slit size. Under these conditions, we are distinctly detecting the photoelectrons emitted by the single bunch, therefore the resolution is only limited by the X-ray pulse duration.

1 A. Shavorskiy, S. Neppl, D. S. Slaughter, J. P. Cryan, K. R. Siefermann, F. Weise, M.-F. Lin, C. Bacellar, M. P. Ziemkiewicz, I. Zegkinoglou, Matthew W. Fraund, C. Khurmi, M. P. Hertlein, T. W. Wright, N. Huse, R. W. Schoenlein, T. Tyliszczak, G. Coslovich, J. Robinson, R. A. Kaindl, B. S. Rude, A. Ölsner, S. Mähl, H. Bluhm and O. Gessner, *Rev. Sci. Instrum.*, 2014, **85**, 093102.

Mathieu G. Silly suggested: In order to correlate the transient electronic behavior evidenced by time resolved photoemission spectroscopy with phase transition, have you tried to perform time resolved X-ray photoelectron diffraction (XPD) which is sensitive to the surface structure modification?

Martina Dell'Angela answered: Unfortunately, with our set-up we cannot perform time resolved XPD, but it will for sure be a very interesting experiment to directly compare with the ultrafast electron diffraction[1] and ultrafast X-ray absorption experiments[2] recently published, which are performed employing different pump wavelengths.

1 A. Krishnamoorthy, M.-F.u Lin, X. Zhang, C. Weninger, R. Ma, A. Britz, C. S. Tiwary, V. Kochat, A. Apte, J. Yang, S. Park, R. Li, X. Shen, X. Wang, R. Kalia, A. Nakano, F. Shimojo, D. Fritz, U. Bergmann, P. Ajayan and P. Vashishta, *Nano Lett.*, 2019, **19**(8), 4981–4989.
2 A. R. Attar, H.-T. Chang, A. Britz, X. Zhang, M.-F. Lin, A. Krishnamoorthy, T. Linker, D. Fritz, D. M. Neumark, R. K. Kalia, A. Nakano, P. Ajayan, P. Vashishta, U. Bergmann and S. R. Leone, *ACS Nano*, 2020, **14**(11), 15829–15840.

Sefik Suzer continued the general discussion of the paper by Mathieu G. Silly: You have shown a very impressive amount of data, emphasizing mostly the physical changes related with surface photovoltage dynamics. However, I see that you were able to record similar changes for different chemical/charge states of Si. My question is: do you see/observe differences in the dynamics between two different charge states; such as Si^{2+} and Si^{0}? This would be even more interesting since chemical (like bonding type, say Si–Si and Si–O, *etc.*), in addition to the physical information, could also be extracted from your data.

Mathieu G. Silly answered: We have focused our work on the bulk silicon component of Si 2p. The time resolved experiment is performed in fixed acquisition mode. The energy range probed is determined by the pass energy used limited by the central kinetic energy (KE) of the probed photoelectrons (50 eV central KE). This has limited us to an energy window covering only the Si 2p component coming from Si–Si species. To measure at the same time the suboxides, the use of a higher pass energy is needed. The same measurement has to be performed at a higher kinetic energy. Compared to Si–Si Si 2p component, suboxide contributions present overlapping broader features, disentangling the signals coming from the suboxides is more challenging but is of high interest. Different lifetimes can be expected coming from possible charge transfer. The capability of such measurement has already been proved with C60 on top of Cu-phthalocyanine where different photoresponses depending on the material, have been evidenced by measuring the temporal evolution of the C 1s core level of the two species under pulsed laser excitation.[1]

1 T. Arion, S. Neppl, F. Roth, A. Shavorskiy, H. Bluhm, Z. Hussain, O. Gessner and W. Eberhardt, *Appl. Phys. Lett.*, 2015, **106**, 121602.

Olivier Renault asked: Would you record the valence band besides the Si core-levels? I was wondering if interesting features fingerprinting particular states related to the surface termination would be detectable.

Mathieu G. Silly responded: Interesting features can be expected by measuring the valence band of the sample excited at 800 nm corresponding to 1.55 eV in energy. Electrons down to 1.55 eV below the Fermi level can be excited and transferred to the conduction band resulting in a modification in the electron population of the valence band. A modification in the shape of the valence band can be expected. We focused our work on the evolution of the Si 2p core level energy shift under laser excitation and surface photovoltage effect. Measurement of the evolution of the valence band is more challenging as the valence band is composed of small and smooth features. The modification in the shape of the valence band is also superimposed to an energy shift coming from the SPV effect. Such a study is of great interest to determine the transient evolution of the electron population of the valence band under laser excitation and would permit correlating it to the transient SPV shift.

Olivier Renault said: This is a question that may seems naïve given my limited knowledge regarding time-resolved experiments, and which concerns surface photo-voltage: don't you have two SPV effects – one from the pump, another one from the probe? Or are you working in such conditions that one is negligible?

Mathieu G. Silly answered: You are right that both laser and synchrotron can be at the origin of the SPV effect. As mentioned in our paper, we have taken particular care to avoid SPV effects from the synchrotron beam during our measurement. We have performed preliminary tests and have reduced the synchrotron incident flux until no energy shift of the Si 2p core level induced by the synchrotron, is observable. In this way, the band bending is not reduced by the synchrotron and only the laser contributes to the observed SPV effect. If the synchrotron beam had contributed to the SPV, it would have lead to a reduction of the band bending and to an underestimation of the amplitude of the SPV effect caused by the laser excitation, but it should not affect the general evolution SPV time response.

Damien Aureau commented: As you are looking at band bending of a semiconductor, did you work at different excitation energies in order to get information from various depths, or did you work so far only at 150 eV? What would you expect by changing energies?

Mathieu G. Silly answered: As you mentioned, by changing the photon energy, we change the kinetic energy of the photoelectrons which results in a different probing depth. To compare the results obtained for the different surfaces, we have only used one photon energy corresponding to the most surface sensitive kinetic energy of the photoelectrons of about 50 eV. This permits us to measure the highest surface photovoltage (SPV) energy shift closest to the maximum SPV effect. By increasing the kinetic energy, we will probe deeper in the depletion layer and we will reduce the observable SPV shift. This could be interesting to probe the excited states beyond the depletion region by working at very high kinetic energy to measure phenomena independent of band bending variations.

Georg Held said: Beam-induced dissociation is a known problem in experiments in water. Can we exclude that the effects you see for water adsorption are beam-induced?

Mathieu G. Silly replied: Water-saturated Si(001)-2×1 surface is a very sensitive surface, especially to beam damage. Even if we tried to minimize the photon flux to avoid too much beam damage, it cannot be completely avoided. Fig. S1 in the ESI of our article (https://doi.org/10.1039/d1fd00107h) shows the evolution of the Si 2p core level of the water saturated Si surface before and after time resolved experiments. Even if there is no appearance of the Si^{2+} oxidation state which shows that beam damage remains constrained, components appear broader and a shift of 40 meV in higher binding energy is noticed. It is shown that the charge surface has been affected by the interaction with synchrotron and laser beams.

Andrew Evans questioned: The photovoltage is effective in the depletion region of the semiconductor and can be induced by excitation by the X-rays as well as the laser. Is there a match, or mismatch, between the depletion width in the silicon and the penetration depth of the radiation?

Mathieu G. Silly responded: We first checked by reducing the incident flux that the synchrotron radiation does not contribute to the measured surface photovoltage effect. Only the excitation coming from the laser has to be considered in the present study. As shown in our paper, the thickness of the depletion layer is between 3 and 783 nm depending on the doping level of the substrate. The penetration depth of the infrared light in silicon is much higher.[1] Only 10% of the light at 800 nm is adsorbed in the first micron which means that only a few parts of the photoexcited electrons are submitted to the electric field of the depletion layer and contribute to the thermionic deexcitation process. The other photoexcited electrons will deexcite through other deexcitation channels. One process is visible at high fluence (Fig. 5 in our article, https://doi.org/10.1039/d1fd00107h) where a bulk deexcitation process occurs and dominates the time response.

1 Y. R. S. Carvalho da Silva, R. Kuroda and S. Sugawa, *Sensors*, 2019, **19**(12), 2755.

Georg Held addressed Martina Dell'Angela and Mathieu G. Silly: To what extent does the irradiation with an intense laser pulse heat the sample? Could local heating induce additional broadening in the XPS signal?

Mathieu G. Silly responded: Compared to ns pulsed laser excitation which is known to anneal the lattice of silicon by thermal process, fs pulsed laser excitation leads to the creation of a huge amount of electron–hole pairs which do not thermalize immediately.[1] Tenths of picoseconds are needed to thermalize, during that time non-thermal transitions occur disordering the lattice. For very high flux, nonreversible effects can happen such as ablation and evaporation of material due to temperature increase.[2] After tens of ns, the system has already reached the thermal diffusion process. The complete thermalization of the system due to the heat charge can take several minutes in silicon.[3,4] As one photoexcitation dynamic measurement takes about 30 min, the laser excitation leads to a global increase of the temperature of the sample. Compared to previous work[4] where a laser excitation of 100 mW at 1030 nm leads to an increase of less than 20 °C, here we have used laser power two orders of magnitude lower in the thermionic regime. The heat charge appears negligible in agreement with our fitting procedure using room temperature for the system.

Due to the time resolution of 30 ns used during this experiment, we are not sensitive to the first deexcitation steps which could lead to peak deformation. To evidence fast thermalization phenomena, we should look at the lineshape evolution of the Si 2p core level of previous measurements we have performed using low alpha mode and 10 ps of time resolution.[5]

1 S. K. Sundaram and E. Mazur, *Nature Mater.*, 2002, **1**, 217–224.
2 A. A. Ionin, S. I. Kudryashov, L. V. Seleznev, D. V. Sinitsyn, A. F. Bunkin, V. N. Lednev and S. M. Pershin, *J. Exp. Theor. Phys.*, **116**, 347–362.
3 D. V. Tran, Y. C. Lam, B. S. Wong, H. Y. Zheng and D. E. Hardt, 2006, **14**(20), 9261–9268.
4 T. Chen, W. Wang, T. Tao, X. Mei and A. Pan, *SPIE Optical Engineering + Applications*, 2019, 11101.
5 M. G. Silly, T. Ferté, M. A. Tordeux, D. Pierucci, N. Beaulieu, C. Chauvet, F. Pressacco, F. Sirotti, H. Popescu, V. Lopez-Flores, M. Tortarolo, M. Sacchi, N. Jaouen, P. Hollander, J. P. Ricaud, N. Bergeard, C. Boeglin, B. Tudu, R. Delaunay, J. Luning, G. Malinowski, M. Hehn, C. Baumier, F. Fortuna, D. Krizmancic, L. Stebel, R. Sergo and G. Cautero, *J. Synchrotron Radiat.*, 2017, **24**(4), 886–897.

Martina Dell'Angela added: The irradiation with the pulsed laser can indeed heat the sample. By employing the model in ref. 1, we estimate that at the maximum applied fluence the maximum temperature reached by the system is 350 K. By comparing the laser on-laser off spectra, we observe an additional broadening of the Mo 3d line of about 15 meV corresponding to a temperature of about 390 K (as described in my response to Anders Nilsson above) when the laser is on. Such temperature is however not high enough to trigger the phase transition.

1 Y. Xu, R. Wang, S. Ma, Lei Zhou, Y. R. Shen and C. Tian, *J. Appl. Phys*, 2018, **123**, 025301.

Anders Nilsson opened a general discussion of the paper by Loren Ban: There can be a major difference in the space charge of the droplets with femtosecond or nanosecond short pulses. Do you see any indication of space charge?

Loren Ban answered: It is true that space charge is one of the major practical concerns for studies of droplets (as well as other condensed phase samples such as liquid jets) with ultrafast laser sources (see for example ref. 1 and 2). In our case, we do not see significant evidence of space charge. This is because we do not focus neither the nanosecond nor femtosecond laser pulses, so that our peak intensities are in the kW cm^{-2} and GW cm^{-2} range, respectively. At these intensities, we do not produce high enough density of charged particles to observe significant space charge effects on the photoelectron spectra or angular distributions. This condition is not so trivial for the femtosecond case as it requires very low pulse energies to prevent excessive ionization of the droplets. In practice, we tuned the beam diameter and pulse energy to the point where the photoelectron spectra are dominated by the direct (1 + 1) REMPI and we see negligible contributions from the higher-order REMPI processes.

1 C. Corder, P. Zhao, J. Bakalis, X. Li, M. D. Kershis, A. R. Muraca, M. G. White and T. K. Allison, *Struct. Dyn.*, 2018, **5**, 054301.
2 G. Schönhense, D. Kutnyakhov, F. Pressacco, M. Heber, N. Wind, S. Y. Agustsson, S. Babenkov, D. Vasilyev, O. Fedchenko, S. Chernov, L. Rettig, B. Schönhense, L. Wenthaus, G. Brenner, S. Dziarzhytski, S. Palutke, S. K. Mahatha, N. Schirmel, H. Redlin, B. Manschwetus, I. Hartl, Y. Matveyev, A. Gloskovskii, C. Schlueter, V. Shokeen, H. Duerr, T. K. Allison, M. Beye, K. Rossnagel, H. J. Elmers and K. Medjanik, *Rev. Sci. Instrum.*, 2021, **92**, 053703.

Anders Nilsson said: As the droplets are charged, can you use electrostatic plates to direct the droplets? This question is more related to other applications than this particular experiment.

Loren Ban responded: To produce a sample of neutral droplets we basically use exactly what you mentioned – a pair of electrostatic plates. This device is called the electrostatic precipitator. By applying a voltage, we can steer the motion of the charged droplets in the aerosol flow and force them to collide onto the plates. Neutral droplets travel undisturbed through the instrument. As a result, we get a sample of only neutral droplets that we can send into our spectrometer. On the other side, charged droplets are produced by unipolar aerosol chargers (see for example, ref. 1). In short, the droplets are charged by the diffusion of charged gas phase ions generated in a corona-wire discharge. These ions travel onto the droplets by simple diffusion and lead to droplet charge which is proportional to the ion concentration. With this method, we can currently produce positively or negatively charged particles with up to 20–30 charge units. Another approach to studying charged particles you might find interesting is by trapping them in between two electrodes, in an electrodynamic trap. By using AC fields one can determine their charge from the characteristic oscillation frequency. This method was, for example, applied to study the ionization process of single particles using soft X-rays.[2]

1 G. Biskos, K. Reavell and N. Collings, *J. Aerosol Sci.*, 2005, **36**, 247–265.
2 M. Grimm, B. Langer, S. Schlemmer, T. Lischke, U. Becker, W. Widdra, D. Gerlich, R. Flesch and E. Rühl, *Phys. Rev. Lett.*, 2006, **96**, 066801.

Philip R. Davies asked: You show that the larger the droplet the faster the electron decay time and also mention that oxygen could be quenching these states. Is it possible that the different decay rates are linked to different concentrations of oxygen in the different-sized droplets?

Loren Ban replied: That's a very interesting point – if the oxygen quenching mechanism holds, I think this could be possible. And I would still classify such an explanation as a surface effect. At equilibrium, I see no reason for a size-dependent oxygen concentration, it should simply reach the equivalent bulk concentration at certain conditions. However, the uptake of oxygen from air into the droplet is a kinetic problem, determined by a number of surface processes which depend on droplet size. In our experiment, the droplets travel through a wide range of pressure conditions – from atmosphere down to high vacuum. One would need a rather complex model to try and estimate how much oxygen would still be left in the droplets when they reach the interaction region. The parameters that one would need to describe this uptake are size-dependent and typically not known precisely. One example is the mass accommodation coefficient, which was recently investigated in our group by photothermal single-particle spectroscopy.[1]

Since this is a complex problem, we propose that the surface plays a role in the slower electron yield decay only in larger droplets. We did not discuss in detail the possible mechanisms by which the surface alters the measured decay kinetics. Before doing that, I would say the first step is to gain more information on the actual decay mechanism of the long-lived state.

1 M. E. Diveky, M. J. Gleichweit, S. Roy and R. Signorell, *J. Phys. Chem. A*, 2021, **125**, 3528–3548.

Philip R. Davies commented: Do you have any information on the distribution of the molecule DEHP through the droplet? Could there be a change in concentration with depth?

Loren Ban replied: We do not have any information on the depth profile of DEHP concentration. However, for this work the depth profile should not play a major role. This is because we are probing only the molecules contained in a thin layer (≤10 nm) around the droplet surface, and diffusion distance is small on the investigated timescale. Nevertheless, this is a very interesting point especially for comparing reaction rates in droplets and bulk. For example, surface enrichment has been discussed as one of the reasons why reactions can be accelerated in droplets compared to bulk.[1] Both our solute (DEHP) and solvent (DEHS) have non-polar aliphatic chains which could lead to interesting behavior at the surface. Learning more about this point would definitely be an interesting future direction.

1 G. Rovelli, M. I. Jacobs, M. D. Willis, R. J. Rapf, A. M. Prophet and K. R. Wilson, *Chem. Sci.*, 2020, **11**, 13026–13043.

Robert Lindsay said: You mentioned that dissolved oxygen may play a role in your results. Did you try to test this possibility by purging the solution with nitrogen to reduce the dissolved oxygen concentration?

Loren Ban answered: We didn't try removing oxygen from our system. Although we generate the droplets with pure nitrogen gas, the surrounding air is mixed with the aerosol in the size control stage. It is, in principle, possible to modify our instruments to work with only nitrogen flow, but these modifications are not straight forward and were therefore not performed within the scope of this work. I agree with you that this would be a promising next step to understand the decay mechanism in detail.

Rosa Arrigo asked: I find much interest in the opportunity to measure potential at the droplet interface. However, you showed that the decay of the excitation process is influenced by charged droplets with a similar kinetic regardless of the sign. I think this is counter intuitive considering the expectation based on the impact of the final state effect on the relaxation processes. Why would this be the case?

Loren Ban responded: To explain this observation, we propose that the electric field generated by the droplet charge facilitates charge-pair separation that takes place during the ionization process (see also ref. 33 in our paper, https://doi.org/10.1039/d1fd00108f). Since a charge-pair is typically formed by particles of opposite charge, an external electric field would facilitate their separation (*i.e.* higher electron yield) for either field polarity. This mechanism assumes that the electric field has an influence only on the ionization process, but not the excitation of the long-lived state. It is possible that other mechanisms could be responsible for our observations. We assume that the decay involves neutral particles, but if this was not the case one could also observe the effect of the electric field on the charged particle diffusion (ref. 61 and 62 in our paper, https://doi.org/10.1039/d1fd00108f).

Rosa Arrigo added: How is the potential being measured?

Loren Ban replied: We do not measure the potential directly but infer it from the droplet charge state. By assuming that the charge is distributed on the droplet surface we can calculate the electric field at the surface by the equation given in the paper. It is important to note that this is a rather simple model and additional complexity probably exists. A way of measuring this potential with our setup was suggested in ref. 1 where we showed that the shift in the photoelectron kinetic energy onset can be used as a measure of the droplet surface potential (Fig. S9 in ref. 1).

1 L. Ban, T. E. Gartmann, B. L. Yoder and R. Signorell, *Phys. Rev. Lett.*, 2020, **124**, 013402.

Rosa Arrigo commented: In terms of lateral resolution, do you measure a single droplet?

Loren Ban responded: The presented data is averaged over multiple laser shots and we therefore probe an ensemble of droplets. This introduces averaging over different physical properties of our droplets and that is why it is important that we can precisely characterize these properties in parallel to the photoelectron spectroscopy experiment. On the other hand, each laser shot probes a fresh droplet sample, which inherently leads to avoidance of long-term radiation damage.

Georg Held commented: The interference patterns in Fig. 13 in your article (https://doi.org/10.1039/d1fd00108f) are calculated for photon energies of 3.1 eV and 4.7 eV and droplets of 300–1200 nm. Could one expect similar patterns in nanoparticles of a few nanometers if the photon energy is scaled up to the hundred electronvolts range such that the diameter : wavelength ratio is the same?

Loren Ban responded: This is a very nice question! These patterns are typical internal electric field signatures of Mie resonances in spherical particles, which are typically called nanofocusing. These are analogous to whispering gallery modes that are known for larger particles (a few tens of micrometers). In other words, the conditions required to observe nanofocusing are: (i) that the particle size is similar or larger to the light wavelength; and (ii) that light is not strongly absorbed (absorption length is larger than particle size). For equations see ref. 1. If both conditions are satisfied, I would expect similar patterns. However, I think that going to the hundreds of electronvolts range might bring some additional challenges: first, the absorption might be too high in the soft X-ray region; and second, ionizing radiation can lead to transient changes in the refractive index which would prevent build-up of Mie resonances. A nice study was recently published investigating the effect of ionization-induced electronic changes on the scattering response of xenon clusters.[2]

1 R. Signorell, M. Goldmann, B. L. Yoder, A. Bodi, E. Chasovskikh, L. Lang and D. Luckhaus, *Chem. Phys. Lett.*, 2016, **658**, 1–6.
2 D. Rupp., L. Flückiger., M. Adolph, A. Colombo, T. Gorkhover, M. Harmand, Maria Krikunova, J. P. Müller, T. Oelze, Y. Ovcharenko, M. Richter, M. Sauppe, S. Schorb, R. Treusch, D. Wolter, C. Bostedt and Thomas Möller, *Struct. Dyn.*, 2020, 7, 034303.

Thorsten Bartels-Rausch said: Could you comment on the phase state of the droplets during the experiment? Are those solid or maybe viscous? If I understood correctly, the droplets are produced from aqueous solutions that are then dried during the differential pumping. May I ask you to give an outlook on which type of interfaces could be investigated with this approach/set-up?

Loren Ban answered: The droplets are generated from a viscous and non-volatile liquid mixture of DEHP and DEHS. Atomization of such a liquid leads to formation of spherical droplets which do not evaporate and allow us to control and measure their size and charge reliably. The droplets are indeed quite viscous and this is probably one of the reasons that allows us to observe the long-lived states on such long timescales (slow diffusion).

The short answer to your second question would be – any particulate system that you can deliver into vacuum through the aerodynamic lens in high enough concentration! We have previously used this setup to study electron escape across a liquid–vacuum interface.[1] From these experiments we found that slow electrons exhibit very high sensitivity to the interface potential shape through quantum effects and proposed that this can be used to determine the escape barrier (conduction band minimum). Extension to solid particles is in principle straightforward.

One can also study buried interfaces using core–shell aerosol particles. This opens the door to studies of solid–liquid interfaces, as well as liquid–liquid interfaces. For example, a similar setup was previously used to investigate electron scattering by the aerosol overlayer method using core–shell particles.[2]

However, extending this method to volatile substances is a standing challenge. Volatile droplets can be measured in our setup, but it is difficult to precisely control their size and in particular, their charge. As you mentioned, droplets can "dry" during their transfer into vacuum which leads to uncertainties in their size. Also, their size/charge cannot be well controlled or characterized by conventional aerosol instrumentation. I believe that further development of *in situ* droplet characterization methods will be important to extend this method to a wide range of volatile substances.

1 L. Ban, T. E. Gartmann, B. L. Yoder and R. Signorell, *Phys. Rev. Lett.*, 2020, **124**, 013402.
2 S. Amanatidis, B. L. Yoder and R. Signorell, *J. Chem. Phys.*, 2017, **146**, 224204.

Philip R. Davies opened a general discussion of the paper by Evgeniy Redekop: In Fig. 8 in your article (https://doi.org/10.1039/d1fd00120e), the XP data shows a delay in the take up of carbon. Is this a real delay? If so, what do you think is going on?

Evgeniy Redekop replied: Part of this delay is due to uncertainty between the introduction of gas into the large, back-filled analysis chamber and the time frame of the measurement, which can be minimized by the explicit synchronization demonstrated in our recent work.[1] However, the delay is still prominent even after considering the uncertainty in synchronization on the scale of seconds. This "real" delay must be due to insufficient sensitivity of the C 1s XPS signal in the limit of very low carbon coverage at this high temperature. In other words, it take a while before sufficient carbon accumulates on the surface to result in a significant C 1s signal. In this regard, the sensitivity of TAP measurements is superior, as it relies on the very efficient detection of incoming and outgoing gaseous species. When comparing experiments with pure propane and with added hydrogen, the presence of hydrogen clearly increases the delay in carbon accumulation *via* the intrinsic pathways of coke formation.

1 E. A. Redekop, N. Johansson, E. Kokkonen, S. Urpelainen, F. Lopes da Silva, M. Kaipio, H.-E. Nieminen, F. Rehman, V. Miikkulainen, M. Ritala and U. Olsbye, 2021, **92**, 044101.

Philip R. Davies said: Your work is an important first step in bringing together information from the separate TAP and XPS techniques. What are the prospects and main challenges in bringing the two techniques together in one instrument?

Evgeniy Redekop replied: The primary fundamental challenge of combining the two experiments in a single instrument is the opposing requirements of the two techniques when it comes to managing gas molecules. APXPS strives to minimize the layer of gas that the photoelectron traverses on its way to the analyzer by means of the differential pumping of the gas. Gas molecules that are pumped away are mostly discarded and not allowed to return to the sample. TAP, on the other hand, relies on quantitative counting of all the molecules scattered off the sample (for quantitative assessment of the reaction kinetics) and attempts to maximize the possibility of multiple collisions between a gas molecule and the sample surface. There are two promising avenues for implementing an optimal compromise between these conflicting demands. One avenue is to drastically minimize the volume of space available to the gas in front of the sample surface, while ensuring that the gas molecules crossing the boundaries of this volume are properly counted. This requires special instrument design and integration, including miniaturized spectroscopic cells and very close proximity between the gas outlet and the ionization region of the mass spectrometer. The virtual cell concept implemented in the POLARIS instrument provides an excellent example of minimizing the gas volume, although it remains to be seen whether the mass balance of such a virtual cell can be closed with mass spectrometry at relevant sub-second time scales. My co-authors and I are currently developing an alternative concept with an actual miniature cell which we hope to present in the upcoming publications. The other, drastically different possible avenue to achieve a fully coupled TAP-XPS experiment is to use a spectroscopic cell capped with an electron-transparent membrane that would retain gas molecules. Such spectroscopic cells have been in development for a number of years, and I expect that the robustness of membranes will eventually make them more accessible for a wider user community and for more tight integration with transient gas feed systems and mass spectrometers. In both cases, it is likely that sample surfaces will have to be prepared *via* a tailored synthesis route like ALD or CVD/PVD. Overall, I am incredibly excited and optimistic about the prospects of implementing a fully coupled experiment. It should be emphasized that the pressure range for the coupled experiment does not necessarily have to be bridged to industrially-relevant conditions, like APXPS at 1 bar, because the fundamental goal of such an experiment is to achieve the next level of (quantitative) kinetic resolution for reactions on complex materials, rather than completely replicate the environment in an industrial reactor.

Günther Rupprechter commented: You have stressed kinetic measurements. For catalytic reactions one typically determines reaction orders and activation energies. Have you measured them with your approach as well?

Evgeniy Redekop responded: Determining the reaction orders, pre-exponential factors, and activation energies is, indeed, an essential task in a quantitative kinetic study. Steady-state measurements yield these reaction characteristics only for the slowest (*i.e.* rate-determining) step. The advantage of transient kinetic methods, such as Temporal Analysis of Products (TAP) employed in our study, is that the reaction orders and activation parameters can be determined for various individual reaction steps or at least sub-networks of the full catalytic mechanism, not only the slowest ones. Examples of such comprehensive kinetic analyses of catalytic reactions based on TAP data can be found in multiple literature sources, including some of our prior work.[1–3] As for the dataset presented in the current manuscript, we have not performed such detailed data evaluation for the main catalytic cycle of propane dehydrogenation, aside from the dependency of the apparent rate coefficient for propane consumption as a function of carbon coverage at 800 K. In this proof-of-principle study, instead, we focused on the quantitative alignment of the TAP and NAP-XPS datasets in terms of the surface exposure to reactants, hoping to highlight the enormous potential of combining XPS information with rigorous gaseous kinetics. Unfortunately, only a limited amount of NAP-XPS data were collected for a single temperature and propane partial pressure during the relevant synchrotron experiment. Furthermore, post-reaction characterization of spent materials revealed extensive sintering of metal nanoparticles and did not allow us to unambiguously identify the surface structure. Extensive TAP experimentation and analysis on these sub-optimal (from the catalytic point of view) materials would hardly contribute valuable information. However, we believe that the methodological advances exemplified in our report will stimulate the development of better model surfaces for the tandem TAP-XPS experiments, and we hope to report a more detailed spectro-kinetic study of similar, but more suitable ALD-derived materials in the near future.

1 U. Menona, V. Galvita, D. Constales, K. Alexopoulos, G. Yablonsky and G. B. Marin, *Cat. Today*, 2015, **258**, 214–224.
2 E. A. Redekop, A. Lazzarini, S. Bordiga and U. Olsbye, *J. Cat.*, 2020, **385**, 300–312.
3 C. Reece, E. A. Redekop, S. Karakalos, C. M. Friend and R. J. Madix, *Nat. Catal.*, 2018, **1**, 852–859.

Wendy Flavell asked: You have written a great deal (outside this meeting) on synchronisation in gas pulsing experiments. I wonder if you have a view on whether it is best to synchronise to the gas sources, or to attempt image recognition and identify spectral changes associated with the ALD process? Could you comment on the positives and negatives?

Evgeniy Redekop responded: Image recognition, *i.e.* detecting an abrupt apparent shift in the core-level spectra of gas-phase species related to the change of the surface work function, is a valuable strategy for post-factum alignment of time frames during a transient XPS measurement. Clearly, the main benefit of this approach is its simplicity in terms of software and hardware, as it can readily be implemented for most experimental setups with time-resolved data acquisition with no extra modifications. ALD processes in particular are well suited for the use of image recognition for time-frame alignment, as ALD precursors often require extra time to passivate the inlet lines before impacting the sample. However, there are considerable advantages of the alternative strategy, *i.e.*, explicit synchronization of the gas perturbation and data acquisition time frames. Such explicit synchronization does not rely on a fortuitous capture of the gas-phase species within the same image (spectral range) as surface species. Furthermore, explicit synchronization allows systematic studies of gas transport phenomena within the device, including transport upstream from the sample surface. Other potential advantages include: (1) the ability to automate complex experimental sequences with programmable injections of various gases, with possible data acquisition in different spectra ranges during different segments of the sequence; (2) simplicity of the data analysis workflow, since the time frames are already aligned at the acquisition stage; and (3) the ability to multiplex several detection modes in the same experiment, *e.g.* XPS and mass spectrometry. It is my opinion that both synchronization schemes should be employed more widely among the user community and have their application scopes, but I believe that explicit synchronization will play an important role in combined spectro-kinetic experiments on short (subsecond) time scales.

Georg Held commented: Given the time scales of gas pulses and the reaction of catalyst surfaces, what is the time resolution one should aim for in time-resolved catalytic experiments?

Evgeniy Redekop responded: It is hardly possible to provide a general answer to this question, given that the characteristic time scales of catalytic reactions varies over many orders of magnitude. The kinetic studies of most technologically-relevant catalytic reactions require time resolution on the order of milliseconds, at least from the point of view of a microkinetic modeling framework, although less time resolved data acquisition can still be very useful for monitoring the kinetics of slower reactions (*e.g.* methanol synthesis) and for tracking the restructuring of the catalyst material itself (*e.g.* segregation of bimetallic alloys or long-term catalyst deactivation). However, many reactions at the frontier of catalytic science will require faster time scales, such as monitoring photo-excited states in photocatalysis. Please note that I specifically limit my answer here to kinetics of reactions as opposed to dynamics of reactions.

Philip R. Davies opened a general discussion of the Concluding Remarks article by Anders Nilsson: In your study of methanol formation over the Cu/ZnO catalyst you see a carbon species with a C(1s) peak at high binding energy that you assign to formate, but could it be due to carbonate?

Anders Nilsson replied: We have considered carbonate but the binding energy of the peak is too high. Usually carbonate is at around $\approx$ 289 eV and the peak that we associate to formate is at $\approx$ 290 eV. Since the spectra are quite noisy there could be a small amount of carbonate.

Wendy Flavell asked: You have commented on the need for higher repetition rate but lower peak flux sources for spectroscopy – and I wholeheartedly agree with you. I wondered if there are any elements of your work that you have been able to carry out at low energy at FLASH, as it now has available pulse trains of 500 1 MHz pulses with 400 matching laser pump pulses?

Anders Nilsson replied: Experiments under UHV using ultrafast photoelectron spectroscopy could be done at FLASH. The best is to have a facility delivering a more even pulse pattern such as the upcoming LCLS-II that will have 1 MHz pulses evenly spaced. We tried ultrafast photoelectron spectroscopy at FLASH using the POLARIS instrument and could measure spectra of carbon on Ni(100) without space charge but we could not pump any laser induced changes due to the pump laser being unable to deliver high enough intensity.

Rosa Arrigo commented: I reckon that the dynamics of the chemical species during a catalytic reaction as observed by ambient pressure XPS provide insight into the reaction mechanism, however these can be misinterpreted without comprehensive experimental screening. The study on the ammonia synthesis you presented is a beautiful example that shows how powerful this technique will be now that you have demonstrated the possibility to operate at pressures as high as 500 mbar and higher. You showed that a stepped surface leads to an increased abundance of the Fe–N species in the N 1s if compared to a flat ordered surface, but the information about the reaction mechanism was evidenced during the temperature dependent study, in which the decrease of the abundance of the Fe–N species in the N 1s (if I remember correctly) consisted of the Fe–N dissociation as the RDS. My question is about what you think in general about using XPS peaks for microkinetic modelling and if you have used them to obtain kinetic parameters such as the activation energy. What are the limitations?

Anders Nilsson answered: An excellent point since we have been working on input to microkinetic modelling from the XPS data. The idea is to modify transition state energies and adsorbate energies where adsorbate–adsorbate interactions are included to modify the output from DFT calculations. From such simulations a deeper understanding of the mechanism can be obtained.

Georg Held remarked: We had a discussion earlier about whether a pressure of several mbar should be called "ambient pressure" or rather "near-ambient pressure", given that the ambient pressure is 1 bar. It was pointed out that the pressure requirements depend very much on the specific gas-surface interaction under investigation. In the context of the Haber–Bosch process, the pressures used in the experiments presented in your talk are in the 1 bar range, which is again about two orders of magnitude lower than actual reaction conditions. Is it possible to establish a rule of thumb which fraction of the typical process pressure would be needed to model realistic conditions?

Anders Nilsson replied: It all depends on specific cases. For many syn-gas reactions including Haber Bosch, the high pressure requirement is mostly based on thermodynamics and not kinetics. That is, to enable a high conversion the gas phase equilibrium needs to be shifted towards the products. The conditions at 1 bar therefore represent the process at low conversion. What is important is to have high enough pressures to allow adsorbate to cover the surface and reactions to proceed at the relevant temperatures during XPS studies. One way to experimentally simulate higher conversion conditions that usually would have required higher pressures is to mix in the reactant gases with the products.

I would distinguish the names: ambient XPS, atmospheric XPS, high pressure XPS and *operando* XPS for the cases with pressures relevant to environmental surface conditions, atmospheric pressures, above one bar and when a real catalytic reaction is rapidly turning over, respectively.

James J. C. Counter asked: With the POLARIS instrument you mention that sample temperature is controlled. Are you also able to control the gas temperature? If so, how does this affect your observations?

Anders Nilsson responded: In the original plan we intended to also heat the gas but the design became rather complicated and we decided to maybe implement that in the next upgrade of the front cone when we had gained experience.

Conflicts of interest

There are no conflicts to declare.

Faraday Discussions

PAPER

Concluding remarks: Photoelectron spectroscopy and the future of surface analysis

Anders Nilsson

Received 17th June 2022, Accepted 24th June 2022
DOI: 10.1039/d2fd90028a

The Faraday Discussion on Photoelectron Spectroscopy and the Future of Surface Analysis was held both in person and virtually in London, April 20–22, 2022. In my concluding lecture I have highlighted three different recent directions in the field; (i) *in situ* and *operando*, (ii) buried interfaces and (iii) time resolved measurements. I have based my lecture on most recent studies conducted in my research group in all these 3 directions. In particular, the new *operando* instrument POLARIS, located at the brightest hard X-ray beamline at DESY for photoelectron spectroscopy, has opened new avenues to address important problems in catalysis that are essential for society to become fossil free.

Introduction

First of all, I would like to thank the organizers for putting together an excellent *Faraday Discussions* meeting on photoelectron spectroscopy in the middle of London. For many of us, it was a unique time to meet in person after 2 years of pandemic. The meeting was stimulated by long discussions after each presented paper. I am also pleased to have been able to deliver the concluding remarks lecture – in particular, since I was the last PhD student that was recruited by the Nobel Laureate Kai Siegbahn in Uppsala, although he never was my advisor. He had then in 1981 received the Nobel Prize in Physics for X-ray Photoelectron Spectroscopy (XPS) or as it was called then ESCA (electron spectroscopy for chemical analysis).

Let me with one single figure capture the essential two aspects of XPS that make it unique for surface chemical analysis. Fig. 1 shows the XPS spectra of the Yb 4f level as it grows layer-by-layer on a Mo(110) single crystal surface.[1] Although the 4f electron levels have low binding energy the electron orbit is relatively close to the nucleus and can be regarded as core electrons. We observe a number of peaks of double nature reflecting the existence of two spin–orbit components, $4f_{3/}$

Department of Physics, Stockholm University, AlbaNova University Center, 106 91 Stockholm, Sweden. E-mail: andersn@fysik.su.se

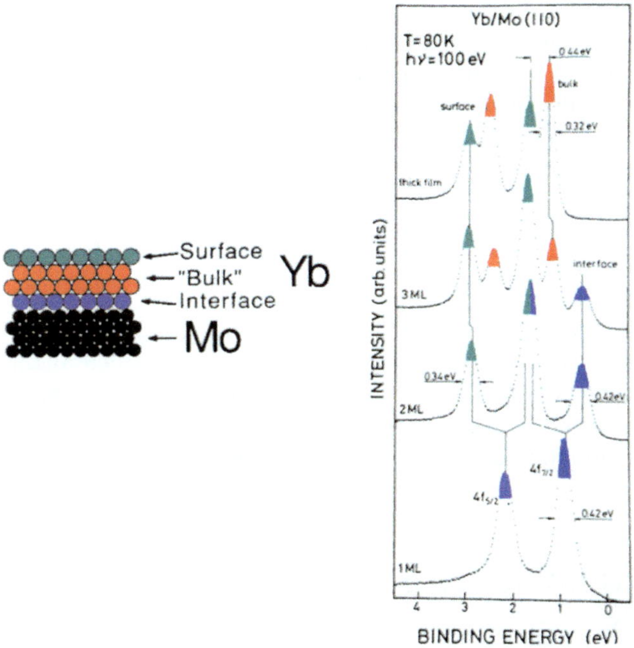

Fig. 1 Yb 4f photoelectron spectra for 1, 2, 3 layer and thick layer films of Yb grown on Mo(110). Reproduced from ref. 1.

$_2$ and $4f_{1/2}$. The existence of several doublets reflects the fact that there are Yb atoms in different environments. This year it is the 60th anniversary of the discovery of chemical shift in photoelectron spectroscopy by Siegbahn and co-workers, who discovered two chemical shifted sulphur peaks for thiosulphate.[2,3] In particular, that discovery set out the first unique aspect of XPS, the ability to perform chemical analysis. In the 3 ML case in Fig. 1 we can identify three different chemically shifted 4f components: the atomic layer at the interface, the middle layer and the top-most layer.[1] As the number of layers increase the interface peak diminishes, the middle layer becomes the bulk and the top layer becomes surface and its shift with respect to the bulk has been denoted the surface core level shift.[4] There are many approaches to interpret chemical shifts but today the standard approach is based on difference in total energy between the ground state and the core ionized state.[5] In the case of Yb on Mo(110) the difference in total energy allowed for the determination of adhesion, interfacial and segregation energies.[1]

The second aspect is related to the surface sensitivity of using electron detection. The electrons scatter strongly against other atoms and undergo inelastic losses. The elastic electrons therefore are emitted from the surface region where the depths depend on the electron mean free path. The latter is strongly dependent on the electron kinetic energy and has a minimum around 100 eV. The photon energy used in Fig. 1 was 100 eV and since the 4f is almost at the Fermi level the kinetic energy is at the minimum in the escape mean free path. The 3 ML case in Fig. 1 nicely demonstrates the surface aspect of XPS. We observe that the intensity of the middle layer is almost half of the first layer and then further half for the interface layer. For each layer the intensity dropped by a factor of 2. These two aspects "surface sensitivity" and "chemical sensitivity" led to the revolution of XPS for surface chemical analysis and the field has developed strongly during the last 5 decades.

In the concluding remarks I like to highlight three different directions that have appeared during the last decade and these were the main focus during the meeting:

- *In situ* measurements in non-vacuum environments
- Probing interfaces
- Time resolved measurements

To illustrate these three different areas of XPS I will draw recent results from our own activities. At the end I will point towards a set-up where all three activities can be built into a single instrument at a new dream facility. However, we should first discuss the challenge the world faces where surface catalysis can be part of the solution and XPS can become the main characterization tool.

Opportunity for XPS studies in fossil-free based catalysis

Climate change makes it essential that we develop alternatives to fossil-based resources for our energy and chemical needs. Production of fuels and chemicals from renewable sources can reduce or even eliminate CO_2 emissions, but it is necessary to find efficient ways to store and transmit energy. Synthetic fuels offer a promising alternative to fossil fuels as they provide the highest energy density of all energy storage media and can be stored cheaply over long periods. Today, fossil resources also form the basis of our current chemical industry with polymers, adhesives, lightweight carbon composites, detergents, and pharmaceuticals as a few examples, all consisting of carbon backbones. The chemical industry uses more than 10% of all fossil resources in Europe. Therefore, replacing the fossil feedstocks currently used in the chemical industry with sustainably produced base chemicals would significantly decrease net CO_2 emissions. The conversion of abundant molecules (in particular, water, CO_2, and N_2) into fuels and base chemicals (*e.g.*, H_2, methanol, light olefins, aromatics, and ammonia) using renewable electricity and electrocatalysis, photocatalysis, or thermal catalysis is an essential component in a sustainable energy and chemical production system, as illustrated in Fig. 2.[6]

There currently exist scientific means to produce fuels and base chemicals from renewable electricity based on electrolysis to generate hydrogen combined with traditional thermal catalysis. However, no existing technology can do it sufficiently well to be economically competitive with fossil resource-based processes. The scientific challenge is centered around molecular and interfacial catalysis. For thermal catalysis, the hydrogen feedstock derived *via* electrolysis would require processes at lower pressures (below 10 bar) and lower temperatures. It will be necessary to develop a new set of catalysts to reduce CO, CO_2, and N_2 into base chemicals and fuels under milder reaction conditions. Furthermore, we need new catalytic materials for electrocatalysis that can be selective towards specific products operating at low overpotentials.

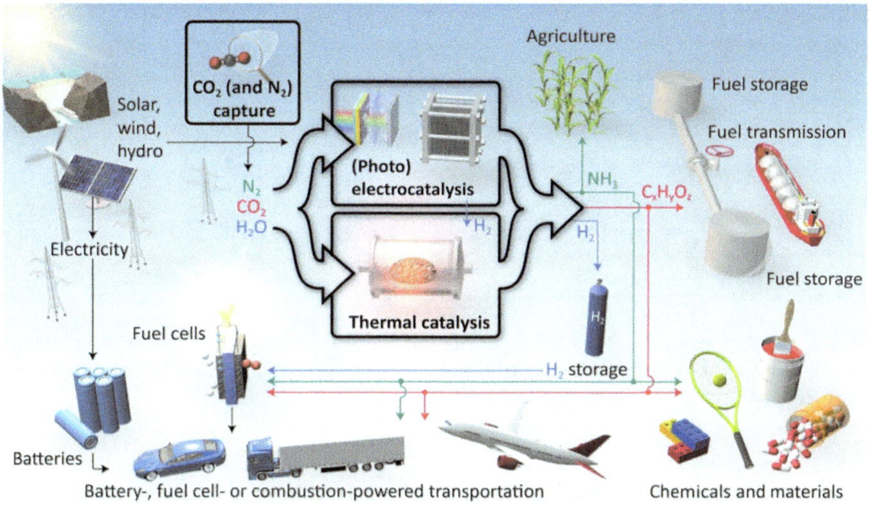

Fig. 2 Illustration of a sustainable energy system to produce base chemicals for industry and transportation fuel based on catalysis. Key processes are electrochemical water splitting and thermal and electrochemical CO_2 and N_2 reduction—graphics courtesy of Jakob Kibsgaard, DTU. An earlier version of this figure appeared in ref. 6.

To develop new catalytic materials, it is essential to derive a fundamental understanding of the active chemical phase and various reaction intermediates as the reaction occurs (*operando*) that can couple with theoretical simulations. This Faraday Discussion has highlighted how X-ray photoelectron spectroscopy (XPS) can be used to derive chemical information with high surface sensitivity. There are already many applications of differentially pumped near ambient pressure XPS (NAP-XPS) where surfaces can be probed under pressure of 0.1 mbar, in exceptional cases, pushed towards 10 mbar.[7] For investigating catalytic surfaces during CO, CO_2, and N_2 hydrogenation, at least 100–1000 times higher pressures are necessary since the reaction occurs at a temperature where otherwise the adsorbate population would be negligible. In order to meet this goal, we decided to develop the POLARIS instrument at Stockholm University, an actual *operando* XPS technique for hydrogenation reactions.[8]

The high-pressure *operando* XPS instrument POLARIS

The principle of the high pressure XPS instrument POLARIS is to have a virtual high-pressure gas cell in between the sample surface and electron spectrometer.[8] If the thickness of the cell is only a few tenths of μm and using high kinetic electron energies above 3000 eV, the electron scattering in the gas phase is not 100% and a detectable signal can be obtained for real hydrogenation reaction systems up to around atmospheric pressures.[8] A photograph of the instrument is shown in Fig. 3a where one can see that it comprises a Scienta R4000 hemispherical analyzer equipped with a HiPP-2 differentially pumped pre-lens.[8] The virtual cell operates *via* that the gas is blown straight onto the sample *via* a gas

Fig. 3 (a) Picture of the POLARIS instrument on-site at beamline P22 at DESY. (b) Image illustrating a cut-through drawing of the front cone, which schematically illustrates the virtual cell concept. Reproduced from ref. 8.

stream integrated into the front cone of the electron lens, as illustrated in Fig. 3b. The virtual cell creates a 2.5 mm diameter and ∼15–30 μm-thin cushion of pressurized gas just between the sample, and the analyzer inlet, where the sample is irradiated with X-rays and electrons are emitted. A rough vacuum around 5–8 mbar is maintained in the rest of the chamber. This cell development allows to create overpressures beyond 1 bar in an otherwise UHV chamber in a simple way. Furthermore, since most of the gas entering the virtual cell is pumped away, the net flow of gases is away from the sample. This directionality results in low susceptibility to contaminants from the chamber walls, and with higher gas flows at the surface, catalysts can be studied in the kinetic regime, not being mass-transfer limited. Moreover, the small volume of the virtual cell means that pressure or gas composition can be changed very fast (*i.e.*, from 2 bar to zero), resulting in an instantaneous effect that opens opportunities for out-of-equilibrium experiments probed time-resolved with a time-resolution of 60 ms, limited by the 17 Hz frame rate of the detector camera.

The beamline P22 at Petra III, Hamburg, Germany, where currently the instrument is located, complements the features of POLARIS extremely well.[9] It is a hard X-ray beamline delivering a photon energy from 2.4 keV to 15 keV with a narrow beam focus (10 × 10 micrometer) and low divergence. Owing to the source's brilliance the beamline is able to achieve the required foci with a horizontal divergence below 0.03° and flux in excess of 10^{12} photons per s at typical setups. The photon resolution is selectable down to 150 meV using a combination of double crystal monochromator crystal and post monochromator.

In order to demonstrate the high-pressure capability, we generated a most favorable case with a Cu surface in contact with He atmosphere that has low cross section for electron scattering. Fig. 4a shows the Cu 2p core level spectra in varying pressures up to 2.5 bar.[8] In the figure, three peaks are identified. Peak (b) and (c) are the spin–orbit split components, Cu $2p_{1/2}$ and Cu $2p_{3/2}$, respectively. With higher pressure, the signal decreases as expected due to attenuation by the gas phase. Peak (a) is related to a peculiar kind of phenomenon where the photoelectron excites the electronic state of the He atom resulting in a discreet but inelastic energy loss. In this case, we observe a 1s to 2p transition. We have observed similar unique molecular fingerprints of electron losses for every gas

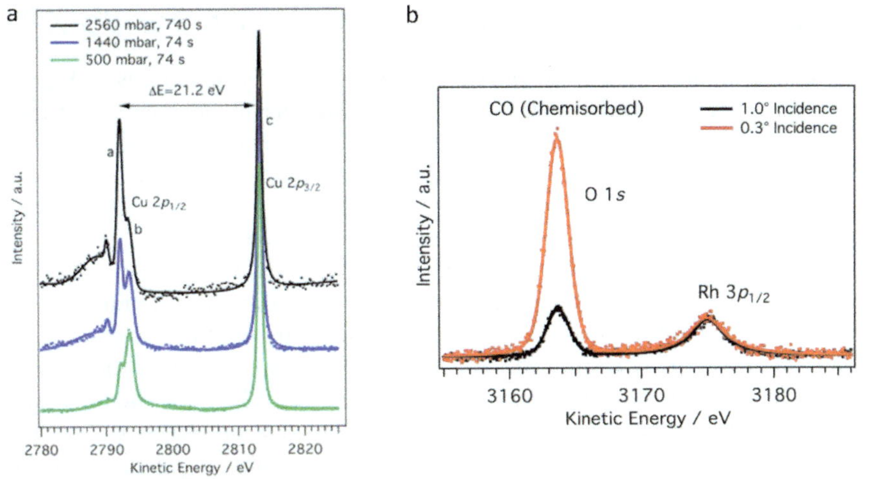

Fig. 4 (a) Cu 2p XPS spectra at different pressures up to 2.5 kbar with indicated measurement times at an incident angle of 1.3° and photon energy of 3.7 keV. (b) Grazing incidence XP spectra of the O 1s and Rh $3p_{1/2}$ region taken at 1° (black) and 0.3° (red) grazing incidence angle on CO covered Rh(111). The peaks have been background-subtracted and normalized to the same peak height of Rh $3p_{1/2}$. The photon energy was tuned to 3.7 keV. Reproduced from ref. 8.

that we have used in the set-up so far. In reactive mixtures we have been able to follow CO oxidation at a total pressure of 1 bar.[10]

Although hard X-rays, generating high kinetic energies of the photoelectrons, are being used, in order to maintain surface sensitivity it is essential to conduct experiments below the critical angle for total external reflection (typically well below one degree). Such measurements allow the effective probing depth to be limited not by the scattering of electrons in the solid but rather by the penetration depth of the evanescent X-ray wave.[11] With typical probing depths of ~15 Å, the coveted surface sensitivity can be maintained even with harder X-rays. The whole instrument, as is seen in Fig. 3a, has been placed on top of a hexapod manipulation stage, which allows for translational and rotational movement in all degrees of freedom with sub-micrometer and microradian precision.[8] This ability to precisely change the incidence angle around total reflection allows the possibility to tune from surface to bulk sensitivity. Fig. 4b shows an example of Rh 3d and O 1s spectra taken from a CO covered surface of Rh(111) at 1° and 0.3° incidence angle, *i.e.* above and below the critical angle of total reflection, respectively.[8] The spectra were normalized to the intensity of the Rh $3p_{1/2}$ component. The O 1s signal increased by a factor of 5 for the surface sensitive measurement. The effect is mainly due to intensity increase in the CO signal, as the bulk signal is affected very little by the change in angle. This measurement demonstrates the surface sensitivity of grazing incidence XPS and that adsorbates on the surface can be detected despite working with hard X-rays.

Probing active phase of catalyst by *operando* XPS

Let us take an example of how the new development of *operando* XPS with pressures that are approaching 1 bar can be used to solve surface chemistry questions in catalysis that can become important in the transformation of the chemical industry. Methanol is currently one of the most important petrochemical basic chemicals, with an annual production of 110 million tones, and can be converted into tens of thousands of different products and used for the manufacture of, for example, plastics, detergents, pharmaceuticals and fuels. Methanol also has the potential to become a future energy carrier where, for example, aviation fuel can be produced using captured carbon dioxide and hydrogen from electrolysis of water instead of using natural gas. The leading industrial catalyst for the CO/CO_2 hydrogenation toward methanol consists of a mixture of Cu, ZnO, and Al_2O_3 and the process is carried out between 150 °C and 300 °C at pressures of 50 to 100 bar.[12] The reaction can also occur at substantially lower pressures of a few millibar.[13] Despite this catalytic process being a century old, years of intense research has aimed at determining the mechanism and active sites. The catalytic nature of the Zn in the catalyst is still highly debated,[12] and the mechanism is not yet fully experimentally verified. The active chemical state of Zn in a Zn–Cu catalyst during CO_2/CO hydrogenation has been debated to be either Zn oxide nanoparticles, metallic Zn, or a Zn–Cu surface alloy. Another mechanistic issue is whether the Zn-promoted mechanism propagates through formate and methoxy species as stable intermediates with different propensities depending on the reactants being either CO_2, CO, or a mixture of both.[14–16] It has been determined from isotope labeling experiments that the reaction proceeds more efficiently from CO_2 than CO,[17] but it remained unresolved how the mixture of the two gases yields the highest production rate of methanol.[18]

In order to enable a direct comparison with theoretical calculations[14] we selected an identical model system with a stepped Cu(211) single crystal promoted by Zn. The Zn/ZnO/Cu(211) surface was prepared by evaporating metallic Zn and subsequent thermal annealing, generating around 15 or 35% surface Zn inside the vacuum chamber without exposure to air.[19] Fig. 5a shows Zn 3d spectra measured under reaction conditions of 180 mbar, 140 °C, 2 : 1 H_2 : CO and 3 : 1 H_2 : CO_2 gas fractions.[19] The spectra were consistently characterized by four peaks at 10.9, 10.3, 9.8, and 9.5 eV throughout the accumulated dataset. The low binding energy peaks at 9.5 and 9.8 eV (green) corresponded to the spin–orbit split states of metallic Zn $3d_{3/2}$ and Zn $3d_{5/2}$. The broad structure at 10.9 eV (gray) was associated with Zn in the +2 oxidation state related to bulk-like ZnO. The component at ~10.3 eV (light blue) was assigned to a Zn-(δ+) oxidation state species related to Zn interacting with formate and methoxy as well as some ZnOH. Thus, we identified the preferential states of Zn directly from the spectra under gas phase composition-dependent reaction conditions. Under CO-rich conditions, the metallic Zn signal became more dominant, whereas under CO_2-rich conditions, ZnO was strongly enhanced. Fig. 5b shows that when the pressure was increased to more extreme conditions at 500 mbar and 230 °C, more noisy spectra were obtained, caused by increased gas phase electron scattering still indicating that, in pure CO, almost only metallic Zn and in CO_2 only ZnO were observed.[19] The relative amount of Zn in at least two different redox states changes

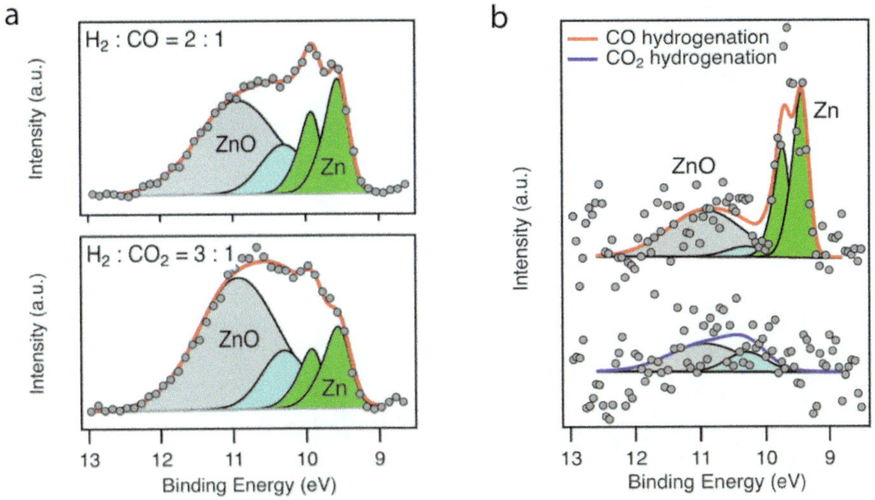

Fig. 5 (a) Zn 3d XPS spectra obtained at 180 mbar, 140 °C, and stoichiometric gas composition using 4.75 keV photon energy. The experiment was conducted at ~35% surface Zn. The relative amount of ZnO to metallic Zn shows a strong dependency on gas composition. (b) Zn 3d XPS spectra obtained at 500 mbar, 230 °C, and $H_2 : CO = 2.6 : 1$ and $H_2 : CO_2 = 2.6 : 1$ using 4600 eV photon energy at a Zn coverage of ~15%. Reproduced from ref. 19.

continuously in different gas mixtures from pure CO to pure CO_2 at 180 °C and 280 mbar. To distinguish the surface from bulk effects, separate measurements were conducted at two different incidence angles 0.3° below and at 0.9° above the critical angle.[19] The results showed that the reduction of ZnO to metallic Zn occurred in close proximity to the surface or directly at the surface, whereas the main fraction of ZnO_x should represent larger assemblies containing substantial bulk ZnO contribution. The valence character of the Zn 3d electronic state was used to address the nature of the metallic Zn, distinguishing metallic Zn islands on the surface, or a surface Cu–Zn alloy.[19] The Zn 3d spectra in Fig. 5a showed a spin–orbit split metallic feature that was enhanced under CO-rich conditions, which was indicative of a decrease in the Zn–Zn interaction. It was thereby possible to identify that under CO-rich conditions there is a surface Cu–Zn alloy.

We next consider the reaction mechanism and the identification of stable intermediates. Density functional theory (DFT) predicts deep minima in the free energy surface with formation of formate and methoxy in the case of CO_2 hydrogenation and formation of only methoxy in the case of CO hydrogenation.[14–16] Fig. 6a shows the C 1s spectra during CO_2 reduction and indeed both formate and methoxy are identified whereas Fig. 6b shows only methoxy from CO.[19] When the temperature is high the coverage of the intermediate adsorbates diminished indicating fast turn-over of the reaction. The reaction mechanism for methanol formation is the reduction of CO_2 that proceeds *via* formate and methoxy. However, the CO_2 is oxidizing Zn into ZnO but since CO is mixed into the reaction mixtures during operational conditions it is from these data possible to infer that the presence of CO makes the Zn stay metallic and alloyed into the surface.[19]

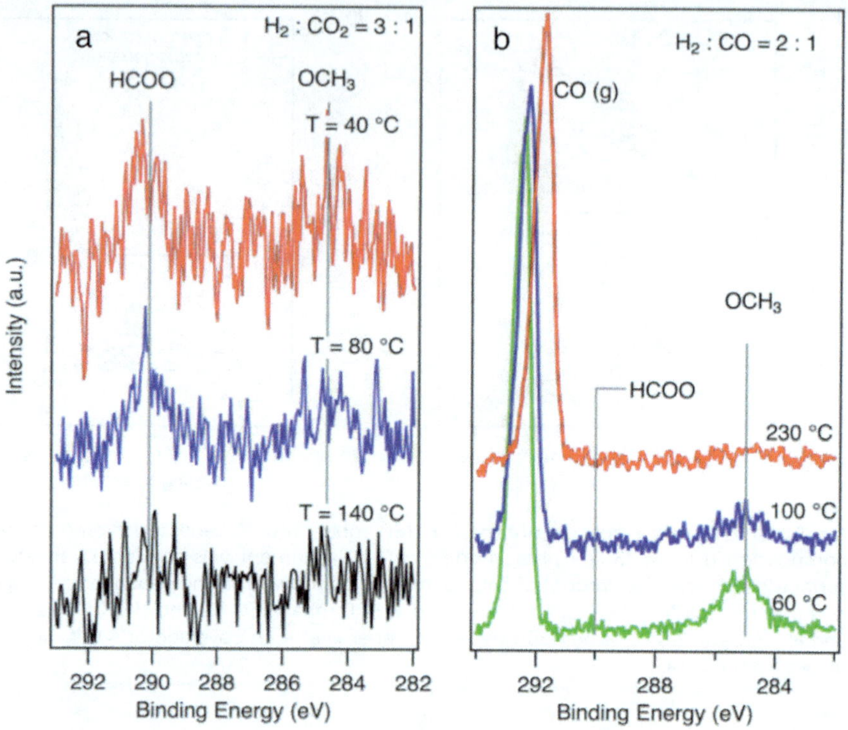

Fig. 6 (a) C 1s XPS spectra during CO_2 hydrogenation obtained at a pressure of 180 mbar and various temperatures and conducted at ~35% surface Zn. (b) C 1s XPS spectra during CO hydrogenation obtained at a pressure of 500 mbar and various temperatures and conducted at ~15% surface Zn. Reproduced from ref. 19.

Operando XPS probing the surface of Rh and Fe during CO hydrogenation reactions in the pressure regime 150–700 mbar has recently revealed important insights on ethanol synthesis[20] and Fischer–Tropsch synthesis of hydrocarbons.[21] The *operando* XPS will offer to provide insights into the nature of the catalyst and reaction intermediates for new catalyst development as is necessary for a non-fossil based chemical industry

Probing interfaces

There is an increasing ability by XPS to probe buried interfaces through the usage of hard X-rays where the electron mean free path becomes longer with the high kinetic energy electrons. Here I will demonstrate how we can obtain depth and morphology information using two approaches by either utilizing different electron kinetic energies of the emitted photoelectrons and thereby varying the electron scattering or the angle dependence of the reflectivity of the incoming X-rays.

The latter was used to derive the long-range structure of a Pd catalyst during CO oxidation by utilizing *operando* XPS.[22] The surface structure is of particular interest in Pd due to the possibility of surface reconstructions and oxide formation during the catalytic reaction.[23] The question is the nature of the active phase: (i) the oxide on top of metallic Pd, (ii) metallic islands are formed on top of the oxide, (iii) the oxide is reduced and is exposing the bulk Pd in terms of pillars or (iv) a combination of (ii) and (iii). Here grazing incidence XPS (GI-XPS) at different angles during reaction conditions can distinguish the different scenarios.[22]

GI-XPS operates on the principle of total external reflection, wherein below the critical angle, the light source is reflected off the surface with a low penetration depth. At low angles, the X-ray light penetrates less deeply into the material; as the angle is increased, the penetration depth of the light increases. By calculating the penetration depth of the light and escape likelihood of the photoelectrons, the different models could be evaluated to determine the structure of the surface.[22]

Fig. 7a shows an example of an angle-resolved Pd $3d_{5/2}$ spectrum during CO oxidation at a total pressure of 300 mbar and a gas ratio of 50 : 4 O_2 to CO.[22] The metallic and oxide phases are well-resolved and are determined well through curve fitting. Fig. 7b shows the measured Pd/PdO intensity ratio as the incident angle of the X-rays relative to the surface is changed. When the incident angle of the photons approaches the critical angle, 0.7° for Pd at this photon energy, the X-ray field intensity within the sample changes. We clearly see a substantial variation in the ratio as the angle is changed. At low angles, below 0.2°, there is a strong metal signal that rapidly diminishes as the angle increases forming a minimum in the ratio between 0.2° and 0.4°. At higher angles, the metal component then grows more rapidly again as for the very small angles since we are above the critical angle and probe more the bulk metal. This profile should be informative in the nature of the Pd surface during CO oxidation and how the metal and oxide are distributed. By calculating the intensity of the field and including the effects of electron scattering through the different materials as a function of depth, different model can be constructed where the fitting parameters are the thicknesses of the metal and oxide layers. Two of the models are shown in Fig. 7b with either a fully covered surface with PdO or small Pd metal islands on top pillars of PdO. It is the latter model that fits the experimental data in a qualitative manner. Following Fig. 7c, we conclude that small Pt metal islands form on top of the oxide and this is the active phase during CO oxidations at realistic conditions of 300 mbar.[22]

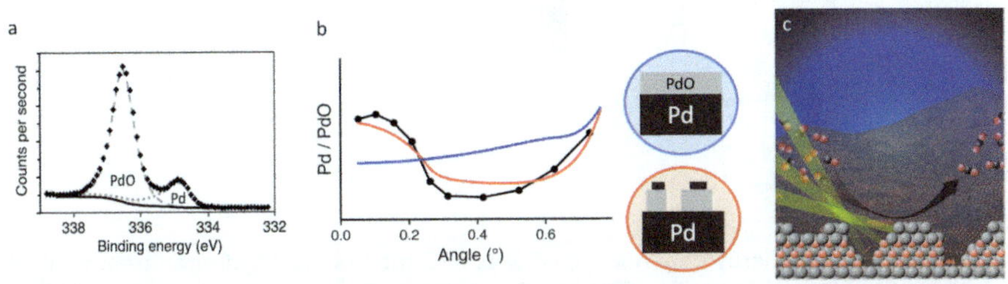

Fig. 7 *Operando* XPS study of CO oxidation on Pd(100): (a) an example of peak fitting for the Pd $3d_{5/2}$ peak with two components, a metal peak based on pure components shown with a dotted line and an oxide with a dashed line, (b) ratio of metal to oxide peak intensity as a function of angle with two calculated models, (c) metallic Pd forms on top of PdO during the reaction. Reproduced from ref. 22.

The other approach to vary the kinetic energy of the photoelectrons is here shown to shed light on the possible oxygen species inside Cu metal after reduction of the oxide. Among various catalysts that have been studied for electrochemical reduction of CO_2, Cu is the only pure metal capable of reducing CO_2 into higher-order hydrocarbons and oxygenates beyond the two-electron transfer products (*e.g.*, formic acid or CO) in comparison to what has been observed on most other metals.[24] It is noteworthy that the activity and selectivity toward multi-carbon products of the CO_2 reduction reaction can be significantly enhanced by fabricating a modified Cu by first oxidizing and subsequently reducing a pristine Cu substrate.[25–27] The origin of such improved performance on oxide-derived Cu has been discussed, but the one here particular related proposal is that the enhanced performance is due to the appearance of subsurface oxygen.[28,29]

The *operando* instrument POLARIS was used to probe if subsurface oxygen could exist in the near-surface region of an oxidized Cu surface that has been reduced *in situ* in H_2 at a pressure of 300 mbar and a temperature of 140 °C.[30] After oxidation, the XPS spectrum showed the presence of Cu_2O and after reduction of Cu metal. Fig. 8a shows the O 1s photon energy dependent XPS spectra at an incidence angle of 1.5° for the reduced Cu sample.[30] Since different photon energies will generate photoelectrons with different kinetic energies, the sampling depth will vary. First, we directly observe that there is oxygen remaining in the sample. There are three different spectral features that vary with photon energy, and their intensities have been determined using curve fitting. There are two distinctly different subsurface oxygen peaks, with one related to lattice defects/Cu vacancy (henceforth: $O_{Cu,vac}$) at 531.6 eV and the other related to interstitial oxygen (O_{int}) intercalated in the metallic Cu structure at 533.7 eV. There is also the dominating Cu_2O feature at 530.6 eV that still exists in the reduced sample.

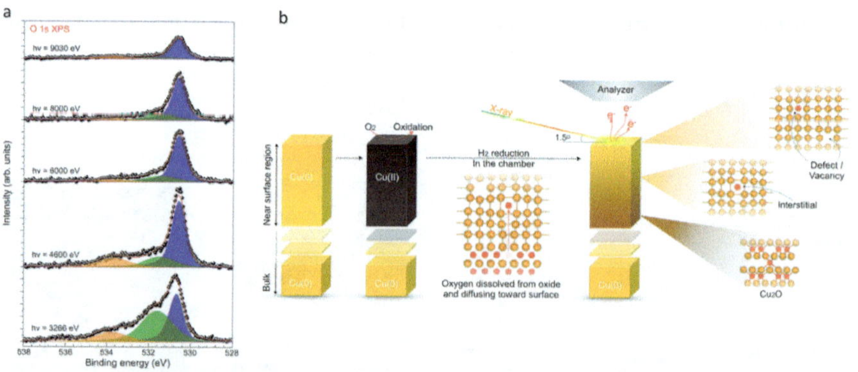

Fig. 8 (a) Photon energy dependence of O 1s XPS spectra after hydrogen treatment of oxidized Cu. The fitting results with blue: Cu_2O, green: $O_{Cu,vac}$, and orange: O_{int}. (b) The illustration of two different subsurface oxygen species, $O_{Cu,vac}$ and O_{int}, formation during oxide derived Cu preparation, where a defect/vacancy rich surface is generated by Cu atom rearrangement under H_2 treatment in the chamber, and the lattice oxygen is dissolved from the oxide and then diffuses toward the surface as O_{int}. Reproduced from ref. 30.

The relative intensities show that at the most surface sensitive probing, photon energy of 3266 eV, the $O_{Cu,vac}$ is significantly stronger than O_{int}, but this difference is effectively removed when probing with 4600 eV photons. This strongly indicates that $O_{Cu,vac}$ species are substantially more concentrated at the near surface region than O_{int}, suggesting interstitial oxygen is not stable near the surface in copper. During the reduction process, the metallic copper lattice gradually formed on top of Cu_2O will be defect rich, and it will not order in a perfect fcc lattice structure. The O_{int} species is here thought of as the oxygen dissolved from Cu oxide that, upon interaction with copper lattice vacancy sites, may form $O_{Cu,vac}$. If it reaches the top-most surface region during reduction conditions, it will be reacted away. There is a Cu_2O layer buried underneath a reduced copper film, and this is based on the relative intensity changes as a function of photon energy and the dominating metallic copper signal in Cu spectra. The results here, summarized in Fig. 8b, demonstrate that subsurface oxygen can exist in the near-surface region after a substantial reduction of the oxide and potentially affects the CO_2 electrochemical reduction by binding the intermediate CO much stronger.[30]

Time resolved measurements

Let me bring in the last topic, that is yet to be fully utilized with photoelectron spectroscopy, allowing measurements of the chemical nature of surfaces on ultrafast time scales. The most obvious case is to follow core level shifts during catalysis as the reaction proceeds in real time.

Surface chemical reactions proceed through several elementary reaction steps that control the overall reaction rate, see Fig. 9a. The kinetic rate of each step in

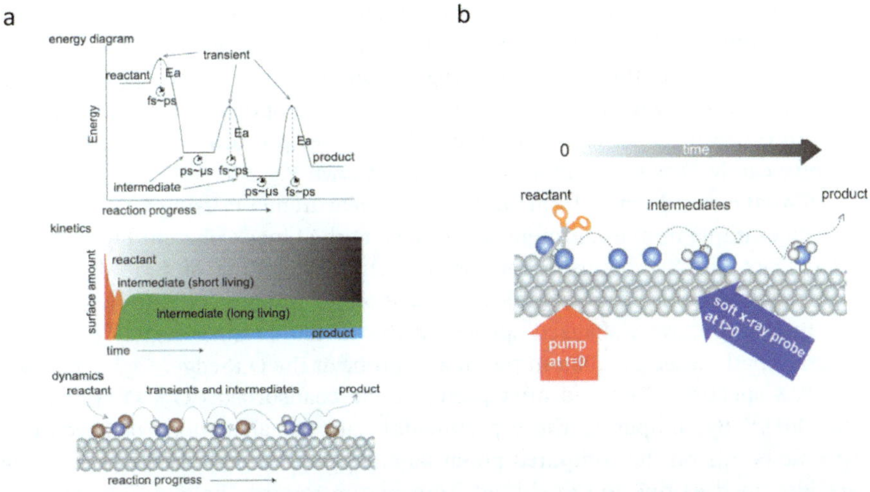

Fig. 9 (a) The top figure shows a schematic energy reaction diagram indicating transition states and intermediates of different elementary reactions. The middle illustrates the time development from reactant to product with both a transient short-lived and stable long-lived intermediate. The bottom picture shows an example of molecular fragments during the various elementary steps. (b) Schematic picture of how the reaction is initiated by an optical laser pump and probed with a soft X-ray spectroscopy pulse. The concentration of reactants, various intermediates and products can be followed in real time.

the Arrhenius description is related to how often a reaction is attempted and how often it is successful in overcoming the barrier to the reaction step. Therefore, this kinetic rate is not a representation of the time scale of the reaction event, but more its reaction probability. The time scales involved in the reaction steps involve atomic and molecular motions related to bond formation and breakage and usually take place on a 100s of femtoseconds (fs) to a few picoseconds (ps) time scale. These processes are usually referred to as molecular dynamics in contrast to reaction kinetics.[31] The outcome of most elementary steps gives rise to transient species which then react further. Some are simply short-lived intermediates while others produce long-lived intermediate states that become a reaction sink. Typically, the overall reaction is a sequence of elementary steps with ps time scales producing the wanted products or a stable intermediate that remains on the surface from nanoseconds to seconds. In order to provide a fundamental predictive picture of catalytic chemistry, we need to both understand the overall path of the reaction with its net kinetic rate and the real time molecular dynamics of the different elementary reaction steps. For the latter, we clearly need fast time resolved techniques to probe short lived-transient species and even the transition state regions of the reaction steps.

From an experimental point of view observing the individual reaction steps in the overall catalytic process is challenging. The catalytic transformation from reactants to products is a rare stochastic event (typical turnover frequencies of a good catalyst are 1 per active site per second). Thus, if we monitor the various species on the surface using *in situ* techniques under steady-state conditions we can only observe the adsorbed reactants, products and maybe some stable intermediates, like formate and methoxy during methanol synthesis.[19] However, other important intermediates will only be populated transiently so that their overall concentration under steady state conditions will be well below detection limits. Fig. 9b shows that we can drive catalytic reactions with an ultrashort (femtosecond) optical laser pulse and this provides a new window into these processes.[32] We can then probe the progression of the reaction using soft X-rays from X-ray lasers with femtosecond resolution. The optical laser pump and X-ray probe can be conducted at different delay times providing the possibility to observe catalysis as it occurs in real time, as shown Fig. 9b.

Of what can be learned from pump–probe measurements using X-ray lasers we give an example from experiments conducted at the Linac Coherent Light Source (LCLS), not with XPS but using the complementary technique of X-ray absorption spectroscopy (XAS).[33] This example is the case of CO oxidation on Ru where we, for the first time, have visualized species in the transition state region along the reaction path based on 400 nm pump and probe at the O K-edge. Fig. 10a shows the XAS spectra before and after pump of the coadsorbed CO + O system on Ru(0001).[33] By comparing the experimental data to calculations of spectra at specific points on the computed potential energy surface these new resonances are identified as due to CO shifted from on-top towards hollow sites, O from hollow sites towards bridge sites and to the formation of a bond between CO and O that is substantially elongated in comparison to in the CO_2 final product. From the time dependent measurement, a molecular movie can be generated as illustrated in Fig. 10b.[33] The O becomes activated with motions parallel to the surface on a time scale below 300 fs whereas CO is activated on a 500 fs time scale. They will start to collide on the surface since the CO is caged by O atoms in neighboring

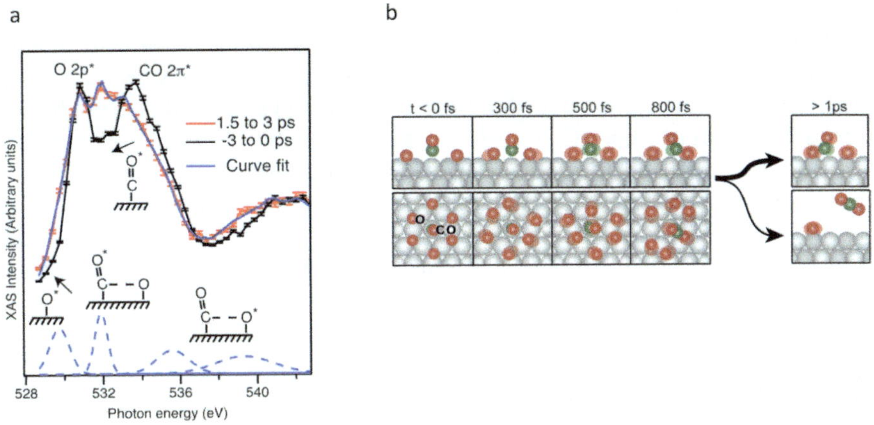

Fig. 10 (a) O K-edge XAS spectra of CO + O coadsorption system before (black) and after (red) pump. Blue line is a curve fit of pumped data using the unpumped spectrum and calculated features (dashed) after activation. (b) Schematic of the reaction steps including characteristic time scales. Reproduced from ref. 33.

sites in the coadsorbed honeycomb structure. These collisions generate a transient population of species in the transition state region that appears on a time scale slightly longer than the initial motion of the CO and O species. Most collisions lead to dissociation back to CO and O with subsequent further collisions into the transition state while only a small fraction of the attempts result in formation of CO_2. A number of different chemical reactions has been investigated at LCLS[33] and FERMI[34,35] using these ultrafast techniques based on X-ray spectroscopy opening-up a new era in catalytic research.

There is an opportunity to utilize XPS as the probe technique for ultrafast measurements of surface chemical reactions to complement X-ray spectroscopy data. The challenge, in comparison to photon-based techniques, is that electron emission during short time scales causes space charge that both broadens and shifts the core level spectra.[36] This can be circumvented by attenuating the incoming X-ray beam in such a way that the number of emitted electrons during the pulse is low. Recently it was demonstrated that indeed important information during CO oxidation over TiO_2 regarding the oxygen activation could be derived.[37] With the development of high repetitions X-ray laser sources, such as LCLS-II, XPS will become an important tool. The number of photons per pulse needs to be low but with a repetition rate of 100 kHz to 1 MHz then a spectrum can be obtained in a short time.

Potential future XPS facility

I have presented here that XPS has been developing in many different directions since the initial discovery of the chemical shift 60 years ago. I have highlighted with examples from my own research three different important recent directions, *operando*, probing interfaces and time resolved measurements. With the development of the POLARIS instrument that opens up measurements at atmospheric pressures using small divergent hard X-ray beams, allowing to make use of large sensitivity around the critical angle of total reflection, we could envision to also add the third component of ultrafast measurements into a single instrument. What would be required is a femtosecond X-ray source of tender X-rays (1–5 keV) with small divergence. The question is if a facility could be built that is not based on an X-ray laser concept. What is needed is only spontaneous X-rays generated in a traditional undulator from a short pulse electron source ($\approx$100 femtoseconds) but with high repetition rate, $\approx$100 kHz to 1 MHz. We could then envision to probe surface chemistry during *operando* at various depths from surface to bulk sensitive during ultrafast time scales. This could provide a new era in the development of Photoelectron Spectroscopy for Surface Analysis.

Conflicts of interest

There are no conflicts to declare.

Acknowledgements

First, I wish to thank the organizing committee for putting together a most exciting and stimulating Faraday Discussion on Photoelectron Spectroscopy. Secondly, I would like to thank the POLARIS team for years of instrumental development and many night shifts at the DESY facility. Thirdly, I thank Hirohito Ogasawara for many years of collaboration and for Fig. 9. Many scientists and technicians have provided outstanding work but I would like to in particular put forward Peter Amann, Hirohito Ogasawara, David Degerman, Patrick Lömker, Chris Goodwin, Mikhail Shipilin, Markus Soldemo, Hsin-Yi Wang and Christoph Schlueter. I would also like to acknowledge the support from the Swedish Research Council (Vetenskapsrådet, VR) with award no 2013-8823 and the Knut and Alice Wallenberg (KAW) Foundation with award no. 2016.0042.

References

1 N. Mårtensson, A. Stenborg, O. Björneholm, A. Nilsson and J. N. Andersen, *Phys. Rev. Lett.*, 1988, **60**, 1731–1734.

2 S. Hagström, C. Nordling and K. Siegbahn, *Arkiv för Fysik*, 1962, **23**, 145–151.

3 S. Hagström, C. Nordling and K. Siegbahn, *Z. Phys.*, 1964, **178**, 439–444.

4 J. F. van der Veen, F. J. Himpsel and D. E. Eastman, *Phys. Rev. Lett.*, 1980, **44**, 189–192.

5 N. Mårtensson and A. Nilsson, *J. Electron Spectrosc. Relat. Phenom.*, 1995, **75**, 209–223.

6 Z. W. Seh, J. Kibsgaard, C. F. Dickens, I. Chorkendorff, J. K. Nørskov and T. F. Jaramillo, *Science*, 2017, **355**, eaad4998.

7 J. Schnadt, J. Knudsen and N. Johansson, *J. Phys.: Condens. Matter*, 2020, **32**, 413003.

8 P. Amann, D. Degerman, M. T. Lee, J. D. Alexander, M. Shipilin, H. Y. Wang, F. Cavalca, M. Weston, J. Gladh, M. Blom, M. Björkhage, P. Löfgren, C. Schlueter, P. Loemker, K. Ederer, W. Drube, H. Noei, J. Zehetner, H. Wentzel, J. Åhlund and A. Nilsson, *Rev. Sci. Instrum.*, 2019, **90**, 103102.

9 C. Schlueter, A. Gloskovskii, K. Ederer, S. Piec, M. Sing, R. Claessen, C. Wiemann, C. M. Schneider, K. Medjanik, G. Schönhense, P. Amann, A. Nilsson and W. Drube, *Synchrotron Radiat. News*, 2018, **31**, 29.
10 S. Blomberg, U. Hejral, M. Shipilin, S. Albertin, H. Karlsson, C. Hulteberg, P. Lömker, C. Goodwin, D. Degerman, J. Gustafson, C. Schlueter, A. Nilsson, E. Lundgren and P. Amann, *ACS Catal.*, 2021, **11**, 9128.
11 L. G. Parratt, *Phys. Rev.*, 1954, **95**, 359.
12 J. Sehested, *J. Catal.*, 2019, **371**, 368–375.
13 S. Kuld, M. Thorhauge, H. Falsig, C. F. Elkjær, S. Helveg, I. Chorkendorff and J. Sehested, *Science*, 2016, **352**, 969–974.
14 M. Behrens, F. Studt, I. Kasatkin, S. Kühl, M. Hävecker, F. Abild-Pedersen, S. Zander, F. Girgsdies, P. Kurr, B. Kniep, M. Tovar, R. W. Fischer, J. K. Nørskov and R. Schlögl, *Science*, 2012, **336**, 893–897.
15 S. Kattel, P. J. Ramírez, J. G. Chen, J. A. Rodriguez and P. Liu, *Science*, 2017, **355**, 1296–1299.
16 L. C. Grabow and M. Mavrikakis, *ACS Catal.*, 2011, **1**, 365–384.
17 G. C. Chinchen, P. J. Denny, J. R. Jennings, M. S. Spencer and K. C. Waugh, *Appl. Catal.*, 1988, **36**, 1–65.
18 G. Ertl, H. Knözinger and J. Weitkamp, *Handbook of Heterogeneous Catalysis*, Wiley, 2008, vol. 1–5.
19 P. Amann, B. Klötzer, D. Degerman, N. Köpfle, T. Götsch, P. Lömker, C. Rameshan, K. Ploner, D. Bikaljevic, H.-Y. Wang, M. Soldemo, M. Shipilin, C. M. Goodwin, J. Gladh, J. H. Stenlid, M. Börner, C. Schlueter and A. Nilsson, *Science*, 2022, **376**, 603–608.
20 D. Degerman, M. Shipilin, P. Lömker, C. M. Goodwin, S. M. Gericke, U. Hejral, J. Gladh, H.-Y. Wang, C. Schlueter, A. Nilsson and P. Amann, *J. Am. Chem. Soc.*, 2022, **144**, 7038.
21 M. Shipilin, D. Degerman, P. Lömker, C. M. Goodwin, G. L. S. Rodrigues, M. Wagstaffe, J. Gladh, H. Y. Wang, A. Stierle, C. Schlueter, L. G. M. Pettersson, A. Nilsson and P. Amann, *ACS Catal.*, 2022, **12**, 7609.
22 C. M. Goodwin, M. Shipilin, S. Albertin, U. Hejral, P. Lömker, H. Y. Wang, S. Blomberg, D. Degerman, C. Schlueter, A. Nilsson, E. Lundgren and P. Amann, *J. Phys. Chem. Lett.*, 2021, **12**, 4461.
23 M. Shipilin, J. Gustafson, C. Zhang, L. R. Merte, A. Stierle, U. Hejral, U. Ruett, O. Gutowski, M. Skoglundh, P. A. Carlsson and E. Lundgren, *J. Phys. Chem. C*, 2015, **119**, 15469.
24 Y. Hori, H. H. I. Wakebe, T. Tsukamoto and O. Koga, *Electrochim. Acta*, 1994, **39**, 1833.
25 F. S. Roberts, K. P. Kuhl and A. Nilsson, *Angew. Chem., Int. Ed.*, 2015, **54**, 5179.
26 A. Eilert, F. S. Roberts, D. Friebel and A. Nilsson, *J. Phys. Chem. Lett.*, 2016, **7**, 1466.
27 C. W. Li, J. Ciston and M. W. Kanan, *Nature*, 2014, **508**, 504.
28 F. Cavalca, R. Ferragut, S. Aghion, A. Eilert, O. Diaz-Morales, C. Liu, A. L. Koh, T. W. Hansen, L. G. M. Pettersson and A. Nilsson, *J. Phys. Chem. C*, 2017, **121**, 25003.
29 A. Eilert, F. Cavalca, F. S. Roberts, J. Osterwalder, C. Liu, M. Favaro, E. J. Crumlin, H. Ogasawara, D. Friebel, L. G. M. Pettersson and A. Nilsson, *J. Phys. Chem. Lett.*, 2017, **8**, 285.

30 H. Y. Wang, M. Soldemo, D. Degerman, P. Lömker, C. Schlueter, A. Nilsson and P. Amann, *Angew. Chem., Int. Ed.*, 2022, **61**, e202111021.
31 A. C. Luntz, The dynamics of making and breaking bonds at surfaces, in *Chemical Bonding at Surfaces and Interfaces*, Elsevier, 2008.
32 A. Nilsson, J. LaRue, H. Öberg, H. Ogasawara, M. Dell'Angela, M. Beye, H. Öström, J. Gladh, J. K. Nørskov, W. Wurth, F. Abild-Pedersen and L. G. M. Pettersson, *Chem. Phys. Lett.*, 2017, **675**, 145.
33 H. Öström, H. Öberg, H. Xin, J. LaRue, M. Beye, M. Dell'Angela, J. Gladh, M. L. Ng, J. A. Sellberg, S. Kaya, G. Mercurio, D. Nordlund, M. Hantschmann, F. Hieke, D. Kühn, W. F. Schlotter, G. L. Dakovski, J. J. Turner, M. P. Minitti, A. Mitra, S. P. Moeller, A. Föhlisch, M. Wolf, W. Wurth, M. Persson, J. K. Nørskov, F. Abild-Pedersen, H. Ogasawara, L. G. M Pettersson and A. Nilsson, *Science*, 2015, **347**, 978.
34 H. Y. Wang, S. Schreck, M. Weston, C. Liu, H. Ogasawara, J. LaRue, F. Perakis, M. Dell'Angelo, F. Capotondi, L. Giannessi, E. Pedersoli, D. Naumenko, I. Nikolov, L. Raimondi, C. Spezzani, M. Beye, F. Cavalca, B. Liu, J. Gladh, S. Koroidov, P. S. Miedema, R. Costantini, L. G. M. Pettersson and A. Nilsson, *Phys. Chem. Chem. Phys.*, 2020, **22**, 2677.
35 E. Diesen, H.-Y. Wang, S. Schreck, M. Weston, H. Ogasawara, J. LaRue, F. Perakis, M. Dell'Angela, F. Capotondi, L. Giannessi, E. Pedersoli, D. Naumenko, I. Nikolov, L. Raimondi, C. Spezzani, M. Beye, F. Cavalca, B. Liu, J. Gladh, S. Koroidov, P. S. Miedema, R. Costantini, T. F. Heinz, F. Abild-Pedersen, J. Voss, A. C. Luntz and A. Nilsson, *Phys. Rev. Lett.*, 2021, **127**, 016802.
36 M. Dell'Angela, T. Anniyev, M. Beye, R. Coffee, A. Föhlisch, J. Gladh, S. Kaya, T. Katayama, O. Krupin, A. Nilsson, D. Nordlund, W. F. Schlotter, J. A. Sellberg, F. Sorgenfrei, J. J. Turner, H. Öström, H. Ogasawara, M. Wolf and W. Wurth, *Struct. Dyn.*, 2015, **2**, 02101.
37 M. Wagstaffe, L. Wenthaus, A. Dominguez-Castro, S. Chung, G. D. L. Semione, S. Palutke, G. Mercurio, S. Dziarzhytski, H. Redlin, N. Klemke, Y. Yang, T. Frauenheim, A. Dominguez, F. Kärtner, A. Rubio, W. Wurth, A. Stierle and H. Noei, *ACS Catal.*, 2020, **10**, 13650–13658.

Poster titles

Real-time measurements of desorbed hydroxyl terminated $(\bar{2}01)$ β-Ga_2O_3 surfaces and band-bending effects, **S. Astley, K. Hazeldine, J. J. Tomes, M. W. Allen, R. Cross and D. A. Evans,** *Aberystwyth University, Wales*

In-situ oxidation and modification of diamond and nanodiamond surfaces at near-ambient pressures, **S. Astley, K. Hazeldine, J. Durk, A. Walton, J. J. Tomes, G. Held, F. Venturini, R. Cross and D. A. Evans,** *Aberystwyth University, Wales*

The role of water's hydrogen bonding network in nucleation and chemistry at environmental surfaces, **T. Bartels-Rausch, H. Yang, J. P. Gabathuler, Y. Manoharan, A. Boucly, L. Artiglia and M. Ammann,** *Paul Scherrer Institute (PSI), Switzerland*

NAP-XPS study of Mo/HZSM-5 under methane dehydroaromatization conditions, **S. Bartling, S. Peters, A. M. Abdel-Mageed, S. Wohlrab, and A Brückner,** *Leibniz Institute for Catalysis, Germany*

Geometric and electronic structure of a model ionic liquid – oxide interface delineated by resonant photoemission and normal incidence x-ray standing waves, **J. M. Seymour, E. Gousseva, D. A. Duncan, R. Grau-Crespo, K. R. J. Lovelock and R. A. Bennett,** *University of Reading, UK*

Lab-scale HAXPES, a new tool for buried interface analysis in an industrial environment, **T. Conard, F. Mascarenhas, A. Vanleenhove and I. Hoflijk,** *Imec, Belguim*

Fundamentals of atmospheric corrosion: elucidating the zinc/water interface as a function of relative humidity, **M. Dowhyj, K. Kousar, F. P. Lydiatt, D. Chekulaev, M. S. Walczak, R. Temperton, J. N. O'Shea, S. Walters, A. G. Thomas and R. Lindsay,** *University of Manchester, UK*

Quantification of the shell thickness of SnO@Au core-shell nanoparticles by x-ray photoelectron spectroscopy, **F. J. G. Zapata, M. Khan, R. Arbi and A. Turak,** *McMaster University, Canada and Tecnológico de Monterrey (ITESM), Mexico*

Electronic behaviour of ionic liquids: theory and experiment, **E. Gousseva, S. D. Midgley, J. M. Seymour, R. Grau-Crespo and K. R. J. Lovelock,** *University of Reading, UK*

Discover the 2D world of elements with XPS, **S. Guan, G. Rossi, P. R. Davies, D. J. Willock, J. Winfield and D. Lennon,** *Cardiff University, UK*

XPS and combinatorial analysis in assisting nanomaterials design, **M. A. Isaacs,** *University College London and Harwell XPS, UK*

Challenges in photoelectron spectroscopy to address relevant issues in catalysis research, **P. Lömker, C. Goodwin, D. Degerman, M. Shipilin, H.-Y Wang, A. Holm, M. Soldemo, R. Rameshan, C. Schlueter, P. Amann and A. Nilsson,** *Stockholm University, Sweden*

Revisiting polymer degradation in x-ray photoelectron spectroscopy, **D. J. Morgan,** *Cardiff University and Harwell XPS, UK*

Chemical information from XPS shake-up and TD-DFT, **G. Ososki, A. Paris, D. J. Morgan, D. J. Willock and P. R. Davies,** *Cardiff University, UK*

Identifying active sites on NiS electrodes for alkaline hydrogen evolution reaction, **X. Ding, F. E. Oropeza, G. Gorni, V. de la Peña O'Shea and K. H. L. Zhang,** *IMDEA Energy, Spain*

Lewis acidity in liquid phase organoboron compounds using XPS, **L. G. Parker, A. McGrogan, J. M. Seymour, R. Seidel, T. J. Penfold, M. Swadźba-Kwaśny and K. R. J. Lovelock,** *University of Reading, UK*

Argon cluster sputtering reveals internal chemical distribution in polymeric core-shell nanoparticles, **Y. Pei, D. Cant, C. Minelli and A. G. Shard,** *National Physical Laboratory, UK*

Metrology developments in XPS and HAXPES, **B. P. Reed, D. J. H. Cant and A. G. Shard,** *National Physical Laboratory, UK*

Interaction of oxygen with haemoglobin studied by ambient pressure x-ray photoelectron spectroscopy, **W. S. J. Skinner, G. Kerherve, P. T. P. Ryan, J. Åhlund, I. Mangialomini and D. J. Payne,** *Imperial College London, UK*

Successes and challenges in applications of a laboratory-based scanning XPS/HAXPES instrument, **K. Artyushkova, A. Vanleenhove, T. Conard, P. -M. Deleuze and O. Renault,** *Physical Electronics Inc, USA*

X-ray spectroscopy of halozincate anions in the liquid phase, **N. O. Alblewi, J. M. Seymour, C. J. Clarke, N. A. Besley, S. Hayama, R. G. Palgrave and K. R. J. Lovelock,** *University of Reading, UK*

Activating 2D materials by ionic bombardment, **J. Zribi, M. Frégnaux and D. Aureau,** *Institut Lavoisier de Versailles (ILV), UniversitéPari-Saclay and UVSQ, France*

First principles design and investigation of 2-dimensional Si doped carbon lattices for anode in Na ion batteries, **N. Yadav and T. J. D. Kumar,** *Indian Institute of Technology, India*

Considering XPS characterisation of ultra-thin films, **J. D. P. Counsell, A. J. Roberts, S. J. Coultas, C. Tupei and Y. Li,** *Kratos Analytical Ltd, UK*

Spurious charge transfer artefacts in core-level constrained x-ray spectroscopy simulations of metal-organic interfaces, **S. J. Hall, B. P. Klein and R. J. Maurer,** *University of Warwick, UK*

Unravelling the chemical state of ball-milled catalysts under operando methane complete combustion conditions, **N. J. Divins, L. Pascua-Solé, M. Danielis,**

A. Mussio, S. Colussi, A. Trovarelli and J. Llorca, *Technical University of Catalonia, Spain*

Surface properties of nickel electrodes during alkaline water electrolysis, **F. Longo, E. Billeter and A. Borgschulte,** *Laboratory for Advanced Analytical Technologies, Empa and University of Zurich, Switzerland*

Composition assessment of vanadium oxides by XPS using the active and static method, comparatively, **L. M. Marmolejo-Rivas, M. Bravo-Sanchez, F. J. González,** *Universidad de Guadalajara, Mexico*

Predicting x-ray photoelectron peak shapes: the effect of electronic structure, **C. R. O'Connor, M. A. van Spronsen, M. Karatok, J. Boscoboinik, C. M. Friend and M. M. Montemore,** *Harvard University and Pacific Northwest National Laboratory, USA*

In-situ imaging of multi-states in a catalytic surface reaction, **P. Winkler, J. Zeininger, M. Raab, Y. Suchorski, A. Steiger-Thirsfeld, M. Stöger-Pollach, M. Amati, L. Gregoratti, H. Grönbeck and G. Rupprechter,** *TU Wien, Austria*

Depth dependence of salt ions at the liquid/vapor interface studied by liquid jet APXPS measurements, **A. Siebert, K. Goodman and M. Blum,** *Lawrence Berkeley National Laboratory, USA*

XPS depth profiling of polymer brush layers, **D. Hammond and J. Alkorbi,** *Sheffield University, UK*

Light and water acceleration of defect mediated photodecomposition in metal halide perovskite thin films by environmental XPS, **K. X. Steirer, A. Ruth, A. Donakowski, E. Sanehira, J. Berry, A. Rockett and M. Irwin,** *Colorado School of Mines, USA*

The Hiden Analytical Poster Prize for the best poster was awarded to Mr Francisco Javier Garza Zapata of McMaster University, Canada and Tecnológico de Monterrey (ITESM), Mexico, for his poster on Quantification of the shell thickness of SnO@Au core-shell nanoparticles by x-ray photoelectron spectroscopy.

List of participants

Ms Najaat Alblewi, *University of Reading, United Kingdom*
Ms Marina Armengol Profitós, *Universitat Politècnica de Catalunya, Spain*
Dr Gavin Armstrong, *Nature Chemistry, United Kingdom*
Dr Rosa Arrigo, *University of Salford, United Kingdom*
Mr Kateryna Artyushkova, *Physical Electronics, USA*
Mr Albert Assonkeng, *Cape Breton University, Canada*
Dr Simon Astley, *Aberystwyth University, United Kingdom*
Dr Damien Aureau, *Institut Lavoisier de Versailles (ILV), Universite Pari-Saclay and UVSQ, France*
Mr Loren Ban, *ETH Zurich, Switzerland*
Dr Yossef Bar, *SRC, Israel*
Dr Iain Barlow, *Evonetix, United Kingdom*
Dr Thorsten Bartels-Rausch, *Paul Scherrer Institute (PSI), Switzerland*
Dr Stephan Bartling, *Leibniz Institute for Catalysis, Germany*
Dr Roger Bennett, *University Of Reading, United Kingdom*
Miss Prajna Bhatt, *University College London, United Kingdom*
Dr Monika Blum, *Lawrence Berkeley National Lab, USA*
Mr Andrea Braga, *Universitat Politècnica de Catalunya, Spain*
Dr Mark A. Buckingham, *The University of Manchester, United Kingdom*
Dr Adam Bushell, *Thermo Fisher Scientific, United Kingdom*
Dr David Cant, *National Physical Laboratory, United Kingdom*
Ms Rachel Cole, *Royal Society of Chemistry, United Kingdom*
Dr Thierry Conard, *IMEC, Belgium*
Dr Jonathan Counsell, *Kratos Analytical Ltd, United Kingdom*
Mr James J.C. Counter, *University of Reading, United Kingdom*
Mr Giulio D'Acunto, *Lund University, Sweden*
Professor Philip R. Davies, *Cardiff University & Harwell XPS, National Facility, United Kingdom*
Dr Martina Dell'Angela, *CNR-IOM, Italy*
Dr Desire Di Silvio, *Centro de Investigación Cooperativa en Biomateriales CIC bio-maGUNE, Spain*
Miss Michael Dowhyj, *University of Manchester, United Kingdom*
Dr Carlos Escudero Rodríguez, *ALBA-CELLS, Spain*
Dr Ad Ettema, *Scienta Omicron, Germany*
Professor Andrew Evans, *Aberystwyth University, United Kingdom*
Dr Samuel Eyley, *KU Leuven, Belgium*
Professor Wendy Flavell, *The University of Manchester, United Kingdom*
Dr Matthew Forster, *Johnson Matthey, United Kingdom*
Professor Giulia Galli, *University of Chicago, USA*
Mr Francisco Javier Garza Zapata, *Tecnológico de Monterrey (ITESM), Mexico*
Dr Joshua S. Gibson, *University of Oxford, United Kingdom*
Ms Ekaterina Gousseva, *University of Reading, United Kingdom*
Mr Stuart Govan, *Royal Society of Chemistry, United Kingdom*

Dr Shaoliang Guan, *Cardiff University*, *United Kingdom*
Mr Samuel Hall, *University of Warwick*, *United Kingdom*
Miss Charlotte Hall, *Diamond Light Source*, *United Kingdom*
Dr Debbie Hammond, *Sheffield University*, *United Kingdom*
Professor Dwayne Heard, *University of Leeds*, *United Kingdom*
Professor Georg Held, *Diamond Light Source & University of Reading*, *United Kingdom*
Miss Elif Hüsam, *Helmholtz-Zentrum Berlin für Materialien und Energie GmbH*, *Germany*
Dr Mark Isaacs, *University College London*, *United Kingdom*
Dr Mohammed Islam, *Thermo Fisher Scientific*, *United Kingdom*
Dr Nuria J. Divins, *Technical University of Catalonia*, *Spain*
Mr Edward Jones, *Loughborough University*, *United Kingdom*
Dr Juhan Matthias Kahk, *University of Tartu, Estonia*, *Estonia*
Mr Claus F. P. Kastorp, *Aarhus University*, *Denmark*
Dr Heath Kersell, *Oregon State University*, *USA*
Mr Markus Knaebbeler-Buss, *Fraunhofer Institute for Solar Energy Systems ISE*, *Germany*
Mr Oltion Kodra, *National Research Council*, *Canada*
Miss Alenka Krizan, *National Institute of Chemistry Slovenia*, *Slovenia*
Dr Alexander I. Large, *Diamond Light Source*, *United Kingdom*
Miss Roxy Lee, *University College London*, *United Kingdom*
Dr Robert Lindsay, *University of Manchester*, *United Kingdom*
Dr Johannes Lischner, *Imperial College London*, *United Kingdom*
Mr Filippo Longo, *Empa*, *Switzerland*
Dr Kevin Lovelock, *University of Reading*, *United Kingdom*
Dr Patrick Lömker, *Stockholm University*, *Sweden*
Miss Lourdes Michel Marmolejo-Rivas, *Universidad de Guadalajara*, *Mexico*
Dr Jette Mathiesen, *Stockhom University/Technical University of Denmark*, *Sweden*
Professor Reinhard Maurer, *University of Warwick*, *United Kingdom*
Dr Siobhan Mckeown Walker, *Laboratory of Advanced Technology, University of Geneva*, *Switzerland*
Dr Rotraut Merkle, *Max Plank Institute for Solid State Research*, *Germany*
Dr David Morgan, *Cardiff University*, *United Kingdom*
Dr Valery Muravev, *Eindhoven University of Technology*, *Netherlands*
Miss Fiona Nalden, *Royal Society of Chemistry*, *United Kingdom*
Dr Slavomir Nemsak, *Advanced Light Source, LBNL*, *USA*
Professor Anders Nilsson, *Stockholm University/Department of Physics*, *Sweden*
Dr Christopher O'Connor, *Pacific Northwest National Laboratory*, *USA*
Miss Sophie Orchard, *Royal Society of Chemistry*, *United Kingdom*
Dr Freddy Enrique Oropeza Palacio, *IMDEA energy*, *Spain*
Ms Genevieve Ososki, *Cardiff University*, *United Kingdom*
Mr Ayomide Osundare, *University of Manchester*, *United Kingdom*
Professor Robert Palgrave, *University College London & Harwell XPS, National Facility*, *United Kingdom*
Mr Lewis Parker, *University of Reading*, *United Kingdom*
Miss Laia Pascua Solé, *Universitat Politècnica de Catalunya*, *Spain*
Professor David Payne, *Imperial College London*, *United Kingdom*
Dr Yiwen Pei, *National Physical Laboratory*, *United Kingdom*
Dr Mirko Prato, *Istituto Italiano di Tecnologia*, *Italy*
Miss Faith Pritchard, *Durham university*, *United Kingdom*
Dr Evgeniy Redekop, *Centre for Materials Science and Nanotechnology (SMN), University of Oslo*, *Norway*

Dr Benjamen P. Reed, *National Physical Laboratory*, *United Kingdom*
Dr Olivier Renault, *CEA-Leti*, *France*
Dr Adam Roberts, *Kratos Analytical*, *United Kingdom*
Professor Günther Rupprechter, *TU Wien*, *Austria*
Mr Oleksandr Selyshchev, *Technische Universität Chemnitz, Institut für Physik*, *Germany*
Dr Zijie Sha, *Imperial College London*, *United Kingdom*
Dr Alexander Shard, *National Physical Laboratory*, *United Kingdom*
Dr Mzamo Shozi, *University of KwaZulu-Natal*, *South Africa*
Dr Andreas Siebert, *Lawrence Berkeley National Laboratory*, *USA*
Dr Mathieu G. Silly, *Synchrotron SOLEIL*, *France*
Mr William S.J. Skinner, *Imperial College London*, *United Kingdom*
Dr Charlotte Slade, *Thermo Fisher Scientific*, *United Kingdom*
Professor Francine Solal, *IPR-Université de Rennes*, *France*
Dr Ben Spencer, *University of Manchester*, *United Kingdom*
Miss Claire Springett, *Royal Society of Chemistry*, *United Kingdom*
Dr Xerxes Steirer, *Colorado School of Mines*, *USA*
Professor Kelsey A. Stoerzinger, *Oregon State University*, *USA*
Professor Sefik Suzer, *Bilkent University, Chemistry Department, Ankara, Turkey*, *Turkey*
Mrs Sarah Thirkell, *Royal Society of Chemistry*, *United Kingdom*
Dr Richard Thompson, *Royal Society of Chemistry*, *United Kingdom*
Miss Frances Towers Tompkins, *University of Reading*, *United Kingdom*
Dr Takashi Toyao, *Hokkaido University*, *Japan*
Dr Juan Jesús Velasco Vélez, *Fritz-Haber-Institut der Max-Planck-Gesellschaft*, *Germany*
Dr Marc Walker, *University of Warwick*, *United Kingdom*
Dr Susan Weatherby, *Royal Society of Chemistry*, *United Kingdom*
Dr Robert Weatherup, *University of Oxford*, *United Kingdom*
Ms Neha Yadav, *Indian Institute of Technology Ropar*, *India*

